Algebra 1

Fourth Edition

bju press
Greenville, South Carolina

Note:
The fact that materials produced by other publishers may be referred to in this volume does not constitute an endorsement of the content or theological position of materials produced by such publishers. Any references and ancillary materials are listed as an aid to the student or the teacher and in an attempt to maintain the accepted academic standards of the publishing industry.

ALGEBRA 1 Student Edition
Fourth Edition

Writers
Timothy King
Gene Bucholtz, MEd
Matthew Rollins, MA

Contributing Writers
Jeffrey A. Jaeger, MEd
Kathy Pilger, EdD
Mark Wetzel, MEd

Biblical Worldview
Ben Adams, MEd
Kyle J. Keesling Jr., ThM

Academic Integrity
Jeff Heath, EdD

Instructional Design
Rachel Santopietro, MEd
Danny Wright, DMin

Educational Technology
Ken Matesevac

Editor
Emily N. House

Designers
Drew Fields
Alina Zhupikov

Cover & Concept Design
Troy Renaux

Illustration, Design & Production
QBS Learning

Production Designers
Lydia Thompson
Clarissa Wisener

Permissions
Ruth Bartholomew
Sharon Belknap
Jennifer Walton

Project Coordinators
Andrea Huerta
Gary Santiago
Kyla J. Smith

Postproduction Liaison
Peggy Hargis

Indexer
Golden Rule Indexing

Photo credits appear on pages 684–86.

The text for this book is set in Adobe Minion Pro, Adobe Myriad Pro, Arial, Calibri by Monotype, Cambria Math, Good Pro by Lukasz Dziedzic, Helvetica, Minion Math by Typoma GmbH, STIX fonts, Symbol, TI84PlusCE Keys by Texas Instruments, and Times New Roman PSMT.

Printed in the United States of America

ISBN 978-1-64626-490-2

15 14 13 12 11 10 9 8 7 6 5 4 3 2 1

CONTENTS

Chapter 12: Rational Expressions & Equations 562

Appendix Lessons

FEATURES

Mind over Math

Technology Corner

Careers in Math

Math in History

Application Problems

STEM Preview

WELCOME TO *ALGEBRA 1*!

"It was interesting to think that the very first liquid ever poured on the moon, and the first food eaten there, were communion elements."
— Buzz Aldrin

In July 1969, Apollo 11 became the first spaceflight to land human beings on the moon. Edwin "Buzz" Aldrin, pilot of the Lunar Module Eagle, marked the occasion by reciting John 15:5 and partaking in the Lord's Supper. As he recalled the atoning death of the one who created the very moon on which the Eagle sat, Aldrin hoped taking communion could symbolize "the thought that God was revealing Himself there too, as man reached out into the universe."

To make this moment happen, the entire Apollo 11 team relied upon the consistency and logical order within mathematics. Math, including algebra, is used extensively by spaceflight teams to calculate every detail for a mission, including supply needs, trajectories, and air pressure changes inside spacecraft.

As you make your own voyage through *Algebra 1*, you will consider God's grand design of an orderly universe and develop a biblical worldview of math itself: why it works, why it is an effective tool, and how it should be used. Each chapter focuses on one of five themes to help you form this biblical view: foundations, modeling, reasoning, design, and ethics.

What sort of math will you be learning? *Algebra* is taken from the Arabic word *al-jabr*, which carries the idea of the reunion or restoration of broken parts. The concepts and skills you will learn from this book will help you take the fragmented information of a problem (unknown quantities) and organize it into a useful mathematical form by manipulating numbers and variables. You will be challenged to develop skill in the following essential mathematical practices.

MP1. Persevere in problem solving.

MP2. Use abstract reasoning.

MP3. Construct logical arguments.

MP4. Use mathematical models.

MP5. Use appropriate tools.

MP6. Use precise language.

MP7. See and apply structure.

MP8. Generalize patterns.

This text provides instruction and exercises created to assist you in developing these skills. Each section comes with clearly stated learning targets, step-by-step examples, Skill Checks, QR codes for tutorial videos, and cumulative review. *Algebra 1* has been written and designed with student success in mind.

Enjoy making the most of this year as you learn algebra. The problem-solving skills you develop will be useful to you whether you go on to be an entrepreneur, an architect, or even an astronaut. Shoot for the moon!

Chapter Introduction

Preview the basic outline of the chapter.

Math in History & Careers in Math

Read biblical perspectives on the lives of famous mathematicians and math-related careers.

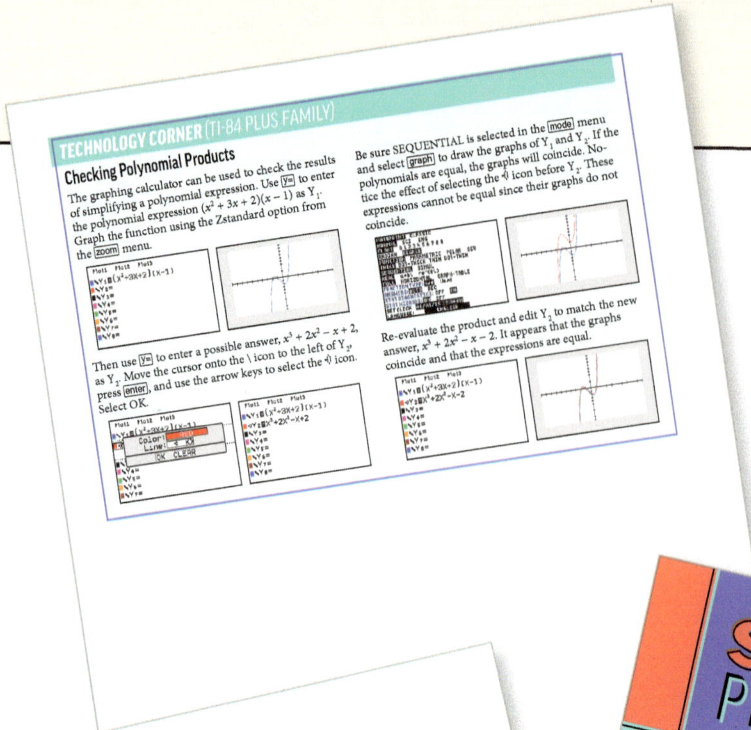

TECHNOLOGY CORNER (TI-84 PLUS FAMILY)

Checking Polynomial Products

The graphing calculator can be used to check the results of simplifying a polynomial expression. Use [Y=] to enter the polynomial expression $(x^2 + 3x + 2)(x - 1)$ as Y_1. Graph the function using the Zstandard option from the [zoom] menu.

Then use [Y=] to enter a possible answer, $x^3 + 2x^2 - x + 2$, as Y_2. Move the cursor onto the \ icon to the left of Y_2, press [enter], and use the arrow keys to select the icon. Select OK.

Be sure SEQUENTIAL is selected in the [mode] menu and select [graph] to draw the graphs of Y_1 and Y_2. If the polynomials are equal, the graphs will coincide. Notice the effect of selecting the icon before Y_2. These expressions cannot be equal since their graphs do not coincide.

Re-evaluate the product and edit Y_2 to match the new answer, $x^3 + 2x^2 - x - 2$. It appears that the graphs coincide and that the expressions are equal.

Technology Corner

Follow step-by-step instructions to apply basic uses of the graphing calculator related to each chapter's content.

APPLICATION PROBLEMS
CALCULATING INTEREST

The stockholders of the First National Bank of Newton have a vested interest in the financial plans that the bank president, Mr. Lee, presents at the annual board meeting. Mr. Lee knows that he must gain their confidence in his plans for investing their money. He summarizes investments that will earn $3.4 million in the coming year. He needs to know how much money he must invest in municipal bonds that yield 8% annually so that the total interest income for the bank will be at least an additional $2.1 million.

The Bible has much to say about the everyday details of life, including such things as borrowing and lending (Prov. 22:7). As long as a person is in debt, he is morally obligated to the person who is lending him money. For this reason, a Christian should be very careful about borrowing money.

One way to prepare is to know about different types of loans, three of which are discussed below. Recall the interest formula $I = Prt$.

1. Simple interest loan: A single payment of interest and principal combined at the end of the specified time.
2. Add-on loan: Simple interest plus principal divided by the number of payments (usually monthly). Each payment applies the same set amount to the principal and the same amount to interest.
3. Discount loan: The interest on the desired amount to be borrowed is calculated and added to the amount that is borrowed at the beginning of the loan. This new amount owed is divided into equal payments throughout the life of the loan.

Compare these types of loans for $5000 borrowed at 6% interest for 2 yr in the 3 ways described above.

1. Simple interest loan: Find the amount of interest and the entire amount due at the end of the loan.
2. Add-on loan: Find the amount of interest, the total amount to repay, and the monthly payment.
3. Discount loan: Calculate the actual amount loaned for the borrower to receive $5000 using the formula $S = \frac{P}{1 - dt}$, where S = the amount that must be borrowed, P = the amount the borrower receives, d = the interest rate, and t = the time in years. Then find the amount of interest and the monthly payment.

If money must be borrowed, it is wise to find the best rate. Interest rates are expressed in 2 ways: annual percentage rate (APR) and annual percentage yield (APY). The APR expresses annual interest without considering compound interest. The APY expresses annual interest, including compound interest (interest earned on both principal *and* previous interest).

When the lender receives monthly compounded interest on the remaining principal throughout the term of a loan, the actual interest rate is not equal to the simple interest rate (APR). Calculating the APY reveals that borrowers are actually paying a higher rate than the APR.

$APY = \left(1 + \frac{APR}{n}\right)^n - 1$, where n is the number of compounding periods per year.

Find the APY of a 5 yr $8000 loan with a 7% … compounded monthly.

Using $n = 12$, $APY = \left(1 + \frac{0.07}{12}\right)^{12} - 1$ $= (1.005833)^{12} - 1 = 0.0723 = 7.2\%$.

The actual interest rate is about 7.2%, … than the 7% APR.

To provide protection for borrowers, … Lending Act requires creditors to … rower a document listing the total … the annual percentage rate (APR) … percentage yield (APY).

Assuming monthly payments, … the following loans. Round to the … percent.

4. Mr. Ellison was quoted an interest rate … APR for a loan to expand his business.

146 CHAPTER 3 Solving Inequalities

STEM PROJECT

THE WATER WHEEL

What if you needed to power something and you could not simply "plug it in"? Water wheels were one of the first machines that converted the power of water flow in a stream or river into energy. God created the earth with an expanse of water and gravity, which causes water to flow downhill. This natural flow of water is a sustainable resource that can produce great amounts of energy. In this STEM project you will build a small water wheel and assess its effectiveness by lifting various weights using the wheel.

STEM PREVIEW 261

STEM Project

Preview the semester STEM projects located in Activities.

Application Problems

Apply learned math skills to solve real-world problems.

Essential Question & Learning Targets

Start each section with a question about the key idea and a list of the skills you should expect to learn.

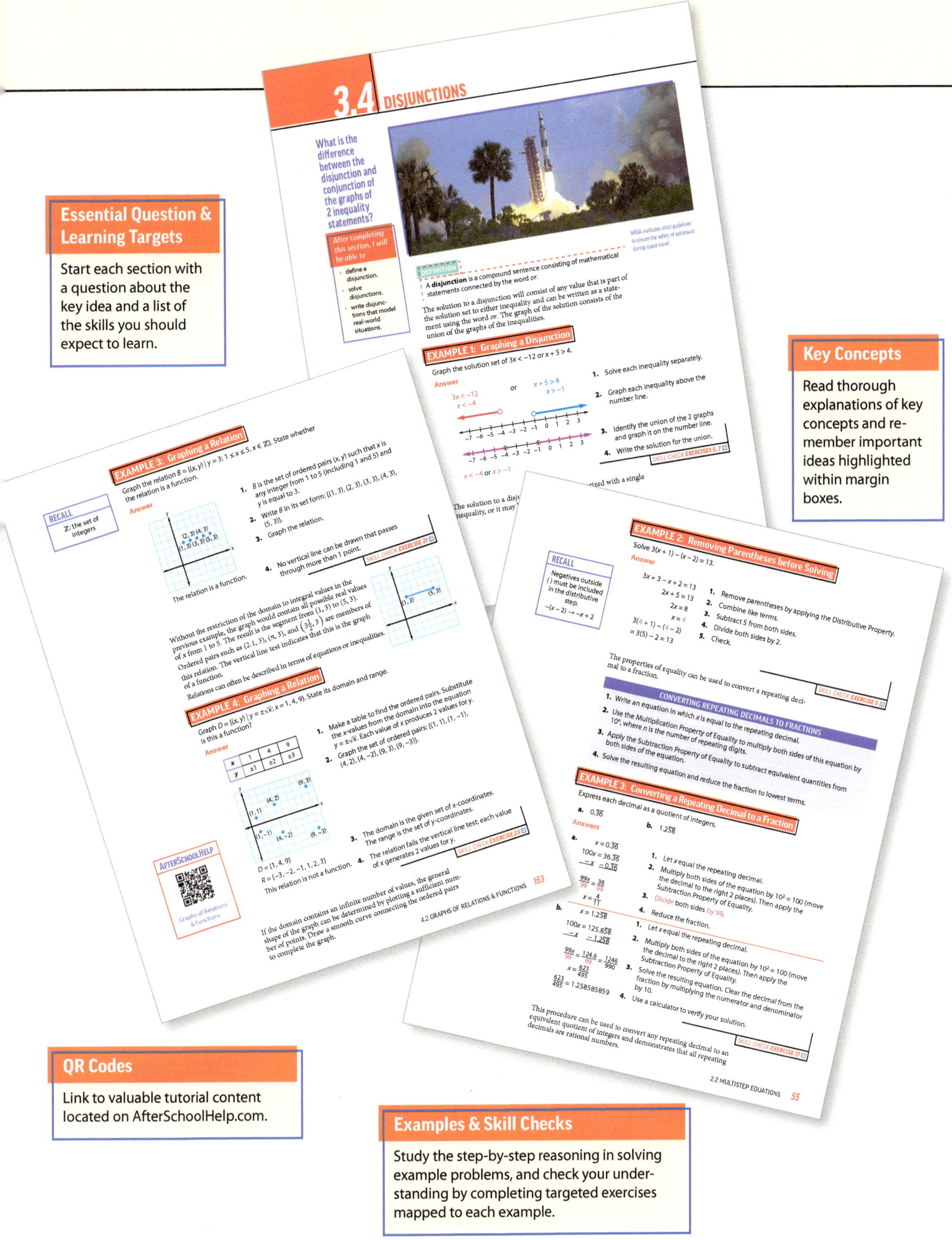

Key Concepts

Read thorough explanations of key concepts and remember important ideas highlighted within margin boxes.

QR Codes

Link to valuable tutorial content located on AfterSchoolHelp.com.

Examples & Skill Checks

Study the step-by-step reasoning in solving example problems, and check your understanding by completing targeted exercises mapped to each example.

Exercises

Build and maintain your skills with carefully sequenced exercises that emphasize the essential question and essential mathematical practices (**MP**s).

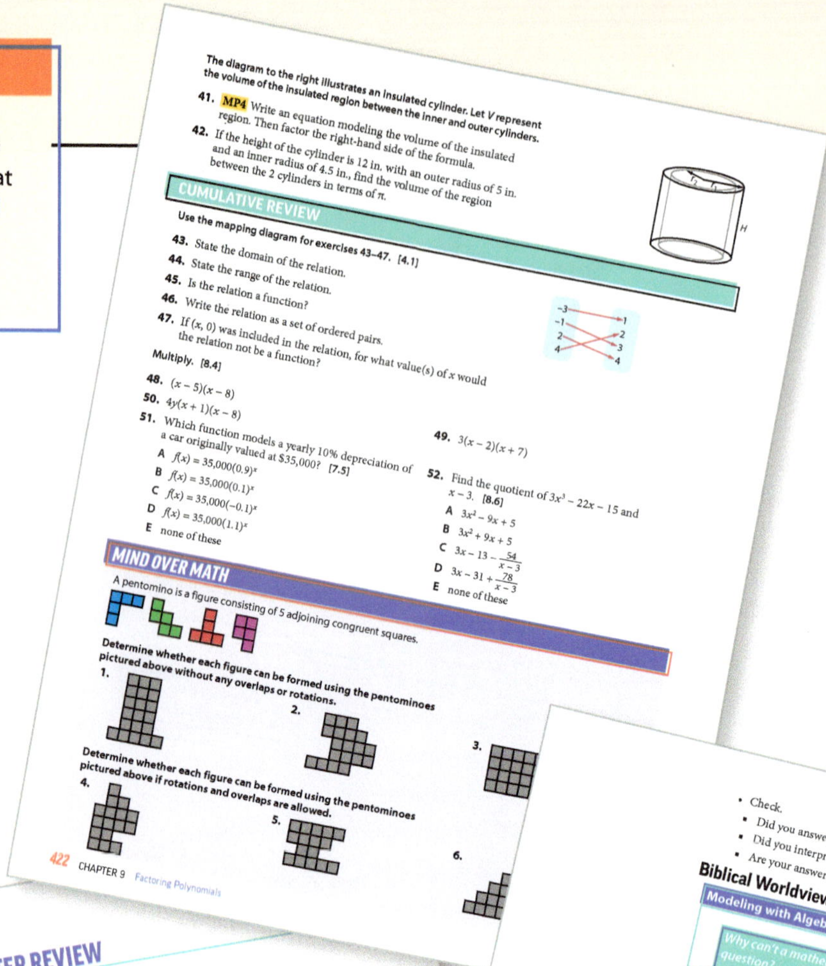
The diagram to the right illustrates an insulated cylinder. Let V represent the volume of the insulated region between the inner and outer cylinders.

41. MP4 Write an equation modeling the volume of the insulated region. Then factor the right-hand side of the formula.
42. If the height of the cylinder is 12 in. with an outer radius of 5 in. and an inner radius of 4.5 in., find the volume of the region between the 2 cylinders in terms of π.

CUMULATIVE REVIEW

Use the mapping diagram for exercises 43–47. [4.1]

43. State the domain of the relation.
44. State the range of the relation.
45. Is the relation a function?
46. Write the relation as a set of ordered pairs.
47. If $(x, 0)$ was included in the relation, for what value(s) of x would the relation not be a function?

Multiply. [8.4]

48. $(x - 5)(x - 8)$
49. $3(x - 2)(x + 7)$
50. $4y(x + 1)(x - 8)$
51. Which function models a yearly 10% depreciation of a car originally valued at \$35,000? [7.5]
A $f(x) = 35{,}000(0.9)^x$
B $f(x) = 35{,}000(0.1)^x$
C $f(x) = 35{,}000(-0.1)^x$
D $f(x) = 35{,}000(1.1)^x$
E none of these
52. Find the quotient of $3x^3 - 22x - 15$ and $x - 3$. [8.6]
A $3x^2 - 9x + 5$
B $3x^2 + 9x + 5$
C $3x - 13 - \frac{54}{x-3}$
D $3x - 31 + \frac{78}{x-3}$
E none of these

MIND OVER MATH

A pentomino is a figure consisting of 5 adjoining congruent squares.

Determine whether each figure can be formed using the pentominoes pictured above without any overlaps or rotations.

1. 2. 3.

Determine whether each figure can be formed using the pentominoes pictured above if rotations and overlaps are allowed.

4. 5. 6.

422 CHAPTER 9 Factoring Polynomials

Cumulative Review

Systematically review key concepts and practice strategies for standardized testing.

Mind over Math

Challenge your critical thinking with this unique feature found in each chapter.

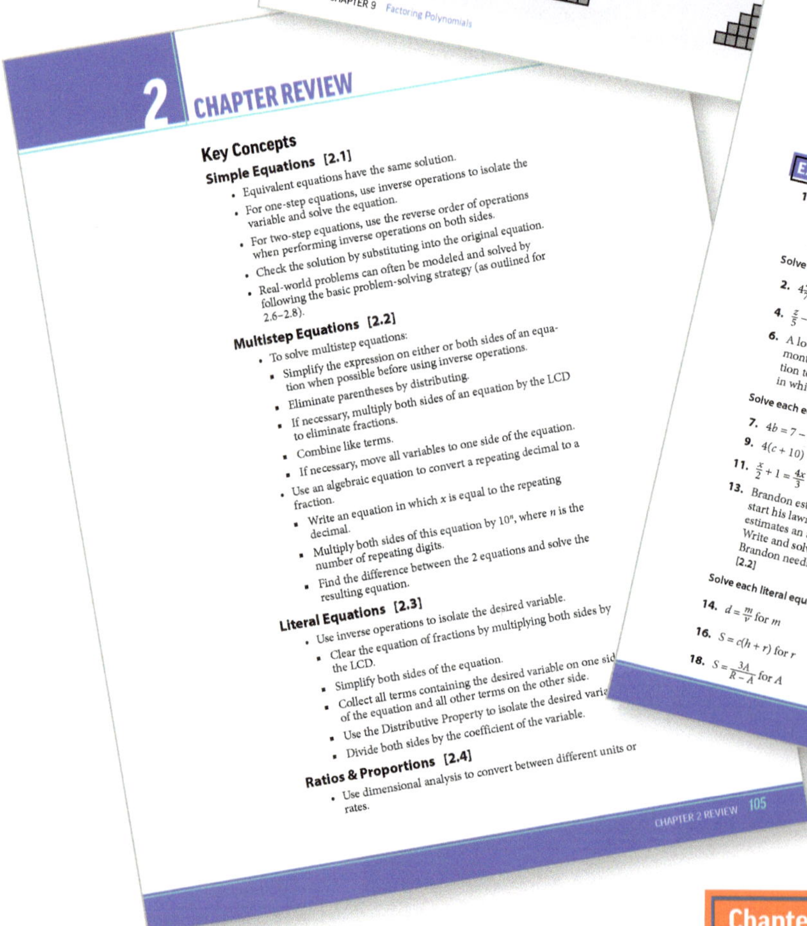
2 CHAPTER REVIEW

Key Concepts

Simple Equations [2.1]

- Equivalent equations have the same solution.
- For one-step equations, use inverse operations to isolate the variable and solve the equation.
- For two-step equations, use the reverse order of operations when performing inverse operations on both sides.
- Check the solution by substituting into the original equation.
- Real-world problems can often be modeled and solved by following the basic problem-solving strategy (as outlined for 2.6–2.8).

Multistep Equations [2.2]

- To solve multistep equations:
  - Simplify the expression on either or both sides of an equation when possible before using inverse operations.
  - Eliminate parentheses by distributing.
  - If necessary, multiply both sides of an equation by the LCD to eliminate fractions.
  - Combine like terms.
  - If necessary, move all variables to one side of the equation.
- Use an algebraic equation to convert a repeating decimal to a fraction.
  - Write an equation in which x is equal to the repeating decimal.
  - Multiply both sides of this equation by 10^n, where n is the number of repeating digits.
  - Find the difference between the 2 equations and solve the resulting equation.

Literal Equations [2.3]

- Use inverse operations to isolate the desired variable.
  - Clear the equation of fractions by multiplying both sides by the LCD.
  - Simplify both sides of the equation.
  - Collect all terms containing the desired variable on one sid[e] of the equation and all other terms on the other side.
  - Use the Distributive Property to isolate the desired varia[ble].
  - Divide both sides by the coefficient of the variable.

Ratios & Proportions [2.4]

- Use dimensional analysis to convert between different units or rates.

CHAPTER 2 REVIEW 105

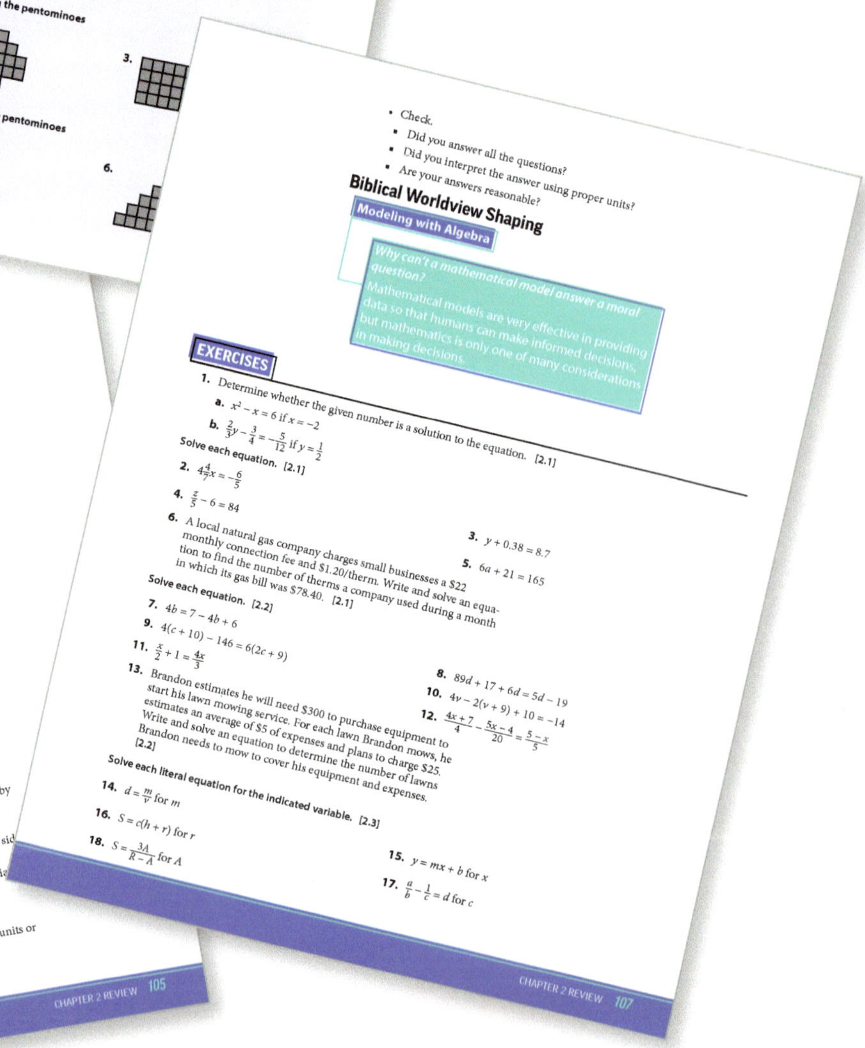
- Check.
  - Did you answer all the questions?
  - Did you interpret the answer using proper units?
  - Are your answers reasonable?

Biblical Worldview Shaping

Modeling with Algebra

Why can't a mathematical model answer a moral question?

Mathematical models are very effective in providing data so that humans can make informed decisions, but mathematics is only one of many considerations in making decisions.

EXERCISES

1. Determine whether the given number is a solution to the equation. [2.1]
a. $x^2 - x = 6$ if $x = -2$
b. $\frac{2}{3}y - \frac{3}{4} = -\frac{5}{12}$ if $y = \frac{1}{2}$

Solve each equation. [2.1]

2. $4\frac{4}{7}x = -\frac{6}{5}$
3. $y + 0.38 = 8.7$
4. $\frac{z}{5} - 6 = 84$
5. $6a + 21 = 165$
6. A local natural gas company charges small businesses a \$22 monthly connection fee and \$1.20/therm. Write and solve an equation to find the number of therms a company used during a month in which its gas bill was \$78.40. [2.1]

Solve each equation. [2.2]

7. $4b = 7 - 4b + 6$
8. $89d + 17 + 6d = 5d - 19$
9. $4(c + 10) - 146 = 6(2c + 9)$
10. $4v - 2(v + 9) + 10 = -14$
11. $\frac{x}{2} + 1 = \frac{4x}{3}$
12. $\frac{4x+7}{4} - \frac{5x-4}{20} = \frac{5-x}{5}$
13. Brandon estimates he will need \$300 to purchase equipment to start his lawn mowing service. For each lawn Brandon mows, he estimates an average of \$5 of expenses and plans to charge \$25. Write and solve an equation to determine the number of lawns Brandon needs to mow to cover his equipment and expenses. [2.2]

Solve each literal equation for the indicated variable. [2.3]

14. $d = \frac{m}{V}$ for m
15. $y = mx + b$ for x
16. $S = c(h + r)$ for r
17. $\frac{a}{b} - \frac{1}{c} = d$ for c
18. $S = \frac{3A}{R - A}$ for A

CHAPTER 2 REVIEW 107

Chapter Review

Prepare for the chapter test with a summary of key concepts, a review of the biblical worldview theme, and review exercises.

Algebra 1

Fourth Edition

CHAPTER 01
EXPRESSIONS

CHAPTER OVERVIEW

1.1 ADDING & SUBTRACTING RATIONAL NUMBERS

How is subtraction the same as addition?

After completing this section, I will be able to

- add rational numbers.
- identify the properties of addition.
- subtract rational numbers.
- evaluate sums and differences to solve real-world problems.
- explain the unity and diversity of rational numbers.

RECALL

8 + (−3)
\ /
addends

The value of stocks on a given day can be modeled by the addition and subtraction of rational numbers.

The ability to add both positive and negative rational numbers is essential to success in algebra. Using the absolute value of the addends simplifies the addition process.

DEFINITION

The **absolute value** of a number, denoted $|x|$, is the number of units (distance) between 0 and the location of the number on a number line.

ADDING POSITIVE & NEGATIVE NUMBERS	
Addends with the Same Sign	Addends with Different Signs
1. Add the absolute values of the addends.	**1.** Subtract the smaller absolute value from the larger absolute value.
2. Use the same sign as the 2 addends.	**2.** Use the sign of the addend with the larger absolute value.

EXAMPLE 1: Adding Integers

Find each sum.

a. $-1 + (-3)$ **b.** $8 + (-3)$ **c.** $4 + (-10)$

Answers

a.	$|-1| + |-3| = 1 + 3 = 4$	**1.** When addends have the same sign, add their absolute values.
	$-1 + (-3) = -4$	**2.** The sum of 2 negatives must be negative.
b.	$|8| - |-3| = 8 - 3 = 5$	**1.** When addends have different signs, subtract the smaller absolute value from the larger.
	$8 + (-3) = 5$	**2.** The sum is positive because $|8| > |-3|$.
c.	$|-10| - |4| = 10 - 4 = 6$	**1.** When addends have different signs, subtract the smaller absolute value from the larger.
	$4 + (-10) = -6$	**2.** The sum is negative because $|-10| > |4|$.

SKILL CHECK **EXERCISE 5** ☑

Mathematicians have created special names for the rules governing operations on numbers.

DEFINITION

A **mathematical property** or *identity* is an equation or statement that is true for any value of the variable.

PROPERTIES OF ADDITION	
Commutative Property of Addition $a + b = b + a$	When adding 2 numbers together in different orders, the sum is the same. $3 + 5 = 5 + 3$
Associative Property of Addition $(a + b) + c = a + (b + c)$	When grouping numbers together differently while adding, the sum is the same. $(2 + 9) + 4 = 2 + (9 + 4)$
Additive Identity Property $a + 0 = 0 + a = a$	The sum of any number and 0 is that number. $3 + 0 = 0 + 3 = 3$
Additive Inverse Property $a + (-a) = 0$	The sum of any number and its additive inverse (opposite) is 0. $3 + (-3) = 0$

Applying these properties can often simplify sums. The Commutative and Associative Properties establish that numbers can be added in any order. Rearranging the order of the addends produces an equivalent expression that can simplify the adding process.

EXAMPLE 2: Adding Multiple Integers

Find each sum.

a. $-1 + 5 + (-3) + (-7) + 4 + (-6)$

b. $8 + (-3) + 15 + (-8)$

Answers

a. $-1 + 5 + (-3) + (-7) + 4 + (-6)$

$= 5 + 4 + (-1) + (-3) + (-7) + (-6)$

1. Apply the Commutative Property of Addition.

$= 9 + (-17)$

2. Add all the positive numbers and add all the negative numbers.

$= -8$

3. Add the results.

b. $8 + (-3) + 15 + (-8)$

$= 8 + (-8) + (-3) + 15$

1. Apply the Commutative Property of Addition.

$= 0 + 12$

2. Apply the Additive Inverse Property.

$= 12$

3. Apply the Additive Identity Property.

SKILL CHECK EXERCISE 17

The process of subtracting signed numbers can often be simplified by rewriting a difference as an equivalent sum.

DEFINITION

Subtraction is defined as adding the opposite. For any real numbers a and b, $a - b = a + (-b)$.

This equation is read "*a* minus *b* equals *a* plus the opposite of *b*." When subtracting a smaller positive number from a larger positive number $(8 - 3)$, it is unnecessary to use the definition to rewrite the expression as addition $[8 + (-3)]$. However, more complicated subtractions such as $4 - (-2)$ and $-3 - 2$ are more easily understood by using the definition to add the opposite.

EXAMPLE 3: Subtracting Integers

Find each difference.

a. $4 - (-2)$ **b.** $-3 - 2$ **c.** $2 - 5$

Answers

a. $4 - (-2) = 4 + 2 = 6$ Subtract −2 by adding its opposite, 2.

b. $-3 - 2 = -3 + (-2) = -5$ Subtract 2 by adding its opposite, −2.

c. $2 - 5 = 2 + (-5) = -3$ Subtract 5 by adding its opposite, −5.

SKILL CHECK **EXERCISE 11**

The unity and diversity we find in mathematics reflects the unity and diversity found in our triune God. Without the revelation of Scripture, we would have no adequate explanation for why our world functions this way. Rational numbers, for example, may be very different, yet they can all be expressed as a quotient of integers.

Our number system reflects diversity in the many different forms of numbers (integers, fractions, decimal numbers, radicals, etc.). At the same time, our number system represents unity since the principles of adding and subtracting integers also apply to all real numbers.

RECALL

Align decimals when adding or subtracting.

AFTERSCHOOLHELP

Adding & Subtracting Rational Numbers

EXAMPLE 4: Adding or Subtracting Rational Numbers

Find each sum or difference.

a. $-112.6 + 39.27$ **b.** $2.39 - (-14.682)$ **c.** $\left(-\frac{1}{12}\right) + \frac{3}{4}$ **d.** $-\frac{5}{12} - \frac{5}{8}$

Answers

a.
$$\begin{array}{r} 112.6 \\ \underline{-\ 39.27} \\ 73.33 \end{array}$$

1. When addends have different signs, find the difference of their absolute values.

$-112.6 + 39.27 = -73.33$

2. The sum is negative because $|-112.6| > |39.27|$.

b. $2.39 - (-14.682) = 2.39 + 14.682$
$= 17.072$

1. To subtract, add the opposite.
2. The sum is positive since both addends are positive.

c. $\left(-\frac{1}{12}\right) + \frac{3}{4}\left(\frac{3}{3}\right) = \left(-\frac{1}{12}\right) + \frac{9}{12}$

1. Rename so the fractions have a common denominator.

$\frac{9}{12} - \frac{1}{12} = \frac{8}{12} = \frac{2}{3}$

2. When addends have different signs, find the difference of their absolute values.

$\left(-\frac{1}{12}\right) + \frac{3}{4} = \frac{2}{3}$

3. The sum is positive because $\left|\frac{3}{4}\right| > \left|-\frac{1}{12}\right|$.

d. $-\frac{5}{12} - \frac{5}{8} = -\frac{5}{12} + \left(-\frac{5}{8}\right)$

1. To subtract, add the opposite.

$= -\frac{5}{12}\left(\frac{2}{2}\right) + \left(-\frac{5}{8}\right)\left(\frac{3}{3}\right)$

$= \left|-\frac{10}{24}\right| + \left|-\frac{15}{24}\right| = \frac{25}{24}$

2. Add the absolute values.

$-\frac{5}{12} + \left(-\frac{5}{8}\right) = -\frac{25}{24}$

3. The sum is negative since both addends are negative.

SKILL CHECK ***EXERCISES 23, 33***

EXAMPLE 5: Solving a Real-World Problem

On Monday the price of stock in an electronics company opened at \$15.75, fell \$0.19 to its daily low before climbing \$0.56 to its daily high, and then fell \$0.20 to its closing price. What was the value of the stock at the end of the day on Monday?

Answer

$15.75 - 0.19 + 0.56 - 0.20$

1. Write an expression modeling the price of the stock.

$= 15.75 + (-0.19) + 0.56 + (-0.20)$

2. To subtract, add the opposite.

$= 15.75 + 0.56 + (-0.19) + (-0.20)$

3. Apply the Commutative Property to rearrange the order of the numbers.

$= 16.31 + (-0.39)$

4. Find the sum of the positives and the sum of the negatives.

$= |16.31| - |-0.39|$

$= 16.31 - 0.39 = 15.92$

5. Subtract the smaller absolute value from the larger. The answer is positive since the larger absolute value is positive.

The value of the stock at the end of the day was \$15.92.

SKILL CHECK ***EXERCISE 41***

VOCABULARY

- absolute value
- mathematical property
- subtraction

A. EXERCISES

Fill in the blank.

1. The ______ of a number, n, is the distance between n and 0 on the number line.
2. A ______ is a statement that is true for any value of the variable in the statement.
3. **Essential Question:** Subtraction is defined as ______ the opposite.

Find each sum.

4. $3 + 21$
5. $-13 + 32$
6. $26 + (-28)$
7. $-14 + (-7)$
8. $-126 + 35$
9. $7 + (-3)$

Find each difference.

10. $3 - (-2)$
11. $4 - 5$
12. $-12 - 5$
13. $-8 - (-2)$
14. $6 - 8$
15. $-16 - (-31)$

B. EXERCISES

Simplify each expression.

16. $-7 + (-8) + 18 + (-5)$
17. $14 + (-9) + 6 + (-28)$
18. $-1 + 2 - (-4) - 3$
19. $5 - 3 - [-(-9)] + 13$
20. $|-5| + (-4) - 3 + (-6) - (-8)$
21. $-8 + |-10| - (-9) - |17|$
22. $-527.69 + (-89.41)$
23. $0.23 + (-624.4)$
24. $-25.6 + 11.23$
25. $621.4 - 0.23$
26. $3.896 - (-11.42)$
27. $-13.716 - (-7.07)$
28. $\frac{1}{5} + \frac{8}{15}$
29. $\frac{4}{7} + \left(-\frac{11}{7}\right)$
30. $-\frac{2}{15} + \frac{3}{5}$
31. $-\frac{12}{21} + \left(-\frac{10}{3}\right)$
32. $-\frac{4}{9} + \frac{5}{2}$
33. $\frac{2}{9} - \left(-\frac{5}{9}\right)$
34. $5 - \frac{3}{4} - \left(-\frac{8}{9}\right)$
35. $-\frac{7}{10} + \frac{3}{5} - \frac{7}{2}$
36. $\frac{8}{9} - \frac{14}{9} + \left(-\frac{5}{3}\right)$
37. $5\frac{3}{7} - 9\frac{1}{5} + \left(-3\frac{13}{35}\right)$

Write and evaluate an expression for each exercise.

38. The temperature on a September day is 68°F at 7:00 AM, increases 14° by noon, and then decreases 20° from noon to 10:00 PM. What is the temperature at 10:00 PM?
39. A motorboat travels at an average speed of 17 knots (nautical miles per hour) in still water. Determine the actual speed of the boat when it is traveling against a current of 5 knots.
40. If the temperature on January 5 increased from −3°F to 14°F, how much did it increase?
41. Find the difference in the height of Mt. Whitney (14,494 ft elevation) and the depth of Death Valley (282 ft below sea level).
42. Mathilda is sewing a top treatment for her bedroom window that requires a special self-enclosed French seam. The first part of the seam measures $\frac{1}{4}$ in., and the second part enclosing the first seam measures $\frac{3}{8}$ in. What is the total amount of fabric that Mathilda should allow for the top treatment seam?

43. Bob is putting gutters on the last home in a new subdivision. The individual measurements (in feet) for each length of gutter needed are 20, $3\frac{3}{4}$, $30\frac{1}{2}$, $7\frac{1}{6}$, $20\frac{5}{6}$, $7\frac{1}{6}$, and $25\frac{1}{4}$. Find the total length of gutters that Bob needs to order if he has 3 pieces measuring $5\frac{1}{2}$ ft, $7\frac{7}{12}$ ft, and $8\frac{2}{3}$ ft left over from another job site.

State the property of addition that justifies each numbered step in the following simplification.

$(-3 + 10) + (-7)$

44. $= [10 + (-3)] + (-7)$ ______

45. $= 10 + [-3 + (-7)]$ ______

$= 10 + (-10)$ addition

46. $= 0$ ______

47. **BWS Essential Question:** What is one way that rational numbers exhibit unity and diversity?

C. EXERCISES

Simplify each expression.

48. $-|3 - 15| + (-9)$

49. $2.9 + 13.05 + (-26.472)$

50. $|-8| - |7| - |-8 - 7|$

51. $4 - (-5) + |2 - 19| - (-6) - 4$

52. $\frac{3}{7} + \frac{5}{14} - \frac{1}{3} + \frac{9}{28}$

53. MP3 Simplify the expression $-\frac{3}{4} - \left(-\frac{2}{3}\right)$ and compare your work with Emily's work.

a. Did Emily get the correct answer if she put $-\frac{1}{12}$ on her answer sheet?

b. Determine which steps in Emily's work contain errors and describe each error that was made.

Step	Emily's Work
	$-\frac{3}{4} - \left(-\frac{2}{3}\right)$
1.	$= -\frac{3}{4} + \left(-\frac{2}{3}\right)$
2.	$= -\frac{9}{12} + \left(-\frac{8}{12}\right)$
3.	$= -\frac{1}{12}$

1.2 MULTIPLYING & DIVIDING RATIONAL NUMBERS

How can I use a property of multiplication to solve a division problem?

After completing this section, I will be able to

- multiply rational numbers.
- divide rational numbers.
- identify the properties of multiplication.
- evaluate products and quotients to solve real-world problems.

RECALL

$25 \cdot 10 = 250$

factors product

Lola manages a food truck in the park during the summer. After the first 5 days, she notices that her expenses exceeded her sales, and she has lost \$7 each day. What is her net income so far?

$$-7 + (-7) + (-7) + (-7) + (-7) = -\$35$$

Lola can save time if the repeated addition is replaced with multiplication.

$$5(-7) = -\$35$$

Recall the following rules for multiplying signed numbers.

MULTIPLICATION RULES FOR THE PRODUCT OF 2 NUMBERS

1. If the factors are both positive or both negative, the product is positive.
2. If 1 factor is positive and the other is negative, the product is negative.

Note how these rules are applied to find each of the following products.

$$2 \cdot 3 = 6 \qquad -2 \cdot 3 = -6 \qquad 2 \cdot (-3) = -6 \qquad -2 \cdot (-3) = 6$$

The following generalizations can be made when finding the product of more than 2 factors since the product of every pair of negative factors is positive.

MULTIPLICATION RULES FOR THE PRODUCT OF MULTIPLE NUMBERS

1. If there is an even number of negative factors, the product is positive.
2. If there is an odd number of negative factors, the product is negative.

EXAMPLE 1: Multiplying Integers

Calculate each product.

a. $-4(-6)$ **b.** $-2(-3)(-5)$

Answers

a.	$-4(-6) = 24$		A product of 2 factors with the same sign is positive.
b.	$2 \cdot 3 \cdot 5 = 30$	**1.**	Multiply the absolute values.
	$-2(-3)(-5) = -30$	**2.**	A product of 3 negative factors is negative.

SKILL CHECK **EXERCISE 7**

The rules for multiplying integers apply to all real numbers.

EXAMPLE 2: Multiplying Rational Numbers

Find each product.

a. $-2\frac{1}{5}\cdot 4\frac{2}{7}$ **b.** $-\frac{4}{15}\left(\frac{5}{14}\right)\left(-\frac{21}{20}\right)$ **c.** $4.2\times(-5.3)$

Answers

a. $-2\frac{1}{5}\cdot 4\frac{2}{7}=-\frac{11}{5}\cdot\frac{30}{7}$ — **1.** Rewrite the mixed numbers as improper fractions.

$=-\frac{11}{\cancel{5}}\cdot\frac{\overset{6}{\cancel{30}}}{7}=-\frac{66}{7}$ — **2.** Cancel before multiplying.

b. $\frac{\cancel{4}}{\underset{3}{\cancel{15}}}\cdot\frac{\cancel{5}}{\underset{2}{\cancel{14}}}\cdot\frac{\overset{3}{\cancel{21}}}{\underset{5}{\cancel{20}}}=\frac{1}{10}$ — **1.** Find the product of the absolute values.

$-\frac{4}{15}\left(\frac{5}{14}\right)\left(-\frac{21}{20}\right)=\frac{1}{10}$ — **2.** A product of 2 negative factors is positive.

c. $4.2\times(-5.3)$ — **1.** Find the product of the absolute values.

$$\begin{array}{r} 4.2 \\ \times\ 5.3 \\ \hline 126 \\ 2100 \\ \hline 22.26 \end{array}$$

$4.2\times(-5.3)=-22.26$ — **2.** A product with 1 negative factor is negative.

SKILL CHECK **EXERCISES 9, 23**

The following properties are often used when simplifying products.

PROPERTIES OF MULTIPLICATION	
Commutative Property of Multiplication $ab=ba$	When multiplying 2 numbers in different orders, the product is the same. $3\cdot 5=5\cdot 3$
Associative Property of Multiplication $a(bc)=(ab)c$	When grouping the numbers differently while multiplying, the product is the same. $2(3\cdot 5)=(2\cdot 3)5$
Zero Property of Multiplication $a\cdot 0=0$	The product of any number and 0 is 0. $4\cdot 0=0$
Multiplicative Identity Property $a\cdot 1=a$	The product of any number and 1 is that number. $13\cdot 1=13$
Multiplicative Inverse Property $a\cdot\frac{1}{a}=1$ for $a\neq 0$	The product of a nonzero number and its reciprocal is 1, the multiplicative identity. $\frac{4}{5}\cdot\frac{1}{\frac{4}{5}}=\frac{4}{5}\cdot\frac{5}{4}=1$

EXAMPLE 3: Applying Properties of Multiplication

Simplify the product $\left(\frac{1}{9}\cdot\frac{13}{22}\right)\cdot 9$, and identify the property used in each step.

Answer

$$\left(\frac{1}{9}\cdot\frac{13}{22}\right)\cdot 9=\left(\frac{13}{22}\cdot\frac{1}{9}\right)\cdot 9$$

1. Apply the Commutative Property of Multiplication.

$$=\frac{13}{22}\cdot\left(\frac{1}{9}\cdot 9\right)$$

2. Apply the Associative Property of Multiplication.

$$=\frac{13}{22}\cdot 1$$

3. Apply the Multiplicative Inverse Property.

$$=\frac{13}{22}$$

4. Apply the Multiplicative Identity Property.

SKILL CHECK **EXERCISE 17**

Recall that division and multiplication are inverse operations of each other.

$3\cdot 2=6$, so $6\div 2=3$ and $6\div 3=2$.
$-3\cdot(-2)=6$, so $6\div(-3)=-2$ and $6\div(-2)=-3$.
$-3\cdot 2=-6$, so $-6\div 2=-3$ and $-6\div(-3)=2$.

Notice that the rules for the division of signed numbers correspond to the rules for multiplication.

RECALL

$161\div 7=23$

dividend | divisor | quotient

DIVIDING POSITIVE & NEGATIVE NUMBERS

1. If the dividend and divisor have like signs, then the quotient is positive.
2. If the dividend and divisor have unlike signs, then the quotient is negative.

$0\div 18=0$ because $0\cdot 18=0$. What is $18\div 0$? Since the product of any number and 0 is 0, there is no number whose product with 0 is 18. Therefore, division by 0 is undefined.

Two numbers whose product is 1 are **reciprocals** (or **multiplicative inverses**). Note that 5 and $\frac{1}{5}$ are reciprocals since $5\cdot\frac{1}{5}=1$. Similarly, $-\frac{2}{3}$ and $-\frac{3}{2}$ are reciprocals since $-\frac{2}{3}\cdot\left(-\frac{3}{2}\right)=1$.

DEFINITION

Division is defined as multiplication by the reciprocal of the divisor. For any real numbers a and b, where $b\neq 0$, $a\div b=\frac{a}{b}=a\cdot\frac{1}{b}$.

EXAMPLE 4: Dividing Rational Numbers

Find each quotient.

a. $\frac{4}{5}\div\frac{9}{10}$ **b.** $-1\frac{7}{12}\div 7\frac{3}{5}$ **c.** $-1.26\div(-0.3)$

Answers

a. $\frac{4}{5}\div\frac{9}{10}=\frac{4}{\not 5}\times\frac{\not{10}^{\,2}}{9}$

1. Rewrite the division as multiplication by the reciprocal.

$$=\frac{8}{9}$$

2. The product of 2 factors with the same sign is positive.

b. $-1\frac{7}{12}\div 7\frac{3}{5}=-\frac{19}{12}\div\frac{38}{5}$

1. Convert the mixed numbers to improper fractions before dividing.

$$=-\frac{\not{19}}{12}\times\frac{5}{\not{38}_{\,2}}$$

2. Rewrite the division as multiplication by the reciprocal.

$$=-\frac{5}{24}$$

3. The product of 2 factors with different signs is negative.

c. $-1.26 \div (-0.3)$

$$\begin{array}{r} 4.2 \\ 0.3\overline{)\,1.26} \\ \underline{12} \\ 06 \\ \underline{6} \\ 0 \end{array}$$

1. Divide the absolute values.

$-1.26 \div (-0.3) = 4.2$

2. The quotient of 2 negatives is positive.

SKILL CHECK **EXERCISES 13, 15** ☑

A **complex fraction** is a fraction whose numerator and/or denominator contains a fraction. A complex fraction can be simplified by first rewriting the division as multiplication.

EXAMPLE 5: Dividing Rational Numbers

Divide.

a. $\dfrac{-\frac{4}{5}}{-\frac{1}{2}}$ **b.** $\dfrac{\frac{4}{3}}{-2}$

Answers

a. $-\frac{4}{5} \div \left(-\frac{1}{2}\right) = -\frac{4}{5}\left(-\frac{2}{1}\right)$

$= \frac{8}{5}$

1. Rewrite the division as multiplication by the reciprocal.
2. The product of 2 factors with the same sign is positive.

b. $\frac{4}{3} \div (-2) = \frac{\overset{2}{\cancel{4}}}{3}\left(-\frac{1}{\cancel{2}}\right)$

$= -\frac{2}{3}$

1. Rewrite the division as multiplication by the reciprocal.
2. The product of 2 factors with different signs is negative.

SKILL CHECK **EXERCISE 21** ☑

AFTERSCHOOLHELP

Multiplying & Dividing Rational Numbers

EXAMPLE 6: Solving a Real-World Problem

The Rochester long-distance relay race covers $18\frac{3}{5}$ mi. How far must each of the 6 team members run if each runner is required to run the same distance?

Answer

$18\frac{3}{5} \div 6 = \frac{93}{5} \div 6$

$= \frac{\overset{31}{\cancel{93}}}{5} \times \frac{1}{\underset{2}{\cancel{6}}} = \frac{31}{10} = 3.1 \text{ mi}$

1. Divide the total by 6 since there are 6 runners.
2. Rewrite the division as multiplication by the reciprocal.

SKILL CHECK **EXERCISE 35** ☑

VOCABULARY

- complex fraction
- division
- multiplicative inverse
- reciprocal

A. EXERCISES

Fill in the blank.

1. If 2 factors have different signs, the product is ______.
2. The ______ Property tells us that the product of a number and its reciprocal is 1.

State the missing number and name the illustrated property.

3. $5 \cdot (\_\_\_\_\_\_ \cdot 7) = (5 \cdot 4) \cdot 7$
4. $(5 \cdot 4) \cdot 7 = (\_\_\_\_\_\_ \cdot 5) \cdot 7$

Find each product.

5. $3(-2)$
6. $-7(-6)$
7. $-8 \cdot 6$
8. $\frac{4}{9} \times \frac{3}{5}$
9. $\frac{-1}{3}\left(-\frac{9}{10}\right) \cdot 3$
10. $-2(-8)(-10)$

Find each quotient.

11. $-27 \div (-3)$
12. $-56 \div 8$
13. $54 \div -0.6$
14. $-\frac{4}{7} \div (-2)$
15. $\frac{1}{4} \div \frac{9}{16}$
16. $\frac{0}{-37}$

B. EXERCISES

17. State the property of multiplication that justifies each step in the following simplification.

$\frac{5}{12} \cdot (-9) \cdot \frac{12}{5}$

a. $= \frac{5}{12} \cdot \frac{12}{5} \cdot (-9)$ ______

b. $= 1(-9)$ ______

c. $= (-9)$ ______

Perform each indicated operation.

18. $-\frac{4}{11} \cdot \frac{7}{12}$
19. $\frac{\frac{2}{17}}{8}$
20. $61 \div 12.2$
21. $-3.9(-8.42)$
22. $37(-24)$
23. $-\frac{1}{10}\left(-\frac{15}{4}\right)$
24. $-6.09 \div (-7)$
25. $-3\frac{7}{9}\left(-\frac{3}{17}\right)$
26. $2\frac{16}{19} \div \frac{18}{7}$
27. $-5\frac{3}{5} \div 2\frac{1}{3}$
28. $\frac{\frac{2}{3}}{-\frac{5}{4}}$
29. $-2\frac{5}{6} \div \left(-3\frac{2}{5}\right)$
30. $-\frac{4}{7}\left(-\frac{16}{5}\right)\left(\frac{3}{11}\right)\left(-\frac{5}{14}\right)\left(\frac{7}{12}\right)$
31. $-2\frac{5}{8}\left(-2\frac{1}{3}\right)\left(-\frac{6}{7}\right)$

32. Kevon's bucket contains $2\frac{1}{2}$ gal of water. If he gives his horse $\frac{2}{3}$ of the water, how much water does the horse get?

33. Mason, Marisol, and Matt each picked $1\frac{3}{5}$ pt of strawberries. How many pints of strawberries did they pick altogether?

34. Ben and Max caught a total of $9\frac{1}{2}$ lb of fish and divided the fish evenly between them. How many pounds of fish did each get?

35. **Essential Question:** State the missing property of multiplication to show how to rewrite $6 \div \frac{3}{4}$ as a product.

Work	Reasoning
$6 \div \frac{3}{4} = \frac{6}{\frac{3}{4}}$	definition of division
$= \frac{6}{\frac{3}{4}} \times 1 = \frac{6 \times \frac{4}{3}}{\frac{3}{4} \times \frac{4}{3}}$	______
$= \frac{6 \times \frac{4}{3}}{1} = 6 \times \frac{4}{3}$	Multiplicative Inverse Property

C. EXERCISES

Calculate each product and write your answer as a fraction.

36. $-2\frac{11}{19} \cdot \frac{5}{7} \cdot 3.04$

37. $-3\frac{6}{11}(3.85)\left(-8\frac{1}{3}\right)$

Answer each question. When possible, express your answer as a mixed number.

38. The product of 2 numbers is 7904, and the absolute value of one of the numbers is 247. Find 2 possible solutions for the other number.

39. Is division commutative? Explain and give an example.

40. McKenna used fencing to create a square play enclosure for her rabbits with side lengths of $8\frac{5}{8}$ ft. If she wanted to use the same amount of material to create an enclosure in the shape of an equilateral triangle, how long would each side be?

41. MP3 Describe any errors that Roger made in his work or reasoning to simplify $\left(-\frac{1}{7} \cdot 19\frac{4}{9}\right) \cdot 7$.

Step	Roger's Work	Reasoning
1.	$\left(\frac{1}{7} \cdot 19\frac{4}{9}\right) \cdot 7$	Ignore the negative sign to start.
2.	$= \left(19\frac{4}{9} \cdot \frac{1}{7}\right) \cdot 7$	Associative Property
3.	$= 19\frac{4}{9} \cdot \left(\frac{1}{7} \cdot 7\right)$	Commutative Property
4.	$= 19\frac{4}{9} \cdot 1$	Multiplicative Inverse Property
5.	$= 19\frac{4}{9}$	Multiplicative Identity Property
6.	$= -19\frac{4}{9}$	An odd number of negative factors results in a negative product.

CUMULATIVE REVIEW

State the property of addition that justifies each statement. [1.1]

42. $(56 + 0) + (79 + 44) = 56 + (79 + 44)$

43. $56 + (79 + 44) = 56 + (44 + 79)$

44. $56 + (44 + 79) = (56 + 44) + 79$

Simplify each expression. [1.1]

45. $-9.07 + (-5.96)$

46. $\frac{7}{24} + \left(-\frac{5}{8}\right)$

47. $|-21 - 5| - (-9) + |3 - 8|$

48. $|-4.7 + 1.2| + (-6) - |5.3 - 9.1|$

49. $-\frac{3}{4} + \frac{5}{6} - \left(-\frac{2}{3}\right)$

Choose the addition property that is illustrated. [1.1]

50. $\left(\frac{2}{3} + \frac{4}{5}\right) + \frac{1}{5} = \frac{2}{3} + \left(\frac{4}{5} + \frac{1}{5}\right)$

A Associative
B Additive Identity
C Additive Inverse
D Commutative
E none of these

51. $22 + (-5 + 5) = 22 + 0$

A Associative
B Additive Identity
C Additive Inverse
D Commutative
E none of these

MIND OVER MATH

Find the digit that will properly fill in each box in the following multiplication problems.

1.

$$\begin{array}{r} \square\square\square \\ \times \quad 39 \\ \hline 47\square 3 \\ 1\square\square 1 \\ \hline 20{,}\square\square 3 \end{array}$$

2.

$$\begin{array}{r} 7\square\square\square \\ \times \quad \square 8 \\ \hline \square 81\square 2 \\ 14\square\square 8 \\ \hline 20\square{,}\square 92 \end{array}$$

3.

$$\begin{array}{r} \square 2\square \\ \times \square 81 \\ \hline \square\square 6 \\ \square 4\square 8 \\ \square 278 \\ \hline 16\square{,}\square 06 \end{array}$$

4.

$$\begin{array}{r} 5\square 4\square \\ \times \quad \square 3 \\ \hline 17\square\square 8 \\ \square\square\square\square 2 \\ \hline 43\square{,}0\square 8 \end{array}$$

CAREERS IN MATH
MATHEMATICIAN

What do you imagine are some of the top jobs in the US? Professional athlete? Media star? For years the Wall Street Journal has published annual evaluations of the country's top jobs by CareerCast.com. The rankings are based on work environment, income, future outlook, physical demands, and stress. Perennially in the top 5 (at times #1) is . . . mathematician! (In fact, 6 of the top 10 jobs in 2021 were mathematical in nature; we will discuss the #1 job later.)

You probably are asking, "What does a mathematician do?" According to the US Department of Labor, "Mathematicians use mathematical theory, computational techniques, algorithms, and the latest computer technology to solve economic, scientific, engineering, physics, and business problems." In other words, they are problem solvers.

Theoretical mathematicians focus on creating pure, abstract mathematical knowledge. They study unsolved problems and formulate new relationships from known mathematical ideas. On the other hand, applied mathematicians focus on applying known theories and techniques. They solve practical problems using mathematical modeling and computational methods. Without the discovery of calculus by Isaac Newton and Gottfried Leibniz, the physics of classical mechanics would not be possible. The development of fractals and chaos theory by Benoit Mandelbrot and others initiated current research of apparently random events in the physical, natural, and even social sciences.

If God has given you ability and interest in mathematics, prayerfully consider mathematics as a college major and future career choice. Whether your skills are used to benefit others directly or indirectly by increasing mathematical knowledge, you can use your work to glorify God and show your love for others.

1.3 EXPONENTS & ROOTS

Why is it possible to define consistent properties in mathematics?

After completing this section, I will be able to

- define *exponent*, *base*, *exponential form*, *square root*, and *cube root*.
- simplify expressions with integral exponents.
- evaluate the square root and cube root of a number.

Traffic engineers use exponents and roots when designing roads and intersections for efficient traffic flow. Exponents provide a way for very large or very small numbers to be represented in a concise way. Just as multiplication is a shorter way to write repeated addition, exponential form is a shorter way to represent repeated multiplication.

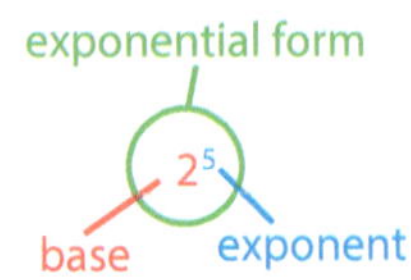

$$2 + 2 + 2 + 2 + 2 = 5(2) \qquad 2 \cdot 2 \cdot 2 \cdot 2 \cdot 2 = 2^5$$

$$-3 + (-3) + (-3) + (-3) = 4(-3) \qquad -3 \cdot (-3) \cdot (-3) \cdot (-3) = (-3)^4$$

RECALL

$\in$: "is an element of"

$\mathbb{N}$: the set of natural numbers

DEFINITIONS

For $x \in \mathbb{N}$, b^x is the product of x factors of b. The repeated factor, b, is the **base**, and x, which indicates the number of times b is used as a factor, is the **exponent**. When numbers are written in this way, they are said to be in **exponential form** or **exponential notation**.

The example above illustrates an important distinction between $(-3)^4$ and -3^4. The expression -3^4 represents the opposite of three to the fourth power:

$$-3^4 = -1 \cdot 3^4 = -1(3 \cdot 3 \cdot 3 \cdot 3) = -81.$$

However,

$$(-3)^4 = -3 \cdot (-3) \cdot (-3) \cdot (-3) = 81.$$

EXAMPLE 1: Evaluating Exponential Form

Evaluate each of the following exponential forms.

a. 9^3 **b.** $(-7)^4$ **c.** $-\left(\frac{4}{5}\right)^2$

Answers

a. $9^3 = 9 \cdot 9 \cdot 9 = 729$

b. $(-7)^4 = -7 \cdot (-7) \cdot (-7) \cdot (-7) = 2401$

c. $-\left(\frac{4}{5}\right)^2 = -\left(\frac{4}{5}\right) \cdot \left(\frac{4}{5}\right) = -\frac{16}{25}$

SKILL CHECK **EXERCISES 7, 15**

Consider what happens when exponential numbers with the same base are multiplied or divided.

$$8^3 \cdot 8^2 = (8 \cdot 8 \cdot 8)(8 \cdot 8) = 8 \cdot 8 \cdot 8 \cdot 8 \cdot 8 = 8^5$$

The resulting exponent (5) is the sum of the exponents (3 and 2) of the original base numbers.

$$5^6 \div 5^4 = \frac{5 \cdot 5 \cdot \cancel{5} \cdot \cancel{5} \cdot \cancel{5} \cdot \cancel{5}}{\cancel{5} \cdot \cancel{5} \cdot \cancel{5} \cdot \cancel{5}} = 5 \cdot 5 = 5^2$$

The resulting exponent (2) is the difference of the exponents (6 and 4) of the original base numbers.

What happens when an exponential expression is raised to a power?

$$(3^2)^4 = 3^2 \cdot 3^2 \cdot 3^2 \cdot 3^2 = 3^{2+2+2+2} = 3^8$$

The resulting exponent (8) is the product of the exponents (2 and 4) in the original expression. In all 3 examples above, we observe patterns that will consistently hold for all exponential expressions. As in Sections 1.1 and 1.2, we call these patterns *properties*.

PROPERTIES OF EXPONENTS		
Product Property	$x^a \cdot x^b = x^{a+b}$	To multiply like bases, add the exponents.
Quotient Property	$\frac{x^a}{x^b} = x^{a-b}$ for $x \neq 0$	To divide like bases, subtract the exponents.
Power Property	$(x^a)^b = x^{ab}$	To raise a power to a power, multiply the exponents.

These properties provide a foundation for simplifying expressions involving exponents.

EXAMPLE 2: Simplifying Exponential Form

Simplify, leaving each answer in exponential form.

a. $5^2 \cdot 5^{10}$ **b.** $3^2 \cdot 3^4 \cdot 3^9$ **c.** $2^4 \div 2$

d. $\frac{10^5}{10^3}$ **e.** $(5^3)^6$

Answers

a. $5^2 \cdot 5^{10} = 5^{2+10} = 5^{12}$ Apply the Product Property.

b. $3^2 \cdot 3^4 \cdot 3^9 = 3^{2+4+9} = 3^{15}$ Apply the Product Property.

c. $2^4 \div 2 = 2^{4-1} = 2^3$ Apply the Quotient Property.

d. $\frac{10^5}{10^3} = 10^{5-3} = 10^2$ Apply the Quotient Property.

e. $(5^3)^6 = 5^{3(6)} = 5^{18}$ Apply the Power Property.

SKILL CHECK **EXERCISES 17, 19**

The properties listed hold consistently and do not change randomly. This is because they are part of God's orderly creation. They also help us understand other mathematical principles.

The Quotient Property gives meaning to the expression x^0. Note that $\frac{x^2}{x^2} = x^{2-2} = x^0$.

Anything (except 0) divided by itself equals 1. Therefore, $x^0 = 1$ when $x \neq 0$.

The Quotient Property also gives meaning to exponential expressions with negative exponents. $\frac{x^2}{x^3} = \frac{x \cdot x}{x \cdot x \cdot x} = \frac{1}{x}$ and $\frac{x^2}{x^3} = x^{2-3} = x^{-1}$ by the Quotient Property. Therefore, $\frac{1}{x} = x^{-1}$. In general, $x^{-a} = \frac{1}{x^a}$ since $\frac{1}{x^a} = \frac{x^0}{x^a} = x^{0-a} = x^{-a}$.

DEFINITIONS

Any nonzero number to the **zero power** equals 1. That is, when $x \neq 0$, $x^0 = 1$.

Negative exponent: For any nonzero number x and integer a, $x^{-a} = \frac{1}{x^a}$.

CONNECT

If $a^{-1} = \frac{1}{a}$, then $\frac{1}{a^{-1}} = a$.

AFTERSCHOOLHELP

Exponents & Roots

EXAMPLE 3: Evaluating Expressions with Negative Exponents

Evaluate the following exponential forms.

a. $(-2)^{-4}$ **b.** -3^{-2} **c.** $\frac{1}{6^{-3}}$

Answers

a. $(-2)^{-4} = \frac{1}{(-2)^4} = \frac{1}{-2(-2)(-2)(-2)} = \frac{1}{16}$

b. $-3^{-2} = -\frac{1}{3^2} = -\frac{1}{3 \times 3} = -\frac{1}{9}$

c. $\frac{1}{6^{-3}} = 6^3 = 6 \cdot 6 \cdot 6 = 216$

SKILL CHECK **EXERCISES 9, 11**

The properties of exponents apply to expressions containing negative exponents as well.

EXAMPLE 4: Simplifying Expressions with Negative Exponents

Simplify. Express answers in positive exponential form.

a. $3^{-9} \cdot 3^7$ **b.** $2^{-3} \cdot 2^{-2}$ **c.** $4^{-2} \div 4^{-3}$ **d.** $(5^{-3})^4$

Answers

a. $3^{-9} \cdot 3^7 = 3^{-9+7} = 3^{-2} = \frac{1}{3^2}$

b. $2^{-3} \cdot 2^{-2} = 2^{-3+(-2)} = 2^{-5} = \frac{1}{2^5}$

c. $4^{-2} \div 4^{-3} = 4^{-2-(-3)} = 4^{-2+3} = 4^1 = 4$

d. $(5^{-3})^4 = 5^{-3(4)} = 5^{-12} = \frac{1}{5^{12}}$

SKILL CHECK **EXERCISES 25, 27**

RADICALS
$\sqrt{n} = x$, if $x^2 = n$ $\sqrt[3]{n} = x$, if $x^3 = n$

The inverse operation for exponents is finding the root, denoted with a radical sign $\sqrt{\ }$. The **square root** is one of a number's 2 equal factors. Since 3^2 and $(-3)^2$ equal 9, then the 2 square roots of 9 are 3 and −3 (or ±3). The positive root of a number ($\sqrt{9} = 3$) is called the **principal root**. To indicate the negative square root, place a negative sign in front of the radical ($-\sqrt{9} = -3$).

Similarly, the **cube root** is one of a number's 3 equal factors. Since $2^3 = 8$, the cube root of 8, denoted $\sqrt[3]{8}$, is 2. However, unlike square roots, there exists only 1 real cube root of any real number.

VOCABULARY

- base
- cube root
- exponent
- exponential form
- exponential notation
- negative exponent
- principal root
- square root
- zero power

EXAMPLE 5: Evaluating Roots

Find the indicated roots.

a. $\sqrt{4}$ **b.** $\sqrt[3]{-64}$ **c.** $-\sqrt{\frac{4}{9}}$

Answers

a. $\sqrt{4} = \sqrt{2^2} = 2$

b. $\sqrt[3]{-64} = \sqrt[3]{(-4)^3} = -4$

c. $-\sqrt{\frac{4}{9}} = -\frac{\sqrt{4}}{\sqrt{9}} = -\frac{\sqrt{2^2}}{\sqrt{3^2}} = -\frac{2}{3}$

SKILL CHECK **EXERCISES 31, 33** ☑

A. EXERCISES

Correct the statement.

1. $5^0 = 0$ because anything to the zero power is 0.

Match the property demonstrated by each equation.

2. $3^7 \div 3^4 = 3^3$

3. $(3^7)^3 = 3^{21}$

4. $3^7 \cdot 3^3 = 3^{10}$

A Power Property

B Product Property

C Quotient Property

Evaluate each expression.

5. 4^3

6. $\frac{1}{2^{-4}}$

7. $(-5)^3$

8. 4^{-2}

9. $(-3)^{-3}$

10. -2^{-4}

11. $\frac{1}{8^{-2}}$

12. $\sqrt{49}$

13. $\sqrt[3]{8}$

14. $-\sqrt{25}$

15. $-\left(\frac{1}{2}\right)^2$

16. $\sqrt{81}$

B. EXERCISES

Simplify each expression. Express answers in exponential form.

17. $(-12)^3(-12)^2$

18. $3^3 \cdot 3$

19. $8^7 \div 8^3$

20. $(8^{-3})^5$

21. $(3^{11})^0$

22. $12^2 \cdot 12^{-7}$

23. $(10^{10})^5$

24. $4^{16} \div 4^{-2}$

25. $10^{12} \cdot 10^{-3}$

26. $7^{-2} \div 7^9$

27. $\frac{4^{-2}}{4^{-9}}$

28. $2^2 \cdot 2^3 \cdot 3^4 \cdot 5^2 \cdot 5$

Evaluate each expression.

29. $\left(\frac{7}{10}\right)^2$

30. $\left(\frac{4}{3}\right)^3$

31. $\sqrt[3]{-125}$

32. $-\sqrt{\frac{4}{25}}$

33. $\sqrt{\frac{256}{121}}$

34. $\sqrt[3]{\frac{27}{64}}$

35. **MP3** State the property that verifies each step.

$(2^2)^3 \cdot 2^4 \div 2^7$

a. $= 2^6 \cdot 2^4 \div 2^7$ ______

b. $= 2^{10} \div 2^7$ ______

c. $= 2^3$ ______

36. **Essential Question:** Why can we define and use consistent properties in mathematics?

C. EXERCISES

Simplify. Express answers using positive exponents.

37. $\dfrac{2^3 \cdot 3^{10} \cdot 2^7 \cdot 5^3}{2^4 \cdot 5^4 \cdot (3^5)^2}$

38. $\dfrac{2^4 \cdot 7^8 \cdot (11^0)^2}{2^3 \cdot (3^2)^3 \cdot 3^5}$

39. $[3 \cdot 2^{-2}(2^5)]^{-4}$

40. $(-2 \cdot 5^{-3} \cdot 7^4)^{-2}$

41. In exponential form, what is the product of $y^5 \cdot y^0$? Use that product and the Multiplicative Identity Property to show that $y^0 = 1$.

Evaluate the cube roots.

42. $\sqrt[3]{-\frac{343}{1000}}$

43. $\sqrt[3]{\frac{128}{250}}$

CUMULATIVE REVIEW

Simplify each expression. [1.1–1.2]

44. $37 - 85$

45. $126 + 395$

46. $-146 - 29 - (-35)$

47. $-\frac{135}{5}$

48. $-38 + (-27)$

49. $23 + (-82)$

50. $-83(-24)$

51. $2(-10)(-82)$

52. Which expression is equivalent to $7 - (-10) - |-8|$? [1.1]

A $7 + 10 + 8$

B $7 - 10 + 8$

C $17 - 8$

D $-3 + 8$

E none of these

53. Which expression is not equivalent to $-18 - 37$? [1.1]

A -45

B 45

C $-18 + (37)$

D all of these

E none of these

1.4 ORDER OF OPERATIONS

Why do I need to follow a specific order of operations?

After completing this section, I will be able to

- state the order of operations.
- evaluate numerical expressions by using the order of operations.
- translate word phrases into numerical expressions.

Evaluating the expression $3 + 2 \cdot 4 - 6 \div 3$ can yield many different answers ($\frac{14}{3}$, $\frac{5}{3}$, $-\frac{10}{3}$, 9, or 18) even though each step is performed correctly. Order is just as important in simplifying expressions as in performing the individual operations. Mathematicians have agreed that numerical expressions without grouping symbols are evaluated in the following order.

G	Grouping Symbols	
E	Exponents (& Square Roots)	
M	Multiplication	*In order from left to right* →
D	Division	
A	Addition	*In order from left to right* →
S	Subtraction	

1. Evaluate all exponential expressions.
2. Evaluate all multiplication and division from left to right.
3. Evaluate all addition and subtraction from left to right.

Following the established order provides consistent results. Since the above expression has no grouping symbols and no exponential expressions, multiplication and division should be performed first, followed by addition and subtraction.

$3 + 2 \cdot 4 - 6 \div 3$
$= 3 + 8 - 2$
$= 11 - 2$
$= 9$

EXAMPLE 1: Evaluating Expressions

Evaluate each expression.

a. $2 + 3^2 - 4 \div 2 \cdot 3$

b. $2^3 - 1^2 + 5 \cdot 4 - 3^2 + 9 \div 3$

Answers

a. $2 + 3^2 - 4 \div 2 \cdot 3$ — **1.** Evaluate the exponential expression first.

$2 + 9 - 4 \div 2 \cdot 3$

$2 + 9 - 2 \cdot 3$ — **2.** Evaluate all multiplication and division from left to right.

$2 + 9 - 6$

5 — **3.** Evaluate all addition and subtraction from left to right.

b. $2^3 - 1^2 + 5 \cdot 4 - 3^2 + 9 \div 3$ — **1.** Evaluate the exponential expressions first.

$8 - 1 + 5 \cdot 4 - 9 + 9 \div 3$

$8 - 1 + 20 - 9 + 3$ — **2.** Evaluate all multiplication and division from left to right.

21 — **3.** Evaluate all addition and subtraction from left to right.

SKILL CHECK **EXERCISES 11, 13**

Sometimes it may be necessary to perform a sequence of operations in an order that is different from the one prescribed by the order of operations. For instance, 2 numbers might need to be added before being divided. Grouping symbols indicate which operation is to be performed first.

Parentheses ( ), brackets [ ], and braces { } are used to indicate that the operations enclosed by them are to be performed prior to any other operations. Absolute value symbols | |, the fraction bar, and the radical sign are also used as grouping symbols.

ORDER OF OPERATIONS
1. Evaluate within the innermost grouping symbols first.
2. Evaluate all exponential expressions.
3. Evaluate all multiplication and division from left to right.
4. Evaluate all addition and subtraction from left to right.

EXAMPLE 2: Evaluating Expressions with Grouping Symbols

Evaluate each expression.

a. $2+\sqrt{3(5+7)}\div 2$ **b.** $\dfrac{2(-3+7)-(7-5)}{5-3(1-4)}$ **c.** $9-3|2+7\cdot 2^3(3-5)|-4$

Answers

a. $2+\sqrt{3(5+7)}\div 2$	**1.** Evaluate within the grouping symbols first.
$2+\sqrt{3(12)}\div 2$	
$2+\sqrt{36}\div 2$	
$2+6\div 2$	
$2+3$	**2.** Divide.
5	**3.** Add.
b. $\dfrac{2(-3+7)-(7-5)}{5-3(1-4)}$	**1.** Evaluate within the grouping symbols first.
$\dfrac{2(4)-2}{5-3(-3)}$	
$\dfrac{8-2}{5+9}$	**2.** Evaluate the multiplication indicated in both the numerator and the denominator.
$\dfrac{6}{14}$	**3.** Evaluate the addition and subtraction.
$\dfrac{3}{7}$	**4.** Reduce the resulting fraction.
c. $9-3\|2+7\cdot 2^3(3-5)\|-4$	**1.** Evaluate within the innermost grouping symbols first.
$9-3\|2+7\cdot 2^3(-2)\|-4$	
$9-3\|2+7\cdot 8(-2)\|-4$	**2.** Evaluate within the grouping symbols by evaluating the exponential expression, multiplication, and then addition.
$9-3\|2-112\|-4$	
$9-3\|-110\|-4$	
$9-3(110)-4$	**3.** Evaluate the absolute value, which functions as a grouping symbol.
$9-330-4$	**4.** Evaluate the resulting expression by performing the multiplication and then addition and subtraction.
-325	

SKILL CHECK **EXERCISES 23, 29, 31**

AFTERSCHOOLHELP

Order of Operations

Translating simple numerical expressions such as the difference of 4 and 3 $(4 - 3)$ or the quotient of 5 and 6 $(5 \div 6)$ is a relatively simple task. However, combining several of these operations into one translated expression and including squares and roots becomes more challenging.

When translating word phrases into numerical expressions, it is important to look for key words that will indicate the operation being performed.

ADDITION	SUBTRACTION	MULTIPLICATION	DIVISION
added to	subtracted from	multiplied by	divided by
sum	difference	product	quotient
increased by	decreased by	times	ratio of
more than	less than	twice (×2)	halved (÷2)
plus	minus	doubled (×2)	thirds (÷3)
gained	lost	tripled (×3)	sixths (÷6)

EXAMPLE 3: Translating a Word Phrase

Write the word phrase as a numerical expression.

a. the difference of 5 times the square root of 9 and 10

b. 3 more than the absolute value of the difference of 5 and 2 cubed

c. 2 squared times the difference of 4 and 3 plus 12

Answers

a. the difference of 5 times the square root of 9 and 10
$5 \times \sqrt{9} - 10$

b. 3 more than the absolute value of the difference of 5 and 2 cubed
$|5 - 2^3| + 3$

c. 2 squared times the difference of 4 and 3 plus 12
$2^2 \times (4 - 3) + 12$

RECALL

"3 squared" → 3^2
"4 cubed" → 4^3

SKILL CHECK **EXERCISES 37, 39**

A. EXERCISES

State the first 2 operations performed to evaluate each expression.

1. $18 \div 3 \cdot 9 + 11$
2. $1 + (3 - 7)^2 \cdot 8$
3. $2 + \sqrt{100} \div 5$

Evaluate each expression.

4. $5 + 3 \cdot 4$
5. $10 + 7 - 3 \cdot 2$
6. $8 + 10 \div 5$
7. $1 + 18 \div 3(-2)$
8. $-10 - 2 \cdot 4 \div 2 + 3$
9. $1 + (-7)^2 \cdot 2 - 50$
10. $27 - 3^2 + 8 \div (-2)$
11. $-81 - 14 \cdot 3 \div 2 \cdot 3$
12. $3 + 4 - 2 - 6 \cdot 3$
13. $7 - 3^2 - 1 + 4^3 \div 2$
14. $5 + 4(-3 + 7)$
15. $6 \cdot 3 \div (5 + 4) + 1 - 6$

B. EXERCISES

Rewrite each expression using the definition of subtraction from Section 1.1. Evaluate.

16. $1 - (-3) - 11 + 17 - 3$
17. $11^2 - 5 - (-4)$
18. Based on the rewritten expressions from exercises 16–17, why can addition and subtraction be evaluated from left to right rather than doing all adding first?

Rewrite each expression using the definition of division from Section 1.2. Evaluate.

19. $12 \div 6 \cdot 4 \div 8$
20. $3 \cdot 5 \div \frac{3}{4} \div -2$
21. Based on the rewritten expressions from exercises 19–20, why can multiplication and division be evaluated from left to right rather than doing all multiplying first?
22. **Essential Question:** Why is it important to follow a specific order of operations?

Evaluate each expression.

23. $-2|5 - 3(2)| + 5^0$
24. $-3^2 - 5|4 + 2(-3)^2|$
25. $3(-1)^2 - 2(-1) - 5$
26. $-\frac{2}{3} + \frac{1}{2} - \frac{3}{4} \cdot \frac{5}{6}$
27. $6 + 8(2 - 5) - [3 + (4 - 7)3] + 6$
28. $4 - 5 + 6(8 + 2 \cdot 4 - 5)$
29. $\sqrt{10^2 - 4(-3)(-3)}$
30. $\sqrt{(2 - 7)^2 + [3 - (-9)]^2}$
31. $\dfrac{6 + (2 - 5)8}{-7 + [4 + (4 - 7)3]}$
32. $\dfrac{2^2 + (2 - 3)6}{2^3 - 9(3 - 2)}$
33. $2\{[4 - (5 - 2^2)](-3 - 9) \div 9 - 3\}$
34. $3 - |2(5 - 9) \div 4 + (6 - 4) \div 2|^5$

Translate each phrase into a numerical expression. Then simplify your expression.

35. the product of five and the sum of six and seven
36. the square of the difference of twenty and thirteen

37. the difference of the squares of twenty and thirteen

38. five more than twice the cube root of eight

39. the product of five and six increased by the quotient of twenty and five

40. seven less than the product of eleven and the reciprocal of nine

C. EXERCISES

Evaluate each expression.

41. $\frac{7 - \sqrt{(-7)^2 - 4(6)(2)}}{2(6)}$

42. $5[3 + 5 \div 2^2 - (4 - 7)^2(-3 + 6)^{-2} - 2]$

43. $[-2(3^{-2}) + 6^{-2} \div 4^{-2}]\frac{1}{9^{-3}}$

44. $-\sqrt{\frac{[2 - (-3)]5}{(1 - 7)^2}} + \sqrt[3]{\frac{10^2 + 5^2}{-4 + 4\sqrt{9}}}$

CUMULATIVE REVIEW

Evaluate each expression. [1.1–1.2]

45. $-8 + |-10 + (-9)| + 17$

46. $-\{-[-(-2)]\} - |-8| - 6 - (-3)$

47. $-\frac{3}{4} - \left(-\frac{8}{9}\right)$

48. $\frac{-\frac{6}{11}}{\frac{2}{7}}$

Simplify. Express each answer in exponential form. [1.3]

49. $8^{15} \div 8^5$

50. $\frac{1}{3^2} \cdot \frac{1}{3^3}$

51. $(28^{-4})^{-2}$

52. $8^{-3} \cdot \frac{1}{8^5}$

53. Which property gives the result in the second line of the following simplification? **[1.1–1.2]**
$-4[11 + (-11)]$
$= -4(0)$
$= 0$

- A Additive Inverse Property
- B Zero Property of Multiplication
- C Additive Identity Property
- D Multiplicative Inverse Property
- E none of these

54. Which of the following is true? **[1.1–1.2]**

- A $|4 + (-7)| = |4| + |-7|$
- B $|4 - (-7)| = |4| - |-7|$
- C $|4(-7)| = |4| \cdot |-7|$
- D all of these
- E none of these

TECHNOLOGY CORNER (TI-84 PLUS FAMILY)

Calculator Settings & Evaluating Expressions

To evaluate an expression using your graphing calculator, enter the expression as you would when writing it on paper. Pressing [enter] (regardless of the cursor location) will evaluate the expression according to the order of operations.

Calculate $\pi(5^2) - 3(2)$.

[2nd][^]5[x²][-]3[(]2[)][enter]

The calculator recognizes implied multiplication in expressions like 2π. Be sure to use the [-] key for subtraction and the [(-)] key for negation.

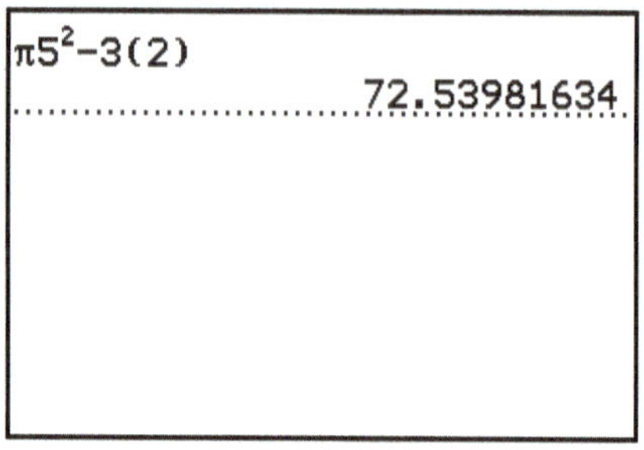

Mode Settings

Modes affect the way numbers are displayed and expressions are evaluated. To change the modes on your calculator, press [mode], use the cursor keys to highlight the desired setting, and press [enter].

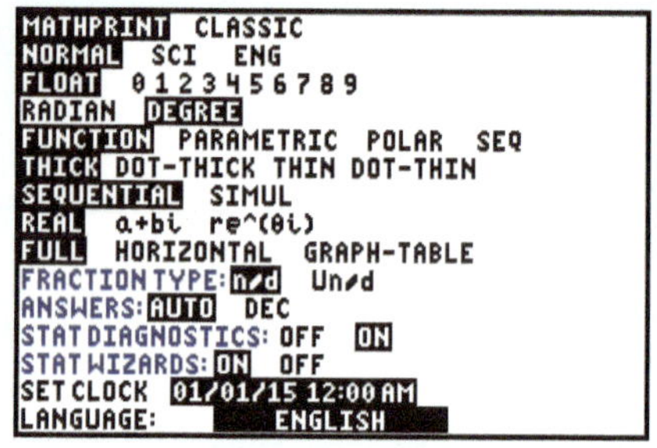

You can choose between *normal*, *scientific*, or *engineering* notation. Decimal numbers can be displayed with a *fixed* or *floating* number of decimal places. For now, be sure your calculator has *degree* mode selected for angles and is set to use *real* numbers. You can choose whether improper fractions are displayed as mixed numbers (*Un/d*) and whether all fractional results should be displayed as decimals (*dec*).

Function mode will typically be used for graphing, and you can choose whether to plot multiple graphs *sequentially* or *simultaneously*. You can also choose whether graphed points are connected (*thick*, *thin*) or dotted (*dot-thick*, *dot-thin*). *Horizontal* or *graph-table* split-screen modes can be chosen instead of the normal full-screen mode. Be sure to keep *stat diagnostics* and *stat wizards* on.

1.5 VARIABLES & ALGEBRAIC EXPRESSIONS

Why is it no surprise that we can describe the world mathematically?

After completing this section, I will be able to

- evaluate an algebraic expression with given variable values.
- translate word phrases into algebraic expressions.
- apply formulas to solve real-world problems.
- explain how a biblical view of creation accounts for the effectiveness of mathematics.

The previous section reviewed the simplifying of numerical expressions using the order of operations. Numerical expressions become algebraic expressions when at least 1 of the numbers is replaced by a variable.

NUMERICAL		ALGEBRAIC
$2(3) + 4^2$	→	$2x + 4^2$

DEFINITIONS

A **variable** is a symbol used to represent any element of a given set of numbers.
The set of numbers is the **domain** of the variable.
A **constant** is a fixed number.
An **algebraic expression** is a mathematical expression containing numbers, variables, and operational signs.

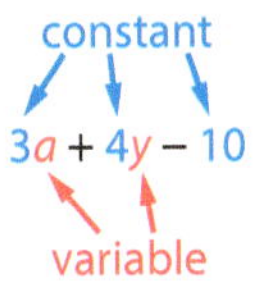

Suppose an amusement park offers ticket prices as displayed in the table. If a represents the number of regular tickets sold, the expression $35a$ would represent the amount of money collected from the regular ticket sales. To evaluate an algebraic expression, substitute values for the variable and simplify the result.

Amusement Park Ticket Prices	
regular (10–55 yr old)	\$35
senior adult (>55 yr old)	\$20
child (<10 yr old)	\$15

EXAMPLE 1: Evaluating Algebraic Expressions

Evaluate the expression $35a$ for each member of the domain $\{2, 5, 8\}$.

Answer

If $a = 2$, then $35a = 35(2) = 70$.

If $a = 5$, then $35a = 35(5) = 175$.

If $a = 8$, then $35a = 35(8) = 280$.

The corresponding set of values is $\{70, 175, 280\}$.

SKILL CHECK **EXERCISE 5**

Evaluating algebraic expressions that contain more than 1 variable means substituting values for all variables and simplifying the result.

EVALUATING AN ALGEBRAIC EXPRESSION

1. Substitute the given value(s) for the variable(s).
2. Evaluate by following the order of operations.

EXAMPLE 2: Evaluating Algebraic Expressions

Evaluate the expression $35a + 20b + 15c$ to determine the amusement park's ticket sales for 1 hr if $a = 250$ (regular tickets), $b = 18$ (senior adult tickets), and $c = 48$ (child tickets).

Answer

$$\begin{aligned} 35a + 20b + 15c &= 35(250) + 20(18) + 15(48) \\ &= 8750 + 360 + 720 \\ &= 9830 \end{aligned}$$

The total ticket sales for 1 hr is \$9830.

SKILL CHECK **EXERCISE 15**

The ability to translate phrases into algebraic expressions is foundational to success in algebra. Note that "a number" indicates an unknown value to be represented by a variable.

EXAMPLE 3: Writing Algebraic Expressions

Translate each word phrase into an algebraic expression.

a. one-third of the square of a number

b. the sum of three and eight times a number

c. eight less than three times a number

d. the quotient of the difference of a number and eight, and twice the number

Answers

a. $\frac{1}{3}n^2$ or $\frac{n^2}{3}$

b. $3 + 8n$

c. $3n - 8$

d. $\frac{n-8}{2n}$

SKILL CHECK **EXERCISES 11, 21**

Algebraic expressions provide an effective way to represent relationships found in God's orderly creation.

If the first number in a sequence of consecutive integers is n, then the following consecutive integers would be $n + 1$, then $n + 2$, and so on. The pattern for consecutive even or odd integers would be n, $n + 2$, $n + 4$, etc.

EXAMPLE 4: Writing Algebraic Expressions

Write an expression for the sum of 3 consecutive integers.

Answer

n = the first integer
$n + 1$ = the second integer
$n + 2$ = the third integer

1. Represent each of the numbers with an expression.

$n + (n + 1) + (n + 2)$

2. The sum is the result of adding the numbers.

SKILL CHECK **EXERCISE 25**

AFTERSCHOOLHELP

Variables & Algebraic Expressions

Geometric relationships can also be modeled algebraically. Consider the supplement and complement of an angle. Recall that supplementary angles are 2 angles whose sum is 180° and complementary angles are 2 angles whose sum is 90°. If an angle has a measure of a degrees, then its supplement measures $(180 - a)$ degrees and its complement measures $(90 - a)$ degrees.

EXAMPLE 5: Writing Algebraic Expressions

Write an expression for the difference of twice an angle and its complement.

Answer

a = the measure of the angle
$90 - a$ = the measure of its complement

1. Represent the angle and its complement with expressions.

$2a - (90 - a)$

2. Translate the word phrase into an algebraic expression.

SKILL CHECK **EXERCISES 27, 29**

Evaluating expressions provides valuable training in working with real-world applications and formulas. God created an orderly world in a spatial and quantifiable way. Since He also created us with the ability to observe and measure, it is no surprise that mathematics is very effective at modeling real-world relationships. Consistent relationships are often modeled with a formula.

DEFINITION

A **formula** is an equation that expresses the relationship between related quantities.

COMMON PERIMETER & AREA FORMULAS		
rectangle	$P = 2l + 2w$	$A = lw$
square	$P = 4s$	$A = s^2$
circle	$C = 2\pi r$	$A = \pi r^2$
triangle	$P = a + b + c$	$A = \frac{1}{2}bh$

EXAMPLE 6: Applying a Formula

What is the total area Gladys needs to till if she wants the radius of her garden to be 12 ft?

Answer

$A = \pi r^2$	**1.** Write the formula for the area of a circle.
$= \pi(12)^2$	**2.** Substitute the length of the radius into the formula.
$= 144\pi \approx 144(3.14)$	**3.** Evaluate, using 3.14 to estimate π.
$= 452.16 \text{ ft}^2$	

Gladys will need to till approximately 452.16 ft^2 of the yard for her garden.

SKILL CHECK **EXERCISES 33, 35**

VOCABULARY

- algebraic expression
- constant
- domain of the variable
- formula
- variable

USING FORMULAS

1. Write the correct formula.
2. Substitute the given values for the variables in the formula.
3. Evaluate the expression.

A. EXERCISES

List the variables and constants in each expression.

1. $3x - 6$

2. $3ab$

3. πr

Evaluate each expression for each member of the domain {−4, 1, 3, 8}.

4. $x - 4$

5. $2a + 4$

6. $|b^2 - 10|$

Evaluate when $x = 4$, $y = -2$, and $h = -1$.

7. $\frac{x + y}{h}$

8. $x^2 + h - 3$

9. $\frac{x}{y} + \frac{4}{5}h$

10. $\frac{3x^2 - y}{h}$

Translate each word phrase into an algebraic expression.

11. the product of eight and a number

12. a number less ten

13. twenty-four greater than a number

14. the sum of twice a number and six

B. EXERCISES

Evaluate xy^2z for the following values.

15. $x = 10, y = \frac{2}{7}$, and $z = \frac{1}{5}$

16. $x = a, y = a$, and $z = a$

17. $x = c, y = c^3$, and $z = c^{-2}$

Evaluate when $x = \frac{1}{2}, y = -\frac{1}{4}, z = 1$, and $h = \frac{1}{3}$.

18. $(x + y)(x - y)$

19. $h^2 + y^2 - x^3$

20. $3x^2y - \frac{z}{5}$

Translate each word phrase into an algebraic expression.

21. a number cubed plus the same number squared

22. the reciprocal of the absolute value of a number

23. the average of the three numbers p, q, and r

24. the sum of three consecutive odd numbers

25. 126 diminished by the product of two consecutive integers

26. three less than the product of two consecutive even integers

27. three times the complement of an angle

28. the sum of an angle's complement and its supplement

29. the difference of an angle's supplement and the measure of the angle itself

30. the perimeter of a rectangle 24 units long and w units wide

31. the amount Evelyn makes the week she gives x haircuts at \$25 each and y perms at \$90 each

32. the number of points a basketball player gets in one game when he makes x three-point shots, y two-point shots, and z free throws (at one point each)

Solve using the formula $R = xp$, where R is revenue (business income), x is the number of items sold, and p is the price per item.

33. If 48 cans of Soda's Peak are sold at \$0.79 per can, how much revenue is collected?

34. The senior class washed 110 cars and charged \$6.50 per car. How much money did they raise?

35. Cycle Path sold 33 bicycles for \$315 each. How much revenue did the company receive?

Use the given formula to answer each question.

36. The formula $d = rt$ calculates distance as the rate multiplied by the time. If a bus carrying spectators to a soccer game travels 47 mi/hr for 2 hr, how far does the bus travel?

37. The surface area of mailbox doors in Mrs. Bailey's neighborhood is represented by $A = w^2 + \frac{1}{2}\pi\left(\frac{w}{2}\right)^2$, where w represents the width of the mailbox. Mrs. Bailey wants to paint the door of her 8-inch-wide mailbox yellow to match her Dutch Master Trumpet Daffodils. How many square inches of surface area will she paint? Use $\pi \approx 3.14$.

38. **Essential Question:** Why is it no surprise that we can describe the world mathematically?

C. EXERCISES

Evaluate when $a = -\frac{2}{3}$, $b = \frac{4}{7}$, and $c = -3$.

39. $|a - b| - \frac{c^2}{b}$

40. $\sqrt{14abc} + \left(\frac{1}{a}\right)^2$

Use the given formula to answer each question.

41. The amount of simple interest Shanda earns on the money she saves from her babysitting jobs is expressed by $I = Prt$. In this formula, I is simple interest, P is the principal (or amount invested), r is the percentage rate of the investment per year in decimal form, and t is the time or length of the investment in years. If Shanda invests $450 at 5% interest, how much interest will she earn in 1 yr?

42. A formula for ideal weight in pounds is $w = 5b + \frac{b}{5}(h - 60)$, where b represents target BMI and h represents height in inches. If Tyler is 5 ft 10 in. tall and is aiming for a BMI of 20, what will his ideal weight be?

43. Chris plans to paint a cylindrical gasoline tank. He finds the number of square feet of surface area on the cylinder using the formula $S = 2\pi rh + 2\pi r^2$, where r is the radius of the cylinder and h is its height. If the height of the cylinder is 10 ft and its diameter is 4 ft, what is the total surface area Chris will paint? Use $\pi \approx 3.14$.

44. If a circle has radius r, find the effect on the circumference and the area when the radius is increased by 50%.

CUMULATIVE REVIEW

Name the property each statement illustrates. [1.1–1.2]

45. $t(sq) = (ts)q$

46. $n\left(\frac{1}{n}\right) = 1$

47. $p + v = v + p$

Simplify each expression. Express answers using positive exponents. [1.3]

48. $\frac{2^7}{2^{-3}}$

49. $(3^7)^{-3}$

50. $\frac{(5^3)^5 5^{-2}}{5^4}$

Evaluate each expression. [1.4]

51. $3 + \{[(2^5 - 2) + 2 + 7 \cdot 4] - 7 \cdot 2\} - 8$

52. $1 + [2(4 + 2)^2 - 6 + (3 - 5)^3] - 8 \cdot 6$

53. Which of these numbers has a multiplicative inverse that is different from the number itself? **[1.2]**

A -1

B 1

C 2

D all of these

E none of these

54. Which expression is equivalent to $\left(\frac{1}{11^4}\right)^3$? **[1.3]**

A $-(11^4 \cdot 11^4 \cdot 11^4)$

B $11^{-4 + (-4) + (-4)}$

C $\frac{1}{11^4} + \frac{1}{11^4} + \frac{1}{11^4}$

D $3 \cdot 11^4$

E none of these

1.6 USING THE DISTRIBUTIVE PROPERTY

How is the Distributive Property useful in algebra?

After completing this section, I will be able to

- apply the Distributive Property to simplify algebraic expressions.

When purchasing new volleyball uniforms, Coach Jones found jerseys for \$27 and shorts for \$18. He calculated the cost of purchasing 12 uniforms in 2 different ways.

12(27 + 18)	and	12(27) + 12(18)
= 12(45)		= 324 + 216
= \$540		= \$540

The fact that both methods produce the same answer illustrates the Distributive Property of Multiplication over Addition.

DISTRIBUTIVE PROPERTY

For any real numbers a, b, and c, $a(b + c) = ab + ac$.

The Distributive Property is often used to simplify algebraic expressions.

EXAMPLE 1: Removing Parentheses

Simplify each expression.

a. $4(a + 3b)$ **b.** $-2(3x^2 - 4x + 5)$

Answers

a. $4(a + 3b)$ — **1.** Apply the Distributive Property.

$= 4(a) + 4(3b)$

$= 4a + 12b$ — **2.** Simplify.

b. $-2(3x^2 - 4x + 5)$ — **1.** Apply the Distributive Property.

$= -2(3x^2) - (-2)(4x) + (-2)(5)$

$= -6x^2 + 8x - 10$ — **2.** Simplify.

SKILL CHECK **EXERCISE 5**

When simplifying an algebraic expression, any *like terms* should be combined. A **term** is a constant, a variable, or the product of a constant and 1 or more variables. **Like terms** are terms that have the same variable(s) with corresponding variables having the same exponents. Each of the following expressions is an example of a term.

$4 \qquad x \qquad x^3 \qquad 4x^3 \qquad \frac{x^3}{4}$

Terms within an algebraic expression are separated by addition or subtraction. The expression $3x - 4y + 5x$ contains 3 terms. The terms $3x$ and $5x$ are like terms because they have the same variable, x, with the same exponent, 1. Combining the like terms produces a simpler equivalent expression, $8x - 4y$.

Combining like terms is actually an application of the Distributive Property, where $ax + bx = (a + b)x$.

EXAMPLE 2: Combining Like Terms

Simplify $7x^2 + 8x - 4x^2 + 10x$.

Answer

$7x^2 - 4x^2 + 8x + 10x$	**1.** Rearrange the terms.
$= (7 - 4)x^2 + (8 + 10)x$	**2.** Apply the Distributive Property.
$= 3x^2 + 18x$	

SKILL CHECK **EXERCISE 13**

When terms contain different powers, they are typically arranged in descending powers of the variable. Expressions with several variables are typically written with the terms in alphabetical order. Multiple variables within a single term are also written in alphabetical order. The constant term is usually listed as the last term.

A variable term's **coefficient** is its numerical factor. In the expression $3x^2 - 7x^2 + x^2$, the terms have coefficients of 3, −7, and 1. The Distributive Property implies that like terms can be combined by adding their coefficients:

$$3x^2 - 7x^2 + x^2 = (3 - 7 + 1)x^2 = -3x^2.$$

EXAMPLE 3: Combining Like Terms

Simplify each expression.

a. $8x^3 - x^2 + x^3 + 4x^3 - x^2 + 6x^3$

b. $4a^2b + 5ab - 10a^2b + 22ab - 8a + 3$

c. $4p^2qr + 2p^2q + pq^2 - 3pqr$

Answers

a.	$8x^3 - x^2 + x^3 + 4x^3 - x^2 + 6x^3$	**1.** Identify the like terms.
	$= 19x^3 - 2x^2$	**2.** Combine the like terms.
b.	$4a^2b + 5ab - 10a^2b + 22ab - 8a + 3$	**1.** Identify the like terms.
	$= -6a^2b + 27ab - 8a + 3$	**2.** Combine the like terms.

c. $4p^2qr + 2p^2q + pq^2 - 3pqr$
The expression is already simplified since there are no like terms.

SKILL CHECK **EXERCISE 21**

AFTERSCHOOLHELP

Using the Distributive Property

Some expressions require applying the Distributive Property to remove parentheses before combining like terms. An algebraic expression is in its **simplest form** when there are no parentheses and all like terms are combined.

EXAMPLE 4: Simplifying Expressions

Simplify each expression.

a. $(2x + 5) + (3x - 9) - (x - 6)$

b. $3k(k + 2) - k(k - 1)$

c. $7g^2 - [3g + 4h(2g - 8h)] + 5gh$

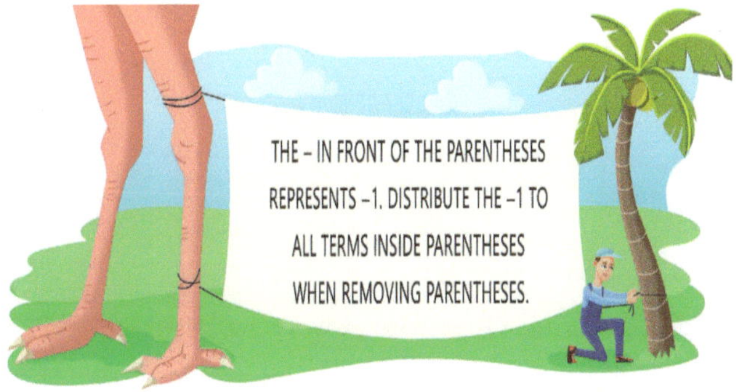

Answers

a. $(2x + 5) + (3x - 9) - (x - 6)$
$= 2x + 5 + 3x - 9 - x + 6$
$= 4x + 2$

1. Remove parentheses by distributing the coefficients of 1 and −1.
2. Combine the like terms.

b. $3k(k + 2) - k(k - 1)$
$= 3k^2 + 6k - k^2 + k$
$= 2k^2 + 7k$

1. Apply the Distributive Property to remove parentheses.
2. Combine the like terms.

c. $7g^2 - [3g + 4h(2g - 8h)] + 5gh$
$= 7g^2 - [3g + 8gh - 32h^2] + 5gh$
$= 7g^2 - 3g - 8gh + 32h^2 + 5gh$
$= 7g^2 - 3g - 3gh + 32h^2$

1. Apply the Distributive Property to remove the innermost parentheses.
2. Remove the brackets by distributing the coefficient of −1.
3. Combine the like terms.

SKILL CHECK **EXERCISE 23** ☑

VOCABULARY

- coefficient
- like terms
- simplest form
- term

A. EXERCISES

For each algebraic expression, list all numerical coefficients and then identify like terms.

1. $3x + 4y - 5y$

2. $8x^2 + 3x + 6x^2$

3. $xy - 2xy - 9xy$

4. $x^2yz - xy^2z + 3x^2yz$

Use the Distributive Property to eliminate the parentheses.

5. $4(x + 2y)$

6. $-2(3z + 7)$

7. $5x(x + y)$

8. $\frac{2}{3}x\left(\frac{1}{2}y - \frac{9}{4}\right)$

Combine like terms to simplify each expression.

9. $3a + 7a$

10. $19k - 13k$

11. $12xy + 4xy$

12. $12ab + 9ab - 7ab$

13. $y - 9y + 3y^2$

14. $6x^2 + 3x^2y - 4x + 8x^2$

B. EXERCISES

Simplify each expression.

15. $4a + (2a + 9)$

16. $5 - (2x - 18)$

17. $2x - (8y + x)$

18. $(x + y) - (x - y)$

19. $(x + y) + (x - y)$

20. $(7b - 8c) - (3b + 2c)$

21. $3x + 2y + (5x - 7y)$

22. $-5a - (3a + b) + (2a - 7b)$

23. $(2a - b) - (3a + 4b) + (6a + b)$

24. $-4(x + 3) + 5(x - 7)$

25. $2.46m + 1.6n - 8.4m$

26. $2.9x(x - 4) + 3.2x(x + 2)$

27. $4a - (2a^2 + 3z) + \frac{4}{5}(a^2 - 2a + 8)$

28. $x(x + y) - y(x + y)$

29. $x(x + y) + y(x + y)$

30. $4 - 3p - 6(p + 5) + 9$

31. $2r^2 + 5r + 3r - r(6r + 2)$

32. $2x^2 - 5x(4x - 3) - x(x^2 - 5x + 3)$

33. $-6[8x + 3(x - 4)]$

34. $7y + 3[y - 2(y + 1)] - 5$

35. **Essential Question:** How is the Distributive Property used in algebra?

C. EXERCISES

Find the missing expression to obtain the final expression.

36. $x^2 + x + 7 + \_\_\_\_\_ = 5x^2 + 3x - 8$

37. $3(a^2 + 4a - 8) + \_\_\_\_\_ = 4a^2 + 12a - 30$

Use the Distributive Property to rewrite the following formulas.

38. $A = \frac{1}{2}h(b_1 + b_2)$

39. $P = 2l + 2w$

40. $Ft = mv_1 - mv_2$

41. $C = \frac{5}{9}(F + 40) - 40$

CUMULATIVE REVIEW

Translate each word phrase into an algebraic expression. [1.5]

42. x increased by y

43. five degrees less than the outside temperature t

Simplify and then evaluate when $x = 2$ and $y = -3.7$. [1.5–1.6]

44. $4x + 3y - x + 6y - 8y$

45. $x^2 + 2x - 7x + 4$

State the property that justifies each statement. [1.1, 1.2, 1.6]

46. $3x^2 + x + 2x^2 - 3x = 3x^2 + 2x^2 + x - 3x$

47. $3x^2 + 2x^2 + x - 3x = 3x^2 + 2x^2 + 1x - 3x$

48. $3x^2 + 2x^2 + 1x - 3x = (3x^2 + 2x^2) + (1x - 3x)$

49. $(3x^2 + 2x^2) + (1x - 3x) = (3 + 2)x^2 + (1 - 3)x$

$(3 + 2)x^2 + (1 - 3)x = 5x^2 - 2x$ addition

50. Which expression is equivalent to $6 \cdot \frac{4}{7}$? **[1.2]**

A $\frac{4}{7} \div \frac{1}{6}$

B $\dfrac{6 \cdot \frac{4}{7}}{\frac{7}{4} \cdot \frac{4}{7}}$

C $(6 \cdot 4)\frac{1}{7}$

D all of these

E none of these

51. The expression $a^3 - 15$ can be translated as **[1.5]**

A fifteen less than the cube of a number.

B the difference of fifteen and a number cubed.

C thrice a number diminished by fifteen.

D a number cubed minus the opposite of fifteen.

E none of these

APPLICATION PROBLEMS
ENERGY COSTS

"Turn the lights off!" is a phrase you have likely heard over and over again. The Creation Mandate found in Genesis 1:28 demands that we not waste the resources God has provided for our use. Christ's example of gathering what was left after feeding the crowd in John 6:12–13 illustrates wise management of God's resources.

How much does it cost to leave that light on? Should you purchase the more expensive, energy-efficient appliance? We should seek to conserve energy when possible and carefully analyze marketers' claims to discern the best solutions to everyday problems.

An electric device is often labeled with the amount of power (measured in watts, W) required to operate it. The appliance's wattage indicates the rate at which energy is consumed. Power companies typically charge consumers for the number of kilowatt-hours (kWh) used. A wise manager of God's resources will be aware of the cost of operating an electric device.

Suppose we want to determine the daily, monthly, and yearly cost of operating a 65 W incandescent light bulb turned on for 8 hr a day if the electric company charges \$0.12/kWh.

$$65\text{ W} \times \frac{1\text{ kW}}{1000\text{ W}} \times 8\text{ hr} = 0.520\text{ kWh each day}$$

$$0.520\text{ kWh/day} \times \frac{\$0.12}{\text{kWh}} = \$0.0624 \approx \$0.06\text{/day}$$

$$\$0.0624\text{/day} \times 30\,\frac{\text{days}}{\text{month}} = \$1.872 \approx \$1.87\text{/month}$$

$$\$0.0624\text{/day} \times 365\,\frac{\text{days}}{\text{year}} = \$22.776 \approx \$22.78\text{/yr}$$

Assume the power company charges \$0.12/kWh.

1. Determine the daily, monthly, and yearly cost of operating a 15 W LED bulb that produces the same amount of light as a 65 W incandescent bulb if it is on for 8 hr a day.
2. Determine the daily, monthly, and yearly cost of running an 80 W television for 4 hr a day.
3. Determine the daily, monthly, and yearly cost of continuously running a 35 W security camera video recorder.
4. Many electronic devices maintain a standby mode while turned off but still plugged in. Determine the cost of keeping a device plugged in for 1 yr if its standby mode uses 1 W.

5. Calculate the annual cost of running a 65 W computer and a 35 W monitor for 8 hr a day and keeping them in sleep mode (using 5 W each) for the rest of the day. What would the annual cost be for a company with 2000 computers?

While many energy-conservation strategies cost nothing to implement, others require an initial investment. The payback period is a simple measure of the cost-effectiveness of the investment.

Suppose a company decides to replace a 65 W incandescent bulb costing \$22.78/yr to operate. Determine the payback period if an extra \$2 is invested in a 15 W LED bulb costing \$5.26/yr to operate.

$$\text{savings} = \$22.78 - \$5.26 = \$17.52\text{/yr}$$

$$\text{payback period} = \frac{\$2}{\$17.52\text{/yr}} \approx 0.114\text{ yr} \approx 42\text{ days}$$

6. Use the example to write a generalized formula for the payback period.
7. A 4 W LED light bulb produces the same amount of light as a 25 W incandescent bulb.
   a. Determine the annual cost for running each bulb continuously, assuming the electric company charges \$0.12/kWh.
   b. If the LED bulb costs \$2 more than the incandescent bulb, calculate the payback period for this investment.

8. A household spends about \$300 annually to heat water. Research shows that a tankless water heater could save 30% of this cost.
   a. How much would they save annually by installing a tankless water heater?
   b. If the tankless water heater costs an extra \$550, what is the payback period for investing in a tankless water heater instead of a storage tank water heater when the water heater needs to be replaced?

Return on investment (ROI) is another measure used to evaluate the effectiveness of an investment.

Find the first-year ROI of spending an extra \$2 on a LED light bulb that saves \$17.52 in electricity each year.

$$\text{ROI} = \frac{\$17.52}{\$2} = 8.76 = 876\%$$

The wisdom of using LED bulbs to conserve energy is apparent when this ROI is compared to the typical ROI of other investments, which is generally well under 10%.

9. Use the example to write a generalized formula for ROI.
10. Using your answers from exercise 7, determine the first-year ROI of installing a 4 W LED bulb instead of a 25 W incandescent bulb.
11. Using your answers from exercise 8, determine the first-year ROI of installing a tankless water heater instead of a storage tank water heater.
12. A homeowner needing to replace the water heater in his home researched his options at the home-improvement store. He can purchase a \$479 electric water heater, estimated to use 3500 kWh of electricity annually, or a \$679 natural gas water heater, estimated to use 275 therms annually.
   a. Determine the annual cost of electricity for the electric water heater if the power company charges \$0.12/kWh.
   b. Determine the annual cost of natural gas for the gas water heater if the gas company charges \$1.10/therm.
   c. Determine the payback period for investing in the gas water heater instead of the electric water heater.
   d. Determine the first-year ROI of this investment.
13. When faced with a \$3000 repair bill for their central air conditioner, a family considers replacing the unit with a new \$6200 energy-efficient unit that would qualify for a \$1500 tax credit (a government rebate). They estimate that the 40% increase in efficiency would translate into an average savings of \$500/yr.
   a. Determine the amount invested by replacing the air conditioner instead of repairing it.
   b. Determine the payback period for replacing instead of repairing the air conditioner.
   c. List several other factors that might influence the family's decision to repair or replace the air conditioner.

1 CHAPTER REVIEW

Key Concepts

Adding & Subtracting Rational Numbers [1.1]

- To add rational numbers with the same sign:
  - add the absolute values, and
  - use the sign of the addends.
- To add rational numbers with different signs:
  - subtract the lesser absolute value from the greater absolute value, and
  - use the sign of the greater absolute value.
- To subtract rational numbers, add the opposite of the subtrahend.

Multiplying & Dividing Rational Numbers [1.2]

- To multiply or divide 2 rational numbers, multiply or divide the absolute values. The sign of the product or quotient is
  - positive when the signs are the same.
  - negative when the signs are opposite.

Properties of Addition, Multiplication, Exponents & Roots [1.1–1.3]

ADDITION	MULTIPLICATION	EXPONENTS	ROOTS
Commutative		Definitions	
$a + b = b + a$	$ab = ba$	$x^0 = 1\ (x \neq 0); x^{-n} = \frac{1}{x^n}$	$\sqrt{n} = x$, if $x^2 = n$ $\sqrt[3]{n} = x$, if $x^3 = n$
Associative		Product of Powers	
$(a + b) + c = a + (b + c)$	$(ab)c = a(bc)$	$x^a \cdot x^b = x^{a+b}$	
Identity		Quotient of Powers	
$a + 0 = 0 + a = a$	$a \cdot 1 = 1 \cdot a = a$	$\frac{x^a}{x^b} = x^{a-b}$ for $x \neq 0$	
Inverse		Power of a Power	
$a + (-a) = -a + a = 0$	$a \cdot \frac{1}{a} = \frac{1}{a} \cdot a = 1$	$(x^a)^b = x^{ab}$	

Order of Operations [1.4]

- The order of operations is defined to make the value of an expression consistent.

G	Grouping Symbols	
E	Exponents (& Square Roots)	
M	Multiplication	In order from left to right →
D	Division	
A	Addition	In order from left to right →
S	Subtraction	

Variables & Algebraic Expressions [1.5]

- Substitute the value(s) for the variable(s) in the expression and evaluate the resulting expression.

Using the Distributive Property [1.6]

- For any real numbers a, b, and c, $a(b + c) = a(b) + a(c)$.
  - The Distributive Property is applied to write equivalent expressions.
  - When reversed, $a(b) + a(c) = (b + c)a$, the Distributive Property is used to combine like terms.

Biblical Worldview Shaping

Foundations of Algebra

Why do we need Scripture to understand math?

Scripture is the foundation of all things. Understanding the unity and diversity we find in mathematics requires a biblical worldview.

EXERCISES

Simplify each expression. [1.1]

1. $5 + (-7)$

2. $-12 - 4$

3. $17 + (-12) - 11 - (-5)$

4. $15 - |-3| + 8 - |12|$

5. $-\frac{7}{12} + \frac{3}{4}$

6. $3\frac{5}{8} - \left(-5\frac{7}{10}\right)$

7. $-\frac{5}{6} + \frac{7}{15}$

8. Cameron spent \$33.35 on gasoline and \$4.79 on snacks and had \$4.11 left over. Write and evaluate an expression modeling the amount Cameron had to begin with.

Perform each indicated operation. [1.2]

9. $98(4)(0)(-5)$

10. $-56.4 \div 0.2$

11. $\frac{-\frac{6}{5}}{9}$

12. $-\frac{18}{35} \cdot \frac{7}{16}$

13. $5\frac{5}{6} \cdot 2\frac{2}{5}$

14. $-6\frac{8}{9} \div \left(-1\frac{7}{24}\right)$

15. Basham Demolition Crew needs to haul away $11\frac{4}{7}$ tn of debris away from a demolition site. How much weight will each truck carry if all the debris is divided evenly among 6 trucks?

16. Determine Crystal's typing speed in words per minute if she typed 72 words in $2\frac{1}{4}$ min.

Simplify, leaving each answer in positive exponential form. [1.3]

17. $8^9 \cdot 8^6$

18. $4^{-8} \cdot 4^3 \div 4^2$

19. $(3^{-4})^6$

Evaluate each expression. [1.3]

20. $(-3)^5$

21. $\left(\frac{3}{4}\right)^2$

22. $\frac{1}{6^{-3}}$

Evaluate each root. [1.3]

23. $-\sqrt{\frac{100}{64}}$

24. $\sqrt{\frac{45}{245}}$

25. $\sqrt[3]{-512}$

Evaluate each expression. [1.4]

26. $4 + 6 \cdot 8 \div 2 - 14$

27. $\frac{5}{6} \div \frac{4}{3} + \frac{1}{3} \cdot \frac{2}{5}$

28. $2^3 + 2[4(3 + 9) - 7] - 15 + 3 \cdot 6$

29. $\sqrt{9} + |-9|(2 - 6) + 4 \cdot 5^0$

Translate each phrase into a numerical expression. Then simplify your expression. [1.4]

30. nine reduced by the absolute value of the sum of two and negative fifteen

31. the product of fifteen and negative three decreased by twelve

32. the sum of three cubed and four

33. Explain the error that Jeremy made in simplifying the expression $10 \div 5 \cdot 6 + \sqrt[3]{(1 - 9)^2}$.

Step	Jeremy's Work
	$10 \div 5 \cdot 6 + \sqrt[3]{(1 - 9)^2}$
1.	$10 \div 5 \cdot 6 + \sqrt[3]{(-8)^2}$
2.	$10 \div 5 \cdot 6 + \sqrt[3]{-64}$
3.	$2 \cdot 6 + (-4)$
4.	$12 + (-4)$
5.	8

Evaluate when $x = 3$, $y = -4$, and $z = 2$. [1.5]

34. $2x + 3y - 6z$

35. $x^2 + yz$

36. $5x^2 + 2x + 7$

Translate each word phrase into an algebraic expression. [1.5]

37. one less than the square root of a number

38. the product of three consecutive odd integers

39. the difference of an angle's supplement and its complement

40. the number of animal legs on a farm containing x horses, y cows, and z chickens

Use the formula $d = \frac{1}{2}gt^2$ to find the distance each object falls in 3 sec when dropped. [1.5]

41. on the earth, where $g = 32$ ft/sec^2

42. on the moon, where $g = 5\frac{1}{3}$ ft/sec^2

Simplify each expression. [1.6]

43. $-3\left(a - 2b - 5c + \frac{7}{6}d\right)$

44. $2x^2 + 3x + 5x$

45. $\frac{1}{2}x - \left(-\frac{3}{2}x\right)$

46. $\frac{1}{4}x(100x^2 + 16xy - 72y)$

47. $d - 4d + 3f + 7f$

48. $7b - 3(2b + 6)$

49. $2a - (4a + 3b) - 6a$

50. $x + 3y - [4x + 2y - (8x - 9y)]$

51. **BWS Essential Question:** How does Scripture give insight into the unity and diversity we find in mathematics?

CHAPTER 02
SOLVING EQUATIONS

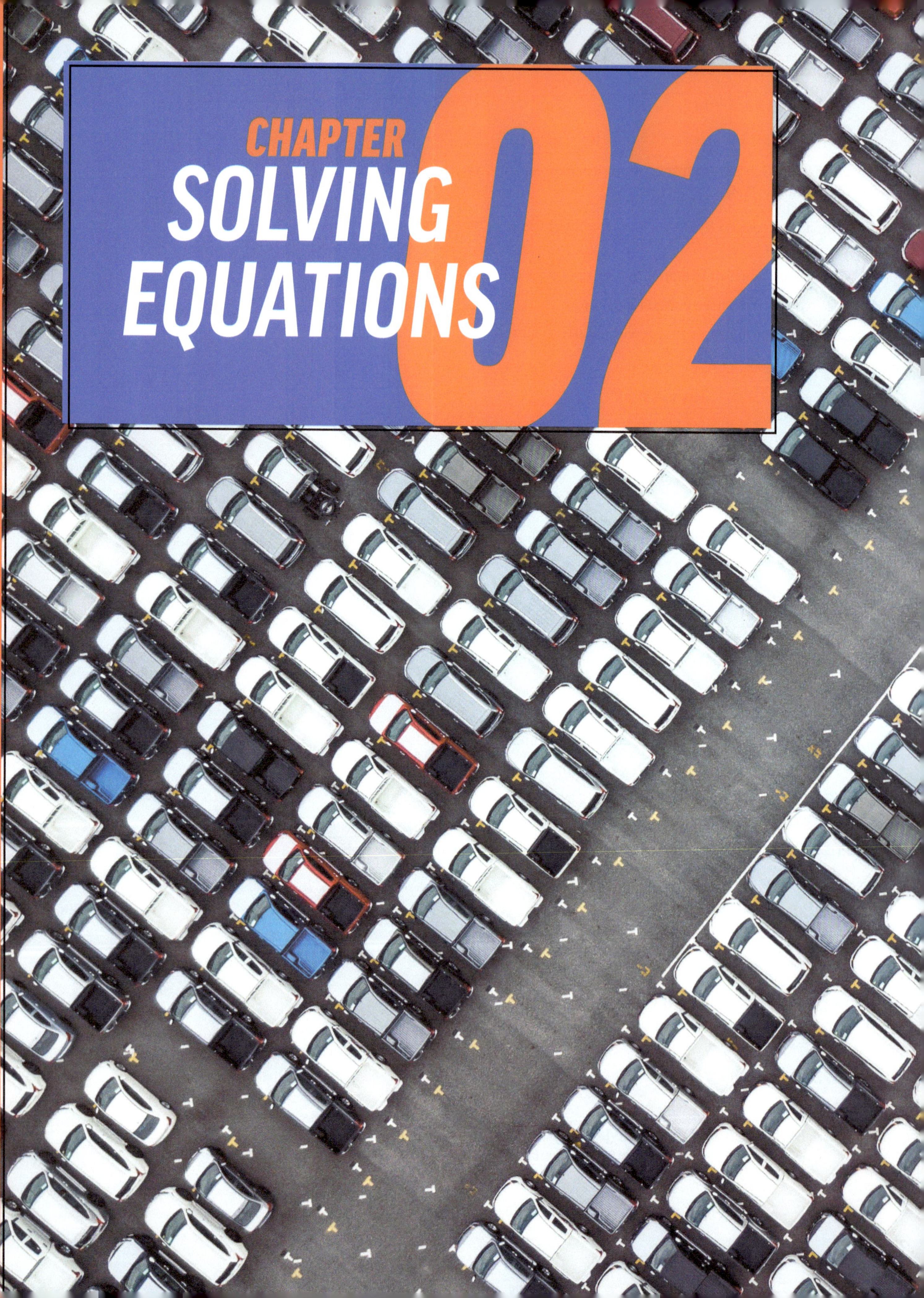

CHAPTER OVERVIEW

2.1 SIMPLE EQUATIONS

How is the order of operations used in solving equations?

After completing this section, I will be able to

- identify the properties of equality.
- apply the properties of equality to solve one-step equations and two-step equations in the form $ax + b = c$.
- solve real-world problems by writing and solving simple linear equations.

Simple equations can be used to determine the cost of heavy payloads.

Algebraic equations are useful in describing God's created world and have many real-world applications. An **equation** is a mathematical sentence stating that 2 expressions are equal. A number that makes an equation true is a **solution** of the equation. Equations can be solved by undoing an operation using its **inverse operation**.

The following properties of equality require the same operation to be performed on both sides of the equation. Equations with the variable on the right side of the equals sign can be rewritten with the variable on the left side by applying the Symmetric Property of Equality.

PROPERTIES OF EQUALITY		
Property	**Symbolization**	**Example**
Addition	If $a = b$, then $a + c = b + c$.	$x - 17 = -35$ $x - 17 + 17 = -35 + 17$ $x = -18$
Subtraction	If $a = b$, then $a - c = b - c$.	$x + 10 = 142$ $x + 10 - 10 = 142 - 10$ $x = 132$
Multiplication	If $a = b$, then $ac = bc$.	$\frac{x}{5} = 21$ $\cancel{5}\left(\frac{x}{\cancel{5}}\right) = 5(21)$ $x = 105$
Division	If $a = b$ and $c \neq 0$, then $\frac{a}{c} = \frac{b}{c}$.	$3x = 9$ $\frac{\cancel{3}x}{\cancel{3}} = \frac{\cancel{9}^{3}}{\cancel{3}}$ $x = 3$
Symmetric	If $a = b$, then $b = a$.	$4 = x + 10 \rightarrow x + 10 = 4$

Each of the properties produced an equivalent equation. **Equivalent equations** are equations that result in the same solution. Note that $x - 17 = -35$ and $x = -18$ are equivalent equations.

KEYWORD SEARCH

solving equations applet

EXAMPLE 1: Solving One-Step Equations

Solve each of the following.

a. $x - 12 = 15$ **b.** $x + \frac{2}{3} = -\frac{7}{12}$ **c.** $-8 = \frac{x}{7}$

Answers

a.

$$\begin{aligned} x - 12 &= 15 \\ \underline{+12} \quad & \underline{+12} \\ x &= 27 \\ 27 - 12 &= 15 \end{aligned}$$

1. Add 12 to both sides of the equation.
2. Simplify both sides.
3. Check the solution.

b.

$$\begin{aligned} x + \frac{2}{3} &= -\frac{7}{12} \\ \underline{-\frac{2}{3}} \quad & \underline{-\frac{2}{3}} \\ x + \frac{2}{3} &= -\frac{7}{12} \\ \underline{-\frac{2}{3}} \quad & \underline{-\frac{8}{12}} \\ x &= -\frac{5}{4} \\ -\frac{5}{4} + \frac{2}{3} = -\frac{15}{12} + \frac{8}{12} &= -\frac{7}{12} \end{aligned}$$

1. Subtract $\frac{2}{3}$ from both sides.
2. Rename to a common denominator.
3. Simplify.
4. Check the solution.

c.

$$\begin{aligned} -8 &= \frac{x}{7} \\ 7(-8) &= 7\left(\frac{x}{7}\right) \\ -56 &= x \\ x &= -56 \end{aligned}$$

1. Multiply both sides by 7.
2. Simplify.
3. Apply the Symmetric Property.

SKILL CHECK **EXERCISES 1, 3**

AFTERSCHOOLHELP

Simple Equations

Since the solution can be checked mentally, show the check only when directed.

SOLVING EQUATIONS

1. Determine the operation performed on the variable in the equation.
2. Perform the inverse operation on both sides of the equation.
3. Simplify both sides of the equation.
4. Check your answer.

Equations in the form of $ax + b = c$, where a, b, and c represent real numbers and x is an unknown value, require 2 steps to be solved completely. Use inverse operations to isolate the variable by undoing any addition or subtraction and then undoing any multiplication or division. The steps to solve equations in the form $ax + b = c$ will reflect the order of operations performed in reverse order.

SOLVING EQUATIONS IN THE FORM $ax + b = c$

1. Undo the addition or subtraction of the constant term.
2. Undo the multiplication or division done to the variable.

EXAMPLE 2: Solving Two-Step Equations

Solve each equation.

a. $21 = 2x + 6$ **b.** $\frac{a}{1.5} - 3.6 = -4.4$

Answers

a.

$$21 = 2x + 6$$
$$-6 \quad -6$$

1. Subtract 6 from both sides.

$$\frac{15}{2} = \frac{2x}{2}$$

2. Divide both sides by 2.

$$x = \frac{15}{2}$$

3. Simplify and apply the Symmetric Property.

$$2\left(\frac{15}{2}\right) + 6 = 15 + 6 = 21$$

4. Check.

b.

$$\frac{a}{1.5} - 3.6 = -4.4$$
$$+3.6 \quad +3.6$$

1. Add 3.6 to both sides.

$$1.5\left(\frac{a}{1.5}\right) = 1.5(-0.8)$$

2. Multiply both sides by 1.5.

$$a = -1.2$$

3. Simplify.

$$\frac{-1.2}{1.5} - 3.6 = -0.8 - 3.6 = -4.4$$

4. Check.

SKILL CHECK **EXERCISES 11, 13**

Equations are very useful in modeling and solving real-world problems. This process often begins with translating a sentence into an equation.

EXAMPLE 3: Writing an Equation

Translate the sentence into an equation and find the missing number.

a. The difference of a number and 10 is 37.

b. Ninety-seven is twenty-five more than four times a number.

Answers

a. The difference of a number and 10 is 37.

$$n - 10 = 37$$

1. Translate the sentence into an equation.

$$+10 \quad +10$$
$$n = 47$$

2. Solve.

b. Ninety-seven is twenty-five more than four times a number.

$97 = 4n + 25$	**1.** Translate the sentence into an equation.
$\underline{-25} \quad \underline{\quad -25}$ $\frac{72}{4} = \frac{4n}{4}$ $18 = n$	**2.** Solve.
$n = 18$	**3.** Apply the Symmetric Property.
$4(18) + 25 = 72 + 25 = 97$	**4.** Check.

SKILL CHECK **EXERCISES 15, 35**

Applying the following steps will assist in solving real-world problems at any level of mathematics.

SOLVING WORD PROBLEMS

1. **Read** the problem, looking for operational words.
   a. *Draw a picture* (if possible) to be sure you understand the problem.
   b. *Assign a variable* to the main unknown.
2. **Plan** a method of attack.
   a. *Make a table* (if possible) to organize relevant information.
   b. *Translate word phrases into expressions* using the variable. Express each unknown quantity in terms of the variable.
3. **Solve** an equation.
   a. *Write an equation.* Find the mathematical verb and identify 2 quantities that are equal.
   b. *Solve the equation.*
4. **Check** your solution. Ask yourself these 3 questions.
   a. *Have I answered all the questions in the problem?*
   b. *What do the numbers represent?* Interpret your answers (including units) in the context of the problem.
   c. *Are the answers reasonable?* Impossible and improbable answers are clues that you made a mistake (e.g., a negative distance is impossible, and a car's speed of 100 mi/hr is unlikely and usually illegal).

EXAMPLE 4: Solving a Real-World Problem

How many quarters must be removed from \$2.64 to leave a total of \$1.89?

Answer

Let q = the number of quarters.	**1. Read** the problem carefully, noting that we are looking for the number of quarters.
$0.25q$ = the value of the quarters	**2. Plan** a strategy. The total value of the quarters is the number of quarters times the value of a quarter.
$2.64 - 0.25q = 1.89$	**3. Solve.** Write an equation modeling the situation.
$\underline{-2.64 \qquad\quad} \quad \underline{-2.64}$	a. Undo addition.
$\frac{-0.25q}{-0.25} = \frac{-0.75}{-0.25}$ $q = 3$ quarters	b. Undo multiplication.
$2.64 - 0.25(3) = 2.64 - 0.75 = 1.89$	**4. Check.**

SKILL CHECK **EXERCISE 37**

EXAMPLE 5: Solving a Real-World Problem

The grain elevator weighed Mr. Young's truck at 17,760 lb before he dumped his wheat. The weight of the empty truck is 6180 lb. If the wheat weighed 60 lb/bu, how many bushels of wheat did he deliver to the elevator?

Answer

Let b = bushels of wheat.

1. **Read** the problem carefully, noting that the total weight of 17,760 lb consists of the weight of the wheat and the weight of Mr. Young's truck.

$60b$ = the weight of the wheat

2. **Plan** a strategy. The question asks for the number of bushels of wheat, b, and $60b$ represents the total weight of the wheat.

$$60b + 6180 = 17{,}760$$

3. **Solve.** Write an equation modeling the situation.

$$\underline{\quad -6180 \quad -6180}$$

a. Undo addition.

$$\frac{60b}{60} = \frac{11{,}580}{60}$$

b. Undo multiplication.

$$b = 193 \text{ bu}$$

$$60(200) + 6000 = 12{,}000 + 6000 = 18{,}000 \approx 17{,}760$$

4. **Check.** Estimation shows that the answer is reasonable.

SKILL CHECK **EXERCISE 39**

VOCABULARY

- equation
- equivalent equations
- inverse operation
- solution

A. EXERCISES

State the operation needed to solve the equation and name the property of equality illustrated by this action.

1. $x + 7 = 15$
2. $x - 3 = 57$
3. $20 = 4x$
4. $\frac{x}{5} = \frac{2}{3}$

Solve each equation and check your answer.

5. $x + 3 = 12$
6. $x - 4 = -23$
7. $\frac{y}{5} = -2$
8. $\frac{c}{-9} = -8$
9. $4x = 164$
10. $3x - 10 = 14$
11. $-4c + 9 = 41$
12. $\frac{p}{4} - 9 = 15$
13. $4 + \frac{y}{6} = 9$

Use the problem-solving plan to write and solve equations modeling each problem.

14. A number multiplied by 12 is 60. Find the number.

15. The difference of a number and 72 is 12. Find the number.

16. The sum of a number and 27 is −14. Find the number.

17. When Laura is six times as old as she is now, she will be 84. How old is she now?

18. Deedra canned 24 qt of beans, which was four times as much as Ava canned. How many quarts did Ava can?

19. Find the length of the third side of a triangle with a perimeter of 21 in. if the sum of the other 2 sides is 15 in.

B. EXERCISES

Solve each equation and check your answer.

20. $82 = x - 24$

21. $a + 0.93 = 1.42$

22. $2w = \frac{1}{9}$

23. $\frac{4}{5}y = 24$

24. $\frac{z}{\frac{11}{2}} = 2$

25. $x - \frac{18}{7} = \frac{4}{5}$

26. $\frac{t}{8} + 11 = 4$

27. $17 - w = -13$

28. $6.8x + 4.7 = -8.22$

29. $\frac{y}{3.7} - 1.5 = -2.8$

30. $-\frac{2}{3}z - \frac{1}{2} = \frac{4}{5}$

31. $\frac{d}{3\frac{1}{2}} + \frac{1}{3} = -\frac{2}{5}$

32. **Essential Question:** How are the methods for evaluating expressions and solving two-step equations related?

Write and solve an equation to model each problem.

33. Four times a number is 20. Find the number.

34. The quotient of a number and 15 is 120. Find the number.

35. The sum of 47 and three times a number is 68. What is the number?

36. Five reduced by twice a number is two. Find the number.

37. Emma made $37.80 for babysitting 6 hr. How much did she earn per hour?

38. Find the average high temperature in Madrid in October if Hong Kong's average high temperature is 84° and is 17° higher than Madrid's average high.

39. Of the $395 that Carlos collected for the fundraiser, $310 was in tens and twenties with the rest in fives. How many fives did he collect?

40. The class decided to purchase T-shirts from an online company that charged $11.95 for each shirt and a shipping and handling fee of $6.95. The bill was $174.25. How many T-shirts did the class order?

41. The horizontal position of an object in uniform motion is given by the formula $d_f = vt + d_i$. How many seconds, t, does it take for an object moving at a velocity, v, of 88 ft/sec to move from an initial position, d_i, of −128 ft to a final position, d_f, of 532 ft?

C. EXERCISES

42. Show that $u = -4$ is a solution to $\dfrac{\sqrt[3]{-u^2 - 2u} - \sqrt{121}}{5u - 3u} = 2(4^{-2}) + \sqrt{\dfrac{3^4}{-9u}}$.

Solve.

43. A trucking company quoted the Fernandez family a rate of \$1782 to use 19 linear feet of trailer space for their cross-country move. An adjustment price of ±\$54/ft for extra or fewer feet used was also agreed upon. How many linear feet did the family use if the trucking bill was \$2052?

44. Miss Miller teaches three times as many students during first hour as she does during third hour. Her second-hour class has twice as many as her third-hour class. If she has a total of 72 students in her 3 morning classes, how many does she have in each class?

45. A triangular lot has a perimeter of 590 ft. The road frontage is twice as long as the second side and 50 ft shorter than the third side. Find the lot's dimensions.

CUMULATIVE REVIEW

Evaluate each expression. [1.1–1.4]

46. $-12 - (-4) - (-3)$

47. $-12(-4)(-3)$

48. $\dfrac{1}{-2^{-2}} - (-3)^2$

49. $\dfrac{3 - 5(-2 + 6)}{|9 - 1|}$

Simplify each expression. [1.6]

50. $7 - (4x - 3)$

51. $6y(x - 2y)$

52. $-7(x + 5) + 3(x - 4)$

53. $7x^2 - 4x^2y - x^2 + 8x^2y$

54. Which expression has a value different from the others when $g = 5$? **[1.5]**

A $9g - 3$

B $8(g + 2)$

C $\dfrac{200}{g} + 2$

D $-7 + (g + 2)^2$

E Every expression has the same value.

55. Simplifying which of the following expressions will not require applying the Distributive Property? **[1.6]**

A $\frac{2}{3}(39a - 24b + 10)$

B $5pq + 4q^2 - 3p + 11pq$

C $15p + 17v + 20s$

D $x^2 - (2x^2 + 8y - 3)$

E none of these

MIND OVER MATH

Here are 5 incorrect equations. To correct each equation, reposition just 1 match.

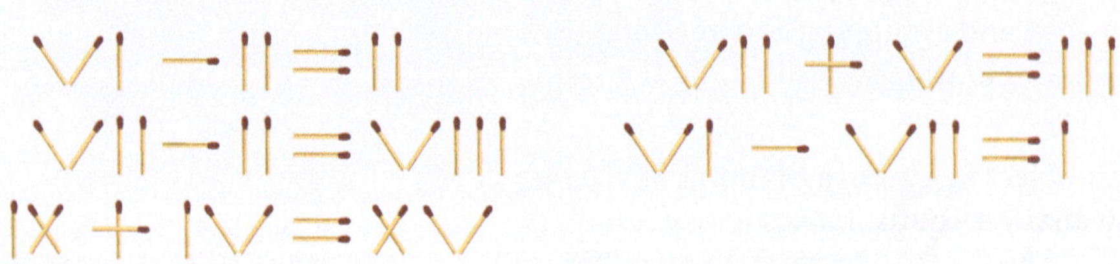

TECHNOLOGY CORNER (TI-84 PLUS FAMILY)

Using Stored Values & Entries

To create any letter A–Z, simply press [alpha] followed by the key with a given letter above it. For example, type A using [alpha] [math].

Storing Numerical Values

Any numerical value, including the result of an evaluated expression, can be stored as a variable A–Z.

Assign $-\frac{1}{2}$ to the variable A and $\frac{4}{5}$ to B. Then display the value of the variable A.

[(-)] [1] [÷] [2] [sto→] [alpha] A [enter]

[4] [÷] [5] [sto→] [alpha] B [enter]

[alpha] A [enter]

```
-1/2→A
                    -0.5
4/5→B
                     0.8
A
                    -0.5
```

Expressions containing the variable can be easily evaluated.

Evaluate AB and $B - A^2$.

[alpha] A [alpha] B [enter]

[alpha] B [−] [alpha] A [x²] [enter]

```
AB
                    -0.4
B-A²
                    0.55
```

Automatic Storage

[entry] is a storage area containing the most recently entered expression or instruction. Selecting [entry] ([2nd] [enter]) places the last instruction or expression at the current cursor location, and it can then be edited and evaluated. Repeatedly selecting [entry] scrolls through recently entered expressions.

[ans] is a storage area containing the most recent result. It can be used to extend a calculation or to store the result. Pressing [ans] ([2nd] [(-)]) inserts the last result at the current cursor location.

Change the value of A to 2 and reevaluate the expression $B - A^2$.

[2] [sto→] [alpha] A [enter]

[entry] [entry] [enter]

Evaluate the expression $5(2 + B - A^2)$ and store the result as C.

[5] [(] [2] [+] [2nd] [(-)] [)] [sto→] [alpha] C [enter]

```
2→A
                       2
B-A²
                    -3.2
5(2+Ans)→C
                      -6
```

2.2 MULTISTEP EQUATIONS

How are properties of real numbers helpful in solving equations?

After completing this section, I will be able to

- apply mathematical properties to solve multistep equations.
- apply mathematical properties to solve equations with variables on both sides of the equals sign.
- solve real-world problems by writing and solving multistep equations.

Solving mathematical problems in the real world often requires making use of equations more complicated than those from Section 2.1. This section presents methods of solving more complex equations.

The Distributive Property is a foundational concept of mathematics that is often used to combine like terms before solving equations. The properties of equality demand that the same operations be performed on both sides of the equation. These steps can be done mentally from now on, as shown in the following example.

EXAMPLE 1: Combining Like Terms before Solving

Sammy's mother is four times as old as Sammy, and Sammy's brother is twice as old as Sammy. If you find the sum of their ages and add 3, you get 80. What are the ages of all 3 people?

Answer

Let x = Sammy's age.

1. **Read** the problem carefully to find the main unknown. All ages are described based on Sammy's age.

$4x$ = Sammy's mother's age
$2x$ = Sammy's brother's age

2. **Plan.** Define the other ages in terms of x.

$$4x + 2x + x + 3 = 80$$
$$7x + 3 = 80$$
$$7x = 77$$
$$x = 11$$

3. **Solve.** Write an equation and apply the Distributive Property and the properties of equality to find the solution.

mother's age: $4(11) = 44$
brother's age: $2(11) = 22$

4. **Check.** Find the other ages and note that $44 + 22 + 11 + 3 = 80$.

Sammy is 11 yr old, his mother is 44 yr old, and his brother is 22 yr old.

SKILL CHECK **EXERCISE 1**

If parentheses appear in the equation, apply the Distributive Property.

EXAMPLE 2: Removing Parentheses before Solving

Solve $3(x + 1) - (x - 2) = 13$.

Answer

RECALL

Negatives outside ( ) must be included in the distributive step.

$-(x - 2) \rightarrow -x + 2$

$3x + 3 - x + 2 = 13$	**1.** Remove parentheses by applying the Distributive Property.
$2x + 5 = 13$	**2.** Combine like terms.
$2x = 8$	**3.** Subtract 5 from both sides.
$x = 4$	**4.** Divide both sides by 2.
$3(4 + 1) - (4 - 2)$ $= 3(5) - 2 = 13$	**5.** Check.

SKILL CHECK **EXERCISE 5**

The properties of equality can be used to convert a repeating decimal to a fraction.

CONVERTING REPEATING DECIMALS TO FRACTIONS

1. Write an equation in which x is equal to the repeating decimal.
2. Use the Multiplication Property of Equality to multiply both sides of this equation by 10^n, where n is the number of repeating digits.
3. Apply the Subtraction Property of Equality to subtract equivalent quantities from both sides of the equation.
4. Solve the resulting equation and reduce the fraction to lowest terms.

EXAMPLE 3: Converting a Repeating Decimal to a Fraction

Express each decimal as a quotient of integers.

a. $0.\overline{36}$ **b.** $1.2\overline{58}$

Answers

a.	$x = 0.\overline{36}$	**1.** Let x equal the repeating decimal.
	$100x = 36.\overline{36}$ $\underline{-x \quad -0.\overline{36}}$	**2.** Multiply both sides of the equation by $10^2 = 100$ (move the decimal to the right 2 places). Then apply the Subtraction Property of Equality.
	$\frac{99x}{99} = \frac{36}{99}$	**3.** Divide both sides by 99.
	$x = \frac{4}{11}$	**4.** Reduce the fraction.
b.	$x = 1.2\overline{58}$	**1.** Let x equal the repeating decimal.
	$100x = 125.8\overline{58}$ $\underline{-x \quad -1.2\overline{58}}$	**2.** Multiply both sides of the equation by $10^2 = 100$ (move the decimal to the right 2 places). Then apply the Subtraction Property of Equality.
	$\frac{99x}{99} = \frac{124.6}{99} = \frac{1246}{990}$ $x = \frac{623}{495}$	**3.** Solve the resulting equation. Clear the decimal from the fraction by multiplying the numerator and denominator by 10.
	$\frac{623}{495} = 1.258585859$	**4.** Use a calculator to verify your solution.

SKILL CHECK **EXERCISE 17**

This procedure can be used to convert any repeating decimal to an equivalent quotient of integers and demonstrates that all repeating decimals are rational numbers.

For equations with variables on both sides, use the properties of equality to move all variables to one side of the equation. In more complicated equations, simplify both sides of the equation first.

EXAMPLE 4: Solving an Equation with Variables on Both Sides

Solve each equation.

a. $2x+5=x-2$ **b.** $3(x-4)=4(x-6)$

Answers

a.

$$\begin{aligned} 2x+5&=x-2 \\ -x\quad&\quad -x \\ x+5&=-2 \\ x&=-7 \\ 2(-7)+5&=-7-2 \end{aligned}$$

1. To eliminate x from the right side, subtract x from both sides.
2. Subtract 5 from both sides.
3. Check.

b.

$$\begin{aligned} 3(x-4)&=4(x-6) \\ 3x-12&=4x-24 \\ -3x\quad&\quad -3x \\ -12&=x-24 \\ 12&=x \\ 3(12-4)&=4(12-6) \\ 3(8)&=4(6) \end{aligned}$$

1. Remove the parentheses by applying the Distributive Property.
2. To eliminate x from the left side, subtract $3x$ from both sides.
3. Add 24 to both sides.
4. Check.

SKILL CHECK **EXERCISE 7**

SOLVING MULTISTEP EQUATIONS

1. Simplify both sides by removing any parentheses and combining like terms.
2. Move the variable terms to one side of the equation.
3. Undo the addition or subtraction of the constant term.
4. Undo the multiplication or division done to the variable.

Equations containing variables on both sides of the equation sometimes produce unexpected results.

EXAMPLE 5: Determining Solutions of Special Equations

Solve each equation.

a. $3(4x-5)+2=2(6x+3)$ **b.** $8-(x-4)=-2(x-6)+x$

Answers

a.

$$\begin{aligned} 12x-15+2&=12x+6 \\ 12x-13&=12x+6 \\ -13&=6 \end{aligned}$$

no solution

1. Distribute to eliminate the parentheses.
2. Combine like terms.
3. Subtract $12x$ from both sides.
4. No value of x will make the original equation true.

b.

$$\begin{aligned} 8-x+4&=-2x+12+x \\ -x+12&=-x+12 \end{aligned}$$

The equation is an identity and the solution is $\mathbb{R}$, the set of real numbers.

1. Distribute to eliminate the parentheses.
2. Combine like terms.
3. Since the equation contains 2 equal expressions, it must be true for all values of x.

SKILL CHECK **EXERCISES 23, 25**

Equations containing fractions can be simplified by multiplying both sides of the equation by the least common multiple of the denominators (least common denominator or LCD).

EXAMPLE 6: Solving Equations by Eliminating Fractions

Solve $\frac{x+2}{6}-\frac{x-1}{3}=\frac{1}{9}$.

Answer

$$18\left(\frac{x+2}{6}-\frac{x-1}{3}\right)=18\left(\frac{1}{9}\right)$$
$$3(x+2)-6(x-1)=2$$

1. Multiply both sides of the equation by 18, the LCD, and apply the Distributive Property.

$$3x+6-6x+6=2$$
$$-3x+12=2$$
$$-3x=-10$$
$$x=\frac{10}{3}$$

2. Solve.

SKILL CHECK **EXERCISE 29**

AFTERSCHOOLHELP

Multistep Equations

SOLVING EQUATIONS CONTAINING FRACTIONS

1. Find the LCD.
2. Eliminate fractions by multiplying both sides by the LCD.
3. Solve the resulting equation.
4. Check the solution in the original equation.

EXAMPLE 7: Solving a Real-World Problem

The coach has developed a drill that requires a rectangular field with the width being $\frac{1}{3}$ of the length. The perimeter of the field is 280 ft. What length and width should the field liner mark off?

Answer

Let x = the length of the field and $\frac{1}{3}x$ = the width.

$$x+x+\frac{1}{3}x+\frac{1}{3}x=280$$

1. Write the equation.

$$3\left(x+x+\frac{1}{3}x+\frac{1}{3}x\right)=3(280)$$
$$3x+3x+x+x=840$$

2. Clear the fractions by multiplying both sides by 3 and apply the Distributive Property.

$$8x=840$$
$$x=105$$
$$\frac{1}{3}x=35$$

3. Solve.

The dimensions are 35 ft by 105 ft.

SKILL CHECK **EXERCISE 35**

A. EXERCISES

Solve each equation.

1. $4a + 8a = 12$

2. $-2c + 2 - c = 20$

3. $4(w - 7) = 8$

4. $-7(y + 3) = 77$

5. $2n - 4(n - 3) = 26$

6. $3x - 7 = 2x$

7. $3x + 5 = 8x - 5$

8. $3(x + 1) = 8x - 12$

9. $7(x - 9) = 3(x - 2)$

10. $6x - 1 = x + 4$

11. $7n - 13 = 4n - 9$

12. $\frac{x}{3} - \frac{x}{4} = 1$

13. $y = \frac{3y}{5} + 12$

14. $\frac{x}{3} + \frac{x}{8} = \frac{11}{24}$

15. $\frac{b}{3} - \frac{8b}{7} = \frac{68}{21}$

B. EXERCISES

Express each of the following repeating decimals as a quotient of integers.

16. $0.151515\ldots$

17. $0.\overline{81}$

18. $1.\overline{785}$

19. $2.4\overline{396}$

Solve. Identify equations that are identities or that have no solution.

20. $(t + 3) + 2(t + 5) = 34$

21. $3.2k + 4(k + 0.7) = 38.8$

22. $-2(j - 1) + 3(j + 4) = 8$

23. $9(2.8a - 1.3) = 18.4a - 11.7 + 6.8a$

24. $8(2t + 1) - (t - 9) = -52 - 8t$

25. $11x - 8 = 12x + 9 - x$

26. $4(m - 6) - m(3 + 4) = 8m - 2$

27. $\frac{t + 1}{2} = \frac{2t - 3}{6}$

28. $\frac{4y - 4}{8} + \frac{26}{2} = \frac{1}{4}$

29. $\frac{107}{6} = \frac{7m}{2} - \frac{3m - 1}{3}$

30. $\frac{a - 3}{2} - \frac{a}{8} = \frac{a}{4} + \frac{a - 12}{8}$

31. $7\left(\frac{6}{5} - b\right) + \frac{3}{5}b = \frac{2}{5}(b + 5)$

32. **Essential Question:** Name 2 properties of real numbers you would use to solve $\frac{2(1 - x) + 5x}{6} + \frac{4(2x - 9)}{3} = \frac{21}{2}$.

Write an equation modeling each problem and solve.

33. The sum of 3 consecutive integers is 84. Find the integers.

34. Stan wants to make a rectangular garden. He has 34 ft of fencing to put around the perimeter. If the length is to be 5 ft more than the width, find the dimensions of the planned garden.

35. If three subtracted from twice a number is equal to 31 more than three times the number, what is the number?

36. The perimeters of 2 rectangles are equal. The width and length of one of the rectangles are x and $x + 14$. The other rectangle has dimensions of $x + 2$ by $7x$. Find the dimensions of each rectangle.

37. **MP1** Find 2 consecutive even integers such that four times the smaller is equal to four more than three times the larger.

38. Ciara has organized an online fundraiser for a friend in need. Four donors helped her collect more than her initial goal when two each chipped in $\frac{1}{6}$ of the total ask amount, another gave $\frac{2}{3}$, and the last gave $\frac{1}{4}$ of the goal amount. If Ciara collected \$375, what was the original goal amount?

C. EXERCISES

Write an equation modeling each problem and solve.

39. The length of a rectangle is three times its width. The perimeter of the rectangle is equal to the perimeter of a triangle that has 1 side equal to the length of the rectangle. The sum of the other 2 sides of the triangle is 85. What are the dimensions of the rectangle?

40. Find 2 consecutive even integers such that three times the smaller is four more than eight times the larger.

41. **MP2** Seema needs a certain length of ribbon to put around the outer edge of her craft project. She has several pieces of equal length. She discovers that if she uses 1 piece, it is 2 in. too short; and if she uses 2 pieces, they are 6 in. too long. What is the length that she needs for her project?

There is often more than one way to solve a problem. Consider the task of finding 2 numbers whose sum is 85 and whose difference is 29.

42. Let x represent the larger number and use the fact that the difference of the numbers is 29 to represent the smaller number. Then write an equation using the fact that their sum is 85. Solve the equation to find the numbers.

43. Let x represent the first number and use the fact that the sum of the numbers is 85 to represent the second number. Then write an equation using the fact that the numbers differ by 29. Solve the equation to find the numbers.

44. **MP6** Compare the 2 methods used to solve the problem, state which method you prefer, and explain why you liked that method better.

CUMULATIVE REVIEW

Evaluate each expression. [1.4]

45. $-3^2 + (-3)^{-2}$

46. $-2 - 2^3(-2)^{-3}$

Evaluate when $x = 3$ and $y = 6$. [1.5]

47. $3x^2 + 6xy$

48. $|x - y|$

49. $\frac{x + y}{2}$

50. $\sqrt{(x + 1)^2 + (y - 3)^2}$

Use the formula $t = pC$ to calculate the sales tax (t) given the tax rate percentage (p) and cost (C). [1.5]

51. Find the tax on a \$23.95 shirt with a 5% tax rate.

52. How much sales tax is there on a \$2300 computer in a state with a 7% tax rate?

53. Which property justifies rewriting $2x = 8$ as $\frac{1}{2}(2x) = \frac{1}{2}(8)$? **[1.2, 2.1]**

A Multiplicative Identity Property

B Multiplicative Inverse Property

C Multiplication Property of Equality

D multiplication

E none of these

54. The solution to $5 - 6x = 2$ is **[2.1]**

A -2

B $\frac{1}{2}$

C $\frac{5}{8}$

D 2

E none of these

2.3 LITERAL EQUATIONS

What makes a model useful?

After completing this section, I will be able to

- solve literal equations for the indicated variable.
- find an unknown quantity within a real-world context by evaluating a solved literal equation.
- explain what makes a mathematical model useful.

Electricians use Ohm's law, $V = IR$, to determine voltage (V), current (I), or resistance (R).

An equation containing more than one variable is called a **literal equation**. Formulas used to find measurements such as distance, perimeter, and volume are examples of common literal equations. Ohm's law ($V = IR$) is a fundamental formula that relates current, voltage, and resistance in electronics. The steps for solving $V = IR$ for the current (I) are the same as solving a simple equation such as $5x = 10$. Just as you would divide both sides by 5 to isolate x in the simple equation, you would divide both sides by R to isolate I in the literal equation, resulting in $\frac{V}{R} = I$.

EXAMPLE 1: Solving a Literal Equation

Solve $F = \frac{WH}{L}$ for H.

Answer

$L \cdot F = \cancel{L}\left(\frac{WH}{\cancel{L}}\right)$ — 1. Multiply both sides by L to clear the fraction.

$\frac{LF}{W} = \frac{\cancel{W}H}{\cancel{W}}$ — 2. Divide both sides by W.

$H = \frac{LF}{W}$ — 3. Use the Symmetric Property of Equality to write the equation with the desired variable on the left side.

SKILL CHECK **EXERCISES 1, 5**

It is often best to solve a formula for the desired variable before substituting values for the other variables.

EXAMPLE 2: Solving a Perimeter Formula

Solve $P = 2l + 2w$ for l. Then find l when $P = 58$ and $w = 6$.

Answer

$P - 2w = 2l$ — 1. Subtract $2w$ from both sides.

$\frac{P - 2w}{2} = l$ — 2. Divide both sides by 2.

$l = \frac{58 - 2(6)}{2} = \frac{58 - 12}{2} = \frac{46}{2} = 23$ — 3. Evaluate to find the numerical value of l.

SKILL CHECK **EXERCISE 19**

The Distributive Property allows the combining of like terms in rewriting more complicated literal equations.

EXAMPLE 3: Using the Distributive Property

Solve $ax - 5b = 7x + 18b$ for x.

Answer

$ax - 5b - 7x = 18b$	**1.** Subtract $7x$ from both sides.
$ax - 7x = 23b$	**2.** Add $5b$ to both sides.
$(a - 7)x = 23b$	**3.** Apply the Distributive Property.
$x = \frac{23b}{a - 7}$	**4.** Divide both sides by $a - 7$.

SKILL CHECK **EXERCISE 21**

When solving literal equations, variables like W in Example 1 and expressions like $a - 7$ in Example 3 are treated like constants.

SOLVING LITERAL EQUATIONS

1. If necessary, use the Multiplicative Property of Equality to clear the equation of fractions by multiplying both sides by the LCD.
2. Simplify both sides of the equation.
3. Collect all terms containing the desired variable on one side of the equation and all other terms on the other side.
4. Use the Distributive Property to isolate the desired variable.
5. Divide both sides by the coefficient of the variable.

EXAMPLE 4: Solving by Clearing Fractions

Solve $\frac{a + 2b}{c} = \frac{cd - a}{b}$ for a.

Answer

$b\cancel{c}\left(\frac{a + 2b}{\cancel{c}}\right) = \cancel{b}c\left(\frac{cd - a}{\cancel{b}}\right)$	**1.** Multiply both sides by bc to clear the equation of fractions.
$b(a + 2b) = c(cd - a)$	
$ab + 2b^2 = c^2d - ac$	**2.** Apply the Distributive Property.
$ab + ac + 2b^2 = c^2d$	**3.** Add ac to both sides to collect all terms containing the desired variable, a, on the left side.
$ab + ac = c^2d - 2b^2$	**4.** Subtract $2b^2$ from both sides.
$(b + c)a = c^2d - 2b^2$	**5.** Use the Distributive Property to isolate the desired variable.
$a = \frac{c^2d - 2b^2}{b + c}$	**6.** Divide both sides by $b + c$.

SKILL CHECK **EXERCISE 13**

AFTERSCHOOLHELP

Literal Equations

Literal equations are used to model real-world processes in many disciplines, including finance, the physical sciences, and even the social sciences. Useful models use clearly defined processes that produce reliable results. Establishing a firm foundation of solving equations will provide the preparation necessary for success in a variety of studies and occupations.

Chemists have noticed that the boiling point of a solution increases according to the number of particles dissolved in it. In order to raise the boiling point to a desired temperature, the chemist must determine the molality, or concentration, of dissolved particles. The following example is a typical application of a literal equation. Subscripted variables, such as B_s and B_e, are often used to distinguish variables with values of the same type, such as the boiling points of 2 different solutions.

EXAMPLE 5: Solving a Real-World Problem

The boiling point of an ethanol solution, B_s, rises as sugar is added according to the formula $B_s = B_e + k_e m$, where B_e is the boiling point of pure ethanol and k_e is a special constant related to ethanol. Solve the formula for m. Then find the molality, m, of a solution that has a boiling point (B_s) of 82.37°C if $k_e = 1.22 \frac{°C}{\text{molal}}$ and $B_e = 78.26°C$.

Answer

$B_s = B_e + k_e m$ — 1. Subtract B_e from both sides.

$B_s - B_e = k_e m$

$\frac{B_s - B_e}{k_e} = m$ — 2. Divide both sides by k_e to isolate m.

$m = \frac{82.37°C - 78.26°C}{1.22 \frac{°C}{\text{molal}}} \approx 3.37 \text{ molal}$ — 3. Evaluate to find the numerical value of m.

SKILL CHECK **EXERCISE 25**

VOCABULARY

- literal equation

A. EXERCISES

Solve each literal equation for the indicated variable.

1. $a = bx$ for x

2. $\frac{x}{n} = q$ for x

3. $\frac{x}{r} = r$ for x

4. $amx = 3a$ for x

5. $y = mx + b$ for b

6. $b - x = d$ for x

7. $I = Prt$ for t

8. $p = a + b + c$ for c

9. $d = rt$ for t

10. $C = 2\pi r$ for r

11. $p = 3s$ for s

12. $V = \pi r^2 h$ for h

B. EXERCISES

Solve each literal equation for the indicated variable.

13. $\frac{P}{6rn} = \frac{7}{9nx}$ for P

14. $f + \frac{y}{3} = \frac{y}{8}$ for f

15. $y = mx + b$ for x

16. $e = \frac{x - a}{a}$ for x

17. $m = \frac{y_2 - y_1}{x_2 - x_1}$ for x_1

18. $\frac{P_1 V_1}{T_1} = \frac{P_2 V_2}{T_2}$ for P_2

19. $S = 2\pi rh + 2\pi r^2$ for h

20. $ax + by = c$ for y

21. $ax + b + a(x + b) = b$ for x

22. $2y - 3x = ay - 2x$ for y

23. $\frac{nx}{b} - n = x$ for n

24. $Pnx = qx - r$ for x

Solve each literal equation for the indicated variable. Then evaluate the expression using the given values.

25. Solve $d_f = vt + d_i$ for v. Given: $d_f = 326$ ft, $t = 2$ sec, and $d_i = 150$ ft.

26. Solve $V = \frac{1}{3}\pi r^2 h$ for h. Given: $V = 288\pi$ cm^3 and $r = 6$ cm.

27. Solve $S = c(h + r)$ for h. Given: $S = 72\pi$ in.^{2}, $c = 6\pi$ in., and $r = 3$ in.

28. Solve $m = \frac{y_2 - y_1}{x_2 - x_1}$ for x_2. Given: $m = -\frac{2}{3}$, $x_1 = -1$, $y_1 = 4$, and $y_2 = 2$.

29. **BWS Essential Question:** Which of these are characteristics of a useful mathematical model?

A provides reliable results

B uses a process that can be explained

C is never wrong

C. EXERCISES

Solve each literal equation for the indicated variable.

30. $\frac{nx}{b} - n = x$ for x

31. $\frac{2p}{m} - \frac{r}{s} = \frac{pr + q}{ms}$ for p

32. $\frac{x}{b} - \frac{x}{c} = x$ for b

33. $a(x + c) = 2ad + c$ for a

34. **MP2** Solve $\frac{P_1 V_1}{T_1} = \frac{P_2 V_2}{T_2}$ for T_2. Then evaluate the resulting expression to find T_2 when $P_1 = 800$ torr, $V_1 = 2.5$ L, $T_1 = 300$ K, $P_2 = 760$ torr, and $V_2 = 1.5$ L.

35. **MP2** Solve $E = mgh + \frac{1}{2}mv^2$ for m. Then evaluate the resulting expression to find m if $E = 580$ kg(m^2/sec^2), $g = 9.8$ m/sec^2, $h = 10$ m, and $v = 6$ m/sec.

CUMULATIVE REVIEW

Simplify each expression. [1.6]

36. $-7(8a^3 - 9a^2 - \frac{1}{14}a + 4)$

37. $xy(x^3y - xy^2 + x^{-1}) - y$

Translate each sentence into a literal equation. [2.1]

38. The net force on an object (F) is equal to the product of its mass (m) and its resulting acceleration (a).

39. A business's net income (I) is calculated by subtracting its expenses (E) from its revenue (R).

Solve each equation. [2.2]

40. $7 - 4(p + 3) = -3$

41. $(t + 5) - 2(t + 3) = 1$

42. $2b - 9b + 6 = 58b + 19$

43. $9x - 20 = 5(x - 4) - 4x$

44. For which equation is $z = -3$ a solution? **[2.1]**

A $5 - (z - 8) = 2^4$

B $\frac{10 - z}{14} = \frac{1}{2}$

C $3z + 1 = -10$

D $5z - 2 = 4z + 1$

E none of these

45. What is the first step in simplifying $2 + \{[(9 \cdot 3 - 2) + 5 + 7 \cdot 4] - 7 \cdot 2\} - 6$? **[1.4]**

A add 5 and 7

B subtract 2 from 3

C subtract 6 from 2

D add 2 and 27

E none of these

2.4 RATIOS & PROPORTIONS

When can I model a problem by using a proportion?

After completing this section, I will be able to

- convert given rates into equivalent rates.
- solve proportions.
- use proportions to solve real-world problems.
- find corresponding lengths and areas of similar figures.

A **ratio** is a comparison of 2 numbers, *a* and *b*, and is commonly expressed in 3 ways—$\frac{a}{b}$, *a* to *b*, or $a : b$. If a student gets 45 out of 60 questions correct on a test, the ratio of correct answers to test questions can be expressed as $\frac{45}{60}$, which should be reduced to $\frac{3}{4}$. Other ratios deduced from this statement include the ratio of incorrect answers to test questions, $\frac{15}{60}$ or $\frac{1}{4}$, and the ratio of correct answers to incorrect answers, $\frac{45}{15}$ or $\frac{3}{1}$.

If the student was given 45 min to take the sixty-question test, the ratio $\frac{60 \text{ questions}}{45 \text{ min}}$, or $\frac{4 \text{ questions}}{3 \text{ min}}$, indicates the rate at which the test must be completed. A **rate** is a ratio comparing 2 quantities with different units.

The rate is more easily understood when expressed as a unit rate: $1\frac{1}{3}$ questions/min.

RECALL

A **unit rate** is a rate in which the denominator equals 1.

EXAMPLE 1: Writing Ratios

A school of 312 students has 168 girls and 20 teachers. Find the following ratios.

a. girls to students

b. boys to students

c. girls to boys

d. unit rate of students per teacher

Answers

a. $\frac{\text{girls}}{\text{students}} = \frac{168}{312} = \frac{7}{13}$

b. $\frac{\text{boys}}{\text{students}} = \frac{144}{312} = \frac{6}{13}$ There are $312 - 168 = 144$ boys.

c. $\frac{\text{girls}}{\text{boys}} = \frac{168}{144} = \frac{7}{6}$

d. $\frac{\text{students}}{\text{teachers}} = \frac{312}{20} = \frac{15.6}{1} = 15.6$ students per teacher Complete the division to get a denominator of 1.

SKILL CHECK **EXERCISES 1, 3**

A rate can be converted to an equivalent rate with different units by a process called **dimensional analysis** (or **unit analysis**). **Unit multipliers**, such as $\frac{1\text{ mi}}{5280\text{ ft}}$, are fractions equal to 1, with different units representing the same quantity in the numerator and the denominator. These unit multipliers can be used to convert between different units of measure.

EXAMPLE 2: Using a Unit Multiplier

Use a unit multiplier to convert to the desired unit.

a. 28 ft to yards

b. 62 in. to centimeters

Answers

a. $28\cancel{\text{ft}}\left(\frac{1\text{ yd}}{3\cancel{\text{ft}}}\right)=\frac{28}{3}\approx 9.3\text{ yd}$

Recall that 3 ft = 1 yd to make a fraction equal to 1. Be sure that the quantity with the unit to be canceled is in the denominator.

b. $62\cancel{\text{in.}}\left(\frac{2.54\text{ cm}}{1\cancel{\text{in.}}}\right)=62(2.54)$

$=157.48\text{ cm}$

Recall that 2.54 cm = 1 in. to make a fraction equal to 1. Be sure that the units cancel.

SKILL CHECK **EXERCISE 15**

Unit multipliers are also used to convert between equivalent rates. Multiplying a rate by one or more unit multipliers will result in an equivalent ratio with different units.

EXAMPLE 3: Solving a Real-World Problem

About 8 min after liftoff, a space shuttle is traveling 17,000 mi/hr. Express this rate in feet per second.

Answer

$17{,}000\text{ mi/hr}=\frac{17{,}000\text{ mi}}{1\text{ hr}}$

1. Write the rate as a unit rate.

$1\text{ mi}=5280\text{ ft}$

$1\text{ hr}=60\text{ min}$

$1\text{ min}=60\text{ sec}$

2. Compare the units in the original rate to those in the desired rate. Miles must be converted to feet, and hours to seconds.

$\frac{17{,}000\cancel{\text{mi}}}{1\cancel{\text{hr}}}\cdot\frac{5280\text{ ft}}{1\cancel{\text{mi}}}\cdot\frac{1\cancel{\text{hr}}}{60\cancel{\text{min}}}\cdot\frac{1\cancel{\text{min}}}{60\text{ sec}}$

3. Multiply the given rate by unit multipliers made by considering the units that need to be canceled.

$\frac{17{,}000(5280)\text{ ft}}{60^2\text{ sec}}=24{,}933.\overline{3}\text{ ft/sec}$

$\approx 24{,}933\text{ ft/sec}$

4. Notice that the dimensions that are left in the problem are feet per second. Use a calculator to convert to decimal form.

SKILL CHECK **EXERCISE 19**

RECALL

extremes means

$\frac{a}{b}=\frac{c}{d}$

$ad=bc$

A **proportion** is a statement of equality between 2 ratios. The proportion $\frac{a}{b}=\frac{c}{d}$ is read "a is to b as c is to d." The first and last numbers, a and d, are the extremes of the proportion, and the 2 middle numbers, b and c, are the means. When solving a proportion, multiply both sides of the equation by the common denominator to clear the equation of fractions.

$$\frac{a}{b} = \frac{c}{d}$$

$$\cancel{b}d\left(\frac{a}{\cancel{b}}\right) = b\cancel{d}\left(\frac{c}{\cancel{d}}\right)$$

$$ad = bc$$

PROPERTY OF CROSS PRODUCTS

In a proportion, the product of the extremes equals the product of the means.

EXAMPLE 4: Solving Proportions

Solve each proportion.

a. $\frac{7}{8} = \frac{x}{4}$ **b.** $\frac{x+9}{5} = \frac{x+15}{7}$

Answers

a.

$\frac{7}{8} = \frac{x}{4}$

$8x = 7(4)$

$8x = 28$ — **1.** Find the product of the extremes and the product of the means.

$x = \frac{28}{8} = 3.5$ — **2.** Divide both sides by 8.

b.

$\frac{x+9}{5} = \frac{x+15}{7}$ — **1.** Apply the Property of Cross Products.

$7(x+9) = 5(x+15)$

$7x + 63 = 5x + 75$ — **2.** Distribute to remove the parentheses.

$2x + 63 = 75$ — **3.** Collect all variable terms on the left side.

$2x = 12$ — **4.** Undo addition by subtracting 63 from both sides.

$x = 6$ — **5.** Undo multiplication by dividing both sides by 2.

SKILL CHECK **EXERCISES 21, 23**

It may be necessary to choose the correct ratio when solving problems involving proportions.

EXAMPLE 5: Solving a Real-World Problem

A car dealership typically sells new and used cars in a ratio of 3 new cars to 2 used cars. How many used cars need to be sold if the dealership wants to sell a total of 325 cars next quarter?

Answer

$\frac{3 \text{ new}}{2 \text{ used}}, \frac{3 \text{ new}}{5 \text{ total}},$ and $\frac{2 \text{ used}}{5 \text{ total}}$ — **1.** Write the 3 stated and implied ratios.

$\frac{2}{5} = \frac{u}{325}$ — **2.** Use the ratio relating the known quantity to the unknown quantity to write a proportion.

$5u = 2(325)$ — **3.** Apply the Property of Cross Products.

$5u = 650$

$u = \frac{650}{5} = 130$ used cars — **4.** Divide both sides by 5.

SKILL CHECK **EXERCISE 29**

Proportions can also be applied to solve unknown lengths in similar figures. **Similar figures** are figures that have the same shape, but not necessarily the same size. Triangle ABC is similar to triangle XYZ, notated $\triangle ABC \sim \triangle XYZ$.

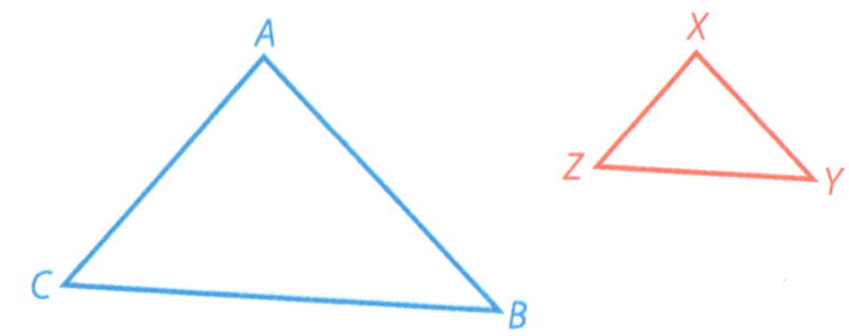

DEFINITION

Similar polygons are polygons in which corresponding angles are congruent and corresponding sides are proportional.

KEYWORD SEARCH

similar triangles

The order in which the notation is written is important. Corresponding vertices must be written in corresponding order. From the notation $\triangle ABC \sim \triangle XYZ$, we know the following statement is true.

$$\frac{AB}{XY} = \frac{BC}{YZ} = \frac{CA}{ZX}$$ Corresponding sides are proportional.

EXAMPLE 6: Finding a Missing Side Length

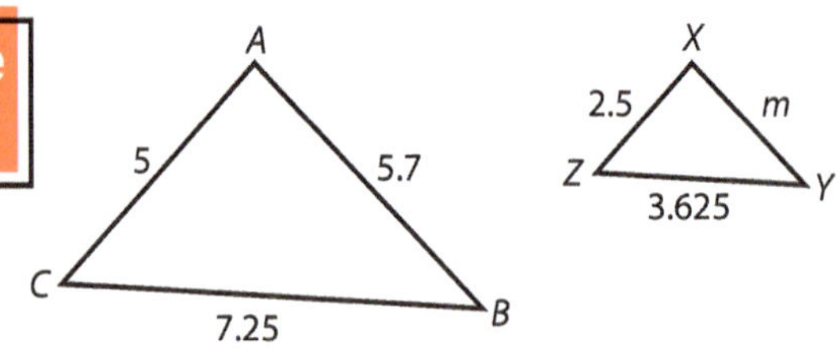

Find XY, given that $\triangle ABC \sim \triangle XYZ$.

Answer

$\frac{CA}{ZX} = \frac{AB}{XY}$, so $\frac{5}{2.5} = \frac{5.7}{m}$.

1. Set up a proportion of corresponding sides of the triangles, as given in the diagram.

$5m = 2.5(5.7)$

$5m = 14.25$

$m = \frac{14.25}{5} = 2.85$

2. Apply the Property of Cross Products and solve for m.

SKILL CHECK **EXERCISE 31**

AFTERSCHOOLHELP

Ratios & Proportions

It can be shown that any ratio of corresponding lengths of similar figures, including their perimeters, is equal to the ratio of their corresponding sides.

$$\frac{AB}{XY} = \frac{BC}{YZ} = \frac{CA}{ZX} = \frac{AB + BC + CA}{XY + YZ + ZX}$$

$$\frac{5.7}{2.85} = \frac{2}{1}; \frac{7.25}{3.625} = \frac{2}{1}; \frac{5}{2.5} = \frac{2}{1}$$

$$\frac{5.7 + 7.25 + 5}{2.85 + 3.625 + 2.5} = \frac{17.95}{8.975} = \frac{2}{1}$$

Not only do the perimeters of similar figures reflect a specific relationship, but the areas of similar figures do as well. Consider 2 circles of radius a and radius b. Since the area of a circle is found by the formula $A = \pi r^2$, the areas of the respective circles are πa^2 and πb^2. The ratio of their areas is $\frac{\pi a^2}{\pi b^2} = \frac{a^2}{b^2} = \left(\frac{a}{b}\right)^2$.

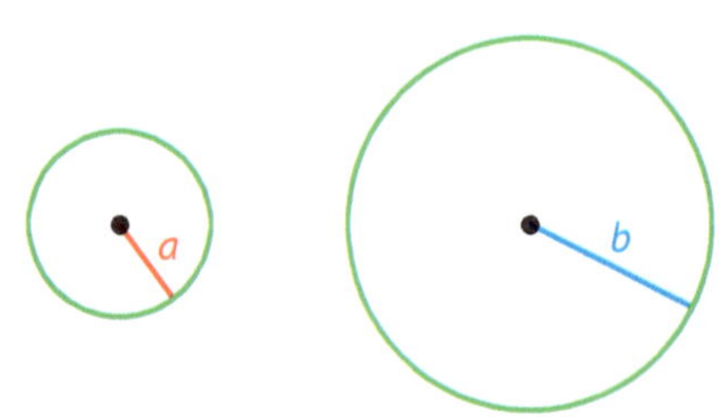

That is, the ratio of the areas of any 2 circles is equal to the square of the ratio of their radii. The same relationship can be shown to hold true for any 2 similar polygons: The ratio of their areas is equal to the square of the ratio of any 2 corresponding sides. Suppose 2 squares have sides in the ratio 5 : 4. Their actual side lengths can be represented by $5x$ and $4x$, and their areas are $(5x)^2$ and $(4x)^2$. So their areas are in the ratio $\frac{(5x)^2}{(4x)^2} = \frac{25x^2}{16x^2} = \frac{25}{16} = \frac{5^2}{4^2} = \left(\frac{5}{4}\right)^2$.

EXAMPLE 7: Finding a Missing Area

VOCABULARY

- dimensional analysis
- proportion
- rate
- ratio
- similar figure
- similar polygon
- unit analysis
- unit multiplier
- unit rate

Two similar rectangles have corresponding sides in the ratio 2 to 3. If the area of the smaller rectangle (A_S) is 36 units^2, what is the area of the larger rectangle?

Answer

$\frac{A_S}{A_L} = \left(\frac{2}{3}\right)^2 = \frac{4}{9}$

1. The ratio of the areas is the ratio of corresponding edges squared.

$\frac{36}{A_L} = \frac{4}{9}$

2. Write a proportion for the areas of the bases.

$4A_L = 324$

$A_L = 81 \text{ units}^2$

3. Apply the Property of Cross Products and solve.

SKILL CHECK **EXERCISE 37**

A. EXERCISES

On a recent homework assignment, 8 students received an A, 10 received a B, 6 received a C, and there were no Ds or Fs. Write the following ratios in lowest terms, using the word "to."

1. the ratio of As to Bs
2. the ratio of As to Cs
3. the ratio of As to students in the class
4. the ratio of Cs to students in the class

Find the unit rate.

5. 85 points scored in 5 games
6. \$27.50 for 11 gal of gas
7. \$370 for 40 hr of work
8. 20,000 bu of wheat from 800 acres
9. Given that $\triangle RSB \sim \triangle ZQM$, list the 3 equal ratios of corresponding sides.
10. Given that $\triangle ABC \sim \triangle DEF$, find x and y.
11. Using the ratio of the sides and the perimeter of $\triangle ABC$, write and solve a proportion to find the perimeter of $\triangle DEF$. Check your work using your answers from exercise 10.

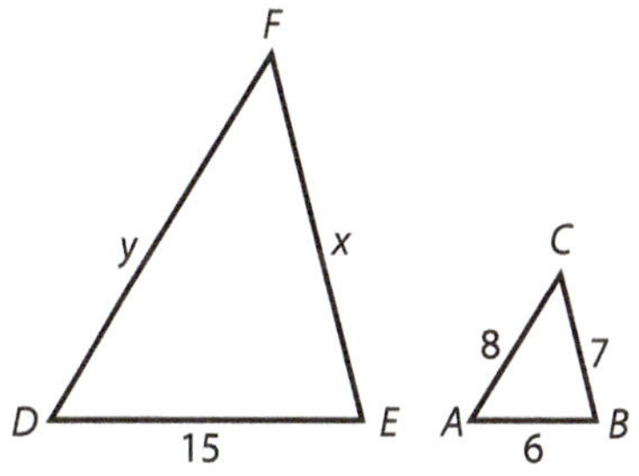

Use proportions to solve.

12. The distance between Nashville and Knoxville is about 160 mi. How far apart are the cities on a map that has a scale of 1 in. : 10 mi?
13. A map of Kentucky has a scale of 1 cm : 10 km. If Lexington and Louisville are 13.2 cm apart on the map, how far apart are the 2 cities?
14. A map of the state of Washington shows the cities of Spokane and Kennewick separated by 4 in. when they are actually about 120 mi apart. Determine the scale of the map.

B. EXERCISES

Convert the given rate to an equivalent rate.

15. 53.6 kg to g

16. 12.88 km to mi

17. 5¢/day to dollars per year

18. Tiffany is buying sliced turkey for \$4/lb but wants less than a full pound. What is the cost of the deli meat in cents per ounce?

19. Stephon is an engineer studying the stopping distance (in feet) of a car traveling 55 mi/hr. What is this speed in feet per second?

Solve each proportion.

20. $\frac{5}{2w} = \frac{45}{36}$

21. $\frac{3x}{25} = \frac{63}{175}$

22. $\frac{d-1}{18} = \frac{12}{27}$

23. $\frac{n+1}{15} = \frac{n-1}{12}$

24. $\frac{2p-3}{6} = \frac{3p+3}{4}$

25. $\frac{6r-5}{9r-3} = -\frac{1}{3}$

26. **Essential Question:** When can I use proportions to solve problems involving figures?

Use proportions to solve.

27. If Jeremy earned \$76 in 8 hr, how much will he earn in 20 hr?

28. Ashima drove the first 348 mi of her trip in 6 hr. How long will it take her, if she drives at the same rate, to drive the remaining 145 mi?

29. A random survey of 30 students found that 17 played a musical instrument. If the school has 450 students, predict how many do not play an instrument.

30. The ratio of soccer players to volleyball players on the bus was 5 : 3. If there were a total of 48 players on the bus, how many were volleyball players?

31. Given that $ABCDE \sim FGHIJ$, find DE and BC.

32. State the ratio of the area of $ABCDE$ to the area of $FGHIJ$. Given that the area of $FGHIJ$ is 44 units^2, write and solve a proportion to find the area of $ABCDE$.

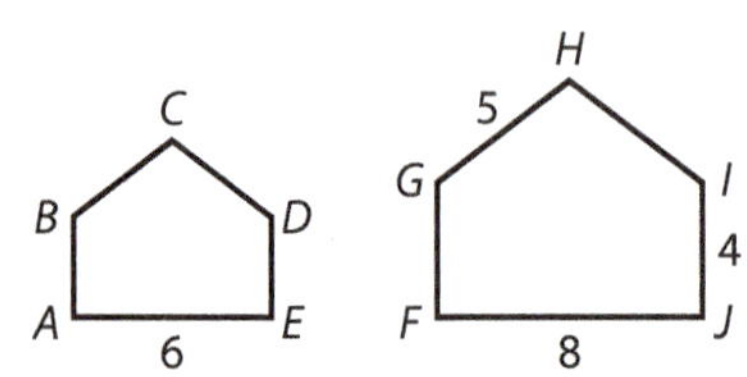

33. The areas of 2 similar plane figures are in the ratio 36 : 25. What is the ratio of their perimeters?

34. MP1 A triangle has sides of 5, 9, and 11 units. A similar triangle's shortest side is 12 units. To the nearest tenth, find the length of the second triangle's 2 other sides and perimeter.

35. The ratio of the sides of 2 similar squares is 2 : 5. What is the ratio of their corresponding areas?

36. The ratio of the sides of 2 similar pentagons is 5 : 7, and the area of the smaller one is 75 ft^2. Find the area of the larger one.

37. The ratio of the sides of 2 similar rectangles is 5 : 2, and the area of the larger one is 400 ft^2. What is the area of the smaller rectangle?

C. EXERCISES

38. If there are b boys in a graduating class of s students, write the following ratios in terms of b and s.

a. boys to girls

b. girls to students

39. MP4 Jorge is trying to save money on his phone plan. For 6 months, Plan A costs \$42 plus \$2 per gigabyte of data used. Plan B offers 6 months of data at a cost of \$15 plus \$5 per GB of data used. Jorge estimates, based on his data usage, that the ratio of Plan A's total cost to Plan B's is 5 to 8. How much money will he save each month using Plan A?

The "golden ratio" described in exercise 40 has fascinated humans for centuries. This special quantity in God's creation has been observed in great works of art and architecture, as well as in nature, from the spatial arrangements of leaves and seeds of plants to the spiral patterns of snail shells and galaxies!

40. The sides of a golden rectangle are in the width to length ratio $1 : \frac{1 + \sqrt{5}}{2}$. Find the length of a golden rectangle whose width is 2.5 units. Round to the nearest tenth.

41. Find the width of a golden rectangle whose length is 8 units. Round to the nearest tenth.

CUMULATIVE REVIEW

Perform each indicated operation. [1.2]

42. $-\frac{14}{15}\left(-\frac{10}{21}\right)$

43. $-\frac{15}{14} \div \frac{10}{21}$

44. $\dfrac{\frac{15}{14}}{\frac{21}{10}}$

Solve each equation. [2.1–2.2]

45. $0.4x + 3.1 = 2.5$

46. $a - (180 - a) = 90$

47. $-2(10 - q) = -32$

Solve each literal equation for the indicated variable. [2.3]

48. $mx + n = 2mx - n$ for x

49. $w = \frac{11(h - 40)}{2}$ for h

50. For which equation is $y = 6$ a solution? **[2.1]**

A $10 - y = -4$

B $\frac{y}{12} + 19 = 21$

C $-8(y - 1) = 7y - 2$

D $-31 = -5y + 1$

E none of these

51. Which equation is an identity? **[2.2]**

A $5w + 4 = 2(3w + 2)$

B $11x - 19 - 2x + 3 = 9x - 22$

C $-2(-6y + 8) = 3(4y - 6)$

D $-3z + 7(4z + 5) = 5(5z + 7)$

E none of these

MATH IN HISTORY
ISAAC NEWTON (1642–1727)

Isaac Newton is the world's foremost mathematician and physicist. Math historians consider him one of the top three mathematicians of all time, along with Archimedes and Gauss. In an end-of-millennium poll of one hundred leading physicists, he was listed with Einstein and Maxwell as one of the top three physicists.

Isaac Newton was born on Christmas Day in 1642. Raised by his grandmother, he found more enjoyment from inventing things and from his personal studies than from his schoolwork. After graduating from Cambridge University, he served as its chair of mathematics for over twenty years. Newton was elected to the famed Royal Society, England's oldest scientific society, at age thirty and was annually reelected as its president from 1703 until his death over twenty years later. In 1701 he was elected to Parliament, and in 1705 he was knighted by Queen Anne.

Newton combined Galileo's advances in terrestrial motion (such as the earth rotating on an axis) and the results of Kepler's laws of celestial motion (orbits) to formulate the universal law of gravitation and his three laws of motion. Despite his fear of criticism that made him reluctant to make his work public, his monumental scientific work *Mathematical Principles of Natural Philosophy* (or *Principia*) was published in 1687 with the encouragement and financial assistance of Edmund Halley. This work confirmed the heliocentric theory of the universe and sparked the scientific revolution. The development of differential and integral calculus, the expansion of infinite series, and the binomial theorem top a long list of Newton's mathematical achievements.

Newton's religious beliefs were complex but important for Christians to consider. Many scientists today claim that theism (a belief in a personal God) is incompatible with good science. Newton would have strenuously disagreed. Commenting on the universe that he studied, he said, "This most beautiful system . . . could only proceed from the dominion of an intelligent and powerful Being." Nevertheless, in Newton there were already the seeds of unbelief that have flowered among modern scientists. Newton thought it was unreasonable to believe in the deity of Christ and the Trinity; he therefore rejected these doctrines. Newton's life demonstrates that belief in God is not incompatible with great achievements in mathematics and science. But his life also presents a warning against granting human reason authority above the Word of God (1 Cor. 1:20–25).

2.5 PERCENT EQUATIONS

How can percent change versus amount change be used to mislead?

After completing this section, I will be able to

- apply the percent formula to find the part, the percent, or the whole.
- solve real-world problems using the percent formula.
- find the percent change when a quantity increases or decreases.

Percent attendance at sporting events can determine whether they are televised in local markets.

Percentages are one of the most commonly used mathematical concepts. Going beyond the classroom, percentages are extensively used in the scientific and business communities as well as politics, the medical field, and sports.

DEFINITION

A **percent** is a ratio of a number to one hundred. The symbol for percent is %.

Percent can be expressed with the common percent symbol (40%), as a fraction $\left(\frac{40}{100}=\frac{2}{5}\right)$, or as a decimal (0.4).

EXAMPLE 1: Converting Percents

a. Write 82% as a fraction and a decimal.

b. Write $\frac{3}{5}$ as a decimal and a percent.

c. Write 1.55 as a percent and a fraction.

Answers

a. $82\% = \frac{82}{100} = \frac{41}{50}$ — **1.** Express percent as a ratio per hundred and reduce to lowest terms.

$82\% = 0.82$ — **2.** Move the decimal point 2 places to the left.

b. $\frac{3}{5} = 0.6$ — **1.** Convert $\frac{3}{5}$ to its decimal form by dividing.

$0.6(100\%) = 60\%$ — **2.** Convert 0.6 to its percent form by multiplying by 1 in the form of 100%.

c. $1.55(100\%) = 155\%$ — **1.** Convert 1.55 to its percent form by multiplying by 1 in the form of 100%.

$155\% = \frac{155}{100} = \frac{31}{20}$ — **2.** Express percent as a ratio per hundred and reduce to lowest terms.

SKILL CHECK **EXERCISES 1, 5**

By representing a percent as a fraction, a proportion can be used to solve a percent problem.

The sentence "a is p percent of b" can be written as the proportion $\frac{a}{b} = \frac{p}{100}$.

EXAMPLE 2: Using a Proportion

25 is what percent of 40?

Answer

$\frac{25}{40} = \frac{p}{100}$	**1.** Write the proportion.
$40p = 2500$	**2.** Apply the Property of Cross Products.
$p = 62.5$	**3.** Solve.
25 is 62.5% of 40.	

SKILL CHECK **EXERCISE 7**

Problems involving percents can also be translated into the percent equation. This single equation can be used to find a percent, a part, or the whole of an original amount.

THE PERCENT EQUATION

The percent of the whole is equal to the part: *percent* × *whole* = *part*.

EXAMPLE 3: Using a Percent Equation

RECALL

Of means "×."
Is means "=."

Write the appropriate percent equation and solve.

a. What percent of 85 is 17?

b. What is 45% of 236?

c. 78 is 130% of what number?

Answers

a. What percent of 85 is 17?	**1.** Translate the question into the percent equation.
$p \times 85 = 17$	
$p = \frac{17}{85} = \frac{1}{5} = 0.2$	**2.** Solve.
$p = 0.2(100\%) = 20\%$	**3.** Convert the decimal to a percent.
$\frac{1}{5} \times 85 = 17$	**4.** Check whether the answer is reasonable.
b. What is 45% of 236?	**1.** Translate the question into the percent equation.
$r = 0.45(236)$	
$r = 106.2$	**2.** Evaluate.
106 is a little less than half of 236.	**3.** Use estimation to check whether the answer is reasonable.
c. 78 is 130% of what number?	**1.** Translate the question into the percent equation.
$1.3w = 78$	
$w = \frac{78}{1.3} = 60$	**2.** Solve.
$33\% \approx \frac{1}{3}$ and $\frac{1}{3}(60) = 20$, so 133% would be about 80.	**3.** Use estimation to check whether the answer is reasonable.

SKILL CHECK **EXERCISES 9, 11**

Remember to convert a percent to an equivalent decimal or fraction when writing a percent equation. This strategy will enable you to successfully solve real-world problems.

EXAMPLE 4: Solving a Real-World Problem

How many gallons of 2% milk does it take to yield 1 gal of butterfat?

Answer

2% of the milk is butterfat. — **1.** Carefully read the problem and identify the key facts.

$0.02m = 1$ — **2.** Write an equation.

$m = \frac{1}{0.02} = 50$ gal — **3.** Solve.

50 gal of 2% milk contains 1 gal of butterfat.

SKILL CHECK **EXERCISE 21**

Often prices are increased or decreased by a certain percentage. The amount of increase or decrease is determined by the percentage as well as the original price of the item. The change in a quantity is often described using the *percent change*, the ratio of the change in value to the original value expressed as a percent.

DEFINITION

Percent change $= \frac{\text{amount of change}}{\text{original amount}} \times 100\%$

The percent change can provide greater insight on how much amounts are affected by a change. Consider a \$10 increase on a \$1000 computer compared to a \$10 increase on a \$5 mouse. The percent change for the computer is 1%, while the percent change for the mouse is 200%.

EXAMPLE 5: Finding Percent Change

A baker decided to raise the price of a loaf of her regular bread from \$3 to \$4. This new price resulted in a large drop in sales, so she reduced the price back to \$3. Determine both the percent increase and the percent decrease in the price of bread during these changes.

Answer

a \$1 increase and a \$1 decrease — **1.** Find the amount of change.

percent increase $= \frac{1}{3} \times 100\% \approx 33\%$ — **2.** Divide the change by the original amount and convert to a percent.

percent decrease $= \frac{1}{4} \times 100\% \approx 25\%$ — **3.** Divide the change by the original amount and convert to a percent.

Notice that the percents are not the same since they are based on different original numbers.

SKILL CHECK **EXERCISES 13, 15**

AFTERSCHOOLHELP

Percent Equations

EXAMPLE 6: Finding Percent Change

Last year 1056 people voted in the city election. This year only 912 voted. Find the percent decrease in votes cast to the nearest tenth of a percent.

Answer

amount of change = 1056 − 912 = 144 — **1.** Subtract to find the amount of change.

percent decrease = $\frac{144}{1056} \times 100\% \approx 13.6\%$ — **2.** Divide the change by the original amount and convert to a percent.

13.6% fewer votes were cast this year than last year.

SKILL CHECK **EXERCISES 27, 29**

When a quantity increases by a given percent, the new amount can be calculated by adding the increase to the original amount. When a $50 pair of shoes is purchased, a 5% sales tax adds $2.50 to the cost, for a total bill of $52.50. The total can be calculated directly using the fact that the total cost is 105% of the cost of the item.

CHANGE	MULTIPLICATION FACTOR
increase of x%	$(100 + x)\%$
decrease of x%	$(100 - x)\%$

EXAMPLE 7: Applying Percent Change

Through dedicated training, a runner was able to decrease her time in the marathon by 13% to a time of 2 hr 47 min. What was her previous time?

Answer

2 hr 47 min = 167 min — **1.** Convert the time to a consistent unit.

$167 = 0.87r$ — **2.** A 13% decrease means that her new time is $(100 - 13)\% = 87\%$ of her original time.

$r = \frac{167}{0.87} \approx 192$ — **3.** Solve.

Her previous time was 192 min or 3 hr 12 min.

SKILL CHECK **EXERCISE 35**

VOCABULARY

- percent
- percent change

A. EXERCISES

For each percent, fraction, or decimal, state the equivalent expressions in the other 2 forms.

1. 75%
2. 187%
3. 0.25
4. 3.5
5. $\frac{3}{5}$
6. $\frac{1}{8}$

Write the appropriate percent equation and solve. Round decimals to the nearest hundredth and percents to the nearest whole percent.

7. What is 84% of 150?
8. 18 is what percent of 76?
9. 58% of what number is 48?
10. 482 is what percent of 24?
11. 175% of 18,900 is what number?
12. 10% of what number is 50?

Find the percent change.

13. from 80 to 100
14. from 200 to 100
15. from 60 to 450

Find the new amount.

16. decrease 600 by 55%
17. decrease 1400 by 9%
18. increase 90 by 250%

B. EXERCISES

Write the appropriate percent equation and solve.

19. Mrs. Gardenghi notices that there are 12 boys in her class of 30. What percent of her class is boys? What percent of her class is girls?
20. If 28% of the 842 respondents to a survey preferred candidate Conley over his opponent candidate Owens, how many chose Conley?
21. Mrs. Chang spends $180 for groceries for her family. How much is the family's weekly income if 12% of the income is spent on groceries?
22. After 4 weeks of training at his new job, Brian will get a 7% increase in his hourly wage. If his increase amounts to $0.60/hr, how much per hour is he getting paid during training?
23. **Essential Question:** Give an example, using values, illustrating that the use of percent change versus amount change (or vice versa) can produce a significantly different impression.

The American Council on Exercise categorizes men's fitness levels according to their percent body fat in the following way.

Description	Percent Body Fat
Essential Fat	2–5%
Athletes	6–13%
Fitness	14–17%
Acceptable	18–24%
Obesity	25% or more

24. If Marcos falls into the Acceptable category and weighs 185 lb, how much of his weight is body fat?
25. The exercise physiologist determines that Dave, who weighs 164 lb, has 20 lb of body fat. Determine his percentage of body fat (to the nearest tenth of a percent) and the category in which he falls.
26. A person's lean body mass (LBM) is his weight minus the weight of his body fat. What is Jesse's LBM if he weighs 195 lb and falls at the lower end of the Fitness category?

Complete the following table. Round percents to the nearest tenth.

	Original Amount	Change	Percent Change	New Amount
27.	68	add 15		
28.	590	subtract 40		
29.	125			86
30.	163			210
31.	274		53% increase	
32.	70		22% decrease	
33.		add 0.3	125% increase	

34. **MP2** Find the result of decreasing 600 by 105%. When would the result not be reasonable?

35. Last year, Raquel experienced an 11% decrease in revenue earned from ridesharing from the previous year. If she made $32,040 this last year, how much money did she make the previous year?

C. EXERCISES

36. In the 2016 national election approximately 136,754,000 Americans voted, representing 54.7% of America's voting-age population. In 2020, the percentage increased to 61.5%, or about 158,408,000 Americans. What was the difference (to the nearest thousand) in the voting-age population of America between these 4 years?

37. The butcher sells 80% lean ground beef at $2.80/lb, 90% lean ground beef at $3.00/lb, and 95% lean ground beef at $3.25/lb. What is the cost per pound of the lean portion of the beef in each case?

38. **MP3** If a number x is increased by 400%, is the new number $4x$? Explain.

39. The Bruins played the Falcons in a 3-game series. If the attendance at the second game increased 50% from the first game and the third game's attendance increased 50% from the second game, what was the percent increase from the first game to the third game?

40. What percent change results from the following?

a. increasing a number by 50% and then decreasing the result by 50%

b. decreasing a number by 50% and then increasing the result by 50%

41. **MP6** If a number is increased by 25%, by what percent must the result be decreased in order to equal the original number?

CUMULATIVE REVIEW

Solve each literal equation for *x*. [2.3]

42. $pq + \frac{2x}{d} = 4b$

43. $ax - \frac{2b}{c} = 3$

44. $\frac{ax + b}{3} = \frac{a - bx}{4}$

Solve each proportion. [2.4]

45. $\frac{7}{12} = \frac{x}{9}$

46. $\frac{7}{9} = \frac{x}{12}$

Use the similar pentagons for the following exercises. [2.4]

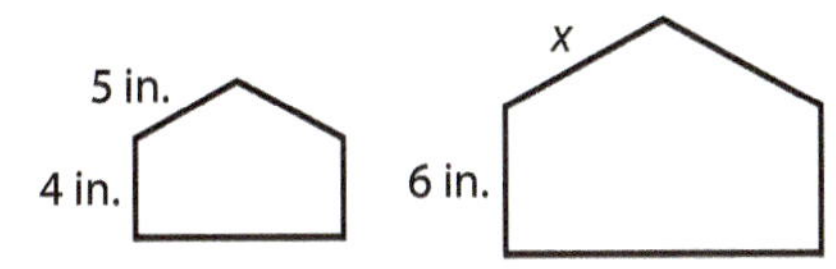

47. Write and solve a proportion to find x.

48. Find the area of the larger pentagon if the area of the smaller pentagon is 44 in.^{2}

49. Two common sizes of index cards are 3" × 5" and 4" × 6". Are cards of these dimensions similar? Explain.

50. Will the sum $-\frac{3}{4} + \frac{1}{6}$ be positive or negative? **[1.1]**

- **A** negative, since $\frac{3}{4} > \frac{1}{6}$
- **B** negative, since $-\frac{3}{4} > \frac{1}{6}$
- **C** positive, since $\frac{3}{4} < \frac{1}{6}$
- **D** positive, since $-\frac{3}{4} < \frac{1}{6}$
- **E** none of these

51. Which expression is equivalent to $\frac{3^{-9}}{3^{-3}}$? **[1.3]**

- **A** $\frac{1}{3^{12}}$
- **B** $\frac{1}{3^{6}}$
- **C** 3^3
- **D** 3^{12}
- **E** none of these

2.6 MONEY PROBLEMS

Why can't mathematical models be used exclusively to make decisions?

After completing this section, I will be able to

- apply percent equations to solve real-world problems involving tips, commission, markups, and discounts.
- solve real-world problems involving simple interest.
- explain the limitations of mathematical models in making decisions.
- compare the benefits and limitations of mathematical models.

Mathematical formulas are very useful in modeling real-world problems involving money. They provide essential information to help us make decisions, but they can never tell us the right thing to do.

Percentages are often used to determine retail and sale prices. After purchasing merchandise from a manufacturer at a given price (*cost*), a store owner must determine the additional amount (*markup*) he must add to arrive at the retail price. This can be expressed mathematically with the equation

$$\textit{cost} + \textit{markup} = \textit{retail price}.$$

The markup is a dollar amount and is often expressed as a percent of the cost, or *markup rate*, using the percent equation

$$\textit{markup} = \textit{markup rate} \times \textit{cost}.$$

DEFINITIONS

The **cost** is the amount a merchant pays for the merchandise that will be sold.
The **markup** is the amount the merchant adds to the cost to arrive at the retail price.
The **markup rate** is the amount of markup on an item as a percent of its cost.
The **retail price** is the merchant's cost plus markup; it is the regular amount a merchant asks customers to pay for merchandise.

EXAMPLE 1: Applying Markup

Tom & Harry's Sporting Goods marks up all clothing 75%. If the merchant paid $15 for a shirt, what is the retail price?

Answer

$$\textit{markup} = \textit{markup rate} \times \textit{cost}$$
$$= 0.75(15) = 11.25$$

1. Use the appropriate formula to find the markup.

$$\textit{cost} + \textit{markup} = \textit{retail price}$$
$$15 + 11.25 = \$26.25$$

2. Use the appropriate formula to find the retail price.

SKILL CHECK **EXERCISES 1, 3**

By noting that the markup rate is the percent increase in the cost of the item, we can use a more direct calculation to solve the problem above.

$$\begin{aligned} \textit{retail price} &= \textit{cost} + 0.75(\textit{cost}) \\ &= (1 + 0.75)(\textit{cost}) \\ &= 1.75(\textit{cost}) \\ &= 1.75(15) = \$26.25 \end{aligned}$$

Merchants frequently offer customers a *discount* off the retail price as an incentive for them to purchase goods. Retailers often advertise the *discount rate* (the "percent off"), leaving it to the customer to calculate the discount and the *sale price* using the following formulas:

$$\textit{discount} = \textit{discount rate} \times \textit{retail price}$$

and

$$\textit{retail price} - \textit{discount} = \textit{sale price}.$$

DEFINITIONS

The **discount rate** is the percent by which the retail price is reduced.
The **discount** is the amount by which the retail price is reduced.
The **sale price** is the customer's cost after the retail price has been discounted.

EXAMPLE 2: Applying a Discount

Determine the retail price of a tennis racquet if Paulo saved \$52.50 by buying it during the 30% off sale. How much did he pay for the racquet?

Answer

$\textit{discount rate} \times \textit{retail price} = \textit{discount}$

1. Use the appropriate formula to find the retail price.

$$0.3r = 52.50$$
$$r = \frac{52.5}{0.3} = \$175$$

2. Write an equation and solve. The dollar amount he saved is the discount.

$\textit{sale price} = 175 - 52.50 = \$122.50$

3. Find the sale price by subtracting the discount from the retail price.

SKILL CHECK **EXERCISES 5, 7**

In Example 2, how would the retail price be found given the discount rate (30%) and the sale price (\$122.50)?

$$\begin{aligned} \textit{sale price} &= \textit{retail price} - \textit{discount} \\ 122.50 &= \textit{retail price} - (\textit{discount rate})(\textit{retail price}) \\ &= \textit{retail price} - 0.3(\textit{retail price}) \\ &= (1 - 0.3)(\textit{retail price}) \\ &= 0.7(\textit{retail price}) \end{aligned}$$

A discount of 30% means a sale price of 70% of the retail price.

EXAMPLE 3: Determining the Amount of Discount

Determine the retail price of a pair of running shoes and the discount if they are marked 40% off and have a sale price of \$54.

Answer

sale price = 60% of the *retail price*

1. Saving 40% is the same as paying 60% of the retail price.

$$54 = 0.6r$$
$$r = \frac{54}{0.6} = \$90$$

2. Write an equation and solve for the retail price.

retail price − *discount* = *sale price*

$$90 - d = 54$$
$$-d = -36$$
$$d = \$36$$

3. Use the appropriate formula to write an equation and solve for the discount.

The retail price of the shoes is \$90, and they were discounted \$36.

SKILL CHECK **EXERCISE 23**

The discount in Example 3 step 3 above could also be calculated using the formula:

discount = *discount rate* × *retail price* = 0.4(90) = \$36.

Waiters and waitresses are usually paid a minimal hourly wage by their employer, but they rely on tips from their customers for the majority of their wages. The tip typically ranges from fifteen to twenty percent of the total bill and can be calculated with another variation of the percent equation:

tip rate × *bill* = *tip*.

A company often pays its salespeople a certain percentage of their total sales in addition to, or instead of, a regular salary. While the term "commission" can refer to either the percent or the amount earned, in this text *commission* will refer to the amount earned, and the percent will be referred to as the *commission rate* or simply the *rate*. The following variation of the percent equation is used to find the commission earned:

commission rate × *sales* = *commission*.

DEFINITIONS

A **tip** is the amount paid for a personal service.
Commission is the dollar amount an employee receives as a result of the sales made.
The **rate** is the percent used to determine a tip or commission based on the amount of the bill or an employee's sales.

Recognize that discount, tip, and commission are specific applications of the percent equation (*percent* × *whole* = *part*) from Section 2.5.

EXAMPLE 4: Finding Commission Rate

Find Jeff's commission rate if his total sales of electronic equipment for the month were \$8250 and his gross income for the month was \$3660, including a monthly salary of \$2175.

Answer

salary + *commission rate* × *sales* = *commission*

1. Choose the appropriate formula for this problem.

$$2175 + 8250r = 3660$$
$$8250r = 1485$$
$$r = \frac{1485}{8250} = 0.18$$
$$= 0.18(100\%) = 18\%$$

2. Write an equation and solve.

SKILL CHECK **EXERCISE 29**

Another variation of the percent formula is used to calculate simple interest on a bank account or a loan:

$$I = Prt,$$

where P = the principal (the amount invested or borrowed), r = the annual interest rate, and t = the time (in years). This formula can help you organize a table for simple interest finance problems.

AFTERSCHOOLHELP

Money Problems

EXAMPLE 5: Applying Simple Interest

Christopher invests the same amount of money in each of 2 accounts and receives a total of \$36 interest at the end of 1 yr. If the accounts earn 5% and 7% interest, how much was invested in each account?

Answer

Let x = the amount in each account (principal).

1. **Read** and assign a variable to the unknown.

P	r	t	=	I
x	0.05	1	=	$0.05x$
x	0.07	1	=	$0.07x$

2. **Plan.** Make a table using $Prt = I$. Fill in the first 3 columns and then multiply across to fill in the last column.

$$0.05x + 0.07x = 36$$
$$0.12x = 36$$
$$x = \$300$$

3. **Solve.** The total interest is \$36.

$0.05(300) = 15$ and $0.07(300) = 21$

$15 + 21 = 36$

Christopher invested \$300 in each account.

4. **Check.**

SKILL CHECK **EXERCISE 31**

The equation in Example 5 step 3 could have been simplified by multiplying both sides by a power of 10 to eliminate the decimals.

$100(0.05x + 0.07x) = 100(36)$	Multiply both sides by 100 to clear the decimals.
$5x + 7x = 3600$	Solve.
$12x = 3600$	
$x = \$300$	

EXAMPLE 6: Applying Simple Interest

Robert borrowed $525 at a simple interest rate of 6%. Three months later he borrowed an additional $950 at 8%. If he paid off both loans at the same time and paid $196 in interest, for how many months did he borrow each amount of money?

Answer

Let x = the time of the longer loan.

1. Assign a variable to the unknown.

P	r	t	=	I
525	0.06	x	=	$525(0.06)x$
950	0.08	$x - 0.25$	=	$950(0.08)(x - 0.25)$

2. Make a table. Three months can be written as $\frac{3}{12} = \frac{1}{4} = 0.25$ yr.

$$525(0.06)x + 950(0.08)(x - 0.25) = 196$$

3. The total interest is $196.

$$525(6)x + 950(8)(x - 0.25) = 19{,}600$$
$$3150x + 7600x - 1900 = 19{,}600$$
$$10{,}750x = 21{,}500$$
$$x = \frac{21{,}500}{10{,}750} = 2 \text{ yr}$$
$$= 24 \text{ months}$$

4. Multiply both sides of the equation by 100 to clear the decimals and then solve.

The $525 loan was for 24 months, and the $950 loan was for 21 months.

SKILL CHECK **EXERCISE 33**

VOCABULARY

- commission
- cost
- discount
- discount rate
- markup
- markup rate
- rate
- retail price
- sale price
- tip

APPLICATIONS OF THE PERCENT FORMULA	
Percent Equation	*percent × whole = part*
Markup	*markup rate × cost = markup*
Discount	*discount rate × retail price = discount*
Tip	*tip rate × bill = tip*
Commission	*commission rate × sales = commission*
Simple Interest	*principal × rate × time = interest*

A. EXERCISES

Complete the following table. Round rates to the nearest whole percent.

	Cost	Markup Rate	Markup	Retail Price
1.	$75	20%		
2.	$24.50			$34.30
3.		15%	$18	
4.			$2.36	$9.11

Complete the following table. Round rates to the nearest whole percent.

	Retail Price	Discount Rate	Discount	Sale Price
5.	$77	30%		
6.			$1.01	$5.74
7.	$30,000			$28,200
8.		33%	$247.17	

Find the following tips.

	Bill	Tip Rate	Round up to the Next	Tip Amount
9.	$26.41	15%	quarter	
10.	$17.25	15%	nickel	
11.	$64.32	20%	dollar	

Complete the following table.

	Sales	Rate	Commission
12.	$29,450	14%	
13.		17.5%	$23,108.75
14.	$6912		$483.84

Complete the following table.

	Principal	Annual Interest Rate	Time	Interest Earned
15.	$2500	4%	3 yr	
16.	$15,000	5%		$3750
17.		8%	10 yr	$4000
18.	$100,000		6 months	$1250

B. EXERCISES

Write an equation and solve.

19. After purchasing soccer shoes at a cost of $30 and marking them up by 125%, the owner of Kicks for Kicks sells them at a 50% off clearance sale. Find the retail price and the sale price.

20. A clothing retailer marks up his merchandise at a rate of 150% so that he can regularly offer a 25% off sale. If the retail price of a sweater is $60, determine the cost to the retailer and the sale price.

21. The owner of Brewed Strength sold coffee beans for $11.89/lb after roasting them, a markup of $4.64/lb. What was the merchant cost for a pound of green coffee beans and the markup rate after roasting them?

22. A suit listed at $699 was reduced by 25% and then by an additional 75%. Find the sale price and the total percent reduction based on the original retail price.

23. A dress priced at $299 was reduced 60% and then an additional 75%. Find the sale price and the total percent reduction based on the original retail price.

24. Emily bought a $45 hoodie at 40% off, a $25 scarf at 35% off, and $44 jeans discounted by 30%. How much money did she save in all?

25. The Hamiltons celebrated Mother's Day by eating out. The pretax bill was $51.79. Determine the tip and the total amount spent if the tax rate is 8% and Mr. Hamilton tipped 18% of the total bill. Round all numbers to the nearest cent.

26. Jennifer, Leigh Anne, and Karyn went out to eat. Jennifer bought an entrée for $12.95 and split a $4.95 dessert with Karyn, who bought a sandwich for $7.95. Leigh Anne bought soup for $3.95, a salad for $6.95, and coffee for $1.70. Determine the total amount each should pay if tax is 6% and each one tips 15% of her individual bill rounded up to the next quarter.

27. A salesman earns $750 per week plus a 5% commission rate on his sales. Determine his total earnings for a week in which he sells $2020 worth of products.

28. A prospective employee is offered a commission rate of 8%. If he wants to make $7000 per month in commission, how much must he sell each month?

29. A prospective employee wants to earn at least $4000 per month in commission. He estimates he can sell $25,000 worth of merchandise each month. What is the minimum commission rate that he needs?

30. A real-estate agent gets a 6% commission on lots he lists and sells. He gets 3% on lots he sells that were listed by someone else. One week he sold a $90,000 lot listed with another company and a $65,500 lot he had listed. Find his commission.

31. Mrs. Jones invested $10,000, divided between 2 separate accounts. One pays 5% interest and the other, 6%. Her combined annual interest income is $575. How much does she have invested in each account?

32. An investor has $7500 invested at one rate and $6200 invested at a different rate. If the second investment is at a rate 2% lower than the first, find the rate for each investment if his combined annual interest income is $698.

33. The Halls borrowed $1800 at 12% and took out a second loan for $2500 at 9%. The time period for the second loan was twice as long as that for the first, and the total interest paid for both combined was $499.50. Find the length of each loan.

34. **Essential Question:** Consider the Halls' decision to borrow money in exercise 33. What is 1 benefit and 1 limitation of the simple interest model in this situation?

35. **MP7** An investor placed her money in a venture paying 7% interest. If she had had another $1500, she would have been able to invest her money at 10% and would have gained an additional $255 in annual interest income. How much did she invest?

C. EXERCISES

36. Write a formula for the total amount of money after simple interest is added to the principal. Use the Distributive Property to write the formula as a product.

37. Mariah invested money in an account earning 8% simple interest. Sukhi invested in an account earning 5% simple interest. How much longer will it take for her to double her investment than Mariah?

38. A borrower is offered a loan of \$10,080 at 10% and a loan of \$7000 at 12%. If the larger loan is 6 months shorter but the total interest paid is the same, find the length of each loan.

39. An automobile buyer is trying to decide whether to buy a new car financed for 5 yr at 7% interest or a used car costing \$11,500 less at 11% for 2 yr. The new car loan will cost \$5130 more in interest than the used car loan. Find the cost of each car.

40. MP2 What discount rate can be offered by a retailer who marks up his merchandise by 125% so that the sale price is the same as the cost of the merchandise?

CUMULATIVE REVIEW

Evaluate each expression for the given values. [1.5]

41. $x^2 - x^3y$ when $x = -2$ and $y = 3$

42. $|c + d|d - c^0$ when $c = -4$ and $d = -1$

43. $a^0 - b(a - b^2)$ when $a = \frac{2}{3}$ and $b = \frac{1}{2}$

44. $-v(z - vz) + |z^2 - v|$ when $v = 2\frac{1}{3}$ and $z = \frac{3}{4}$

Simplify each expression. [1.6]

45. $a^2b - ab^2 + 3ab^2 - 2a^2b$

46. $2x(x - y) - y(x - 2y)$

Exercises 47–48 refer to 2 similar triangles in which the ratio of the perimeters is 25 : 9. [2.4]

47. If a side of the smaller triangle is 45 in. long, what is the length of the corresponding side of the larger triangle?

48. If the area of the larger triangle is 3750 in.^{2}, find the area of the smaller triangle.

49. Which question is not the same as the others? **[2.5]**

A What number is 30% of 75?

B 75 is 30% of what number?

C 30% of 75 is what number?

D 0.3(75) is what number?

E none of these

50. Which equation has no solution? **[2.2]**

A $16 - 5a = -5a + 9$

B $5b + 3 = b + 3$

C $14{,}641 = 1331c$

D $d = 42$

E none of these

2.7 MOTION PROBLEMS

How can I use equations to model real-world motion problems?

After completing this section, I will be able to

- solve real-world problems about related distances by using $d = rt$.

Air traffic controllers direct planes from the tower.

A jet leaves O'Hare Airport in Chicago at 7:00 AM and flies south toward Atlanta, 606 air miles away, at an average speed of 420 mi/hr. The distance formula, $d = rt$, is used in technology to determine the approximate arrival time in Atlanta. The distance formula calculation becomes more difficult when 2 objects are in motion, traveling in the same or opposite directions.

Making a diagram and using a table to organize data will often simplify more complicated motion problems.

EXAMPLE 1: Using Equal Distances

The junior-high Sunday school class leaves the church on a hike and travels east at 3 mi/hr. The high-school class leaves an hour later traveling at 5 mi/hr on the same route. How long will it take the high-school class to catch up with the junior-high class?

Answer

Let t = the time required for the high-school class to catch up.

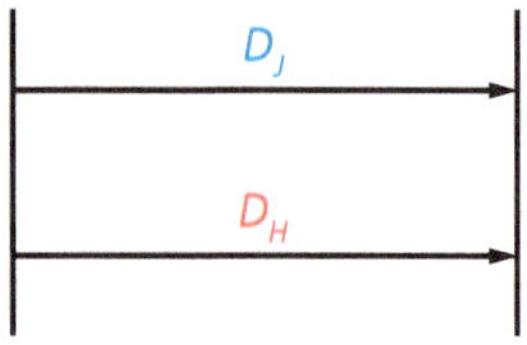

1. **Read** the problem carefully.
   a. Identify the main unknown in the problem.
   b. Draw a diagram to represent the problem.

	r	·	t	=	d
Junior High	3	·	$t + 1$	=	$3(t + 1)$
High School	5	·	t	=	$5t$

2. **Plan** your strategy.
   a. Since the junior-high class left 1 hr earlier, they traveled $t + 1$ hr.
   b. Fill in the rate column using the numbers from the problem.
   c. Using $rt = d$, fill in the distance column.

$3(t+1)=5t$
$3t+3=5t$
$3=2t$
$t=\frac{3}{2}$

3. Solve. Since both groups start and end at the same place, the distances must be equal. Write an equation and solve.

The high-school class will take $1\frac{1}{2}$ hr to catch up with the junior-high class.

4. Check. Does this answer the question in the problem? In this problem, *t* represents the time the high-school class hikes. It takes them 1.5 hr to hike the 7.5 mi that the junior-high class hikes in 2.5 hr.

SKILL CHECK **EXERCISES 13, 19**

Use a diagram to help clarify the relationship between the distances in a uniform motion problem. These problems can be categorized according to the relationship between the distances.

KEYWORD SEARCH

uniform motion problems

1. The distances are the same when a person takes the same trip to and from a destination (roundtrip) or when a person leaves later to catch up to someone who left the same place earlier.

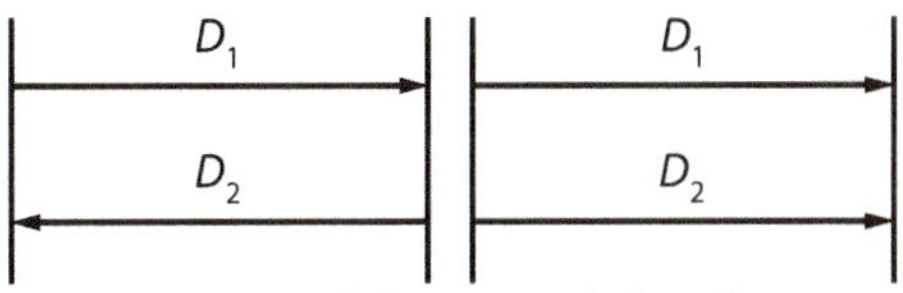

roundtrip or pursuit: $D_1 = D_2$

2. The sum of the distances is known when individuals travel from separate locations to meet or when they travel in opposite directions from a common location.

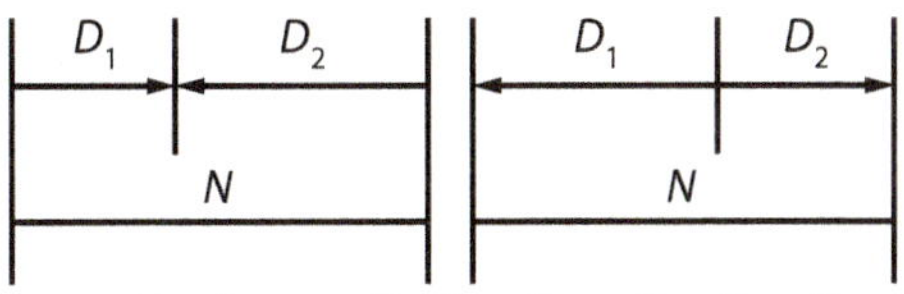

motion in opposite directions: $D_1 + D_2 = N$

3. A given distance must be added to a shorter distance to equal the longer distance.

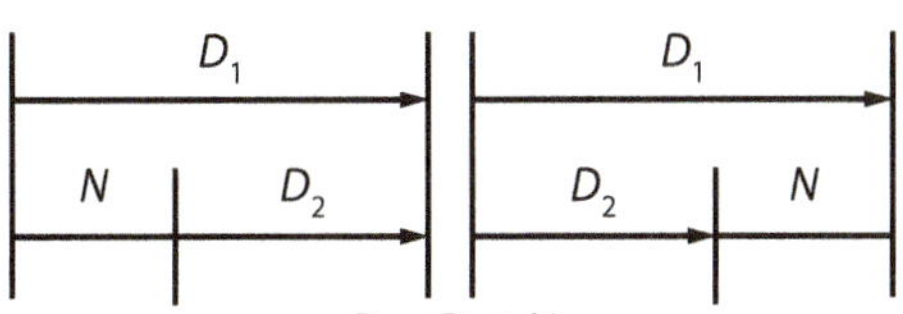

$D_1 = D_2 + N$

SOLVING UNIFORM MOTION PROBLEMS

1. **Read** the problem carefully.
   a. Identify the main unknown in the problem and let the variable stand for this unknown.
   b. Draw a diagram to represent the problem.
2. **Plan** a strategy.
   a. Make a table and fill in the rate and time columns. One column will be specific numbers from the problem, and the other will involve expressions using the variable.
   b. Using $rt = d$, fill in the distance column.
3. **Solve** an equation.
   a. Use your diagram to help you write an equation relating the distances.
   b. Solve the equation.
4. **Check.**
   a. Did you answer all the questions?
   b. Did you interpret the answer using proper units?
   c. Are your answers reasonable?

EXAMPLE 2: Using Total Distance

Two trains leave Denver at the same time. The northbound train travels 5 mi/hr faster than the southbound train. What is the rate of each train if after 3 hr they are 333 mi apart?

Answer

Let r = the rate of the southbound train and $r + 5$ = the rate of the northbound train.

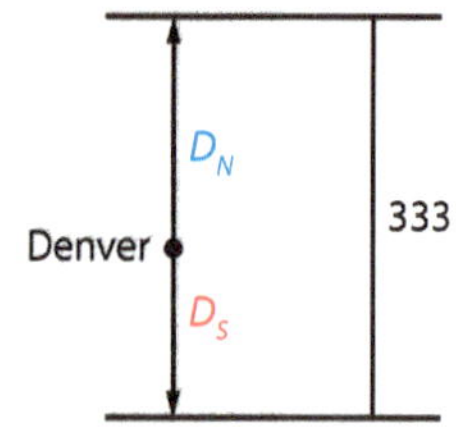

1. **Read** the problem carefully.
   a. Identify the main unknown in the problem.
   b. Draw a diagram to represent the problem.

	r	·	t	=	d
Northbound	$r + 5$	·	3	=	$3(r + 5)$
Southbound	r	·	3	=	$3r$

2. **Plan** your strategy.
   a. Fill in the rate and time columns.
   b. Using $rt = d$, fill in the distance column.

$$3(r + 5) + 3r = 333$$
$$3r + 15 + 3r = 333$$
$$6r + 15 = 333$$
$$6r = 318$$
$$r = 53$$

3. **Solve.** The sum of the trains' distances is 333 mi. Write an equation and solve.

The southbound train travels 53 mi/hr, and the northbound train travels $r + 5$ = 58 mi/hr.

4. **Check.** Be sure to give the rate of each train and to check your solution: $58(3) + 53(3) = 333$.

SKILL CHECK **EXERCISES 5, 17**

EXAMPLE 3: Using Distance Apart

Jon, a senior on the track team, runs at a training rate of 320 m/min, while Paul, a freshman, trains at 240 m/min. How long does it take for Jon to lap Paul if they are training on a 400 m track? How many laps has each run during this time?

Answer

Let t = the time of both Jon and Paul.

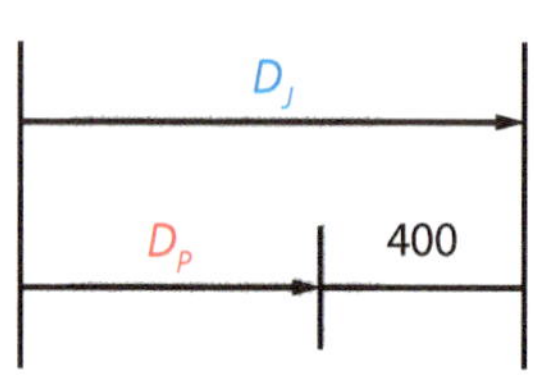

1. **Read** the problem carefully.
   a. Identify the main unknown in the problem.
   b. Draw a diagram to represent the problem. Lapping the slower runner requires the faster runner to run an extra 400 m.

AfterSchoolHelp

Motion Problems

	r	·	t	=	d
Jon	320	·	t	=	$320t$
Paul	240	·	t	=	$240t$

2. **Plan** your strategy. A table will help you organize the information.
   a. Fill in the rate and time columns.
   b. Using $rt = d$, fill in the distance column.

$320t = 240t + 400$

$80t = 400$

$t = 5$ min

$320(5) \div 400 = 4$ laps for Jon

$240(5) \div 400 = 3$ laps for Paul

3. **Solve.** The diagram illustrates that the longer distance equals the sum of the shorter distance and 400. Write an equation and solve.

4. **Check.** Be sure to answer all the questions posed in the problem. Find the distances run and the corresponding number of laps. Jon has lapped Paul in 5 min by running exactly 1 additional lap.

SKILL CHECK **EXERCISES 11, 23** ☑

A. EXERCISES

Complete the following table.

	Rate	Time	Distance
1.	3	18	
2.		2	132
3.	19		d
4.	35	$r + 2$	

Use the information provided in each exercise to complete each table.

	r	t	d
Salvador		30	
Shayne	r		

5. Salvador is running 2 ft/sec faster than Shayne during a 30 sec sprint. Combined, they run 380 yd during this time.

	r	t	d
Russells	50		
Rollinses		t	

6. The Russells leave an hour and a half before the Rollinses at an average travel speed of 50 mi/hr. The Rollinses travel 10 mi/hr faster.

In exercises 7–15, use the tables to write an equation modeling each exercise. Do not solve.

	r	t	d
Freight	50	t	$50t$
Passenger	72	t	$72t$

7. A freight train is heading east to Schenectady from Rochester and a passenger train is heading west to Rochester from Schenectady. If they are 210 mi apart, how long will it take for the trains to meet?

8. If the freight train and passenger train leave the same station traveling in opposite directions, how long will it take for them to be 183 mi apart?

9. How long will it take the passenger train to travel 110 mi farther than the freight train if they leave the station at the same time?

	r	t	d
Kashif	$\frac{3}{2}r$	3	$\frac{9}{2}r$
Steve	r	3	$3r$

10. Kashif and Steve drive home from the same location in opposite directions. How fast are they traveling if they are 330 mi apart after 3 hr?

11. What is Steve's rate in yd/min if Kashif runs 315 yd farther in 3 min?

12. Kashif and Steve are swimming at the same time. What is Kashif's rate in mi/min if they have swum a combined distance of $\frac{3}{20}$ mi in 3 min?

	r	t	d
Jen	12	$t - 1$	$12(t - 1)$
Tiana	9	t	$9t$

13. Jen left 1 hr after Tiana on her bicycle traveling in the same direction. How long did it take for her to catch up to her friend?

14. Jen and Tiana are 23 mi apart and ride their bikes to meet in the middle. If Tiana leaves 1 hr earlier than Jen, how far will she travel?

15. Tiana began her ride 1 hr earlier than Jen, but Jen rode 3 mi farther. How long was Tiana bicycling?

B. EXERCISES

Write and solve an equation for each exercise.

16. Sadie leaves Lewiston for Clarksville 388 mi away and drives 52 mi/hr. Monica leaves Clarksville for Lewiston at the same time and travels 7 mi/hr slower. How soon will they meet?

17. Mr. Taylor rides his moped down the country road in front of his house. His wife and their 2 sons ride bicycles in the opposite direction on the same road. If they leave at the same time and Mr. Taylor travels 24 mi/hr while Mrs. Taylor and the boys travel 8 mi/hr, how much time will pass before they are 24 mi apart?

18. Brantley and Shawna walk to Grandma's house at a rate of 4 mi/hr. They ride their bicycles back at 8 mi/hr over the same route that they walked. It takes 1 hr longer to walk than ride. How long did it take them to walk to Grandma's?

19. Veronica leaves the corner of First and Maple on her bicycle and travels an average speed of 9 mi/hr. Two hours later, her sister Sage leaves from First and Maple, following her on a moped traveling 27 mi/hr. How long will it take Sage to catch up with Veronica?

20. Ted runs two-thirds as fast as Frank. In 2 hr Frank runs 8 mi farther than Ted. How fast does each run?

21. A fishing boat leaves Tampa Bay at 4:00 AM and travels at 12 knots. At 5:00 AM a second boat leaves the same dock for the same destination and travels at 14 knots. How long will it take the second boat to catch up with the first one?

22. A passenger train leaves Charleston, WV, heading east at the same time a loaded coal train leaves Charleston heading west. The passenger train travels 23 mi/hr faster than the coal train, and in 4 hr the 2 trains are 412 mi apart. What is the speed of each train?

23. A roller coaster goes down the first grade 8 times faster than it goes up the other side. The distance down is 100 ft more than the distance up, and it takes 70 sec to go up and 10 sec to come down. What is the speed in feet per second of the roller coaster when it goes up and when it comes down?

24. **MP7** The Martins spend 5 hr traveling to Grandma's house and back home. How far does Grandma live from the Martins if they drove 48 mi/hr on the way to Grandma's house and 52 mi/hr on the way back home?

25. Mr. Thomas drove his old truck to the city at a rate of 40 mi/hr and drove back home on the interstate at a rate of 60 mi/hr. The total trip took 6 hr. How far is Mr. Thomas's home from the city?

26. A freight train leaves Centralia for Chicago at the same time a passenger train leaves Chicago for Centralia. The freight train travels at a speed of 45 mi/hr, and the passenger train travels at a speed of 64 mi/hr. If Chicago and Centralia are 218 mi apart, how long will it take for the trains to meet?

27. At the auto race one car travels 190 mi/hr while another travels 195 mi/hr. How long will it take the faster car to gain 2 laps on the slower car if the speedway track is $2\frac{1}{2}$ mi long?

28. **Essential Question:** Describe how to write an equation used to model the distances described in each situation. *Hint*: Refer back to diagrams in this section.

    a. the total distance of 2 objects moving toward each other to meet

    b. One object is pursuing another and they travel the same distance.

    c. One object is traveling farther than another, and the 2 objects are traveling the same direction for the same amount of time.

    d. The total distance describes 2 objects moving in opposite directions from each other.

C. EXERCISES

Write and solve an equation for each exercise.

29. **MP2** Zeke and Jay ran the 220 yd dash, and Jay won the race by 10 yd. If Jay ran the race in 30 sec, what was the rate of each boy in yards per second? How long did it take Zeke to run the 220 yd dash? Round answers to the nearest tenth.

30. Joseph and Peter bicycled at 12 mi/hr to a lake and then canoed at 4 mi/hr across it. The entire trip took $4\frac{1}{2}$ hr. If they bicycled 6 mi more than three times the distance they canoed, how far did they travel by each method?

31. Makoa is training for a triathlon. He swam at a rate of 1.5 mi/hr followed by running at a 10 min/mi pace for a total of 1 hr and 40 min. If he ran 16 times as far as he swam, what was Makoa's total distance?

32. **MP6** Anna does a 5 km rowing workout on an ergometer. Her workout includes rowing at 12 km/hr and increasing her speed by 50% for short bursts. How much of her 24 min workout was spent at each speed?

CUMULATIVE REVIEW

Solve each equation. [2.1–2.2]

33. $\frac{w}{\frac{2}{3}} = 6$

34. $0.2x - 3 = -3.14$

35. $-2(y - 3) + 4y = 24$

36. $3(2z - 1) = 4z - 2$

37. The formula $v_f = at + v_i$ relates the initial and final velocities (v_i and v_f) of an object experiencing an acceleration (a) for a given amount of time (t). Solve the formula for t. Then determine the amount of time it takes to accelerate at a rate of 32 ft/sec^2 from 40 ft/sec to 520 ft/sec. **[2.3]**

Write an equation and solve. [2.5–2.6]

38. If the population of Bridgetown increased 8% over the last decade to a population of 54,000, what was its population 10 yr ago?

39. If the annual profit of $22,440 for Morgan's Painting, Inc, represents a decrease of 12% from the previous year, what was the company's annual profit during the previous year?

40. A $200 coat was reduced 40% and then an additional 20% for a clearance sale. What is the clearance price of the coat?

41. Solve $5(ax + b) = 2(a - bx)$ for x. **[2.3]**

A $x = \frac{a}{b}$

B $x = \frac{7b}{3a}$

C $x = \frac{2a + 5b}{5a - 2b}$

D $x = \frac{5a + 2b}{2a - 5b}$

E none of these

42. A 20% discount after a markup of 30% on $45 of merchandise is calculated as **[2.6]**

A 45(0.20)(0.30)

B 45(1.30)(1.20)

C 45(1.30)(0.80)

D 45(0.70)(1.20)

E none of these

2.8 MIXTURE PROBLEMS

How can I use algebra to make a desired mixture?

After completing this section, I will be able to

- solve real-world problems involving mixtures.

Cattle feed must contain the correct mixture of nutrients.

Mixtures have many common everyday uses. Chemicals are often mixed for household cleaning purposes, treating a lawn, or spraying for bugs. Even the feed for pets or cattle is carefully mixed to ensure a properly balanced diet. No matter the application, algebra plays an important role in determining the proper balance of ingredients to maximize the effectiveness of the mixture.

Mixture problems involve the combination of different substances. Making a table containing the given information will help organize the data similarly to the method used for interest and motion problems. The problem-solving plan outlined below can be used as a guide for solving mixture problems.

SOLVING MIXTURE PROBLEMS

1. **Read** the problem carefully to find the unknown quantities. Translate each into an algebraic expression.
2. **Plan.**
   a. Make a table to organize the information in the problem. Fill in all columns except the last one.
   b. Use the information from the table to fill in the last column.
3. **Solve.**
   a. Write an equation using the information in the last column.
   b. Solve the equation.
4. **Check.**
   a. Did you answer all the questions?
   b. Did you interpret the answer using proper units?
   c. Are your answers reasonable?

EXAMPLE 1: Creating Animal Feed

A feed store owner wants to mix corn and barley to make 150 bu of silage additive for cattle feed. If corn costs \$3.80/bu and barley costs \$2.60/bu, how many bushels of each grain would he mix to yield an additive worth \$3.00/bu?

Answer

Let c = the number of bushels of corn and $150 - c$ = the number of bushels of barley.

1. Read carefully to find the unknown quantities. Since the combined total is 150 bu, assign a variable to one part and subtract from 150 to get the other part.

	Number of Bushels	Price per Bushel	Total Cost
Corn	c	3.80	$3.8c$
Barley	$150 - c$	2.60	$2.6(150 - c)$
Mix	150	3.00	450

2. **Plan.**
   a. Make a table and fill in the first 2 columns.
   b. Fill in the third column by multiplying the number of bushels by the price per bushel to get the total cost of each component.

$$3.8c + 2.6(150 - c) = 450$$
$$38c + 26(150 - c) = 4500$$
$$38c + 3900 - 26c = 4500$$
$$12c = 600$$
$$c = 50 \text{ bu of corn}$$

3. **Solve.**
   a. The total cost of the corn plus the total cost of the barley equals the total cost of the mixture. Write an equation.
   b. Multiply each term by 10 to clear the equation of decimals. Solve.

The additive should contain 50 bu of corn and $150 - c = 100$ bu of barley.

4. **Check:** $50(3.80) + 100(2.60) = 190 + 260 = 450$

SKILL CHECK **EXERCISES 7, 11** ☑

Notice there are 2 unknown amounts in the previous example but only 1 variable. It is important to represent both unknowns using the same variable to avoid introducing 2 variables.

EXAMPLE 2: Solving for an Alloy

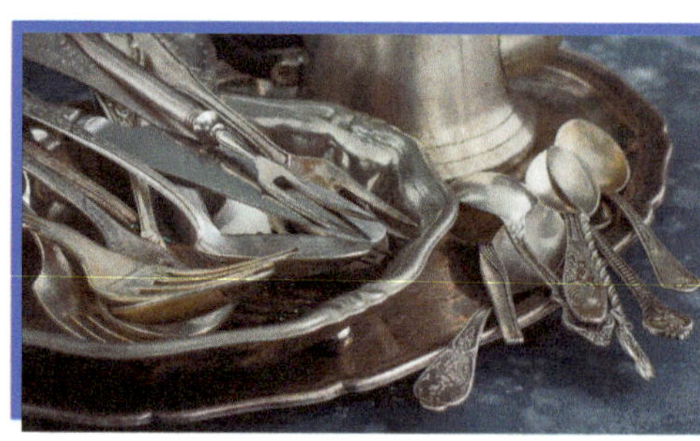

Nickel silver, often used for coating tableware, is an alloy containing 15% nickel along with copper and zinc. How much pure nickel must be melted with 70 lb of nickel silver to make an alloy that is 20% nickel?

Answer

Let p = the number of pounds of pure nickel.

1. **Read** carefully to find the unknown quantity.

	Pounds of Metal	Percent of Nickel	Pounds of Nickel
Nickel Silver	70	15%	0.15(70)
Pure Nickel	p	100%	$1p$
Alloy	$70 + p$	20%	$0.2(70 + p)$

2. **Plan.**
   a. Make a table and fill in the first 2 columns.
   b. Use the percent equation to fill in the third column.

$$0.15(70) + 1p = 0.2(70 + p)$$
$$15(70) + 100p = 20(70 + p)$$
$$1050 + 100p = 1400 + 20p$$
$$80p = 350$$
$$p = 4.375 \text{ lb of pure nickel}$$

3. **Solve.**
   a. Write an equation.
   b. Multiply each term by 100 to clear the equation of decimals. Solve.

$$0.15(70) + 4.375 = 0.2(70 + 4.375)$$
$$10.5 + 4.375 = 14.875$$

4. **Check.**

SKILL CHECK **EXERCISES 9, 13**

A simplified version of the Solving Mixture Problems procedure can be used for solving coin problems. When checking a coin problem, multiply each coin's value by the number of coins and verify that the sum equals the given total.

Coin	Value per Coin	Number of Coins	Value of Coins
pennies	\$0.01	p	$0.01p$
nickels	\$0.05	n	$0.05n$
dimes	\$0.10	d	$0.10d$
quarters	\$0.25	q	$0.25q$

EXAMPLE 3: Solving a Mixture of Coins

Chris has 17 coins, all dimes and quarters. If he has a total of \$2.45, how many of each coin does he have?

Answer

Let q = the number of quarters and $17 - q$ = the number of dimes.

1. **Read** carefully to find the unknown quantities.

Coin	Number of Coins	Value of Coins
quarters	q	$0.25q$
dimes	$17 - q$	$0.10(17 - q)$

2. **Plan.**
   Make a table to organize expressions for the numbers and values of the coins.

$$0.25q + 0.10(17 - q) = 2.45$$
$$100[0.25q + 0.10(17 - q)] = 100(2.45)$$
$$25q + 10(17 - q) = 245$$
$$25q + 170 - 10q = 245$$
$$15q + 170 = 245$$
$$15q = 75$$
$$q = \frac{75}{15} = 5$$

3. **Solve.**
   a. Write an equation expressing the total value.
   b. Multiply both sides by 100 to clear the decimals.
   c. Distribute.
   d. Combine like terms.
   e. Solve.

There are 5 quarters and 12 dimes.

$0.25(5) + 0.1(12) = 1.25 + 1.2 = \$2.45$

4. **Check.**

SKILL CHECK **EXERCISE 15**

AFTERSCHOOLHELP

Mixture Problems

EXAMPLE 4: Solving a Mixture of Several Coins

A coin bank contains four more quarters than nickels, twice as many dimes as nickels, and five more than three times as many pennies as nickels. How many of each coin is in the bank if it contains $22.25?

Answer

Let n = the number of nickels,

$n + 4$ = the number of quarters,

$2n$ = the number of dimes, and

$3n + 5$ = the number of pennies.

1. **Read** carefully to find the unknown quantities.

Coin	Number of Coins	Value of Coins
quarters	$n+4$	$0.25(n+4)$
dimes	$2n$	$0.10(2n)$
nickels	n	$0.05n$
pennies	$3n+5$	$0.01(3n+5)$

2. **Plan.** Make a table to organize expressions for the numbers and values of the coins.

$$0.25(n+4) + 0.10(2n) + 0.05n + 0.01(3n+5) = 22.25$$
$$25(n+4) + 10(2n) + 5n + (3n+5) = 2225$$
$$25n + 100 + 20n + 5n + 3n + 5 = 2225$$
$$53n + 105 = 2225$$
$$53n = 2120$$
$$n = \frac{2120}{53} = 40$$

3. **Solve.**
   a. Write an equation.
   b. Multiply both sides by 100.
   c. Distribute.
   d. Combine like terms.
   e. Solve.

$40 + 4 = 44$, $2(40) = 80$, and $3(40) + 5 = 125$

There are 44 quarters, 80 dimes, 40 nickels, and 125 pennies.

$44(0.25) + 80(0.10) + 40(0.05) + 125(0.01) = 22.25$

4. **Check.**

SKILL CHECK **EXERCISE 23** ☑

A. EXERCISES

Complete the following tables.

	Amount of Mixture	Percent of Salt	Amount of Salt
1.	90 oz	8%	
2.	15 g		4.95 g
3.		18%	14.4 L

	Amount of Mixture	Percent of Ingredient	Amount of Ingredient
4.	80 L	20%	
5.	x	50%	
6.	$x + 80$	25%	

Complete each table based on the information provided in exercises 7–10.

7. Mr. Harper, the candy store manager, wants to mix butterscotch candies and cinnamon balls to make a deluxe 20 lb mix to sell for \$3/lb. Butterscotch candy sells for \$3.40/lb and the cinnamon balls sell for \$2.90/lb.

	Pounds of Candy	Price/ Pound	Total Price
	c		$3.4c$
Cinnamon		2.90	
Mix	20		

8. Udder Brilliance is mixing cream that is 30% butterfat with milk that is 3% butterfat to make 45 gal of cream that is 12% butterfat.

	Gallons of Solution	Percent of Butterfat	Gallons of Butterfat
Cream		0.3	
	$45 - b$	0.03	
Mix	45		

9. A laboratory is adding pure water to 10 gal of 30% salt water to make a solution that is 25% salt water.

	Gallons of Solution	Percent of Salt	Gallons of Salt
Salt Water	10		
Water	w		
Mix		0.25	

10. Katie is counting her cash. She has twice as many quarters as \$1 bills and four fewer \$5 bills than \$1 bills.

Coin or Bill	Number of Coins or Bills	Value of Coins or Bills
		$0.25(2x)$
\$1 bills	x	
\$5 bills		

Choose the equation that will answer each question.

	Pounds of Tea	Price/Pound	Total Price
Indian	t	3.25	$3.25t$
Chinese	$35 - t$	4.00	$4(35 - t)$
Mix	35	3.70	129.5

11. Mr. Li owns a tea company and wants to make a blend of Chinese and Indian teas. If he wants to make 35 lb that will sell for \$3.70/lb, how many pounds of each tea should he mix?

A $4(35 - t) - 3.25t = 129.5$

B $3.25t = 4(35 - t) + 129.5$

C $3.25t + 4(35 - t) = 129.5$

D none of these

Bill	Number of Bills	Value of Bills
ones	$3x$	$1(3x)$
fives	$x + 2$	$5(x + 2)$
tens	x	$10x$

12. A cashier has two more five-dollar bills than tens, and three times as many ones as tens. If she has \$100 in her register drawer, how many of each bill is there?

A $3x + (x + 2) + x = 100$

B $1(3x) + 5(x + 2) = 10x$

C $1(3x) + 10x = 5(x + 2) + 100$

D none of these

Make a table and answer each question.

13. Determine the total number of gallons of solution if 4 gal of pure ethanol are added to 46 gal of 2% ethanol. How much ethanol is contained in the mixture? What is the strength (percentage) of the ethanol in the mixture?

14. If 3 gal of 10% salt water are boiled until there are only 2 gal left, what is the percent of salt in the remaining 2 gal?

15. Lyuba has 18 coins totaling \$3.45. If the coins consist only of dimes and quarters, how many of each coin does Lyuba have?

16. Hugh earned \$72 in tips at the car wash. He received 4 times as many one-dollar bills as five-dollar bills. How many of each bill did he receive?

B. EXERCISES

Write and solve an equation.

17. Kim sells Pistachio Pleasure, containing 80% pistachio nuts, and Pistachio Surprise, containing 30% pistachio nuts. How much of each should she mix to make 12 lb of a new mix that is 40% pistachio nuts?

18. A nurse wants 4 L of a 5% iodine solution. She already has 3 L of a 3% iodine solution. What percentage of iodine should the additional liter contain?

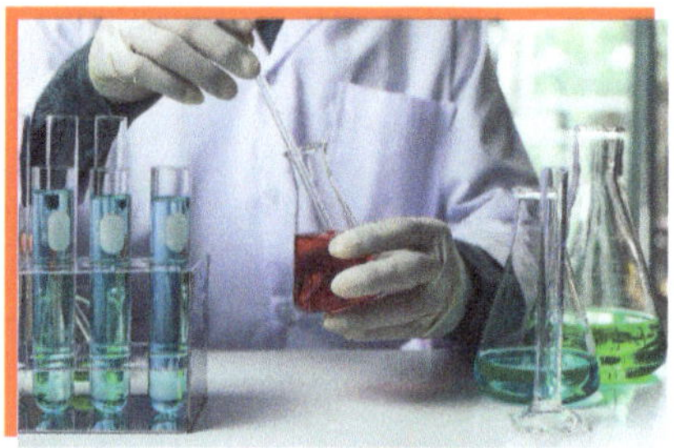

19. A chemist has 2 L of a solution that is 10% hydrochloric acid, and she wants a solution that is 25% hydrochloric acid. How much concentrated (37%) acid must she add?

20. Lynn has $8.58 in quarters and pennies. If she has eight times as many pennies as quarters, how many of each does she have?

21. A bank teller knows she received 34 dimes and nickels combined from a customer. If the change totaled $2.05, how many of each did she receive?

22. Jill has $9.96 in nickels, pennies, and dollar bills. She has twelve more pennies than nickels and five fewer dollar bills than nickels. How many of each does she have?

23. One Saturday Omar collected from his newspaper customers twice as many one-dollar bills as fives and one fewer ten than fives. If Omar collected $58, how many tens, fives, and ones did he get?

24. **MP3** Clair made the table below to help determine how much pure water to add to 5 L of 15% salt solution in order to create a 12% salt solution. Describe and correct Clair's error.

	Liters of Solution	Percent of Salt	Liters of Salt
Salt Water	5	0.15	0.75
Pure Water	w	1	$1w$
Mix	$5 + w$	0.12	$0.12(5 + w)$

25. A butcher makes a meat loaf by mixing hamburger, which sells for $1.89/lb, and sausage, which sells for $2.99/lb. How many pounds of each should he use if he is to make 30 lb of the mix and sell it for $2.33/lb?

26. Jo makes fruit salad with grapefruit and pineapple. Grapefruit costs $1.25 per 12 oz can, and pineapple costs $0.83 per 12 oz can. How many cans should she mix to obtain 84 oz of a mixture that will sell for $1.01 for 12 oz?

27. How many gallons of pure alcohol must you add to 25 gal of a 28% alcohol solution to obtain a solution that is 40% alcohol?

28. **MP2** How much water should you boil off of a 30 gal solution of 10% salt to increase it to 15% salt?

29. Josef has a jar of dimes, nickels, and pennies. He has one dime more than five times the number of nickels and four times as many pennies as nickels. How many of each coin does he have if their total value is $7.77?

30. In his pocket Chip has four more dimes than quarters. If the dimes and quarters total $2.85, how many of each coin does Chip have?

31. Sam has five more dimes than quarters and four fewer nickels than quarters. If he has a total of $4.70, how many of each does he have?

32. A teller has 36 rolls of coins in his drawer, consisting of quarters ($10 per roll), dimes ($5 per roll), and nickels ($2 per roll). He has three times as many rolls of quarters as nickels and twice as many rolls of dimes as nickels. The total is $252. How many rolls of each does he have?

33. **MP3** There are 3 more pennies than nickels and twice as many dimes as nickels for a total of $1.85. Describe and correct the error in Caden's work when determining the number of nickels.

Step	Caden's Work
1.	$0.01(n + 3) + 0.05n + 0.1(2n) = 1.85$
2.	$10(n + 3) + 5n + 1(2n) = 185$
3.	$10n + 30 + 5n + 2n = 185$
4.	$17n + 30 = 185$
5.	$17n = 155$
6.	$n \approx 9.1$ nickels

34. **Essential Question:** Explain how algebra can be used to create a desired mixture.

C. EXERCISES

Write and solve an equation.

35. Babbitt metal (named for its developer, Isaac Babbitt, who first made the metal in 1839) is used primarily for machine gearings. This metal alloy is composed of 90% tin, 7% antimony, and 3% copper. How many ounces of pure tin must be added to the alloy to make 6 oz of a metal that is 94% tin?

36. One 1700 lb load of feed mixture contains 25% corn and 40% bran; the rest is roughage. How many pounds of corn should be added to a load of feed mixture to make the mixture 40% corn?

37. **MP6** Layla counts up the spare change she has accumulated and finds that the combined value of the nickels and dimes matches that of the pennies and quarters. She has 9 more nickels than dimes, 5 fewer quarters than dimes, and twice as many pennies as quarters. How many of each coin does she have?

38. Hassan accepts payment in quarters, ones, fives, and tens at his food cart. At the end of the day, he has 15 more quarters than one-dollar bills, 6 more fives than ones, and the number of tens is $\frac{1}{3}$ the number of ones. The combined value of the tens and $\frac{1}{3}$ of the fives is equal to that of all the other bills and coins, including the other $\frac{2}{3}$ of the fives. How many of each has he collected?

CUMULATIVE REVIEW

Use the similar triangles for exercises 39–40. [2.4]

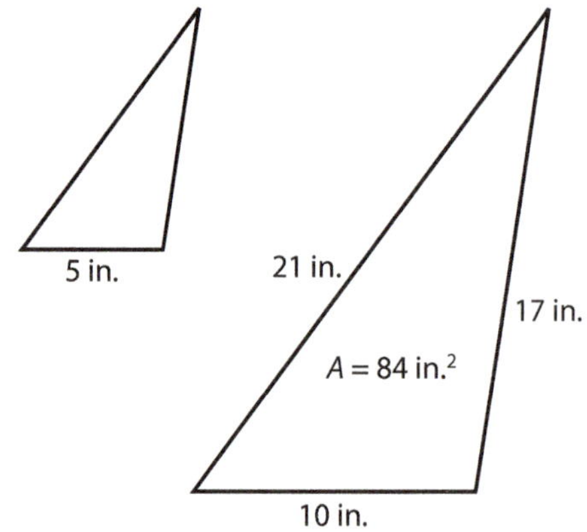

39. Find the perimeter of the smaller triangle.

40. Find the area of the smaller triangle.

Write an equation and solve.

41. A train going 80 mi/hr moves 10 times as fast as Marla runs. How fast can Marla run? **[2.1]**

42. The sum of twice a number and 7 less than 4 times the number is 15. Find the number. **[2.2]**

43. The basketball team won $\frac{4}{7}$ of the games it played. If the team lost 9 games in the season, how many did it win? **[2.4]**

44. Find the average speed of a world-class sprinter (to the nearest tenth of a mile per hour) if he runs 100 yd in 10 sec. **[2.4]**

45. The school had set a goal of having 65% of the student body involved in the fine arts program. If 100 of the school's 300 students are currently not participating in the program, by how many students did the school exceed or fall short of its goal? **[2.5]**

46. A merchant purchased a car for $20,000 and marked it up 25%. When it did not sell, he reduced it 20%. How much did he make if the car sold at this discounted price? **[2.6]**

47. Two trains simultaneously leave the same station, traveling in opposite directions at 51 and 65 mi/hr respectively. Which equation models the time at which the trains will be 290 mi apart? **[2.7]**

A $51t + 65t = 290$

B $65t = 51t + 290$

C $51t = 65t + 290$

D $65t - 51t = 290$

E none of these

48. Ms. Enplace teaches 15 students in her cooking class. The ratio of the number of students receiving an A to those not receiving an A is 2 : 3. How many students received an A? **[2.4]**

A 10

B 6

C 9

D 5

E none of these

APPLICATION PROBLEMS
TRANSPORTATION COSTS

Mr. Hill drives a work van that averages 10 mi/gal. The compact car his wife drives gets 30 mi/gal. They can afford either to replace Mr. Hill's work van with an SUV that gets 13 mi/gal or to replace Mrs. Hill's car with a hybrid that gets 60 mi/gal. Assuming they drive the same number of miles annually, which trade-in would save the most gasoline?

1. Complete the first 6 calculations in the table below describing the gasoline consumption (to the nearest gallon) and savings for the trade-ins of Mr. and Mrs. Hill's vehicles. Assume that the Hills' vehicles will each be driven 15,000 mi annually.
2. Fill in the rest of the table by writing general expressions for the calculations. Use m to represent the number of miles driven each year, d for the gas mileage (in mi/gal) of the old vehicle, and n for the gas mileage of the new vehicle.

	Vehicle	Gallons Consumed	Gallons Saved
Mr. Hill	work van		
	SUV		
Mrs. Hill	compact car		
	hybrid		
General Expressions	old vehicle		
	new vehicle		

The expression for the number of gallons saved annually can by rewritten as $m\left(\frac{n-d}{nd}\right)$.

3. Use this expression to confirm the number of gallons saved for each trade-in mentioned in the table above.
   a. SUV for work van
   b. hybrid for compact car
4. Determine the number of gallons of gasoline saved if someone who drives an old sedan getting 18 mi/gal a distance of 15,000 mi/yr replaces the vehicle with a new sedan that gets 22 mi/gal. Round to the nearest gallon.

The possible gasoline savings for 2 vehicle trade-ins can be compared by examining the ratio $\frac{n-d}{nd}$ (which gives the number of gallons per mile saved) for each trade-in.

5. What fraction of the original gasoline consumption is saved for each proposed trade-in?
   a. SUV (13 mi/gal) for work van (10 mi/gal)
   b. hybrid (60 mi/gal) for compact car (30 mi/gal)
   c. new sedan (22 mi/gal) for old sedan (18 mi/gal)
6. Compare these fractions to determine which trade-in will result in the greatest fuel savings and which will result in the least fuel savings.

A proportion can be used to find the gas mileage of a new vehicle that will produce the same gasoline savings as another trade-in.

Suppose we want to find the gas mileage of a replacement sedan for the old sedan (18 mi/gal) that would produce the *same* gasoline savings as trading in the work van for an SUV.

Solve the proportion $\frac{n-18}{18n}=\frac{13-10}{13(10)}$.

$$\frac{n-18}{18n}=\frac{3}{130}$$
$$(n-18)(130)=18n(3)$$
$$130n-2340=54n$$
$$76n=2340$$
$$n\approx 30.8\text{ mi/gal}$$

7. Determine the gas mileage of a replacement vehicle for the compact car that would produce the same gasoline savings as trading in the work van for an SUV.
8. If we want to match the fuel savings obtained by trading in the compact car for a hybrid, what must the gas mileage be for a car replacing the (a) work van and (b) old sedan?

9. If we want to match the fuel savings obtained by trading in the old sedan for a new sedan, what must the gas mileage be for a car replacing the (a) work van and (b) compact car?

Summarize your results from exercises 7–9 in the table below.

Gas Mileage of Vehicle Replacing the Following			Producing the Same Savings as the Following
Work Van	Old Sedan	Compact Car	
NA	30.8		SUV for work van
		NA	hybrid for compact car
	NA		new sedan for old sedan

10. While a hybrid may not be the best vehicle for an owner's needs, the greatest fuel savings possible would be obtained by trading in the work van getting 10 mi/gal for a hybrid getting 60 mi/gal. Determine the number of gallons saved over 15,000 mi by making this trade-in.

11. For each proposed trade-in, assuming a yearly total of 15,000 mi, use current gas prices to determine the average *monthly* savings.
   a. hybrid (60 mi/gal) for work van (10 mi/gal)
   b. SUV (13 mi/gal) for work van (10 mi/gal)
   c. hybrid (60 mi/gal) for compact car (30 mi/gal)
   d. newer sedan (22 mi/gal) for older sedan (18 mi/gal)

In the previous chapter, the return on investment (ROI) and payback period were used to analyze the benefits of energy-efficient investments. Recall the following formulas:

$$\text{payback period} = \frac{\text{amount of investment}}{\text{savings per period}}$$

$$\text{first-year ROI} = \frac{\text{savings per year}}{\text{amount of investment}}$$

Using the monthly fuel cost savings calculated in exercise 11 and the vehicle costs listed in the table below, determine the net cost, the payback period, and the first-year ROI (to the nearest tenth of a percent) for each of the following trade-ins.

Current Vehicle (age; miles)		Possible Trades			
		Used Vehicle (age; miles)		New Vehicle	
work van (3 yr; 45,000 mi)	$17,000	small SUV (3 yr; 45,000 mi)	$18,500	small SUV	$30,000
compact car (3 yr; 45,000 mi)	$16,500	hybrid (3 yr; 45,000 mi)	$21,500	hybrid	$30,000
sedan (5 yr; 75,000 mi)	$12,500			sedan	$35,000

	Trade-in	Net Cost	Payback Period	ROI
12.	work van for a small used SUV			
13.	compact car for a used hybrid			
14.	work van for a new SUV			
15.	compact car for a new hybrid			
16.	older sedan for a new sedan			
17.	work van for a new hybrid			

18. Which of the proposed trade-ins are justifiable on the basis of fuel cost savings? Explain your reasoning.

19. The table below lists typical prices of 2 different car models over the course of 6 yr. Use a calculator or a spreadsheet to complete the rows of the chart showing the yearly percent change in value for each car.

		New	1 yr old	2 yr old	3 yr old	4 yr old	5 yr old	6 yr old
Car A	Value	$27,370	$19,506	$16,453	$10,918	$8755	$7329	$6551
	Percent Change		29%					
Car B	Value	$29,045	$22,499	$18,829	$14,509	$14,031	$12,731	$11,112
	Percent Change							

20. Which car maintains its value better? Justify your answer mathematically.

2 CHAPTER REVIEW

Key Concepts

Simple Equations [2.1]

- Equivalent equations have the same solution.
- For one-step equations, use inverse operations to isolate the variable and solve the equation.
- For two-step equations, use the reverse order of operations when performing inverse operations on both sides.
- Check the solution by substituting into the original equation.
- Real-world problems can often be modeled and solved by following the basic problem-solving strategy (as outlined for 2.6–2.8).

Multistep Equations [2.2]

- To solve multistep equations:
  - Simplify the expression on either or both sides of an equation when possible before using inverse operations.
  - Eliminate parentheses by distributing.
  - If necessary, multiply both sides of an equation by the LCD to eliminate fractions.
  - Combine like terms.
  - If necessary, move all variables to one side of the equation.
- Use an algebraic equation to convert a repeating decimal to a fraction.
  - Write an equation in which x is equal to the repeating decimal.
  - Multiply both sides of this equation by 10^n, where n is the number of repeating digits.
  - Find the difference between the 2 equations and solve the resulting equation.

Literal Equations [2.3]

- Use inverse operations to isolate the desired variable.
  - Clear the equation of fractions by multiplying both sides by the LCD.
  - Simplify both sides of the equation.
  - Collect all terms containing the desired variable on one side of the equation and all other terms on the other side.
  - Use the Distributive Property to isolate the desired variable.
  - Divide both sides by the coefficient of the variable.

Ratios & Proportions [2.4]

- Use dimensional analysis to convert between different units or rates.

- Proportions can be solved using the Property of Cross Products.

  $\frac{a}{b} = \frac{c}{d} \rightarrow ad = bc$
- Use proportions to solve for a missing corresponding length, perimeter, or area in similar figures.
  - The ratio of the areas of similar figures equals the square of the ratio of any 2 corresponding sides.

Percent Equations [2.5]

- Use a proportion: $\frac{x}{100} = \frac{part}{whole}$ or the percent formula: $percent \times whole = part$.
- Percent change $= \frac{\text{amount of change}}{\text{original amount}} \times 100\%$

Money Problems [2.6]

APPLICATIONS OF THE PERCENT FORMULA	
Percent Equation	*percent × whole = part*
Markup	*markup rate × cost = markup*
Discount	*discount rate × retail price = discount*
Tip	*tip rate × bill = tip*
Commission	*commission rate × sales = commission*
Simple Interest	*principal × rate × time = interest*

Motion Problems [2.7]

THREE COMMON DISTANCE RELATIONSHIPS FOR MOTION PROBLEMS

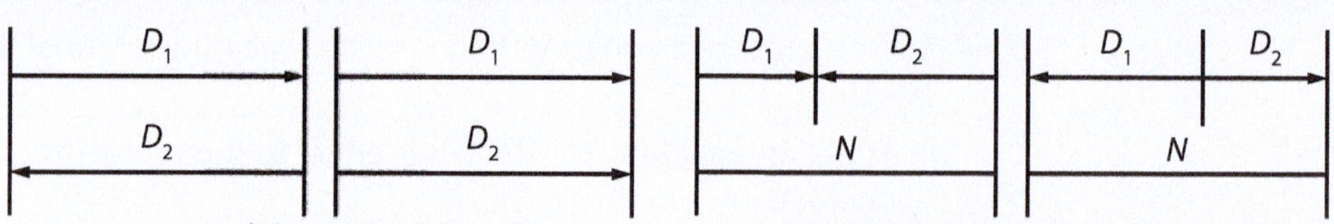

roundtrip or pursuit: $D_1 = D_2$

motion in opposite directions: $D_1 + D_2 = N$

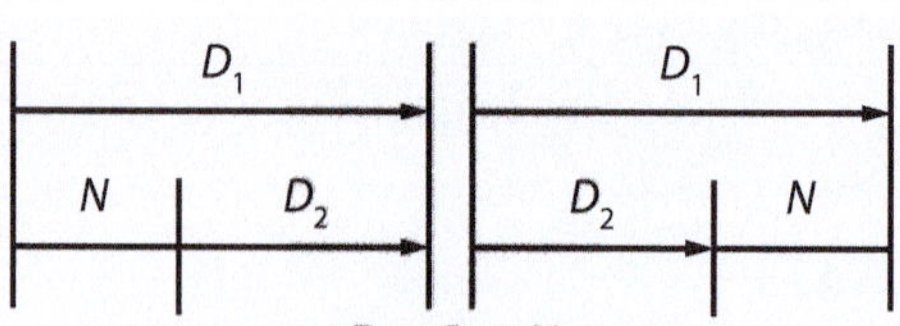

$D_1 = D_2 + N$

Solving Simple Interest, Motion & Mixture Problems [2.6–2.8]

- Read the problem carefully.
  - Find the unknown quantities.
  - Assign a variable to one unknown quantity and represent the others in terms of that variable.
- Plan.
  - Make a table to organize the information in the problem. Fill in all columns except the last one.
  - Use the information from the table to fill in the last column.
- Solve.
  - Write an equation using the information in the last column.
  - Solve the equation.

- Check.
  - Did you answer all the questions?
  - Did you interpret the answer using proper units?
  - Are your answers reasonable?

Biblical Worldview Shaping

Modeling with Algebra

Why can't a mathematical model answer a moral question?

Mathematical models are very effective in providing data so that humans can make informed decisions, but mathematics is only one of many considerations in making decisions.

EXERCISES

1. Determine whether the given number is a solution to the equation. **[2.1]**

a. $x^2 - x = 6$ if $x = -2$

b. $\frac{2}{3}y - \frac{3}{4} = -\frac{5}{12}$ if $y = \frac{1}{2}$

Solve each equation. [2.1]

2. $4\frac{4}{7}x = -\frac{6}{5}$

3. $y + 0.38 = 8.7$

4. $\frac{z}{5} - 6 = 84$

5. $6a + 21 = 165$

6. A local natural gas company charges small businesses a \$22 monthly connection fee and \$1.20/therm. Write and solve an equation to find the number of therms a company used during a month in which its gas bill was \$78.40. **[2.1]**

Solve each equation. [2.2]

7. $4b = 7 - 4b + 6$

8. $89d + 17 + 6d = 5d - 19$

9. $4(c + 10) - 146 = 6(2c + 9)$

10. $4v - 2(v + 9) + 10 = -14$

11. $\frac{x}{2} + 1 = \frac{4x}{3}$

12. $\frac{4x + 7}{4} - \frac{5x - 4}{20} = \frac{5 - x}{5}$

13. Brandon estimates he will need \$300 to purchase equipment to start his lawn mowing service. For each lawn Brandon mows, he estimates an average of \$5 of expenses and plans to charge \$25. Write and solve an equation to determine the number of lawns Brandon needs to mow to cover his equipment and expenses. **[2.2]**

Solve each literal equation for the indicated variable. [2.3]

14. $d = \frac{m}{v}$ for m

15. $y = mx + b$ for x

16. $S = c(h + r)$ for r

17. $\frac{a}{b} - \frac{1}{c} = d$ for c

18. $S = \frac{3A}{R - A}$ for A

19. Solve $a = \frac{v_2 - v_1}{t_2 - t_1}$ for v_2. Then evaluate the resulting expression to find v_2 when $a = 9.8$ m/sec², $t_2 = 5$ sec, $t_1 = 2$ sec, and $v_1 = 19.6$ m/sec. **[2.3]**

Convert the given rate to an equivalent rate. [2.4]

20. 88 ft/sec to miles per hour

21. 100 km/hr to meters per second

Solve each proportion. [2.4]

22. $\frac{12}{25} = \frac{20}{y}$

23. $\frac{x - 1}{6} = \frac{x + 1}{9}$

24. If the ratio of the masses of hydrogen and oxygen in water is 1 : 8, how many grams of each are in 90 g of water? **[2.4]**

25. $\triangle ABC \sim \triangle DEF$; the area of $\triangle ABC$ is 210 cm². **[2.4]**

a. Write and solve a proportion to find x.

b. Find the perimeter and area of $\triangle DEF$.

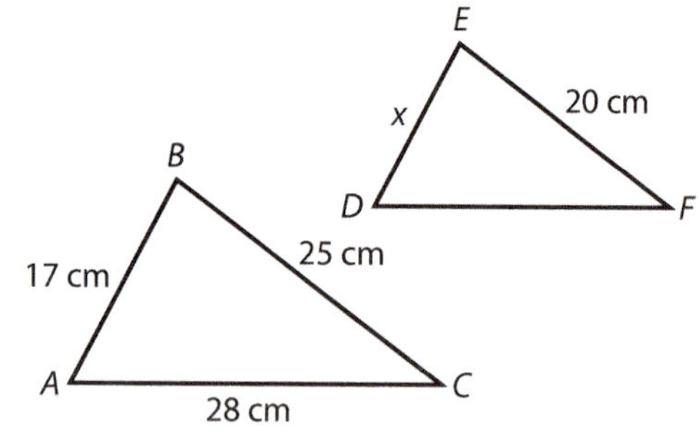

26. For each percent, fraction, or decimal, state the equivalent expressions in the other 2 forms. **[2.5]**

a. 0.5%

b. $\frac{1}{5}$

Write the appropriate percent equation and solve. [2.5]

27. 36% of what number is 54?

28. What percent of 16 is 76?

Complete the following table. [2.5]

	Original Amount	Change	Percent Change	New Amount
29.	900	add 36		
30.		subtract 35	20% decrease	

Write the appropriate percent equation and solve. Round to the nearest tenth of a percent. [2.5]

31. Thirty percent of those in attendance had never seen a lacrosse match before. How many of the 400 in attendance were watching their first game of lacrosse? How many had previously seen a game of lacrosse?

32. The price of a stock rose from a yearly low of \$9.68 to \$18.61. What was the stock's percent increase from its low price that year?

Write an equation and solve. Round to the nearest cent or to the nearest tenth of a percent. [2.6]

33. A merchant bought a boat for \$8000 and marked it up 15.5%. What is the retail price? What would the buyer have to pay if there is a 5% sales tax?

34. A traveler priced an airplane ticket at \$650. He later was able to get it for \$400. What percent did he save?

35. The retail price of a suit was \$719. It was reduced 33% and later an additional 67%. What was the price after both reductions? What was the total percent reduction of the original retail price?

36. Omar earns a 15% commission on his sales in addition to a $1500 monthly salary. Determine his monthly sales goal if he budgets $3750 as his monthly income.

37. Timothy invested some of his savings in an account yielding 4% interest and twice as much in an account yielding 6% interest. If Timothy earned $800 in interest last year, how much did he invest in each account?

Write an equation and solve. [2.7–2.8]

38. Julia drives the 120 mi from Greenup to Newton at 46 mi/hr, while Beth drives from Newton to Greenup at 34 mi/hr. How long will it take for Julia and Beth to meet each other if they leave at the same time?

39. Jorge jogged the trail to the waterfall at 120 yd/min and walked back at 80 yd/min. If the entire trip took 2 hr, how long is the trail (in yards and to the nearest tenth of a mile)?

40. Brad is 15 mi down the road when Carl leaves the rest area driving at 65 mi/hr. How long will it take Carl to catch up if Brad is driving at 60 mi/hr?

41. Nate leaves home at 6 AM and drives 55 mi/hr to Grandma's house. His brother, Caleb, leaves at 8 AM driving 65 mi/hr and arrives at Grandma's house at the same time as his brother. How far is Grandma's house? What time did the 2 brothers arrive?

42. Janessa has a total of $3.40 in nickels and quarters. If she has twenty coins, how many are nickels and how many are quarters?

43. Dan opened up his piggy bank and found a total of thirty-five coins, made up of nickels, dimes, and quarters. If he has five more nickels than dimes and a total of $6, how many of each coin were in his bank?

44. A coffee merchant wants to make a mixture of 50 lb of Colombian and Brazilian coffees that will sell for $3.19/lb. If Colombian coffee sells for $3.35/lb and Brazilian coffee sells for $2.95/lb, how much of each type of coffee should be mixed together?

45. Tickets to the play cost $5 for children and $8 for adults. If the auditorium, which holds 350 people, was sold out for all 3 performances and the net income from ticket sales was $7455, how many children and how many adults attended the play?

46. How much water should be added to 20 L of a 20% salt solution to dilute it to a 15% salt solution?

47. How much water should be evaporated from 20 L of a 20% salt solution to increase it to a 25% salt solution?

48. **BWS Essential Question:** What is the role of mathematical models in making moral decisions?

CHAPTER 03

SOLVING INEQUALITIES

CHAPTER OVERVIEW

3.1 SIMPLE INEQUALITIES

How is solving an inequality different from solving an equation?

After completing this section, I will be able to

- solve simple linear inequalities by applying the properties of inequality.
- interpret negated inequalities by using the Trichotomy Property.
- solve real-world problems by writing and solving simple linear inequalities.
- explain why valid reasoning is important in solving inequalities.

The previous chapter covered solving equations where 2 expressions are equal. However, many applications in mathematics require expressions to be compared using an inequality instead.

DEFINITION

An **inequality** is a mathematical sentence stating that 2 quantities are not always equal.

Mathematical inequalities include many phrases that signify that the 2 quantities are not equal. The following table shows the 5 basic inequality symbols as well as some common corresponding phrases.

SYMBOL	MEANING	EXAMPLE
$<$	is less than	$4 < 6$
$>$	is greater than	$5 > 3$
$\neq$	is not equal to	$-2 \neq 4$
$\leq$	is less than or equal to	$-1 \leq 8$
$\geq$	is greater than or equal to	$6 \geq 0$

Any comparison of 2 real numbers results in 1 of 3 possible outcomes.

TRICHOTOMY PROPERTY

If a and b represent 2 real numbers, then $a = b$, $a > b$, or $a < b$.

Placing a slash through the symbols $<$, $>$, $\leq$, and $\geq$ results in their negation, just as the slash through = means "is not equal to." The Trichotomy Property implies that if $x \not> 5$, then either $x = 5$ or $x < 5$ (that is, $x \leq 5$). The inequality $x < 5$ can also be expressed as $x \not\geq 5$.

Just like equations, inequalities can be true or false. Any value of the variable that makes the inequality true is a solution to the inequality.

EXAMPLE 1: Identifying Solutions to an Inequality

For each inequality, determine which of the following are solutions: {2, 3, 4}.

a. $a + 3 > 6$ **b.** $6 \geq 2b$ **c.** $1 - c \nless -2$

Answers

a. $a + 3 > 6$

$(2) + 3 \overset{?}{>} 6$

$5 \overset{?}{>} 6$; false

$(3) + 3 \overset{?}{>} 6$

$6 \overset{?}{>} 6$; false

$(4) + 3 \overset{?}{>} 6$

$7 \overset{?}{>} 6$; true

4 is a solution.

b. $6 \geq 2b$

$6 \overset{?}{\geq} 2(2)$

$6 \overset{?}{\geq} 4$; true

$6 \overset{?}{\geq} 2(3)$

$6 \overset{?}{\geq} 6$; true

$6 \overset{?}{\geq} 2(4)$

$6 \overset{?}{\geq} 8$; false

2 and 3 are solutions.

c. Rewrite as $1 - c \geq -2$.

$1 - (2) \overset{?}{\geq} -2$

$-1 \overset{?}{\geq} -2$; true

$1 - (3) \overset{?}{\geq} -2$

$-2 \overset{?}{\geq} -2$; true

$1 - (4) \overset{?}{\geq} -2$

$-3 \overset{?}{\geq} -2$; false

2 and 3 are solutions.

SKILL CHECK **EXERCISE 3**

Inequalities have properties similar to the properties of equality that we use to solve equations. Adding or subtracting the same number on both sides of an inequality yields an equivalent inequality.

PROPERTIES OF INEQUALITY	
Addition	If $a < b$, then $a + c < b + c$.
Subtraction	If $a < b$, then $a - c < b - c$.

Even though < is used in stating the properties above, the properties of addition and subtraction hold true for any of the 5 inequality symbols.

Usually, it is not possible to list all the solutions to an inequality. Therefore, the solutions to inequalities are often graphed on a number line.

EXAMPLE 2: Solving Addition & Subtraction Inequalities

Solve and graph the solution to each of the following inequalities.

a. $x+7<8$ **b.** $x-5\neq 3$ **c.** $-0.6\leq x-0.4$

Answers

a. $x+7<8$ Subtract 7 from both sides.

$\underline{-7}\ \ \underline{-7}$

$x<1$

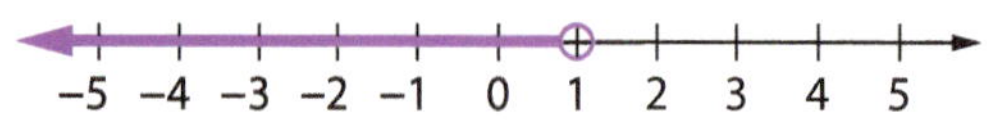

b. $x-5\neq 3$ Add 5 to both sides.

$\underline{+5}\ \ \underline{+5}$

$x\neq 8$

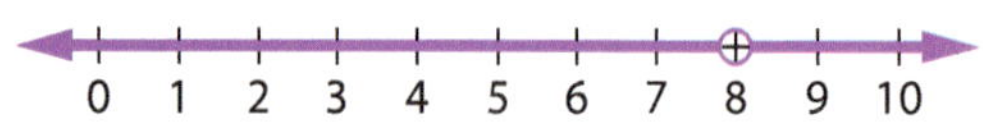

c. $-0.6\leq x-0.4$ **1.** Add 0.4 to both sides.

$\underline{+0.4}\ \ \underline{+0.4}$

$-0.2\leq x$

$x\geq -0.2$ **2.** Rewrite the inequality with the variable on the left side.

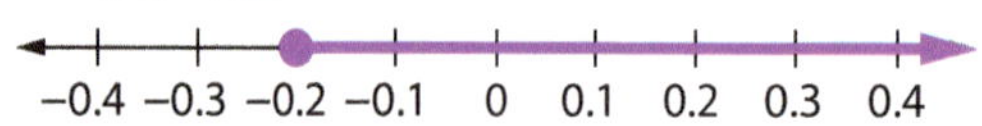

SKILL CHECK **EXERCISE 7**

TIP

If the variable is on the left side, the inequality symbol will indicate the direction the arrow should point on the graph.

As with addition and subtraction inequalities, an equivalent inequality can be created by multiplying or dividing both sides by the same number. However, when multiplying or dividing by a negative number, the symbol must be reversed to create an equivalent inequality.

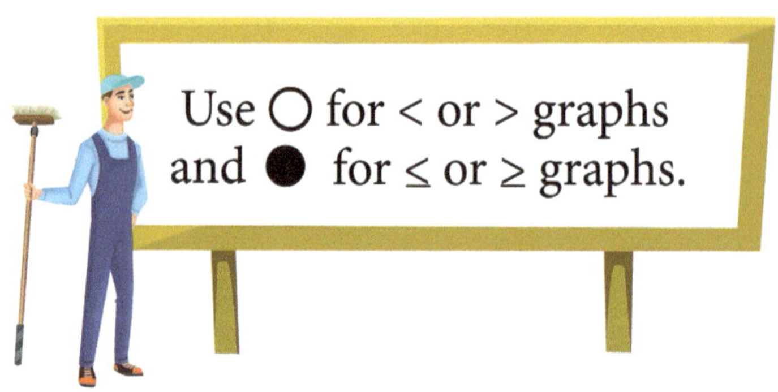

Case 1	$3<5$	Is this statement true?
	$2\cdot 3<2\cdot 5$	Multiply both sides by 2.
	$6<10$	The inequality is still true.
Case 2	$4<6$	Is this statement true?
	$-2\cdot 4<-2\cdot 6$	Multiply both sides by −2.
	$-8<-12$	The inequality is no longer true.
	$-8>-12$	The inequality must be reversed to make the statement true.

PROPERTIES OF INEQUALITY	
Multiplication	If $a<b$ and c is a positive number, then $ac<bc$. If $a<b$ and c is a negative number, then $ac>bc$.
Division	If $a<b$ and c is a positive number, then $\frac{a}{c}<\frac{b}{c}$. If $a<b$ and c is a negative number, then $\frac{a}{c}>\frac{b}{c}$.

These properties provide a valid reason for each step when solving inequalities that require multiplication or division. The properties of multiplication and division also hold true for the other 5 inequality symbols.

EXAMPLE 3: Solving Multiplication & Division Inequalities

Solve and graph the solution to each of the following inequalities.

a. $-3x < -9$ **b.** $\frac{x}{12} > -4$ **c.** $25 \le -\frac{5}{2}x$

Answers

a. $\frac{-3x}{-3} > \frac{-9}{-3}$

$x > 3$

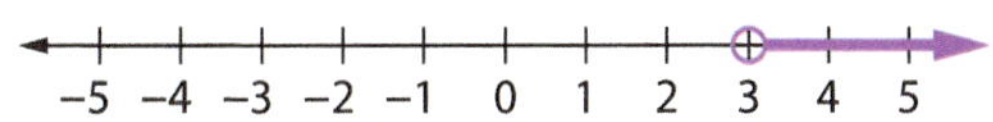

Divide both sides by −3 and reverse the inequality symbol.

b. $12\left(\frac{x}{12}\right) > 12(-4)$

$x > -48$

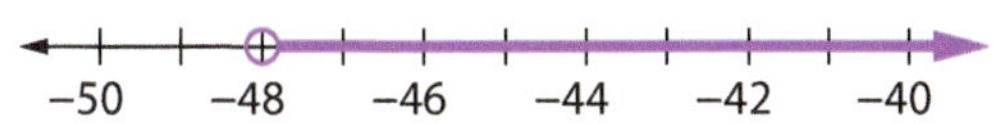

Multiply both sides by 12.

c. $-\frac{2}{5}(25) \ge -\frac{2}{5}\left(-\frac{5}{2}\right)x$

$-10 \ge x$

$x \le -10$

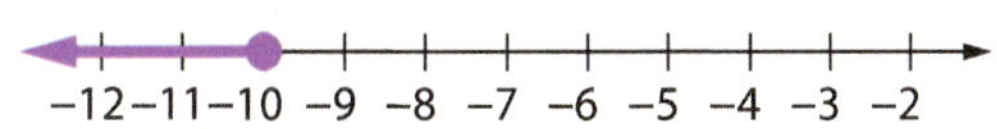

1. Multiply both sides by $-\frac{2}{5}$ and reverse the inequality symbol.
2. Express the inequality with the variable on the left side.

SKILL CHECK EXERCISE 9

Inequalities can be used to model and solve questions from everyday life.

EXAMPLE 4: Solving a Real-World Problem

How much money must be invested at 7% simple interest to produce a yearly income of at least $875?

Answer

Let P = the amount invested.

P	r	t	=	I
P	0.07	1	=	$0.07P$

$0.07P \ge 875$

$7P \ge 87{,}500$

$P \ge 12{,}500$

$12,500 or more must be invested.

$12,500(0.07)(1) = $875

1. **Read** the problem carefully. Note that the phrase "at least" signifies ≥.
2. **Plan:** Write the formula for simple interest: $I = Prt$. Make a table or substitute into the formula.
3. **Solve:** Write an equation. The interest earned must be $875.
4. **Check:** Since this product is exactly $875, any amount larger than $12,500 will yield more than $875 in interest. The answer is reasonable.

SKILL CHECK EXERCISE 29

AFTERSCHOOLHELP

Simple Inequalities

The properties of inequality can be easily applied to solve two-step inequalities. Solving two-step inequalities similar to $ax + b < c$ (where a, b, and c are real numbers and x is an unknown value) requires the same two-step process as solving two-step equations of the same form. Use inverse operations to undo any addition or subtraction and then to undo any multiplication or division.

EXAMPLE 5: Solving Two-Step Inequalities

Solve each inequality.

a. $3x + 5 > 20$ **b.** $-\frac{x}{3} - 8 \geq 12$

Answers

a. $3x + 5 > 20$

$\quad -5 \quad -5$

$\frac{3x}{3} > \frac{15}{3}$

$x > 5$

1. Subtract 5 from both sides.
2. Divide both sides by 3.

b. $-\frac{x}{3} - 8 \geq 12$

$\quad +8 \quad +8$

$-3\left(-\frac{x}{3}\right) \leq -3(20)$

$x \leq -60$

1. Add 8 on both sides.
2. Multiply both sides by -3 and reverse the inequality symbol.

SKILL CHECK **EXERCISE 15**

VOCABULARY

- inequality

A. EXERCISES

1. According to the Trichotomy Property, what 3 comparisons are possible between any 2 numbers a and b?
2. Group the following symbols in pairs of opposites: $=, <, >, \leq, \geq, \neq$.

Determine whether each number is a solution to the inequality.

3. $2m < 4$
- **a.** -2
- **b.** 2
- **c.** 4

4. $n - 7 \geq -3$
- **a.** -1
- **b.** 4
- **c.** 8

5. $2x + 5 \not> -1$
- **a.** -4
- **b.** -3
- **c.** -2

6. $8 - 3s \not\leq 13$
- **a.** -2
- **b.** 0
- **c.** 7

Graph each inequality.

7. $x + 2 > -5$ **8.** $x - 6 \neq -9$ **9.** $5x \leq -10$

10. $2x \not> 4$ **11.** $7 + z \not< 11$ **12.** $8x \not\geq 0$

Solve each inequality.

13. $\frac{1}{2}y - 13 > 41$ **14.** $-\frac{y}{4} + 5 > 8$

15. $2w + 73 \leq -23$ **16.** $3z - 5 \neq 85$

B. EXERCISES

Solve and graph each inequality.

17. $2r + 7 < -5$

18. $29 \leq 2 + 3v$

19. $-2x + 8 > 20$

20. $-4 < 11 - 3m$

21. $7x + 2 \ngeq 51$

22. $-\frac{3}{4}t < -\frac{9}{10}$

Solve each inequality.

23. $12 - 24n > -6$

24. $\frac{3}{4}y + 5 \nless -\frac{1}{4}$

25. $\frac{4}{15} > \frac{1}{10} - \frac{3}{30}w$

26. $\frac{11}{8}y + \frac{5}{6} \nless -\frac{1}{4}$

Write an inequality and solve; then use a word phrase to state your answer.

27. Find all the numbers such that the sum of the number and 25 is less than 82.

28. Find all the numbers such that one-third of the number is at most five.

29. An elevator is rated with a maximum weight capacity of 750 lb. How many teenagers with an average weight of 150 lb can safely ride the elevator?

30. A season pass to the amusement park costs \$117, and daily admission costs \$39. Determine the number of trips to the park for which paying daily admission is more expensive than purchasing a season pass.

31. **MP4** The current leader in the super combined skiing event had a downhill time of 84.96 sec and a slalom time of 45.12 sec. The final competitor had a downhill time of 84.49 sec in her first run. If the winner is determined by the total of the times in the 2 races, what slalom times will allow the final competitor to win the gold medal?

32. Find the least positive integer such that 4 more than 7 times the number is greater than 95.

33. Mr. Sanders prefers to golf when it is above 59°F. Use the formula $F = \frac{9}{5}C + 32$ to determine the temperatures in Celsius at which he will enjoy golfing as he vacations in Europe.

34. Mr. Daniels is paid a \$250 weekly salary and receives a 15% commission on any sales he makes. What is his weekly sales goal if he needs to make at least \$700 a week?

35. **BWS Essential Question:** State a valid reason for reversing the inequality symbol when solving $-3x - 2 < 4$. Explain the importance of valid reasoning when solving inequalities.

36. **MP3** John has made \$237 in sales. Describe and correct his error in determining how much more he needs to make in sales to meet or exceed his goal of \$500.

Step	John's Work
1.	$237 + s \geq 500$
2.	$s \geq 263$
3.	John needs at most \$263 in sales.

37. MP3 Describe and correct Clara's error in solving $-\frac{5}{2} \leq 3h + 5$.

Step	Clara's Work
1.	$-\frac{15}{2} \leq 3h$
2.	$-\frac{5}{2} \geq h$
3.	$h \leq -\frac{5}{2}$

38. **Essential Question:** State 2 differences between solving $-2y + 8 = 1$ and $-2y + 8 > 1$.

C. EXERCISES

Solve each inequality.

39. $4 \cdot 3 \nleq 4x - 3x + 12$

40. $7 \ngtr 2m - (3m + 4)$

41. Is -1 a solution to $\frac{x - 3^2}{4x} + \frac{\frac{1}{2} - \frac{x}{2}}{1 - x} \ngtr \sqrt{2^3 - x}$? Show all work.

42. Suppose that $x < 6$ and that $x > -2$ as well. Graph the solutions for x on a number line.

CUMULATIVE REVIEW

Use the Distributive Property to rewrite the following formulas. [1.6]

43. $P = 2l + 2w$

44. $S = 2\pi r(h + r)$

State the property that justifies each step in the solution of the equation $\frac{1}{3}x + \frac{4}{9} = \frac{1}{6}(3 + x)$. [2.1–2.2]

45. $18\left(\frac{1}{3}x + \frac{4}{9}\right) = 3(3 + x)$

46. $6x + 8 = 9 + 3x$

47. $3x = 1$

48. $x = \frac{1}{3}$

Write an equation and solve. [2.7–2.8]

49. Michael and Gabriel bike home in opposite directions after a ball game. If Michael travels 11 mi/hr and Gabriel travels 7 mi/hr, how long will it take for them to be 6 mi apart?

50. How many ounces of water must Lucy add to 15 oz of 20% iodine solution to dilute it to 15% iodine?

51. Which of the algebraic expressions correctly translates "the product of 3 consecutive even numbers"? **[1.5]**

A $2n(2n + 1)(2n + 2)$

B $n(n + 1)(n + 2)$

C $2n(4n)(6n)$

D $n(n + 2)(n + 4)$

E none of these

52. Solve $11x - 5 - (3 - x) = 4(3x + 2)$ for x. **[2.2]**

A $x = -8$

B $x = -5$

C $x = 0$

D identity; all real numbers

E no solution

3.2 MULTISTEP INEQUALITIES

How can I write $-2 < x$ in a more comprehensible form?

After completing this section, I will be able to

- simplify inequalities by applying the Distributive Property.
- solve inequalities containing variables on both sides.
- solve real-world problems by writing and solving multistep inequalities.

Solving multistep inequalities requires you to "undo" the operations done to the variable by using inverse operations in reverse order. Similar to the process of solving multistep equations, applying the Distributive Property is often the first step.

EXAMPLE 1: Applying the Distributive Property

Solve each inequality.

a. $5x - 8x + 6 > 24$ **b.** $3(2 + 2x) - 4x \geq 10$

Answers

a. $5x - 8x + 6 > 24$ — **1.** Combine like terms.

$-3x + 6 > 24$

$-3x > 18$ — **2.** Subtract 6 from both sides.

$x < -6$ — **3.** Divide both sides by −3. Reverse the inequality symbol.

b. $3(2 + 2x) - 4x \geq 10$ — **1.** Apply the Distributive Property.

$6 + 6x - 4x \geq 10$

$6 + 2x \geq 10$ — **2.** Combine like terms.

$2x \geq 4$ — **3.** Subtract 6 from both sides.

$x \geq 2$ — **4.** Divide both sides by 2.

SKILL CHECK **EXERCISES 1, 3**

For inequalities with variables on both sides, use the properties of inequality to move all variables to 1 side of the inequality.

EXAMPLE 2: Solving an Inequality with Variables on Both Sides

Solve $6x + 10 \leq 3x - 2$.

Answer

$3x + 10 \leq -2$ — **1.** Subtract $3x$ from both sides.

$3x \leq -12$ — **2.** Subtract 10 from both sides.

$x \leq -4$ — **3.** Divide both sides by 3.

SKILL CHECK **EXERCISE 7**

More complicated inequalities may require a couple of simplifying steps to both sides of the inequality at the beginning of the solving process. If variables are on both sides of the inequality, division by a negative number can be avoided by adding the opposite of the variable with the smaller coefficient to both sides.

EXAMPLE 3: Solving a Multistep Inequality

Solve $3x - (2x - 4) \geq 5(x + 6) - 2x$.

Answer

$3x - 2x + 4 \geq 5x + 30 - 2x$	**1.** Apply the Distributive Property.
$x + 4 \geq 3x + 30$	**2.** Combine like terms.
$4 \geq 2x + 30$	**3.** Combine the variables by subtracting x from both sides.
$-26 \geq 2x$	**4.** Subtract 30 from both sides.
$-13 \geq x$ or $x \leq -13$	**5.** Divide both sides by 2.

SKILL CHECK **EXERCISE 25**

Many real-world problems require working within constraints. Often these constraints involve working within maximum or minimum limits.

EXAMPLE 4: Solving a Real-World Problem

Geoffrey wants to fence in a rectangular garden and has up to 60 ft of fencing available. If the width must be 5 ft more than half the length, what are the possible dimensions for the garden?

Answer

length $= l$ width $= \frac{l}{2} + 5$	**1.** Choose a variable for the length and then express the width in terms of the length.
$2l + 2w \leq 60$	**2.** Write an equation. The perimeter of the garden (rectangle) is at most 60 ft.
$2l + 2\left(\frac{l}{2} + 5\right) \leq 60$	**3.** Substitute the variable expression used for the width of the rectangle.
$2l + l + 10 \leq 60$	**4.** Use the Distributive Property to remove the parentheses.
$3l + 10 \leq 60$	**5.** Combine like terms on the left side.
$3l \leq 50$	**6.** Subtract 10 from both sides.
$l \leq 16\frac{2}{3}$	**7.** Divide both sides by 3. Note that $\frac{2}{3}$ ft is the same as 8 in.

The length is at most 16 ft 8 in., and the width is at most 13 ft 4 in.

SKILL CHECK **EXERCISE 29**

AfterSchoolHelp

Multistep Inequalities

A. EXERCISES

Solve each inequality.

1. $4b - 12 - 3b < -142$

2. $14y \geq -5 + 13y$

3. $5(c - 2) > -5$

4. $4a - 4 - 6a > 10$

5. $-3(x - 7 - 4) \leq 9$

6. $-\frac{a}{4} - 6 + a \neq 9$

7. $2m - 9 < 3m + 9$

8. $4n + 5 \geq 9n - 2$

Solve each of the following inequalities and match them with the graph of the solution.

9. $2x - x - 1 \neq -7(8)$

10. $2 \cdot 10^2 \leq 3x - x$

11. $6(x + 1) \geq -6$

12. $6 > -2(1 - x)$

13. $2x \neq 4 + x$

14. $9x \nleq 8x - 1$

A
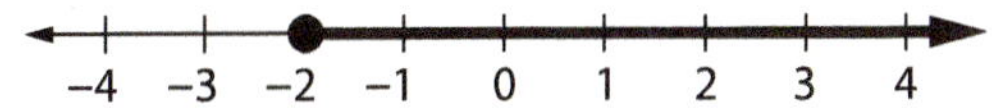

B
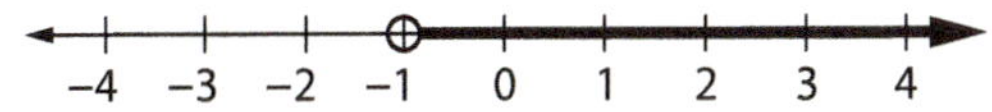

C
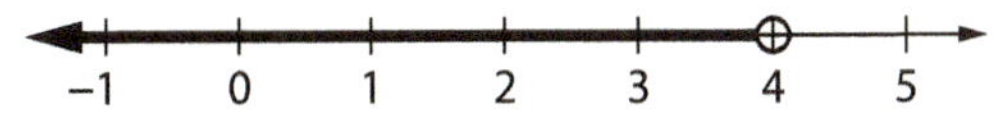

D
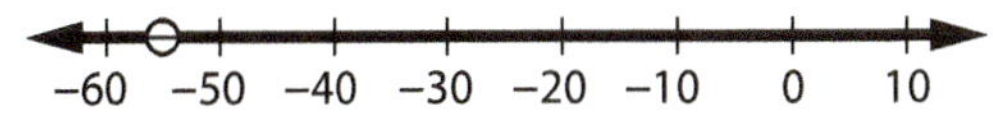

E
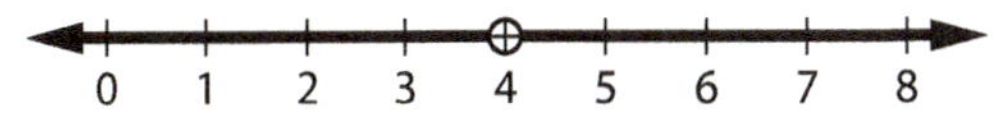

F
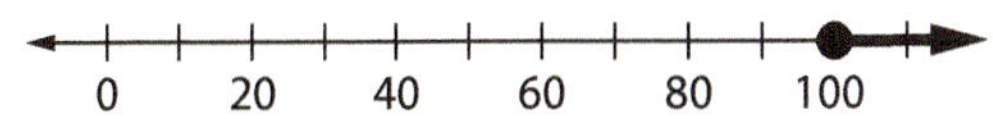

B. EXERCISES

Solve each inequality.

15. $10(2z + 3) \nless 21z$

16. $-6r + 3 \leq -5r - 2r + 6$

17. $-8s + 4s + 6 < 3s + 6$

18. $4 - (3y + 2) > 6 - 5y$

19. $9y - 7 \leq 6y - 3 + 2y$

20. $x - 5 + 3x \neq 2x - 4$

21. $-\frac{8}{3}z + 2z - 6 \leq 3z + 5$

22. $\frac{11}{16}c - \frac{1}{6} \leq \frac{7}{12}c + \frac{5}{8}$

23. $6(-2x + 4) \leq 3 - 5x + 7$

24. $-2(x - 3) \leq 5 - x$

25. $5x - 4x > 3(2x - 4) - 6x$

26. $-7(d - 1) + d > 9 - 2d$

Write an inequality and solve.

27. Dante needs an average test score of at least 90 to get an A in a course. If he has received grades of 95, 78, and 91 on previous tests, what grade does he need to earn on his final test to get an A?

28. Find the greatest 3 consecutive integers such that the sum of the first and third is less than 20.

29. Heston earns twice as much per week as his sister Naomi. If she makes \$30 per week more than her other brother Jake and the sum of all their salaries is at most \$330, what range can their salaries take each week?

30. Pam wants to buy three times as many cans of green beans (62¢ per can) as cans of peas (69¢ per can). If she has no more than \$10 to spend, what is the maximum number of cans of peas she can buy?

31. Each ounce of whole milk contains twice as many calories as an ounce of skim milk. A mixture of 4 oz of whole milk and 2 oz of skim milk contains at least 100 calories. What is the minimum number of calories in each ounce of whole and skim milk?

32. Jed bikes for 3 hr, and his brother bikes for 4 hr. Although Jed bikes 15 mi/hr faster, his total distance is at most twice as far as his brother's. How fast does each bike?

33. The president of First State Bank must invest the bank's funds so that the total interest income for the bank will be at least \$1.5 million annually. If the bank's current investments will earn \$1.3 million in interest this year, how much must he invest in municipal bonds that yield 8% interest annually?

34. **MP3** **Essential Question:** Although Jess got the correct answer, her work in solving $12 < -3(x + 1)$ contains 2 common errors students make when solving inequalities. Which steps contain the errors? Explain what Jess did incorrectly.

Step	Jess's Work	Reasoning
1.	$12 < -3x - 3$	Distributive Property
2.	$15 < -3x$	Addition Property of Inequality
3.	$-5 < x$	Division Property of Inequality
4.	$x < -5$	Rewrite the inequality with the variable on the left side.

C. EXERCISES

Solve each inequality.

35. $3x + 3 - 5x \nleq 7(2x - 3) - (6x - 4)$

36. $\frac{x+8}{3} - \frac{x}{2} \neq \frac{5}{6} - (x + 4)$

37. **MP1** $\frac{q+11}{5} - \frac{q+3}{2} \nless \frac{8-q}{3} - \frac{q+5}{10}$

Write an inequality and solve.

38. Cassie hopes to make money selling her crafts at an upcoming festival. Her revenue is \$5 for each item sold. It costs \$2 to produce each item and \$77 to rent space at the festival. Cassie assumes she'll make 5 more items than she sells. What is the minimum number of items she should craft so that her revenue is not less than her costs?

39. **MP2** Mr. Stevens invests \$2500 between 2 accounts. One account earns 5% interest, and the other earns 8%. Since he cannot afford the early withdrawal penalty in the 8% account, he wants to invest as much as possible at 5% and still earn at least \$150 this year in interest. What is the least amount that he can invest at 8% and still earn the desired amount of interest?

40. Water from the Dead Sea is eight times saltier than ocean water. Together, 50 gal of Dead Sea water and 100 gal of ocean water contain at least 20 gal of salt. What is the percentage of salt in each water source?

CUMULATIVE REVIEW

Solve each equation. [2.1–2.2]

41. $\frac{2}{3}x = \frac{8}{9}$

42. $4y - 3 = 12$

43. $5z - (2z + 3) = 12$

44. $5(w + 3) = 2w - 5$

Solve each literal equation for the indicated variable. [2.3]

45. $a + b + x = c$ for x

46. $3x - 2y = 6$ for y

Clear each equation of fractions or decimals and then solve. [2.2, 2.6]

47. $4.6a - 3.1 = 10.7$

48. $\frac{5}{3}x - \frac{1}{2} = 12$

49. Which number is a solution of $-3n + 1 \geq 11$? **[3.1]**

A -3

B 3

C -2

D all of these

E none of these

50. Which expression is equivalent to $m \ngtr 2$? **[3.1]**

A $m < 2$

B $2 > m$

C $m \geq 2$

D $2 \geq m$

E none of these

3.3 CONJUNCTIONS

How can graphs provide the solution for a conjunction?

After completing this section, I will be able to

- define a conjunction.
- solve conjunctions.
- write conjunctions that model real-world applications.

In mathematics, a *compound sentence* is made by connecting 2 statements using the word *and* or the word *or*. Compound sentences connected with the word *or* are called disjunctions and will be studied in the next section.

DEFINITION

A **conjunction** is a compound sentence consisting of mathematical statements connected by the word *and*.

For a conjunction to be a true statement, both parts of the conjunction must be true. The solution of a conjunction is the intersection of the 2 mathematical statements.

EXAMPLE 1: Graphing a Conjunction

Graph the solution set of $x > -1$ and $x \leq 4$.

Answer

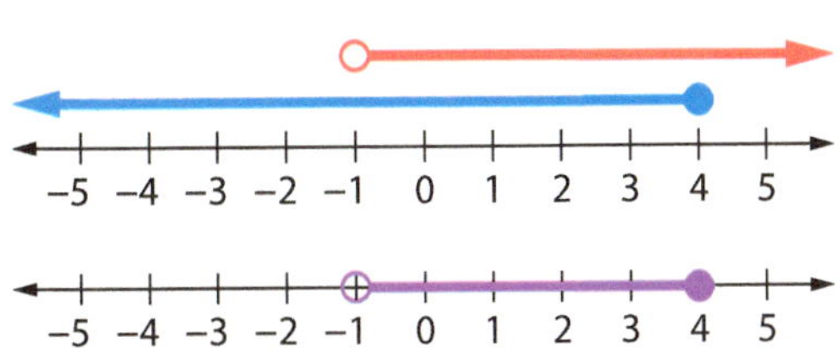

1. Graph each inequality above the number line.

2. Identify the intersection of the 2 graphs and graph it on the number line.

SKILL CHECK **EXERCISE 3**

The intersection of 2 simple inequalities is often written as a single statement. The solution to Example 1 can be expressed as the combined conjunction $-1 < x \leq 4$, which can be read "-1 is less than x, which is less than or equal to 4." Alternatively, we can read the same statement from the middle as "x is greater than -1 and less than or equal to 4."

Just like equations, conjunctions can produce varying results, such as no solution or a solution summarized with just 1 inequality.

EXAMPLE 2: Determining Solutions of Conjunctions

Graph and state the solution to each of the following.

a. $x \geq 5$ and $x < 0$ **b.** $2x + 5 \leq 11$ and $-3x > 18$

Answers

a. $x \geq 5$ and $x < 0$

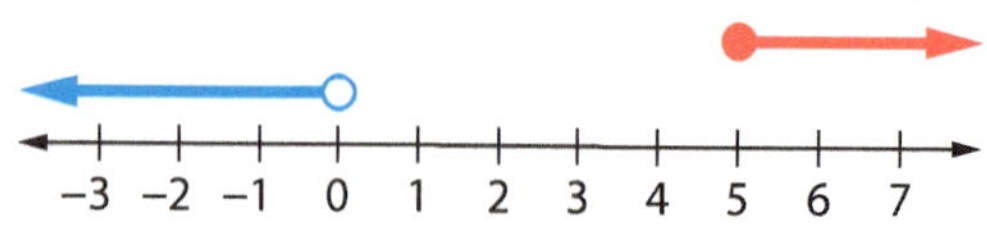

1. Graph each inequality above the number line.

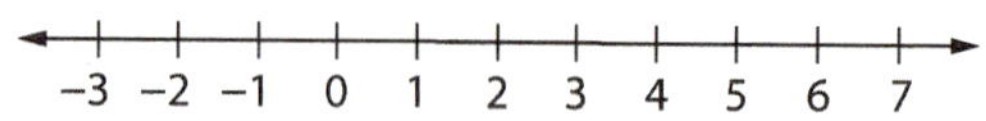

2. The 2 graphs have no points in common, so there are no points in the solution set.

$\varnothing$

3. The solution is the empty set.

b. $2x + 5 \leq 11$ and $-3x > 18$

$2x + 5 \leq 11$ and $-3x > 18$

$2x \leq 6$ $x < -6$

$x \leq 3$

1. Solve each inequality separately. Reverse the inequality symbol when dividing by a negative.

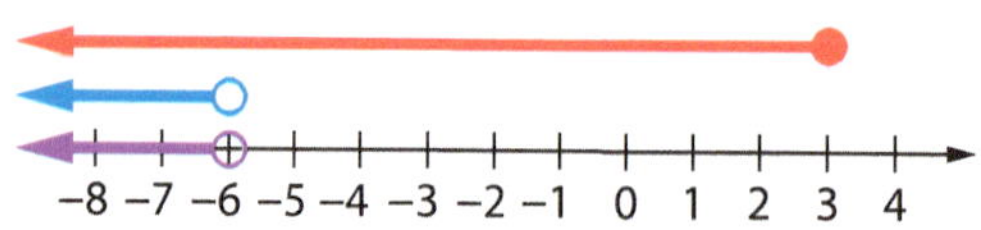

2. Graph each inequality above the number line. Then graph their intersection on the number line.

$x < -6$

3. Write the solution for the intersection.

SKILL CHECK **EXERCISES 5, 9**

The combined conjunction $-1 < 3x + 5 < 8$ can be separated into 2 inequalities joined by the word *and.*

$-1 < 3x + 5$ and $3x + 5 < 8$

Since the same steps would be used to solve each separate inequality, the properties of inequality can be used on each of the 3 parts of the combined conjunction.

EXAMPLE 3: Solving a Combined Conjunction

Solve and graph the solution set for $-1 < 3x + 5 < 8$.

Answer

$-6 < 3x < 3$

1. Subtract 5 from each expression.

$-2 < x < 1$

2. Divide each expression by 3.

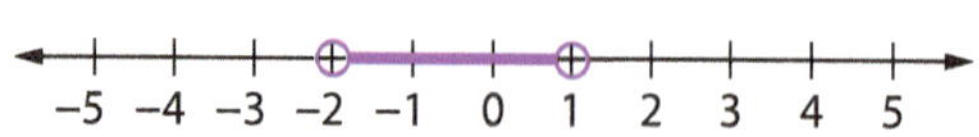

3. Graph the solution.

SKILL CHECK **EXERCISE 23**

When the inequalities that form the combined conjunction require different steps in their solutions, the combined conjunction must be rewritten as 2 simple inequalities separated by the word *and* before solving.

EXAMPLE 4: Solving a Combined Conjunction

Solve and graph the solution set of $x - 3 \le 2x - 9 < 15$.

Answer

$x - 3 \le 2x - 9$ and $2x - 9 < 15$

1. Separate the combined conjunction into 2 inequalities separated by *and*.

$-3 \le x - 9$ and $2x < 24$

$6 \le x$ $\quad$ $x < 12$

$x \ge 6$

2. Solve each inequality.

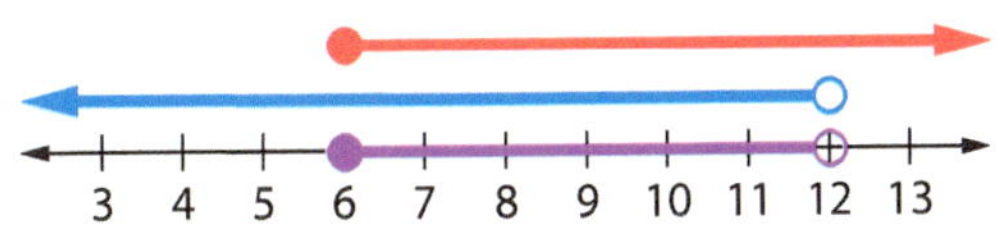

3. Graph each inequality above the number line. Then graph their intersection on the number line.

$6 \le x < 12$

4. Write the solution as a compound inequality.

SKILL CHECK **EXERCISE 15**

AfterSchoolHelp

Conjunctions

EXAMPLE 5: Translating a Compound Inequality Statement

Find the numbers such that 10 more than 3 times a number is more than –8 and no greater than 12.

Answer

10 more than 3 times a number is more than –8 and no greater than 12.

$3n + 10 > -8$ and $3n + 10 \le 12$

$-8 < 3n + 10 \le 12$

1. Rewrite as a compound inequality.

$-18 < 3n \le 2$

$-6 < n \le \frac{2}{3}$

2. Solve.

SKILL CHECK **EXERCISE 33**

Compound inequalities can be used to solve problems related to ranges of values. The grade assigned in a class, the percentage of income that is paid in taxes, and the strength of a hurricane are all based on a range of values.

EXAMPLE 6: Solving a Real-World Problem

If Brittany received a 73 and an 85 on her first 2 tests, what must she earn on her final test to receive a B in the class?

Grade	Test Average
A	90
B	80
C	70
D	60
F	below 60

Answer

Let t = the score on the third test.

1. **Read** the problem carefully. Define a variable for the unknown quantity.

$80 \leq \dfrac{73 + 85 + t}{3} < 90$

2. **Plan:** Write a combined inequality.

$240 \leq 158 + t < 270$

$82 \leq t < 112$

3. **Solve:** Multiply each expression by 3. Then subtract 158 from each expression.

Brittany must get at least an 82. A score of 112 needed to get an A is not possible.

4. **Check:** $82 + 73 + 85 = 240$ and $240 \div 3 = 80$. Any grade below an 82 results in an average less than 80.

SKILL CHECK **EXERCISE 35**

VOCABULARY

- conjunction

A. EXERCISES

1. In a conjunction, inequality statements are joined by the word ______.
2. The solution of a conjunction with no values in common is ______.

Graph the intersection of the given inequalities on a number line.

3. 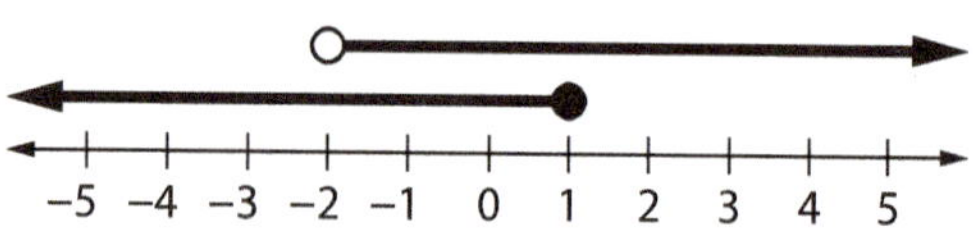

4. 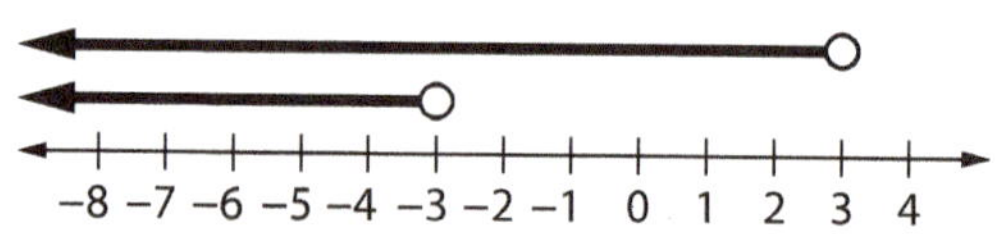

5. 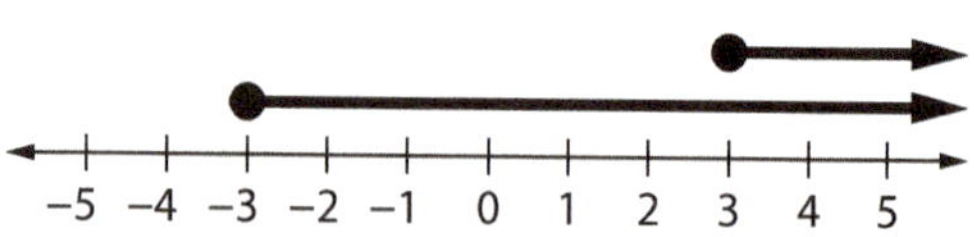

6. 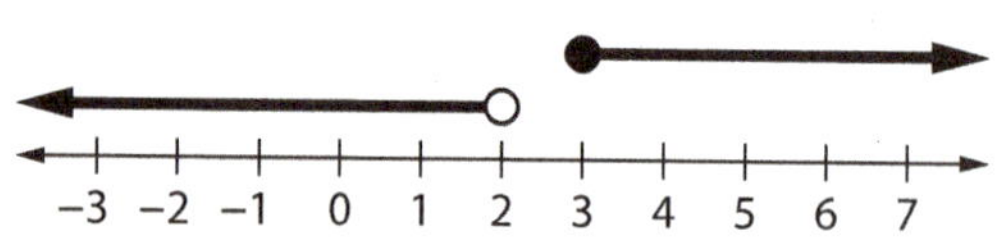

Graph each conjunction.

7. $x > 4$ and $x \geq 2$
8. $x \leq 3$ and $x < 6$
9. $x > -3$ and $x \leq 1$
10. $x > 5$ and $x \leq -2$
11. $x < -1$ and $x \geq 4$
12. $x \leq 5$ and $x \geq 3$

B. EXERCISES

Solve. Express the answer as an inequality statement, an equality, or the empty set.

13. $2x > -6$ and $x \leq 1$
14. $-5x + 13 > -7$ and $4x - 6 \leq 10$
15. $x - 3 \leq 4x$ and $2x + 6 < 18$
16. $9x + 2 < -7$ and $-2x - 8 \leq 12$

17. $2x - 1 < 5$ and $3x + 2 > -4$

18. $4x - 8 < 2x + 5$ and $4x + 4 > -2x - 4$

19. $4x - 2 - 5x > 10$ and $3x - 6 \le 12$

20. $8x - 14 + x - 6 \le 7$ and $x - 16 \ge 0$

21. $x \ne -3$ and $x = -3$

22. $4x \ge 8$ and $x + 6 \le 8$

Solve each combined conjunction. Express the answer as an inequality statement or the empty set.

23. $-2 < x + 4 < 7$

24. $-4 \le x + 3 \le -8$

25. $|-11| < 3x - 1 < 20$

26. $3 < \frac{x}{2} + 5 \le 7$

27. $-6 \le -2x + 4 < -2$

28. $20 > 4 - 6x > -16$

29. **Essential Question:** How can graphs be used to find the solution of a conjunction?

Use the diagram displaying the layered zones of the ocean for exercises 30–31.

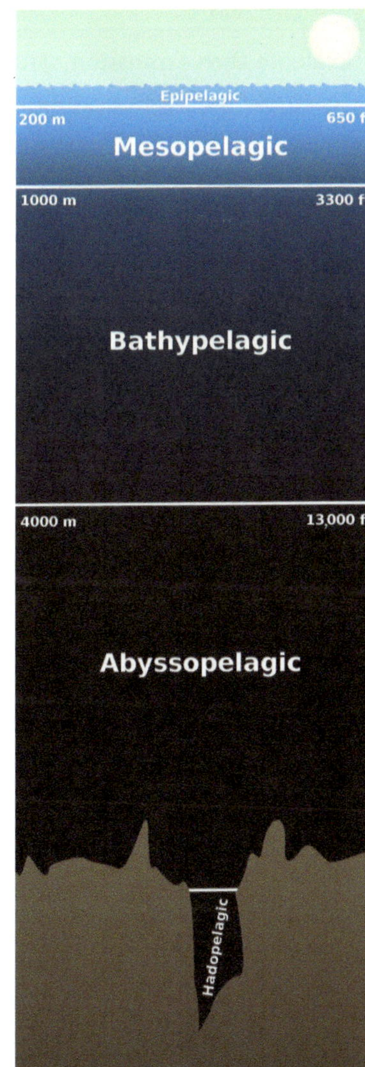

30. God's special design for a small deep-sea lantern shark found in the bathypelagic zone includes a visible light from a chemical reaction (bioluminescence) at the tip of its spine, used as a deterrent for predators. Write an inequality to describe the depth (in feet) at which these sharks live.

31. The lantern shark and other lantern fish are known for their vertical migrations between the mesopelagic zone, where they live during the day, and the epipelagic zone, where they feed at night. Write an inequality describing the depth (in meters) at which lantern fish live during the day.

32. Arches National Park classifies an opening with a span of at least 3 ft as an arch. The longest arch in the park is Landscape Arch, 306 ft long. Write a compound inequality for arch spans in the park.

Write an inequality and solve.

33. Find the numbers such that four less than half the number is at least five and at most ten.

34. Find the angles whose supplements measure more than three times the angle and whose complements measure less than twice the angle.

35. The Saffir-Simpson Hurricane Scale rates a storm's intensity according to its sustained wind speed. By the time a hurricane reached the coast, its sustained winds had decreased by 25 mi/hr, so it was downgraded to a Category 3 hurricane. Determine the possible sustained wind speeds of the hurricane at its peak.

Category	Wind Speed
1	74–95 mi/hr
2	96–110 mi/hr
3	111–30 mi/hr
4	131–55 mi/hr
5	156+ mi/hr

36. Hooke's law relates the distance a spring is stretched, d, to the force, F, applied to the spring by the formula $F = kd$, where k is an experimentally determined constant. An industrial scale has $k = 15$ lb/in. Determine the distances that the spring will be stretched as the scale weighs car parts between 240 lb and 360 lb.

C. EXERCISES

Solve each conjunction.

37. $2x < x - 4 < 5$

38. $x + 7 < 3x - 5 < 2x + 10$

39. $1 - \frac{2x}{3} < -10$ and $2 - \frac{4x}{5} > -6$

40. $3(x - 5) + 7 \leq 4 + x - 20$ and $x - 3(x + 1) \leq 5$

41. A pen whose length is twice its width is to be fenced in with an area of at least 800 ft^2. No more than 150 ft of fencing is to be used. Find the range of widths and lengths that can be used.

42. A plumbing retailer sells three times as many of faucet A as he does of faucet B. Faucet A sells for \$75, and faucet B sells for \$60. How many of faucet B should he order if he wants to sell at least 100 faucets and if he wants to gross at least \$10,000?

CUMULATIVE REVIEW

Evaluate each expression. [1.4]

43. $\frac{175}{36} \cdot \frac{45}{140} - \frac{35}{64}$

44. $\frac{175}{36} \div \frac{45}{140} - \frac{64}{27}$

Solve each proportion. [2.4]

45. $\frac{x}{36} = \frac{7}{8}$

46. $\frac{4}{13} = \frac{10}{x}$

47. $\frac{x + 2}{5} = \frac{5}{12}$

48. $\frac{2x + 1}{9} = \frac{4x - 3}{15}$

Use proportions to solve. [2.4]

49. Twelve of the 46 trees on Mr. Schmidt's property are oak trees, and the rest are pine trees. What is the ratio of pine trees to oak trees?

50. A map has a scale of 1 in. : 80 mi. How far apart on the map are 2 towns that are 260 mi apart?

51. Which equation can be used to determine how many gallons of water, w, to add to 10 gal of 25% salt water to make 15% salt water? **[2.8]**

A $10(0.25) + w = 0.15(10 + w)$

B $10(0.25) = 0.15(10 + w)$

C $10(0.25) = 0.15(10 - w) + w$

D $0.25(10 - w) = 0.15w$

E none of these

52. What is the solution set for $-18 \leq -\frac{2}{3}t$? **[3.1]**

A $t \geq 27$

B $t \leq 27$

C $t \geq 12$

D $t \leq 12$

E none of these

MIND OVER MATH

Use the clues to find the digits to solve each problem. No 2 letters in a problem represent the same number.

1.

four
+ one
five

Clues:
v = 2e
e = f − u
u = 5
$f = 2^3$

2.

soft
+ ball
game

Clues:
b = 2t
$f = a^2$
e = 8
l = 5

3.

four
− two
ten

Clues:
w = 3
n = r + e
o = 8

3.4 DISJUNCTIONS

What is the difference between the disjunction and conjunction of the graphs of 2 inequality statements?

After completing this section, I will be able to

- define a disjunction.
- solve disjunctions.
- write disjunctions that model real-world situations.

NASA institutes strict guidelines to ensure the safety of astronauts during space travel.

DEFINITION

A **disjunction** is a compound sentence consisting of mathematical statements connected by the word *or*.

The solution to a disjunction will consist of any value that is part of the solution set to either inequality and can be written as a statement using the word *or*. The graph of the solution consists of the union of the graphs of the inequalities.

EXAMPLE 1: Graphing a Disjunction

Graph the solution set of $3x < -12$ or $x + 5 > 4$.

Answer

$3x < -12$ or $x + 5 > 4$

$x < -4$ $\qquad$ $x > -1$

1. Solve each inequality separately.

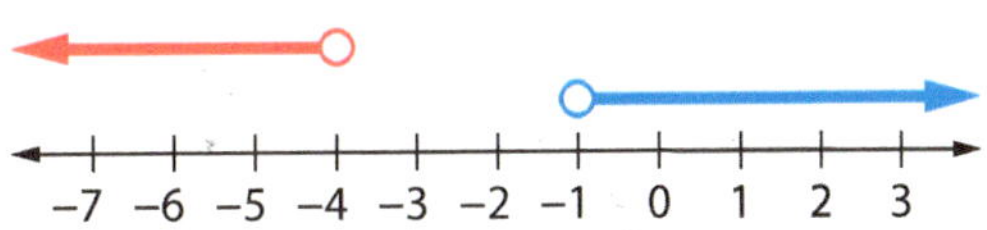

2. Graph each inequality above the number line.

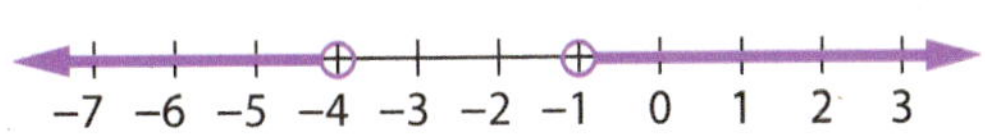

3. Identify the union of the 2 graphs and graph it on the number line.

$x < -4$ or $x > -1$

4. Write the solution for the union.

SKILL CHECK **EXERCISES 5, 7**

The solution to a disjunction may be summarized with a single inequality, or it may be all real numbers.

EXAMPLE 2: Solving Disjunctions

Solve and graph the solution set of each disjunction.

a. $x + 3 < 7$ or $2x - 1 \le 3$ **b.** $3x + 4 > 16$ or $-5x + 7 \le 27$

Answers

a. $x + 3 < 7$ or $2x - 1 \le 3$

$x < 4$ $2x \le 4$

$x \le 2$

1. Solve each inequality separately.

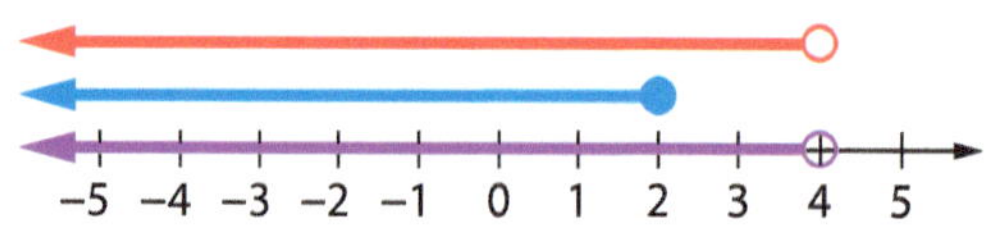

2. Graph each inequality above the number line. Then graph their union on the number line.

$x < 4$

3. Write the solution for the union.

b. $3x + 4 > 16$ or $-5x + 7 \le 27$

$3x > 12$ $-5x \le 20$

$x > 4$ $x \ge -4$

1. Solve each inequality separately.

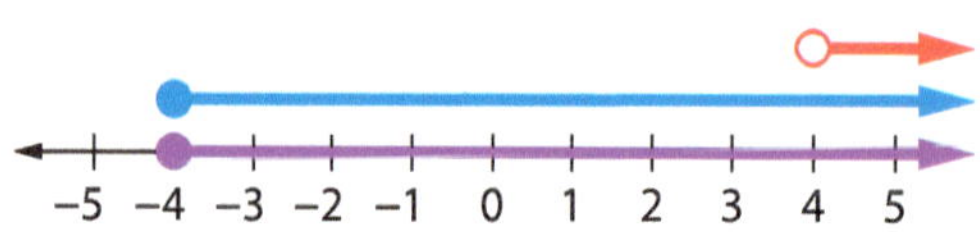

2. Graph each inequality above the number line. Then graph their union on the number line.

$x \ge -4$

3. Write the solution for the union.

SKILL CHECK **EXERCISE 15**

EXAMPLE 3: Determining the Solution of a Special Case

Solve and graph the solution set of $3x - 6 \le 12$ or $3(x + 4) + 2x > -13$.

Answer

$3x - 6 \le 12$ or $3(x + 4) + 2x > -13$

$3x \le 18$ $3x + 12 + 2x > -13$

$x \le 6$ $5x + 12 > -13$

$5x > -25$

$x > -5$

1. Solve each inequality separately.

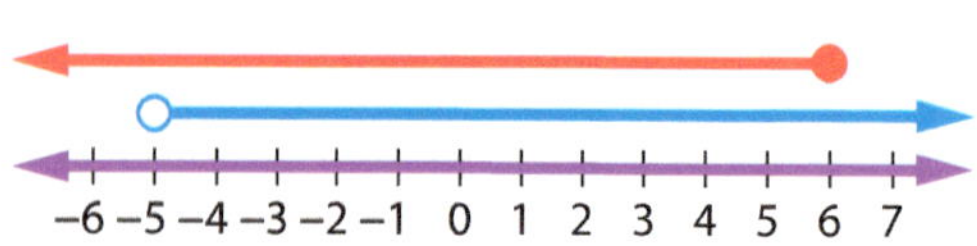

2. Graph each inequality above the number line. Then graph their union on the number line.

$\mathbb{R}$

3. Write the solution for the union. Since the entire number line is shaded, x can be any real number.

SKILL CHECK **EXERCISE 25**

Disjunctions can be used to model the real world.

EXAMPLE 4: Modeling the Real World

NASA determined that the safe reentry angle for the Apollo 13 spacecraft had to be no less than 5.2 degrees and no more than 7.8 degrees. Write a conjunction defining the safe reentry angles. Then write a disjunction to define the angles of reentry that were determined to be unsafe for the Apollo 13 crew.

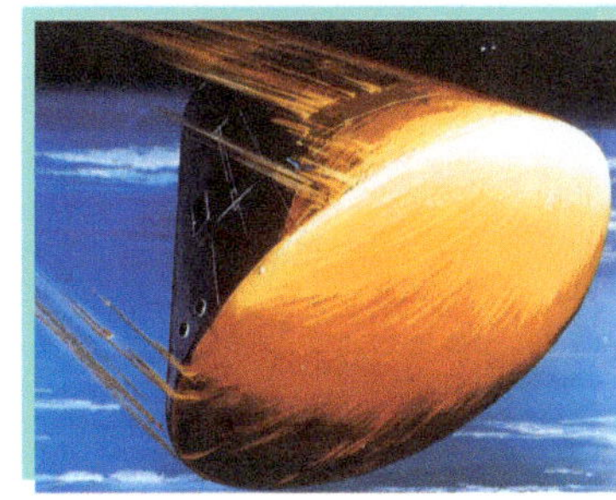

Apollo 13's communication blackout lasted approximately 6 min due to a shallow reentry angle.

Answer

Let s = the safe zone and d = the dangerous zone.

1. Define variables for the unknown quantities.

$5.2 \le s \le 7.8$

$s \ge 5.2$ and $s \le 7.8$

2. Write a combined inequality describing the safe reentry range of angles.

$d \not\ge 5.2$ or $d \not\le 7.8$

$d < 5.2$ or $d > 7.8$

3. After applying the Trichotomy Property, the resulting compound inequalities become a disjunction which represents the unsafe zone.

SKILL CHECK **EXERCISE 33**

AFTERSCHOOLHELP

Disjunctions

VOCABULARY

- disjunction

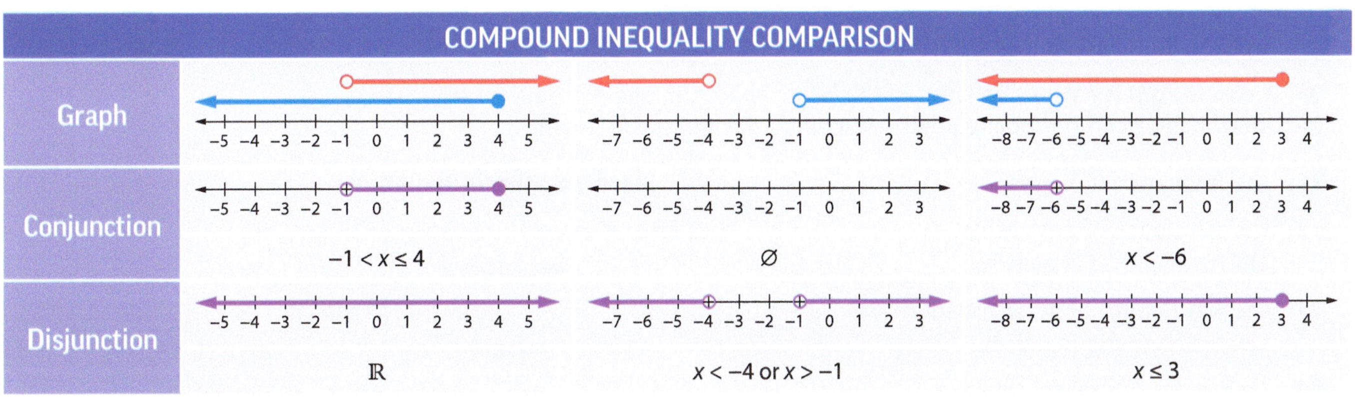

A. EXERCISES

1. In a disjunction, inequality statements are joined by the word _____.

2. The solution to a disjunction is the _____ of the 2 solution sets.

Graph the union of the given inequalities on a number line.

3. 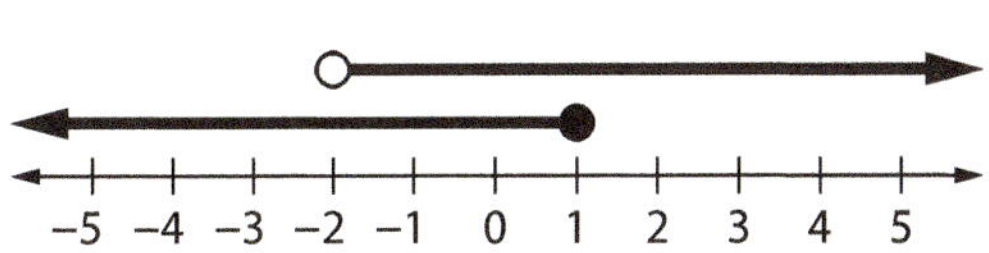

4. 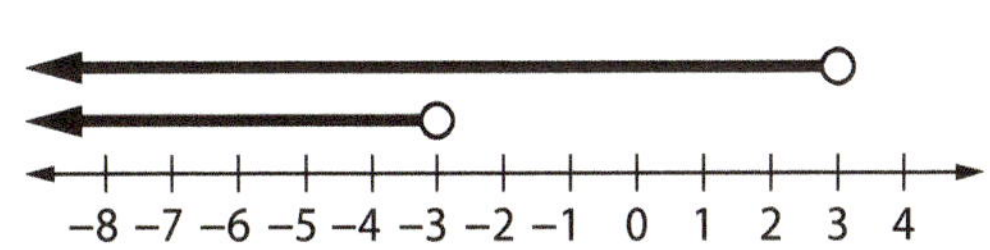

5. 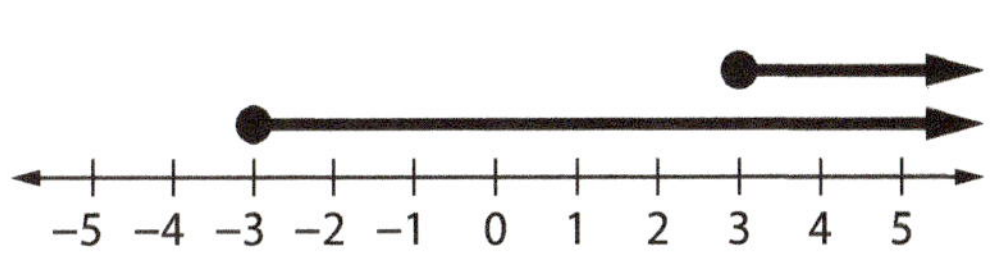

6. 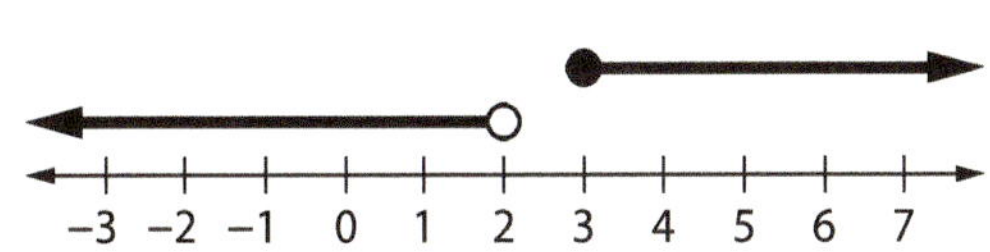

Graph each disjunction.

7. $x < 2$ or $x > 5$

8. $x < 2$ or $x < 5$

9. $x > 2$ or $x > 5$

10. $x > 2$ or $x < 5$

11. $x < -4$ or $x > -4$

12. $x < -3$ or $x \le -2$

B. EXERCISES

Solve. Express the answer as an inequality statement or the set of real numbers.

13. $x \neq 6$ or $x \neq 3$

14. $2a + 5 > 17$ or $a + 2 \leq 8$

15. $y - 6 \leq -6$ or $3y - 5 > 10$

16. $3z + 5 > 2z$ or $6z - 3z + 4 \geq 16$

17. $2b - 6 \geq 14$ or $8b + 2b - 3 > 27$

18. $r - 7 > 0$ or $2r + 16 \leq 2$

19. $3x + 6 \leq -9$ or $x < 6$

20. $2s - 7 > -21$ or $4s + 6 \neq 22$

Solve each compound inequality. Express the answer as an inequality statement, the set of real numbers, or the empty set.

21. $3 - 2x > -2$ or $3x + 6 < 8$

22. $7y > 2y + 10$ and $2y > 5y - 15$

23. $4m - 4 \leq 6m - 8$ and $-3m + 1 \geq 5m - 3$

24. $4x + 6 > 10 - x$ or $9 - 3x \geq -1$

25. $3n - 2 < 4$ or $2n + 8 < 5n + 11$

26. $6 - 3x \geq 4 - 2x > 5 - 4x$

27. $2z + 3 \leq -2$ or $3z - 5 > 4$

28. $2c + 50 < 7c + 2 + 3c$ and $5 - 4c < 53$

29. A chef roasting black angus beef keeps the internal temperature t between 125°F (rare) and 160°F (well done). Write a disjunction describing undesirable internal temperatures for the meat.

30. A section of I-85 in South Carolina has a speed limit of 65 mi/hr and a minimum speed limit of 45 mi/hr. Write a disjunction describing illegal speeds on this portion of I-85.

MP2 **Visible light is a very small part of the larger spectrum of electromagnetic radiation. Write a compound inequality describing each of the following.**

31. the frequency of x-rays

32. the frequency of nonvisible electromagnetic radiation

33. the wavelength of nonultraviolet electromagnetic radiation

34. the wavelength of microwaves

Write a compound inequality describing all the numbers that are not in the solution set of the following compound inequalities.

35. $-3 \leq x \leq 1$

36. $x > 4$ or $x < 2$

37. $x \leq -5$ or $x > 0$

38. $x > -8$ and $x \leq 13$

39. **Essential Question:** Describe the difference between the disjunction and conjunction of the graphs of 2 inequality statements.

Electromagnetic Radiation

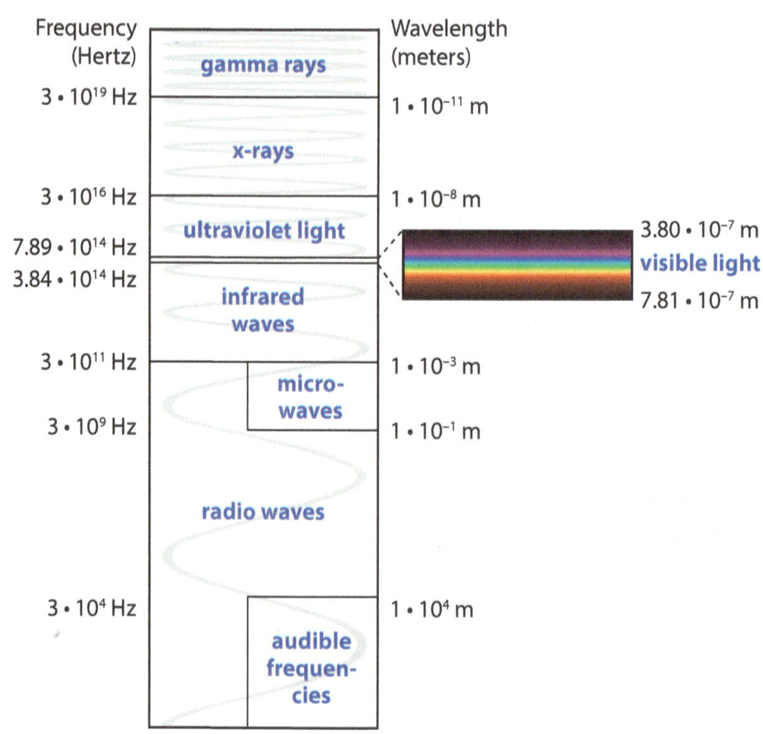

C. EXERCISES

Solve each compound inequality.

40. $5x - 1 < 1$ or $3x - 2 \neq 5$

41. $\frac{5}{7}x + \frac{1}{2} < 2$ or $\frac{3}{4}x - \frac{1}{5} \leq \frac{1}{10}$

42. $3x + 1 \geq 6$ or $9(4 - x) \leq 15$

43. **MP4** A photo studio sells 5" × 7" portraits for $45 and 8" × 10"s for $75. They consistently sell about 150 more 5" × 7"s than 8" × 10"s each month. They want to meet at least one of the following goals next month: sell at least 400 portraits or gross at least $20,000. How many 8" × 10"s do they need to sell to meet one of their goals? both goals?

CUMULATIVE REVIEW

Evaluate each expression. [1.1]

44. $|13 - 58|$

45. $|-45| - |61|$

46. $|-14 - (-7)|$

Evaluate each expression. [1.3]

47. $-\sqrt{100}$

48. $\sqrt[3]{64}$

49. $\frac{1}{3^{-4}}$

Evaluate when $x = -2$, $y = 3$, and $z = -5$. [1.5]

50. $2xy - y^2z$

51. $(2yz - x)^2$

52. Solve $4 - (3x + 2) \leq 3(2 - 3x)$. **[3.2]**

A $x \leq 0$

B $x \geq \frac{2}{5}$

C $x \leq \frac{2}{3}$

D $x \leq \frac{4}{5}$

E none of these

53. Solve the conjunction $-5 > 4x + 3$ and $3x - 3 \geq 3$. **[3.3]**

A $x > -2$

B $-2 < x \leq 2$

C $\mathbb{R}$

D $\varnothing$

E none of these

CAREERS IN MATH
ACTUARY

The actuarial profession has been consistently rated as one of the top careers in the United States. Many colleges and universities offer an actuarial science major in their math department.

An actuary evaluates the likelihood of future events and works to mitigate the impact of undesirable events that occur in our fallen world. Most actuaries are employed by insurance companies to determine risk and set premiums, but their job responsibilities are often much broader than that.

An actuary deals with risk management, especially as it impacts the financial well-being of a business. Actuaries help companies build strategies to improve financial security as they navigate changing marketplaces and interest rates (see Application Problems). Thus, an actuary must have good math and analytical skills as well as business knowledge. An actuary determines the likelihood of uncertain events, such as the probability that a thirty-year-old man will live to be sixty-five.

James reminds Christians of the uncertainty of future events and instructs even businessmen to submit to whatever God wills (James 4:13–15). Nonetheless, the Bible teaches that the way of wisdom includes making prudent plans for the future (Prov. 22:3; 27:12).

Government and insurance industry opportunities for actuaries currently abound and are predicted to increase. In addition to working with insurance plans, pensions, and social welfare programs, actuaries work to design creative ways to reduce the likelihood of undesirable events occurring. Christian actuaries can glorify God by using their God-given abilities to anticipate and plan for the difficult circumstances that may occur in the lives of others.

3.5 ABSOLUTE VALUE EQUATIONS

When do absolute value equations have 2 solutions?

After completing this section, I will be able to

- define an absolute value equation.
- write a disjunction representing an absolute value equation.
- solve absolute value equations.

Turkey's capital city, Ankara, dates back to the time of the Hittites, circa 2000 BC. Its age is approximately $|2023 - (-2000)| = 4023$ yr.

Recall that the absolute value of a number can be thought of as the number's distance from 0.

$$|6| = 6 \text{ and } |-6| = 6$$

The absolute values of 6 and −6 are the same because they are both 6 units away from 0 on the number line. A more precise definition of absolute value is required when more complicated expressions are included within the absolute value symbols.

DEFINITIONS

The **absolute value** of a number, x, is defined as follows:

$$|x| = \begin{cases} x \text{ if } x \geq 0 \\ -x \text{ if } x < 0 \end{cases}.$$

An **absolute value equation** is an equation whose variable expression is contained within the absolute value symbols.

This definition of absolute value states that if x is a nonnegative number, such as 6, then $|x|$ is equal to the number itself. So $|6| = 6$. However, if x is a negative number, such as −6, then $|x|$ equals the opposite of the number. So $|-6| = -(-6) = 6$.

Consider the equation $|x| = 4$. Since we do not know whether x is positive or negative, the definition leads to 2 possible conclusions.

1. If $x \geq 0$, then $|x| = 4$ implies that $x = 4$.
2. If $x < 0$, then $|x| = 4$ implies that $-x = 4$ and that $x = -4$.

Thus, the equation $|x| = 4$ has 2 solutions. The notation $x = \pm 4$ is often used to indicate the positive and negative solutions.

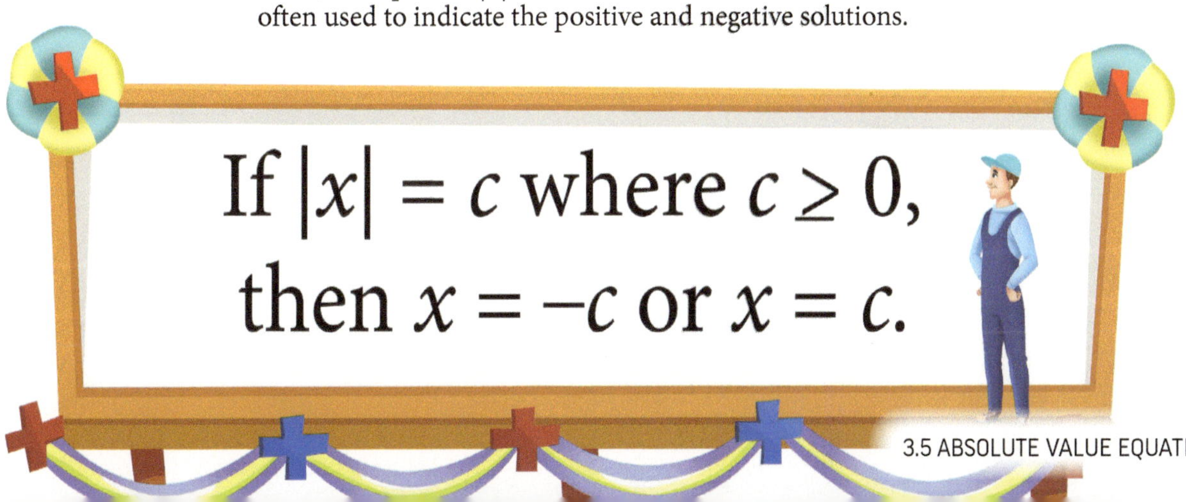

EXAMPLE 1: Solving an Absolute Value Equation

Solve $|2x + 5| = 11$.

Answer

$2x + 5 = 11$ or $2x + 5 = -11$

1. The expression within the absolute value symbols must equal 11 or −11.

$2x = 6 \qquad 2x = -16$

$x = 3 \qquad x = -8$

2. Solve each equation separately.

$|2(3) + 5| = |6 + 5| = |11| = 11$

$|2(-8) + 5| = |-16 + 5| = |-11| = 11$

The values for x are 3 and −8.

3. Check your solutions.

SKILL CHECK **EXERCISES 1, 13**

SOLVING ABSOLUTE VALUE EQUATIONS

1. Simplify within the absolute value symbols.
2. Isolate the absolute value on 1 side of the equation.
3. Use the definition of absolute value to write 2 equations.
4. Solve each equation separately.
5. Check your solutions.

The number the absolute value equation equals will determine the number of solutions. Since the absolute value equation equaled a positive number in Example 1, the result was 2 solutions. Consider the cases when the absolute value equation is not positive.

EXAMPLE 2: Solving Special Absolute Value Equations

Solve each absolute value equation.

a. $|x - 5| = 0$ **b.** $|2x + 8| + 10 = 4$

Answers

a. $|x - 5| = 0$

$x - 5 = 0$

1. Only 1 equation is possible since 0 has no opposite.

$x = 5$

2. Solve.

b. $|2x + 8| + 10 = 4$

$|2x + 8| = -6$

1. Isolate the absolute value on 1 side of the equation.

$\varnothing$

2. Since the absolute value cannot be negative, there is no solution.

SKILL CHECK **EXERCISE 7**

As exemplified in Example 2b, it is important to wait to determine the number of solutions until after the absolute value has been isolated.

EXAMPLE 3: Solving an Absolute Value Equation

Solve $5 - |2y + 4 - 5y| = -11$.

Answer

$5 - \lvert -3y + 4 \rvert = -11$		**1.** Simplify within the absolute value symbols.
$-\lvert -3y + 4 \rvert = -16$		**2.** Subtract 5 from both sides.
$\lvert -3y + 4 \rvert = 16$		**3.** Divide both sides by −1.
$-3y + 4 = 16$ or	$-3y + 4 = -16$	**4.** Solve each problem separately.
$-3y = 12$	$-3y = -20$	
$y = -4$	$y = \frac{20}{3}$	
$5 - \lvert 2(-4) + 4 - 5(-4) \rvert$	$5 - \left\lvert 2\left(\frac{20}{3}\right) + 4 - 5\left(\frac{20}{3}\right) \right\rvert$	**5.** Check your solutions.
$5 - \lvert -8 + 4 - (-20) \rvert$	$5 - \left\lvert \frac{40}{3} + 4 - \frac{100}{3} \right\rvert$	
$5 - \lvert 16 \rvert = -11$	$5 - \lvert -16 \rvert = -11$	
$y = -4$ and $\frac{20}{3}$		

SKILL CHECK **EXERCISE 23**

KEYWORD SEARCH

absolute value equation

DETERMINING THE NUMBER OF SOLUTIONS

$\lvert x \rvert = n$	$n > 0$	2 solutions
$\lvert x \rvert = 0$		1 solution
$\lvert x \rvert = n$	$n < 0$	0 solutions

The absolute value equation $|x - 5| = 3$ has 2 solutions, $x = 2$ and $x = 8$. By graphing these solutions, we can see that both solutions are a distance of 3 units from 5 on the number line.

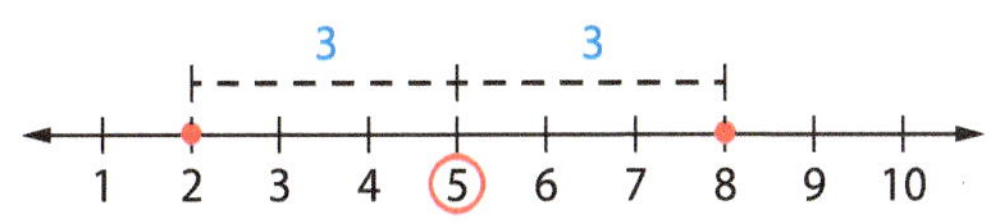

Notice that 5 is the coordinate of the midpoint of the segment with endpoints at 2 and 8, the solutions to the equation that are both 3 units away from the midpoint.

EXAMPLE 4: Writing Absolute Value Equations

Write and solve absolute value equations to find the numbers that are as follows.

a. 4 units from 6 **b.** 2 units from −7

Answers

a.	$\lvert x - 6 \rvert = 4$	**1.** The distance is 4 units, so the absolute value equals 4. Since the distances are measured from 6, 6 is subtracted from x within the absolute value symbols.
	$x - 6 = 4$ or $x - 6 = -4$	**2.** Use the definition of absolute value to write 2 equations.
	$x = 10$ or $x = 2$	**3.** 10 and 2 are both 4 units from 6.
b.	$\lvert x - (-7) \rvert = 2$	**1.** The distance is 2 units, so the absolute value equals 2. Since the distances are measured from −7, −7 is subtracted from x within the absolute value symbols.
	$x + 7 = 2$ or $x + 7 = -2$	**2.** Use the definition of absolute value to write 2 equations.
	$x = -5$ or $x = -9$	**3.** −5 and −9 are both 2 units from −7.

SKILL CHECK **EXERCISE 27**

AFTERSCHOOLHELP

Absolute Value Equations

VOCABULARY

- absolute value
- absolute value equation

A. EXERCISES

Solve. Write ∅ if no solution exists.

1. $|x| = 3$

2. $|x| = 21$

3. $|x| - 6 = -1$

4. $4|x| = 44$

5. $-3|x| = -8$

6. $-5|x| = 35$

7. $|x| + 8 = 5$

8. $|x + 5| = 7$

9. $|3y| = 129$

10. $|7x| = 4$

11. $|x + 2| = -7$

12. $|x - 8| = 14$

13. $|y + 4| = 8.5$

14. $\left|x - \frac{1}{2}\right| = \frac{3}{2}$

B. EXERCISES

Solve. Write ∅ if no solution exists.

15. $|x + 1| + 2 = 8$

16. $|3z + 4 - z| = 16$

17. $|2y - 5| - 9 = 24$

18. $|2x - 3| + 7 = 2$

19. $|5y + 12| = 24$

20. $-1 + |7x + 9| = 50$

21. $19 = |12x - 5|$

22. $|3x - 4 + 4x| - 6 = 3$

23. $|2z - 10 + 5z| - 20 = -2$

24. $|x - 1.2| - 2 = 11$

25. $-36 = 2 - |7x + 3|$

26. $|3y + 4 - 6y| - 14 = 12$

Write an absolute value equation representing the following. Then solve the equation.

27. numbers that are 4 units from 2

28. numbers that are 7 units from −3

29. numbers that are 8 units from −4

30. numbers that are 5 units from 15

31. numbers that are 15 units from 5

32. MP2 Explain why $|x - 4| = -2$ has no meaning in relation to the number line.

33. **Essential Question:** When does $|x - 3| = n$ have 2 solutions?

C. EXERCISES

Use the average of each pair of numbers to write an absolute value equation having the following solutions.

34. 7 and 13

35. −4 and −10

36. −15 and 3

37. −7 and 20

Use the definition of $|x|$ to write similar definitions of the absolute values below.

38. $|x + 1|$

39. MP7 $|2x - 3|$

CUMULATIVE REVIEW

Write an equation and solve.

40. 38 is what percent of 40? **[2.5]**

41. Increase \$250 by 3.5%. **[2.5]**

42. What is the percent change (to the nearest percent) of a stock that dipped from \$83 per share to \$65 per share? **[2.5]**

43. What is the original retail price of a sweater that has been discounted by 40% for a sale price of \$36? **[2.6]**

44. Determine the commission earned by Zach if he receives 15% on last week's sales of \$2500. **[2.6]**

45. Lawrence deposits \$4000 in a CD (certificate of deposit) earning 4% simple interest. Determine the interest earned and the total amount in his CD at the end of 2 yr. **[2.6]**

46. A plane leaves Charlotte at 9:00 AM and flies north at 400 knots. At the same time, another plane flies south from Charlotte at 250 knots. How long is it before they are 1000 nautical miles apart? **[2.7]**

47. Coffee selling for \$10/lb is mixed with coffee selling for \$4/lb. To the nearest pound, how much of each is used to produce 50 lb of a mixture that sells for \$8/lb? **[2.8]**

48. If the ratio of pine trees to other trees on Mr. Locke's property is 2 : 3 and he has a total of 60 trees, how many trees does he have that are not pine trees? **[2.4]**

A 40

B 36

C 24

D 20

E none of these

49. For which value of k will the disjunction $x > 8$ or $x \le k$ have $\mathbb{R}$ as the solution set? **[3.4]**

A 3

B −8

C 11

D −4

E none of these

3.6 ABSOLUTE VALUE INEQUALITIES

Why is an absolute value never less than 0?

After completing this section, I will be able to

- write a conjunction or disjunction representing an absolute value inequality.
- solve absolute value inequalities.
- solve real-world problems by writing and solving absolute value inequalities.
- evaluate the limitations of deductive reasoning in determining truth.

Recall that the absolute value of a number refers to its distance from the origin. This will help us interpret absolute value inequalities. The solutions to $|x| < 3$ will be all the numbers whose distance from the origin is less than 3. For instance, the numbers −2, 1.5, 0, and −0.5 will make this sentence true. The graph illustrates all the solutions to this inequality.

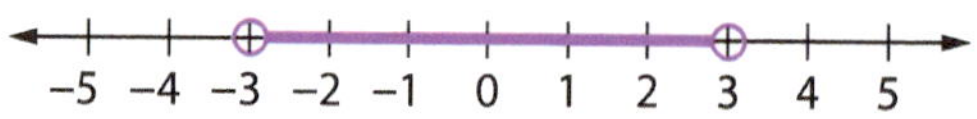

The average human body temperature is 98.6°F.

This inequality can also be written without the absolute value symbols as a conjunction using the word *and*,

$x > -3$ and $x < 3$,

or as the combined conjunction

$-3 < x < 3$.

The solution to the absolute value inequality $|x| \geq 3$ includes all numbers whose distance from the origin is greater than or equal to 3. The graph shows that this includes −3 and all numbers to its left as well as 3 and all numbers to its right.

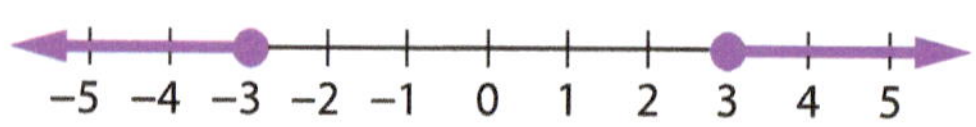

Note that this inequality is written as a disjunction using the word *or*,

$x \leq -3$ or $x \geq 3$.

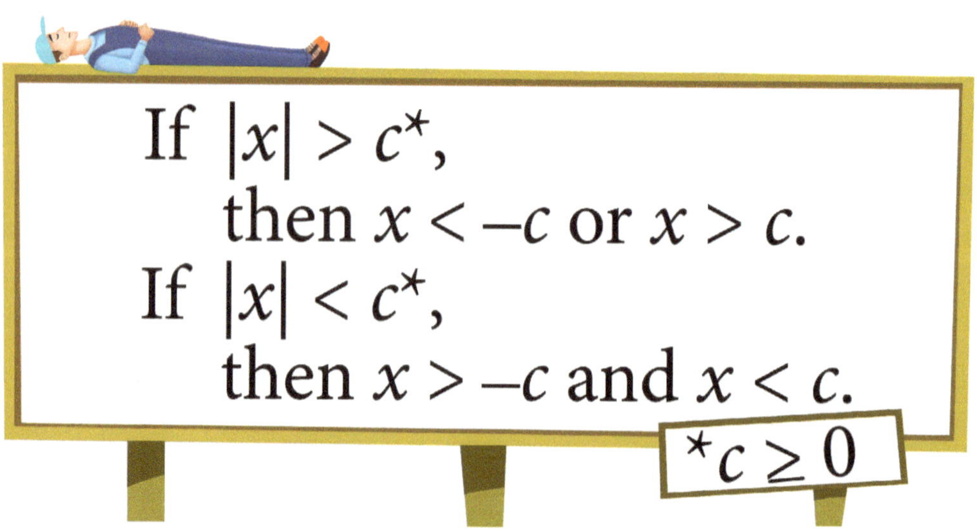

EXAMPLE 1: Solving an Absolute Value Inequality

Solve and graph the solution set of $|x+5| \le 10$.

Answer

$x+5 \ge -10$ and $x+5 \le 10$

$x \ge -15$ $\quad$ $x \le 5$

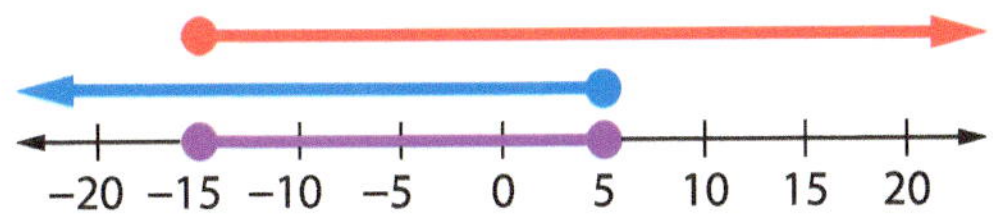

$-15 \le x \le 5$

1. Write a conjunction and solve the 2 inequalities.
2. Graph each inequality above the number line. Then graph their intersection on the number line.
3. Write the solution as a combined conjunction.

SKILL CHECK **EXERCISE 9**

As with absolute value equations, the absolute value must be isolated before writing the 2 separate inequalities.

EXAMPLE 2: Simplifying before Solving an Absolute Value Inequality

Solve and graph the solution set of $|-3x+6| - 24 > -6$.

Answer

$|-3x+6| > 18$

$-3x+6 > 18$ or $-3x+6 < -18$

$-3x > 12$ $\quad$ $-3x < -24$

$x < -4$ $\quad$ $x > 8$

$x < -4$ or $x > 8$

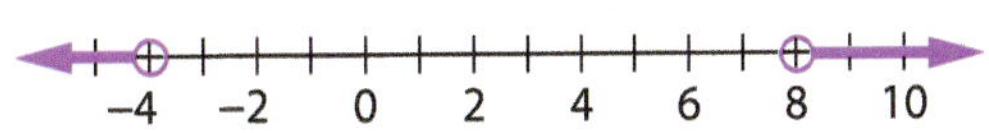

1. Isolate the absolute value expression.
2. Write a disjunction and solve the 2 inequalities.
3. Write the solution for the union.
4. Graph the union on the number line.

SKILL CHECK **EXERCISE 21**

The following table contrasts the key differences in absolute value inequalities and shows their similarities to conjunctions and disjunctions.

	ABSOLUTE VALUE INEQUALITIES			
	Conjunction (*and*)		Disjunction (*or*)	
	$\lvert x\rvert \le n$	$\lvert x\rvert < n$	$\lvert x\rvert \ge n$	$\lvert x\rvert > n$
$n > 0$	$-n \le x \le n$ 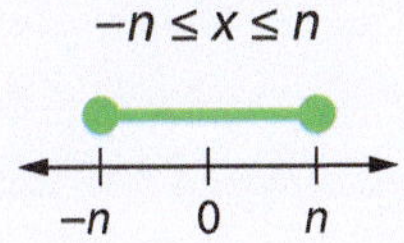	$-n < x < n$ 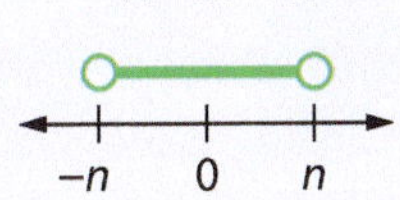	$x \le -n$ or $x \ge n$ 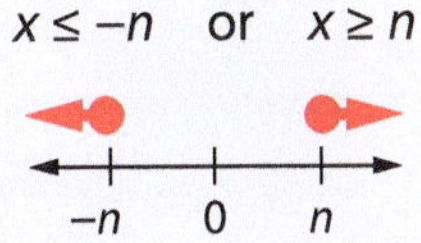	$x < -n$ or $x > n$ 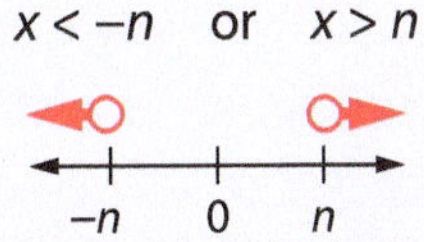
$n = 0$	$x = 0$	$\varnothing$	$x \le 0$ or $x \ge 0$ $\mathbb{R}$	$x < 0$ or $x > 0$ 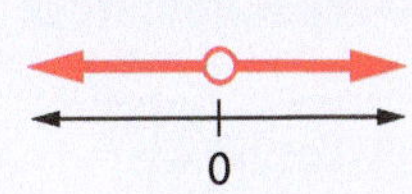 $x \ne 0$
$n < 0$	$\varnothing$	$\varnothing$	$\mathbb{R}$	$\mathbb{R}$

EXAMPLE 3: Solving a Real-World Problem

The ideal overall tank temperature for a ball python is 80°F. Any temperature 15°F or more in either direction for too long is dangerous for the health of the python. Write and solve an absolute value inequality representing the dangerous temperatures for a python.

Answer

$|x - 80| \geq 15$

1. Any temperature 15°F or more from the ideal temperature is dangerous.

$x - 80 \leq -15$ or $x - 80 \geq 15$

$x \leq 65$ $\quad$ $x \geq 95$

2. Write a disjunction and solve the 2 inequalities.

Any temperature ≤ 65°F or ≥ 95°F is dangerous for the python.

SKILL CHECK **EXERCISE 33**

Reasoning is a gift from God that provides great insight, but even valid reasoning can lead to error if the initial assumptions are not correct. If we assumed, for example, that the temperatures in the previous example were Celsius instead of Fahrenheit, our python would not last long even if there were no mistakes in the math. This may be unlikely given the temperatures here, but passenger planes have run out of fuel midair and space probes have crashed due to similar assumption errors.

EXAMPLE 4: Solving a Real-World Problem

Most human muscle cells function well within 2° of their optimal point, 98.6°F. Describe this optimal range of temperature with an absolute value inequality; then solve to find the range.

Answer

$|t - 98.6| \leq 2$

1. The temperature must be within 2° of the optimal point.

$t - 98.6 \leq 2$ and $t - 98.6 \geq -2$

$t \leq 100.6$ $\quad$ $t \geq 96.6$

2. Write a conjunction and solve the 2 inequalities.

$96.6 \leq t \leq 100.6$

3. Write the solution as a combined conjunction.

SKILL CHECK **EXERCISE 35**

AFTERSCHOOLHELP

Absolute Value Inequalities

A. EXERCISES

1. If n is a positive number, then $|x| < n$ can be written as a(n) ______ using the word ______.
2. If n is a positive number, then $|x| > n$ can be written as a(n) ______ using the word ______.

Solve and graph the solution on a number line.

3. $|x| > 3$
4. $|x| < 10$
5. $|x| \leq 5$
6. $|x| \geq 7$
7. $|2x| > 16$
8. $|-3x| > 6$
9. $|x - 3| \leq 10$
10. $|x + 6| \geq 5$

Rewrite each compound inequality as an absolute value inequality.

11. $-15 < x < 15$
12. $x > 7$ or $x < -7$
13. $y \leq -19$ or $y \geq 19$
14. $z \geq -23$ and $z \leq 23$
15. **Essential Question:** Why is an absolute value never negative?

B. EXERCISES

Solve each absolute value inequality.

16. $|2x + 6| \geq 14$
17. $|3x - 10| < 9$
18. $|4x + 17| > 7$
19. $1 + \left|\frac{2x}{3} - 7\right| \leq 5$
20. $\left|\frac{5}{9}x + 6\right| > 5$
21. $|2x - 8| - 7 < 3$
22. $-7 + |4x - 7| \geq 22$
23. $|-4x - 8| > 4$
24. $|3x + 2| - 5 \leq 1$
25. $3|4x - 6| < 72$
26. $|7x + 9 - 6x| > 8$
27. $|2x - 6 + 5x| < 10$
28. Write an absolute value inequality representing all numbers that are at most 2 units from 5.
29. Write an absolute value inequality representing all numbers that are at least 2 units from 5.
30. Machine parts are manufactured with a given tolerance, the permissible limit of variation. Write an absolute value inequality representing the length of a bolt that is to be 3 cm long with a tolerance of ±0.02 cm. Then solve the inequality to state the range of acceptable lengths.
31. The hole for the bolt in exercise 30 is designed to be 5 mm in diameter with a tolerance of ±0.05 mm. Write an absolute value inequality representing the diameter of the hole. Then solve the inequality to state the range of acceptable diameters.
32. The USDA has defined the danger zone for meat, in which bacteria are most likely to grow, as between 40°F and 140°F. Meat should not be kept in this temperature zone for more than 2 hr. Write an absolute value inequality representing the danger zone.
33. Meat should be stored below 40°F or kept above 140°F after cooking, outside the danger zone described in exercise 32. Write an absolute value inequality representing safe temperatures for meat.

34. **MP4** To reduce storage costs, many manufacturers have implemented just-in-time delivery of the components that are assembled for their products. A car company needs parts delivered within 3 min of every half hour. Write an absolute value inequality representing the time of delivery. Then solve the inequality to state the range of acceptable times.

35. The results of a poll claiming a 4% margin of error indicated that a candidate was favored by 48% of the voters. Write an absolute value inequality representing the percent of voters favoring the candidate. Then solve the inequality to state the range of the candidate's support.

36. **BWS Essential Question:** Is it possible to use valid reasoning and reach a wrong conclusion? Explain.

C. EXERCISES

Solve each absolute value inequality.

37. $3 - |-4x| > 5$

38. $|y + 7| \le 0$

39. $|3z - 1| > 0$

40. $|5a + 2| \ge a$

41. **MP7** $|4b + 2| \le -2b + 5$

CUMULATIVE REVIEW

Write an inequality equivalent to each given inequality. [3.1]

42. $q \not> -8$

43. $t \not\le 101$

Solve each literal equation for *r*. [2.3]

44. $\frac{2}{3}r = a + 6$

45. $4r + b = 2c$

46. What is the percent increase from 80 to 100? **[2.5]**

47. A solution heated to 90°C was removed from heat. Five minutes later the temperature of the solution was 78°C. Determine the percent decrease in temperature to the nearest tenth of a percent. **[2.5]**

48. Langston went out for lunch and purchased a \$9 entrée and a beverage for \$3. He paid an 8% sales tax and tipped his server 20% of the pretax bill. What is the total amount he spent on lunch? **[2.6]**

49. Mr. Scurtu wants to invest some of his \$10,000 in an aggressive stock fund that he expects to earn 12% annually and the rest in less risky bonds making 4% annually. If he hopes to earn \$1000 in annual interest, how should he split his investment between the stock fund and the bonds? **[2.6]**

50. Find the solution set for $3 \le 11 - 2x < 29$. **[3.3]**

   A $-20 < x \le -7$

   B $-9 < x \le -4$

   C $7 \le x < 20$

   D $-9 < x \le 4$

   E none of these

51. For which value of k will $|2x| + k = 3$ not have a solution? **[3.5]**

   A 5

   B 3

   C −3

   D −5

   E none of these

TECHNOLOGY CORNER *(TI-84 PLUS FAMILY)*

Graphing Inequality Solutions

The [graph] function can be used to graph the solutions to more complicated inequalities. Begin by selecting FUNCTION from the fourth line of the [mode] screen and then ZStandard from the [zoom] menu.

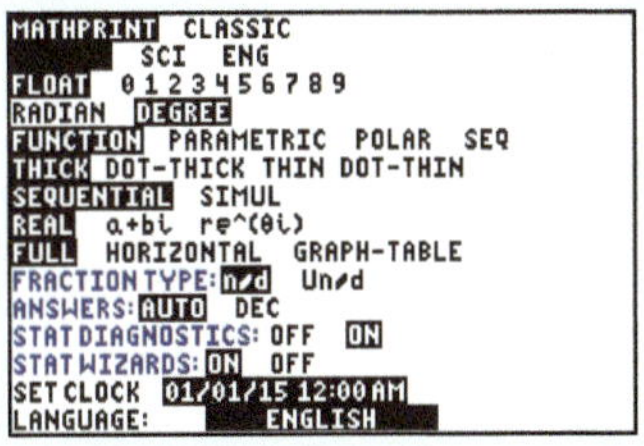

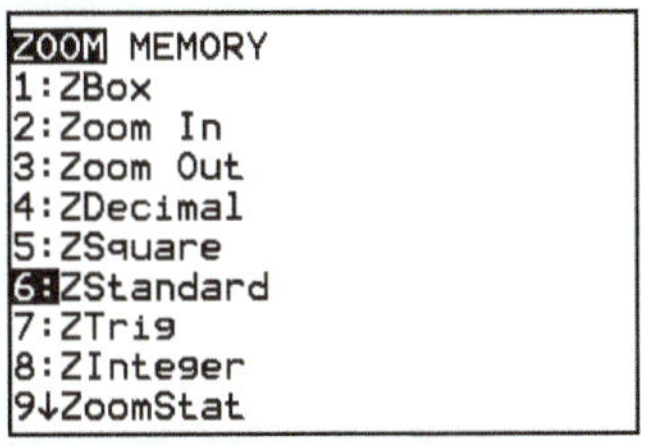

To graph the solution to $2x + 1 > 3$, select [y=] to go to the function editor and enter the inequality as Y_1, using the [test] menu ([2nd] [math]) to enter the inequality symbol.

2 [X,T,θ,n] [+] 1 [test] [3] 3

Use [graph] to plot the solution set; then select [trace] and use the [◀] and [▶] keys to explore the values of x that are graphed. Notice that the function Y_1 equals 1 when the inequality is true and 0 when the inequality is false.

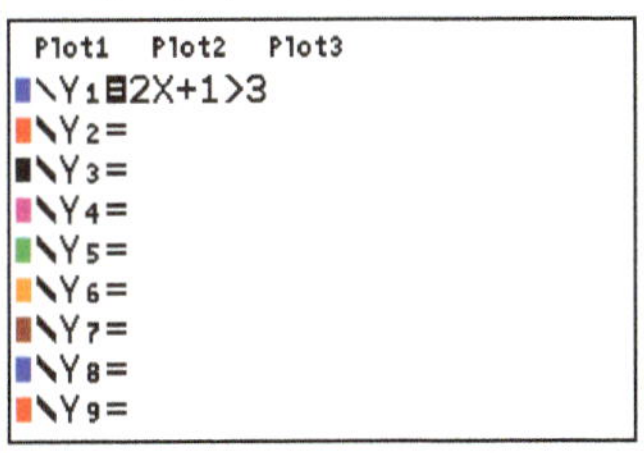

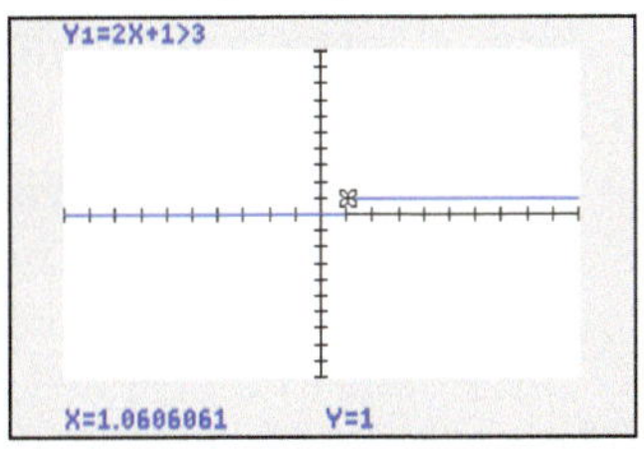

Conjunctions & Disjunctions

The LOGIC menu on the [test] screen can be used to enter disjunctions and conjunctions. Select [y=] to return to the function editor and clear Y_1. Enter the combined conjunction $-3 \le x \le 3$ in Y_1 as 2 inequalities connected by the word *and*.

[X,T,θ,n] [test] [4] [(-)] 3 [test] [▶] [1] [X,T,θ,n] [test] [6] 3

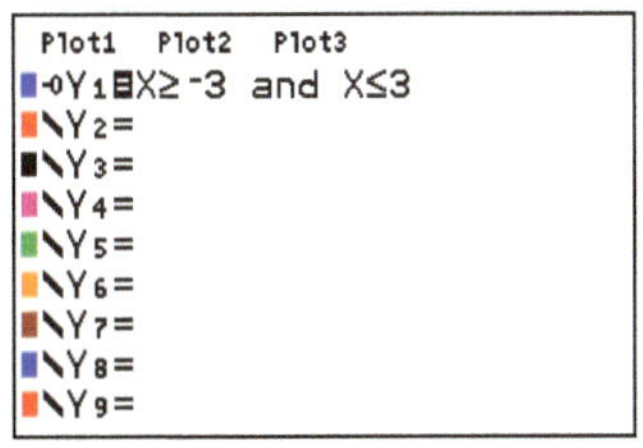

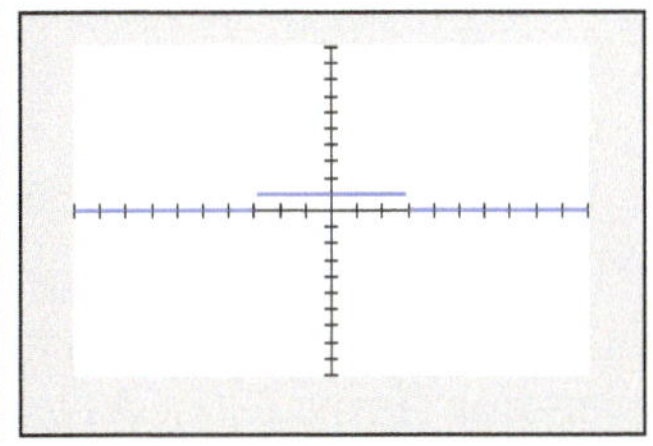

Absolute Value Inequalities

Use the NUM menu from the [math] screen to enter the absolute value function, abs(). To graph the solution set of $|x + 2| > 4$, return to the function editor using [y=] and enter the following in Y_1.

[math] [▶] [1] [X,T,θ,n] [+] 2 [▶] [test] [3] 4

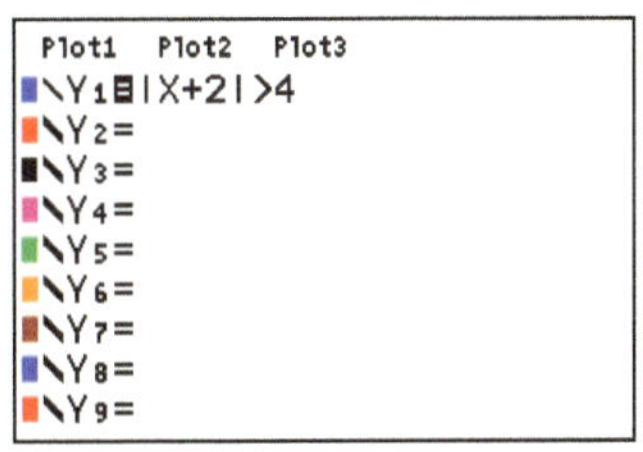

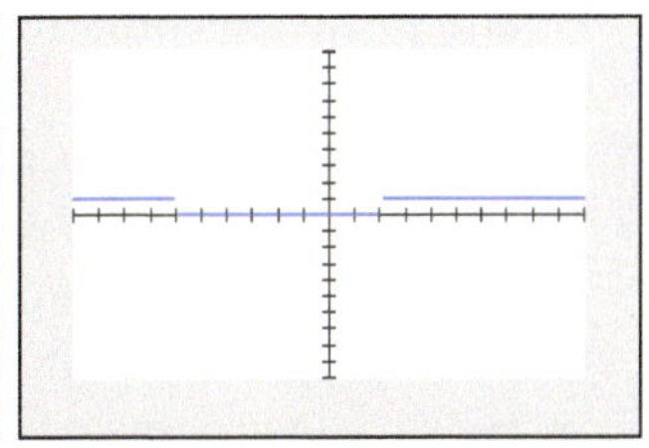

APPLICATION PROBLEMS
CALCULATING INTEREST

The stockholders of the First National Bank of Newton have a vested interest in the financial plans that the bank president, Mr. Lee, presents at the annual board meeting. Mr. Lee knows that he must gain their confidence in his plans for investing their money. He summarizes investments that will earn $3.4 million in the coming year. He needs to know how much money he must invest in municipal bonds that yield 8% annually so that the total interest income for the bank will be at least an additional $2.1 million.

The Bible has much to say about the everyday details of life, including such things as borrowing and lending (Prov. 22:7). As long as a person is in debt, he is morally obligated to the person who is lending him money. For this reason, a Christian should be very careful about borrowing money.

One way to prepare is to know about different types of loans, three of which are discussed below. Recall the interest formula $I = Prt$.

1. Simple interest loan: A single payment of interest and principal combined at the end of the specified time.
2. Add-on loan: Simple interest plus principal divided by the number of payments (usually monthly). Each payment applies the same set amount to the principal and the same amount to interest.
3. Discount loan: The interest on the desired amount to be borrowed is calculated and added to the amount that is borrowed at the beginning of the loan. This new amount owed is divided into equal payments throughout the life of the loan.

Compare these types of loans for $5000 borrowed at 6% interest for 2 yr in the 3 ways described above.

1. Simple interest loan: Find the amount of interest and the entire amount due at the end of the loan.

2. Add-on loan: Find the amount of interest, the total amount to repay, and the monthly payment.

3. Discount loan: Calculate the actual amount loaned for the borrower to receive $5000 using the formula $S = \frac{P}{1 - dt}$, where S = the amount that must be borrowed, P = the amount the borrower receives, d = the interest rate, and t = the time in years. Then find the amount of interest and the monthly payment.

If money must be borrowed, it is wise to find the best rate. Interest rates are expressed in 2 ways: annual percentage rate (APR) and annual percentage yield (APY). The APR expresses annual interest without considering compound interest. The APY expresses annual interest, including compound interest (interest earned on both principal *and* previous interest).

When the lender receives monthly compounded interest on the remaining principal throughout the term of a loan, the actual interest rate is not equal to the simple interest rate (APR). Calculating the APY reveals that borrowers are actually paying a higher rate than the APR.

$APY = \left(1 + \frac{APR}{n}\right)^n - 1$, where n is the number of compounding periods per year.

Find the APY of a 5 yr $8000 loan with a 7% APR compounded monthly.

Using $n = 12$, $APY = \left(1 + \frac{0.07}{12}\right)^{12} - 1$

$\approx (1.005833)^{12} - 1 \approx 0.0723 \approx 7.2\%$.

The actual interest rate is about 7.2%, a higher rate than the 7% APR.

To provide protection for borrowers, the Truth in Lending Act requires creditors to provide the borrower a document listing the total finance charges, the annual percentage rate (APR), and the annual percentage yield (APY).

Assuming monthly payments, find the APY for each of the following loans. Round to the nearest tenth of a percent.

4. Mr. Ellison was quoted an interest rate of 9% APR for a loan to expand his business.

5. A credit card company discloses an 18% APR for outstanding balances.

The APY is often used to advertise the interest rate for investments because it includes the compounding effect. The actual annual percentage rate can be found by solving $APY = \left(1 + \frac{APR}{n}\right)^n - 1$ for the APR, producing the following formula.

$APR = n(APY + 1)^{\frac{1}{n}} - n$, where n is the number of compounding periods per year.

Find the APR of a \$2000 30-month CD with an advertised 3% APY if interest is compounded monthly.

Using $n = 12$, $APR = 12(0.03 + 1)^{\frac{1}{12}} - 12$
$\approx 12.0296 - 12 = 0.0296 = 2.96\%$.

Notice that there is little difference in the interest rates. But once again, the APY is greater because it includes the effect of compounding.

Assuming monthly compounding of interest, find the APR for each of the following investments. Round to the nearest tenth of a percent.

6. \$100,000 deposited into a 3 yr CD at 4.5% APY
7. a \$10,000 investment at 4% APY for 10 yr
8. the same \$10,000 investment at 6% APY for 10 yr

To be faithful with God's resources, you should be knowledgeable about amortized loans since this is the type of loan used to buy a home. An amortized loan is one in which equal periodic payments (usually monthly) are made throughout the life of the loan. Like the add-on loan, each payment is for the same amount. Unlike the add-on loan, the interest is figured on the unpaid balance. At first, most of the money goes to pay interest, with only a small amount going toward the principal. However, as the borrower nears the end of the contracted time, nearly all the payment goes toward principal and only a small amount is interest.

The following 2 formulas are finance formulas. The first finds the monthly payment, A, you would need to make on an amortized loan.

$A = P\left[\frac{i(1+i)^n}{(1+i)^n - 1}\right]$, where P = the principal,
n = the total number of payments (payments per year × number of years), and
i = the periodic interest rate (APR ÷ payments per year).

The second formula determines the monthly deposit, D, you would need to make to accumulate S dollars in an account with compound interest (assuming no withdrawals).

$$D = S\left[\frac{i}{(1+i)^n - 1}\right]$$

9. Find the monthly payment on a 2 yr \$5500 loan with 6% annual interest.
10. Find the monthly deposit you would need to make in order to save \$5500 at 6% interest in 2 yr.
11. Why are the answers to exercises 9 and 10 different? The amounts are the same, the rates are the same, and the time is the same.

Wise investors must distinguish among several investments to determine how they can best manage the resources entrusted to them. Determining the annual percentage yield (APY) allows you to compare different investments.

Which yields the better rate of return, 5% simple interest or 4.7% interest compounded daily?

5% simple interest: $APY = \left(1 + \frac{APR}{n}\right)^n - 1$
$= \left(1 + \frac{0.05}{1}\right)^1 - 1 = 1.05 - 1 = 0.05 = 5\%$

4.7% interest compounded daily: APY
$= \left(1 + \frac{APR}{n}\right)^n - 1 = \left(1 + \frac{0.047}{365}\right)^{365} - 1$
$\approx 1.0481 - 1 = 0.048 = 4.8\%$

The 5% simple interest rate has a higher APY than 4.7% compounded daily.

Find the APY (to the nearest hundredth of a percent) for each of the following pairs of investments; then determine which has the greater yield.

12. 6% simple interest or 5% interest compounded quarterly
13. 4.4% interest compounded quarterly or 4.2% interest compounded monthly
14. 5.6% interest compounded monthly or 5.65% simple interest
15. 3.75% interest compounded daily or 3.8% interest compounded semiannually

3 CHAPTER REVIEW

Key Concepts

Simple Inequalities [3.1]

SYMBOL	MEANING	EXAMPLE GRAPH
$x < n$	is less than	
$x > n$	is greater than	
$x \neq n$	is not equal to	
$x \leq n$	is less than or equal to	
$x \geq n$	is greater than or equal to	

- The Trichotomy Property states that if a and b represent 2 real numbers, then $a = b$, $a > b$, or $a < b$.

PROPERTIES OF INEQUALITY	
Addition	If $a < b$, then $a + c < b + c$.
Subtraction	If $a < b$, then $a - c < b - c$.
Multiplication	If $a < b$ and c is a positive number, then $ac < bc$. If $a < b$ and c is a negative number, then $ac > bc$.
Division	If $a < b$ and c is a positive number, then $\frac{a}{c} < \frac{b}{c}$. If $a < b$ and c is a negative number, then $\frac{a}{c} > \frac{b}{c}$.

- For two-step inequalities, use the reverse order of operations when performing inverse operations on both sides.
- If both sides of an inequality are multiplied or divided by a negative number, the inequality symbol must be reversed.

Multistep Inequalities [3.2]

- To solve multistep inequalities, use the same process as in solving multistep equations.
  - Simplify both sides by removing any parentheses and combining like terms.
  - Move the variable terms to 1 side of the equation.
  - Undo the addition or subtraction of the constant term.
  - Undo the multiplication or division of the variable term.

Conjunctions [3.3]

- Conjunctions are compound inequalities connected by the word *and*.

- To solve a conjunction:
  - Solve each part of the compound inequality by applying inverse operations to isolate each variable.
  - Graph each resulting inequality on the same number line.
  - The final graph is the intersection of the 2 parts.

Disjunctions [3.4]

- Disjunctions are compound inequalities connected by the word *or*.
- To solve a disjunction:
  - Solve each part of the compound inequality by applying inverse operations to isolate each variable.
  - Graph each resulting inequality on the same number line.
  - The final graph is the union of the 2 parts.

COMPOUND INEQUALITY COMPARISON

Graph	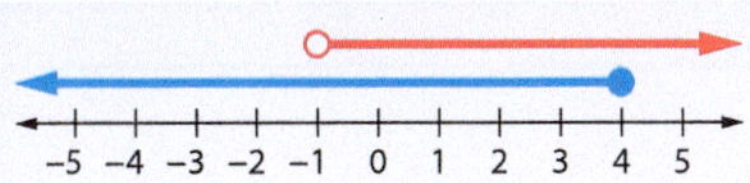	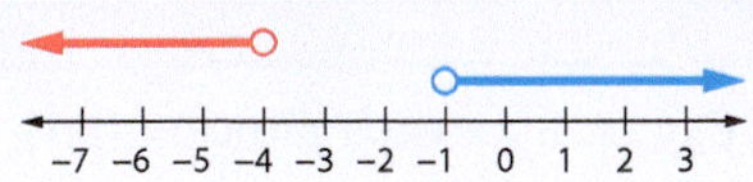	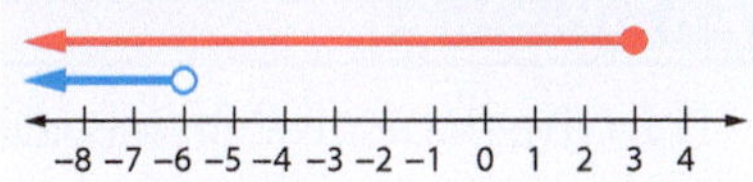
Conjunction	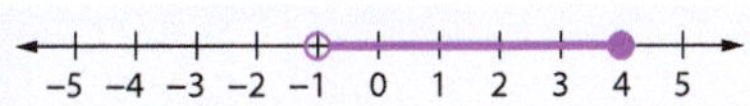 $-1 < x \le 4$	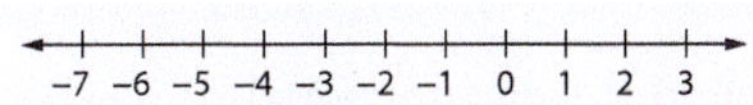 $\varnothing$	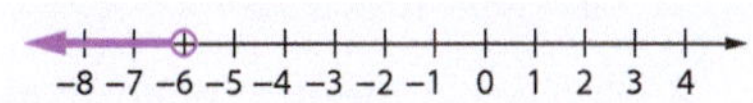 $x < -6$
Disjunction	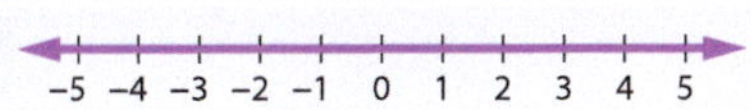 $\mathbb{R}$	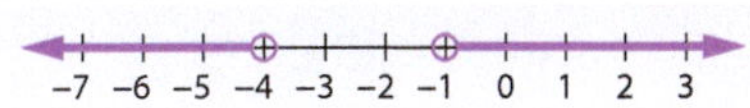 $x < -4$ or $x > -1$	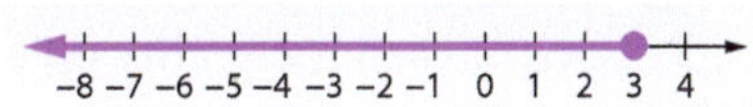 $x \le 3$

Absolute Value Equations [3.5]

- To solve absolute value equations:
  - Simplify within the absolute value symbols.
  - Isolate the absolute value on 1 side of the equation.
  - Use the definition of absolute value to write 2 equations.
  - Solve each equation separately.
  - Check your solutions.

Absolute Value Inequalities [3.6]

ABSOLUTE VALUE INEQUALITIES

	Conjunction (*and*)		Disjunction (*or*)	
	$\lvert x\rvert \le n$	$\lvert x\rvert < n$	$\lvert x\rvert \ge n$	$\lvert x\rvert > n$
$n > 0$	$-n \le x \le n$ 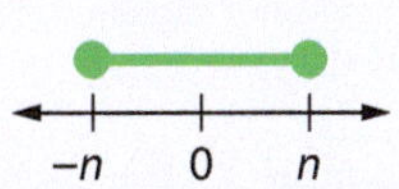	$-n < x < n$ 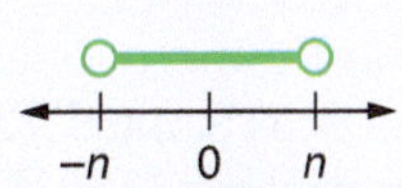	$x \le -n$ or $x \ge n$ 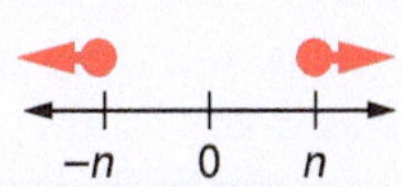	$x < -n$ or $x > n$ 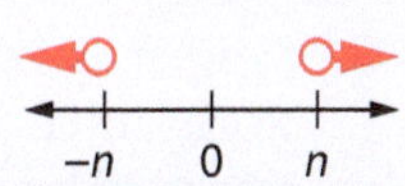
$n = 0$	$x = 0$	$\varnothing$	$x \le 0$ or $x \ge 0$ $\mathbb{R}$	$x < 0$ or $x > 0$ 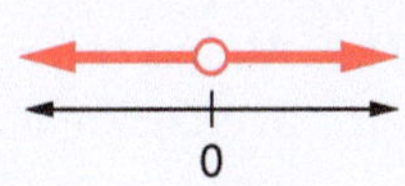 $x \ne 0$
$n < 0$	$\varnothing$	$\varnothing$	$\mathbb{R}$	$\mathbb{R}$

- To solve absolute value inequalities:
  - Isolate the absolute value expression.
  - Write and solve the corresponding compound inequality based on the inequality symbol.

Biblical Worldview Shaping

Reasoning with Algebra

Why does valid reasoning not always lead to truth?

Valid reasoning must start with true assumptions in order to be sure the conclusion is true.

EXERCISES

1. Give the equivalent inequality symbol for each phrase. **[3.1]**

a. is at least **b.** does not exceed **c.** fewer than

2. Write an inequality equivalent to the given inequality. **[3.1]**

a. $y \ngtr 3$ **b.** $z \nless -10$

3. Graph each inequality on a separate number line. **[3.1]**

a. $x \geq -2$ **b.** $y < -3$

Solve each inequality. [3.1–3.2]

4. $x - 17 < -6$

5. $-\frac{y}{6} > 8$

6. $4a - 7 > 5$

7. $-5 \leq -\frac{4}{9}b + 7$

8. $2x - 12 > 5x - 4$

9. $-3(5x - 8) > 5 - 7x + 2$

10. $5(3x - 4) \leq 2(x - 10)$

11. $5(2 - x) \leq x - 4(x + 3)$

Write an inequality and solve. [3.1–3.2]

12. Find all the numbers such that adding 15 to 23 times the number is greater than 429.

13. Find the largest integer such that eight more than six times the integer is less than the opposite of forty-two.

14. Find all the numbers such that the product of seven and the number is more than the sum of seven and the number. What is the smallest integer that is a solution?

15. Samuel makes twice as much per hour as Kimberly, but Kimberly works three times as many hours per week as Samuel. Kimberly works 36 hr, and the sum of their wages exceeds $480. How much does each of them make per hour?

16. Ali has test scores of 95, 78, and 93, and he needs to have a test average of at least 90 to get an A on his report card. What is the lowest score he can get on his final test and still have an A?

17. Two inequality statements joined by the word *and* are called a(n) ______. **[3.3]**

18. The union of 2 solution sets is called a(n) ______. **[3.4]**

Solve each compound inequality and graph its solution on a number line. [3.3–3.4]

19. $x > 3$ or $x > -1$

20. $x > -2$ and $x \leq 4$

21. $x > 3$ and $x < 7$

22. $x < 3$ or $x \geq -1$

23. $x \neq 5$ or $x < 2$

24. $4x - 12 \geq 28$ and $16x + 4x - 6 > 54$

Solve each compound inequality. Express the answer as an inequality statement, the set of real numbers, or the empty set. [3.3–3.4]

25. $2x \leq -10$ or $2x > 8$

26. $6x + 5 < 29$ and $3x > 9$

27. $3x - 5 < 4$ and $3x + 2 > 12$

28. $5x + 6 - 2x \leq 12$ or $8x < 40$

29. $7 < 4x - 9 < 19$

30. $5 - 3x < 10$ and $4x + 6 < 10$

31. $3x + 6 \leq 6$ or $2x + 8 < -3$

32. $5 - 3x < 10$ or $4x + 6 < 10$

Solve each absolute value equation. Write ∅ if no solution exists. [3.5]

33. $-5|x| = -125$

34. $-2|x| + 6 = 20$

35. $|x + 3| + 2 = 12$

36. $|4z - 20 + 10z| - 40 = -4$

37. Write an absolute value equation representing:

a. numbers that are 5 units from −2.

b. numbers that are 4 units from −3.

c. numbers that are 3 units from 4.

Solve and graph the solution on a number line. [3.6]

38. $|x| < 4$

39. $|x + 3| > 7$

40. $|3x - 8| \geq 11$

41. $|3x + 6| < 15$

Write an absolute value inequality and solve. [3.6]

42. The proper pH of the water in swimming pools is 7.4, ±0.2. Write an absolute value inequality representing this range. Then solve the inequality to state the range of the pH.

43. An investor has decided to sell his shares of stock if the stock's price increases or decreases more than $14 from its present value of $73 per share. Write an absolute value inequality representing the price at which he should sell his shares. Then solve the inequality to state the range of prices.

44. A person functions comfortably in temperatures from 16°C to 27°C. Write a compound inequality and an absolute value inequality representing a person's optimal temperature range.

45. **BWS Essential Question:** How can valid reasoning lead to a false conclusion?

CHAPTER 04
FUNCTIONS

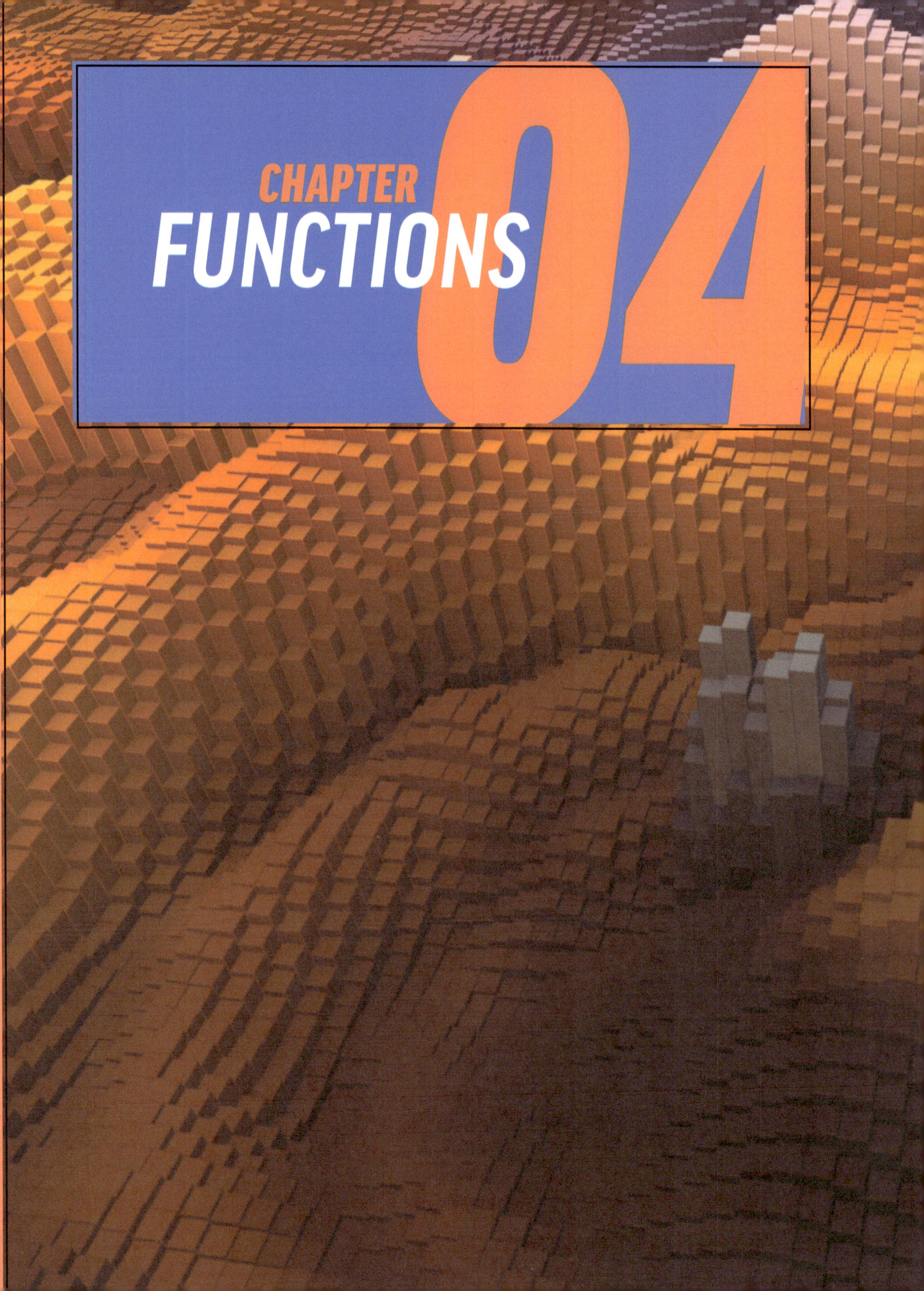

CHAPTER OVERVIEW

4.1 RELATIONS & FUNCTIONS

How does a function compare to a relation?

After completing this section, I will be able to

- represent relations and functions by using sets of ordered pairs, tables, mapping diagrams, and graphs.
- identify the domain and range of relations and functions.
- determine whether a relation is a function.

The term "relation" has a variety of meanings. The word can refer to a relative, someone related to you. It can also mean the connection or relationship between people.

More broadly, a relation can be an association between any 2 (or more) items. A common relation exists in a grocery store between items and their prices.

x	*y*
gum	\$2.58
apple	\$0.84
yogurt	\$1.24

In mathematics, a relation has a specific meaning for an association between 2 sets.

RECALL

A set is a collection of objects generally expressed within brackets, { }.

DEFINITIONS

A **relation** is any set of ordered pairs.

The **domain** of a relation is the set of first elements (*x*-coordinates) of the ordered pairs.

The **range** of a relation is the set of second elements (*y*-coordinates) of the ordered pairs.

Relations are often listed as a set of ordered pairs and named with a capital letter.

$$A = \{(-3, 5), (2, 2), (2, 4), (4, -1)\}$$

A relation can also be illustrated using a table or a graph. The graph of a relation is the set of points in a plane that corresponds to the ordered pairs of the relation.

x	*y*
−3	5
2	2
2	4
4	−1

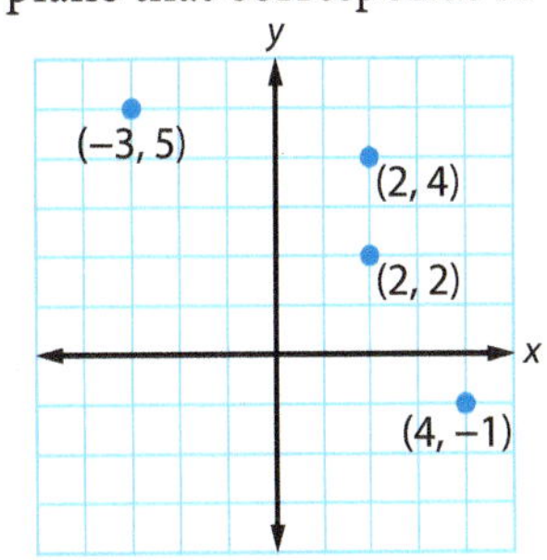

EXAMPLE 1: Identifying the Domain & Range

State the domain, D, and range, R, of relation A shown above.

Answer

$D = \{-3, 2, 4\}$ — List the set of x-coordinate values.

$R = \{-1, 2, 4, 5\}$ — List the set of y-coordinate values.

SKILL CHECK ***EXERCISE 7***

A fourth representation of a relation, a mapping diagram, emphasizes the domain and range. The arrows show the pairing of each number in the domain with its associated number in the range.

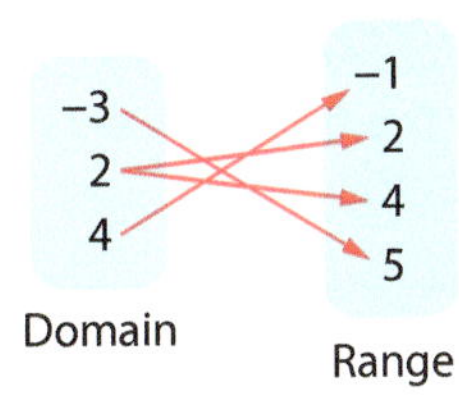

A **function** is a special type of relation in which every element in the domain is paired with 1 and only 1 element in the range. It cannot have 2 or more ordered pairs with the same first coordinate.

Because a function is a relation, it has a domain and a range and can be illustrated using any of the representations of a relation: a set of ordered pairs, a table, a graph, or a mapping diagram. While any function must be a relation, not all relations are functions.

EXAMPLE 2: Representing a Relation

Write a set of ordered pairs and draw a mapping diagram of the relation illustrated by the graph. Is this relation a function?

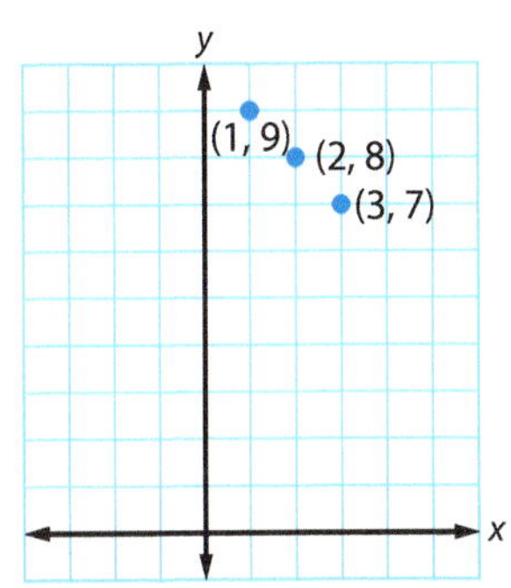

Answer

$B = \{(1, 9), (2, 8), (3, 7)\}$

1. Write the set of ordered pairs.
2. Draw a mapping diagram.

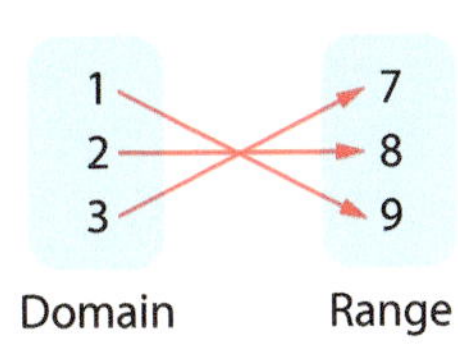

The relation is a function.

3. Each element of the domain is paired with a unique element of the range.

SKILL CHECK ***EXERCISE 23***

EXAMPLE 3: Identifying a Function

Determine whether each relation is a function.

a. $A = \{(-3, 5), (2, 2), (2, 4), (4, -1)\}$ **b.** $B = \{(0, 4), (-1, 2), (3, 4), (-4, -3)\}$

Answers

a. *A* is not a function. The domain element 2 is paired with 2 different members of the range.

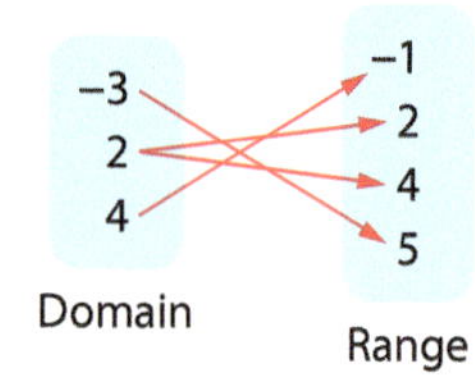

b. *B* is a function. Even though 2 elements of the domain are paired with the same range element, 4, each element of the domain is paired with 1 and only 1 member of the range.

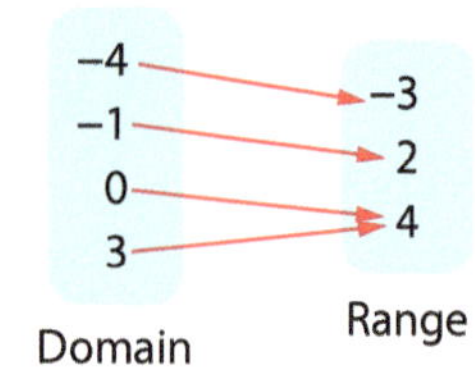

SKILL CHECK **EXERCISE 17**

A scatter plot is a graph on a Cartesian plane that summarizes the relationship between 2 sets of numbers. Tables are often used to summarize the data from a statistical study. Ordered pairs representing the data can be graphed so that a statistician can use the patterns formed by the points to determine the relationship between the variables in the study.

EXAMPLE 4: Modeling the Real World

Make a scatter plot of the heights and weights of students as recorded in the following table and describe the relationship between the variables.

Weight/Height Comparison of Students										
height (in.)	64	60	62	58	63	61	62	60	59	62
weight (lb)	130	124	138	98	159	136	142	126	101	145

Answer

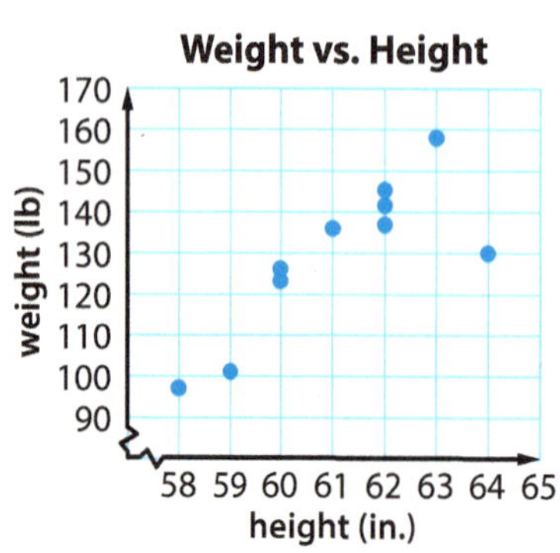

1. Adjust the scale of the axes and their intersection to fit the data.

2. Plot the ordered pairs.

In general, as height increases, weight increases.

3. The graph rises from left to right.

SKILL CHECK **EXERCISE 37**

AFTERSCHOOLHELP

Relations & Functions

VOCABULARY

- dependent variable
- domain
- function
- independent variable
- range
- relation

Lowercase letters, particularly f and g, are often used to name functions.

$$f = \{(6, 9), (4, 12), (0, 4)\}$$

The variable that represents the values of the domain, typically x, is called the **independent variable.** Because the value of the second coordinate, usually y, depends on the value chosen for the first coordinate, it is called the **dependent variable.** When a function is used to model the relationship between an item and its price, the independent variable is assigned to the item since it determines the price, the dependent variable.

x	y
6	9
4	12
0	4

or

x	f(x)
6	9
4	12
0	4

Function notation emphasizes the fact that the value of the y-coordinate depends on the chosen value of the x-coordinate. It symbolizes the y-coordinate as $f(x)$, which is read "f of x." Since $y = f(x)$, the ordered pair (x, y) can be written $(x, f(x))$. Notice that $f(6) = 9$, which means $y = 9$ when x is 6. Likewise, $f(4) = 12$ and $f(0) = 4$.

Functions and relations are often used to model relationships between 2 quantities in everyday life.

EXAMPLE 5: Modeling the Real World

The local grocery store is having a sale on potato chips. The price of 1 bag is \$2, but if you buy 2 bags, you get a third one free. Represent the relationship between the number of bags purchased (from 0 to 12) and their cost using a table and a graph. Are these ordered pairs a relation? What is the domain? What is the range? Are the ordered pairs a function?

Answer

bags, x	0	1	2	3	4	5	6	7	8	9	10	11	12
cost in \$, f(x)	0	2	4	4	6	8	8	10	12	12	14	16	16

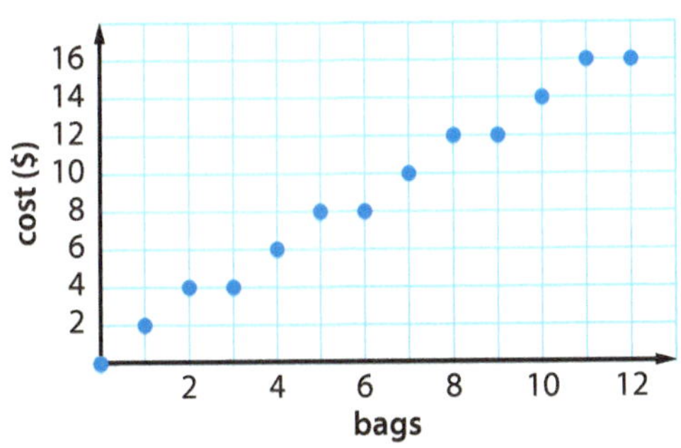

Yes, it is a relation. **1.** Any set of ordered pairs is a relation.

$D = \{0, 1, 2, \ldots, 12\}$ **2.** List the values of the independent variable.

$R = \{0, 2, 4, \ldots, 16\}$ **3.** List the values of the dependent variable.

Yes, it is a function. **4.** No element of the domain relates to more than 1 element of the range.

Using function notation, we can also note that in the function above $f(2) = 4$, $f(4) = 6$, $f(9) = 12$, and so on.

SKILL CHECK **EXERCISE 37**

A. EXERCISES

1. Which of the following cannot be used to represent a relation?

   A graph **C** number line **E** table

   B mapping diagram **D** set of ordered pairs

State whether each set represents a relation.

2. $\{(2, 4), (5, 9), (8, 2)\}$
3. $\{3, 4, 7, 9\}$
4. $\{(-2, 5, 7), (5, -8, 2), (2.8, 6, 7)\}$
5. $\{(1, -2), (1, 3), (1, -4), (1, 5)\}$

State the domain and range of each relation.

6. $\{(6, 10), (-9, 14), (4, 2), (-9, 8)\}$
7. $\{(3, 2), (3, 3), (3, 4), (3, 5), (3, 6)\}$
8. $\{(1.3, 3.7), (1.4, 3.7), (1.5, 3.7), (1.6, 3.7)\}$
9. $\left\{\left(\frac{1}{2}, \frac{3}{5}\right), \left(\frac{2}{3}, \frac{7}{8}\right), \left(\frac{4}{5}, \frac{5}{9}\right), \left(\frac{5}{6}, \frac{3}{5}\right)\right\}$

Draw a mapping diagram of each relation and state whether it is a function.

10. $\{(-5, 1), (-5, 2), (2, 8), (5, 6)\}$
11. $\{(-6, -5), (-3, 4), (1, 7), (4, -3)\}$
12. $\{(1.4, 3.2), (4.2, 9.5), (6.8, 1.4), (8.4, 6.8)\}$
13. $\left\{\left(\frac{4}{3}, \frac{2}{5}\right), \left(\frac{4}{7}, \frac{8}{9}\right), \left(\frac{4}{3}, \frac{3}{8}\right)\right\}$

Choose the letter corresponding to the correct description of each set.

A both a relation and a function **B** a relation that is not a function **C** neither a relation nor a function

14. $\{(2, 3), (4.7, -5)\}$
15. $\{(-2.6, 3), (4.9, 3)\}$
16. $\left\{2, 3, -\frac{4}{5}, 8.1\right\}$
17. $\{(2, 3), (3, 4), (3, 5)\}$
18. $\left\{(2, 3), \left(2, \frac{5}{4}\right)\right\}$
19. $\{(5, 9, 2), (-7, -3, 1), (6, 7, 0)\}$

B. EXERCISES

A relation can be represented by a set of ordered pairs, a table, a graph, or a mapping diagram. For each relation, create the other 3 representations and state whether the relation is a function.

20. $\{(-6, -2), (-1, 0), (3, 4), (4, 3)\}$

21.

x	y
−2	0
−1	4
0	3
1	−1
1	4

22.

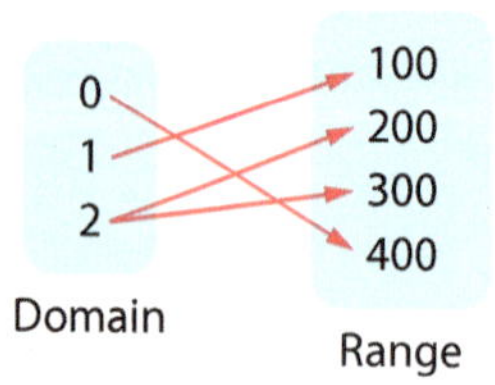

23.

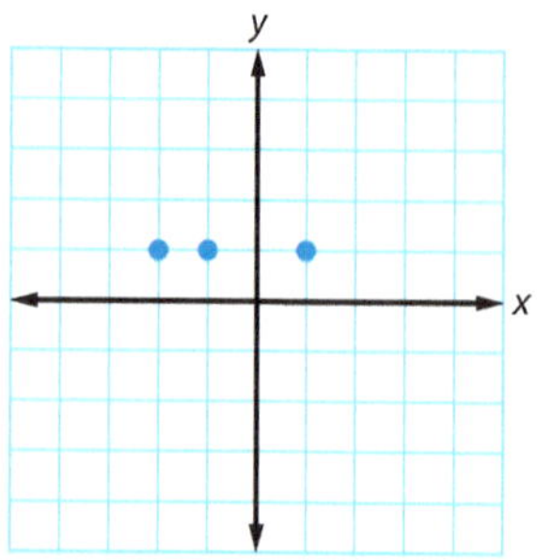

Given the function $f = \{(6, 28), (-7, -4), (3, -9), (0, -4), (-9, 2), (1, -10)\}$, find the following.

24. $f(-7)$
25. $f(0)$
26. $f(6)$
27. $f(1)$

For exercises 28–31, $A = \{1, 6, 9\}$ and $B = \{3, 4\}$.

28. Make a relation in which the domain is A and the range is B. If possible, make this relation a function; if that is not possible, explain why not.

29. Make a relation in which the domain is B and the range is A. If possible, make this relation a function; if that is not possible, explain why not.

30. Make a function in which the domain is B and the range is a subset of A. (Recall that the elements of a subset are contained in another set. If $C = \{4\}$, then C is a subset of B.) Represent the function as a set of ordered pairs, a table, a graph, and a mapping diagram.

31. Make a function in which the domain is A and the range is a subset of B. Represent the function as a set of ordered pairs, a table, a graph, and a mapping diagram.

32. **Essential Question:** How does a function compare to a relation?

For each of the situations described, state which quantity should be represented by the independent variable and which by the dependent variable.

33. modeling the time driven and the distance traveled if the driver maintains a constant speed

34. determining the total earnings of several student workers given the rate of $8/hr and the number of hours each student worked

35. graphing the ages and heights of students in a nutritional study

36. modeling the relationship between body mass index, which measures the level of physical fitness, and the number of minutes spent exercising each day

MP4 Create a scatter plot for each set of data and determine the general relationship between the variables.

37.

daily minutes of exercise (x)	15	25	12	35	20	45	28	60
resting heart rate (y)	72	68	75	66	71	58	70	56

38.

calories per day (x)	weight (y)
2100	152
1750	130
1580	121
2640	165
1900	149
1975	138
2065	144
1490	119

C. EXERCISES

Given the function $g = \{(a, m), (b, n), (c, o), (d, p), (e, q)\}$, find the following.

39. $g(c)$

40. $g(e)$

41. $g(a)$

42. $g(b)$

The greatest integer function is a unique function symbolized by [x]. The domain is any real number x, and the range, [x], is the largest integer less than or equal to x.

43. Complete the following table of values for this function.

x	$4\frac{1}{2}$	5	$5\frac{1}{4}$	$6\frac{3}{4}$	$-\frac{1}{2}$	$-2\frac{1}{2}$	−5.75	$\frac{1}{3}$	$-\frac{1}{4}$
$[x]$									

44. Make a graph for $f(x) = [x]$, using all real numbers as the domain.

CUMULATIVE REVIEW

State the property of equality used to solve each equation. [2.1]

45. $5x = 8$

46. $x - 15 = 39$

Solve each equation. [2.2]

47. $2x + 8 + 4x - 1 = 25$

48. $2x + 5 + 7x + 1 = 42$

49. $3x - 2 = 5x + 8$

50. $3(4x - 1) = 2x + 4$

Solve each absolute value inequality. [3.6]

51. $|5 - x| \leq 3$

52. $|2x + 7| > 11$

53. Jeff and Elaine hope to make at least $2325 in combined sales commission this week, with Jeff estimated to make twice as much as Elaine. Which inequality can be used to find how much each needs to earn? **[3.2]**

A $x + 2x \leq 2325$

B $2x + x \geq 2325$

C $2x \geq 2325 + x$

D $x > 2325 - 2x$

E none of these

54. Which of the following has no solution? **[3.3–3.4]**

A $x > -5$ and $x > 1$

B $x > 0$ or $x < -3$

C $x < 2$ and $x > 2$

D $x \neq -6$ or $x \neq \frac{10}{3}$

E none of these

MIND OVER MATH

Plot the following ordered pairs on a graph and connect them with line segments in the order they are listed, working down each column. Start a new portion of the graph after the word *end*.

(−4, 5)
(−4, 1)
(−1, 1)
(−1, 5)
(−4, 5)
end

(−1, −1)
(−1, −2)
(1, −2)
(1, −1)
(−1, −1)
end

(1, 5)
(4, 5)
(4, 1)
(1, 1)
(1, 5)
end

(−5, 5)
(−3, 7)
(−2, 7)
(−1, 6)
end

(6, 10)
(6, −7)
(5, −8)
(2, −9)
(−2, −9)
(−5, −8)
(−6, −7)
(−6, 10)
end

(4, 2)
(3, 2)
(3, 1)
end

(−4, −3)
(−3, −4)
(−2, −5)
(2, −5)
(3, −4)
(4, −3)
(2, −4)
(−2, −4)
(−4, −3)
end

(−6, 10)
(−5, 11)
(−3, 12)
(3, 12)
(5, 11)
(6, 10)
(5, 9)
(2, 8)
(−2, 8)
(−5, 9)
(−6, 10)
end

(1, 6)
(2, 7)
(3, 7)
(5, 5)
end

(−1, 2)
(−2, 2)
(−2, 1)
end

4.2 GRAPHS OF RELATIONS & FUNCTIONS

How can I graph a function containing an infinite number of points?

After completing this section, I will be able to

- determine the domain and range of a relation by using graphed points.
- determine whether graphed points represent a function.
- graph relations and functions.

SET BUILDER NOTATION

The set notation $\{t \mid 0 \leq t \leq 3\}$ is read "the set of all time (t) such that t is any real number from 0 to 3."

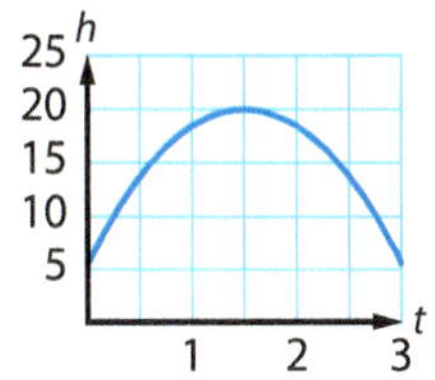

This graph illustrates the relationship between the height of a ball thrown from the outfield and the time that it has been in the air. Since the height of the ball depends on the length of time it has been in the air, t is the independent variable and h is the dependent variable. Since there are an infinite number of points in this relation, the domain and range are best stated using inequalities.

$$D = \{t \mid 0 \leq t \leq 3\} \text{ and } R = \{h \mid 6 \leq h \leq 20\}$$

The **vertical line test** can be used on the graph of a relation to determine whether it is a function. If any vertical line intersects the relation's graph in more than 1 point, the relation is not a function because the intersected points have the same x-coordinate. The fact that no vertical line can be drawn containing more than 1 point of the relation shows that the relation is a function.

EXAMPLE 1: Applying the Vertical Line Test

State the domain and range of each relation. Then use the vertical line test to determine whether the relation is a function.

a.

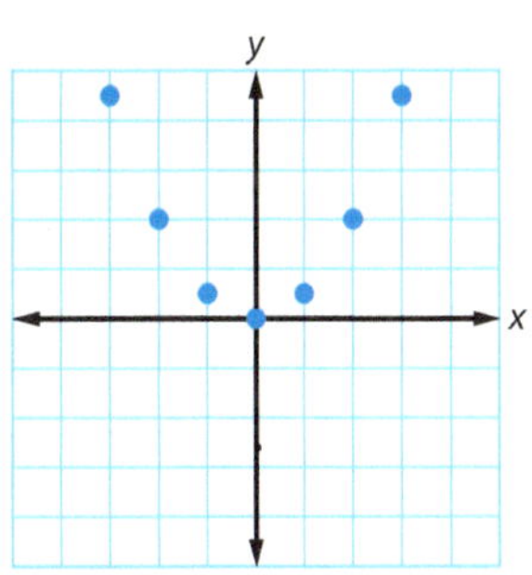

b.

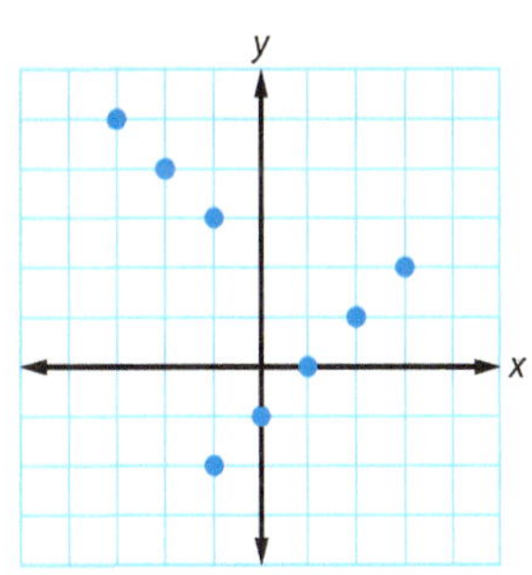

c.

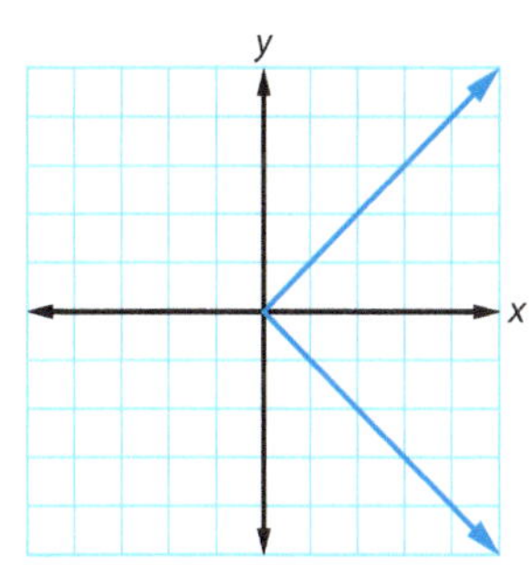

Answers

a. $D = \{0, \pm1, \pm2, \pm3\}$
$R = \{0, 0.5, 2, 4.5\}$
The relation is a function.

No vertical line intersects the graph at more than 1 point.

b. $D = \{0, \pm1, \pm2, \pm3\}$
$R = \{-2, -1, 0, 1, 2, 3, 4, 5\}$
The relation is not a function.

A vertical line crossing the x-axis at -1 contains both points $(-1, -2)$ and $(-1, 3)$.

c. $D = \{x \mid x \geq 0\}$
$R = \mathbb{R}$
The relation is not a function.

1. The domain is any nonnegative number. The range is the real numbers.
2. Any vertical line crossing the positive x-axis crosses the graph at more than 1 point.

SKILL CHECK **EXERCISE 7**

An equation is often used to clearly define the rule that determines the values of the range associated with each member of the domain.

EXAMPLE 2: Graphing a Relation

Given the relation $A = \{(x, y) \mid y = x + 2; x = 1, 2, 3, 4\}$, make a table of ordered pairs to represent the relation. Then graph the relation and state whether it is a function.

Answer

x	$y = x + 2$	Ordered Pair
1	$1 + 2 = 3$	(1, 3)
2	$2 + 2 = 4$	(2, 4)
3	$3 + 2 = 5$	(3, 5)
4	$4 + 2 = 6$	(4, 6)

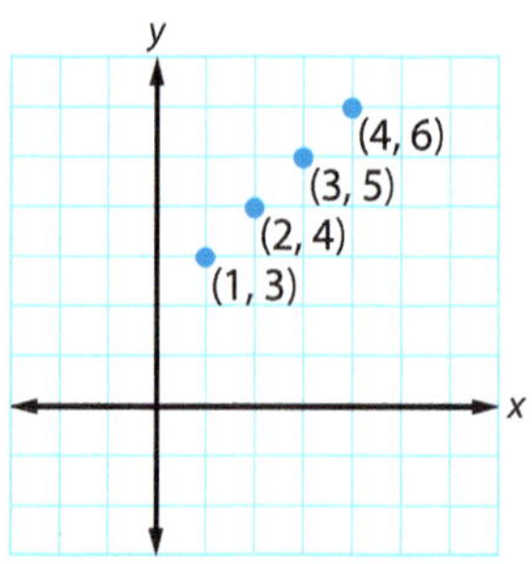

This relation is a function since no vertical line can be drawn that passes through more than 1 point.

SKILL CHECK **EXERCISE 13**

What if the values for x were not stated in the previous example? When no domain is stated, it is assumed to be the real numbers. Therefore, the graph of the function $y = x + 2$ follows the pattern established by the ordered pairs already graphed. Since the coordinates of any point on the graphed line satisfy the equation, the line is the graph of the function.

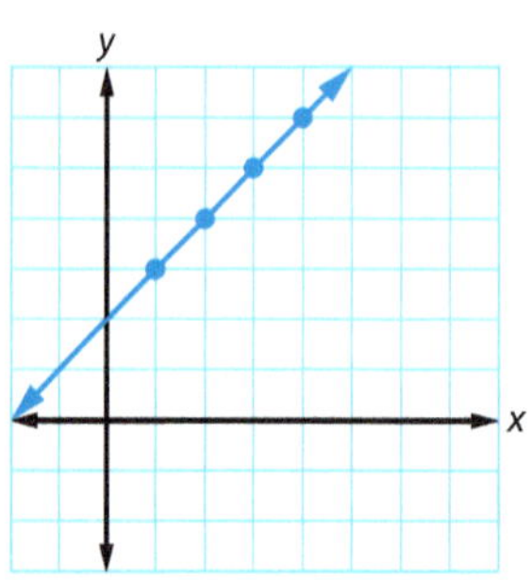

EXAMPLE 3: Graphing a Relation

> **RECALL**
> $\mathbb{Z}$: the set of integers

Graph the relation $B = \{(x, y) \mid y = 3;\ 1 \le x \le 5,\ x \in \mathbb{Z}\}$. State whether the relation is a function.

Answer

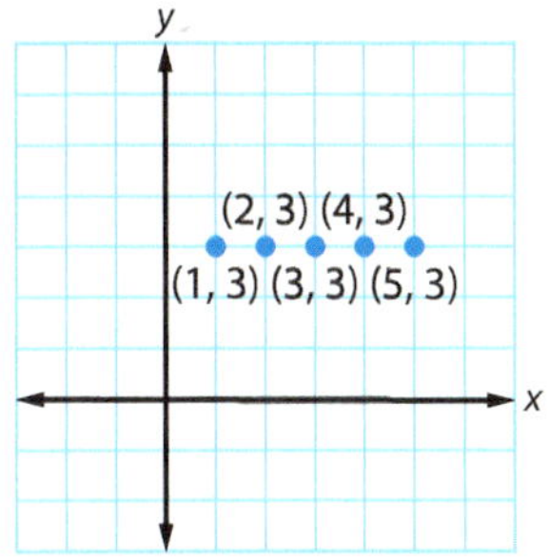

1. B is the set of ordered pairs (x, y) such that x is any integer from 1 to 5 (including 1 and 5) and y is equal to 3.
2. Write B in its set form: $\{(1, 3), (2, 3), (3, 3), (4, 3), (5, 3)\}$.
3. Graph the relation.

The relation is a function.

4. No vertical line can be drawn that passes through more than 1 point.

SKILL CHECK **EXERCISE 21**

Without the restriction of the domain to integral values in the previous example, the graph would contain all possible real values of x from 1 to 5. The result is the segment from (1, 3) to (5, 3). Ordered pairs such as (2.1, 3), (π, 3), and $\left(3\frac{1}{2}, 3\right)$ are members of this relation. The vertical line test indicates that this is the graph of a function.

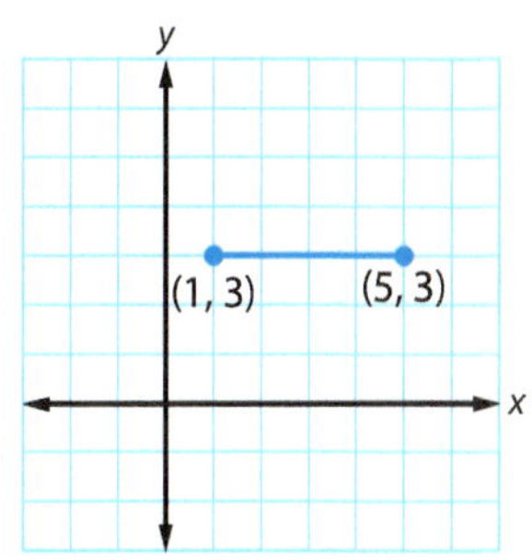

Relations can often be described in terms of equations or inequalities.

EXAMPLE 4: Graphing a Relation

Graph $D = \{(x, y) \mid y = \pm\sqrt{x};\ x = 1, 4, 9\}$. State its domain and range. Is this a function?

Answer

x	1	4	9
y	±1	±2	±3

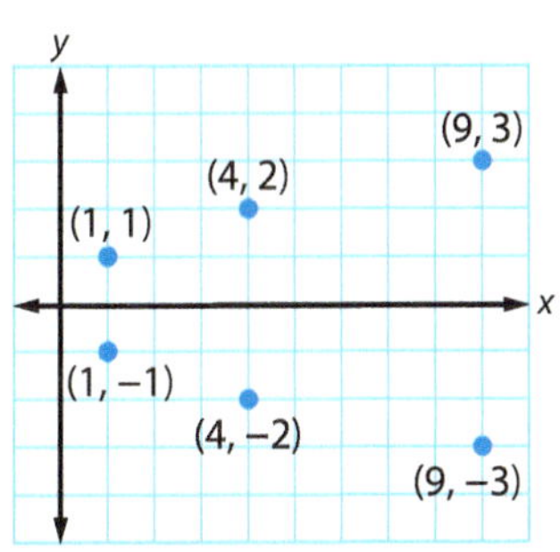

1. Make a table to find the ordered pairs. Substitute the x-values from the domain into the equation $y = \pm\sqrt{x}$. Each value of x produces 2 values for y.
2. Graph the set of ordered pairs: $\{(1, 1), (1, -1), (4, 2), (4, -2), (9, 3), (9, -3)\}$.

$D = \{1, 4, 9\}$

$R = \{-3, -2, -1, 1, 2, 3\}$

3. The domain is the given set of x-coordinates. The range is the set of y-coordinates.

This relation is not a function.

4. The relation fails the vertical line test; each value of x generates 2 values for y.

SKILL CHECK **EXERCISE 23**

> **AFTERSCHOOLHELP**
> Graphs of Relations & Functions

If the domain contains an infinite number of values, the general shape of the graph can be determined by plotting a sufficient number of points. Draw a smooth curve connecting the ordered pairs to complete the graph.

EXAMPLE 5: Graphing a Function

Graph the function $f(x) = -x^2$. Then state its domain and range.

Answer

x	$-x^2$	$f(x)$
−2	$-(-2)^2$	−4
−1	$-(-1)^2$	−1
0	-0^2	0
1	-1^2	−1
2	-2^2	−4

1. Make a table and plot enough ordered pairs so you can see the shape of the graph.

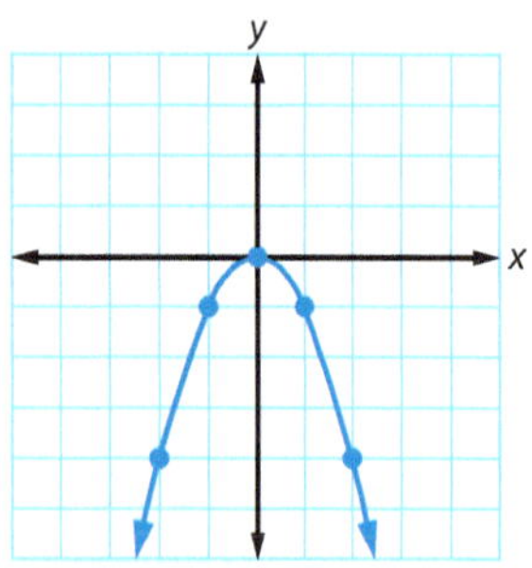

2. Connect the points with a smooth curve.

$D = \mathbb{R}$

$R = \{y \mid y \leq 0\}$

3. Since there is no restriction on the domain, any real number can be used. The range consists of all nonpositive numbers.

SKILL CHECK **EXERCISE 19**

VOCABULARY

- vertical line test

A. EXERCISES

Is the graphed relation a function?

1.

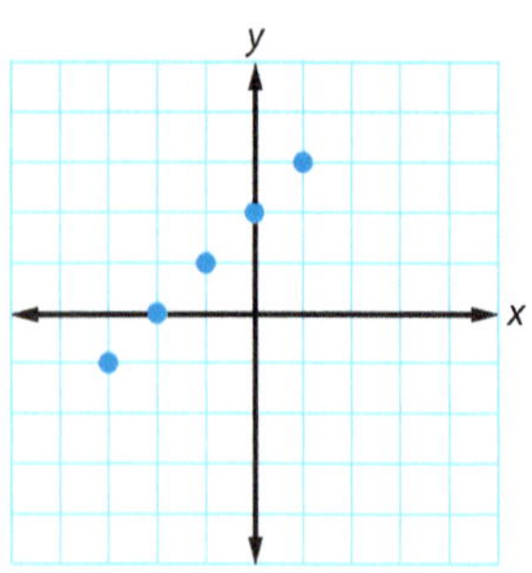

2.

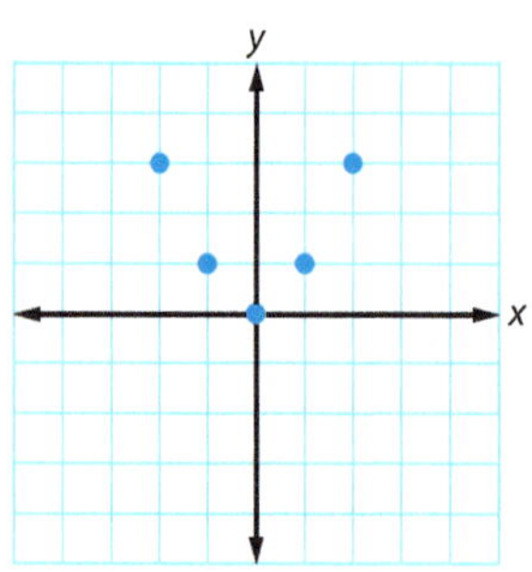

3.

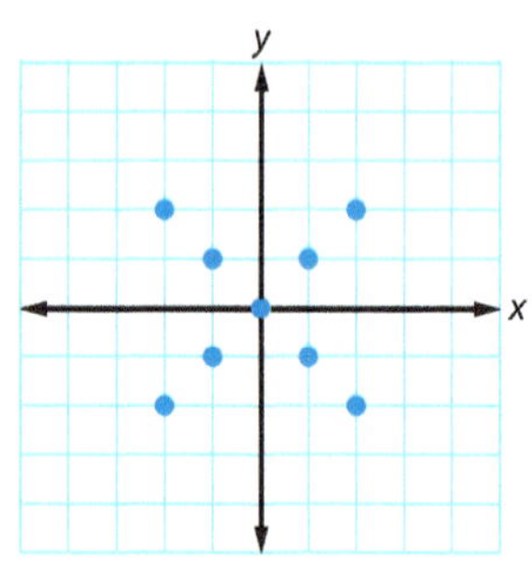

4.

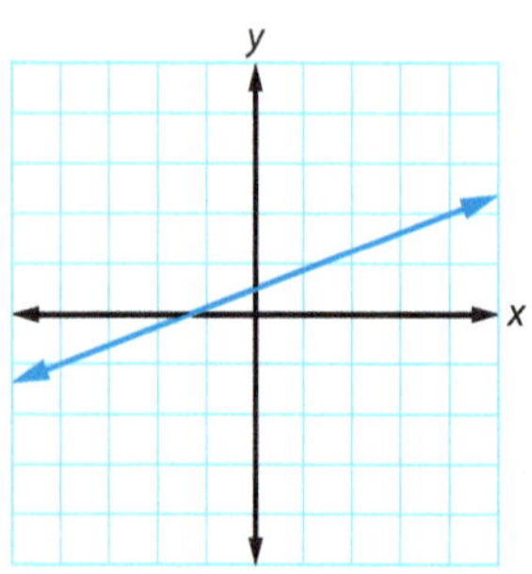

5.

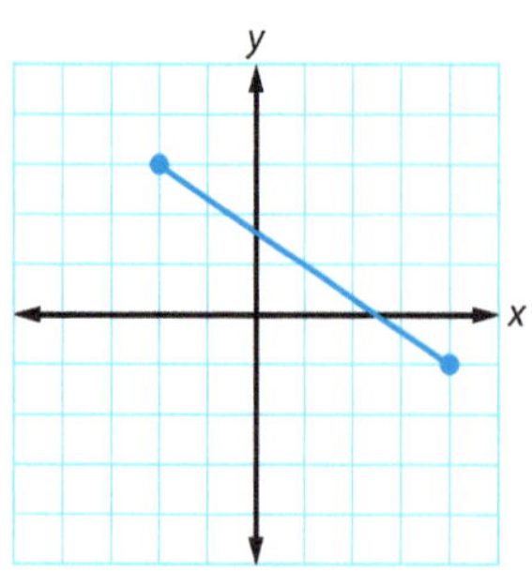

6.

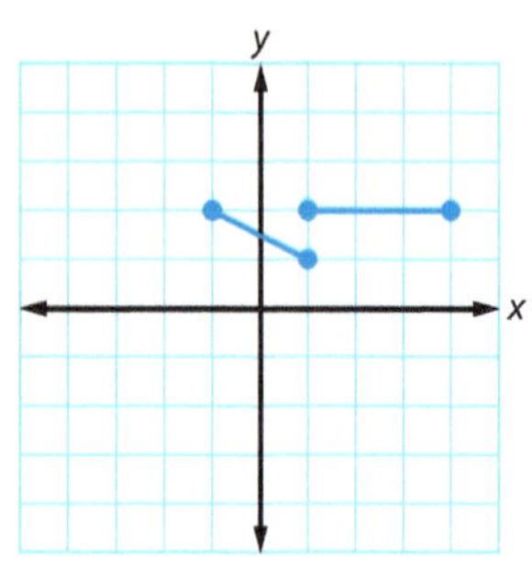

For each graphed relation, state its domain and range and whether it is a function.

7.

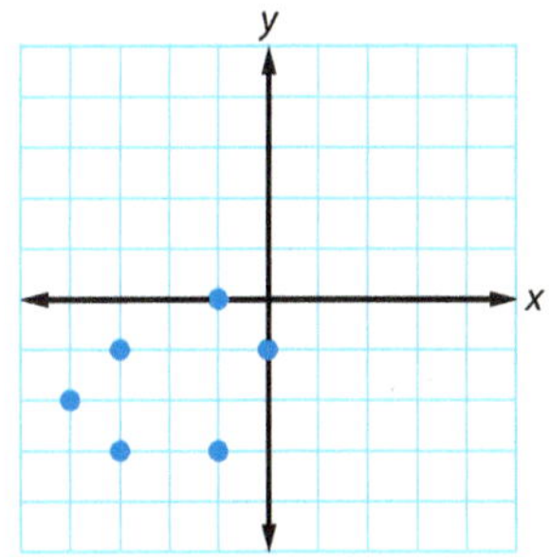

8.

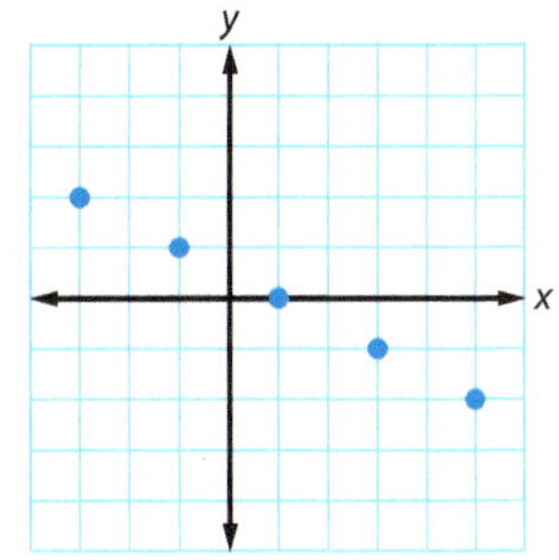

9.

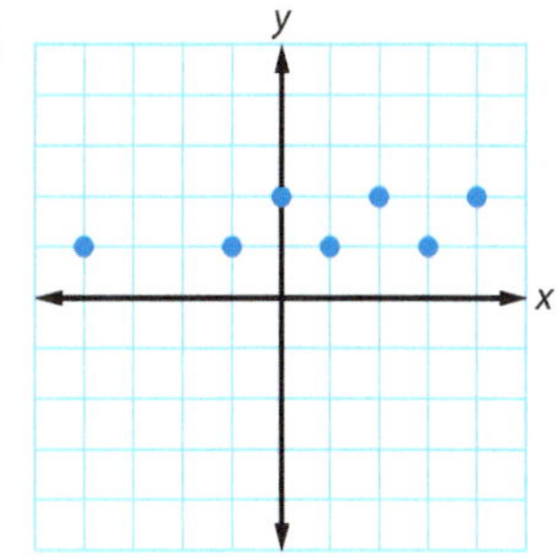

10.

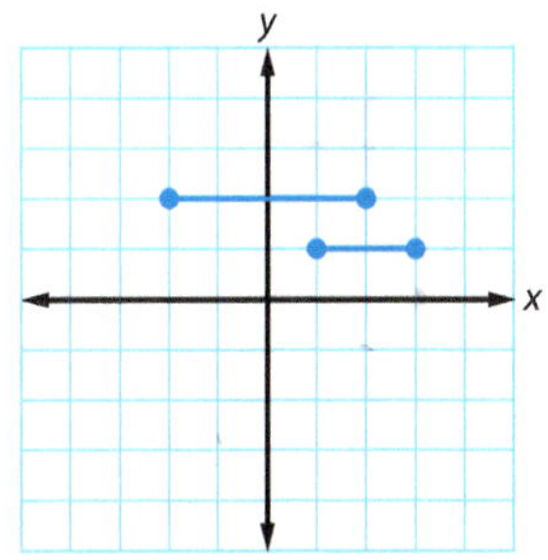

11.

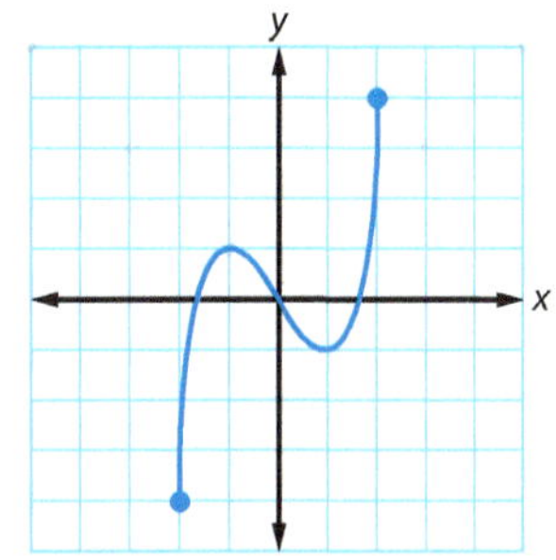

12.

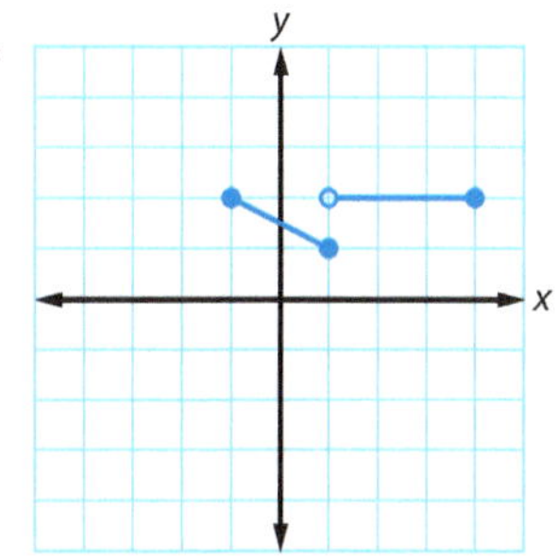

B. EXERCISES

Represent each function as a set of ordered pairs and graph the function.

13. $\{(x, y) \mid y = 3x; x = -1, 0, 1\}$

14. $\{(x, y) \mid y = x^2; x = -1, 0, 1, 3\}$

15. $\{(x, f(x)) \mid f(x) = 2x + 3; x = -2, 0, \frac{1}{2}, 1\}$

16. $\{(x, f(x)) \mid f(x) = -x - 4; x = -2.5, -1, 0, 1.5, 3\}$

Make a table of ordered pairs and graph each function. Then state the function's domain and range.

17. $\{(x, y) \mid y = 3x\}$

18. $\{(x, y) \mid y = x^2\}$

19. $\{(x, f(x)) \mid f(x) = 2x + 3\}$

20. $\{(x, f(x)) \mid f(x) = -x - 4\}$

Graph each relation and state whether it is a function.

21. $\{(x, y) \mid -1 \leq y \leq 3; x = 2\}$

22. $\{(x, y) \mid y = \pm x; x = -4, 2, 5\}$

23. $\{(x, y) \mid y = |x|; x = -5, -3, -1, 0, 3, 5\}$

24. $\{(x, y) \mid y = |x|\}$

25. **Essential Question:** If a function does not include a domain, what is the implied domain? How would you graph such a function?

The conversion of temperature from Celsius to Fahrenheit is given by the equation $F = \frac{9}{5}C + 32$.

26. Complete the table.

C	0	10	20	30
F				

27. Graph the relation.

28. Is the relation described by this formula a function?

29. Determine the Celsius equivalent of 500°F.

The perimeter of a regular hexagon can be expressed by the equation $p = 6s$, where s is the length of a side.

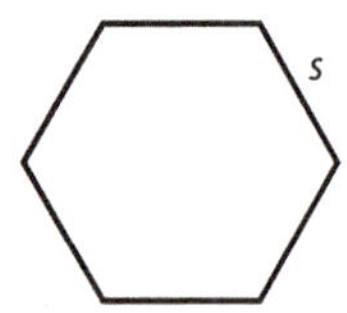

30. State a reasonable domain for this function.

31. Graph the function.

32. Determine the side length of a regular hexagon with a perimeter of 114 m.

MP4 **The cost of replacing the carpet in a home depends on the number of square yards to be covered. If the set fee for labor is \$250 and the carpet (with padding) costs \$15 per square yard, then the total cost represented by C can be found using the following equation: $C = 15x + 250$, where x represents the number of square yards needed.**

33. State a reasonable domain for this function.

34. Graph the function.

35. Determine the cost of carpeting 1800 ft^2 of a home.

A construction company is able to buy 2 in. × 4 in. × 8 ft framing studs at the listed bulk rates.

More Than	Up to	Price per Stud
	100	\$6.00
100	500	\$5.50
500	1000	\$5.00
1000		\$4.50

36. **MP4** Which graph best represents the relationship between the number of studs purchased and the price for each stud?

A

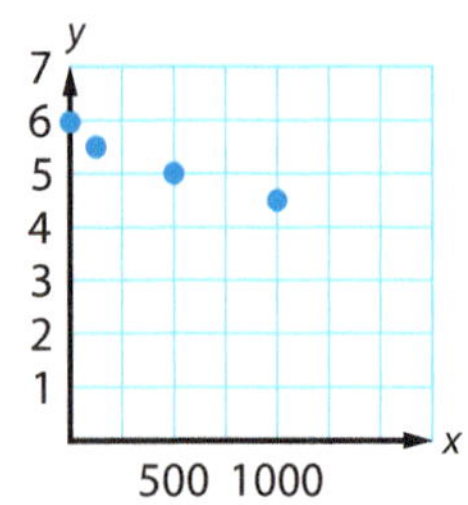

B

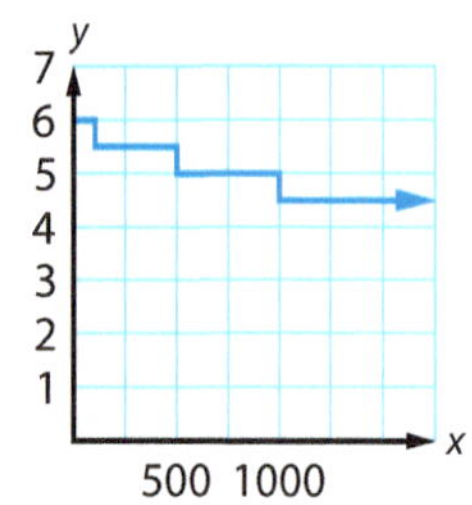

C

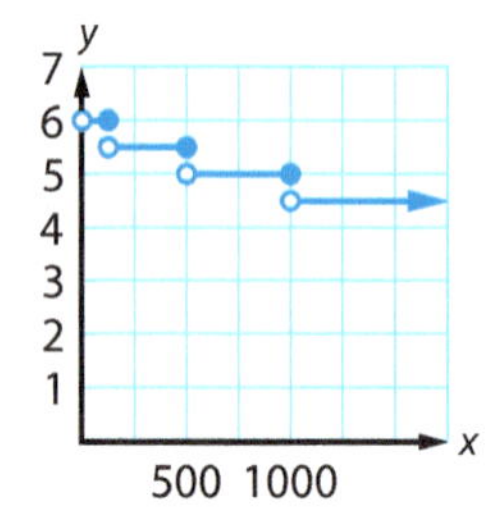

37. Is the relation from exercise 36 a function?

C. EXERCISES

MP2 **Graph each relation and state whether it is a function.**

38. $\{(x, y) \mid -3 \le y \le 2; 0 \le x \le 4, x \in \mathbb{Z}\}$

39. $\{(x, y) \mid -3 \le y \le 2; 0 \le x \le 4\}$

Graph each set of functions on 1 Cartesian plane. Assume that the domain is the set of real numbers.

40. $y = x$
$y = 2x$
$y = 3x$

41. $y = |x|$
$y = |2x|$
$y = |3x|$

42. MP7 How does increasing the number in front of x affect the graphs in exercises 40–41?

CUMULATIVE REVIEW

Clear each equation of fractions or decimals and then solve. [2.2]

43. $\frac{x}{6} - \frac{5}{7} = 2$

44. $\frac{3}{16}x + \frac{2}{3} = \frac{5}{8}$

45. $2.5x - 0.65 = 3.1$

46. $0.2x + 0.78 = 12$

Solve each literal equation for *x*. [2.3]

47. $3axb + c = d$

48. $\frac{dx}{c} = \frac{b}{2a}$

Graph each compound inequality. [3.3–3.4]

49. $x \ge -3$ and $x < 4$

50. $x < 2$ or $x < -1$

51. Shipp Screw manufactures #8 wood screws with a body width within $\frac{1}{64}$ in. of $\frac{5}{32}$ in. Which absolute value inequality represents acceptable widths for these screws? **[3.6]**

A $\left|d - \frac{5}{32}\right| \le \frac{1}{64}$

B $\left|d - \frac{5}{32}\right| \ge \frac{1}{64}$

C $\left|d - \frac{1}{64}\right| \le \frac{5}{32}$

D $\left|d - \frac{1}{64}\right| \ge \frac{5}{32}$

E none of these

52. Which value of m ensures that the relation $\{(3, 2), (1, -1), (m, 3), (0, 5)\}$ is not a function? **[4.1]**

A -1

B 0

C 2

D 5

E none of these

TECHNOLOGY CORNER (TI-84 PLUS FAMILY)

Displaying Functions

Your graphing calculator can be used to display several representations of a function. First, be sure that the calculator is in function mode by selecting FUNCTION on the [mode] menu. Then select [y=] to enter the function editor, and input the function rule $f(x) = 2x - 1$ as Y_1.

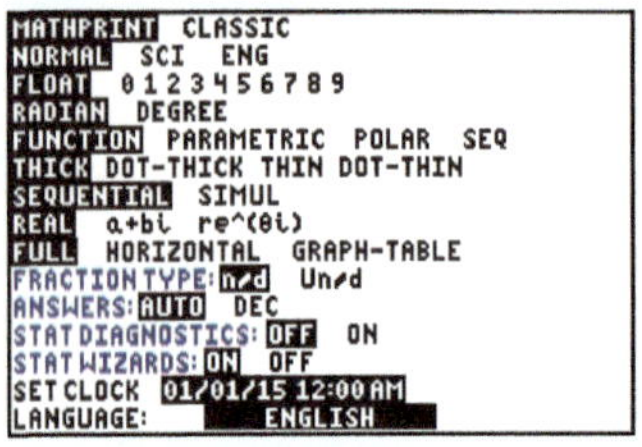

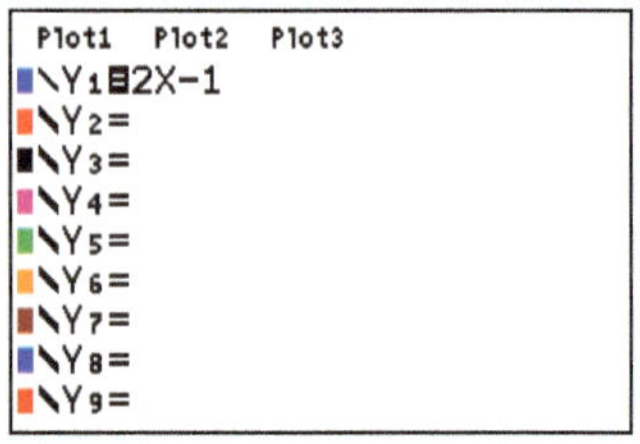

Use the [table] command ([2nd] [graph]) to display a table of ordered pairs for the defined function. The [tablset] command ([2nd] [window]) produces the TABLE SETUP menu. TblStart and ΔTbl are used to change the initial value and increment for the independent variable. (Setting Indpnt to Ask allows you to enter individual values for x into the table.)

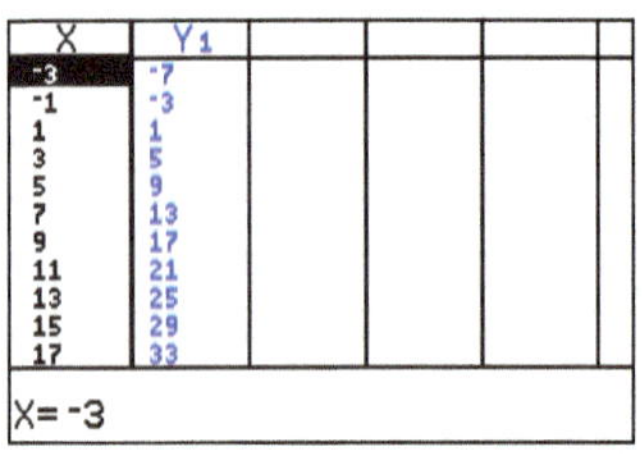

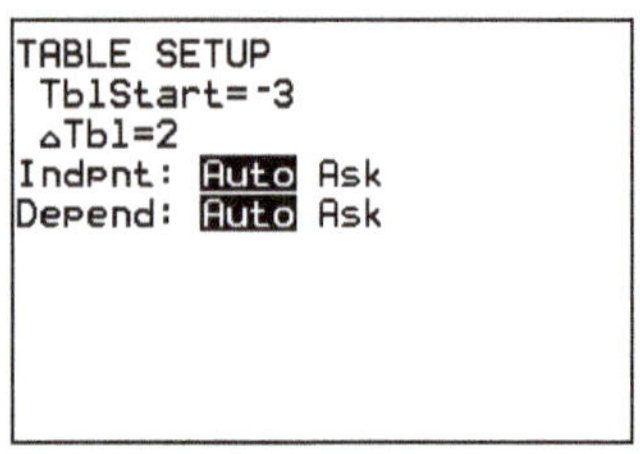

Use the [graph] command to display a graph of the function and use the [zoom] and [window] commands to adjust the portion of the graph that is viewed. Selecting ZStandard from the [zoom] menu causes the function to be graphed in a window with $-10 \le x \le 10$ and $-10 \le y \le 10$. Notice that this causes each axis to have a different scale.

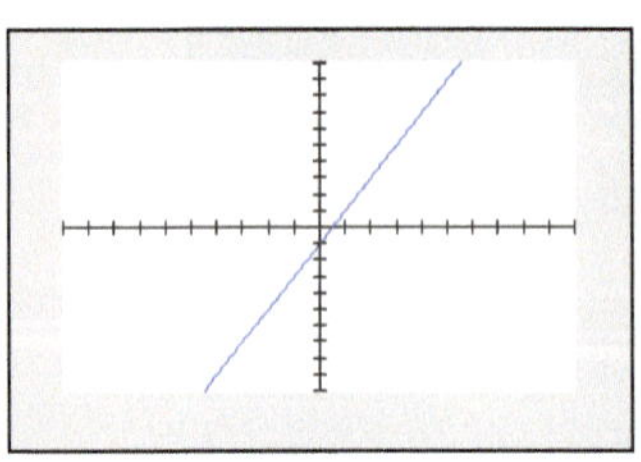

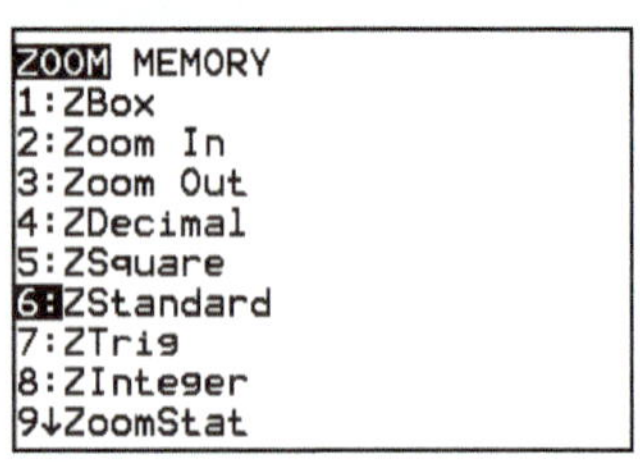

Selecting ZSquare from the [zoom] menu generates an axis with the same scale but different minimum and maximum values.

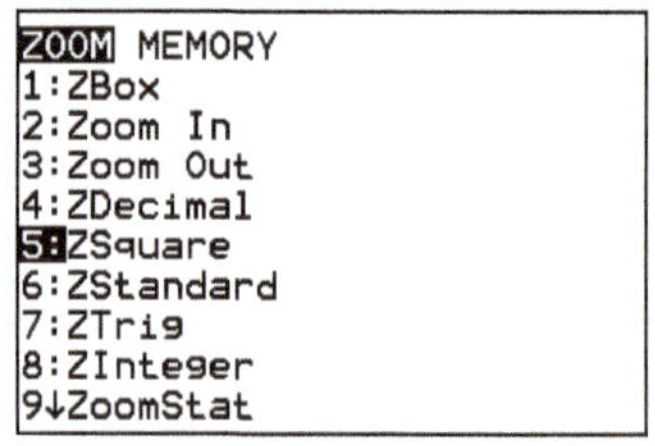

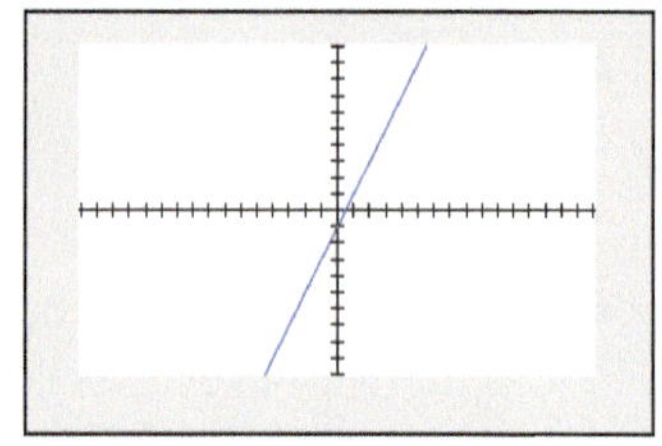

You can enter desired limits and scales for each axis using the [window] menu. The x-values included fall between Xmin and Xmax, and y-values fall between Ymin and Ymax. With experience, you will be able to quickly determine the best window to display the key characteristics of a function's graph.

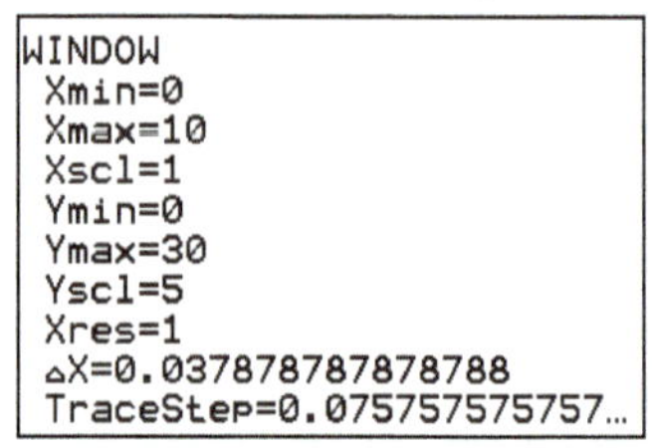

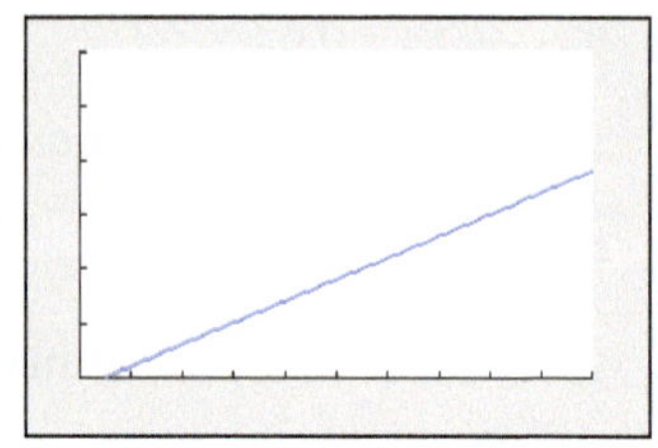

4.3 USING GRAPHS

How can I identify the independent variable in a real-world problem?

After completing this section, I will be able to

- draw graphs to model real-world data.
- interpret graphs representing real-world situations.

Graphs are often used to visualize the relationship between variables. They can illustrate a great variety of phenomena, such as the growth of a fish or the motion of a Ferris wheel. You will often need to draw or interpret graphs that model real-world situations.

Growth of a Largemouth Bass	
age (yr)	**length (in.)**
1	6.5
2	11
3	14
4	16
5	17
6	17.5

The table gives the approximate length of a largemouth bass based on its age. A graph can be used to show the relation between length and age. Since the age of a fish is not affected by the fish's length, it would be graphed on the horizontal axis as the independent variable. The fish's length is the dependent variable since it is affected by the fish's age. It is graphed on the vertical axis.

By drawing a smooth curve that approximates the shape of the plotted points, you can discover general characteristics and make predictions. Rapid growth occurs when the fish is young, but growth slows down as the fish gets older. We could predict that a $1\frac{1}{2}$-yr-old bass would be about 9 in. long and that a 15 in. long bass is about $3\frac{1}{2}$ yr old. We could also estimate the length of an 8-yr-old bass to be almost 18 in. This process of graphing data to analyze the relationship between 2 quantities is an example of **mathematical modeling**.

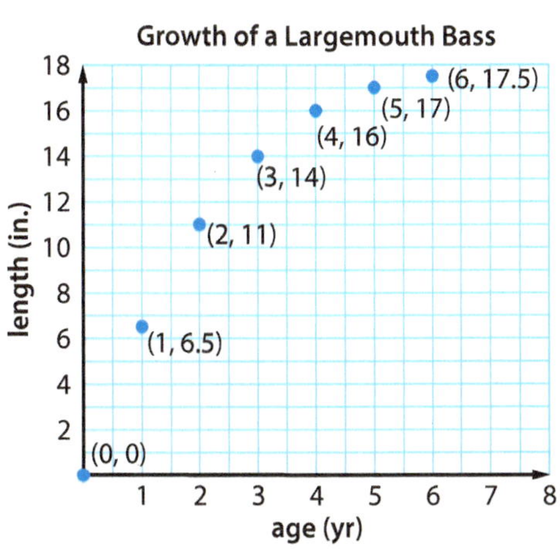

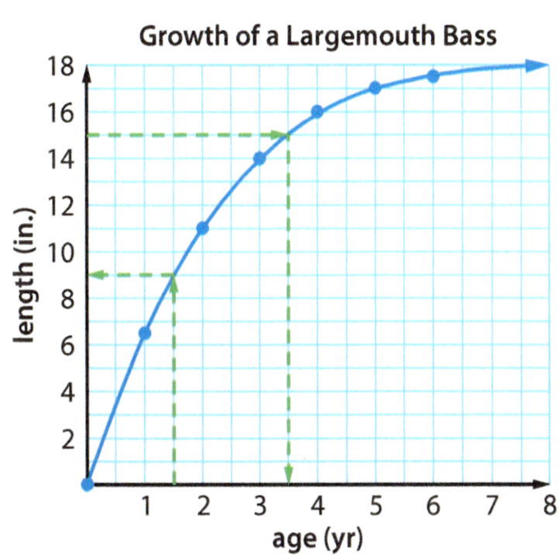

EXAMPLE 1: Using Graphs to Make a Prediction

Model the first-class postage rate data given in the table and state the general characteristics of the rate of increase of postage over the years. Predict what the postage rate will be in 2050.

year	stamp cost (cents)	year	stamp cost (cents)	year	stamp cost (cents)
1865	6	1970	6	2005	37
1885	2	1975	13	2010	44
1900	3	1985	22	2015	49
1930	3	1990	25	2020	55
1960	4	1995	32	2022	58
1965	5	2000	33		

Answer

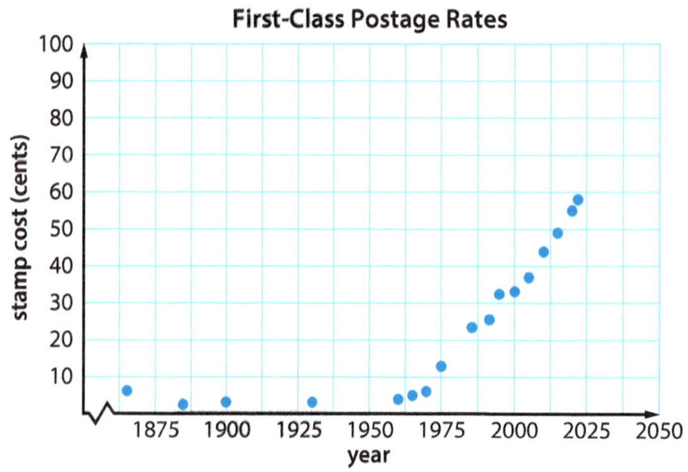

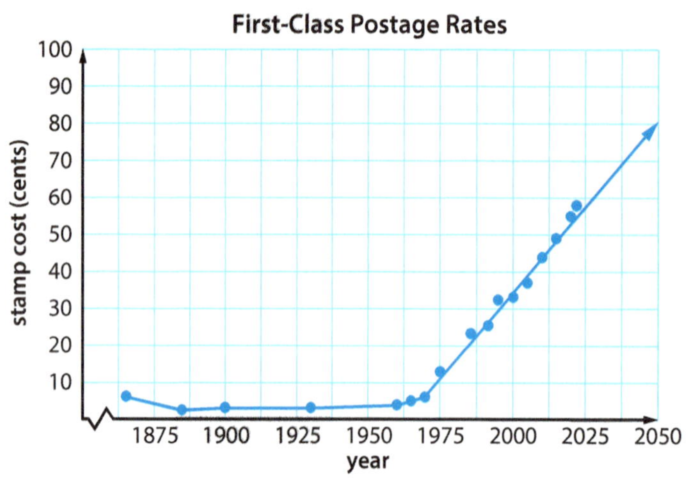

1. The year is the independent variable and is graphed on the horizontal axis. The price is the dependent variable and is graphed on the vertical axis.
2. Early in the history of postage stamps, the rate decreased slightly and then stayed relatively steady for almost 100 yr. Rates have increased dramatically in recent years.
3. Draw a smooth curve that shows the general shape of the postal rate distribution over the time period pictured on the graph. If the price continues to increase at a similar rate, by 2050 it will cost about 80 cents to mail a letter.

SKILL CHECK **EXERCISE 13**

Interpreting the data represented on a graph and analyzing the general relationship between the variables are important skills to master. Consider the graphs modeling the distance fallen and the speed of a skydiver during his fall.

EXAMPLE 2: Interpreting a Graph

Use the distance-time graph to answer the following questions.

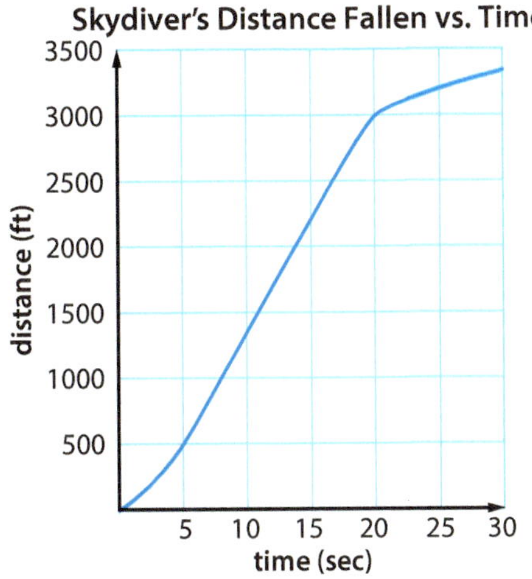

a. How far has the skydiver fallen after 10 sec?

b. How long does it take the skydiver to fall 2000 ft?

c. Describe the fall of the skydiver in the first 5 sec, the next 15 sec, and the final 10 sec illustrated.

Answers

a. At 10 sec he has fallen about 1350 ft.

b. It takes about 14 sec to fall 2000 ft.

c. The skydiver falls greater distances each second for the first 5 sec.

He then falls about the same number of feet each second for the next 15 sec.

He then falls at a much slower constant rate during the last 10 sec.

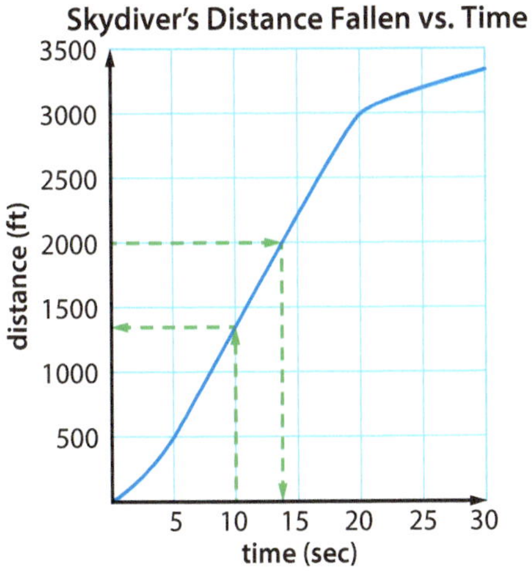

SKILL CHECK **EXERCISE 35**

Graphs without numbered axes can be drawn to illustrate the relationship between variables. The graph illustrates the relative speeds at which the skydiver falls. A more thorough investigation of the distance-time graph in Example 2 would indicate that he is falling approximately 167 ft/sec once he reaches a constant speed and about 33 ft/sec after he deploys his parachute.

Skydiver's Speed vs. Time

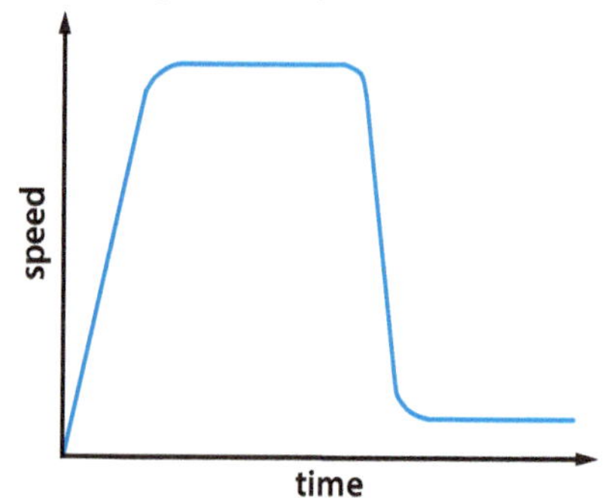

EXAMPLE 3: Interpreting a Graph

The Wadsworths spent a day hiking in the mountains. Mrs. Wadsworth's GPS-enabled sports watch records her heart rate and position, including altitude. Analyze the following graphs produced from their trip to describe their hike during each period of time.

Answer

A Her heart rate is elevated as they ascend the first slope.

B Her heart rate returns to normal while they are at the top of the mountain.

C Her heart rate is elevated slightly as they walk part way down the mountain.

D Her heart rate is elevated greatly as they climb a steeper slope.

E At the top of the second mountain, her heart rate returns to normal.

F As she descends to the trailhead, her heart rate remains slightly elevated.

Mountain Hiking

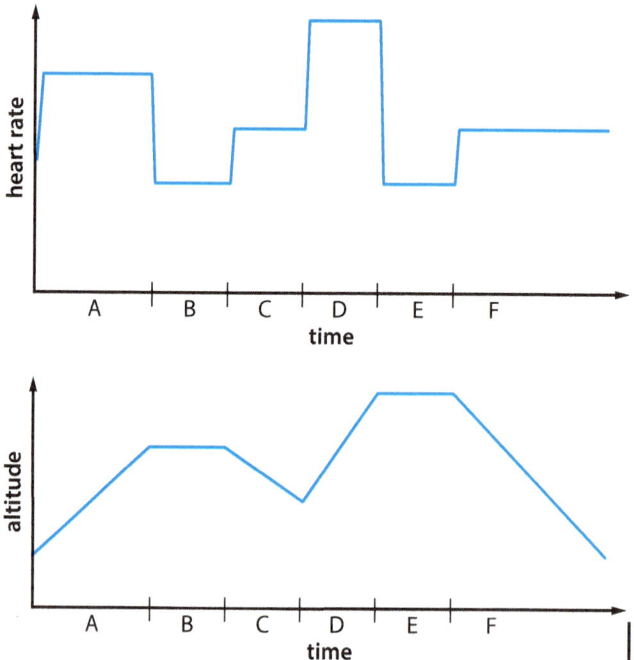

SKILL CHECK **EXERCISE 3**

AFTERSCHOOLHELP

Using Graphs

VOCABULARY

- mathematical modeling

A. EXERCISES

1. Which graph would represent the height of water in each container as water is poured into it at a continuous rate?

a.

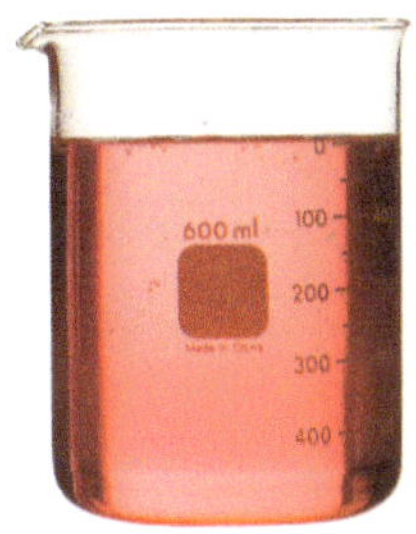

b.

c.

I

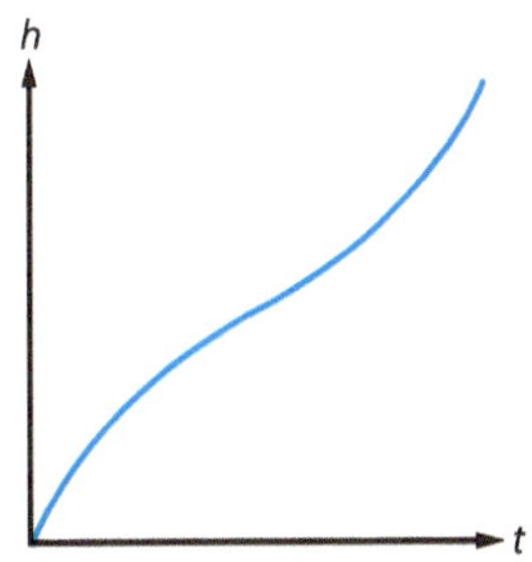

II

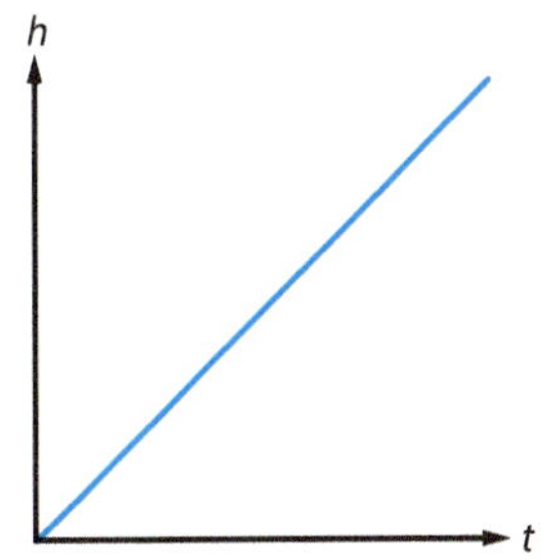

III

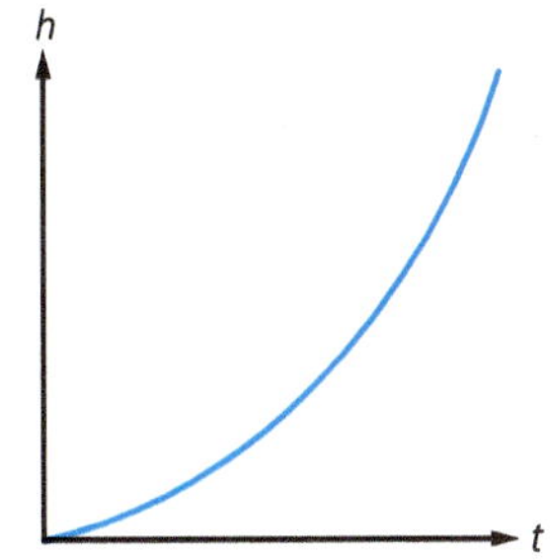

2. Which of these graphs would represent the speed of a ball thrown straight up in the air and then caught by the same person?

A

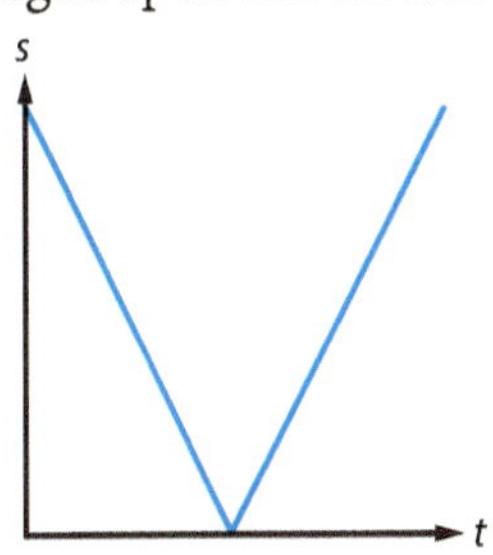

B

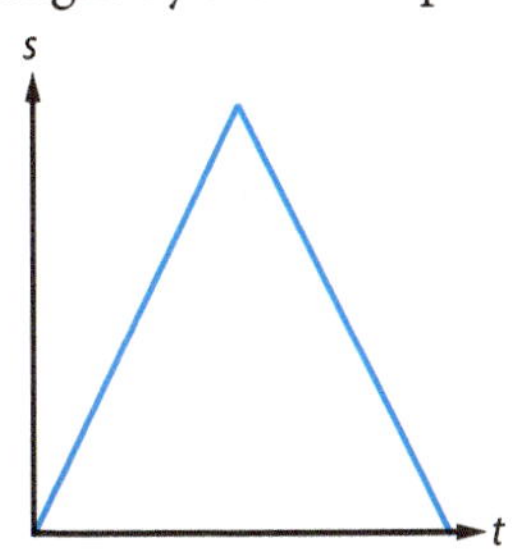

C

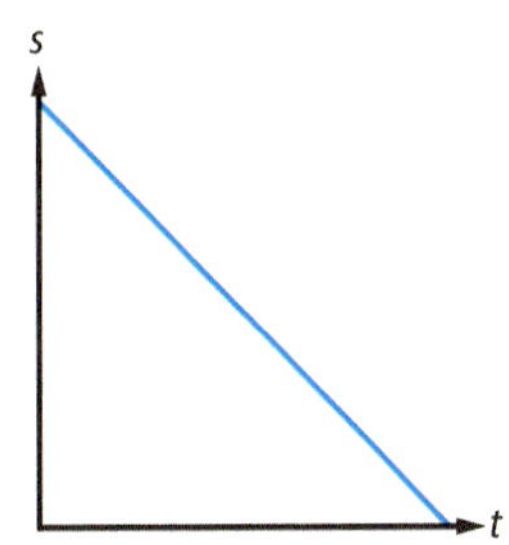

D

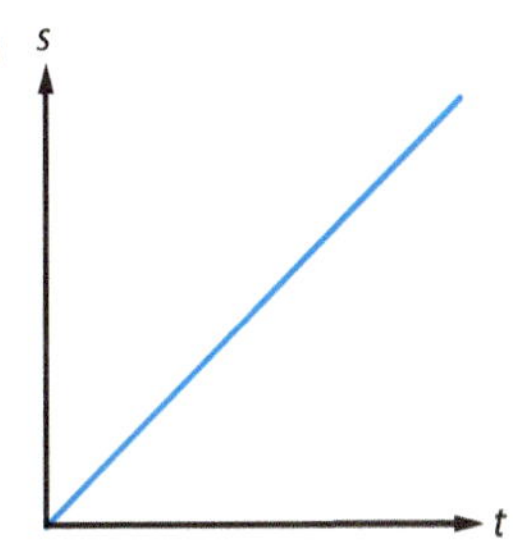

Interpret each graph, explaining each labeled section.

3. Daily Savings Account Balance

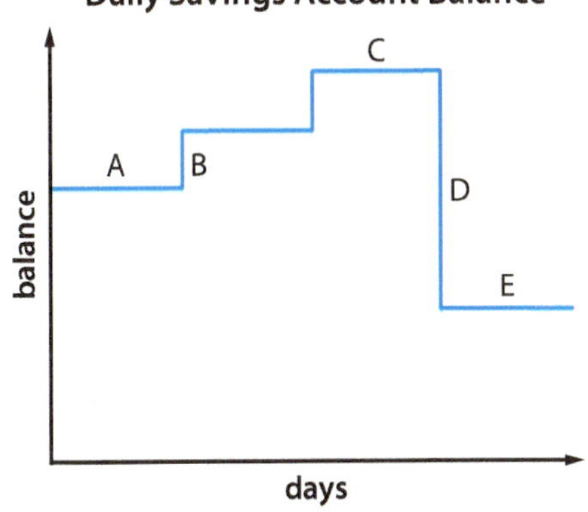

4. Cars Passing through an Intersection

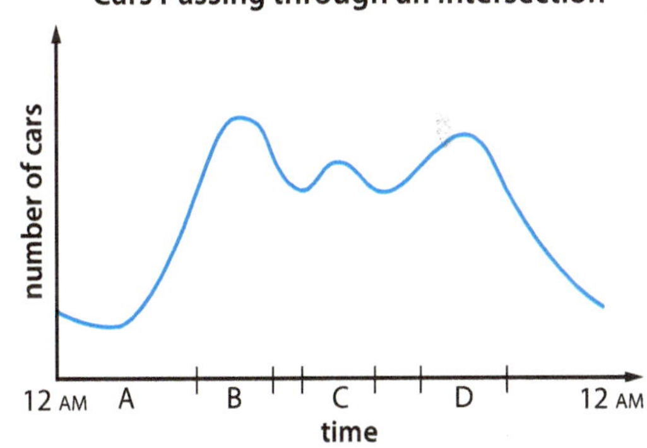

5. Altitude during a Hot Air Balloon Ride

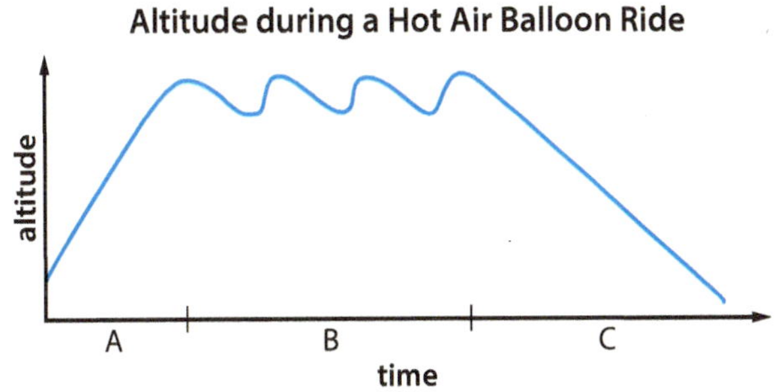

6. Speed of a Bus

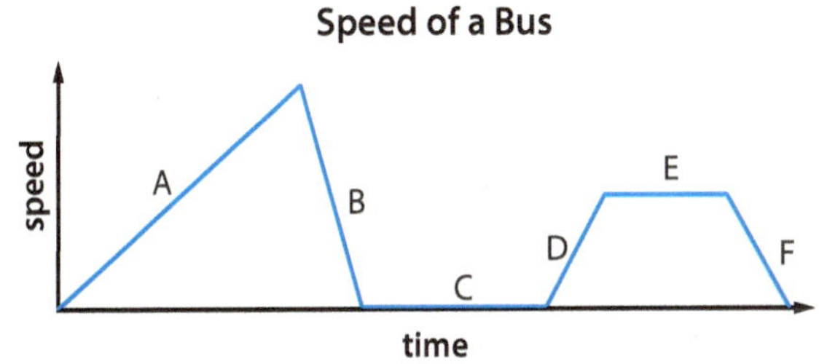

Sketch a graph (without numbered axes) illustrating each of the following situations.

7. Susan's heart rate as she drives to the track, warms up, runs 3 mi, cools down, and then drives home
8. Larry's distance from home as he drives to the grocery store, shops for half an hour, and then returns home
9. Katie's total distance driven as she drives to the bank, deposits a check and withdraws some cash there, and drives home
10. yesterday's outside temperatures
11. the height of a punted football during its flight through the air
12. the speed of a dragster during a quarter-mile race

B. EXERCISES

Use the table on the right for exercises 13–16.

income ($)	offering ($)
50	5
100	10
250	25
500	50
750	75
1000	100

13. Graph the data and draw a smooth curve that models the relationship between Maria's income and her offering.
14. Describe the general trend of the data.
15. Use your graph to determine her offering if she made $300.
16. Use your graph to determine her income if she gave an offering of $12.50.

Use the table on the right for exercises 17–20.

minutes playing basketball	calories burned
10	94
20	187
30	280
40	375
50	470
60	565

17. Graph the data and draw a smooth curve that models the relationship between the number of calories Jonathan burned and the time he spent playing basketball.
18. Describe the general trend of the data.
19. How many minutes does Jonathan need to play basketball in order to burn 150 calories?
20. How many calories does Jonathan burn if he plays basketball for 45 min?

Use the table below for exercises 21–24.

boiling point of water (°C)	101.2	101	100	99.8	97.9	96.2	96.1	95.7
elevation (ft)	0	70	100	350	700	1500	2300	3500

21. Which quantity should be represented by the independent variable? Explain why.
22. Graph the data and draw a smooth curve that models the relationship between elevation and water's boiling point.
23. Describe the general trend of the data.
24. Use your graph to determine water's boiling point at an elevation of 1000 ft.

Use the table on the right for exercises 25–28.

diameter (cm)	area (cm^2)
2	3
4	13
6	28
10	79
16	201
20	314

25. Graph the data and draw a smooth curve that models the relationship between the diameter of a circle and its area.
26. Describe the general trend of the data.

27. Use your graph to determine the area of a circle whose diameter is 5 cm.

28. Use your graph to determine the diameter of a circle with an area of 140 cm^2.

Use the table below for exercises 29–32.

elapsed time (hr)	0	2	4	6	8	10	12	14	16
concentration (mg/L)	0	30	50	42	36	28	20	18	12

29. Graph the data and draw a smooth curve that models the relationship between the concentration of medication in the blood and the amount of time after the medication is taken.

30. Describe the general trend of the data.

31. Use your graph to determine the concentration of the medication after 9 hr.

32. When is the concentration 45 mg/L?

33. **Essential Question:** When graphing data, how do you identify the independent (x) variable? Provide an example by including your reasoning for one of the choices you made in a previous exercise.

MP2 Give a detailed description of the behavior modeled by each graph. Include significant quantities in your description.

34. Describe the growth of bacteria in a Petri dish.

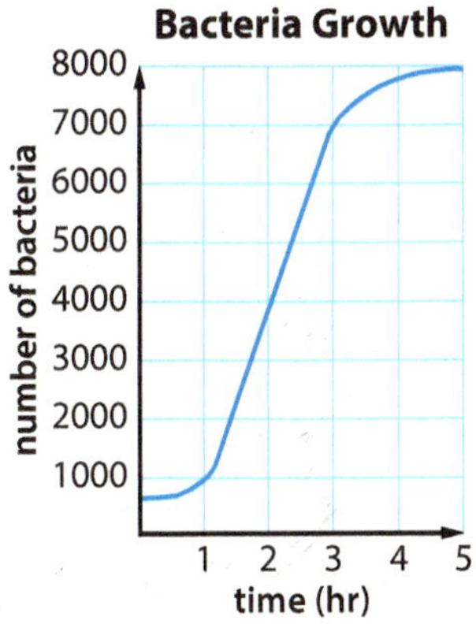

35. Describe the motion of a student.

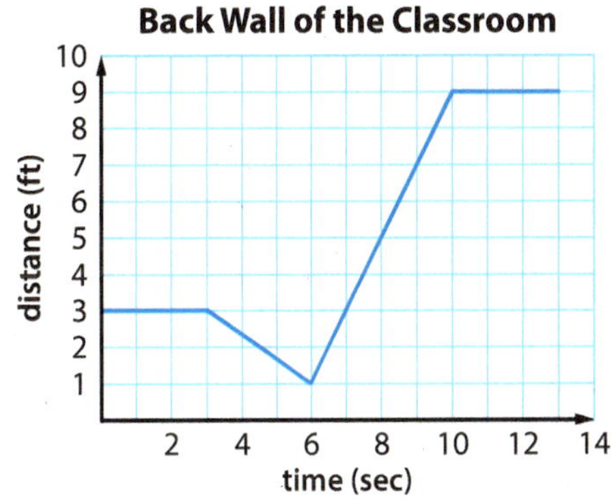

36. Describe the motion of a dropped ball.

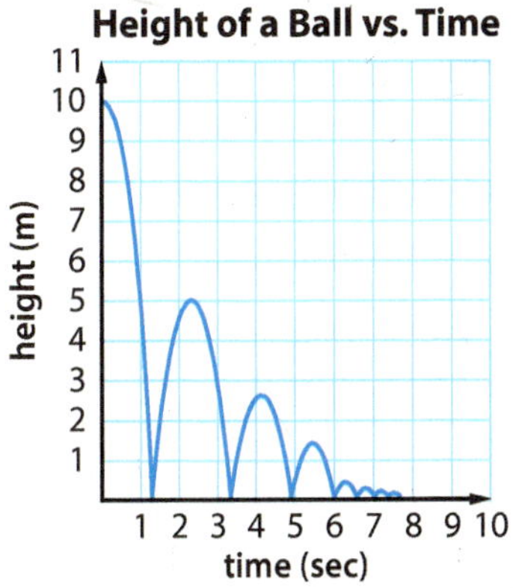

37. Describe the occupancy of the gymnasium on a typical Friday.

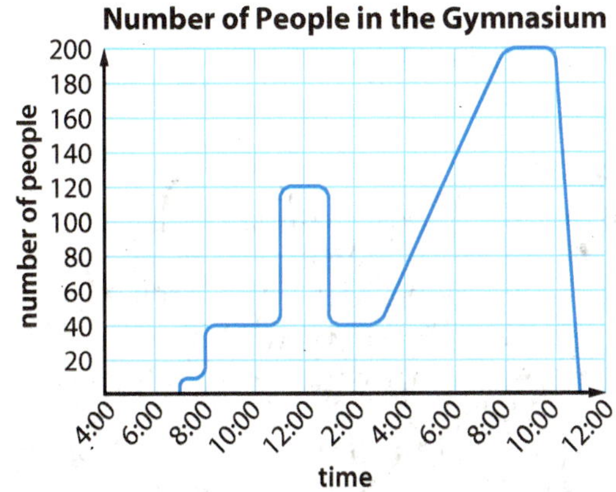

C. EXERCISES

Use the following table listing the height of a Ferris wheel rider for exercises 38–41.

time (sec)	height (ft)	time (sec)	height (ft)
0	5	70	27
10	27	80	5
20	80	90	27
30	133	100	80
40	155	110	133
50	133	120	155
60	80	130	133

38. Graph the data and draw a smooth curve that models the height of a rider on a Ferris wheel and the time of the ride.

39. Describe the general trend of the data.

40. Use your graph to determine the height of the rider 15 sec into the ride.

41. At what times is the rider 100 ft high?

CUMULATIVE REVIEW

Translate each word phrase into an algebraic expression. [1.5]

42. five more than twice a number

43. twenty decreased by three times a number

Solve for *x*. [3.5–3.6]

44. $|x - 3| = 8$

45. $|2x + 5| = 17$

46. $|2x - 3| < 8$

47. $|4x + 3| > 77$

Solve and graph the solution on a number line. [3.6]

48. $|x| > 3$

49. $|x - 4| < 6$

50. In which situation below are the independent (x) and dependent (y) variables correctly identified? **[4.1]**

- A x = earnings, y = hours worked
- B x = number of tree rings, y = tree age
- C x = quiz score, y = hours slept previous night
- D x = time, y = distance traveled
- E none of these

51. The function $f(x) = |-2x|$ has the domain $\{2, 4, 10\}$. What is its range? **[4.2]**

- A $\{4, 8, 20\}$
- B $\{-20, -8, -4, 4, 8, 20\}$
- C $\{1, 2, 5\}$
- D $\{-5, -2, -1, 1, 2, 5\}$
- E none of these

4.4 FUNCTION RULES

What can mathematics tell me about God?

After completing this section, I will be able to

- create sets of ordered pairs, tables, graphs, or mapping diagrams by using given function rules.
- write function rules by using sets of ordered pairs, tables, graphs, or mapping diagrams.
- use function rules to model real-world situations.
- explain how mathematics helps us see design in creation.

The function is a fundamental concept of all higher mathematics. It can be thought of as a machine, consisting of an input, a process, and resulting output, as illustrated. Values are chosen for x, put into the machine, and manipulated by a function rule, and the resulting output is the corresponding y-value.

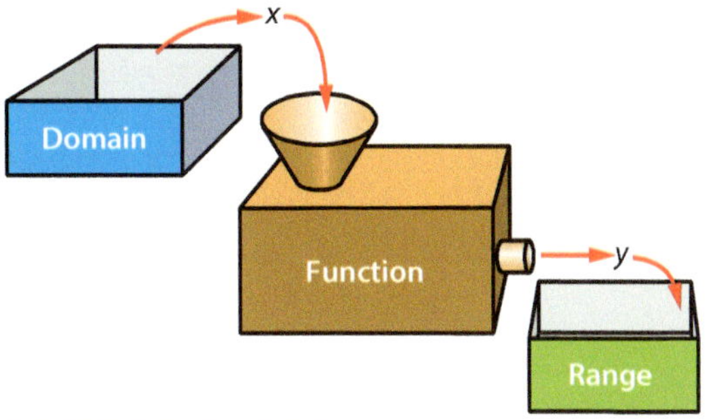

DEFINITION

A **function rule** is an equation that pairs each member of the domain with an element of the range.

If a function rule and the domain are given, then the entire function is defined and can be listed as a set, written as a table, graphed, or illustrated with a mapping diagram.

EXAMPLE 1: Displaying a Function

Given the domain {−3, −1, 0, 2, 4} and the function rule $f(x) = 3x + 2$, state the function as a table, a set of ordered pairs, a graph, and a mapping diagram.

Answer

x	$3x + 2$	$f(x)$
−3	$3(-3) + 2 = -7$	$f(-3) = -7$
−1	$3(-1) + 2 = -1$	$f(-1) = -1$
0	$3(0) + 2 = 2$	$f(0) = 2$
2	$3(2) + 2 = 8$	$f(2) = 8$
4	$3(4) + 2 = 14$	$f(4) = 14$

1. Use the function rule, multiply by 3 and add 2, to find the element of the range associated with each given element of the domain.

$f = \{(-3, -7), (-1, -1), (0, 2), (2, 8), (4, 14)\}$

2. Write the function as a set of ordered pairs.
3. Create a graph from the ordered pairs.
4. Draw a mapping diagram.

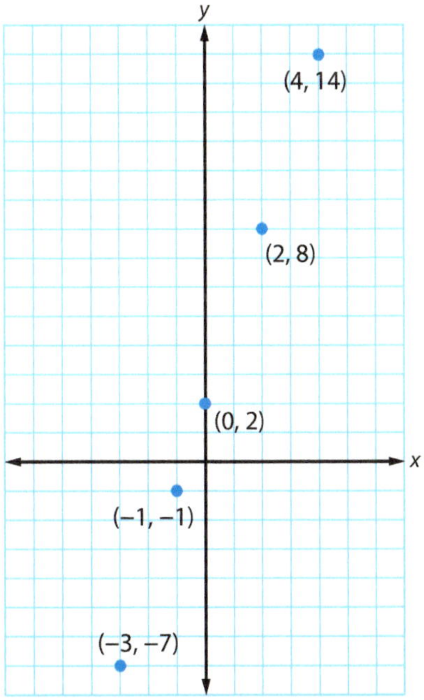

SKILL CHECK EXERCISE 3

Recall from Section 4.1 that $f(x)$ is pronounced "f of x" or "the value of f at x." In Example 1, $f(-3) = -7$, which is read "f of -3 equals -7" and means "the value of f at -3 is -7."

EXAMPLE 2: Graphing a Function

Graph the function $y = |x|$.

Answer

x	$\lvert x\rvert$	y
−3	$\lvert -3\rvert = 3$	3
−1	$\lvert -1\rvert = 1$	1
0	$\lvert 0\rvert = 0$	0
2	$\lvert 2\rvert = 2$	2
5	$\lvert 5\rvert = 5$	5

1. Select several values of the domain to make a table of values for the function.

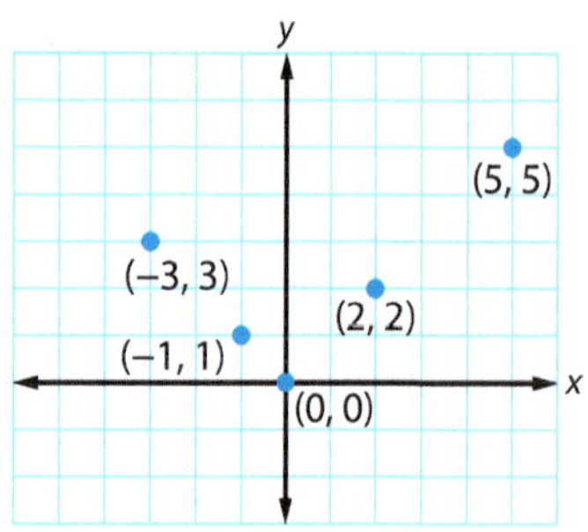

2. Plot enough ordered pairs to determine the general shape of the graph.

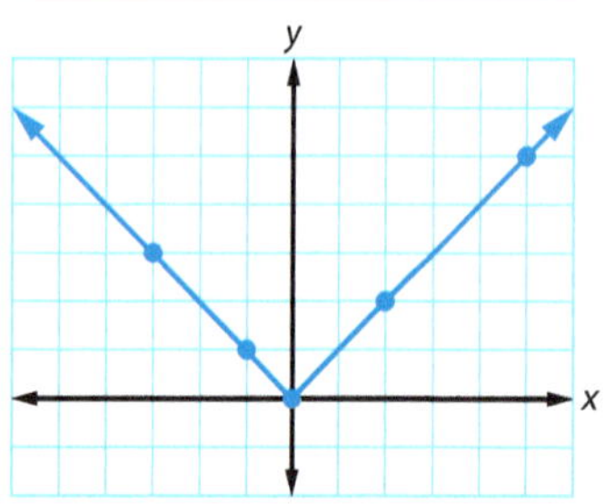

3. Since no domain is given, assume that all real numbers are members of the domain. Connect the graphed points to illustrate the general shape of the function.

SKILL CHECK **EXERCISE 7**

You can find the function rule from the other representations of a function. Carefully examine the domain elements and their related range elements to determine a consistent pattern. Each element of the range is the result of the same function rule.

The table represents a function. By comparing the coordinates, you can observe that each y-coordinate is 4 more than its x-coordinate. Thus, the function rule that describes this pattern is $f(x) = x + 4$. The function rule can be verified by checking to make sure the rule gives the correct range value for each domain value.

x	$f(x)$
−1	3
0	4
2	6
3	7
4	8

EXAMPLE 3: Finding the Function Rule from a Table

Find the function rule for the table.

x	−2	−1	0	1	2
f(x)	−6	−3	0	3	6

Answer

1 1 1 1

x	−2	−1	0	1	2
f(x)	−6	−3	0	3	6

3 3 3 3

1. As the values for x increase by 1, the values for y increase by 3.

$f(x) = 3x$

2. The ratio of y to x remains constant. Check that the function rule is true for all the ordered pairs.

SKILL CHECK ***EXERCISE 9***

EXAMPLE 4: Finding the Function Rule from a Graph

Find the function rule for the graph.

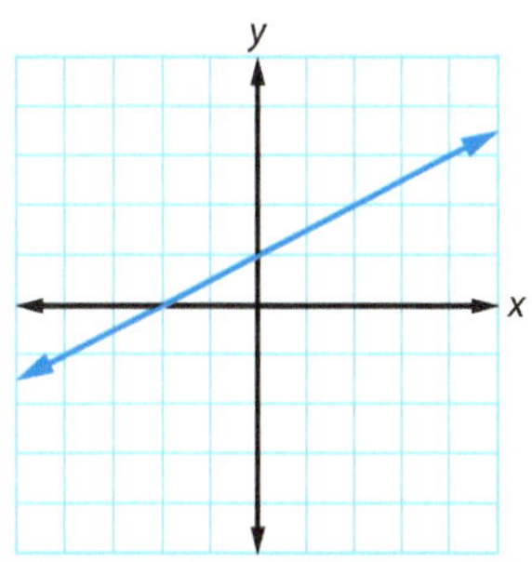

Answer

2 2 2 2

x	−4	−2	0	2	4
y	−1	0	1	2	3

1 1 1 1

1. Make a table using several ordered pairs from the graph. Notice that the x-values differ by 2, while the y-values differ by 1. This suggests that $y = \frac{1}{2}x$ may be the function rule.

x	−4	−2	0	2	4
$\frac{1}{2}x$	−2	−1	0	1	2
y	−1	0	1	2	3

2. A check of this function rule produces values that are always 1 less than y.

$y = \frac{1}{2}x + 1$

3. Revise your function rule by adding 1.

SKILL CHECK ***EXERCISE 19***

EXAMPLE 5: Finding the Function Rule from Ordered Pairs

Find the function rule for the ordered pairs given in the graph.

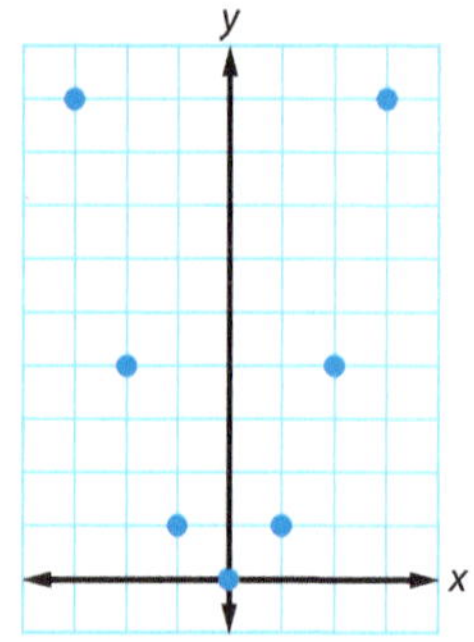

AfterSchoolHelp

Function Rules

Answer

x	−3	−2	−1	0	1	2	3
y	9	4	1	0	1	4	9

$f(x) = x^2$

1. List the ordered pairs in table form.
2. Each value of the domain has been squared to obtain its associated range value.

SKILL CHECK **EXERCISE 17**

Because God has created an orderly world, functions abound in real-world applications. Your experience in translating word problems into equations will help you write function rules that describe the world around us. The design we see in the world helps us learn about the wisdom, creativity, and beauty of its Creator.

More descriptive letters can be used for functions. For instance, $C(w)$ might be used to express the cost, C, of producing w widgets. The function $d(t)$ can be used for distance traveled, where time, t, is the independent variable.

EXAMPLE 6: Applying a Function Rule

Ian bought a lawn mower for $200 to begin his own lawn-mowing business in his neighborhood. He charges $25 per lawn and estimates gas and oil will cost $6 per lawn. Write a function rule defining the profit Ian makes for cutting x lawns. Use the function to determine the number of lawns he needs to cut to reach a goal of $465 for a mission trip.

Answer

$25 - 6 = 19$

$P(x) = 19x - 200$

$465 = 19x - 200$
$665 = 19x$
$35 = x$

Ian needs to cut 35 lawns to receive $465.

1. The profit per lawn is the difference in the amount charged and the cost.
2. Express the profit as a function of the number of lawns (x) mowed.
3. Substitute $465 in for the profit and solve.

SKILL CHECK **EXERCISE 37**

VOCABULARY

- function rule

A. EXERCISES

Find each value using the given function rule.

1. $f(x) = 3x - 8$

a. $f(0)$

b. $f(2)$

c. $f(5)$

2. $v(x) = (x + 1)^2$

a. $v(-3)$

b. $v(2)$

c. $v(9)$

Write a set of ordered pairs and make a mapping diagram representing each function.

3. $f(x) = x - 4; D = \{-2, 0, 3\}$

4. $g(x) = -2x - 3; D = \{-3, 0, 2, 5\}$

Make a table representing each function and then graph the function.

5. $y = -3x; D = \{-1, 0, 2\}$

6. $y = \frac{1}{2}x + 1; D = \{-4, 0, 3, 6\}$

7. $h(x) = 2x + 1$

8. $f(x) = \frac{1}{3}x$

Write the rule for each function using *f*(*x*) notation.

9. $\{(-2, 8), (-1, 4), (0, 0), (1, -4), (5, -20)\}$

10. $\{(-4, 2), (-2, 1), (0, 0), (2, -1), (4, -2)\}$

11. $\{(-3, -5), (-1, -3), \left(\frac{1}{3}, \frac{-5}{3}\right), (2, 0), (5, 3)\}$

12. $\{(-8, -5), (-4, -1), (-3, 0), (2, 5), (4, 7)\}$

13.

x	−5	−2	−1	3	5
y	−15	−6	−3	9	15

14.

x	−3	−1	0	1	3
y	−4	−2	−1	0	2

15.

x	−80	−50	0	20	30
y	82	52	2	−18	−28

16.

x	−30	−20	−10	0.4	1
y	−15	−10	−5	0.2	0.5

17.

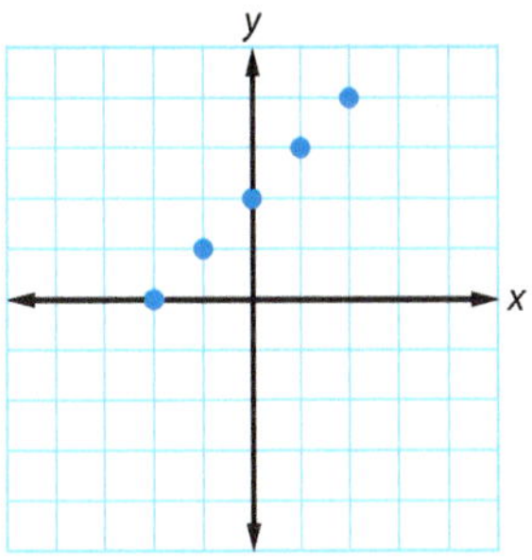

18.

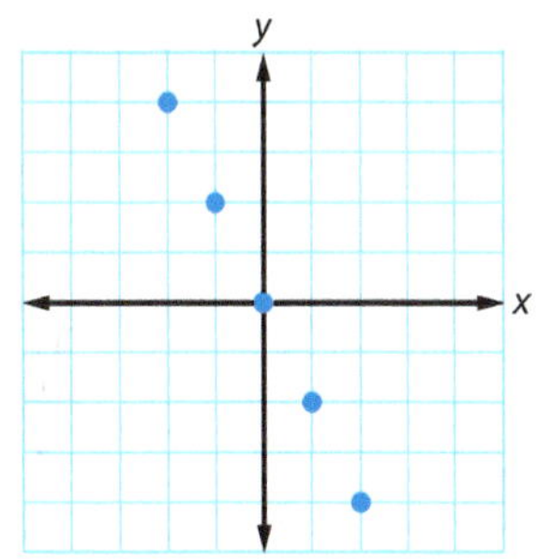

19.

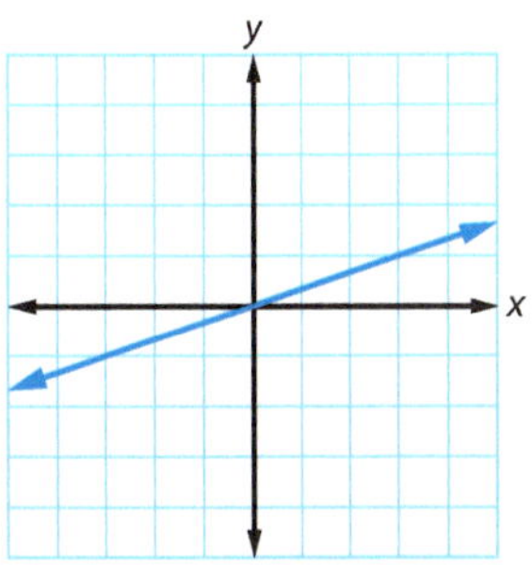

20.

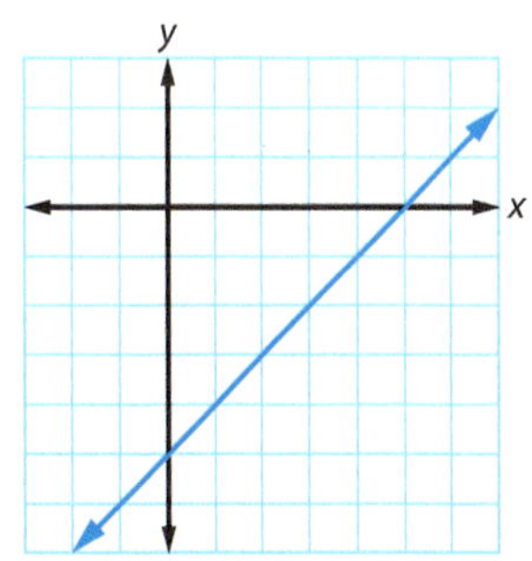

B. EXERCISES

Write the rule for each function using function notation.

21. $f = \{(0, 1), (1, 3), (2, 5), (3, 7), (4, 9)\}$

22. $g = \{(-4, -10), (-2, -4), (0, 2), (2, 8), (4, 14)\}$

23.

x	−3	−1	1	3	5
r(x)	−13	−3	7	17	27

24.

x	−2	−1	0	1	2
s(x)	8	1	0	−1	−8

25.

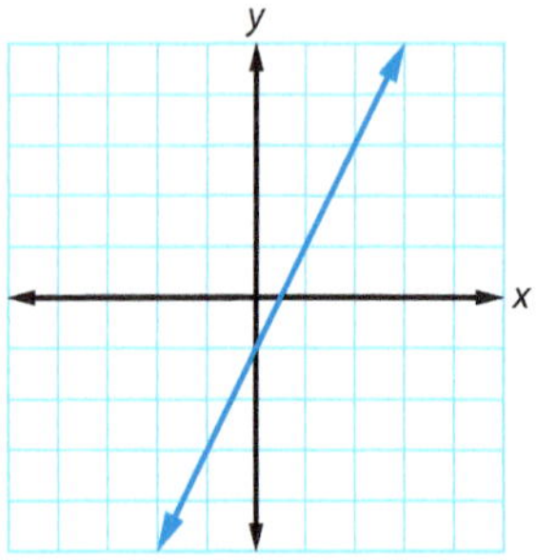

26.

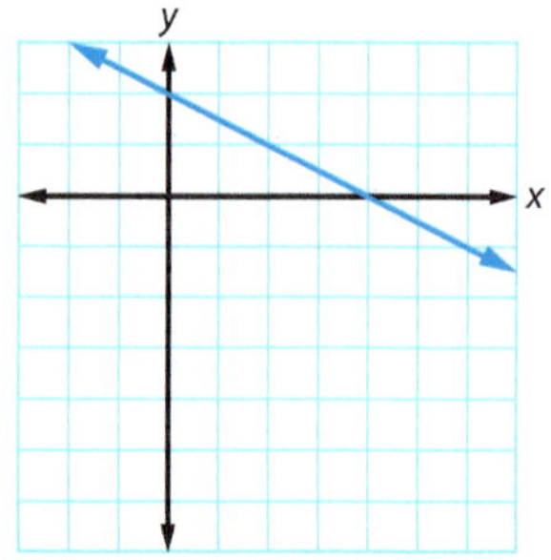

Write a function rule to describe each of the following scenarios.

27. Carissa is using a copier to reduce the size of some of her drawings. She wants them to each be one-fourth of their original size. If d is the length of a side of one of her original drawings, express the length of the reduced side as $f(d)$.

28. If 1 in. of rain is equivalent to 10 in. of snow, express the number of inches of snow that is equivalent to r inches of rain.

29. Noel is 2 yr older than Angelica. Express Noel's age as a function of Angelica's age, a.

30. the perimeter of a regular pentagon with side length s

31. the profit that you make selling c candy bars if each candy bar sold produces $0.25 profit

32. the number of cups of flour required for a triple batch of a recipe that calls for c cups per batch

33. **MP4** the time it takes for Dmitri to travel distance d at 50 mi/hr

Write a function rule for each of the following. Then evaluate the function at the given value.

34. the height of a tree in yards if you know that it is f feet tall; 48 ft

35. the area of a semicircle with radius r; 3 in.

36. the height (h) of a newly planted 3 ft tall Leyland cypress tree after y years if it grows an average of 4 more feet each year; 12 yr

37. the total distance Jimmy traveled if he averaged 52 mi/hr for t hours after having traveled for 40 mi; 3.5 hr

38. the cost of playing r games of racquetball each month at a fitness club that charges a $35 monthly membership fee plus $2 per game of racquetball; 12 games

39. **Essential Question:** As we attempt to model God's world using functions, what can mathematics tell us about God?

C. EXERCISES

Use function rules to answer each of the following exercises.

40. MP4 Evgeni is the manager of a business that manufactures and sells tennis balls for \$1.05 each. The production costs are \$0.48 per ball plus fixed costs of \$1200 per month for the lease and utilities. Write function rules for the revenue, $R(x)$, and the cost, $C(x)$. Graph each function on the same set of axes to determine how many tennis balls must be made and sold each month for the company to break even [that is, $R(x) = C(x)$].

41. MP5 Java Jack's Electronics offers Jamal a base salary of \$150 per week plus \$15 for each computer he sells. Cache Grab offers him \$200 per week plus \$12 for each computer he sells. Write a function rule for each and graph the functions to determine which store pays better under what circumstances.

42. MP2 A Canadian study indicated that the more money a man earns, the less sleep he gets on average. For every \$1000 increase in his annual salary, he sleeps 1 min less. On average, a man who is unemployed and has no income sleeps 485 min (8 hr 5 min) each night. If x represents thousands of dollars of annual income, write a function rule that determines the amount of sleep a man gets based on his income. Graph the function and estimate from it how much sleep he will get if he makes \$60,000 per year and how much he would earn when he is getting no sleep. Will this pattern produce reasonable results indefinitely? Explain.

CUMULATIVE REVIEW

Perform each indicated operation. [1.2]

43. $1\frac{1}{2} \times \frac{3}{4}$

44. $2.751(4.3)$

45. $210 \div 0.005$

46. $\frac{6}{19} \div \frac{8}{3}$

Simplify each expression. [1.5]

47. $3x - (x - 5)$

48. $1.5(3x + 6) + 0.5(3 - x)$

Evaluate when $x = -3$, $y = 2$, and $z = \frac{1}{2}$. [2.2]

49. $\frac{y^3}{xz}$

50. $3x^2yz$

51. $\triangle ABC$ has side lengths of 6 cm and 10 cm while $\triangle XYZ$ has corresponding side lengths of 9 cm and 15 cm. What is the ratio of the area of $\triangle XYZ$ to $\triangle ABC$? **[2.4]**

A $\frac{3}{2}$

B $\frac{9}{4}$

C $\frac{2}{3}$

D $\frac{4}{9}$

E none of these

52. Which equation can be used to solve for the number of dimes if there are 10 more nickels than dimes, 3 fewer quarters than dimes, and \$1.75 total in coins? **[2.6]**

A $0.05(x - 10) + 0.1x + 0.25(x + 3) = 1.75$

B $0.05x + 0.1(x + 10) + 0.25(x - 3) = 1.75$

C $0.05(x + 10) + 0.1x + 0.25(x - 3) = 1.75$

D $10(0.05x) + 0.1x - 3(0.25x) = 1.75$

E none of these

MATH IN HISTORY

RENÉ DESCARTES (1596–1650)

René Descartes was a prominent French mathematician and philosopher. He was formally educated from ages 8 to 16 at a Jesuit school, where he came to view mathematics as the only source of certain knowledge. In mathematics Descartes is most famous for his development of the Cartesian plane and analytic geometry, the study of equations and their graphical representations. You will be graphing many functions in the Cartesian plane in this chapter!

Descartes wrote mathematical papers and books but was initially afraid to publish them, fearing the Catholic church would declare them heretical as it had Galileo's discoveries. In 1637 Descartes's friends convinced him to publish his masterpiece on mathematics, *Discourse on the Method of Reasoning Well and Seeking Truth in the Sciences*, which revealed analytic geometry to the world.

In *Discourse on the Method* Descartes also presented a new method for arriving at truth. He concluded that since even famous teachers disagreed among themselves, relying on what others taught was not a good way to come to truth. He proposed that a person doubt everything that he had ever been taught until he could demonstrate its truth with reason. He believed that, even if he doubted everything, his doubting demonstrated that he was thinking. This is the origin of Descartes's famous saying, "I doubt, therefore I think; I think, therefore I am."

Descartes did not attack Christianity. In fact, he attempted to use his new method to prove the existence of God. But Descartes's method itself is dangerous to Christianity. Reason, instead of being a servant in the quest to understand the revealed Word of God, is made into a judge of what may or may not be true. This makes the individual's mind, rather than revelation from God, the final authority on what is truth.

René Descartes serves as a warning that even highly intelligent teachers who are right on many points may make grave errors. Christians must test the ideas they encounter not by reason alone, but by the Word of God (John 17:17; Acts 17:11). In the end, Descartes's method has not provided a way to certain knowledge. Philosophers still wrangle over the questions that puzzled the Greeks. But the Word of God remains sure and reliable.

4.5 ARITHMETIC SEQUENCES

When is an arithmetic sequence used in the real world?

After completing this section, I will be able to

- define an arithmetic sequence.
- find missing terms in a sequence by determining the sequence's pattern.
- describe arithmetic sequences with multiple representations.
- solve real-world problems involving arithmetic sequences.

Many examples of sequences can be found in God's creation.

Recognizing patterns is foundational in understanding and appreciating God's creation. These patterns can often be displayed using an ordered list of numbers called a **sequence**. Each value in a sequence is called a **term**.

DEFINITION

An **arithmetic sequence** is a sequence in which the difference between any 2 consecutive terms is a constant called the **common difference**.

The sequence displayed in the table has a specific pattern, but it is not an arithmetic sequence since there is not a common difference between consecutive terms.

THE FIBONACCI SEQUENCE
0, 1, 1, 2, 3, 5, 8, 13, . . . a_n
1st term (0), 5th term (3), *n*th term (a_n)

EXAMPLE 1: Identifying an Arithmetic Sequence

Determine whether each sequence is arithmetic. If so, state the common difference and the next 3 terms.

a. 3, 7, 11, 15, . . .

b. 2, −4, 6, −8, . . .

Answers

a. 3 → 7 → 11 → 15 (differences: 4, 4, 4)

1. It is arithmetic with a common difference of 4.

19, 23, 27

2. Add 4 to generate the next 3 terms.

b. 2 → −4 → 6 → −8 (differences: −6, 10, −14)

It is not arithmetic since there is not a common difference.

SKILL CHECK **EXERCISES 3, 5**

An arithmetic sequence can be expressed as a set of ordered pairs using its position in the sequence, n, and its value, a_n. A sequence without an ellipsis (. . .) at the end indicates the sequence has a finite number of terms. Note that n must be a natural number.

3, 5, 7, 9, 11

position (n)	1	2	3	4	5
term (a_n)	3	5	7	9	11

{(1, 3), (2, 5), (3, 7), (4, 9), (5, 11)}

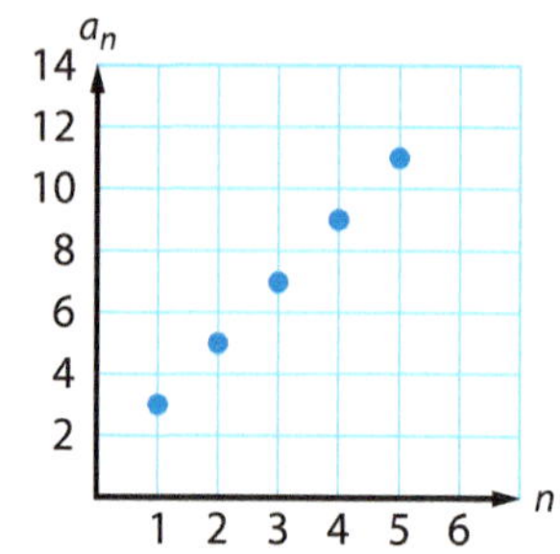

A **recursive formula** can be used to generate successive terms by adding the common difference d to the previous term.

RECURSIVE FORMULA FOR ARITHMETIC SEQUENCES

$$a_n = a_{n-1} + d$$

In the formula a_n denotes the nth term and a_{n-1} denotes the previous term.

EXAMPLE 2: Displaying an Arithmetic Sequence

Use a table, a graph, and a list to represent the first 4 terms in an arithmetic sequence where $a_1 = -4$ and $d = 3$.

Answer

n	$a_{n-1} + 3$	a_n	Ordered Pair
1	given	−4	(1, −4)
2	−4 + 3	−1	(2, −1)
3	−1 + 3	2	(3, 2)
4	2 + 3	5	(4, 5)

1. Make a table displaying the term's position and its corresponding value.

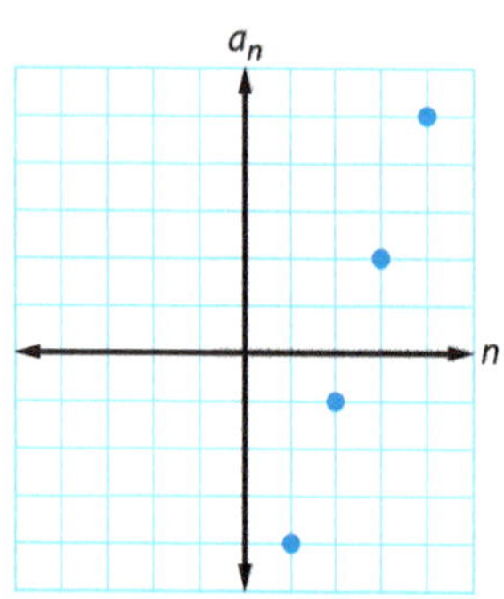

2. Graph the points.

−4, −1, 2, 5

3. List the first 4 terms in the sequence.

SKILL CHECK **EXERCISE 19**

Notice the graphed points of an arithmetic sequence will always lie on a straight line.

The recursive formula works well for finding the first few terms of a sequence, but a more generalized formula is often helpful when finding a specific term in the sequence.

The **explicit formula** or *formula for the nth term* can be derived using the following pattern.

$$
\begin{aligned}
&a_1 \\
&a_2 = a_1 + d \\
&a_3 = a_2 + d &&= (a_1 + d) + d \\
& &&= a_1 + 2d \\
&a_4 = a_3 + d &&= (a_1 + 2d) + d \\
& &&= a_1 + 3d \\
&a_n = a_{n-1} + d &&= a_1 + (n-2)d + d \\
& &&= a_1 + (n-1)d
\end{aligned}
$$

EXPLICIT FORMULA FOR ARITHMETIC SEQUENCES

$$a_n = a_1 + (n-1)d$$

KEYWORD SEARCH

arithmetic sequences

EXAMPLE 3: Finding the nth Term

Write an explicit formula for the *n*th term of the following sequence: 9, 1, –7, ... Then find the 38th term in the sequence.

Answer

$d = 1 - (9) = -8$

1. Determine the common difference.

$a_n = 9 + (n-1)(-8)$
$= 9 - 8n + 8$
$= -8n + 17$

2. Substitute a_1 and d into the explicit formula and simplify.

$a_{38} = -8(38) + 17 = -287$

3. Evaluate the formula for the 38th term.

SKILL CHECK **EXERCISE 27**

The explicit formula can be written using function notation by replacing a_n with $f(n)$.

$$f(n) = a_1 + (n-1)d$$

EXAMPLE 4: Applying Sequences to the Real World

AFTERSCHOOLHELP

Arithmetic Sequences

James began saving up for the ultimate gaming system by setting aside \$150 from his first week's paycheck. He plans to set aside an additional \$50 each week to save the remaining funds. Write a function for the *n*th term and use it to determine which week he will have saved \$1250.

Answer

$d = 50$
$a_1 = 150$

1. The common difference is the amount set aside each week.

$f(n) = 150 + (n-1)50$
$= 150 + 50n - 50$
$= 50n + 100$

2. Substitute a_1 and d into the explicit formula and simplify.

$1250 = 50n + 100$
$1150 = 50n$
$23 = n$

3. Solve for n when $f(n) = 1250$.

James will have saved the needed funds after the 23rd week.

SKILL CHECK **EXERCISE 35**

VOCABULARY

- arithmetic sequence
- common difference
- explicit formula
- recursive formula
- sequence
- term

EXAMPLE 5: Applying Sequences to the Real World

The graph represents a fundraising event where each participant in a 5K race paid an entrance fee to help local military families. Assume the event began with a $500 balance.

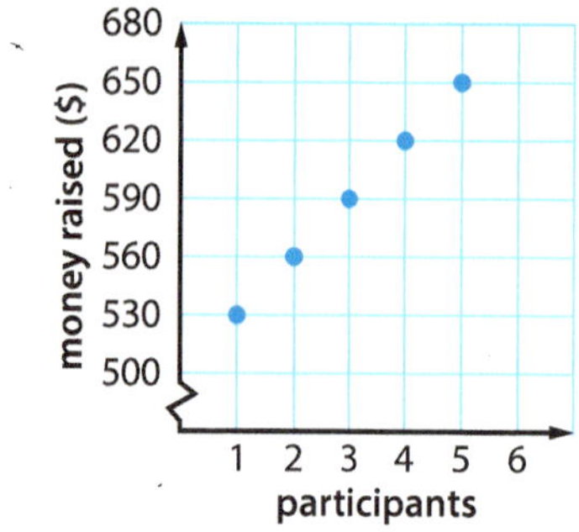

a. Write an explicit formula for the nth term of the sequence.

b. What does the common difference represent?

c. Find the fund's balance if there are 875 race entries.

Answers

a. $d = a_2 - a_1$
$= 560 - 530$
$= 30$

1. Subtract consecutive terms to find the common difference.

$a_n = 530 + (n - 1)30$
$= 530 + 30n - 30$
$= 30n + 500$

2. Substitute a_1 and d into the explicit formula and simplify.

b. The common difference 30 represents the entrance fee paid by each participant.

c. $a_{875} = 30(875) + 500$ Evaluate the formula for the 875th term.

The fundraising balance will be $26,750 after the 875th race entry.

SKILL CHECK **EXERCISE 31**

CONNECT

The a_n formula can also be written in function form.

$f(n) = 30n + 500$

A. EXERCISES

1. Match each vocabulary word with its definition.

a.	an ordered list of numbers	**I**	common difference
b.	constant difference between 2 terms	**II**	explicit formula
c.	used to generate terms by adding d to the previous term	**III**	sequence
d.	used to generate the nth term efficiently	**IV**	recursive formula

2. What is the domain of an arithmetic sequence a_n?

A $\mathbb{R}$ (all real numbers)

B $\mathbb{Z} = \{\ldots, -2, -1, 0, 1, 2, \ldots\}$

C $\mathbb{N} = \{1, 2, 3, 4, \ldots\}$

D none of these

Determine whether each sequence is arithmetic. If so, state the common difference and provide the next 3 terms.

3. 1, 0.5, 0, −0.5, . . .

4. 5, −10, 15, −20, . . .

5.

n	1	2	3	4
a_n	2	4	7	11

6.

n	1	2	3	4
a_n	−5	−1	3	7

7.

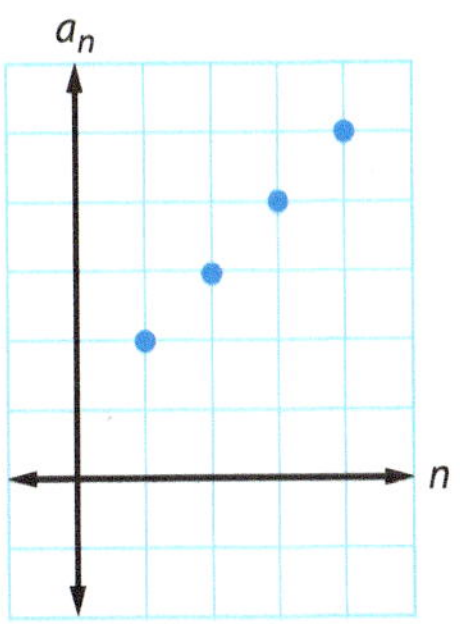

8.

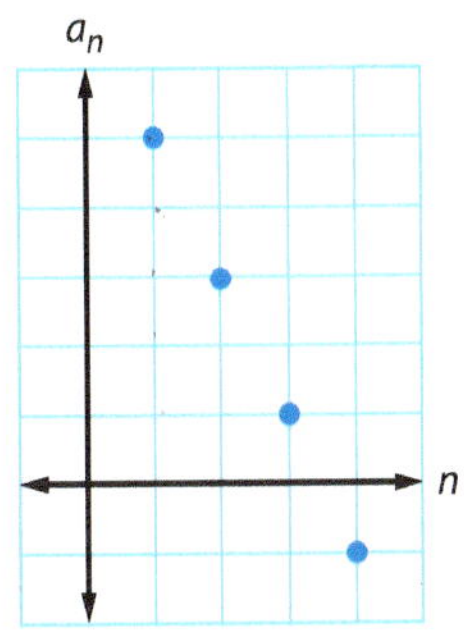

Generate the next 4 terms of each sequence using the given formulas.

9. $a_1 = 5; a_n = a_{n-1} + 2$

10. $a_1 = -12; a_n = a_{n-1} - 7$

11. $a_1 = 17; a_n = 10n + 7$

12. $a_1 = 2; a_n = 3n - 1$

Graph the related ordered pairs of the first 4 terms of each sequence described.

13. first term: −3; common difference: 1

14. $a_1 = 2; d = 2$

15. $a_1 = 6; d = -4$

16. $f(1) = 5; d = -\frac{1}{2}$

B. EXERCISES

17. **MP3** Describe and correct the error Ewan made below in finding the common difference of the sequence 39, 25, 11, −3, ... "The common difference is 14 since 39 − 25 = 14, 25 − 11 = 14, and 11 − (−3) = 14."

Graph the first 4 terms of each arithmetic sequence described below.

18. $a_1 = -5; d = 1.5$

19. $a_1 = 35; d = -20$

20. $f(1) = 20; f(n) = f(n - 1) - 5$

21. $f(1) = -4; f(n) = f(n - 1) + 8$

Find the specified formula for each of the following sequences.

22. 50, 45, 40, 35, 30, . . .; recursive

23. −5, 6, 17, 28, 39, . . .; recursive

24. 14, 24, 34, 44, 54, . . .; explicit

25. 3, 5, 7, 9, 11, . . .; explicit

26. **MP3** Describe and correct the error Rita made when finding a_{20} for the sequence 9, 16, 23, 30, ... She used the formula $a_n = 7 + (n - 1)9 = 9n - 2$ to determine that $a_{20} = 9(20) - 2 = 178$.

Find the explicit formula, and use it to find the given term for each sequence.

27. 12, 27, 42, 57, . . .; a_{52}

28. −30, −17, −4, 9, . . .; $f(11)$

29. 9.6, 9.48, 9.36, 9.24, . . .; $f(81)$

30. **MP1** $\frac{7}{3}, \frac{41}{24}, \frac{13}{12}, \frac{11}{24}, \ldots; a_{17}$

Solve each problem by using an arithmetic sequence formula.

31. Chad and Daisy are assessing the expenses for their upcoming nuptials. The graph represents dining costs based on the number of guests attending.

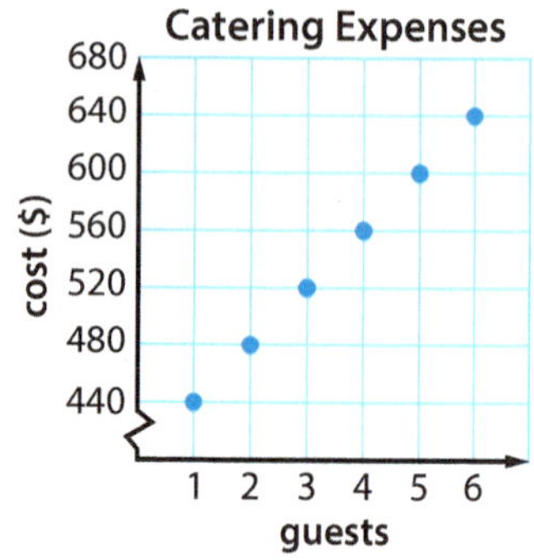

a. What is the cost per plate for each guest? How much will Chad and Daisy pay for service fees upfront?

b. Write an explicit formula for the cost of feeding n guests at the wedding.

c. If 80% of the 120 invited guests show up, how much will the newlyweds pay for catering?

32. Hermann is paying off the remainder of the loan he used to purchase a car. The graph represents how much he still owes after n months.

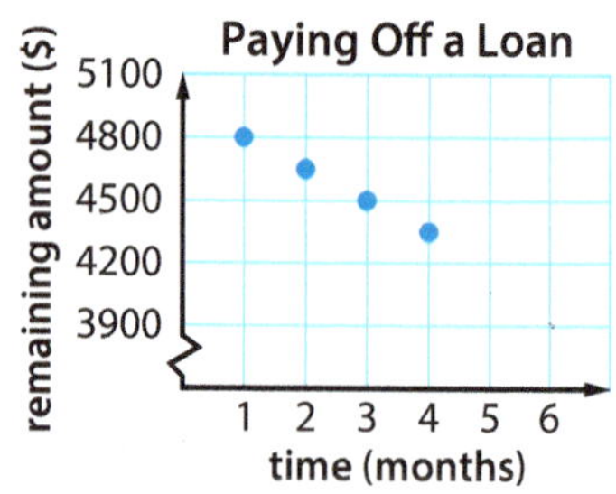

a. Write an explicit formula for the outstanding loan balance. How much is still owed after 12 months?

b. After how many months will he have paid off the entire loan?

33. Essential Question: What do the graphs for exercises 31–32 reveal about when arithmetic sequences are useful in real-world applications?

34. Najara is beginning months of training for a marathon. In the first week, she runs 2 mi during each workout. Each week thereafter she increases the mileage per workout by $\frac{3}{4}$ mi. How many miles will she put into each workout during week 15?

35. The first house on Feuerbach Circle is number 13, followed by number 19. Assuming an arithmetic sequence of address numbers, how many houses are on this street if the last house is number 145?

36. Todd travels roughly the same amount for work each month and earns points based on airline miles. In 1 month, Todd will have 5452 points. He will have 5669 points in 2 months and 5886 in 3 months. How long will it take him to earn a reward worth 15,000 points?

C. EXERCISES

37. Find the common difference of an arithmetic sequence if the first term is 2 and the ninth term is 16.

38. **MP6** Warrick is investing in a retirement account. His initial deposit was \$3000. Warrick contributed a total of \$3300 after 2 months and \$3600 after 4 months. How long after the initial deposit will he have contributed \$8400?

39. **MP8** Given $a_n = a_{n-1} - 87$ and $a_4 = 1759$, find a_{23}.

40. **MP7** Find the first negative term in the sequence 750, 734, 718, 702, ...

41. Describe the pattern of the sequence generated by $a_n = a_{n-1} - a_{n-2}$, where $a_1 = 2$ and $a_2 = 7$. Is the sequence arithmetic?

42. Describe the pattern of the sequence generated by $a_n = a_{n-1} + a_{n-3} - a_{n-2}$, where $a_1 = 3$, $a_2 = 5$, and $a_3 = 7$. Is the sequence arithmetic?

CUMULATIVE REVIEW

Solve using the percent equation. [2.5]

43. What is 28% of 25?

44. Find 165% of 2300.

45. Zane received an 87.5% exam grade after scoring 56 points. How many total points were on the exam?

46. Brittany improved her rebounding numbers by 60% from her sophomore year to her senior year. If this amounts to 2.7 more rebounds per game, how many was she collecting on average during her sophomore year?

Solve each inequality. [3.2]

47. $\frac{1}{10}(x - 11) > 20$

48. $-3(x - 4) + x > 18$

49. $5(x - 9) \neq x + 7$

50. $\frac{1}{3}x - \frac{1}{4} \geq \frac{1}{9}x$

51. Which equation has 2 solutions? **[3.5]**

A $-4|x| = -28$

B $|x| + 5 = 2$

C $|x| - 11 = -11$

D $-2|x| = 10$

E none of these

52. What is the rule for the function below? **[4.4]**

x	−2	−1	0	1	2
p(x)	−2	1	4	7	10

A $p(x) = x^3 + 6$

B $p(x) = -2(x + 1)$

C $p(x) = 3x - 4$

D $p(x) = 3x + 4$

E none of these

4.6 DIRECT & INVERSE VARIATIONS

What is the significance of defining natural laws related to direct and inverse variations?

After completing this section, I will be able to

- classify a function as a direct variation, inverse variation, or neither.
- find the constant of variation for direct and inverse variations.
- write functions modeling direct and inverse variations to solve real-world problems.
- explain the significance of recognizing design in nature.

Jose rode his bike at a constant rate of 22 ft/sec, as indicated by each ordered pair listed in the table. This relationship can be expressed using the equation $y = 22x$. The distance Jose traveled (y) is a function of the amount of time (x) he spent riding. Notice that the ratio $\frac{y}{x} = 22$ for every ordered pair.

time (sec)	distance (ft)
1	22
2	44
5	110
10	220

DEFINITIONS

A **direct variation** is a function in which the ratio of variables is a nonzero constant, k. That is, $\frac{y}{x} = k$, where $k \neq 0$. The quantities are said to *vary directly* or to be **directly proportional**.

The constant k is called the **constant of variation** or the *constant of proportionality*.

A direct variation can also be stated using the general equations $y = kx$ or $f(x) = kx$. This implies 1 variable is a constant multiple of the other. Most applications of direct variation involve a positive constant of proportionality ($k > 0$), in which an increase in one of the variables causes an increase in the other variable.

EXAMPLE 1: Determining a Direct Variation

Does the table represent a direct variation? If so, find the constant of variation and write the function rule.

a.

x	y
1	7
2	14
3	21
4	28

b.

x	y
1	6
2	8
3	10
4	12

Answers

a. The table represents a direct variation since the ratio $\frac{y}{x} = 7$ for each ordered pair. The constant of variation is 7, and the function rule is $f(x) = 7x$.

b. The table does not represent a direct variation since the ratio $\frac{y}{x}$ is not a constant.

SKILL CHECK EXERCISE 17

EXAMPLE 2: Finding the Constant of Variation

Find the constant of variation if y varies directly with x and $y = 6$ when $x = 3$. Write the function rule for the variation.

Answer

$k = \frac{y}{x} = \frac{6}{3} = 2$ — **1.** In the direct variation, the ratio $\frac{y}{x} = k$.

$y = 2x$ — **2.** Write the function rule for the direct variation.

SKILL CHECK **EXERCISE 13**

Whenever 2 different ordered pairs, (x_1, y_1) and (x_2, y_2), represent the same direct variation, their ratios will equal the same constant,

$$\frac{y_1}{x_1} = k \text{ and } \frac{y_2}{x_2} = k.$$

The Transitive Property of Equality states that 2 expressions equal to a third expression must be equal to each other. Therefore, in any direct variation,

$$\frac{y_1}{x_1} = \frac{y_2}{x_2}.$$

This explains why quantities that vary directly are often simply said to be proportional.

EXAMPLE 3: Solving a Real-World Problem

If the distance traveled varies directly with time and Jose rode 12 mi in 45 min, how far can he ride in 2 hr?

CONNECT

Time (t) is the independent variable (x), and distance (d) is the dependent variable (y).

Answer

$\frac{d_1}{t_1} = \frac{d_2}{t_2}$ — **1.** Write a proportion for the quantities that are directly related.

$\frac{12}{45} = \frac{d_2}{120}$ — **2.** Use consistent units in the proportion.

$45d_2 = 1440$ — **3.** Solve the proportion.

$d_2 = 32$ mi

SKILL CHECK **EXERCISE 33**

Example 3 could also be solved by finding the constant of variation. Since Jose rode 12 mi in $\frac{3}{4}$ hr, $\frac{d}{r} = k = 16$ mi/hr. Then use the equation $d = 16t = 16(2) = 32$ mi.

The graph of this function illustrates a key characteristic of a direct variation when $k > 0$. An increase in 1 variable causes a proportional increase in the other variable.

Jose's Bike Ride

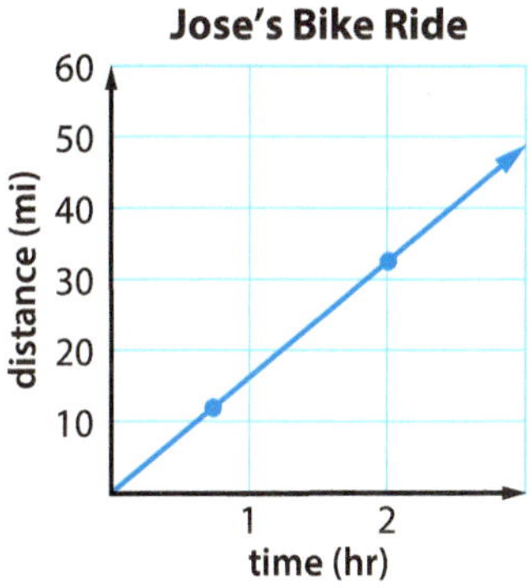

EXAMPLE 4: Solving a Real-World Problem

Hooke's law states that the length of the stretch in a spring is directly proportional to the amount of force applied to the spring. If a spring stretches 8 in. when suspending a 20 lb weight, how far will it stretch when suspending a 50 lb weight?

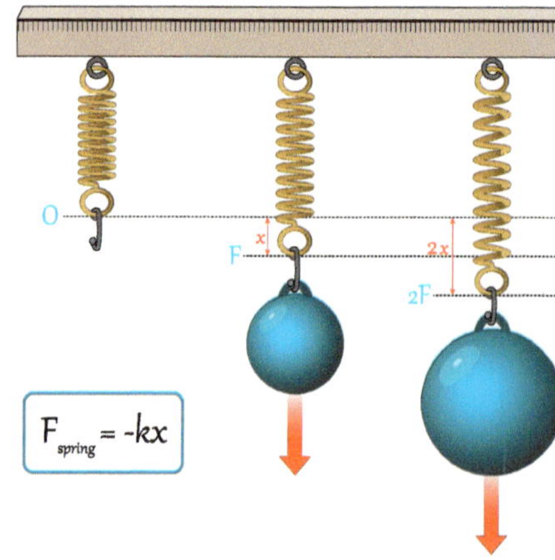

Answer

$$\frac{s_1}{w_1} = \frac{s_2}{w_2}$$

1. Write a proportion for the quantities that are directly proportional.

$$\frac{8}{20} = \frac{s_2}{50}$$
$$20s_2 = 400$$
$$s_2 = 20 \text{ in.}$$

2. Replace the variables with known values and solve.

SKILL CHECK **EXERCISE 35**

Does the following table of base and height measurements of a series of rectangles represent a direct variation?

base (*b*)	1	2	3	4	6	12
height (*h*)	12	6	4	3	2	1

Notice that the ratio of b to h is not constant. Instead, the product of the base and the height is constant: $bh = 12$. The data in this table models an inverse variation.

DEFINITIONS

An **inverse variation** is a relation in which the product of the 2 variables is a nonzero constant, k. That is, $xy = k$, where $k \neq 0$. The quantities are said to *vary inversely* or to be **inversely proportional**.

The constant k is once again called the **constant of variation**.

When the equation $xy = k$ is solved for y, general equations for inverse variations can be stated: $y = \frac{k}{x}$ or $f(x) = \frac{k}{x}$. As with direct variations, most applications of inverse variation involve a positive constant, k. In such inverse variations, an increase in 1 variable causes a decrease in the other.

EXAMPLE 5: Determining an Inverse Variation

Determine whether each table represents a direct variation, an inverse variation, or neither. If it is a variation, find the constant of variation and write the function rule for the variation.

a.

x	y
2	72
4	36
8	18
12	12

b.

x	y
1	5
2	8
4	14
8	26

Answers

a. This is an inverse variation.

$k = 144$

$y = \frac{144}{x}$ or $f(x) = \frac{144}{x}$

1. The ratio $\frac{y}{x}$ is not constant, but the product xy is.
2. For each ordered pair, the product $xy = 144$.
3. Divide both sides of $xy = 144$ by x.

b. This is neither an inverse nor a direct variation.

Neither the ratio $\frac{y}{x}$ nor the product xy is constant.

SKILL CHECK **EXERCISE 19**

Whenever 2 different ordered pairs, (x_1, y_1) and (x_2, y_2), represent the same inverse variation, their products will equal the same constant,

$$x_1y_1 = k \text{ and } x_2y_2 = k.$$

Therefore,

$$x_1y_1 = x_2y_2.$$

This equation can help you solve problems involving inverse variations.

AFTERSCHOOLHELP

Direct & Inverse Variations

EXAMPLE 6: Solving a Real-World Problem

The number of census workers varies inversely with the number of days it takes to complete the census in Pleasantburg. If it takes 24 workers 15 days to complete the census, how many workers are needed to complete the census in 10 days?

Answer

$w_1d_1 = w_2d_2$

1. Write an equation for the quantities that are inversely proportional.

$24(15) = w_2(10)$

$360 = 10w_2$

$w_2 = 36$ workers

2. Replace the variables with known values and solve.

SKILL CHECK **EXERCISE 37**

This solution is equivalent to determining the constant of variation ($k = wd = 360$ workers · days) and then solving the equation $10w = 360$.

	DIRECT VARIATION	INVERSE VARIATION
Definition	Ratio is constant.	Product is constant.
Constant	$\frac{y}{x} = k$	$xy = k$
Implies	$\frac{y_1}{x_1} = \frac{y_2}{x_2}$	$x_1y_1 = x_2y_2$
Function Rule	$y = kx$	$y = \frac{k}{x}$
Graph ($k > 0$)	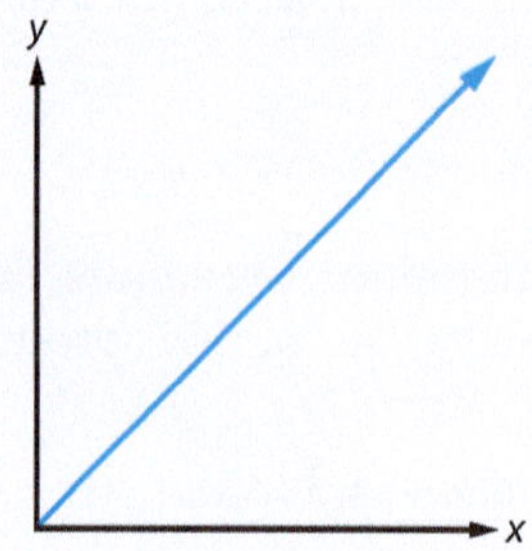	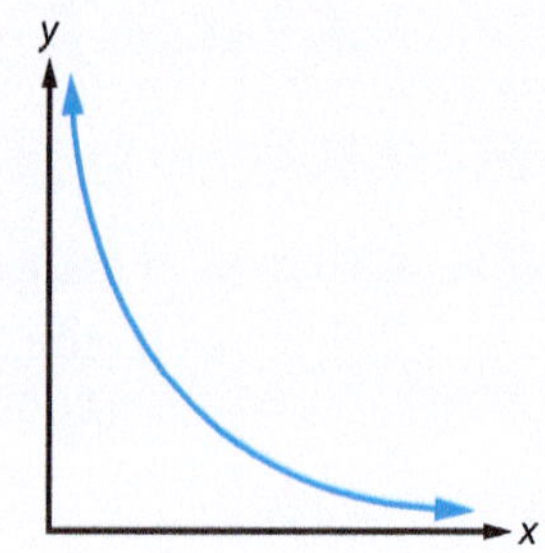
Intersects the Origin	always	never
Real-World Connection	adding more jobs for 1 worker to do	adding more workers to accomplish 1 job

VOCABULARY

- constant of proportionality
- constant of variation
- directly proportional
- direct variation
- inversely proportional
- inverse variation

We are able to consistently describe real-world processes like motion and spring tension with mathematical functions because God has created an orderly world. Not only does math help us fulfill our stewardship role over creation, it also helps us recognize the handiwork of our Creator. While fallen humans suppress this truth (Rom. 1), believers respond with praise, giving glory to God for His wisdom, creativity, and beauty (Ps. 8).

A. EXERCISES

1. Match each algebraic equation or expression with vocabulary from this section.

a. $k = xy$ **I** constant of variation

b. $k = \frac{y}{x}$ **II** direct variation

c. k **III** inverse variation

State whether each of the following represents a direct variation, an inverse variation, or neither.

2. $a = 2b$

3. $y = \frac{7}{x}$

4. $xy = -3$

5. $y = \frac{x}{3}$

6.

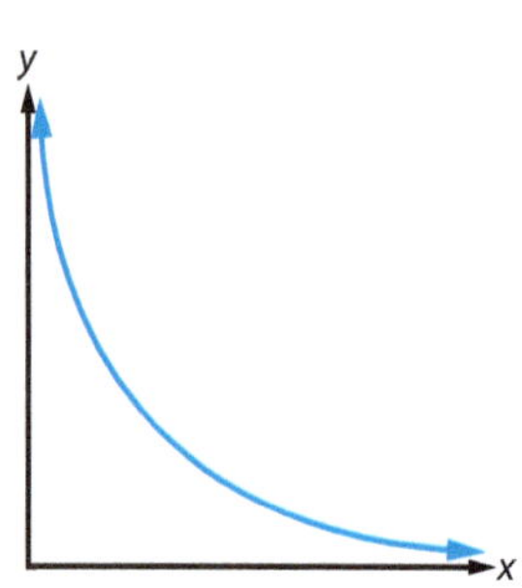

7.

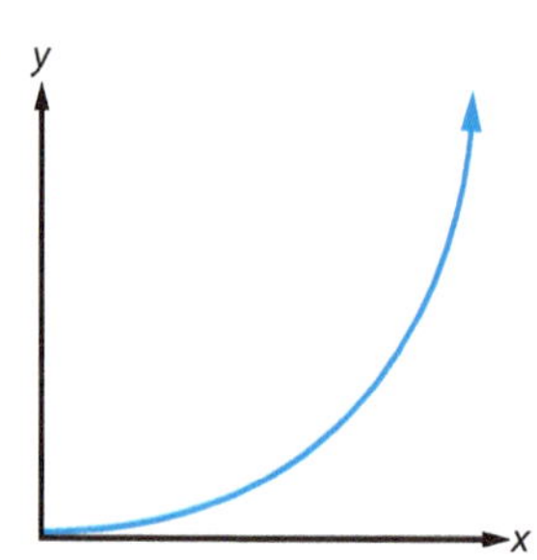

8.

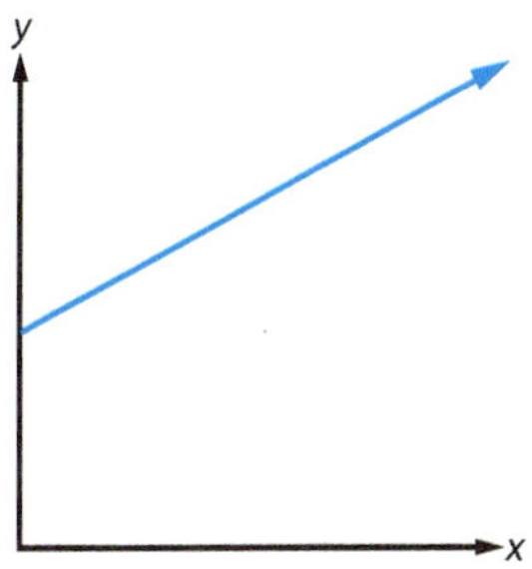

9.

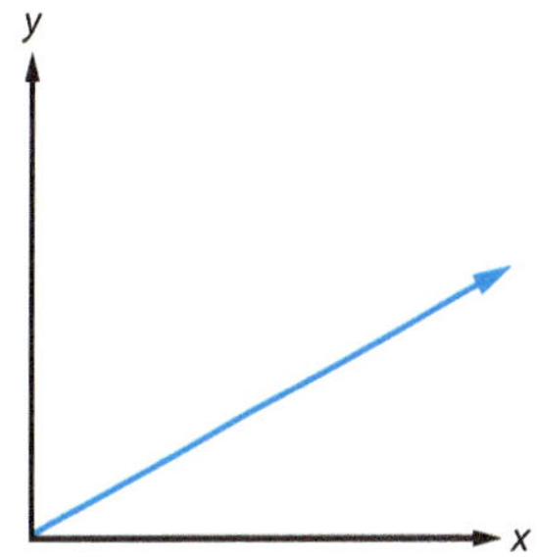

Find the constant of variation and write the function rule for each variation.

10. y varies directly with x and $y = 18$ when $x = 9$.

11. y varies inversely with x and $y = 12$ when $x = 6$.

12. m is inversely proportional to a and $m = 9$ when $a = 4$.

13. c is directly proportional to v and $v = 9$ when $c = 6$.

14. r is inversely proportional to s and $r = 2$ when $s = 2.5$.

15. g varies directly with h and $h = 5$ when $g = 0.5$.

B. EXERCISES

State whether each table represents a direct variation, an inverse variation, or neither. For each variation, write the function rule and graph the first quadrant of the function.

16.

x	y
1	6
2	12
3	18
4	24

17.

x	y
16	64
8	32
4	16
2	8

18.

x	y
2	9
3	6
6	3
9	2

19.

x	y
2	5
3	$\frac{10}{3}$
4	2.5
5	2

20.

x	y
20	30
15	$22\frac{1}{2}$
10	15
5	$7\frac{1}{2}$

21.

x	y
1	25
2	16
3	9
4	4
5	1

22.

x	y
1	2
2	3
3	4
4	5

23.

x	y
0.125	12
0.5	3
2	0.75
6	0.25

Find the missing value for each variation.

24. If y is directly proportional to x and $y = 16$ when $x = 4$, find y when $x = 10$.

25. If y varies inversely with x and $y = 4$ when $x = 2$, find y when $x = 8$.

26. If y is inversely proportional to x and $y = 0.25$ when $x = 4$, find y when $x = 10$.

27. If y varies directly with x and $y = 3$ when $x = 2$, find y when $x = 7$.

State whether each of the following represents a direct or inverse variation and find the constant of variation.

28. The formula for the circumference of a circle is $C = \pi d$, where d is the diameter of the circle.

29. The formula for the additional pressure resulting from being underwater is $P = 0.433d$, where d is the depth below the surface.

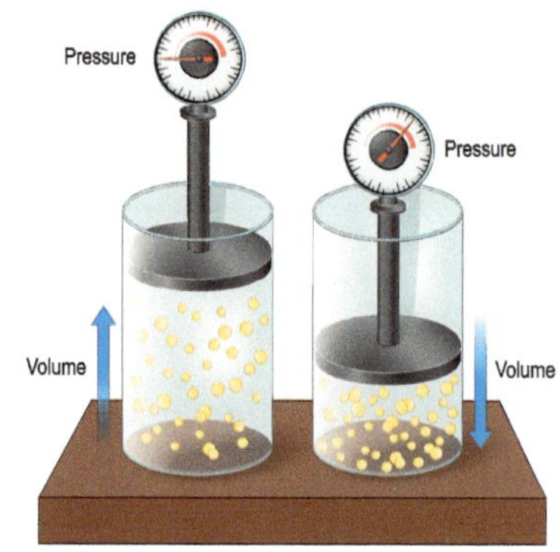

30. Boyle's law states that the product of the pressure, P, exerted by a gas and the gas's volume, V, is constant when there is no change in temperature. Gene was able to determine that 5 m^3 of a gas exerted a pressure of 0.8 kilopascals.

31. Charles's law states that the volume, V, of a gas increases proportionally as the temperature in Kelvin, T, increases if the pressure remains constant. At a temperature of 500 K, a given amount of gas occupies 6 m^3.

32. **Essential Question:** What is the significance of discovering consistent mathematical relationships such as Charles's law and Boyle's law in the real world?

Write an equation expressing the variation described; then use it to solve the problem.

33. Find the distance, d, a car travels at the rate of 52 mi/hr if the distance varies directly with the time, t, in hours. How far will the car travel in 2.5 hr?

34. The width, w, of a rectangle with an area of 20 square units varies inversely with its length, l. Find the width when the length is 8.

35. Find the sales tax, t, paid on purchases at the rate of 7% if tax varies directly with the total purchase price, p. How much sales tax is paid on a $495 computer purchase?

36. The average monthly attendance, a, at an amusement park varies inversely with the number of days it rains during the month, r. If the attendance in July was 104,000 and it rained 3 days, what would you predict the attendance to be for August if it rains 6 days?

37. It takes 1 person 18 hr to paint a fence. If the time, t, that it takes to paint a fence varies inversely with the number of painters, p, how long will it take 4 people to paint the fence?

38. The height, h, of a triangle varies inversely with its base, b. If the base of a triangle is 6 and its height is 6 also, find the height of a triangle whose base is 4.

C. EXERCISES

Use the table on the right for exercises 39–41.

time, t (sec)	distance fallen, d (m)
1	5
2	20
3	45
4	80
5	125

39. Graph the data and determine whether the function is a direct variation, an inverse variation, or neither.

40. Make a table of time squared and distance. Graph your data and determine whether the relation between t^2 and d is a direct variation, an inverse variation, or neither.

41. MP2 Write a function rule for the relationship between t^2 and d and use it to find the following.

a. the distance fallen after 7.5 sec

b. the time it takes to fall 500 m

Write an equation expressing the variation described; then use it to solve the problem.

42. MP6 The weight of an object varies inversely with the square of its distance from the center of the earth, which has a radius of 4000 mi. If a boy weighs 140 lb on the surface of the earth, find his weight 10,000 mi from the center of the earth.

43. MP6 The volume of a ball varies directly with the cube of its diameter. If the volume of a ball 4 in. in diameter is 33.5 in.^{3}, find the volume of a ball with a 6 in. diameter (to the nearest cubic inch).

CUMULATIVE REVIEW

Solve each equation. [2.1–2.2]

44. $2.5x = 10$

45. $x - 25.4 = 57.29$

46. $3(2x - 5) = -(x - 8)$

47. $a - 2(a - 3) = 20$

Write an equation and solve.

48. Find the percent increase (to the nearest tenth of a percent) in attendance from 15,000 for the first game to 17,500 for the second. **[2.5]**

49. Find the discount rate (to the nearest whole percent) of a $42 item that is on sale for $28. **[2.6]**

50. Two cars left the same point at the same time traveling in opposite directions at 60 mi/hr and 46 mi/hr respectively. How long will it be before they are 371 mi apart? **[2.7]**

51. Suppose the cars in exercise 50 were traveling in the same direction and the slower one left 2 hr before the faster one. How long will it take for the faster car to overtake the slower car? **[2.7]**

52. Which graph could represent the complete flight of a baseball after it is hit? **[4.3]**

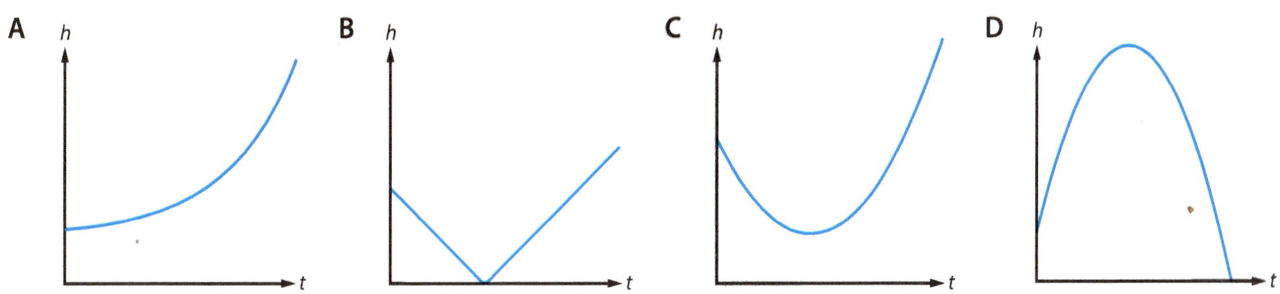

E none of these

53. Which represents the explicit formula for the sequence −3, −9, −15, −21, . . .? **[4.5]**

A $a_n = 6n - 3$

B $a_n = -6n + 3$

C $a_n = 3n - 6$

D $a_n = a_{n-1} - 6; a_1 = -3$

E none of these

4.7 GRAPHING ABSOLUTE VALUE FUNCTIONS

What values will affect the graph of the parent function $f(x) = |x|$?

After completing this section, I will be able to

- create sets of ordered pairs, tables, and graphs representing absolute value functions.
- describe translations of the basic absolute value function $y = |x|$.
- write a function rule for the graph of an absolute value function.

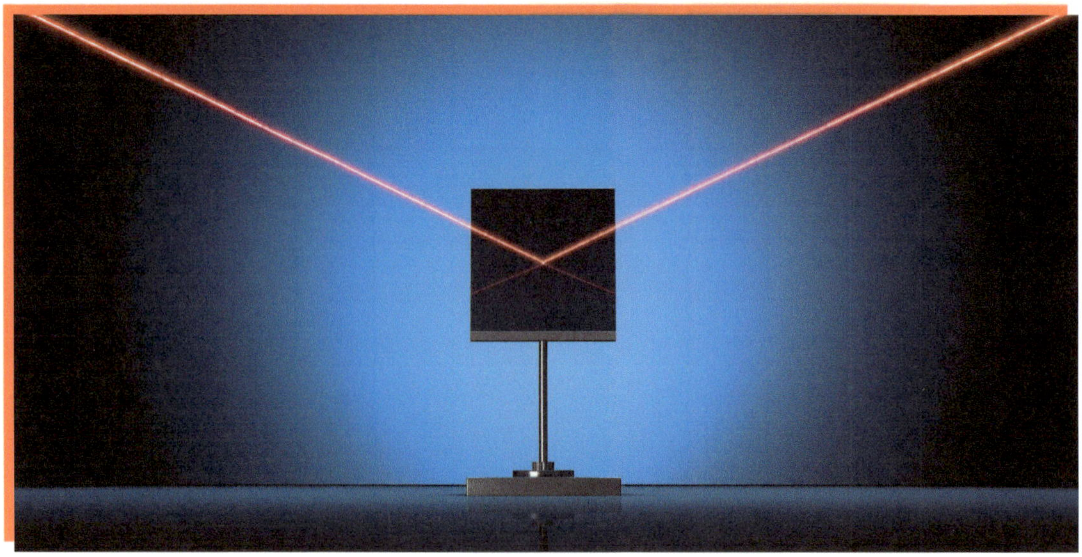

Laser reflections can be modeled by absolute value functions.

Recall the formal definition of absolute value.

$$|x| = \begin{cases} x \text{ if } x \geq 0 \\ -x \text{ if } x < 0 \end{cases}$$

This definition can be used to construct a variety of representations of the basic absolute value function, $f(x) = |x|$.

DEFINITION

An **absolute value function** is a function containing an absolute value expression.

Using the domain $D = \{-3, -2, -1, 0, 1, 2, 3\}$, we can plot a few points to determine the general shape of the graph of $f(x) = |x|$.

x	**f(x) = |x|**	**(x, f(x))**
−3	$f(-3) = \|-3\| = 3$	(−3, 3)
−2	$f(-2) = \|-2\| = 2$	(−2, 2)
−1	$f(-1) = \|-1\| = 1$	(−1, 1)
0	$f(0) = \|0\| = 0$	(0, 0)
1	$f(1) = \|1\| = 1$	(1, 1)
2	$f(2) = \|2\| = 2$	(2, 2)
3	$f(3) = \|3\| = 3$	(3, 3)

Graph the function: $f(x) = |x|$.

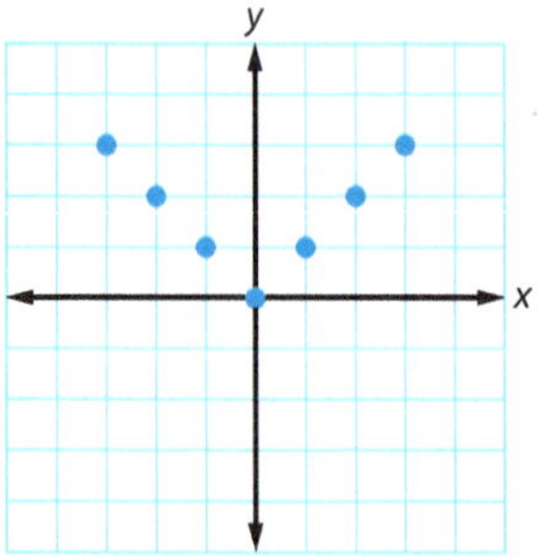

If no specific domain is given, the domain is understood to be all real numbers. The graph of the absolute value function $f(x) = |x|$ forms a V with the turning point at the origin. This turning point is called the **vertex**.

A **parent function** is the simplest form of a family of functions. The basic absolute value function, $f(x) = |x|$, defined above is the parent function for the family of absolute value functions. The graph of this parent function is moved up or down and left or right in the plane when values are added to or subtracted from the function. This movement of the graph is called a **translation**. All possible translations form a family of functions.

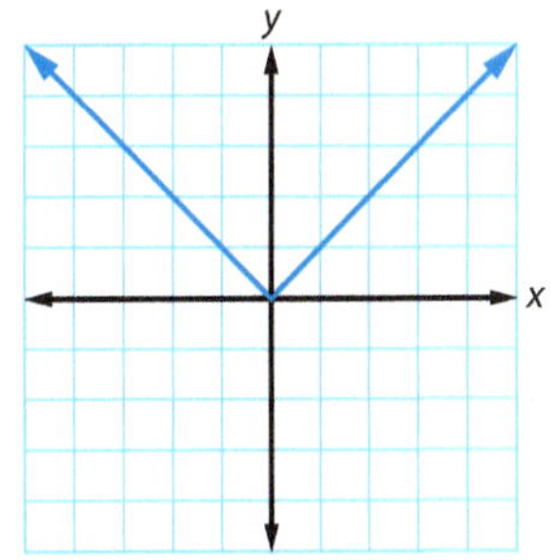

parent absolute value function

EXAMPLE 1: Graphing a Vertical Translation

Graph $f(x) = |x| + 4$. Then compare this graph with the graph of the parent function to determine the effect of adding 4 to the absolute value function rule.

Answer

x	|x| + 4	f(x)
−3	$\|-3\| + 4 = 3 + 4 = 7$	$f(-3) = 7$
−2	$\|-2\| + 4 = 2 + 4 = 6$	$f(-2) = 6$
−1	$\|-1\| + 4 = 1 + 4 = 5$	$f(-1) = 5$
0	$\|0\| + 4 = 0 + 4 = 4$	$f(0) = 4$
1	$\|1\| + 4 = 1 + 4 = 5$	$f(1) = 5$
2	$\|2\| + 4 = 2 + 4 = 6$	$f(2) = 6$
3	$\|3\| + 4 = 3 + 4 = 7$	$f(3) = 7$

1. Evaluate the function rule for several elements of the domain.

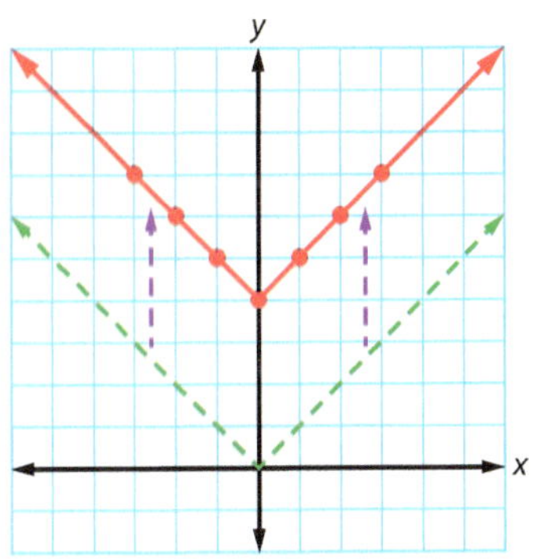

2. Plot enough ordered pairs to determine the shape of the graph, and then connect the points to create a graph of the function.
3. Compare the graph of $f(x) = |x| + 4$ with the graph of the parent function.

The entire graph has moved 4 units up.

4. The vertex of the translated graph is (0, 4).

SKILL CHECK **EXERCISE 7**

When a constant is added or subtracted within the absolute value symbols, a horizontal translation occurs. The vertex is moved to the x-value that causes the expression within the absolute value symbols to equal 0.

EXAMPLE 2: Graphing a Horizontal Translation

Graph $f(x) = |x + 3|$. Then compare this graph with the graph of the parent function to determine the effect of adding 3 within the parent function.

Answer

$$x + 3 = 0$$
$$x = -3$$

1. Set the expression within the absolute value function equal to 0. Solve to find the value of x that is used to find the vertex.

x	$\|x+3\|$	$f(x)$
−6	$\|-6+3\| = \|-3\| = 3$	$f(-6) = 3$
−5	$\|-5+3\| = \|-2\| = 2$	$f(-5) = 2$
−4	$\|-4+3\| = \|-1\| = 1$	$f(-4) = 1$
−3	$\|-3+3\| = \|0\| = 0$	$f(-3) = 0$
−2	$\|-2+3\| = \|1\| = 1$	$f(-2) = 1$
−1	$\|-1+3\| = \|2\| = 2$	$f(-1) = 2$
0	$\|0+3\| = \|3\| = 3$	$f(0) = 3$

2. Evaluate the function rule to find the vertex and several ordered pairs on either side of the vertex.

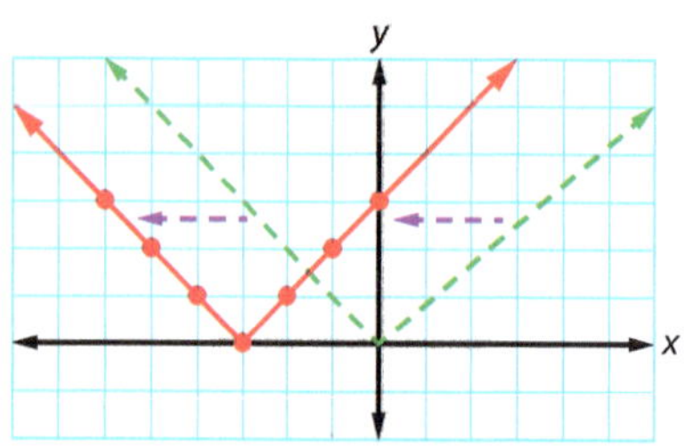

3. Plot enough ordered pairs to determine the shape of the graph, and then connect the points to create a graph of the function.
4. Compare the graph of $f(x) = |x + 3|$ with the graph of the parent function.

The entire graph has moved 3 units left.

5. The vertex of the translated graph is (−3, 0).

SKILL CHECK **EXERCISE 11**

The graphs of $y = |x| + k$ and $y = |x + h|$ are translations of the parent function $f(x) = |x|$. A positive value of k will move the graph up k units and a positive value of h will move the graph to the left h units. If h and k are negative values, their graphs will move in the opposite directions.

FUNCTION	GRAPH	VERTEX
$f(x) = \|x\|$	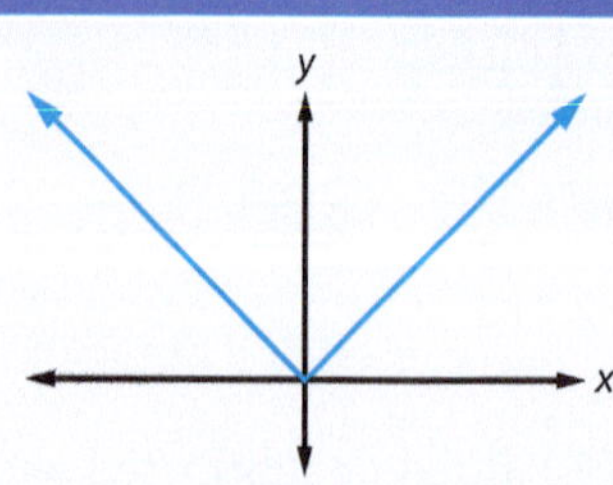	(0, 0)
$f(x) = \|x\| + k$	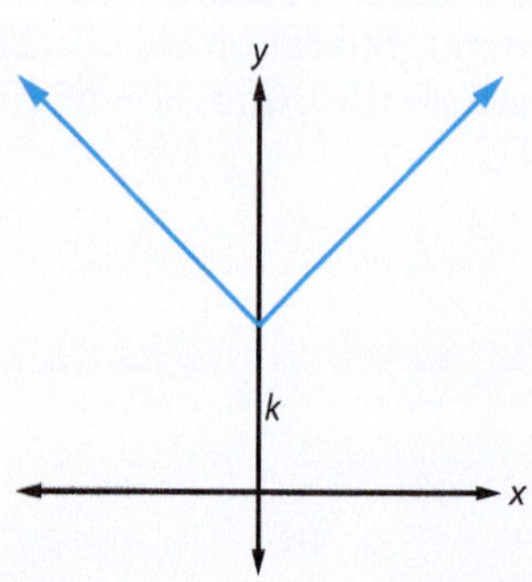	(0, k)

$f(x) = \lvert x \rvert - k$	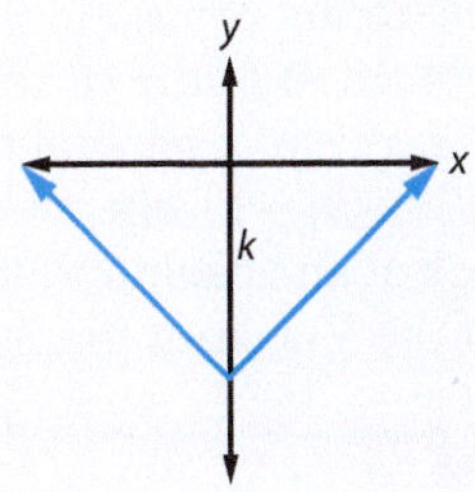	$(0, -k)$
$f(x) = \lvert x + h \rvert$ $= \lvert x - (-h) \rvert$	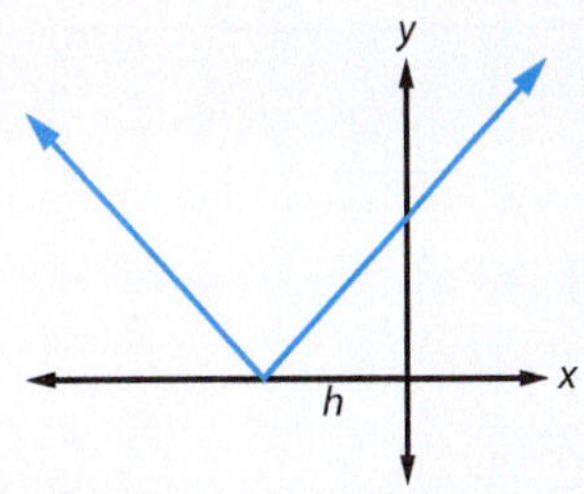	$(-h, 0)$
$f(x) = \lvert x - h \rvert$	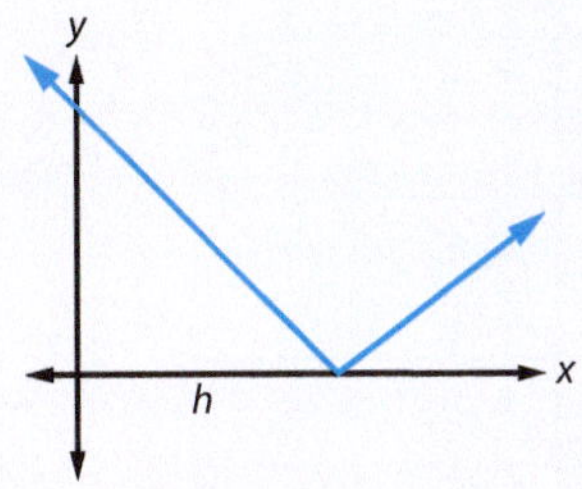	$(h, 0)$

EXAMPLE 3: Finding the Function Rule

Write the function rule that describes each graph.

a. 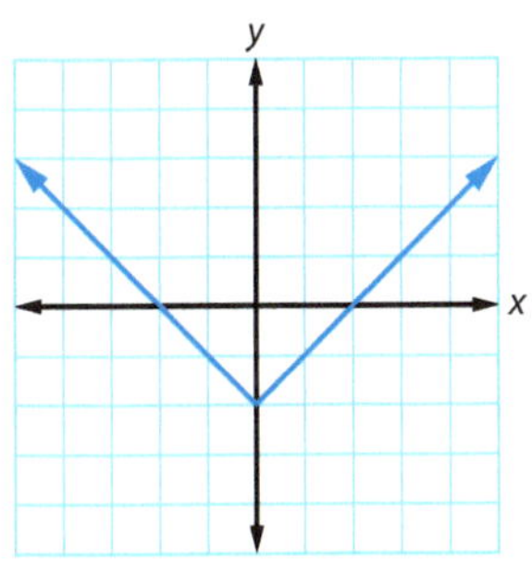

b. 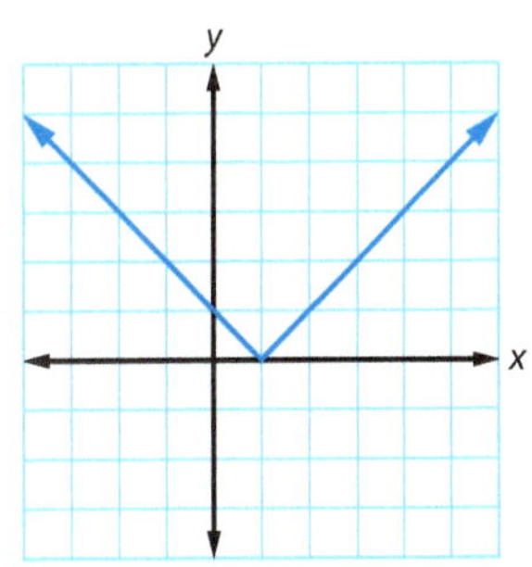

Answers

a. $f(x) = |x|$

1. Notice that the graph has the shape of the parent function. Begin by writing its basic function rule.

$f(x) = |x| - 2$

2. The graph has been translated 2 units down.

$f(-2) = |-2| - 2 = 0$
$f(5) = |5| - 2 = 3$
$(-2, 0)$ and $(5, 3)$ are on the graph.

3. Check the function rule by evaluating a few ordered pairs and making sure that the points are on the graph.

b. $f(x) = |x|$

1. Notice that the graph has the shape of the parent function. Begin by writing its basic function rule.

$f(x) = |x - 1|$

2. The graph has been translated 1 unit right.

$f(-1) = |-1 - 1| = 2$
$f(4) = |4 - 1| = 3$
$(-1, 2)$ and $(4, 3)$ are on the graph.

3. Check the function rule by evaluating a few ordered pairs and making sure that the points are on the graph.

SKILL CHECK **EXERCISES 15, 17**

CONNECT

Translated Vertex Function

$f(x) = |x - h| + k$

Vertex Location

(h, k)

Both vertical and horizontal translations may occur within the same function. The location of the vertex can be determined by examining the function rule. A general form involving both h and k is written as $f(x) = |x - h| + k$, where (h, k) is the vertex. You should be able to look at an absolute value function rule to determine the location of the V-shaped graph.

EXAMPLE 4: Graphing with Multiple Translations

Graph $f(x) = |x - 2| + 1$.

Answer

$f(x) = |x - h| + k$

$f(x) = |x - 2| + 1$

vertex $= (h, k) = (2, 1)$

1. Compare to the general form of the function to identify the vertex.

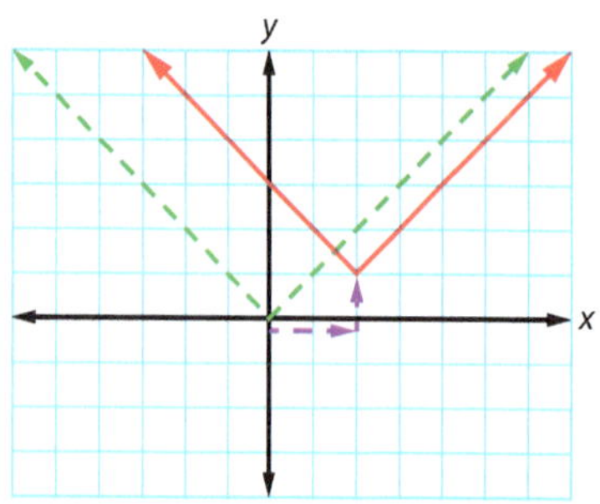

2. Graph the function as a translation of $f(x) = |x|$ with vertex at (2, 1).

$f(0) = |0 - 2| + 1 = 2 + 1 = 3$

$f(5) = |5 - 2| + 1 = 3 + 1 = 4$

(0, 3) and (5, 4) are on the graph.

3. Check the graph by evaluating a few ordered pairs and verifying that the points are on the graph.

SKILL CHECK **EXERCISE 13**

AFTERSCHOOLHELP

Graphing Absolute Value Functions

KEYWORD SEARCH

function graphing tool

The graph of the parent function is also affected by multiplying the function by a constant. The value of a in the function $f(x) = a|x|$ causes a vertical shrink ($|a| < 1$) or stretch ($|a| > 1$) in the parent function.

EXAMPLE 5: Graphing a Vertical Stretch

Graph $f(x) = 2|x|$.

Answer

$f(x) = |x|$

$f(x) = 2|x|$

1. Compare the given function rule with the parent function rule.

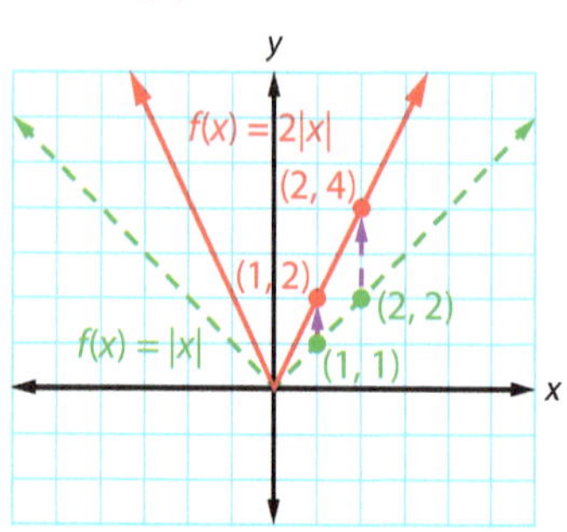

2. Notice the points on the red graph ($f(x) = 2|x|$) are twice as far from the x-axis as the corresponding points on the green graph ($f(x) = |x|$).

$f(1) = 2|1| = 2$

$f(2) = 2|2| = 4$

(1, 2) and (2, 4) are on the graph.

3. Evaluate a few ordered pairs and make sure that the points are on the graph.

SKILL CHECK **EXERCISE 27**

EXAMPLE 6: Graphing a Vertical Shrink

Graph $f(x) = \frac{1}{3}|x|$.

Answer

$f(x) = |x|$

$f(x) = \frac{1}{3}|x|$

1. Compare the given function rule with the parent function rule.

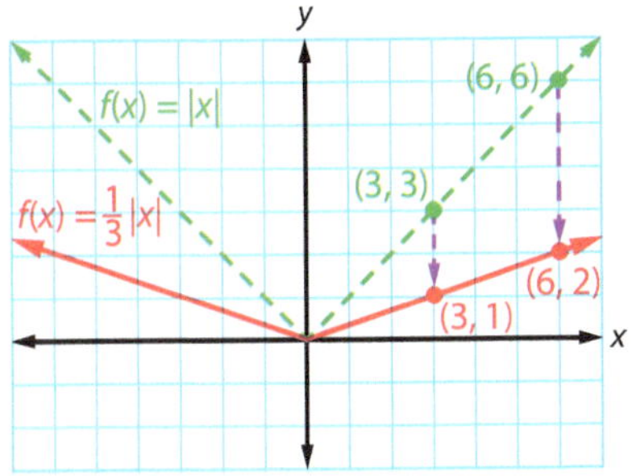

2. Notice the points on the red graph $\left(f(x) = \frac{1}{3}|x|\right)$ are one-third as far from the *x*-axis as the corresponding points on the green graph ($f(x) = |x|$).

$f(3) = \frac{1}{3}|3| = 1$

$f(6) = \frac{1}{3}|6| = 2$

(3, 1) and (6, 2) are on the graph.

3. Evaluate a few ordered pairs and make sure that the points are on the graph.

SKILL CHECK **EXERCISE 25** ☑

VOCABULARY

- absolute value function
- basic absolute value function
- parent function
- translation
- vertex

A. EXERCISES

Match the graph that best models each function. $\{a, b \in \mathbb{R} \mid a, b > 0\}$.

1. $y = |x| - a$
2. $y = |x - a|$
3. $y = 3|x|$
4. $y = |x - a| + b$
5. $y = |x + a| - b$
6. $y = 0.4|x|$

A

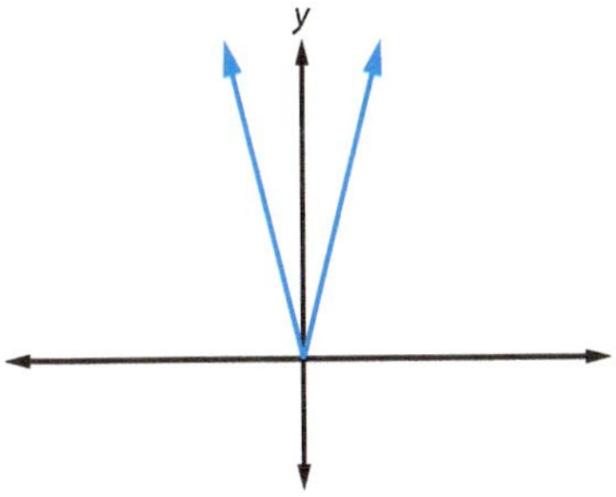

B

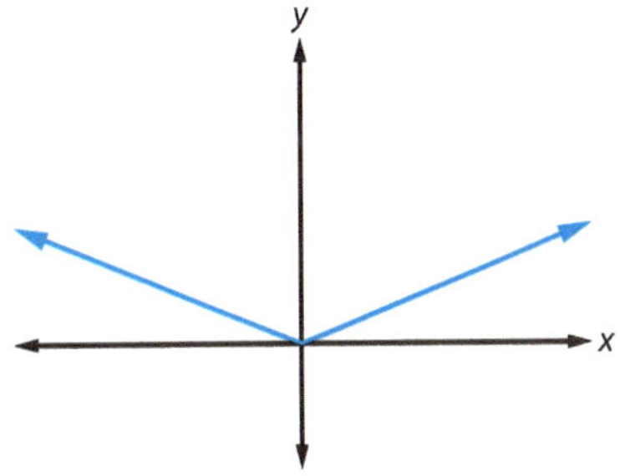

C

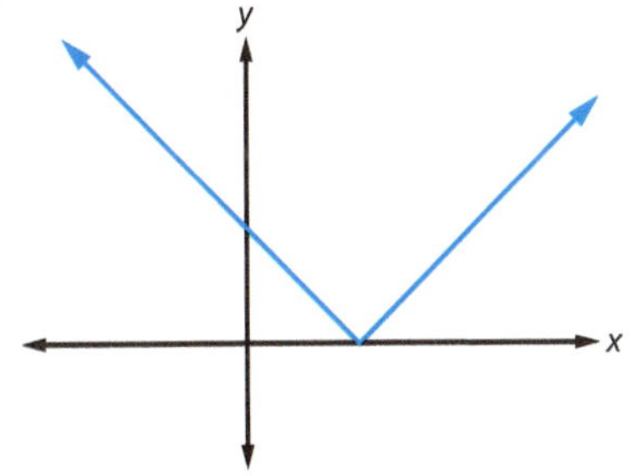

D

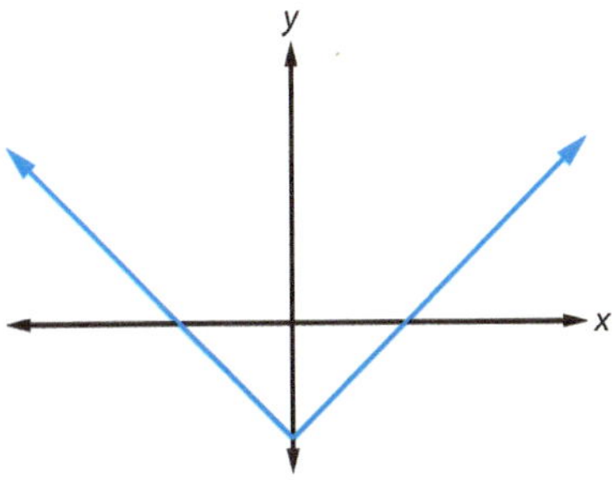

E

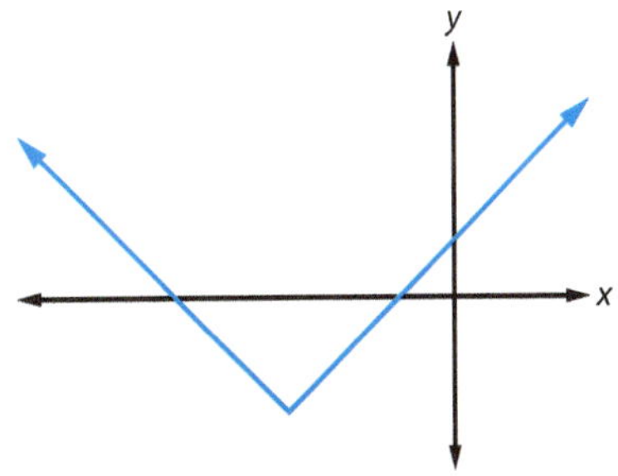

F

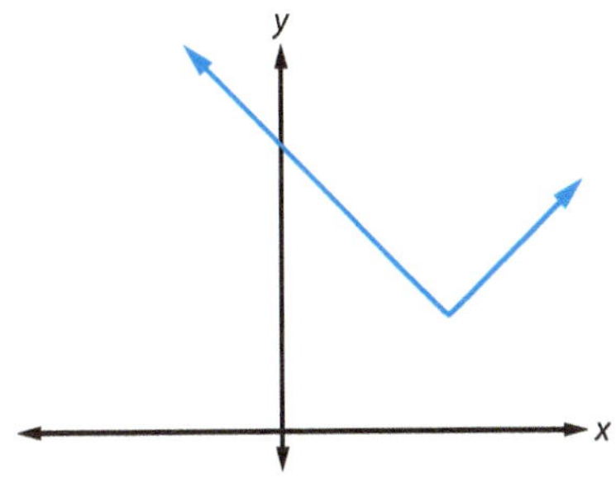

Make a table of at least 7 ordered pairs and graph each function.

7. $f(x) = |x| - 4$
8. $f(x) = |x| + 1$
9. $f(x) = 3|x|$
10. $y = |x + 3|$
11. $f(x) = |x - 5|$
12. $y = \frac{1}{2}|x|$
13. $f(x) = |x - 2| - 3$
14. $y = |x + 2| + 4$

Write a function rule for each translation of $y = |x|$.

15. down 2.5
16. up 3
17. left 4
18. right 7
19. **Essential Question:** How do changes in the variables h, k, and a in $f(x) = a|x - h| + k$ affect the graph of the function?

B. EXERCISES

Describe the translations of the graph of $y = |x|$ that produce the graph of each function.

20. $f(x) = |x - 3| - 4$
21. $y = |x + 2| - 6$
22. $y = |x + 7| + 1$
23. $f(x) = |x - 1| + 3$

Graph each function.

24. $f(x) = 5|x|$
25. $y = \frac{2}{3}|x|$
26. $f(x) = \frac{3}{4}|x|$
27. $y = \frac{7}{3}|x|$

Write the function rule for each translation of $y = |x|$.

28. 2 units left, 6 units up
29. 6 units down, 3 units right
30. 5.9 units up, 6.23 units right
31. $\frac{8}{9}$ units left, $\frac{2}{5}$ units down

Use translations and the location of the vertex to graph each function without evaluating ordered pairs.

32. $f(x) = |x + 3.5|$
33. $f(x) = |x| - 1.5$
34. $y = |x + 3| - 2$
35. $y = |x - 4| - 3$
36. $f(x) = \left|x - \frac{1}{2}\right| + 2$
37. $y = |x + 2| - \frac{2}{3}$

C. EXERCISES

Make a table of at least 7 ordered pairs and graph each function.

38. $f(x) = -|x|$

39. $y = -|x| + 3$

40. $y = -|x - 2|$

41. $f(x) = -|x + 1|$

42. $y = 2|x + 2| - 3$

43. $f(x) = -\frac{1}{2}|x - 3| + 2$

44. MP7 **Essential Question:** From your observations in exercises 38–43, what effect does the negative sign in front of the absolute value expression have on the graph of $f(x) = |x|$?

CUMULATIVE REVIEW

45. State the Trichotomy Property using x and y. **[3.1]**

46. Write an inequality symbol equivalent to $\not>$. **[3.1]**

Solve each inequality. [3.2]

47. $-5x + 2x < 51$

48. $4x - 6 \geq x - 10$

49. $6(x + 2) < 8 - x$

Solve each compound inequality. [3.3–3.4]

50. $3x + 6 < 81$ and $2x > 10$

51. $4x < -8$ or $x + 20 < 30$

52. $7x + 5 \geq -2$ or $9 - 2x < 19$

53. Which expression is the correct translation of "9 more than twice the square root of 16"? **[1.4]**

A $(9 + 2)\sqrt{16}$

B $2 + \sqrt{16} + 9$

C $2\sqrt{16} + 9$

D $2\sqrt{16 + 9}$

E none of these

54. If $4x - 3 < 32$, which of the following are solutions? **[3.1]**

A $\frac{26}{3}$

B 9.1

C −6

D A and B

E A and C

APPLICATION PROBLEMS
FRACTALS

"Clouds are not spheres, mountains are not cones, coastlines are not circles, and bark is not smooth, nor does lightning travel in a straight line," explains Benoit Mandelbrot in his 1982 book, *The Fractal Geometry of Nature*. Traditional geometric shapes are rapidly being replaced by fractals as the scientist's model of the seemingly random processes of the real world. By using fractals, scientists are now just beginning to understand some of the amazing complexity of God's created universe.

Fractals are rough or fragmented geometric shapes that are repeated at an ever-smaller scale. They are generated by repeated applications of a simple procedure or *iteration*. Some of the early fractals were created by Helge von Koch in 1904 and Wacław Sierpiński in 1915.

While much of the research being done today is beyond the scope of this course, many fractals can be easily generated, and their foundational concepts explained using the basic operations of arithmetic.

The Koch curve was one of the earliest fractals studied. It begins with a line segment having a length of 1 unit.

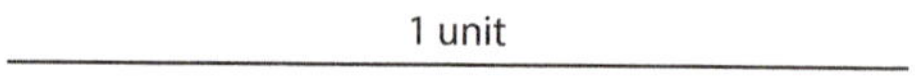

Drawing an equilateral triangle based on the middle third of the segment and then removing the middle third of the segment produces the first iteration.

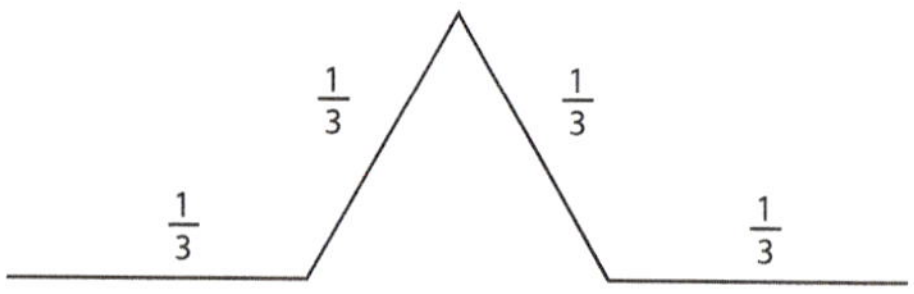

1st iteration

Performing the same steps to each segment produces the next iteration. For each segment draw an equilateral triangle on the middle third of the segment and then remove the middle third of the segment.

2nd iteration

1. What is the total length of the first iteration of Koch's curve?
2. Describe the second iteration of Koch's curve.
   a. How many segments are in the curve?
   b. How long is each segment?
   c. What is the total length of the second iteration of the curve?
3. Draw the next iteration of the fractal and use your sketch to answer the following questions.
   a. How many segments are in the curve?
   b. How long is each segment?
   c. What is the total length of the third iteration of the curve?
4. Make a table similar to the one below and summarize the data for the first 3 iterations of Koch's curve.

Iteration	Number of Segments	Length of Each Segment	Total Length of the Curve	Decimal Equivalent of the Total Length
0	1	1	1	1
1	4	$\frac{1}{3}$	$\frac{4}{3}$	1.33
2	16			
3	64			

5. Analyze each pattern in the table and state the relationships in successive iterations of Koch's curve.

   a. the number of segments in successive iterations

   b. the length of each segment in successive iterations

   c. the total length of the curve in successive iterations

6. Extend the table from exercise 4 to predict the number of segments, the length of each segment, and the total length of the curve for the fourth and fifth iterations of the Koch curve.

7. Mathematicians model patterns using expressions for the general, or *n*th, case. For example, in the *n*th iteration there are 4^n segments.

   a. Continue your table from exercise 4 by filling in expressions for length of each segment and the total length of the curve in the *n*th iteration of Koch's curve.

   b. Use these general expressions and a calculator to complete the missing values for the tenth and the hundredth iterations of Koch's curve.

8. Koch's snowflake is the result of performing the iterations of the Koch curve on an equilateral triangle. The figures illustrate the fourth and fifth iterations of Koch's snowflake. If each side of the original triangle has a length of 1, what is the perimeter of the fifth iteration of Koch's snowflake?

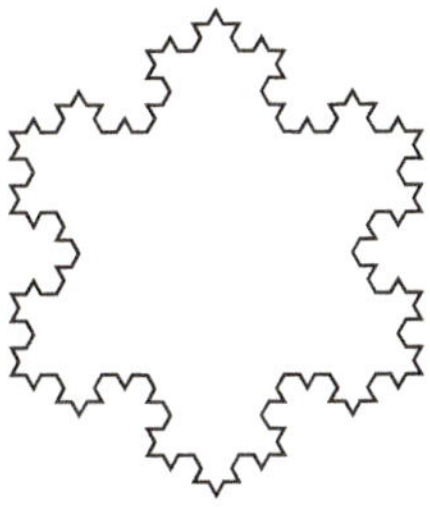

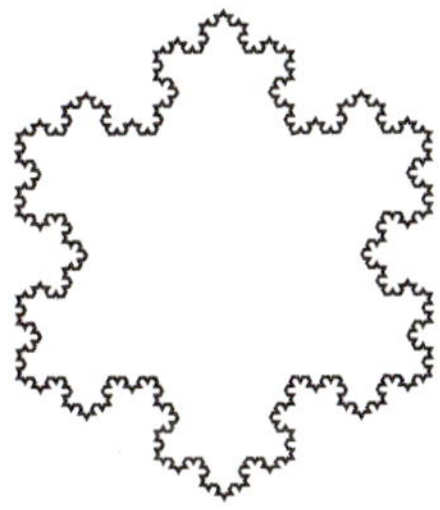

9. If the iterations of the Koch curve continued forever, describe what would happen to each of the following.

   a. the number of sides

   b. the length of each segment

   c. the length of the curve

Mathematically, it can be shown that the lengths of both the Koch curve and the Koch snowflake are infinite when there are an infinite number of iterations. Our finite human minds have difficulty with this concept, but thankfully that does not prevent mathematicians from using fractals as models in the solution of a variety of problems.

KEYWORD SEARCH

fractal applications

The figures below illustrate the first 3 iterations of another famous fractal, the Sierpinski triangle. Each iteration is performed by connecting the midpoints of any upward-pointing triangles.

10. Beginning with your own equilateral triangle, complete the first 4 iterations of the Sierpinski triangle. Shading the downward-pointing triangles created in each iteration with a different color will produce "fractal art."

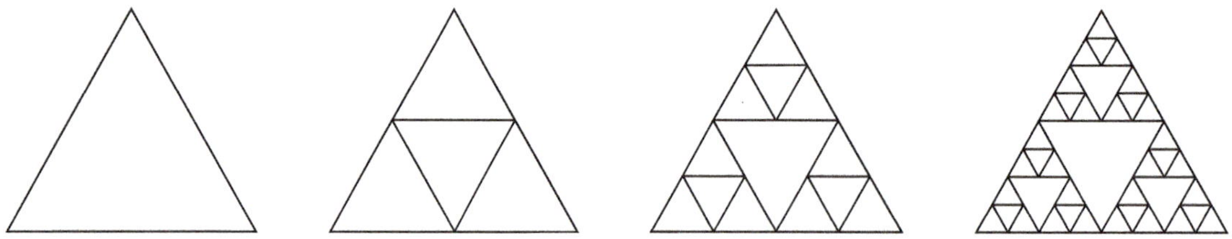

initial triangle *1st iteration* *2nd iteration* *3rd iteration*

11. Notice that each iteration of the Sierpinski triangle produces identical smaller upward-pointing triangles. Use a table like the one below to record the following data for each iteration. Assume the original triangle has an area of 1 square unit.

	Iteration	0	1	2	3	4
a.	number of upward-pointing triangles	1				
b.	area of each upward-pointing triangle	1				
c.	total area of upward-pointing triangles	1				

12. Use your Sierpinski triangles to extend the table to determine expressions for the nth iteration. Then use the nth iteration expressions to complete the table for the tenth and hundredth iterations.

	Iteration	n	10	100
a.	number of upward-pointing triangles			
b.	area of each upward-pointing triangle			
c.	total area of upward-pointing triangles			

Fractals are used to generate lifelike landscapes in computer games as well as a variety of other phenomena in training simulators.

13. Research and briefly describe how the midpoint-displacement algorithm is used to model landscapes.

4 CHAPTER REVIEW

Key Concepts

Relations & Functions [4.1]

- A relation is a set of ordered pairs. Its domain is the set of x-coordinates, and its range is the set of y-coordinates.
- A function is a relation in which each element in the domain is paired with only 1 element in the range.
- The function notation $f(x)$ is read "f of x" and means "the function of f at the value of x." It emphasizes x as the independent variable and $f(x)$ as the dependent variable: $(x, y) = (x, f(x))$.
- Relations and functions can be represented with ordered pairs, tables, mapping diagrams, or graphs.

Graphs of Relations & Functions [4.2–4.3]

- The vertical line test is used to determine whether a graphed relation represents a function.
- After plotting enough points, the function can be graphed by connecting the points with a smooth curve.
- Graphs of relations and functions can be used to model and interpret real-world data.

Function Rules [4.4]

- A function rule is an equation that pairs each member of the domain with an element of the range.
- If no domain is given, assume that $D = \mathbb{R}$.

Arithmetic Sequences [4.5]

- An arithmetic sequence is an ordered list of numbers (terms) where the difference between 2 consecutive terms is called the common difference.
- The nth term of the sequence is denoted a_n (a_1 is the first term, a_2 is the second, and so on).

ARITHMETIC SEQUENCE FORMULAS	
Recursive	Explicit
$a_n = a_{n-1} + d$, where a_1 is given	$a_n = a_1 + (n - 1)d$

Direct & Inverse Variations [4.6]

- A direct variation is a function of the form $y = kx$, where k is the constant of variation.
- An inverse variation is a function of the form $y = \frac{k}{x}$, where k is the constant of variation.

Graphing Absolute Value Functions [4.7]

FUNCTION	GRAPH	VERTEX
$f(x) = \|x\|$	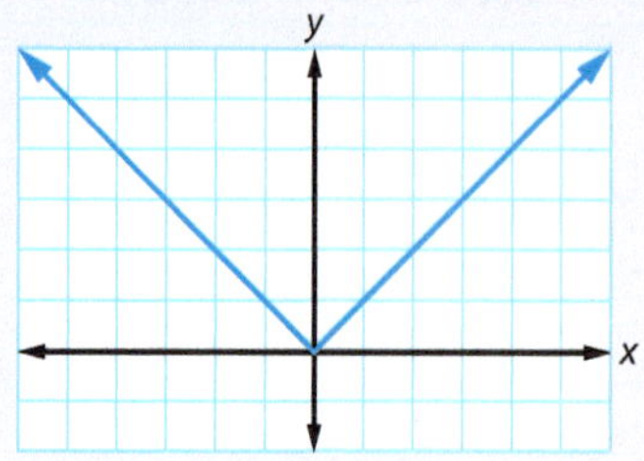	(0, 0)
$f(x) = \|x\| \pm k$	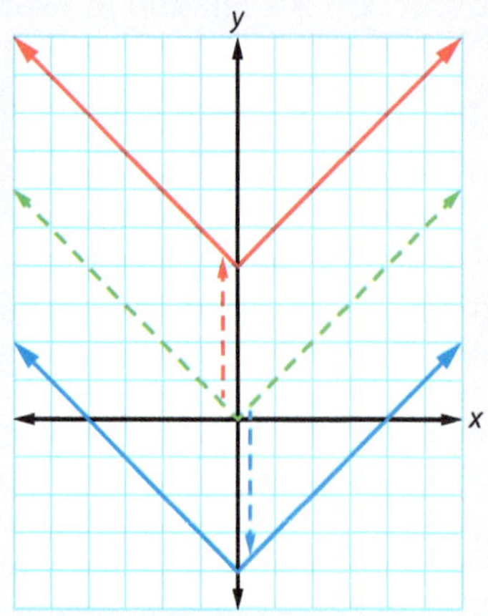	$(0, \pm k)$
$f(x) = \|x \pm h\|$	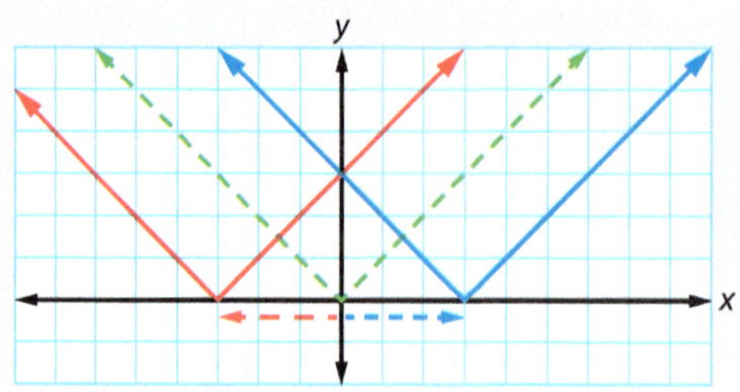	$(\mp h, 0)$
$f(x) = \|x - h\| + k$	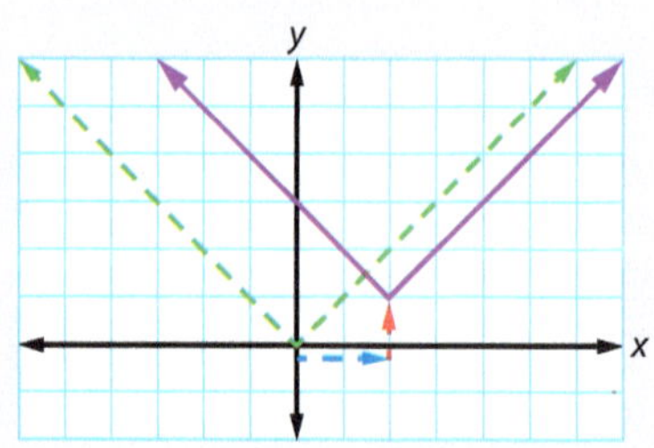	(h, k)
$f(x) = a\|x\|$	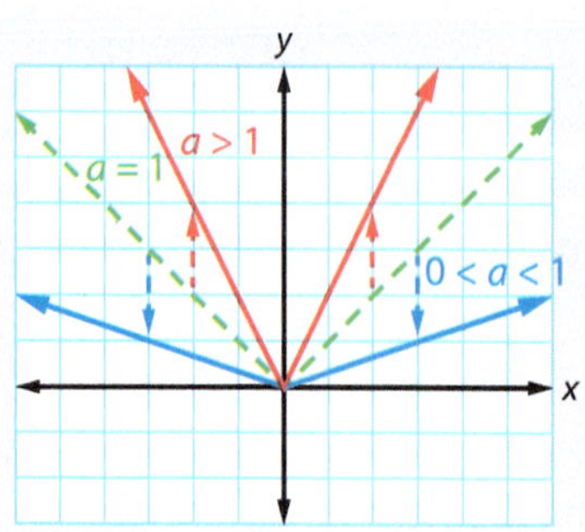	(0, 0)

Biblical Worldview Shaping

Design through Algebra

Why do we see design in the world?

God has created man with the ability to see, recognize, and describe complex patterns in creation that give testimony to His glory.

EXERCISES

1. State the domain and range of the following relation: $G = \{(-7, 5), (-3, 4), (-2, 3), (1, 7), (3, 8), (5, 6), (8, 5)\}$. **[4.1]**
2. Draw a mapping diagram of relation G. **[4.1]**
3. Is relation G a function? **[4.1]**

The following table relates the number of pages in a book to the cost of the book. [4.1]

pages	600	330	430	380	340	670
cost	$42.50	$18.50	$21.50	$20.00	$19.00	$49.00

4. Which quantity should be represented by the independent variable? Explain why.
5. Create a scatter plot of the data in the table.
6. Describe the relationship between the number of pages and the cost of the book.

State the domain and range of each relation and whether it is a function. [4.2]

7. 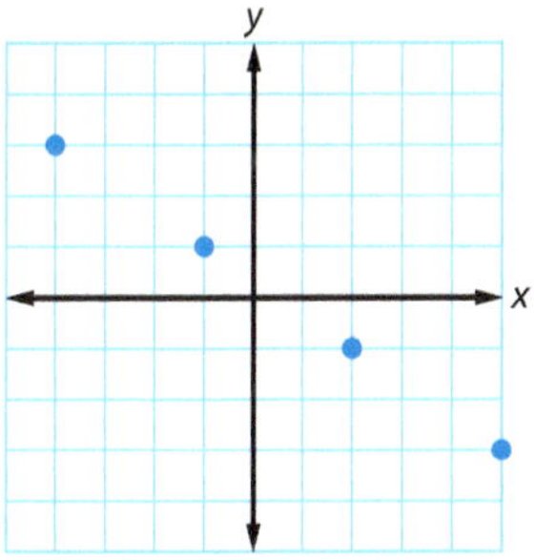

8. 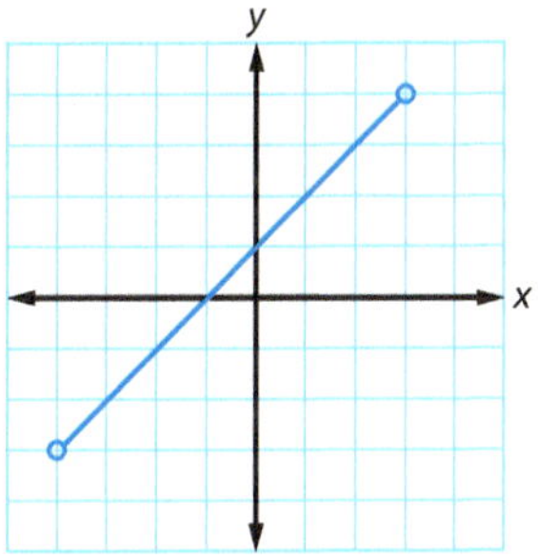

9. 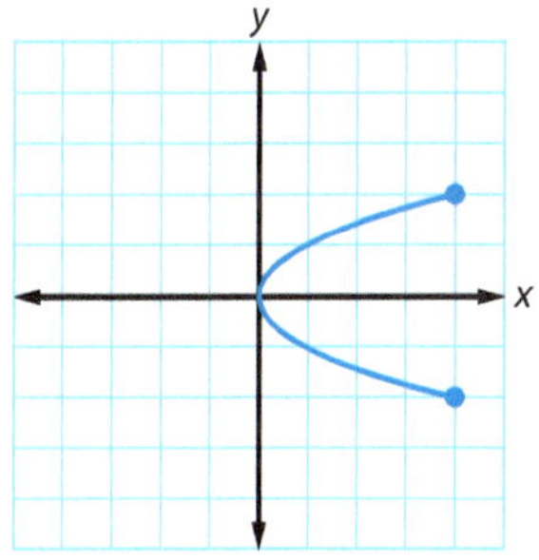

10. Make a table of at least 5 ordered pairs for relation $H = \{(x, y) \mid y = 3x - 5\}$. Is it a function?
11. Is the relation $\{(x, y) \mid y = 3\}$ a function?
12. When no domain is stated for a function, what is the assumed domain?

Graph each relation and state whether it is a function. [4.2]

13. $A = \{(x, y) \mid y = 2x - 4; x = 0, 2, 5\}$
14. $B = \{(x, y) \mid x = 2; 2 \le y \le 5\}$

Interpret each graph, explaining each labeled section. [4.3]

15.

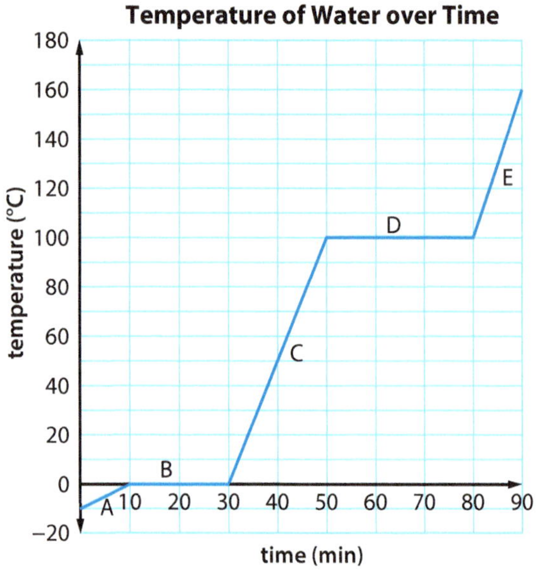

16.

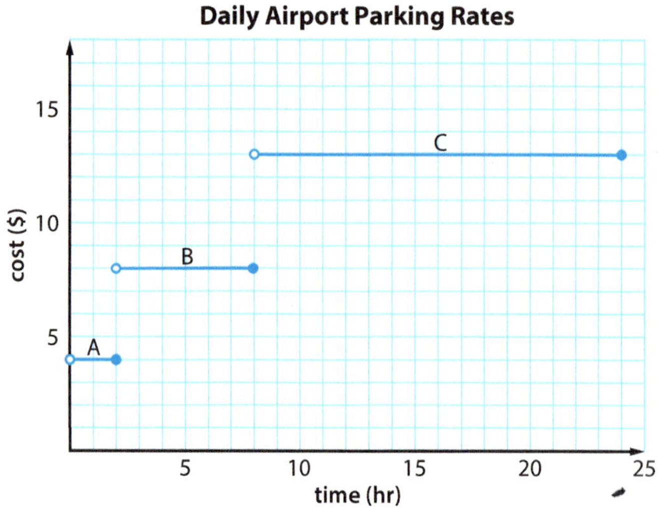

17. Sketch a graph of a man's heart rate over the course of an hour in which he mows the lawn with a push mower for 20 min, rests for 15 min, lifts weights for 10 min, and then reads for the rest of the hour. **[4.3]**

18. Illustrate the relationship between the amount of gas used and the number of miles traveled for Mr. Blake's car by graphing the data in the table and drawing a smooth curve that models the relation. **[4.3]**

miles traveled	gallons of gas used
22	1
44	2
88	4
154	7
330	15

19. Use your graph to estimate the amount of gas used if he travels 200 mi. **[4.3]**

20. How far can Mr. Blake travel using 40 gal of gas? **[4.3]**

21. If $f(x) = 2x - 3$, find: **[4.4]**

a. $f(0)$ **b.** $f(4)$ **c.** $f(63)$ **d.** Graph $f(x)$.

For exercises 22–24, $g(x) = x^2 - 3$; $D = \{-2, -1, 0, 1, 2\}$. [4.4]

22. Make a table to represent the function.

23. Write the function as a set of ordered pairs.

24. Graph $g(x)$.

Write the rule for each function. [4.4]

25.

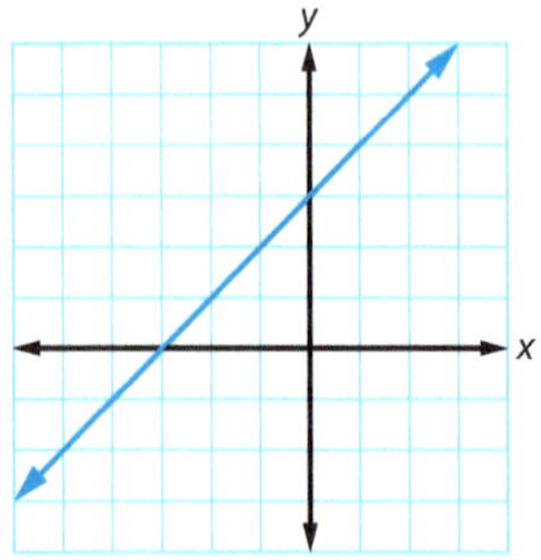

26. $\{(-2, 1), (0, 0), (2, -1), (4, -2), (6, -3)\}$

27.

x	−2	−1	0	1	2
$h(x)$	−8	−5	−2	1	4

28. a rental company charges a \$45 fee for a moving truck plus \$0.35 per mile

Graph the first 4 terms of each sequence. [4.5]

29. $a_1 = -8; a_n = a_{n-1} + 3$

30. $a_1 = 20; a_n = a_{n-1} - 8$

Find the specified formula for each of the following sequences. [4.5]

31. −6, 0, 12, 18, . . .; recursive

32. 30, 18, 6, −6, −18, . . .; recursive

33. −8, −2, 4, 10, 16, . . .; explicit

34. 19, 28, 37, 46, 55, . . .; explicit

Find the 50th and 89th term of each sequence. [4.5]

35. −16, −3, 10, 23, . . .

36. 2, 9, 16, 23, 30, . . .

37. In his first 4 days running his new business, Nate has netted profits of −\$82, −\$59, −\$36, and −\$13. Assuming an arithmetic progression of profits, after how many days can Nate expect to net at least \$100 in profit? **[4.5]**

State whether each table represents a direct variation, an inverse variation, or neither. For each variation, write the function rule. [4.6]

38.

x	2	4	6	8	12
y	9	18	27	36	54

39.

x	3	4	6	8	10
f(x)	12	9	6	4.5	3.6

40. If y varies directly with x and $y = 8$ when $x = 5$, find the constant of variation and write the function rule. **[4.6]**

41. If the amount of paint varies directly with the area to be painted and 2 gal covers 690 ft^2, how much paint (to the nearest tenth of a gallon) will it take to cover the remaining 970 ft^2? **[4.6]**

42. If the volume of a gas, V, varies inversely with pressure, P, and $V = 400$ mL when $P = 1.8$ atm, what pressure is needed to reduce the volume to 375 mL? **[4.6]**

43. Write the function $f(x) = |x|; D = \{-4, -2, 0, 2, 4\}$ as a set of ordered pairs. **[4.7]**

44. Draw a mapping diagram of the function $h(x) = |x| - 2;$ $D = \{-4, -2, 0, 2, 4\}$. **[4.7]**

45. Write a function rule for each translation of $f(x) = |x|$. **[4.7]**

a. down 8 **b.** left 5.3 **c.** up 11.5 and right 4

Graph each function. [4.7]

46. $f(x) = |x| + 1$

47. $f(x) = |x + 2|; D = \{-2, -1, 0, 1, 2\}$

48. $f(x) = |x + 3| + 2.5$

49. $f(x) = 3|x|$

50. $f(x) = \frac{2}{5}|x|$

51. Algebraic functions help us to recognize and describe complex patterns that occur in nature and should result in glory to

A Mother Nature. **B** man. **C** God. **D** mathematics.

CHAPTER 05
LINEAR FUNCTIONS

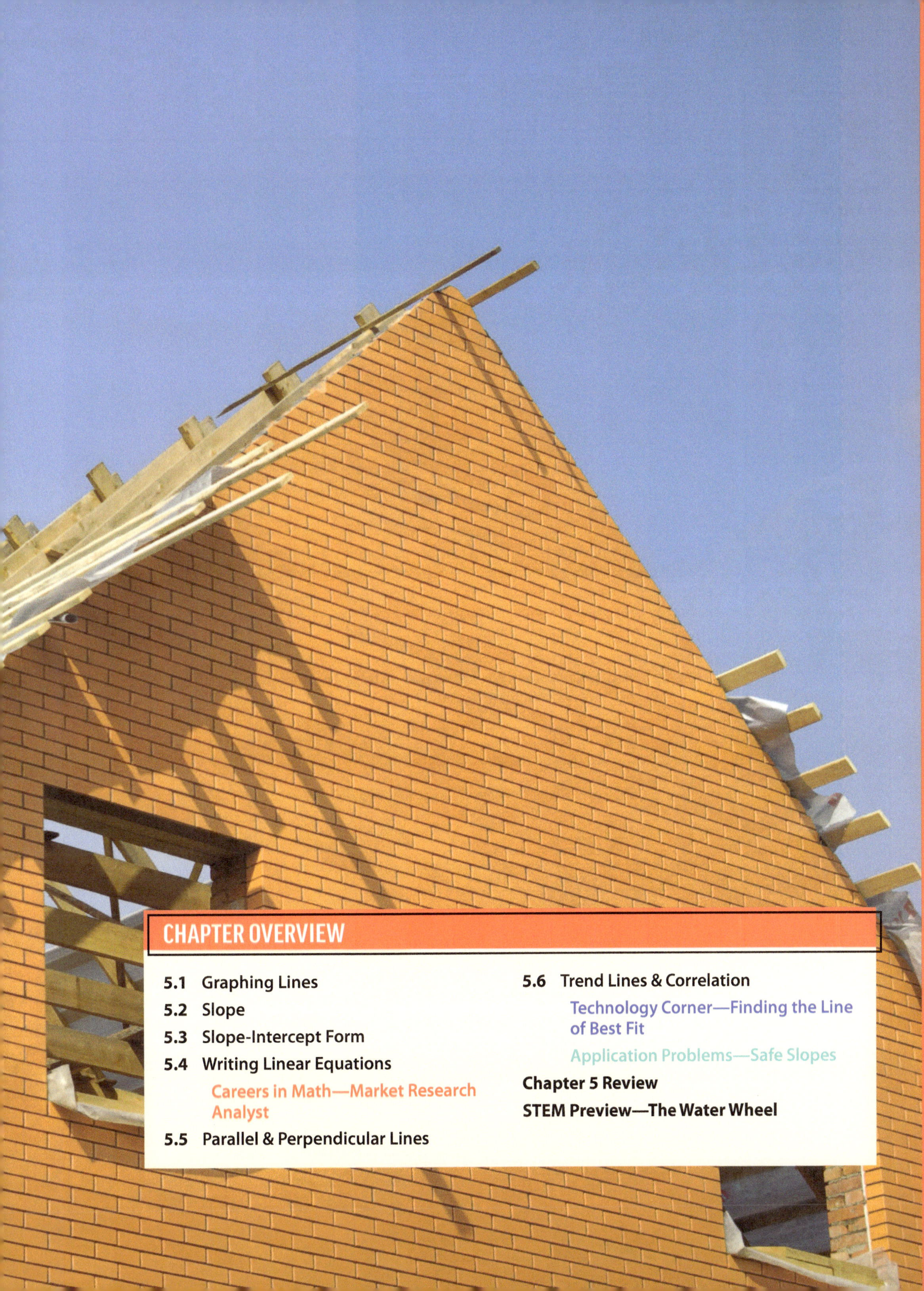

CHAPTER OVERVIEW

5.1 GRAPHING LINES

How can I determine the correct location of the graph of a linear function?

After completing this section, I will be able to

- graph linear equations on the coordinate plane by using ordered pairs.
- convert equations of lines between standard form and function form.
- graph a linear equation by using the x- and y-intercepts.

The graph of a relation or function can be sketched by graphing enough ordered pairs to indicate the shape of the graph and then connecting the points. This is the method used by graphing calculators. By recognizing key characteristics of the function, the shape and location of the graph can be quickly determined.

Functions such as $f(x) = 2x + 7$ can also be written as $y = 2x + 7$. They are called **linear functions** or **linear equations** because all the points that satisfy them form a straight line.

EXAMPLE 1: Using Ordered Pairs

Graph $\{(x, y) \mid y = x + 5\}$.

Answer

x	$x + 5$	(x, y)
−2	$-2 + 5 = 3$	(−2, 3)
−1	$-1 + 5 = 4$	(−1, 4)
0	$0 + 5 = 5$	(0, 5)
1	$1 + 5 = 6$	(1, 6)
2	$2 + 5 = 7$	(2, 7)

1. Make a table of ordered pairs using convenient values for x.

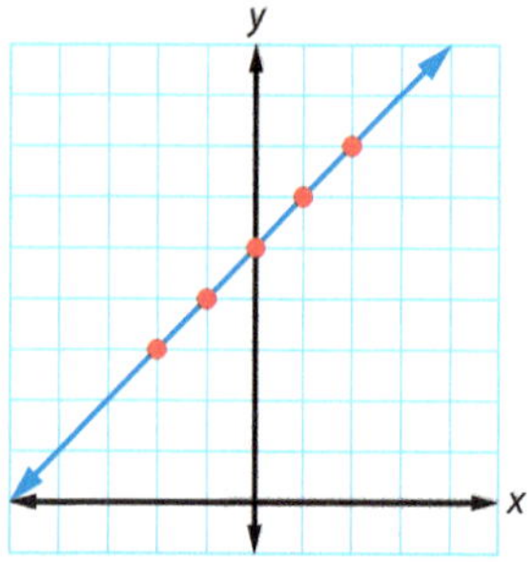

2. Plot the points. Notice that the points are on a straight line.
3. Connect the points with a straight line.

SKILL CHECK **EXERCISE 1**

The **standard form** of a linear equation is $Ax + By = C$, where A, B, and C are real numbers. The equation in Example 1 can be converted to standard form by subtracting x from both sides: $-x + y = 5$. Equivalent standard form equations can be obtained by multiplying both sides by the same number, such as $x - y = -5$ and $2x - 2y = -10$. It is also common for A to be nonnegative.

If a linear equation in standard form contains fractional or decimal coefficients, multiply each term by a common denominator or power of 10 to change A, B, and C to integers.

EXAMPLE 2: Writing an Equation in Standard Form

Write $f(x) = \frac{5}{7}x + 2$ in standard form.

Answer

$y = \frac{5}{7}x + 2$ — 1. Write the equation, replacing $f(x)$ with y.

$-\frac{5}{7}x + y = 2$ — 2. Subtract $\frac{5}{7}x$ from both sides.

$5x - 7y = -14$ — 3. Multiply each term by −7.

SKILL CHECK **EXERCISE 11**

EXAMPLE 3: Graphing an Equation Written in Standard Form

Graph $x + 3y = 4$.

Answer

$$x + 3y = 4$$
$$3y = -x + 4$$
$$y = \frac{-x + 4}{3}$$

1. Solve the equation for y by subtracting x from both sides and then dividing both sides by 3.

x	$\frac{-x+4}{3}$	(x, y)
0	$\frac{0+4}{3} = \frac{4}{3}$	$\left(0, \frac{4}{3}\right)$
1	$\frac{-1+4}{3} = \frac{3}{3} = 1$	$(1, 1)$
4	$\frac{-4+4}{3} = \frac{0}{3} = 0$	$(4, 0)$

2. Make a table of ordered pairs using convenient values for x.

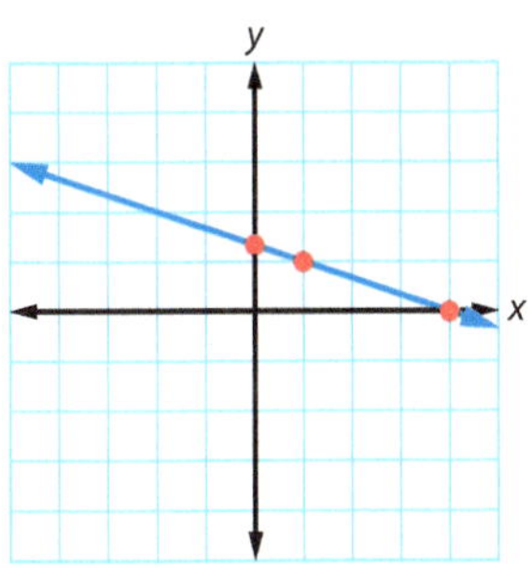

3. Plot the points.
4. Connect the points with a straight line.

SKILL CHECK **EXERCISE 29**

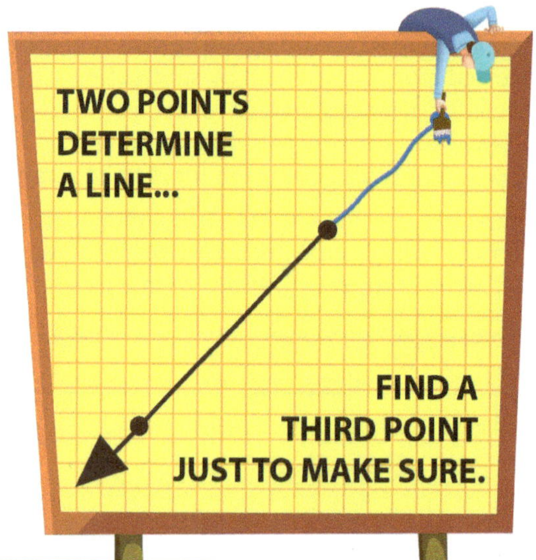

GRAPHING A LINEAR EQUATION

1. Solve the linear equation for y.
2. Make a table of at least 3 ordered pairs using convenient values for x.
3. Graph the ordered pairs on a Cartesian plane.
4. Connect the points with a straight line.

The points on a graph where a line intersects the axes are called **intercepts**. The graphed line from Example 3 indicates the y-intercept is $\left(0, \frac{4}{3}\right)$, and the x-intercept is $(4, 0)$.

DEFINITIONS

The ***y*-intercept** is the point where the line intersects the y-axis.

The ***x*-intercept** is the point where the line intersects the x-axis.

AFTERSCHOOLHELP

Graphing Lines

Each intercept can be determined without graphing by using the equation of the line. Since every point on the y-axis has an x-coordinate of 0, the y-intercept can be determined by substituting 0 for x and solving for the y-coordinate. The y-intercept is commonly expressed as an ordered pair $(0, y)$. Similarly, the x-intercept can be determined by substituting 0 for y in the equation and solving for the x-coordinate. The x-intercept is also commonly expressed as an ordered pair $(x, 0)$.

EXAMPLE 4: Graphing Equations by Using Intercepts

Use the x- and y-intercepts to graph $2x + y = 6$.

Answer

1. Identify the intercepts.

   a. Substitute 0 for y and solve for x.

$$2x + y = 6$$
$$2x + 0 = 6$$
$$2x = 6$$
$$x = 3;\ x\text{-intercept is } (3, 0).$$

   b. Substitute 0 for x and solve for y.

$$2x + y = 6$$
$$2(0) + y = 6$$
$$y = 6;\ y\text{-intercept is } (0, 6).$$

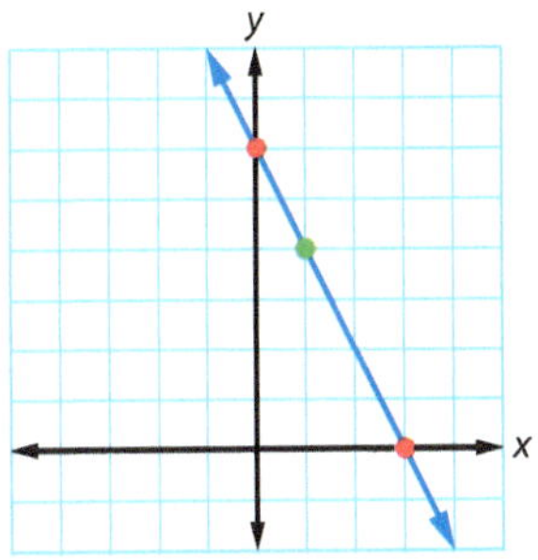

2. Plot the intercepts.
3. Draw the line through the points.
4. Choose a convenient point on the line, (1, 4), and check that it is a solution.

SKILL CHECK **EXERCISE 17**

The equations $y = c$ and $x = c$ are special cases of the standard form $Ax + By = C$ in which either $A = 0$ or $B = 0$. These equations form graphs that are horizontal or vertical lines.

EXAMPLE 5: Graphing Special Cases

Graph each equation; then determine whether it is a function.

a. $y = -2$ **b.** $x = 3$

Answers

a.

x	y
−3	−2
0	−2
4	−2

1. $y = -2$ regardless of the value chosen for x.

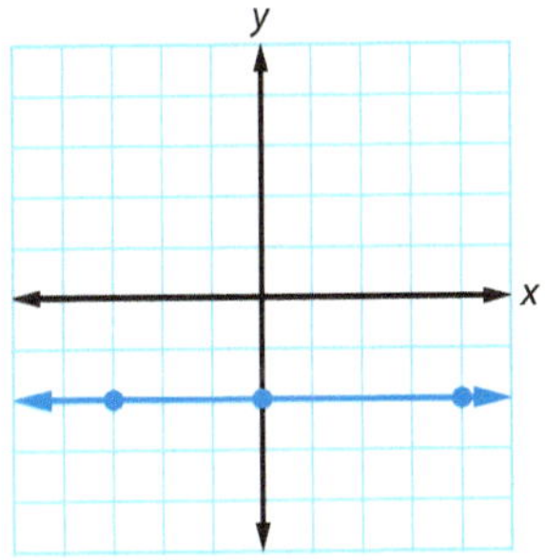

2. Plot the points and draw the line.

$y = -2$ is a function.

3. The graph passes the vertical line test.

b.

x	y
3	−2
3	0
3	4

1. $x = 3$ regardless of the value chosen for y.

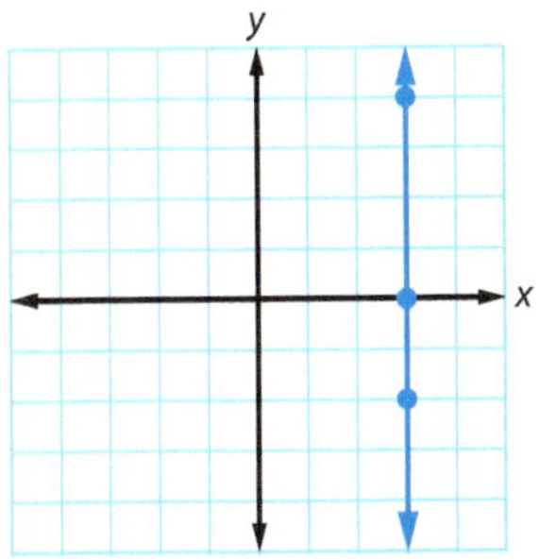

2. Plot the points and draw the line.

$x = 3$ is not a function.

3. The graph fails the vertical line test.

SKILL CHECK **EXERCISE 23**

VOCABULARY

- intercepts
- linear equation
- linear function
- standard form (of a linear equation)
- x-intercept
- y-intercept

A. EXERCISES

Make a table of at least 3 ordered pairs; then graph each linear equation.

1. $y = x - 6$
2. $y = -x + 8$
3. $y = -2x + 5$
4. $y = -4x + 1$
5. $y = 4$
6. $x = -1$
7. $y = \frac{2}{3}x - 2$
8. $y = \frac{x - 3}{2}$

Write each linear equation in standard form.

9. $y = -x + 9$
10. $3x = 18 + y$
11. $y = \frac{4x - 2}{7}$
12. $y = \frac{5}{8}x - 3$

Solve each linear equation for *y*.

13. $5x + y = 20$
14. $3x - y = 8$
15. $8x - 2y = 14$
16. $x + 4y = -12$

B. EXERCISES

Find the *x*- and *y*-intercepts and then graph each equation.

17. $x - 2y = 4$
18. $x - 3y = -3$
19. $y = -x + 4$
20. $x = 3y + 6$
21. $x = 2y - 3$
22. $3x + y = 1$
23. $x = 2$
24. $y = -3$

Graph each linear equation.

25. $y = 2x + 3$
26. $4x + 2y = 8$
27. $y = -\frac{1}{2}x$
28. $x = 3y$
29. $x - y = 8$
30. $y = x - 9$
31. $y = 8x + 12$
32. $7x + 2y = 5$

33. What is the x-coordinate of every y-intercept?

34. What is the y-coordinate of every x-intercept?

35. Is the graph of $y = c$ a horizontal or a vertical line?

36. Which axis does the graph of $x = c$ intersect? Name the point of intersection.

37. **Essential Question:** How do you know where to draw the graph of a linear function?

C. EXERCISES

38. Find general expressions for the x- and y-intercepts of the linear equation $Ax + By = C$.

39. Use the general expressions developed in exercise 38 to find the x- and y-intercepts of $3x + 2y = -12$. Then check your results.

40. Can a function ever have 2 x-intercepts? 2 y-intercepts? Explain.

41. Graph the following equations on the same set of axes. What do the equations have in common? What do the graphs have in common?
$y = x + 1; y = 2x + 1; y = -2x + 1; y = -\frac{1}{2}x + 1$

42. Graph the following equations on the same set of axes. What do the equations have in common? What do the graphs have in common?
$y = 2x; y = 2x - 4; y = 2x + 3; y = 2x + 8$

CUMULATIVE REVIEW

Solve each literal equation for the indicated variable. [2.3]

43. $S = b^2 + 2bl$ for l

44. $C = \frac{5}{9}(F - 32)$ for F

Use proportions to solve. [2.4]

45. A triangle has sides of 3 cm, 5 cm, and 6 cm. A similar triangle's shortest side is 4.5 in. Find the lengths of the other 2 sides.

46. Two similar figures have corresponding sides of 26 mm and 65 mm. What is the ratio of their sides?

Solve each absolute value inequality. [3.6]

47. $|2a + 7| < 3$

48. $|4b - 7| > 11$

49. $|6c - 14| \geq 10$

50. $|15d - 8| \leq 13$

51. What is the function rule if z varies directly with y and $z = 18$ when $y = 3$? **[4.6]**

A $y = \frac{54}{z}$

B $y = 6z$

C $z = \frac{54}{y}$

D $z = 6y$

E none of these

52. Which function rule represents a translation of the graph of $f(x) = |x|$ 2 units up and 8 units right? **[4.7]**

A $g(x) = |x + 8| + 2$

B $g(x) = |x - 8| + 2$

C $g(x) = |x + 2| + 8$

D $g(x) = |x - 2| + 8$

E none of these

MIND OVER MATH

To the right is a 3 × 3 magic square. Use the numbers 1 through 9 to fill in the square so that the sum of each row, column, and diagonal is equal to 15.

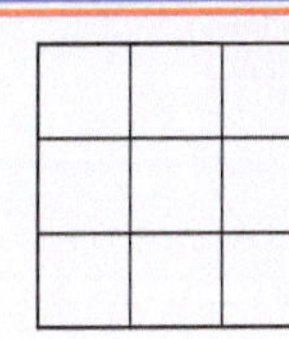

You can make a magic square for any odd number by following the instructions given below. The sum s of each row, column, and diagonal can be determined by using the formula $s = \frac{n^3 + n}{2}$, where n is the length of one side of the square.

Try a 5 × 5 square. For the following instructions think of the 5 × 5 square as a grid that is made up of 25 small squares.

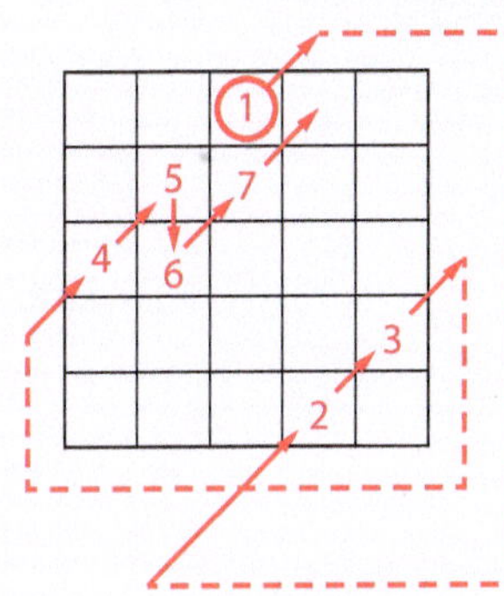

1. Place a 1 in the center square of the top row.
2. Move up diagonally to the right. Since you have moved out of the top of the grid, put 2 in the square at the bottom of that column.
3. Move up diagonally to the right from 2 and put 3 in the square.
4. Move up diagonally to the right again. Since you have moved out of the right side of the grid, put 4 in the square at the left end of that row.
5. Move up diagonally to the right and put 5 in the square.
6. Since you cannot move up diagonally to the right from 5, put 6 in the square below 5. Then proceed with the steps as given above.

Can you finish the magic square (grid)? What is the sum of each row, column, and diagonal? Can you make a 7 × 7 magic square? What is the sum of each row, column, and diagonal of a 7 × 7 magic square? What is the sum of the 4 corners and the center square?

5.2 SLOPE

How can math be applied to slopes to help others?

After completing this section, I will be able to

- determine the slope of a line from a graph.
- find the slope of a line passing through 2 given points.
- apply slope in a real-world context.

The ideal slope for a boat ramp is between 12% and 15% grade.

Slopes are a major safety design consideration for many things in our world, such as stairways, ski slopes, handicap ramps, and road grades.

In mathematics the *slope* of a line describes the incline or steepness of the line.

DEFINITION

The **slope** of a line, m, is the ratio of the rise (vertical change) to the corresponding run (horizontal change).

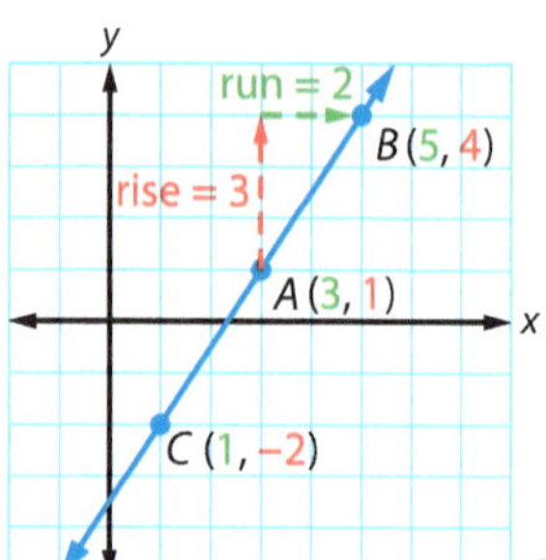

Going from point A to point B on the graph, the line rises up 3 as it runs to the right 2. Therefore, the line's slope is $\frac{3}{2}$ or 1.5. This ratio is the same for any 2 points on the line. To move from B to C, you must go 6 units down and 4 units left. This ratio is $\frac{-6}{-4} = \frac{3}{2}$ or 1.5. Notice that the slope is the vertical change of the line for each unit of horizontal change from left to right.

The coordinates of points can also be used to find the rise and the run. Using A (3, 1) and B (5, 4),

$$m = \frac{\text{rise}}{\text{run}} = \frac{4-1}{5-3} = \frac{3}{2}.$$

EXAMPLE 1: Finding the Slope from a Graph

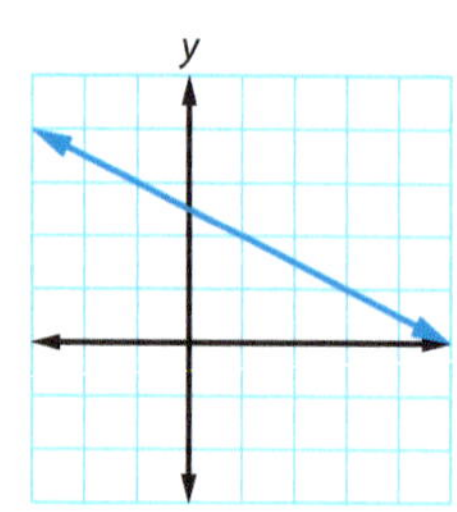

Find the slope of the line.

Answer

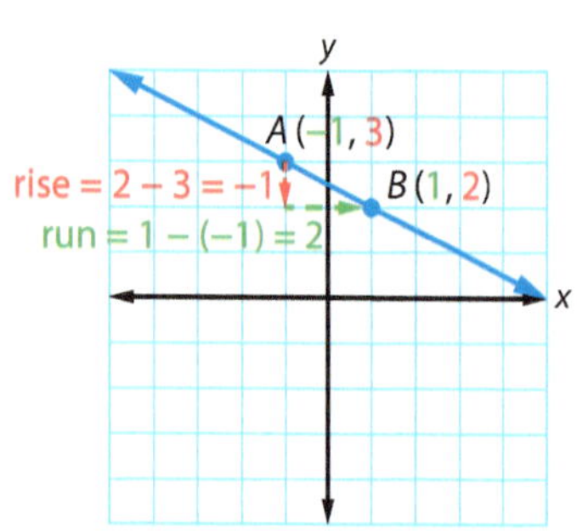

1. Choose 2 convenient points: A (−1, 3) and B (1, 2).
2. Find the rise (change in y). Moving from A to B, you must go 1 unit down.
3. Move 2 units right to point B to find the run (change in x).
4. Find the slope.

$$m = \frac{\text{rise}}{\text{run}} = \frac{-1}{2} = -\frac{1}{2}$$

SKILL CHECK **EXERCISE 7** ☑

If you had started with point B and moved to A, you would have gone 1 unit up and 2 units left, but the slope would be the same.

$$m = \frac{\text{rise}}{\text{run}} = \frac{1}{-2} = -\frac{1}{2}$$

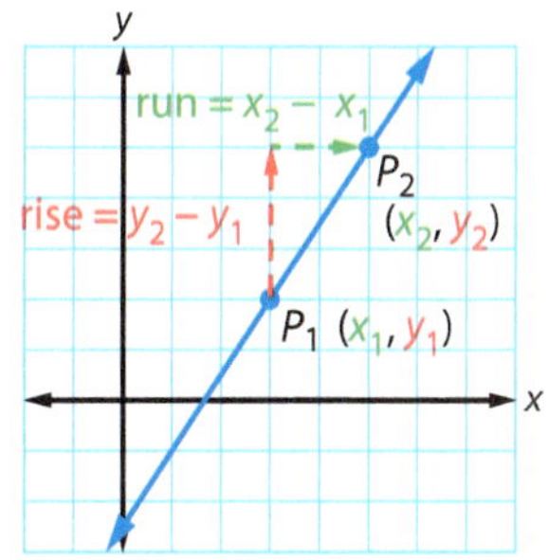

A more generalized example using 2 arbitrary points, P_1 (x_1, y_1) and P_2 (x_2, y_2), gives the general formula for finding the slope of a line through 2 points.

SLOPE FORMULA

$$m = \frac{\text{rise}}{\text{run}} = \frac{y_2 - y_1}{x_2 - x_1}$$

The slope formula can be used to find the slope of a line without its graph if you know the coordinates of any 2 points on the line.

EXAMPLE 2: Finding the Slope between 2 Points

Find the slope of the line passing through (2, –3) and (4, 1).

Answer

P_1 $(x_1, y_1) = (2, -3)$
P_2 $(x_2, y_2) = (4, 1)$

1. Label the points.

$$m = \frac{y_2 - y_1}{x_2 - x_1} = \frac{1 - (-3)}{4 - 2} = \frac{4}{2} = 2$$

2. Substitute values into the slope formula and evaluate.

SKILL CHECK **EXERCISE 13**

When using the slope formula, be sure that the order of subtraction is the same in the numerator and the denominator. In Example 2, choosing (4, 1) to be P_1 and (2, –3) to be P_2 produces the same slope:

$$m = \frac{y_2 - y_1}{x_2 - x_1} = \frac{-3 - 1}{2 - 4} = \frac{-4}{-2} = 2.$$

The slope of this line is 2, but writing it as $\frac{2}{1}$ illustrates that the rise is 2 and the run is 1.

FINDING THE SLOPE OF A LINE

When given the graph, determine the rise and run from one point on the line to another point on the line. Then write the ratio of rise over run and reduce if possible.

When given 2 points, label the points $P_1\ (x_1, y_1)$ and $P_2\ (x_2, y_2)$. Then use the slope formula, $m = \frac{y_2 - y_1}{x_2 - x_1}$.

EXAMPLE 3: Finding the Slope of a Horizontal Line

Find the slope of the line $y = 4$.

Answer

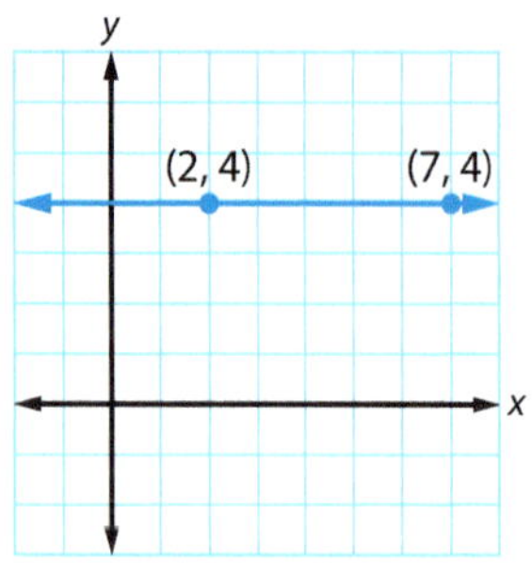

1. Graph the horizontal line $y = 4$.
2. Choose 2 convenient points: (2, 4) and (7, 4).
3. The rise (change in y) is 0.
4. The run (change in x) is 5.

$m = \frac{\text{rise}}{\text{run}} = \frac{0}{5} = 0$

5. Find the slope.

$m = \frac{y_2 - y_1}{x_2 - x_1} = \frac{4 - 4}{7 - 2} = \frac{0}{5} = 0$

6. Check with the slope formula.

SKILL CHECK **EXERCISE 15**

The slope of any horizontal line is always 0 because there is no vertical change.

EXAMPLE 4: Finding the Slope of a Vertical Line

Find the slope of the line $x = 3$.

Answer

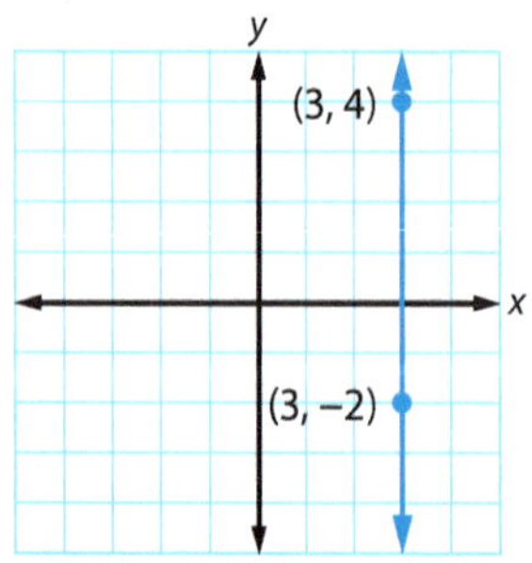

1. Graph the vertical line $x = 3$.
2. Choose 2 convenient points: (3, −2) and (3, 4).
3. The rise (change in y) is 6.
4. The run (change in x) is 0.

$m = \frac{\text{rise}}{\text{run}} = \frac{6}{0}$

5. Find the slope.

$m = \frac{y_2 - y_1}{x_2 - x_1} = \frac{4 - (-2)}{3 - 3} = \frac{6}{0}$

The slope is undefined.

6. Check with the slope formula.

SKILL CHECK **EXERCISE 11**

The slope of any vertical line is always undefined because the denominator of the ratio will always be 0.

EXAMPLE 5: Graphing a Line Using the Slope

Graph the line passing through the point (–1, 2) with a slope of $-\frac{1}{3}$.

Answer

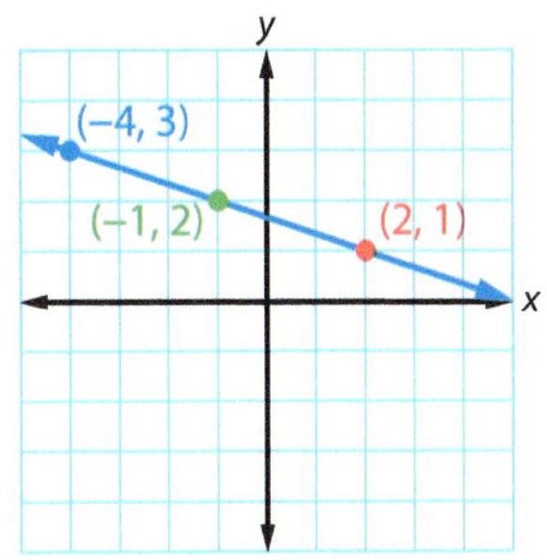

1. Plot the point (–1, 2).
2. Using the slope, $m = -\frac{1}{3} = \frac{-1}{3}$, plot another point on the line. From (–1, 2), move 1 unit down and then 3 units right to plot the point (2, 1).
3. Use an equivalent form of the slope, $m = \frac{1}{-3}$, to plot a third point. From (–1, 2), move 1 unit up and then 3 units left to plot the point (–4, 3).
4. Draw a line through the points.

SKILL CHECK **EXERCISE 25**

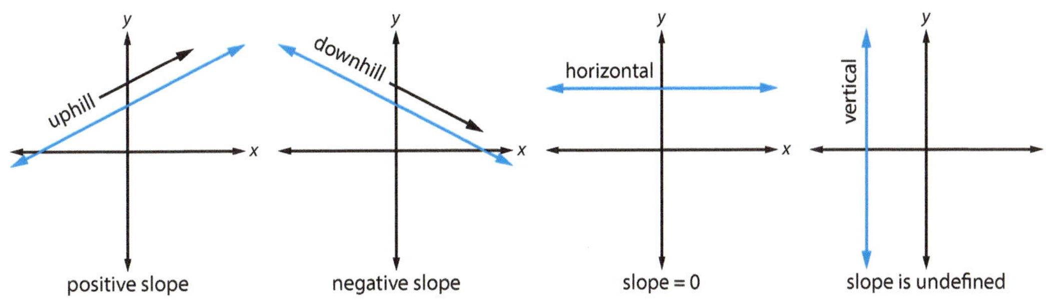

KEYWORD SEARCH

slope line applet

AFTERSCHOOLHELP

Slope

The sign of the slope indicates whether the line slants up or down when moving from left to right. The absolute value of the slope indicates the "steepness" of the line. Slope has been defined as $\frac{\text{rise}}{\text{run}}$ in order that steeper lines would have slopes with larger absolute values.

There are several different notations for slope:

$$m = \frac{\text{rise}}{\text{run}} = \frac{y_2 - y_1}{x_2 - x_1} = \frac{\Delta y}{\Delta x} = \frac{\text{change in } y}{\text{change in } x}.$$

The Greek letter delta, Δ, is used to symbolize the change in a quantity. Using this notation emphasizes that slope measures a rate of change. Rates of change compare 2 changing quantities. For instance, the slope of a distance-time graph indicates the change in distance per unit of time, or the speed of the object.

EXAMPLE 6: Applying the Slope

Determine the speed of the train for each section of the graph.

a. 0–2 hr **b.** 2–4 hr **c.** 4–7 hr

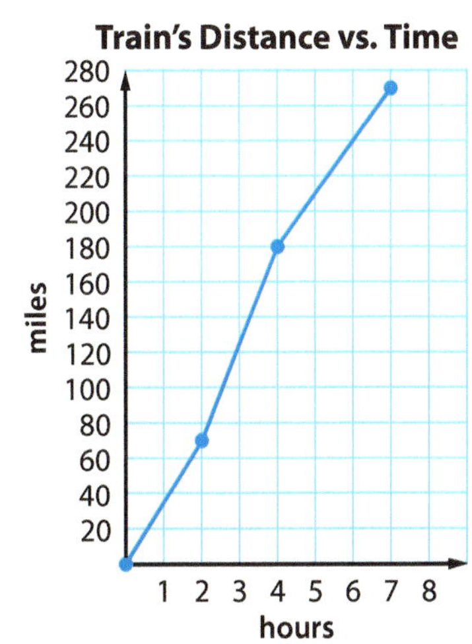

Answers

a. $m = \frac{70 - 0}{2 - 0} = \frac{70}{2} = 35$ mi/hr

b. $m = \frac{180 - 70}{4 - 2} = \frac{110}{2} = 55$ mi/hr

c. $m = \frac{270 - 180}{7 - 4} = \frac{90}{3} = 30$ mi/hr

SKILL CHECK **EXERCISE 29**

VOCABULARY

- slope

A. EXERCISES

Use the following graphs for exercises 1–6.

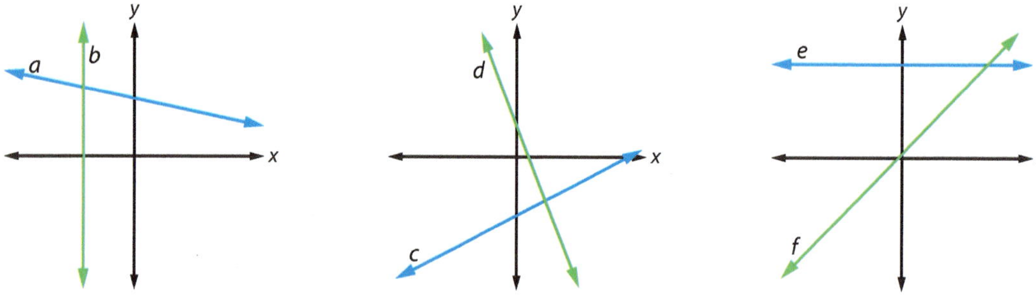

1. Which line(s) have a negative slope?
2. Which line(s) have an undefined slope?
3. Which line(s) have a slope of 0?
4. Which line(s) have a positive slope?
5. Which line has the slope with the greatest absolute value?
6. Which line has the slope with the least nonzero absolute value?

Find the slope of each line.

7. 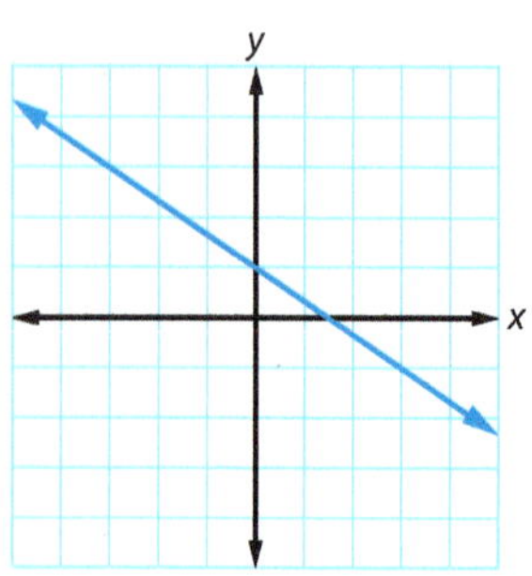

8. 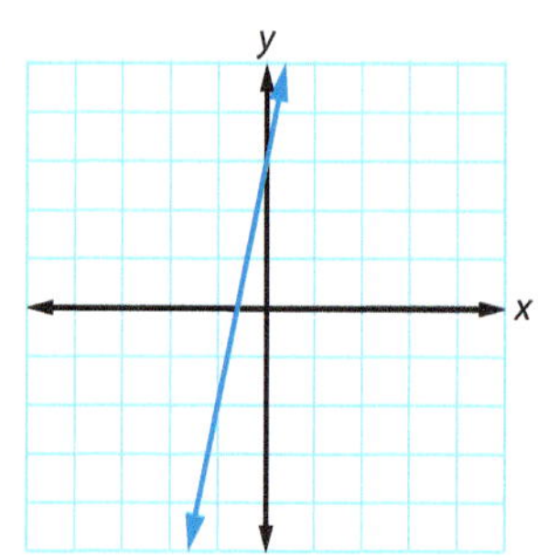

9. 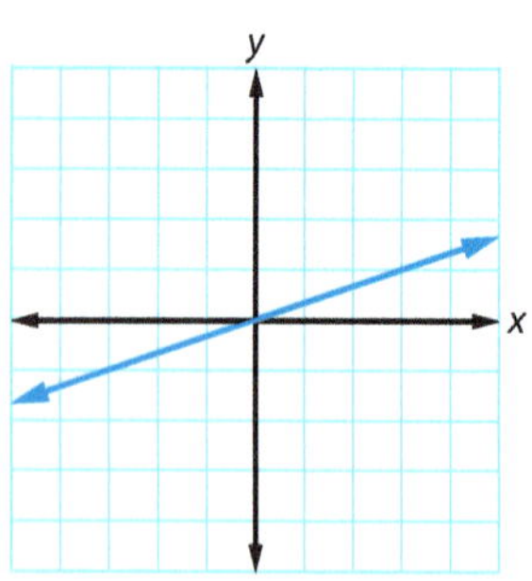

10. 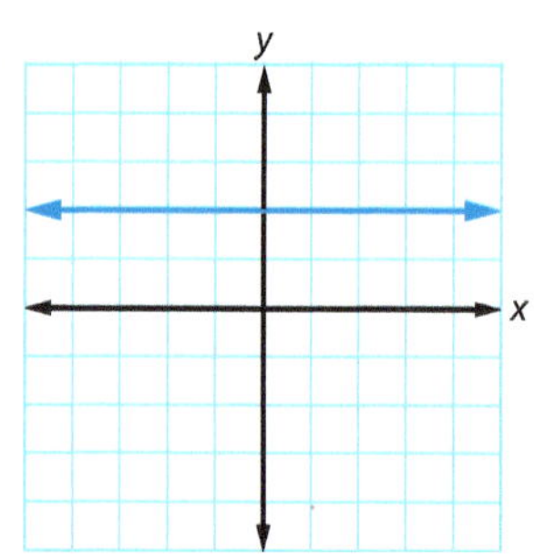

11. 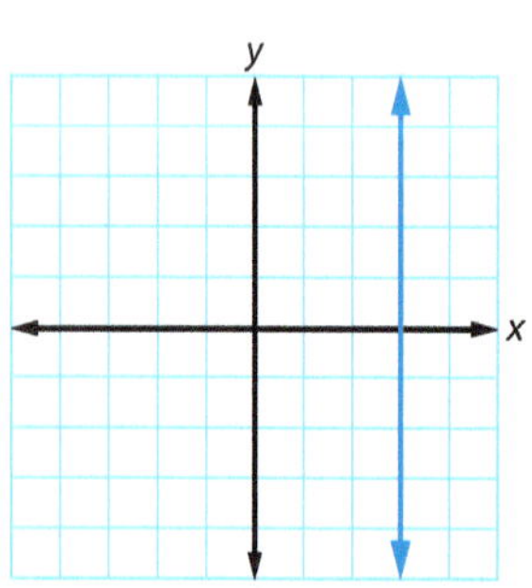

12. 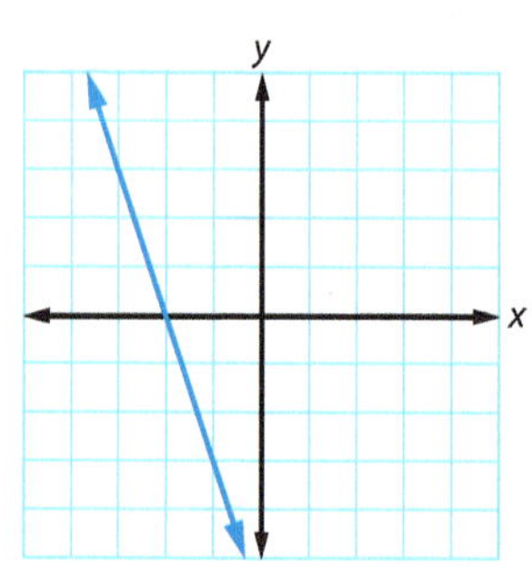

Find the slope of the line passing through the given points.

13. (2, 5), (3, 9)
14. (3, 6), (1, 9)
15. (2, 7), (3, 7)
16. (2, 5), (2, 3)

B. EXERCISES

Find the slope of the line passing through the given points.

17. $(3, 6), (-4, -1)$

18. $(-4, -6), (-3, -2)$

19. $(5, 7), (1, -3)$

20. $(-2, -5), (1, 0)$

21. $(-1, 7), (-1, 2)$

22. $(-4, -1), (-1, 7)$

23. $(4.75, -2.5), (7.25, 4.5)$

24. $\left(\frac{2}{3}, -\frac{3}{2}\right), \left(\frac{7}{3}, -4\right)$

Graph the line passing through the given point with the given slope.

25. $(3, 1), m = -\frac{1}{4}$

26. $(-4, 0), m = 6$

27. $(2, -5), m = \frac{3}{2}$

28. $(0, 1), m = -\frac{5}{3}$

29. Use the distance-time graph to determine the speed of the bicyclist for each section of the graph.

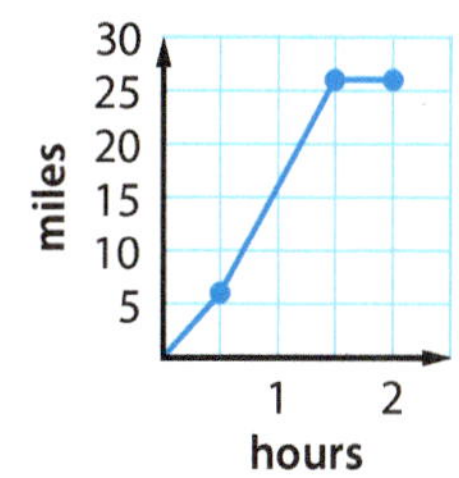

a. from 0 to 0.5 hr

b. from 0.5 to 1.5 hr

c. from 1.5 to 2 hr

d. Describe the motion of the bicyclist from 1.5 to 2 hr.

30. **MP2** The slope of a speed-time graph indicates the change in speed per unit of time, or the *acceleration* of the object. Determine the acceleration (in m/sec^2) of the car for each section of the graph.

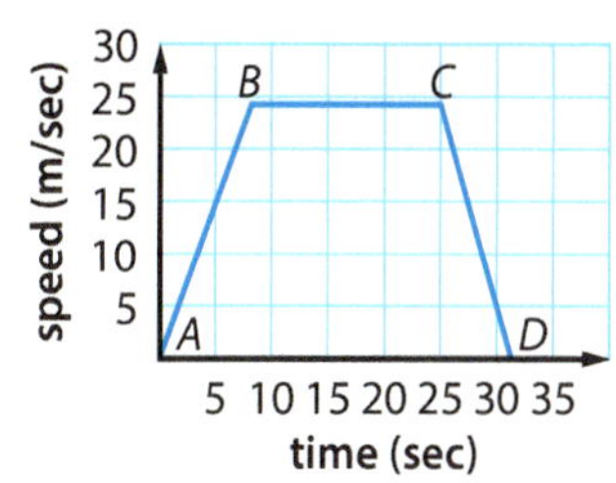

a. from A to B

b. from B to C

c. from C to D

d. Describe the motion of the car from B to C.

31. Pediatricians often track a patient's height. Determine the growth rate (in cm/yr) of a typical girl within each time period.

age (yr)	height (cm)
0	50
2	86
8	134
13	159
18	164

a. 0–2 yr

b. 2–8 yr

c. 8–13 yr

d. 13–18 yr

32. The price of gasoline often fluctuates dramatically. Determine the rate of change (in \$/month) of the cost of a gallon of gasoline for each time period.

a. from \$2.55/gal in January 2020 to \$2.33/gal in January 2021

b. from \$2.33/gal in January 2021 to \$3.14/gal in July 2021

c. from \$3.14/gal in July 2021 to \$3.32/gal in January 2022

Mr. Bateau is overseeing the construction of a new boat ramp at the local marina. The slope of the ramp must be between 12% and 15% grade.

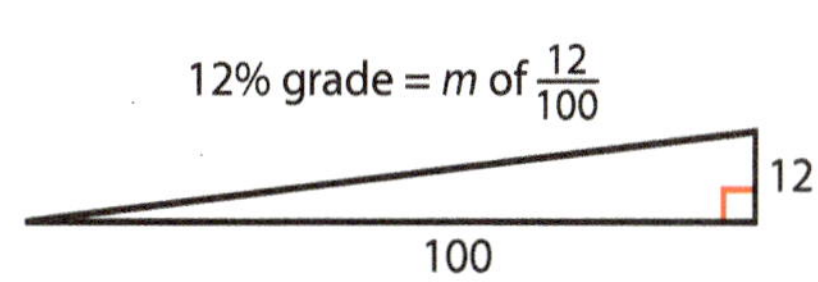

33. Write the percentage grades as fractions in lowest terms.

34. If Mr. Bateau uses the gentlest slope and ensures the bottom of the ramp is 4 ft below the surface of the water, what will the horizontal run of the ramp be?

35. If Mr. Bateau uses the steepest slope and allots 25 ft of horizontal run, at what depth below the surface of the water will the bottom of the ramp be?

36. Ramp Arts is building a wheelchair ramp that needs to rise to a height of 2.5 ft. For safety, the slope must be $\frac{1}{12}$. What is the horizontal run of the ramp?

37. MP1 Determine the rise of a ramp in inches if the ramp has a horizontal run of 18 ft and a slope of $\frac{1}{8}$.

38. **Essential Question:** Measuring the steepness of slopes in real-world contexts contributes to safety and can be an important part of our stewardship. If you know the required slope of a ramp as well as its height (rise), how can you determine the required horizontal length (run)?

C. EXERCISES

The slope of a steep road is often given as a percent to warn motorists. In exercises 39–41 assume that distances measured in miles represent horizontal distances.

39. MP6 If a sign states "6% downgrade next 3 miles," how much elevation will be lost (to the nearest tenth of a foot)?

40. Express the slope of a road that descends 800 ft over 2 mi as a fraction. Then express this downgrade as a percent (to the nearest tenth).

41. MP1 An engineer is planning a road from Barren Pass, elevation 6030 ft, to Fern Creek, elevation 4502 ft. How long (to the nearest tenth of a mile) must the road be to avoid grades over 5%?

CUMULATIVE REVIEW

Write an equation and solve. [2.6, 2.8]

42. Find the sale price of a \$70 item reduced by 40%.

43. Find the sale price of a \$250 item reduced by 30% and then by another 60%.

44. How many pounds of nuts selling for \$4/lb should be mixed with 20 lb of nuts selling for \$8/lb to make a mix selling for \$7/lb?

45. How much pure antifreeze should be mixed with 5 qt of a 30% antifreeze solution to obtain a solution that is 40% antifreeze?

46. How much water must be added to 5 qt of a 45% antifreeze solution to dilute it to a solution that is 35% antifreeze?

47. Express $0.\overline{9}$ as a simplified quotient of integers. **[2.2]**

Solve each equation. [3.5]

48. $|2x - 5| = 10$

49. $|5x + 8| = 12$

50. What is the range of the graphed relation? Is it a function? **[4.2]**

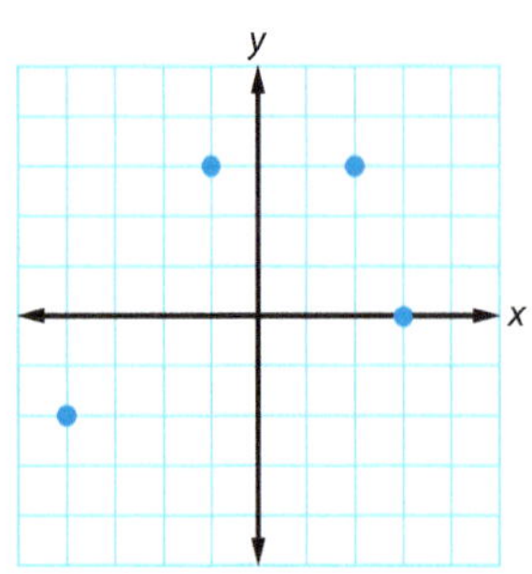

A $\{-4, -1, 2, 3\}$; no

B $\{-4, -1, 2, 3\}$; yes

C $\{-2, 0, 3\}$; no

D $\{-2, 0, 3\}$; yes

E none of these

51. Tori works twice as many hours as Kyle. Which equation expresses the number of hours Kyle works as a function of t hr worked by Tori? **[4.4]**

A $k(t) = t + 2$

B $k(t) = t - 2$

C $k(t) = 2t$

D $k(t) = \frac{t}{2}$

E none of these

5.3 SLOPE-INTERCEPT FORM

What makes the slope-intercept form of an equation useful?

After completing this section, I will be able to

- identify the slope and y-intercept of a linear equation.
- model linear equations in slope-intercept form.
- model real-world situations by using linear equations in slope-intercept form.

Linear functions can be used to determine fuel efficiency.

Equations of lines can be written in different forms by applying the properties of equality. As you saw in Section 5.1, equations can be written in standard form, such as $2x + y = 6$. This equation can also be written as $y = -2x + 6$ or $f(x) = -2x + 6$.

This form of the equation of a line is more commonly known as **slope-intercept form** or **function form**. The slope-intercept form of a linear equation is $y = mx + b$, where m is the slope of the line and b is the y-coordinate of the y-intercept. By substituting $x = 0$ and simplifying to $y = b$, we can see that the y-intercept will be at $(0, b)$.

EXAMPLE 1: Writing in the Slope-Intercept Form

Write $3x + 2y = 6$ in slope-intercept form; then state the slope and y-intercept.

Answer

$$3x + 2y = 6$$
$$2y = 6 - 3x$$
$$\frac{2y}{2} = \frac{6}{2} - \frac{3x}{2}$$
$$y = 3 - \frac{3}{2}x$$

1. Solve for y.

$$y = -\frac{3}{2}x + 3$$

2. Apply the Commutative Property of Addition to rearrange the right side so the x-term is first.

$$m = -\frac{3}{2};\ b = 3$$

3. Compare this equation to $y = mx + b$ to identify the slope, m, and the y-intercept, b.

slope: $-\frac{3}{2}$
y-intercept: $(0, 3)$

4. Name the y-intercept using the ordered pair $(0, b)$.

SKILL CHECK **EXERCISES 1, 15**

Three forms of this equation can be used to describe this line.

$3x + 2y = 6$	standard form
$y = -\frac{3}{2}x + 3$	slope-intercept form
$f(x) = -\frac{3}{2}x + 3$	function form

All linear equations, except for vertical lines, can be written in function notation.

CHANGING A LINEAR EQUATION TO SLOPE-INTERCEPT FORM

1. Solve the linear equation for y.
2. Arrange terms on the right side with the variable term first.

The slope-intercept form of a line can be used to quickly graph the line by plotting the y-intercept $(0, b)$ and then using the slope, m, to plot several other points. While it takes only 2 points to determine a line, graphing extra points serves as a check of your work.

EXAMPLE 2: Graphing from Slope-Intercept Form

Graph $y = \frac{1}{2}x + 4$.

Answer

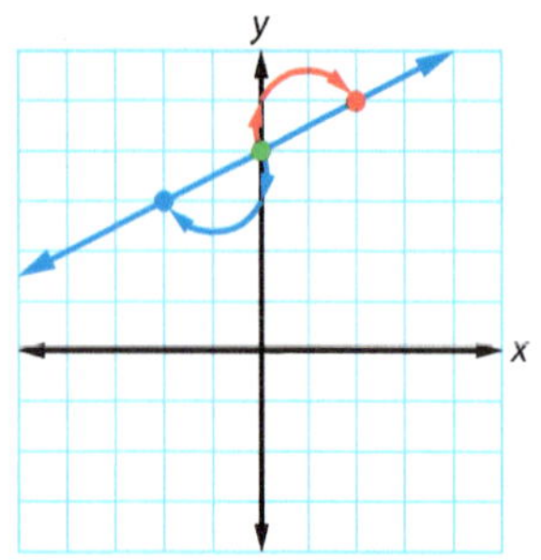

1. Plot the y-intercept. Since $b = 4$, the y-intercept is $(0, 4)$.
2. Use the slope, $m = \frac{1}{2}$, to plot another point on the line. From $(0, 4)$, the y-intercept, move 1 unit up and then 2 units right to plot the point $(2, 5)$.
3. Use an equivalent form of the slope, $m = \frac{-1}{-2}$, to plot a third point. From $(0, 4)$, move 1 unit down and then 2 units left to plot the point $(-2, 3)$.
4. Draw a line through the points.

SKILL CHECK **EXERCISE 9**

USING SLOPE-INTERCEPT FORM TO GRAPH A LINEAR EQUATION

1. Write the equation in slope-intercept form $(y = mx + b)$.
2. Plot the y-intercept, $(0, b)$, on the graph.
3. Starting at the y-intercept, use the slope to plot several other points.
4. Draw a line through the points.

EXAMPLE 3: Graphing from Slope-Intercept Form

Write in slope-intercept form, state the slope and y-intercept, and graph each equation.

a. $2x + 3y = -3$ **b.** $y = 3$

Answers

a. $2x + 3y = -3$

$$3y = -2x - 3$$
$$\frac{3y}{3} = \frac{-2x}{3} - \frac{3}{3}$$
$$y = -\frac{2}{3}x - 1$$

1. Write the equation in slope-intercept form.

$y = mx + b$

$m = -\frac{2}{3}; b = -1$

2. Compare this equation to $y = mx + b$ to identify m and b.

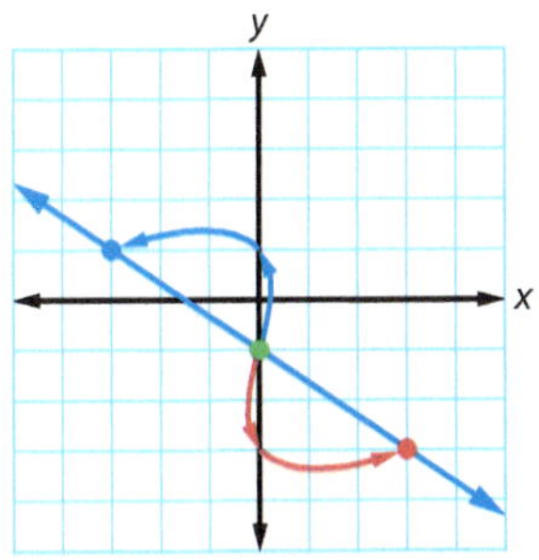

3. Plot the y-intercept, (0, −1).
4. Use the slope, $m = -\frac{2}{3} = \frac{-2}{3}$, to plot another point on the line. From (0, −1), move 2 units down and then 3 units right to plot the point (3, −3).
5. Use an equivalent form of the slope, $m = \frac{2}{-3}$, to plot a third point. From (0, −1), move 2 units up and then 3 units left to plot the point (−3, 1).
6. Draw a line through the points.

b.

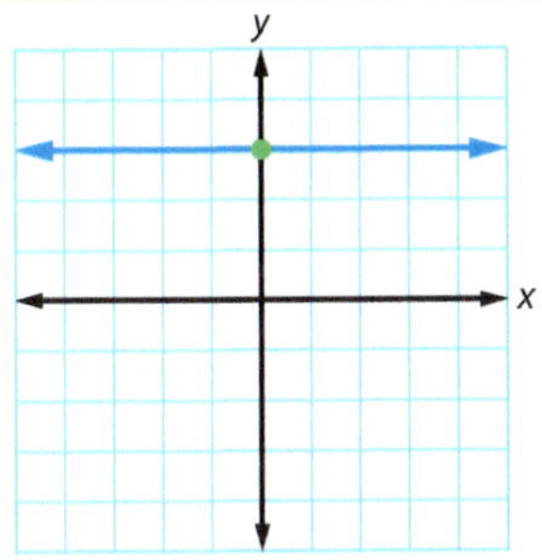

1. Compare $y = 3$ with $y = mx + b$. Since there is no x-term, $y = 3$ can be written as $y = 0x + 3$.
2. Since $b = 3$, the y-intercept is (0, 3), and since $m = 0$, the line is horizontal.
3. Draw the horizontal line through (0, 3).

SKILL CHECK **EXERCISE 25**

Since the graph of $y = 3$ passes the vertical line test, the equation is a function and can be written as $f(x) = 3$. And since this function contains no x-term ($m = 0$), it is called a constant function.

Many of the functions you have studied are linear functions of the form $f(x) = mx + b$. The initial value of the function is b, and the rate of change is m. Using current data to express functions and create graphs allows us to make predictions about future events.

AFTERSCHOOLHELP

Slope-Intercept Form

EXAMPLE 4: Applying the Slope-Intercept Form

An online company produces photo albums from uploaded photo files. If Sarah chooses to pay a membership fee of \$25, a premium photo album will cost her \$1.25 per page with free shipping. Write a linear function rule for the cost and graph the function. Then determine how much it would cost Sarah for a membership and a 23-page photo album.

Answer

Let n = the number of pages.

$C(n) = 1.25n + 25$

1. Identify the independent variable.
2. Write the linear function, where $b = 25$ (the initial fee) and $m = 1.25$, the rate of change (the cost per page).

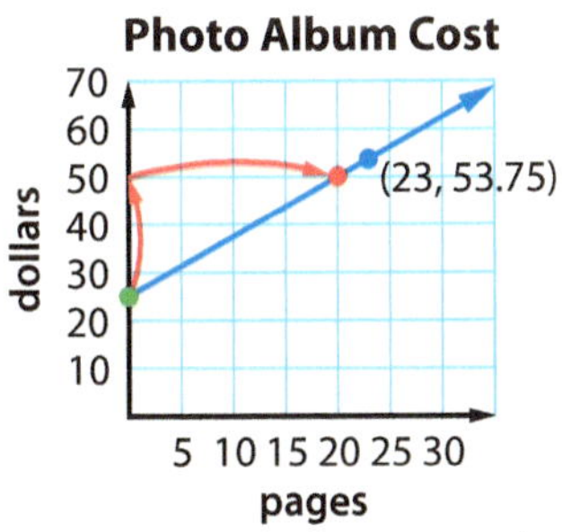

3. Graph the function using the y-intercept, (0, 25), and the slope, $m = \frac{125}{100} = \frac{5}{4}$ or $\frac{25}{20}$.

$C(23) = 1.25(23) + 25$

$= 28.75 + 25 = \$53.75$

4. Substitute 23 for the number of pages in the function rule and simplify to find the cost.

SKILL CHECK **EXERCISE 33**

VOCABULARY

- function form
- slope-intercept form

A. EXERCISES

State the slope and y-intercept of each linear equation.

1. $y = 3x + 4$

2. $y = -\frac{1}{3}x - 2$

3. $y = -x$

4. $y = 6$

Match each equation with its graph.

A

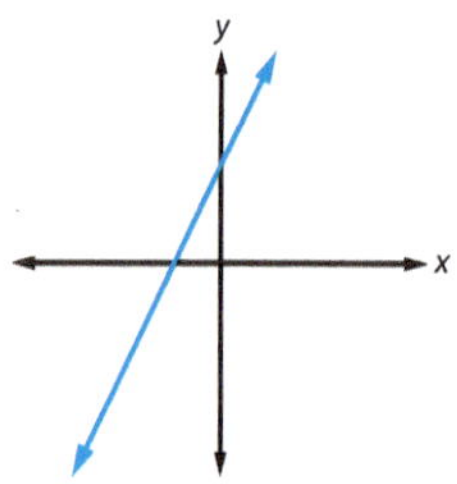

B

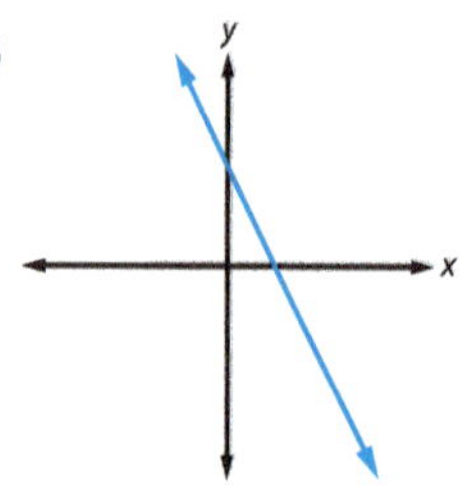

C

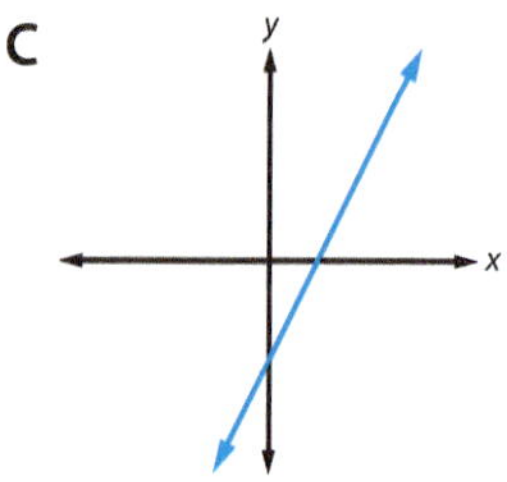

D

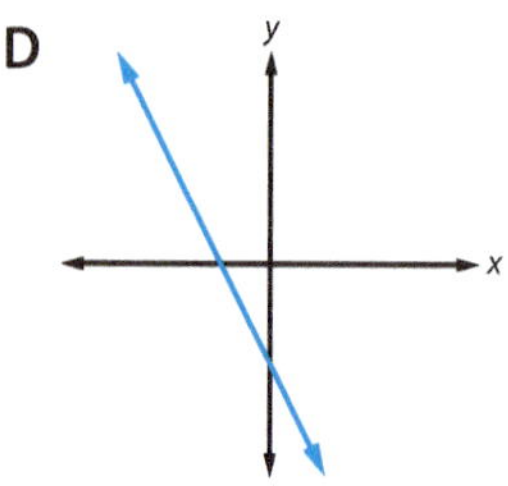

5. $y = 2x - 1$

6. $y = -2x - 1$

7. $y = -2x + 1$

8. $y = 2x + 1$

Graph each equation using its slope and y-intercept.

9. $y = -2x + 4$

10. $y = 4x - 3$

11. $y = -\frac{1}{4}x + 5$

12. $y = -2$

13. $y = \frac{2}{3}x + 2$

14. $y = -\frac{3}{5}x + 1$

B. EXERCISES

Write each equation in slope-intercept form; then state the slope and y-intercept.

15. $3x + y = 5$

16. $x - y = -6$

17. $x + 7y = 14$

18. $3x + 7y = 21$

19. $8x - 3y = 9$

20. $x = 5y$

21. $dx + y = k$

22. MP7 $ry = sx + t$

Write each equation in slope-intercept form; then graph the equation.

23. $2x + y = 5$

24. $x + y = 2$

25. $x + 2y = 8$

26. $\frac{1}{2}y = x - 3$

27. $3x + 4y - 8 = 0$

28. $4x - 5y - 10 = 0$

29. $x = 3y - 9$

30. $3x = -2y + 8$

31. $3x + 2(y - 5) = 7x$

32. MP1 $3(2x - 3) + 1 = 2(y - 1)$

Solve each problem by writing a linear function.

33. The enrollment at Green Mountain Christian School is increasing at a rate of 18 students per year. If the enrollment was 226 in the fall of 2021, write a linear function that models the enrollment. Use your function to predict the number of students enrolled in the fall of 2041.

34. MP4 When a new car is purchased, it immediately begins to lose value, or depreciate. If its depreciation is a linear function and a \$28,550 new car depreciates to a value of \$13,500 over a 7 yr period, what is the rate of change? Write a linear function that models the value of the car over time. Use your function to predict the value of the car when it is 10 yr old.

35. Luke's motorcycle has a 4.5 gal gas tank and uses 1.4 gal to travel 77 mi. At what rate is the gasoline consumed (in gal/mi)? Write a linear function, g, that models the number of gallons left in the tank after he travels m miles, assuming the tank was full when he started. How many gallons are left in the tank after he travels 220 mi?

36. **Essential Question:** What makes the slope-intercept form of an equation useful?

C. EXERCISES

37. MP8 Write $Ax + By = C$ in slope-intercept form. Then find general expressions for the slope and y-intercept of a line written in standard form.

38. Use the general expressions developed in exercise 37 to find the slope and y-intercept of $6x - y = 5$. Then check your results by solving the equation for y.

Write each equation in slope-intercept form; then graph the equation.

39. $4(x - 2) + 7y = 5 - (2x - 3y)$

40. $-2x + 8(3x - 1) = 7x + y - 2(3y + 4)$

CUMULATIVE REVIEW

Perform the indicated operations. [1.1–1.3]

41. $4.3 - 2.95827$

42. $-16 + \frac{2}{3} - \frac{1}{6}$

43. $4\frac{4}{5}\left(\frac{5}{24}\right)$

44. $4\frac{7}{8} \div 2\frac{1}{9}$

45. $\left(\frac{3}{4}\right)^5$

46. $2^3 \cdot 5^{-2}$

Find the slope of the line passing through the given points. [5.2]

47. $(-3, 14)$ and $(2, -6)$

48. $(-1, -44)$ and $(3, 24)$

49. For which expression, represented by □, does $7x - 6y + 4 = \square - 2(3y - 2x + 4)$? [1.6]

A $3x$

B $3x + 12$

C $11x$

D $11x - 4$

E none of these

50. Which of the following is an arithmetic sequence? [4.5]

A $-7, 14, -21, 28, -35, \ldots$

B $2, 4, 8, 16, 32, \ldots$

C $1, 1, 2, 3, 5, \ldots$

D $-6, -1, 4, 7, 10, \ldots$

E none of these

5.4 WRITING LINEAR EQUATIONS

Why are linear models for population growth often inaccurate?

After completing this section, I will be able to

- write the equation of a line by using its slope and a point on the line.
- write the equation of a line by using the point-slope form.
- write the equation of a line given its graph.
- model real-world situations by using linear equations.
- explain why it is important to check the accuracy of a model.

On a steep slope, engineers extend the horizontal distance with a pigtail curve, as on Iron Mountain Road in South Dakota.

Tables or graphs are often used to represent real-world applications involving linear functions. Key characteristics such as the slope or *y*-intercept can be easily determined from these representations. The slope and *y*-intercept can be used to write the function as a linear equation in slope-intercept form.

EXAMPLE 1: Using the Slope & y-Intercept

Write the equation of the line with a slope of $\frac{2}{3}$ and a *y*-intercept of (0, −5).

Answer

$y = \frac{2}{3}x - 5$

Since $m = \frac{2}{3}$ and $b = -5$, substitute these values into the equation $y = mx + b$.

SKILL CHECK **EXERCISE 3**

If the slope and a point other than the *y*-intercept are known, the slope-intercept form can be used to find the *y*-intercept.

EXAMPLE 2: Using the Slope & a Point

Write the equation of the line passing through the given point with the given slope.

a. (1, 5); 3 **b.** (−5, 6); $-\frac{1}{2}$ **c.** (1, −6); 0

Answers

a.

$y = mx + b$

$5 = 3(1) + b$

1. Substitute the slope, $m = 3$, and the point (1, 5) into the slope-intercept form of the equation.

$3 + b = 5$

$b = 2$

2. Solve for *b*.

$y = 3x + 2$

3. Substitute the values for *m* and *b* into the equation $y = mx + b$.

b.

$y = mx + b$

$6 = -\frac{1}{2}(-5) + b$

1. Substitute the slope, $m = -\frac{1}{2}$, and the point (−5, 6) into the slope-intercept form of the equation.

$\frac{5}{2} + b = 6$

$b = \frac{12}{2} - \frac{5}{2} = \frac{7}{2}$

2. Solve for *b*. Be sure to use common denominators when adding or subtracting fractions.

$y = -\frac{1}{2}x + \frac{7}{2}$

3. Substitute the values for *m* and *b* into the equation $y = mx + b$.

c. $y = mx + b$
$-6 = 0(1) + b$
$b = -6$
$y = 0x + (-6)$
$y = -6$

1. Substitute the slope, $m = 0$, and the point $(1, -6)$ into the slope-intercept form of the equation.
2. Substitute the values for m and b into the equation $y = mx + b$ and simplify. The line is a horizontal line.

SKILL CHECK **EXERCISE 15**

Along with slope-intercept and standard forms, a third form for writing equations of lines is the **point-slope form**. This form provides an alternative method for determining the equation of a line from the slope, m, and any given point, (x_1, y_1), on the line.

$$\frac{y - y_1}{x - x_1} = m$$

Multiplying both sides by $x - x_1$ produces the point-slope form.

$$y - y_1 = m(x - x_1)$$

Applying this equation to Example 2a provides a quick transition to the other 2 forms of the equation of a line.

$$y - y_1 = m(x - x_1)$$
$$y - 5 = 3(x - 1)$$

Substitute the slope 3 and the point (1, 5) into the point-slope form.

$$y - 5 = 3x - 3$$
$$y = 3x + 2$$

Solve for y to write the equation in slope-intercept form.

$$-3x + y = 2$$
$$3x - y = -2$$

Convert to standard form.

The slope-intercept form of a line can also be written when only 2 points are known. Once the slope is determined, either point can be used to generate the equation of the line.

EXAMPLE 3: Using 2 Points

Write the equation of the line passing through each set of points.

a. (1, 3) and (4, −2) **b.** (7, 4) and (7, 2)

Answers

a. $m = \frac{y_2 - y_1}{x_2 - x_1} = \frac{-2 - 3}{4 - 1} = \frac{-5}{3} = -\frac{5}{3}$

1. Find the slope of the line.

$y - y_1 = m(x - x_1)$
$y - 3 = -\frac{5}{3}(x - 1)$

2. Substitute the slope $-\frac{5}{3}$ and one of the points, (1, 3), into the point-slope form.

$y - 3 = -\frac{5}{3}x + \frac{5}{3}$
$y = -\frac{5}{3}x + \frac{14}{3}$

3. Solve for the slope-intercept form.

DOING THE MATH

$\frac{5}{3} + 3 = \frac{5}{3} + \frac{9}{3}$
$= \frac{14}{3}$

b. $m = \frac{y_2 - y_1}{x_2 - x_1} = \frac{2 - 4}{7 - 7} = \frac{-2}{0}$

1. Since the slope of the line is undefined, the line must be vertical.

$x = 7$

2. The equation of a vertical line is of the form $x = k$, where k is the value of the x-coordinates.

SKILL CHECK **EXERCISE 19**

If the slope and y-intercept can be easily determined from the graph of a line, substitute the values of m and b into the slope-intercept form to write the equation. If not, use the slope and another convenient point or use the coordinates of 2 convenient points to determine the line's equation, as was shown in Examples 2–3.

EXAMPLE 4: Modeling a Real-World Situation

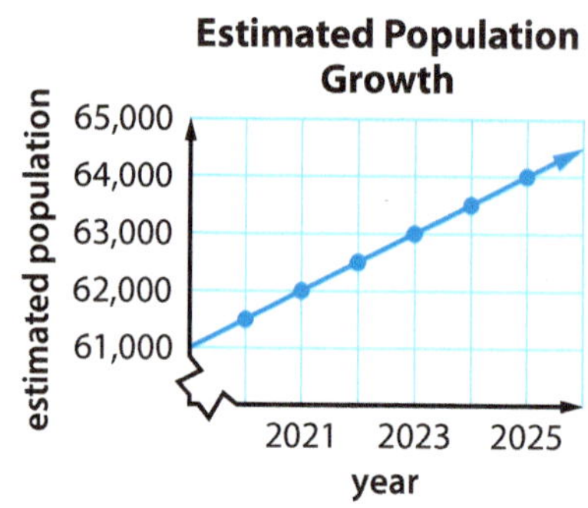

The planning committee for the city of Brownsville used census data to project the growth of the city. Determine the estimated rate of growth and write the function rule describing this population estimate. Then use the function rule to predict the population in 2035.

Answer

P_1 (2020, 61,500) P_2 (2025, 64,000)	**1.** Choose 2 convenient points from the graph.
$m = \frac{64{,}000 - 61{,}500}{2025 - 2020} = \frac{2500}{5} = 500 \frac{\text{people}}{\text{year}}$	**2.** Use the slope formula to determine the estimated rate of growth.
$y - 61{,}500 = 500(x - 2020)$	**3.** Use one of the points, (2020, 61,500), and the slope to write the point-slope form.
$y - 61{,}500 = 500x - 1{,}010{,}000$ $y = 500x - 948{,}500$	**4.** Write in slope-intercept form.
$P(t) = 500t - 948{,}500$	**5.** Rewrite the equation showing population as a function of time.
$P(2035) = 500(2035) - 948{,}500$ $= 69{,}000$	**6.** Evaluate the function rule for the year 2035.

The projected population for Brownsville in 2035 is 69,000 people.

SKILL CHECK **EXERCISE 35**

AFTERSCHOOLHELP

Writing Linear Equations

This function acts as a model of the population growth. While the model may be useful for years close to the data, making predictions well beyond the data points often causes errors. Estimating the founding of the city to be 1897 by setting the population to 0 and solving for t assumes a constant rate of growth over all those years. Predictions of negative population in years prior to 1897 are obviously unrealistic. Unforeseen circumstances may also result in a growth rate different than predicted, so checking the model against actual census results is important to verify its accuracy. Notice that checking the model with census data indicates that we have more faith in the accuracy of the census data than in the model.

VOCABULARY

- point-slope form

A. EXERCISES

Write the equation of the line with the given slope and *y*-intercept.

1. $m = -5$; $(0, 10)$

2. $m = 4$; $(0, -9)$

3. $m = -\frac{1}{3}$; $(0, 3)$

4. $m = 6$; $\left(0, \frac{1}{2}\right)$

5. $m = 0$; $(0, 7)$

6. $m = -\frac{3}{2}$; $(0, 0)$

Write the equation of each line.

7.

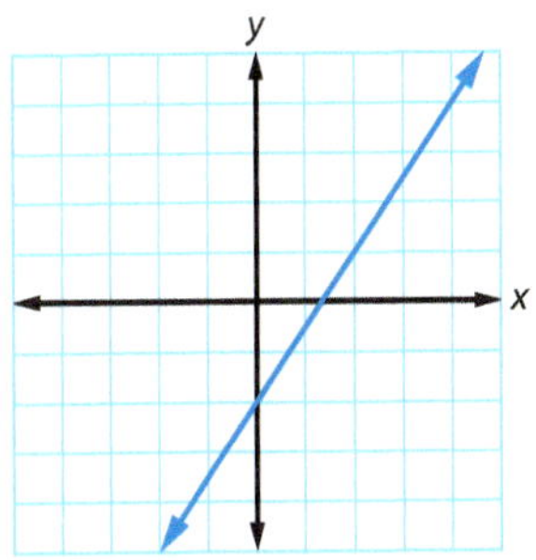

8.

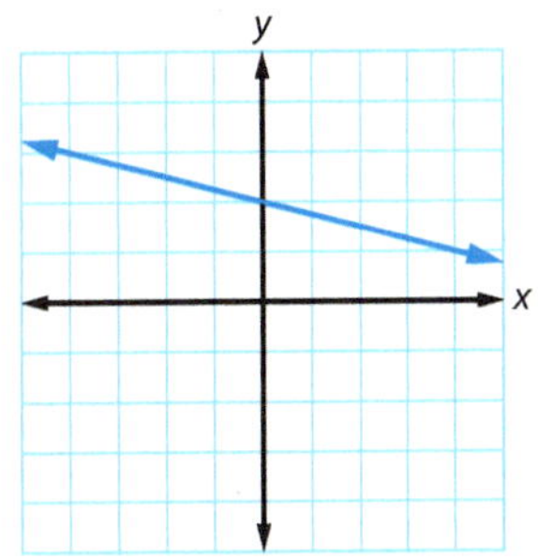

9.

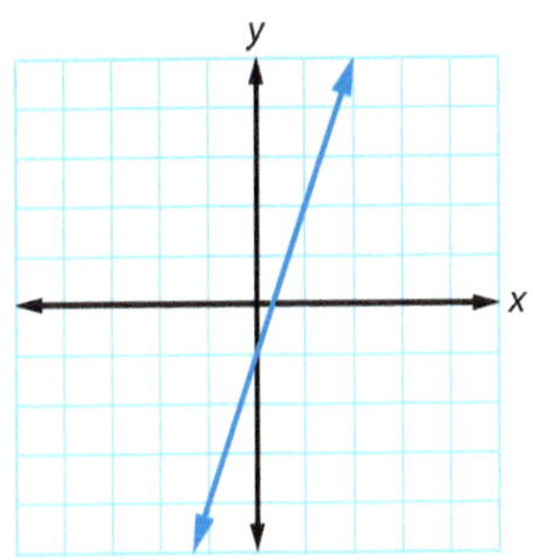

10.

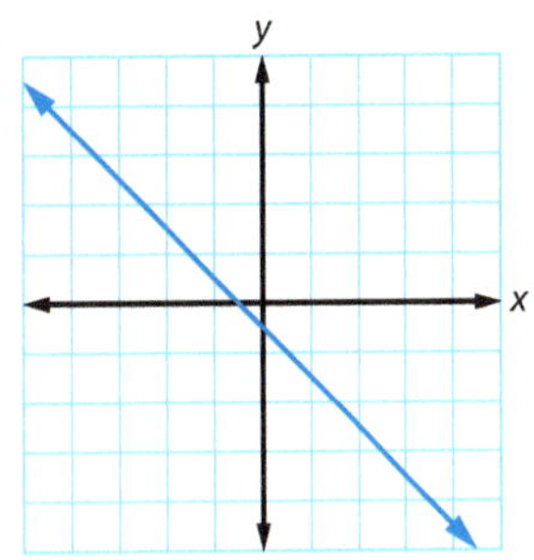

Use the points (−2, 11) and (3, 1) to answer exercises 11–13.

11. Find the slope of the line passing through the 2 points.

12. Write the point-slope form of the line using

a. the point $(-2, 11)$.

b. the point $(3, 1)$.

13. Verify that the equations from exercise 12 are the same by writing them in slope-intercept form.

B. EXERCISES

Use the given information to write the equation of each line in slope-intercept form.

14. $(0, 4)$, $m = \frac{3}{5}$

15. $(2, -5)$, $m = -3$

16. $(-3, -1)$, $m = \frac{5}{3}$

17. $(-5, 2)$, $m = -\frac{4}{5}$

18. $(1, 4)$, $(0, 0)$

19. $(0, 4)$, $(2, 7)$

20. $(4, 2)$, $(-3, 9)$

21. $(2, -3)$, $(2, 4)$

Use the point-slope form to write the equation of the line in slope-intercept form.

22. $(-3, 2), m = 8$

23. $(1, 6), m = -\frac{1}{4}$

24. $(-9, 0), m = -\frac{1}{8}$

25. $(3, 7), m = \frac{1}{4}$

26. $(-5, -4), (3, -4)$

27. $(1, 3), (2, 5)$

28. $(-4, 2), (7, 3)$

29. $(-3, 2), (10, -6)$

Write the equation of each line.

30. a line with a slope of 0.4 passing through the point $(-1, -3)$

31. a line passing through $(1, 3)$ and $(-4, -9)$

32. a line passing through $(2, 0)$ and $(6, -5)$

33. a line passing through the point $(13, 9)$ with a slope of $-\frac{1}{2}$

34. Use the graph of Dr. Jokowski's salary for each year of his practice to write a linear function modeling his salary. Use the function to predict how much he will make during his twentieth year of practice.

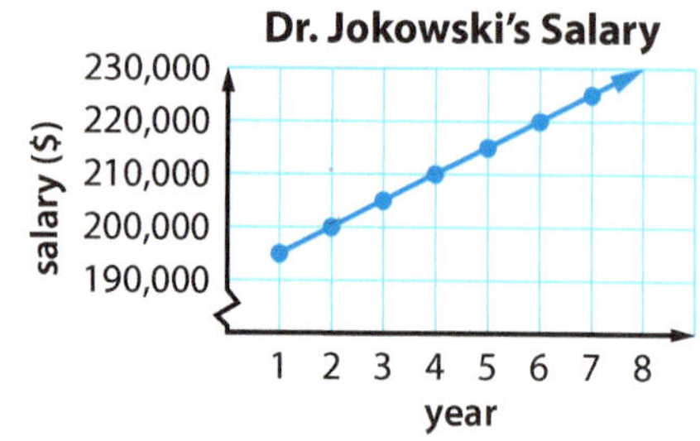

35. A hot-air balloon takes off from Buckingham, PA, and ascends at a linear rate. Use the graph to find its rate of ascent; then write an equation representing the altitude of the balloon as a function of time.

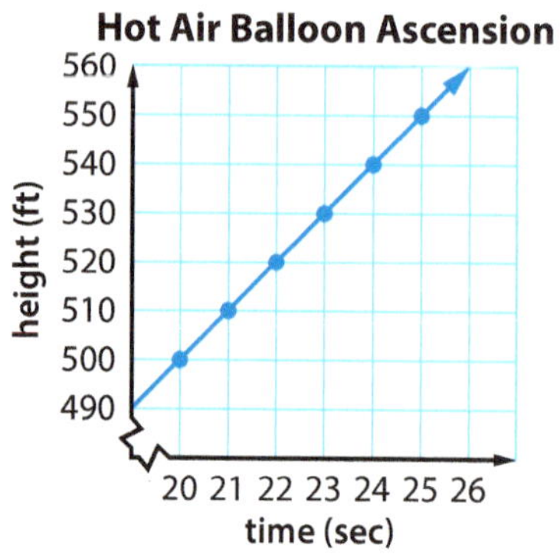

36. Justin is filling a 50 gal aquarium with water. After 5 min the aquarium has 3 gal of water in it, and Justin increases the inflow rate to a constant rate of 4 gal/min. How much water will be in the tank 8 min after he began filling it?

37. **MP4** Two years ago the enrollment of Charis Christian Academy was 520 students; and the students ate, on average, 419 hot lunches per week. Last year the 540 students consumed an average of 444 hot lunches per week. Write a linear function modeling the weekly consumption of hot lunches based on the school's enrollment. Then use this function to predict the number of hot lunches the school should provide each week if the enrollment is 584 students this year.

38. In 2015, Glens Falls had 14,496 residents. In 2020, that number fell to 14,322. Write a linear function modeling the population change of Glens Falls since 2015. Based on this model, what is the population expected to be in 2060?

39. **Essential Question:** Consider the population models in Example 4 and exercise 38. What assumptions are made for predictions well beyond the given data points, such as when the population was (or will be) 0? How might this affect the accuracy of the model?

C. EXERCISES

Write the standard-form equation of the line passing through the given point with the given slope.

40. $(0, 8), m = \frac{1}{5}$

41. $(3, 0), m = \frac{9}{4}$

42. $(6, -1), m = -4.5$

43. $\left(\frac{15}{7}, -\frac{3}{7}\right), m = -\frac{3}{5}$

44. MP7 Mr. Rodriguez works in a clothing store. He notices that when shirts are priced at \$60 each, he sells 20 shirts per week. For each \$10 increase in the price of the shirts, he sells 3 fewer shirts that week. Find the rate of change in the number of shirts sold per dollar increase in price. Then write a linear function, s, modeling the number of shirts sold based on the cost of the shirts, c. Determine the price at which no shirts will sell.

CUMULATIVE REVIEW

Find the x- and y-intercepts of each linear equation; then use them to graph the equation. [5.1]

45. $4x - 3y = 12$

46. $5x + 2y = 10$

Find the slope of the line passing through the given points. [5.2]

47. $(-5, 11)$ and $(7, 11)$

48. $(4, 5)$ and $(3, 13)$

49. $(-6, -2)$ and $(9, 8)$

50. $\left(\frac{8}{3}, 2\right)$ and $\left(\frac{8}{3}, 9\right)$

Write each equation in slope-intercept form; then state the slope and y-intercept. [5.3]

51. $2x + 3y = 15$

52. $3x + 4y = 36$

53. What is the value of $\frac{|q + w|}{q + w}$ if $q = 15$ and $w = -16$? **[1.5]**

A -1

B 1

C -31

D 31

E none of these

54. Which of the following is a function? **[4.1]**

A $\{(0, -1), (1, 4), (3, -1), (-2, 4)\}$

B $\{(2, -5), (0, 0), (-1, 10), (2, 8)\}$

C $\{(-3, 3), (-7, 0), (-3, 9), (10, 2)\}$

D $\{(1, 5), (-1, -5), (0, 12), (0, -18)\}$

E all of these

CAREERS IN MATH
MARKET RESEARCH ANALYST

A 2006 *Money Magazine* article based on an online survey of 26,000 workers ranked the job of a market research analyst sixth in its ranking of top jobs. Like other statisticians, market research analysts gather and analyze data, but they focus on helping businesses market their products to consumers or to other businesses for maximum profit. Christians might wonder whether it would be worthwhile to spend their lives helping businesses make greater profits. However, the successful marketing of beneficial products enables the business to employ more people and to research other means of satisfying consumer needs.

Large corporations and government agencies hire their own market research analysts. Many other analysts are employed by consulting firms and marketing research firms that serve small companies, which often prefer to contract out work rather than support their own marketing department. People with good communicative and quantitative skills will have the best opportunities in the field, and a strong background in math and computer science is especially helpful.

Market research analysts supervise the collection of data through a variety of methods, including surveys, interviews, and focus groups. A market research analyst must know how to calculate an appropriate sample size and how to take a sample from a larger population. After compiling and interpreting the data collected, an analyst provides information and recommendations that will help the business make decisions concerning product design, pricing, and advertising. This information may also lead to other decisions, such as promotional rebates or new store locations. Analysts must also be able to calculate and to state the degree of certainty for their conclusions.

Market research analysts make life easier for the consumer by making the things the consumer wants more easily available. They also increase the business's profit through their advice on which items to promote. A Christian market research analyst should also take into account the ethical questions that arise in such work. Just because something can be marketed does not mean it should be, and just because the market may want a product does not mean a business should supply it. In addressing these ethical issues, the Christian has a great advantage over a secular counterpart.

5.5 PARALLEL & PERPENDICULAR LINES

What are the key characteristics that help us to determine how lines are related?

After completing this section, I will be able to

- classify a pair of equations as parallel, perpendicular, or neither.
- write equations of parallel and perpendicular lines.

We saw in Section 5.3 that an equation of a line written in slope-intercept form allows us to quickly discern the slope and the y-intercept of the graph. By observing these key characteristics, we can determine how lines relate to each other.

Determine the slope of each line in the graph below. Notice that the lines appear to be parallel. **Parallel lines** are lines in the same plane that have no points in common. They have the same slope but different vertical positions.

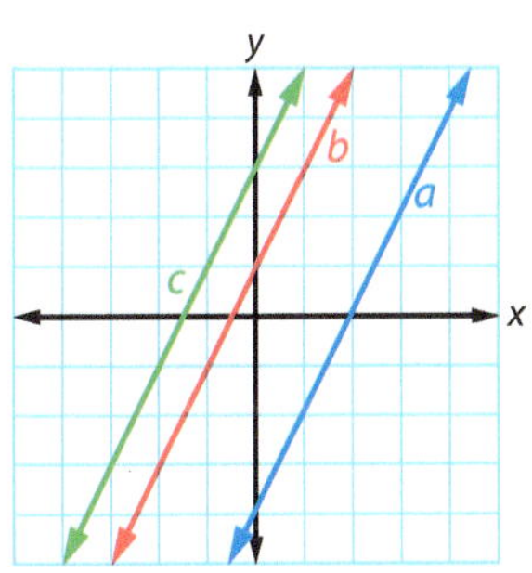

Line	Slope	y-Intercept	Equation
a	2	(0, −4)	$y = 2x - 4$
b	2	(0, 1)	$y = 2x + 1$
c	2	(0, 3)	$y = 2x + 3$

The graphs of 2 linear equations are parallel if they have the same slope but different y-intercepts.

EXAMPLE 1: Identifying Parallel Lines

Determine whether the lines described by $3x - y = 6$ and $-9x + 3y = 12$ are parallel.

Answer

$3x - y = 6$ $-y = -3x + 6$ $y = 3x - 6$	$-9x + 3y = 12$ $3y = 9x + 12$ $y = 3x + 4$	**1.** Write each equation in slope-intercept form.
$m_1 = 3$ y-intercept: (0, −6)	$m_2 = 3$ y-intercept: (0, 4)	**2.** Determine the slope and y-intercept of each line.
The lines are parallel.		**3.** The lines have the same slope but different y-intercepts.

SKILL CHECK **EXERCISE 7**

Notice that the graphs of the equations in Example 1 are parallel.

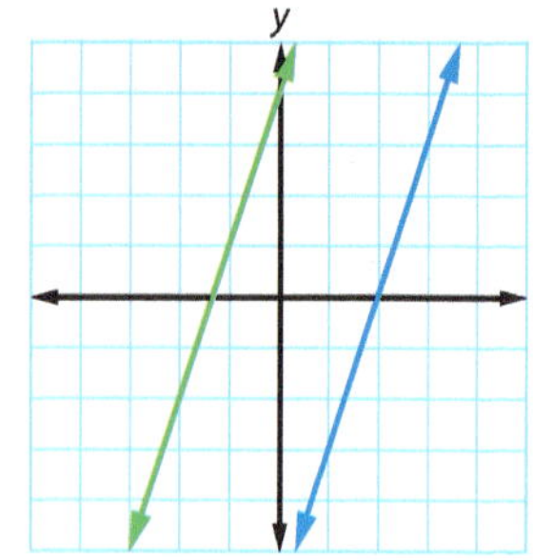

The fact that parallel lines have the same slope can be used to find the equation of a line parallel to a given line and passing through a given point.

EXAMPLE 2: Writing the Equation of a Parallel Line

Write the slope-intercept form of the equation of the line parallel to $4x + 3y = 17$ and passing through $(-5, 9)$.

Answer

$4x + 3y = 17$ $3y = -4x + 17$ $y = -\frac{4}{3}x + \frac{17}{3}$	**1.** Write the equation in slope-intercept form.
$m = -\frac{4}{3}$	**2.** Determine the slope of the given line.
$y - y_1 = m(x - x_1)$ $y - 9 = -\frac{4}{3}[x - (-5)]$	**3.** Substitute the slope $-\frac{4}{3}$ and the point $(-5, 9)$ into the point-slope form of the equation.
$y - 9 = -\frac{4}{3}x - \frac{20}{3}$ $y = -\frac{4}{3}x + \frac{7}{3}$	**4.** Rewrite in slope-intercept form.

DOING THE MATH

$$-\frac{20}{3} + 9 = -\frac{20}{3} + \frac{27}{3} = \frac{7}{3}$$

SKILL CHECK **EXERCISE 17**

Whereas parallel lines do not intersect and their slopes are the same, **perpendicular lines** meet at right angles (90°) and have slopes that are negative reciprocals of each other. The reciprocal of a number is found by interchanging its numerator and denominator. The negative reciprocal is the opposite of the reciprocal. For instance, the negative reciprocal of $\frac{5}{9}$ is $-\frac{9}{5}$, and the negative reciprocal of -8 is $\frac{1}{8}$.

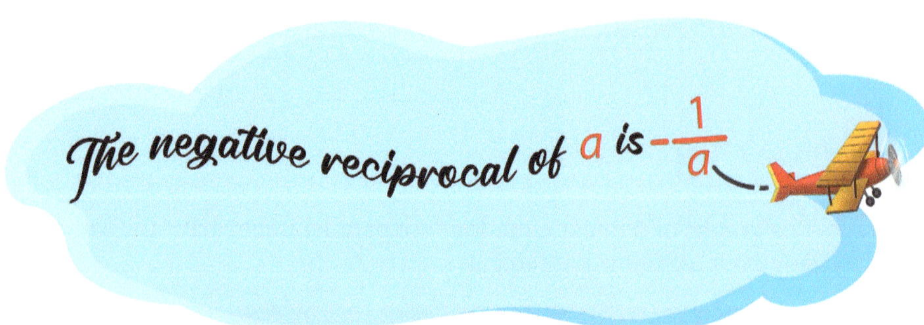

EXAMPLE 3: Finding the Negative Reciprocal

Give the negative reciprocal of each number.

a. $\frac{8}{5}$ **b.** 9 **c.** $-\frac{13}{7}$

Answers

a. The reciprocal of $\frac{8}{5}$ is $\frac{5}{8}$. The negative reciprocal of $\frac{8}{5}$ is $-\frac{5}{8}$.

b. Write 9 as $\frac{9}{1}$. The reciprocal of $\frac{9}{1}$ is $\frac{1}{9}$. The negative reciprocal of 9 is $-\frac{1}{9}$.

c. The reciprocal of $-\frac{13}{7}$ is $-\frac{7}{13}$. The negative reciprocal of $-\frac{13}{7}$ is $\frac{7}{13}$.

SKILL CHECK **EXERCISES 1, 11**

EXAMPLE 4: Writing the Equation of a Perpendicular Line

Write the slope-intercept form of the equation of the line perpendicular to $y = -4x + 9$ and passing through (2, 7).

Answer

$m_1 = -4$
$m_2 = \frac{1}{4}$

1. Determine the slope of the line.
2. The slope of any line perpendicular to the given line is the negative reciprocal of −4.

$y - y_1 = m(x - x_1)$
$y - 7 = \frac{1}{4}(x - 2)$
$y - 7 = \frac{1}{4}x - \frac{1}{2}$
$y = \frac{1}{4}x + \frac{13}{2}$

3. Since the lines are perpendicular, substitute the new slope and the point (2, 7) into the point-slope form of a line.

DOING THE MATH

$-\frac{1}{2} + 7 = -\frac{1}{2} + \frac{14}{2}$
$= \frac{13}{2}$

SKILL CHECK **EXERCISE 21**

SLOPES OF PARALLEL & PERPENDICULAR LINES

Parallel lines have the same slope.

Perpendicular lines have slopes that are negative reciprocals of each other.

Even though horizontal and vertical lines are perpendicular, their slopes are not negative reciprocals of each other. However, knowing the standard form of these equations, $y = a$ and $x = b$ where a and b are constants, enables us to write equations of lines perpendicular to any horizontal or vertical line.

The equation perpendicular to $y = 4$ that passes through the given point (3, −2) is $x = 3$.

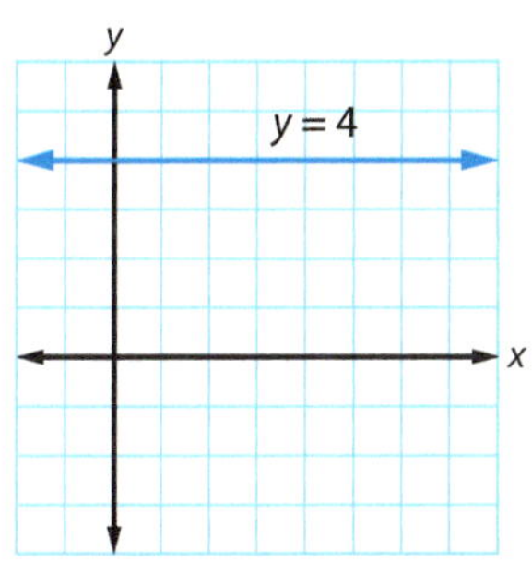

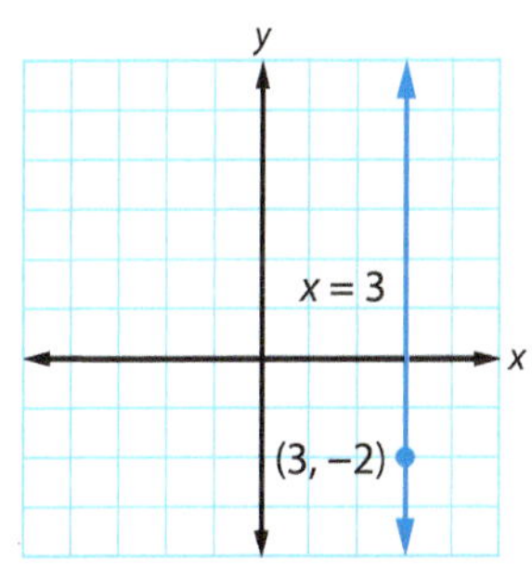

AFTERSCHOOLHELP

Parallel & Perpendicular Lines

EXAMPLE 5: Identifying the Relationship between Lines

Are the graphs of the following lines parallel, perpendicular, or neither?

$2x + 5y = 9$ $\quad\quad$ $4x + y = 1$

Answer

$2x + 5y = 9$
$5y = -2x + 9$
$y = -\frac{2}{5}x + \frac{9}{5}$
$m_1 = -\frac{2}{5}$

$4x + y = 1$
$y = -4x + 1$
$m_2 = -4$

1. Write each equation in slope-intercept form.

The lines are neither parallel nor perpendicular.

2. The slopes are neither equal nor negative reciprocals.

SKILL CHECK **EXERCISE 27**

VOCABULARY

- parallel lines
- perpendicular lines

A. EXERCISES

Give the negative reciprocal of each slope.

1. $m = 7$ **2.** $m = -8$ **3.** $m = -\frac{3}{5}$ **4.** $m = \frac{1}{9}$ **5.** $m = \frac{4}{7}$

State the slope of a line parallel to the given line.

6. $5x - y = 3$ **7.** $-4x + 6y = 7$ **8.** $2x - y = -5$

9. $-8x + 5y = 7$ **10.** $x + 9y = -4$

State the slope of a line perpendicular to the given line.

11. $2x - 10y = 3$ **12.** $x + 9y = -3$ **13.** $4x - y = 13$

14. $7x + 3y = 8$ **15.** $5x - 9y = 1$

B. EXERCISES

Use the point-slope form to write the slope-intercept form of the equation of the line parallel to the given line and passing through the given point.

16. $x + y = -4; (7, 0)$ **17.** $4x + y = 9; (-2, 5)$

18. $2x + 7y = 4; (6, 2)$ **19.** $x - 9y = -1; (-1, -8)$

20. $6x + 5y = -8; (-4, 10)$

Use the point-slope form to write the slope-intercept form of the equation of the line perpendicular to the given line and passing through the given point.

21. $x + y = 8; (0, 4)$ **22.** $3x - y = 7; (-6, 1)$

23. $x = 8; (2, 5)$ **24.** $y = 9x + 2; (8, -3)$

25. $5x + 7y = -12; (3, 0)$

Determine whether the graphs of each pair of equations are parallel, perpendicular, or neither.

26. $-4x + 6y = -24$
$2x - 3y = -15$

27. $x - 5y = -5$
$5x - y = -6$

28. $3x - y = 1$
$x + 3y = -15$

29. $5x - 9y = -9$
$18x + 10y = -40$

30. $18x - 3y = 3$
$6x - y = -5$

Write the equations of each pair of lines and determine whether they are parallel, perpendicular, or neither.

31.

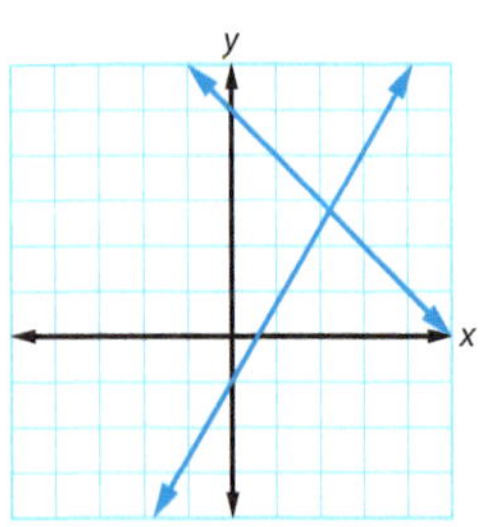

32.

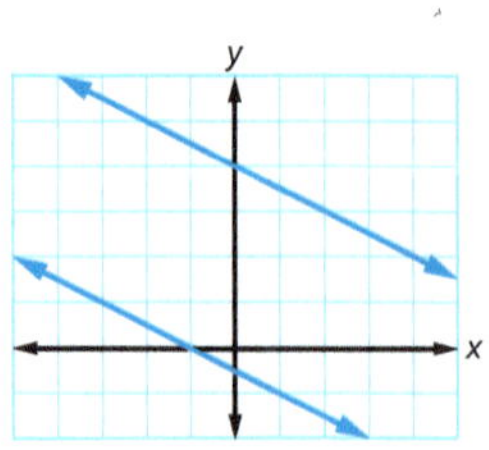

33.

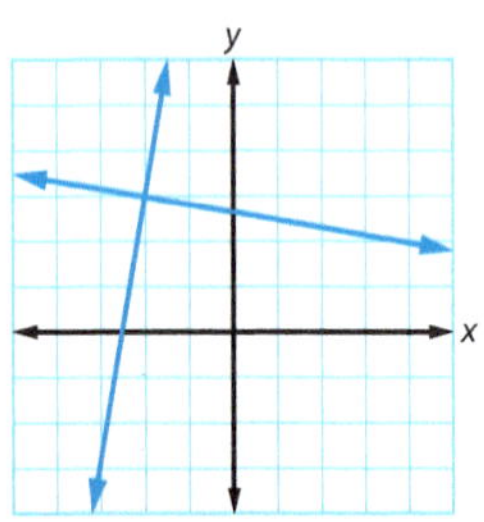

34.

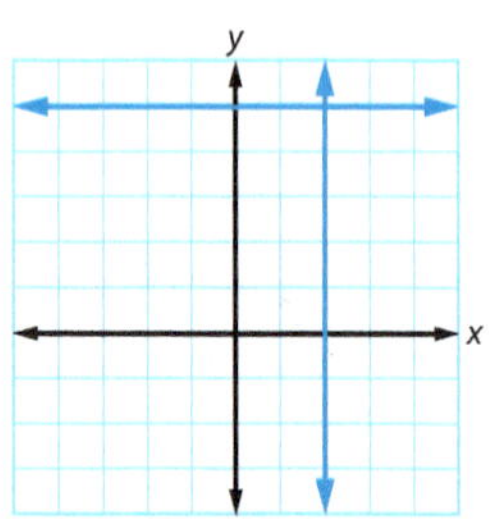

35. **Essential Question:** How can we determine how lines in the same plane relate to one another?

C. EXERCISES

36. Write the standard-form equation of the line perpendicular to $4x + 3y = 1$ and passing through its y-intercept.

37. Write the standard-form equation of the line perpendicular to $x + 7y = 4$ and passing through its x-intercept.

38. If $3x + 7y = 2$ and $Ax + 2y = -9$ are parallel lines, what is the value of A?

39. **MP2** If $6x + By = 3$ and $8x + y = 5$ are perpendicular lines, what is the value of B?

The vertices of a square are A (2, 2), B (−1, −2), C (−5, 1), and D (−2, 5).

40. Find the slope of each side of the square. Use the slopes of the sides to show that the opposite sides are parallel and the consecutive sides are perpendicular.

41. **MP1** Find the slope of each diagonal of the square (the segment connecting opposite vertices). What relationship do the diagonals of the square have?

42. **MP7** A triangle's vertices are R (−6, 2), S (3, −5), and T (4, −1). Use the slopes of the triangle's sides to determine whether it is a right triangle. Justify your conclusion.

CUMULATIVE REVIEW

Write an equation and solve. [2.7]

43. A plane flies west from Wichita at 350 knots at the same time another plane flies east from the same airport at 430 knots. How long (to the nearest minute) will it take for the planes to be 1500 nautical miles apart?

44. Joshua leaves home for Memphis at 8 AM, but his sister Savannah cannot leave until 9:30 AM. If she travels, on average, 10 mi/hr faster than Joshua and catches up to him at 5:30 PM, how fast (to the nearest mile per hour) was each car traveling?

Translate each sentence into a compound inequality. [3.3–3.4]

45. A number is at least −2 and not more than 5.

46. A number is not less than 7, or it is less than 3.

47. A number is no less than −4, or it is at most −10.

48. A number is less than 6 and not less than −1.

Solve and graph the solution on a number line. [3.6]

49. $|0.5x + 3| < 15$

50. $|2x + 3| \geq 12$

51. A store marked up a \$35 sweater by 15%, then reduced that price by 30% during a sale. Which expression provides the sale price? **[2.6]**

   A \$35(1.15)(0.70)

   B \$35(1.15)(0.30)

   C \$35(0.85)(0.70)

   D \$35(0.15)(0.30)

   E none of these

52. What is the function rule for the graph below? **[4.7]**

   A $f(x) = |x - 3| + 2$

   B $f(x) = |x + 3| + 2$

   C $f(x) = |x - 2| - 3$

   D $f(x) = |x + 2| - 3$

   E none of these

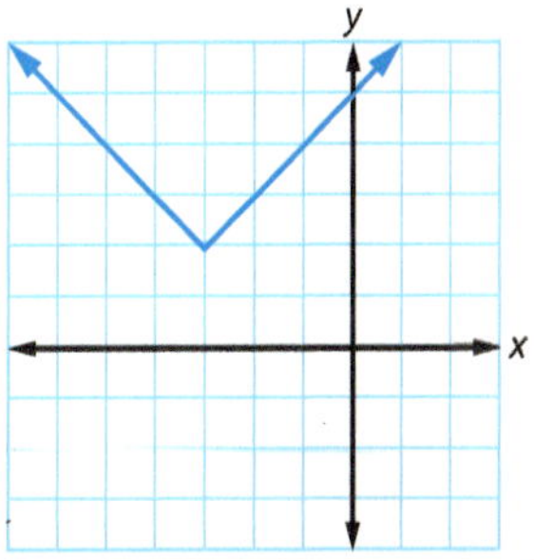

5.6 TREND LINES & CORRELATION

When will I use a line of best fit in the real world?

After completing this section, I will be able to

- graph a trend line from a given scatter plot.
- make interpolations or extrapolations of the data using a trend line.
- describe the linear correlation of a given data set.
- determine the line of best fit from real-world data.

Scatter plots are used to illustrate the relationship between 2 groups of data. While these points rarely lie on a straight line, the relationship can be more clearly seen when a *trend line* is drawn on a scatter plot as close to as many points as possible. Though the exact method of drawing a trend line is quite involved, a carefully drawn estimate can produce useful results. Graphing calculators, spreadsheets, and online tools can be used to find the exact equation of the *line of best fit.*

Trend lines are one of the indicators that help companies determine the best way to limit cost and optimize profit.

DEFINITIONS

A **trend line** is a line that estimates the relationship between data sets.

A **line of best fit** is the trend line that best illustrates the relationship between data sets.

You have already seen how a scatter plot illustrates data, such as that in the table below, which lists the sales revenues of an electronics company. It can also be used to write a linear function modeling the data.

year	sales revenues
1966	$5,000,000
1971	$6,000,000
1976	$5,000,000
1988	$11,000,000
2000	$15,000,000
2001	$16,000,000
2004	$17,000,000

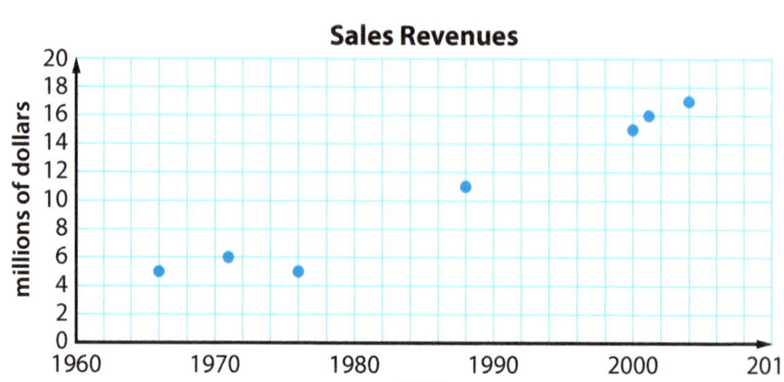

EXAMPLE 1: Finding a Trend Line

Draw and find the equation of a trend line that models the sales revenues.

Answer

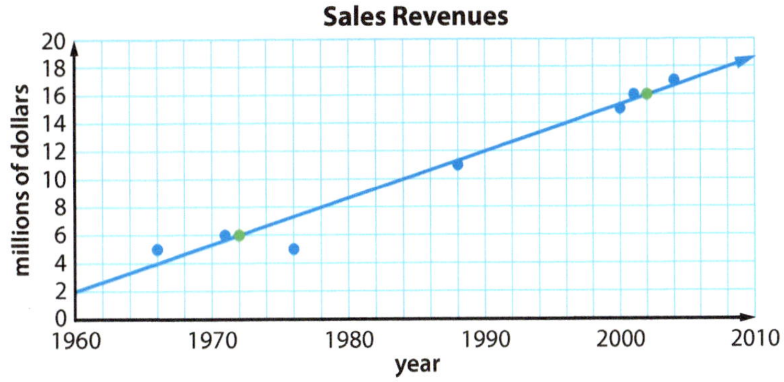

1. Carefully estimate and draw a trend line that by visual inspection minimizes the distances to the data points.

(1972, 6) and (2002, 16)

$$m = \frac{y_2 - y_1}{x_2 - x_1} = \frac{16 - 6}{2002 - 1972} = \frac{10}{30} = \frac{1}{3}$$

$$y - y_1 = m(x - x_1)$$

$$y - 6 = \frac{1}{3}(x - 1972)$$

$$y - 6 = \frac{1}{3}x - \frac{1972}{3}$$

$$y = \frac{1}{3}x - \frac{1954}{3} \text{ or } y = 0.\overline{3}x - 651.\overline{3}$$

2. Choose 2 points on the trend line.
3. Find the equation of the line passing through those 2 points.

DOING THE MATH

$$-\frac{1972}{3} + \frac{18}{3} = -\frac{1954}{3}$$
$$= -651.\overline{3}$$

SKILL CHECK **EXERCISE 7**

You can use the equation of a trend line to make estimates or predictions related to the data. **Interpolation** is the process of making estimates within the range of the existing data points, while **extrapolation** makes predictions (future) and estimations (past) outside this range. Interpolation results tend to be more accurate than extrapolation results.

EXAMPLE 2: Using a Trend-Line Equation

Using the trend-line equation from Example 1, estimate the sales revenue for 1995 and predict the year in which the company will have sales of $24 million. Which is more likely to be accurate?

Answer

$$y = \frac{1}{3}x - \frac{1954}{3}$$

$$y = \frac{1}{3}(1995) - \frac{1954}{3} = \frac{1995 - 1954}{3}$$

$$y = \frac{41}{3} = 13.\overline{6}$$

1. Substitute 1995 for x in the trend-line equation and evaluate.

The model estimates sales of $14 million in 1995.

$$y = \frac{1}{3}x - \frac{1954}{3}$$

$$24 = \frac{1}{3}x - \frac{1954}{3}$$

$$72 = x - 1954$$

$$x = 2026$$

2. Substitute 24 for y in the trend-line equation. Solve for x after clearing the equation of fractions.

The model predicts sales of $24 million in 2026.

Interpolating $14 million in sales in 1995 is likely more accurate than extrapolating $24 million in sales in 2026.

SKILL CHECK **EXERCISES 23–25**

The linear correlation of the data describes how closely the data points are clustered along the trend line. When the majority of the points are close to the line, there is a strong correlation. Data with a trend line from lower left to upper right has a positive correlation, whereas data with a trend line from upper left to lower right has a negative correlation. Trend lines from data with stronger correlations are more likely to produce accurate estimates and predictions.

strong positive correlation

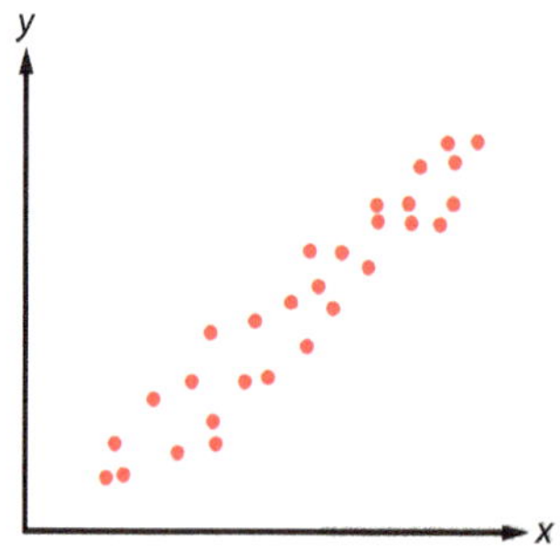

little or no correlation

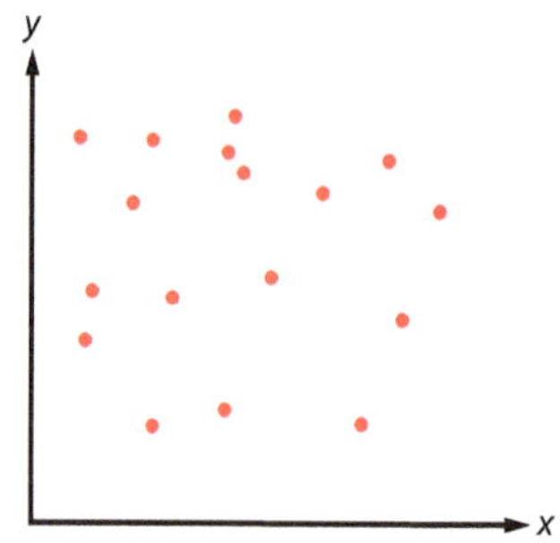

strong negative correlation

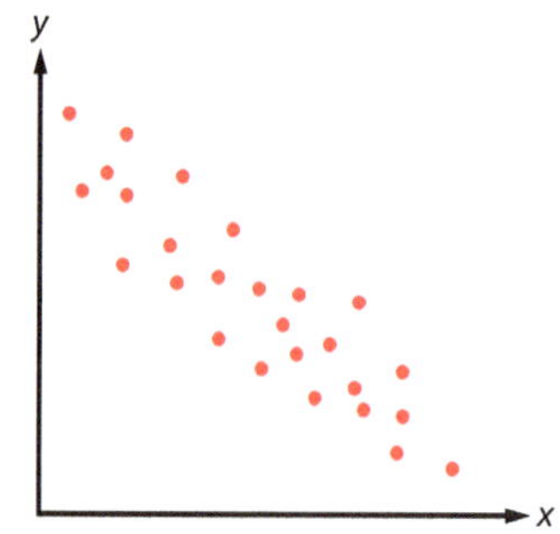

weak positive correlation

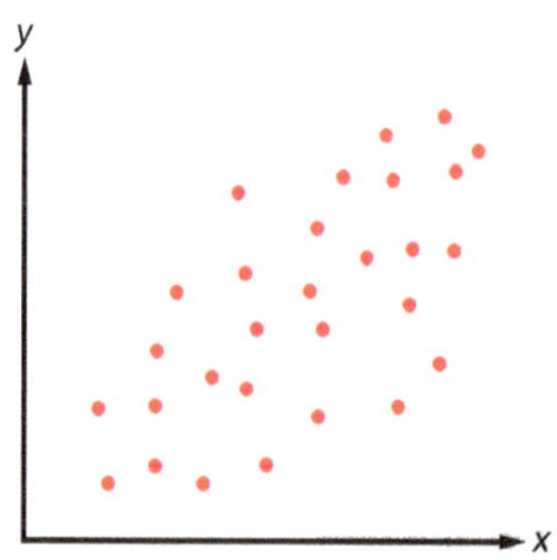

weak negative correlation

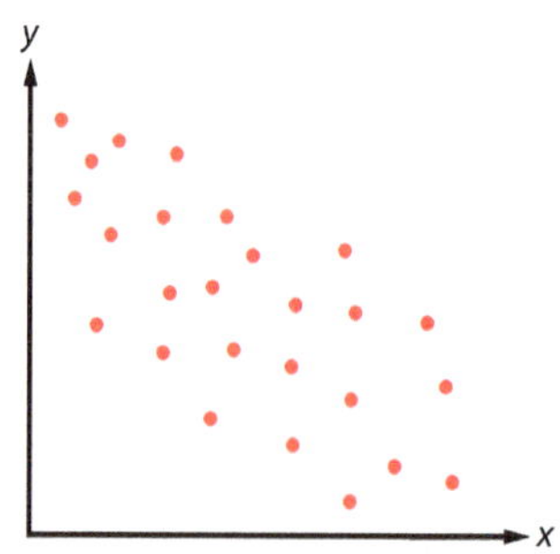

AfterSchoolHelp

Trend Lines & Correlation

VOCABULARY

- extrapolation
- interpolation
- line of best fit
- trend line

TECHNOLOGY CORNER (TI-84 PLUS FAMILY)

Finding the Line of Best Fit

You can use the STAT menu of your graphing calculator to produce a scatter plot of a data set and determine the line of best fit.

Pressing [stat] [enter] selects the first option, Edit, which allows you to enter x-values under L_1 and y-values under L_2. Use [enter] and the arrow keys, [▲], [▼], [◄], and [►], to enter the data from Example 1 in this section.

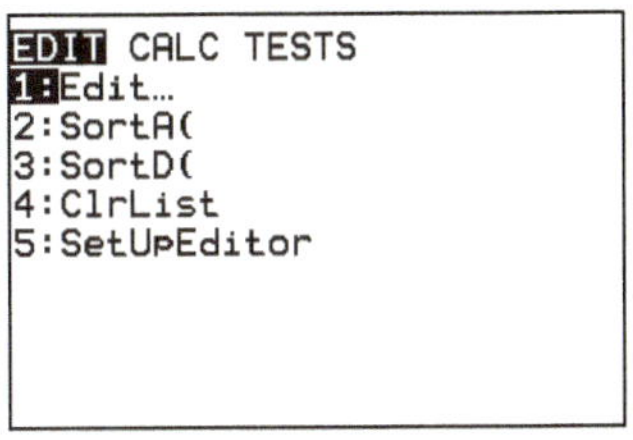

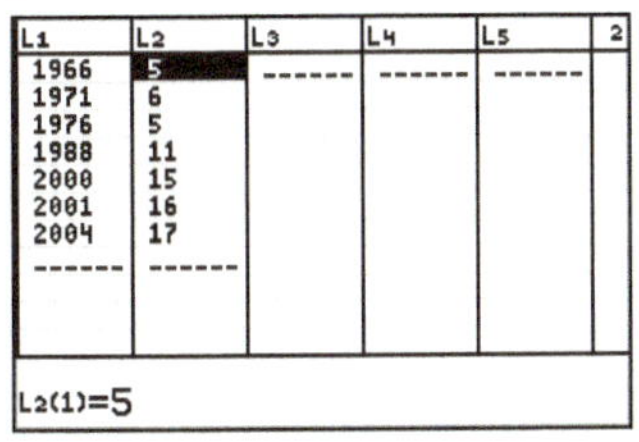

The STAT PLOT menu is used to define the data's scatter plot. Initially press [y=] and use [clear] to erase any functions that have been entered previously. Then make sure all plots are off by selecting [stat plot] ([2nd] [y=]) [4], which returns the user to the home screen. Select [enter].

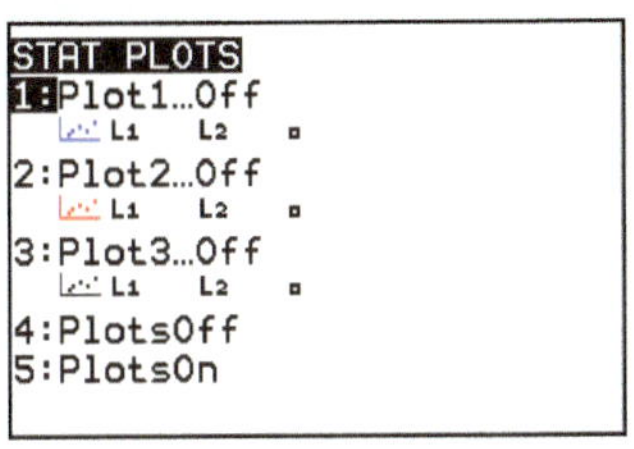

Select [stat plot] again and then hit [enter] to select Plot1. Hit [enter] to turn Plot1 on. The default option for Type is the scatter plot, so leave it as is. Be sure the Xlist and Ylist options are set to their default values of L_1 ([2nd] [1]) and L_2 ([2nd] [2]), respectively. The Mark option allows the user to choose how to display the plotted data points.

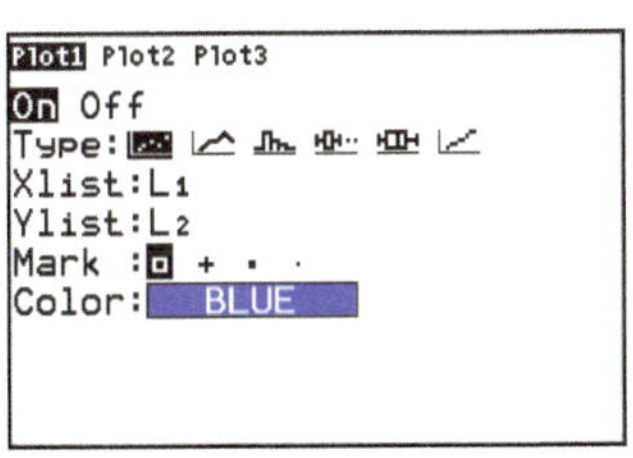

Press [zoom] [9] to select ZoomStat. This function draws the scatter plot and automatically sizes the window to include all the data points. [window] can be used to view and adjust the scale and range of the axes.

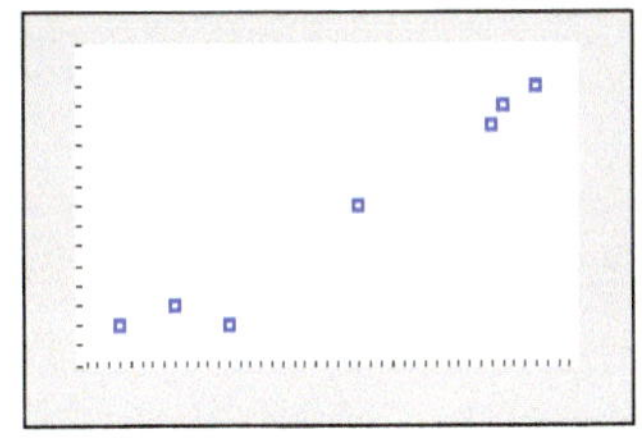

The CALC option of the STAT menu is used to determine the equation of the line of best fit. Select LinReg(ax+b) to place the linear regression command on the home screen. Press [enter] 3 times to highlight the cell next to Store RegEQ. Select Y_1 as the location to store the linear equation by pressing [alpha] [trace] [enter].

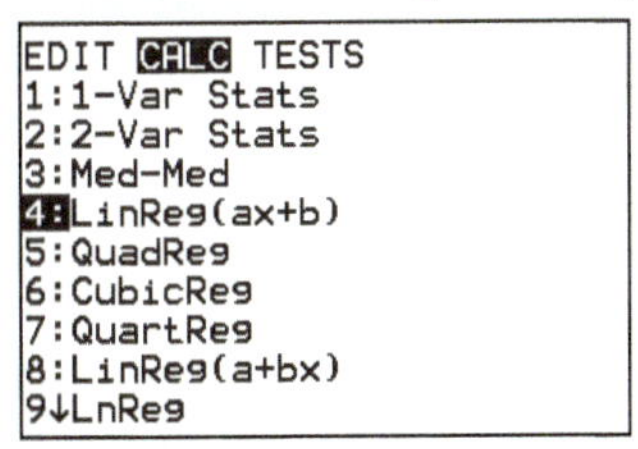

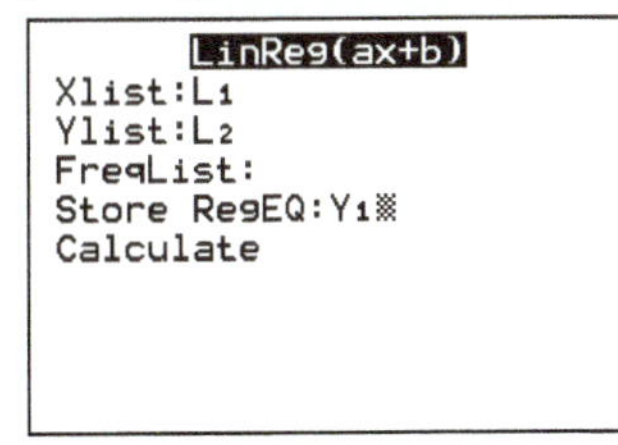

Selecting Calculate displays the values stored in the equation Y_1. The values of r^2 and r will be displayed when Stat Diagnostics is set to ON in the MODE settings. Press [y=] to view the stored equation, and press [graph] to view the line of best fit.

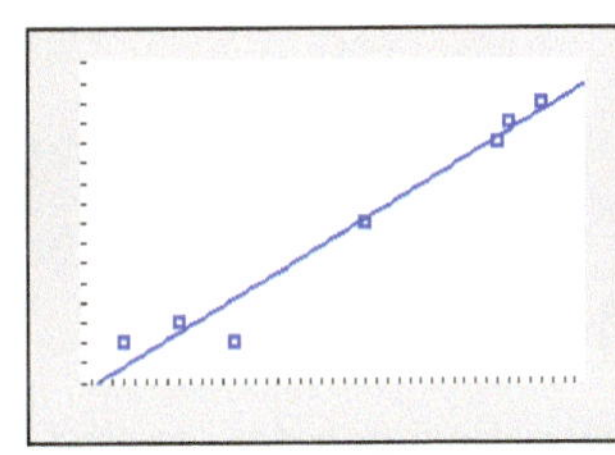

Use the function Y_1 stored in your calculator to estimate the sales revenue from Example 2 more accurately. Enter Y_1 by pressing [vars] and selecting the Y-VARS menu option Function, or by pressing [alpha] [trace] [enter]. To make predictions for the years 1995 and 2026, simply enter $Y_1(1995)$ and $Y_1(2026)$.

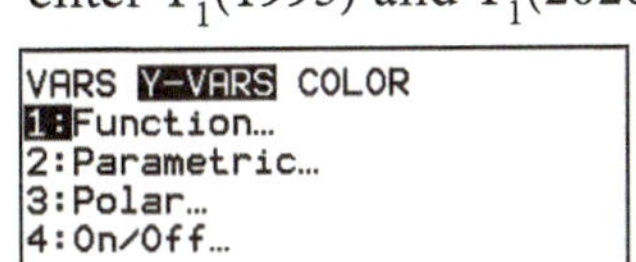

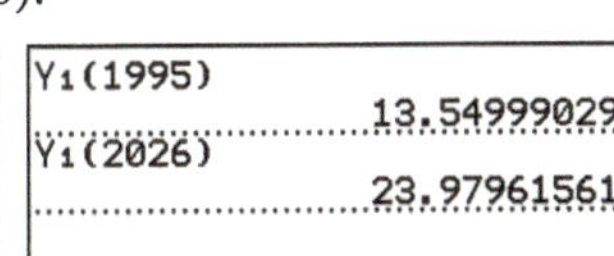

Statisticians more specifically define the correlation between data sets by calculating the correlation coefficient, r. The value of r can vary from -1 to 1 ($-1 \leq r \leq 1$). An r-value of -1 is called a perfect negative correlation, whereas an r-value of 1 is called a perfect positive correlation. Any r-values close to 1 or -1 are considered strong correlations. An r-value near 0 would describe a weak correlation and would mean the graphed dots do not line up at all but appear to be randomly scattered.

A. EXERCISES

Choose the option that describes the linear correlation of each data set.

A strong positive correlation

B weak positive correlation

C little or no correlation

D weak negative correlation

E strong negative correlation

1. 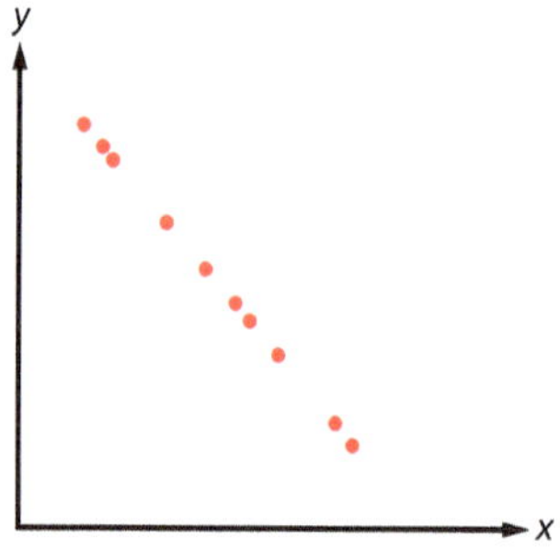

2. 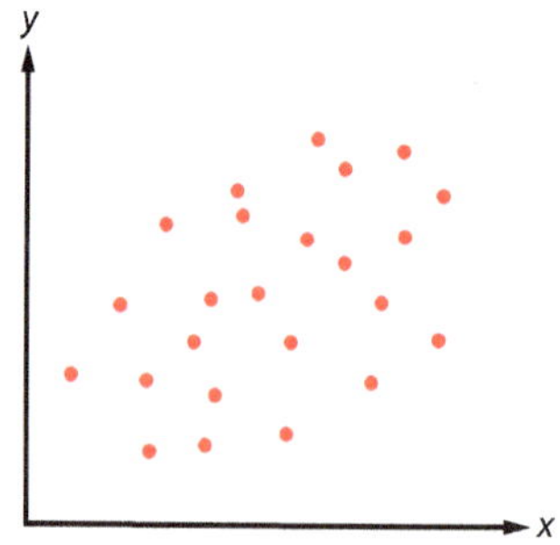

3. 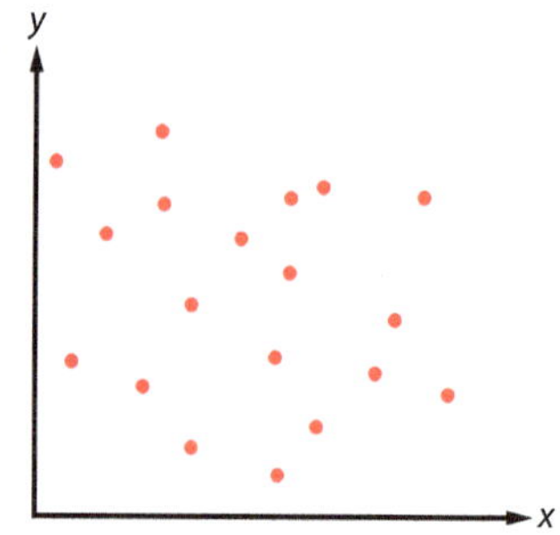

4. 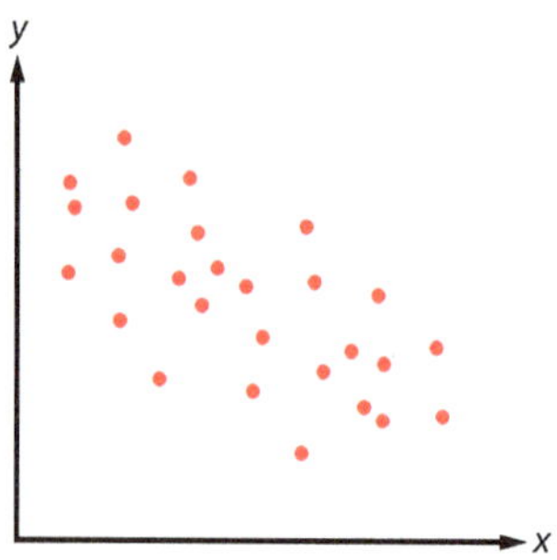

5. 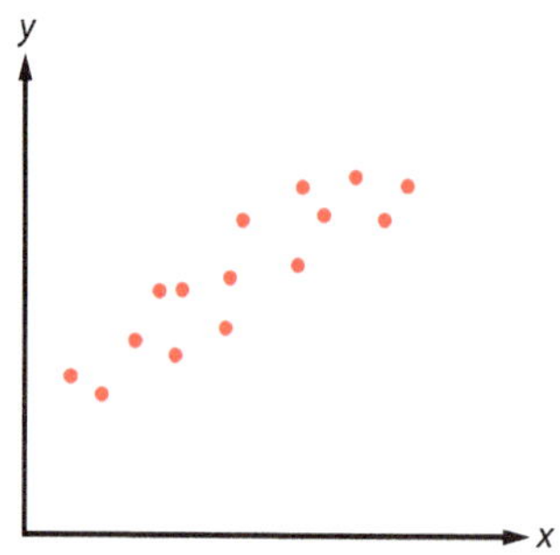

Use the given data table for exercises 6–10.

x	y
1	27.5
4	35
7	42.5
10	50
13	57.5

6. Make a scatter plot of the data and describe the linear correlation of the data.

7. Draw and determine the equation of a trend line that models the data set.

8. Use your equation to estimate the value of y when $x = 3.5$.

9. Use your equation to predict the value of y when $x = 20$.

10. Does the correlation of the data suggest predictable results?

Use the given data table for exercises 11–15.

x	y
1	7
2	8
3	6
4	3
5	3
6	1

11. Make a scatter plot of the data and describe the linear correlation of the data.

12. Draw and determine the equation of a trend line that models the data set.

13. Use your equation to estimate the value of y when $x = 3.5$.

14. Use your equation to predict the value of y when $x = 17$.

15. Does the correlation of the data suggest predictable results?

B. EXERCISES

Use the table listing the final history and English grades of 8 students for exercises 16–22.

history grade	English grade
87	93
75	84
81	78
91	86
69	71
86	88
74	78
95	97

16. Make a scatter plot of the data using the x-axis for history grades and the y-axis for English grades.

17. Draw and determine the equation of a trend line that models the data set.

18. Use your equation to estimate the English grade of a student who earned an 83 in history.

19. Use your equation to estimate the history grade of a student who earned an 83 in English.

20. Describe the correlation of the data. Does a high grade in one class cause a high grade in the other class? Explain.

21. Use a graphing calculator to determine the equation of the line of best fit.

22. To the nearest hundredth, use the equation of the line of best fit to estimate the grades from exercises 18–19.

Use the table giving the population of Rhode Island for exercises 23–29.

year	population of Rhode Island (1000s)
1920	604
1930	687
1940	713
1950	792
1960	859
1970	945
1980	947
1990	1003

23. Make a scatter plot of the data and describe the linear correlation of the data.

24. Draw and determine the equation of a trend line that models the data set.

25. Use your equation to predict the population of Rhode Island in 2020.

26. Use your equation to estimate when the population of Rhode Island was 0.

27. Which of these results is more likely to be accurate? Explain.

28. Use a graphing calculator to determine the equation of the line of best fit. Use it to predict the population of Rhode Island in 2020.

29. **MP5** Determine the r-value of the data to the nearest hundredth. Does this support your answer from exercise 23? Explain.

30. **Essential Question:** How can a line of best fit be useful?

Use the table listing the size (by the radius of the wheels) and the weight of mountain bikes for exercises 31–37.

wheel size (in.)	weight (lb)
19	32
20	30
18	31
21	31
15	29
19	27
17	30

31. Make a scatter plot of the data using the x-axis for wheel size and the y-axis for weight.

32. Draw and determine the equation of a trend line that models the data set.

33. Use your equation to estimate the weight of a 16" bike.

34. Use your equation to predict to the nearest hundredth the wheel size of a bike weighing 31 lb.

35. Describe the correlation of the data. What does this correlation imply about the accuracy of the predictions made by the equation modeling the data?

36. **MP5** Use a graphing calculator to determine the equation of the line of best fit and the correlation coefficient (to the nearest hundredth) for the data.

37. To the nearest hundredth, use the equation of the line of best fit to estimate the weight and wheel size from exercises 33–34.

C. EXERCISES

Use the table listing the speeds and stopping distances of a car for exercises 38–42.

stopping distance (ft)	speed (mi/hr)
16	10
59	25
80	30
154	45
183	50
216	55
251	60

38. Which quantity should be defined as the independent variable? Which is the dependent variable?

39. Make a scatter plot of the data. Would a line or a curve best model the data? By visual inspection, sketch a trend line or curve to model the data.

40. Make a table listing the speed squared for each stopping distance. Then make a scatter plot using the x-axis for speed squared and the y-axis for stopping distance.

41. **MP5** Determine the equation of the line of best fit and draw it on the scatter plot made in exercise 40. Since the x-axis represents speed squared, v^2, and the y-axis represents stopping distance, d, write the final form of this equation as $d = mv^2 + b$ instead of $y = mx + b$.

42. Use your equation to estimate the stopping distances at 40 mi/hr and at 70 mi/hr.

CUMULATIVE REVIEW

Solve each equation. [2.1–2.2]

43. $12x + 11 = 56$

44. $\frac{5}{2}x - 27 = 48$

45. $3(2x - 9) = 4x + 5$

46. $9 - (4x - 29) = 2(x + 7)$

State whether each of the following represents a direct variation, an inverse variation, or neither. [4.6]

47. $y = \frac{2}{x}$

48. $y = 3.5x$

49.

x	y
2	10
3	18
4	28
5	40

50.

x	y
3	8
6	4
12	2
15	1.6

51. What is the solution set for $4 - 3z \nleq 19$? **[3.1]**

A $z < -5$

B $z > -5$

C $z \geq -5$

D $z < 5$

E $z > 5$

52. Which pair of equations is neither parallel nor perpendicular? **[5.5]**

A $y = \frac{2}{3}x - 11$
$y = \frac{2x}{3} + 6$

B $y = 5 - 8x$
$y = \frac{1}{8}x + 3$

C $3x + 4y = 20$
$9x + 12y = 6$

D $x = 10$
$y = -10$

E none of these

APPLICATION PROBLEMS
SAFE SLOPES

Brad, a construction worker, longed to be used by the Lord in his vocation. His desires and gifts in construction, along with his understanding of math and Scripture, led him to serve Christian ministries, students, and individual families by constructing buildings and teaching Scripture and skills that would last for generations. Brad would come to glorify God in the secular workplace as well.

Brad's service for the Lord required a good understanding of safe construction practices, including the careful calculation of slopes. Correctly calculating slopes can help workers and customers avoid serious injuries or even fatalities! These application problems explore the use of math to love our neighbors by creating safe slopes.

Setting up a ladder at the correct angle is crucial to working safely. Emergency rooms in the US attend to more than 150,000 people annually who have been injured while using ladders. While most slopes are described as rise over run, the slope (or pitch) of a ladder is often described using a run-to-rise ratio. The run is the horizontal distance from the legs of a ladder to a building, and the rise is the vertical height at which the ladder contacts the building. Recommended ladder pitch may vary, but a 1 : 4 ratio is common.

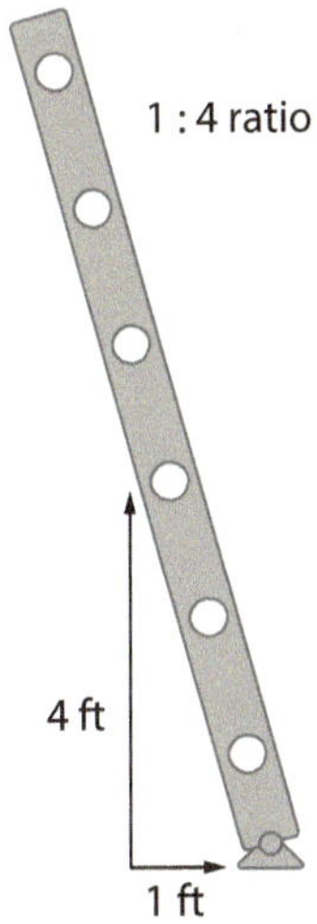

Use the recommended 1 : 4 ratio in exercises 1–2.

1. If Brad's ladder contacts a house at a point 24 ft above the ground, how far should the bottom of the ladder be from the side of the house?
2. If the bottom of a ladder is 4 ft from a wall, at what height on the wall should the ladder make contact?
3. If a ladder contacts a building 15 ft above the ground and the bottom of the ladder is 30 in. from the building, what is the ladder's pitch?
4. What could happen if the distance from the wall to the bottom of the ladder is too short?
5. Name 2 things that could happen if the distance from the wall to the bottom of the ladder is too far.

Brad's construction work involved a lot of time on roofs. Once, he experienced a severe fall from a roof, breaking his back. Although he fully recovered from this fall, many do not. One of the most significant factors to consider when working on a roof is its slope (or pitch). A 4/12 pitch is generally considered to be walkable. This means that the roof rises 4 in. for every foot of horizontal distance. Note that roof pitch ratios, instead of being reduced to lowest terms, are always expressed with a denominator of 12 (inches per foot).

6. On a roof with a 4/12 pitch, how high does the roof rise for every 36 in. of horizontal run?
7. On a 7/12 pitched roof, how high does the roof rise for every 4 ft of run?
8. To determine the slope of his roof, Tyler went into the attic and held a 2 ft level against the roof sheathing. He measured from the other side of the level to the roof. The measurement was exactly 14 in. What is the pitch of Tyler's roof?

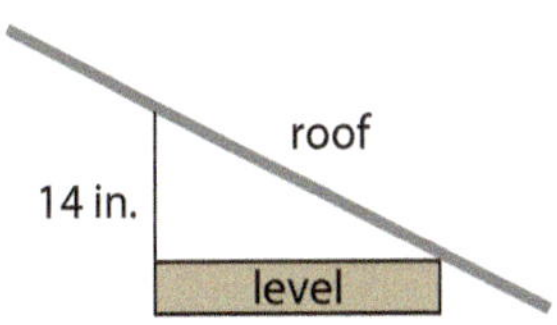

9. Brad's crew built a small church in Alaska with an 8/12 pitched roof. If the width of the building including the eaves is 120 ft, find the height to the ridge, h. Then use the Pythagorean Theorem ($a^2 + b^2 = c^2$) to find the width of the roof from the eaves to the ridge, w, to the nearest inch.

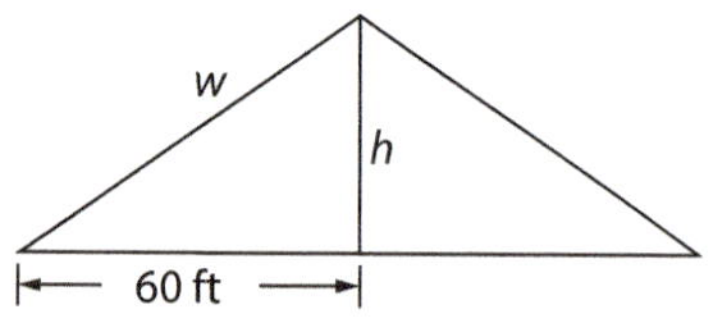

Brad's crew flew to St. Thomas to help rebuild after severe hurricane damage. They constructed concrete stairs to one of the new buildings. The plan proposed a rise of 118 in. and a run of 150 in. It was up to Brad and his crew to plan out the risers and tread of the staircase to meet the current building code safety requirements.

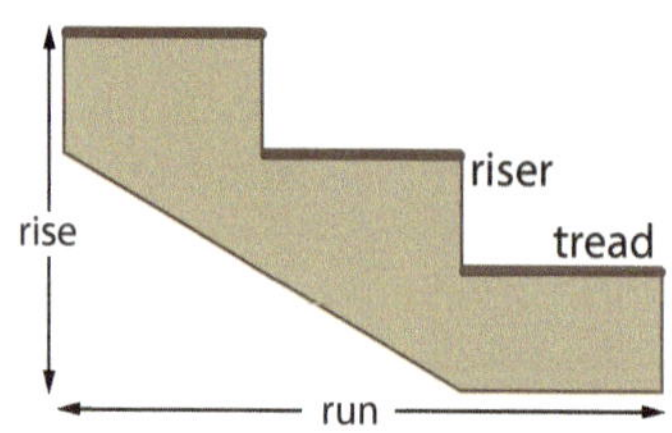

Use the following fixed-stairs building code.

riser: 6.5–9.5 in.
tread: 8–11 in.
slope: maximum 1.2

10. Determine the slope of the planned stairs and state whether the plan meets code.
11. How many risers are needed if Brad chooses a riser height as close to 7.5 in. as possible?
12. The building code requires all risers to be the same height. Using the results of exercise 11, determine the height of each riser.
13. Using the results of exercise 11, determine the width of each tread. Does this width meet code?

The installation of wheelchair ramps helps people in need by allowing greater accessibility for those with limited mobility. Codes commonly limit wheelchair ramps to a maximum slope of 1 : 12.

Brad's crew wants to construct a wheelchair access ramp at the home of an elderly couple in their church. Michael takes measurements and finds that the distance from their front porch to their driveway is 40 ft and the floor of the porch is 32 in. high.

14. Express the code's maximum ramp slope in decimal form.
15. Find the slope ratio of Michael's proposed 40 ft ramp. Would this ramp meet code?
16. What is the shortest ramp they can construct and remain in code?

Brad has depended on his math skills, his knowledge of construction, his crew, and his relationship with God to minister to the needs of many people. Brad and his crew demonstrated the love of God by using the abilities they had received from God to help others (1 Pet. 4:10).

5 CHAPTER REVIEW

Key Concepts

Graphing Lines [5.1]

- Linear equations can be graphed by connecting several plotted points together.
- Linear equations in standard form can be graphed by using their x- and y-intercepts.
- The standard form of a line is $Ax + By = C$, where A is positive and A, B, and C are real numbers.

Slope [5.2]

- The slope of a line can be found by using $m = \frac{\text{rise}}{\text{run}}$ or $m = \frac{y_2 - y_1}{x_2 - x_1}$.
- A horizontal line, $y = b$, has $m = 0$.
- A vertical line, $x = a$, has an undefined slope.
- The slope represents the rate of change.

Slope-Intercept Form [5.3]

- The slope-intercept form of a line is $y = mx + b$, where m is the slope and the y-intercept is $(0, b)$.
- An equation written in slope-intercept form can be graphed by connecting the y-intercept with other points generated using the slope.

Writing Linear Equations [5.4]

- An equation can be written to model a linear function represented by a graph or table.
  - Calculate the slope, m.
  - Determine the value of b from the y-intercept or by substituting the slope and a known point's coordinates into $y = mx + b$ and solving for b.
  - Substitute the values of m and b into $y = mx + b$.
- The point-slope form $y - y_1 = m(x - x_1)$ can be used to easily convert to slope-intercept form or standard form.
  - If 2 points are given, calculate m.
  - Substitute m and one of the given points into point-slope form.
  - Convert to slope-intercept form or standard form.

Parallel & Perpendicular Lines [5.5]

- Parallel lines are lines in the same plane that never intersect.
  - Parallel lines have the same slope but different y-intercepts.
- Perpendicular lines meet at right angles.
  - The slopes of nonvertical perpendicular lines are negative reciprocals of each other.
  - Horizontal lines ($m = 0$) and vertical lines (m is undefined) are perpendicular to each other.

Trend Lines & Correlation [5.6]

- A trend line is a line illustrating the relationship between 2 data sets on a scatter plot.
- Trend lines can be used to make estimates within the known data (interpolation) or predictions beyond the known data (extrapolation).
- A line of best fit is the most accurate trend line illustrating the relationship between data sets.
- The correlation coefficient, r, describes the correlation between data sets.
  - The r-value ranges from -1 to 1 ($-1 \le r \le 1$).
  - Data sets with r-values close to 1 or -1 indicate a strong correlation.
  - Data sets with r-values close to 0 indicate a weak correlation.

Biblical Worldview Shaping

Modeling with Algebra

What is the highest standard for validating any model or idea?

Scripture is the ultimate standard.

EXERCISES

Solve each linear equation for *y*. [5.1]

1. $5x - y = 6$

2. $3x - 5y = 5$

Make a table of at least 3 ordered pairs; then graph each linear equation. [5.1]

3. $\{(x, y) \mid y = x - 3\}$

4. $x - 5y = 6$

Find the *x*- and *y*-intercepts and then graph each equation. [5.1]

5. $x - 4y = 4$

6. $3x - 2y = 8$

Find the slope of each line. [5.2]

7.

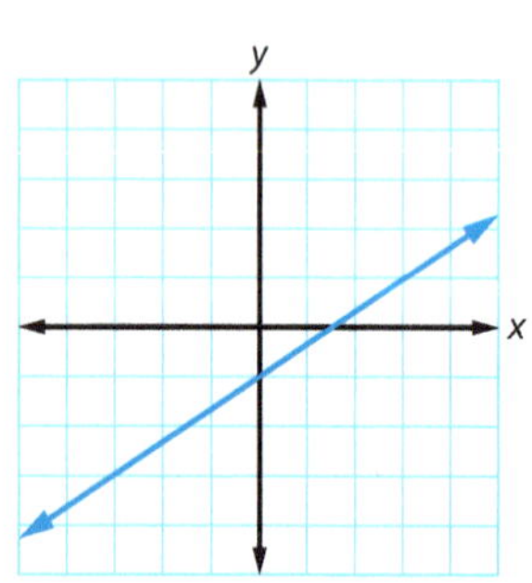

8.

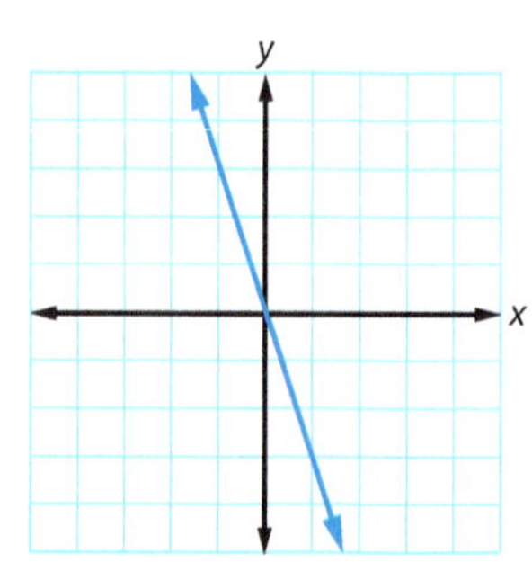

Find the slope of the line passing through the given points. [5.2]

9. (5, 9), (3, 2)

10. (−5, 6), (4, 2)

11. (5, 9), (3, 9)

12. The slope of every vertical line is ______.

13. Graph the line passing through the point (−4, 2) with a slope of $-\frac{3}{4}$.

14. In 1970, the approximate number of US households with TVs was 59,550,000. By 2020, that number had risen to 121,000,000. Determine the growth rate (in households/year) of TVs during this period.

State the slope and y-intercept of each linear equation. [5.3]

15. $y = \frac{5}{6}x - 3$

16. $x = 4$

Graph each equation using its slope and y-intercept. [5.3]

17. $y = \frac{2}{3}x + 2$

18. $3x + y = -6$

Write each equation as a linear function; then graph the function. [5.3]

19. $5x + 4y = 8$

20. $2x - 3y = 9$

21. A spring that is normally 8 in. long stretches $\frac{1}{4}$ in. per pound of applied force. Write a linear function, f, modeling the length of the spring based on the applied force, x. How many pounds of force are required to stretch the spring to a total length of $13\frac{1}{2}$ in.? **[5.3]**

Write the equation of each line in slope-intercept form. [5.4–5.5]

22. slope of $\frac{5}{2}$ and y-intercept of $(0, -1)$

23. slope of 2 and y-intercept of $(0, 3)$

24. passing through $(1, 4)$ with a slope of $-\frac{1}{2}$

25. passing through $(-3, -2)$ with a slope of $\frac{3}{2}$

26. passing through $(-2, 3)$ with a slope of 0

27. passing through $(-4, 6)$ and $(2, 3)$

28. passing through $(3, -7)$ and $(5, -1)$

29. passing through $(4, -3)$ and $(7, -3)$

30. passing through $(0, 0)$ and $(4, 3)$

31. parallel to $2x - 7y = 12$ and passing through $(3, 4)$

32. perpendicular to $4x + y = 7$ and passing through $(5, 4)$

33. Write the standard-form equation of the graphed line.

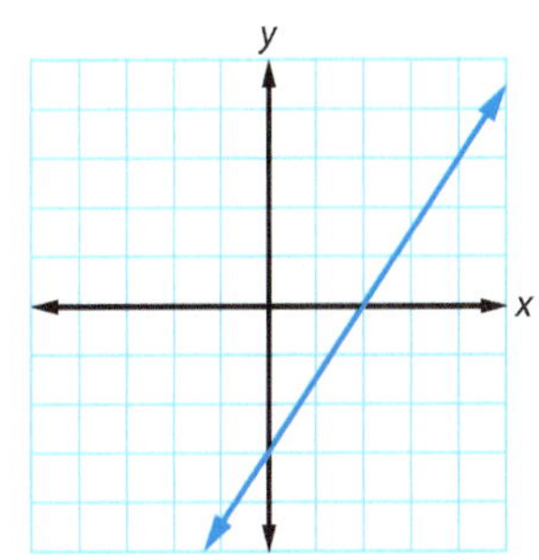

34. Write the standard-form equation of the line parallel to $2x + y = 1$ and passing through $(-2, 6)$.

35. State the negative reciprocals of -2 and $\frac{3}{8}$.

Determine whether the graphs of each pair of equations are parallel, perpendicular, or neither. [5.5]

36. $3x + y = 8$
$-2x + 6y = 18$

37. $4x + 7y = 12$
$5x + 9y = 18$

Find the slopes of each pair of lines and determine whether they are parallel, perpendicular, or neither. [5.5]

38.

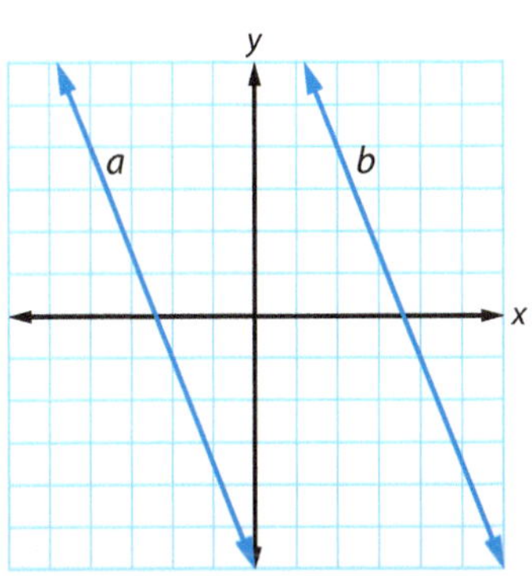

39.

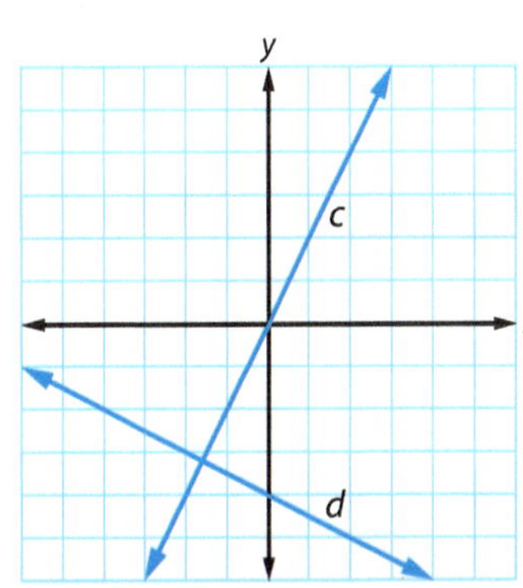

40. The US national debt at the end of the 2008 fiscal year (September 30) was about $10.025 trillion, and at the end of the 2018 fiscal year, it was about $21.516 trillion. Write a linear function, d, to predict the national debt for any year, x. Then predict the debt at the end of 2028. Explain why you think your linear function is or is not a good model for the national debt. **[5.6]**

The fork length of a shark is the distance from the tip of the nose to the fork in the tail. Use the given data table for exercises 41–44. [5.6]

fork length (cm)	weight (kg)
188	92
197	116
169	59
180	77
158	45
117	22
136	39

41. Make a scatter plot of the data using the x-axis for fork length and the y-axis for weight. Describe the linear correlation of the data.

42. Draw and determine the equation of a trend line that models the data set.

43. Use your equation to estimate to the nearest tenth of a kilogram the weight of a 160 cm shark.

44. Use your equation to predict the fork length to the nearest centimeter of a shark weighing 130 kg.

45. The predictions made from trend lines tend to be most accurate when the data exhibits which type of correlation? **[5.6]**

A positive

B negative

C strong

D weak

E little or none

46. True or false: Predictions made by interpolation tend to be more accurate than those made by extrapolation. **[5.6]**

47. **BWS Essential Question:** If you check the accuracy of a population growth model by referring to census data, which one is the standard, the mathematical model or the census?

STEM PROJECT

THE WATER WHEEL

What if you needed to power something and you could not simply "plug it in"? Water wheels were one of the first machines that converted the power of water flow in a stream or river into energy. God created the earth with an expanse of water and gravity, which causes water to flow downhill. This natural flow of water is a sustainable resource that can produce great amounts of energy. In this STEM project you will build a small water wheel and assess its effectiveness by lifting various weights using the wheel.

CHAPTER 06
LINEAR SYSTEMS

CHAPTER OVERVIEW

6.1 SOLVING SYSTEMS BY GRAPHING

What does the graph of a system of linear equations reveal about its number of solutions?

After completing this section, I will be able to

- solve systems of linear equations by graphing.
- describe the 3 possible types of solutions for a system of 2 linear equations.
- solve real-world problems by writing and graphing a system of linear equations.
- explain why the intersection of 2 distinct lines is exactly 1 point.

A **system of linear equations** is made up of 2 or more linear equations with the same variables. Any ordered pair that makes all the equations true is a **solution to the system of equations.**

EXAMPLE 1: Identifying a Solution

Determine whether each ordered pair is a solution to the system below.

$x - y = 9$

$2x + y = 9$

a. (11, 2) **b.** (6, −3)

Answers

a.

$x - y = 9$	$2x + y = 9$	Substitute 11 for x and 2 for y in both equations.
$11 - 2 \stackrel{?}{=} 9$	$2(11) + 2 \stackrel{?}{=} 9$	
$9 \stackrel{?}{=} 9$; true	$24 \stackrel{?}{=} 9$; false	

(11, 2) is not a solution since it does not satisfy both equations.

b.

$x - y = 9$	$2x + y = 9$	Substitute 6 for x and −3 for y in both equations.
$6 - (-3) \stackrel{?}{=} 9$	$2(6) + (-3) \stackrel{?}{=} 9$	
$9 \stackrel{?}{=} 9$; true	$9 \stackrel{?}{=} 9$; true	

(6, −3) is a solution since it satisfies both equations.

SKILL CHECK **EXERCISE 5** ☑

The ordered pair that satisfies both equations might be determined intuitively in a simple system, but most systems will require an organized approach to solve.

Graphing both equations on the same coordinate plane will provide a visual representation of the system. Since the system's solution consists of any ordered pairs that satisfy both equations, the solution of the system is a single point of intersection.

SOLVING A SYSTEM BY GRAPHING

1. Graph both equations on the same coordinate plane.
2. Identify the solution, the coordinates of the point of intersection.
3. Check the solution by substituting it into each equation.

EXAMPLE 2: Solving by Graphing

Solve the system by graphing.

$y = -x - 2$

$y = x - 4$

Answer

$y = -x - 2$ $\quad$ $y = x - 4$

1. Both equations are already in slope-intercept form.

$m = -1$, y-intercept: $(0, -2)$ $\quad$ $m = 1$, y-intercept: $(0, -4)$

2. Identify the slope and y-intercept of each line.

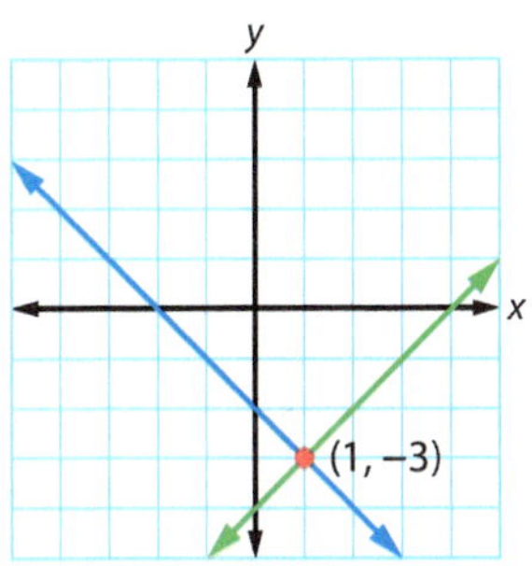

3. Graph both lines on the same coordinate plane.

solution: $(1, -3)$

4. Identify the point of intersection.

$y = -x - 2$ $\quad$ $y = x - 4$

$-3 = -(1) - 2$ $\quad$ $-3 = 1 - 4$

5. The solution checks in both equations.

SKILL CHECK **EXERCISES 1, 15**

Often systems of equations can be easily graphed using the x- and y-intercepts. Solving $Ax + By = C$ for its intercepts provides simple intercept formulas that can be applied to equations written in standard form. When $y = 0$, x will be $\frac{C}{A}$, indicating an x-intercept at $\left(\frac{C}{A}, 0\right)$. When $x = 0$, y will be $\frac{C}{B}$, indicating a y-intercept at $\left(0, \frac{C}{B}\right)$.

EXAMPLE 3: Solving by Graphing

Solve the system by graphing.

$x - y = -6$

$8x + 2y = -8$

Answer

$x - y = -6$	$8x + 2y = -8$	1. Find the intercepts by applying the intercept formulas.
$x\text{-int} = \left(\frac{C}{A}, 0\right) = \left(\frac{-6}{1}, 0\right) = (-6, 0)$	$x\text{-int} = \left(\frac{C}{A}, 0\right) = \left(\frac{-8}{8}, 0\right) = (-1, 0)$	
$y\text{-int} = \left(0, \frac{C}{B}\right) = \left(0, \frac{-6}{-1}\right) = (0, 6)$	$y\text{-int} = \left(0, \frac{C}{B}\right) = \left(0, \frac{-8}{2}\right) = (0, -4)$	

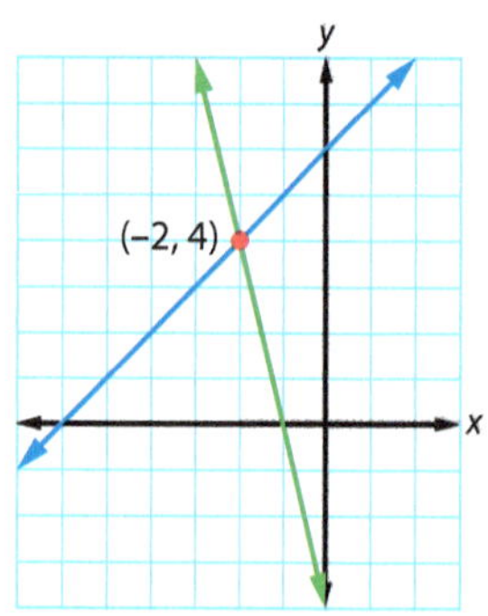

2. Graph both lines on the same coordinate plane.

solution: $(-2, 4)$

3. Identify the point of intersection.

$x - y = -6$	$8x + 2y = -8$	4. Check the solution in each equation.
$-2 - 4 = -6$	$8(-2) + 2(4) = -8$	

SKILL CHECK **EXERCISE 19**

The solution to each system in Examples 2–3 is a single ordered pair where the lines intersect. Some systems, however, have equations that represent parallel lines or the same line. Graphs of parallel lines have no common point; therefore, the system has no solution. When the graphs are the same line, they are said to *coincide*. The solution is the set of all the points on the line.

EXAMPLE 4: Solving Special Systems

Solve each system by graphing.

a. $x + y = 3$
$2y = -2x + 6$

b. $x + y = 3$
$x + y = 7$

Answers

a. $x + y = 3$ → $y = -x + 3$ $2y = -2x + 6$ → $y = -x + 3$

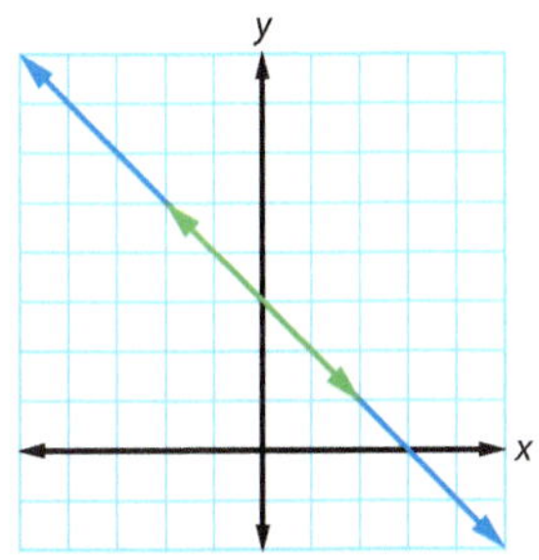

solution: all points on $y = -x + 3$

1. Write each equation in slope-intercept form.
2. Graph both lines on the same coordinate plane.

$m = -1$ y-intercept: (0, 3) $m = -1$ y-intercept: (0, 3)

3. Identify the solution as any point such that $y = -x + 3$.

b. $x + y = 3$ → $y = -x + 3$ $x + y = 7$ → $y = -x + 7$

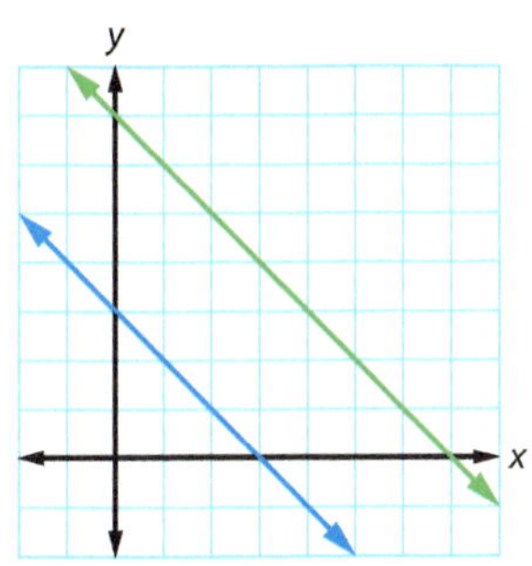

solution: $\varnothing$

1. Write each equation in slope-intercept form.
2. Graph both lines on the same coordinate plane.

$m = -1$ y-intercept: (0, 3) $m = -1$ y-intercept: (0, 7)

3. Identify the solution for a system with no points in common.

SKILL CHECK **EXERCISE 29**

AfterSchoolHelp

Solving Systems by Graphing

POSSIBLE SOLUTIONS TO A SYSTEM			
Graph	intersecting lines 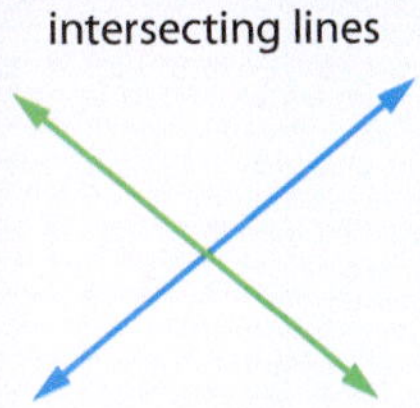	parallel lines 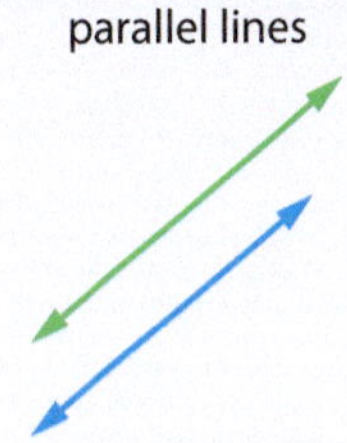	coinciding lines 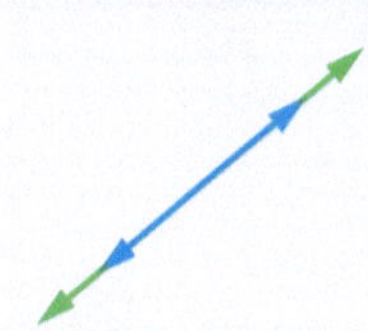
Number of Solutions	1 solution	no solution	infinitely many solutions
Answer	ordered pair	$\varnothing$	all the points on a line

These solutions assume that lines are straight and coplanar.

EXAMPLE 5: Solving a Real-World Problem

Nancy and Drew are each saving for a car. Nancy started with \$3000 and added \$250/month. Drew started with \$2000 and added \$350/month to his savings. Write a function rule for each person's savings. Then graph the system to find when Nancy and Drew will have saved the same amount of money.

Answer

Comparison of Savings

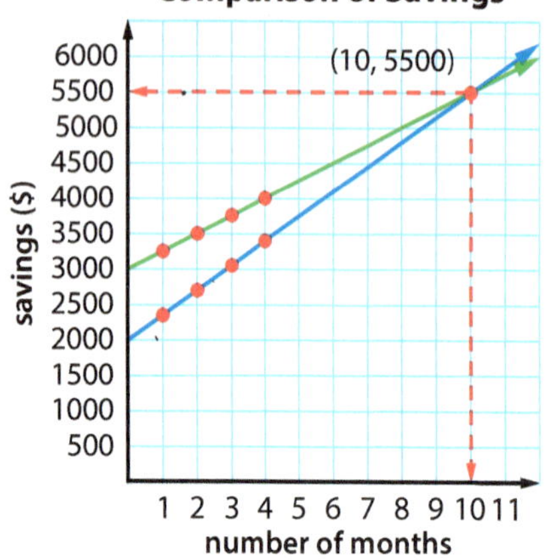

solution: $x = 10$ months

1. Write and graph functions representing each account. Since the number of months and the savings amount are never negative, use only the first quadrant.
   - **a.** Nancy: $n(x) = 250x + 3000$
     Plot the y-intercept, (0, 3000), and use the slope to plot several more points: (1, 3250), (2, 3500), and so on.
   - **b.** Drew: $d(x) = 350x + 2000$
     Plot the y-intercept, (0, 2000), and use the slope to plot several more points: (1, 2350), (2, 2700), and so on.
2. Identify the coordinates of the point of intersection.
3. Nancy and Drew will have saved \$5500 at 10 months.

SKILL CHECK **EXERCISE 35**

VOCABULARY

- solution to a system of equations
- system of linear equations

A. EXERCISES

Identify the solution to each system from its graph. If the solution is a line, state the line's equation.

1.

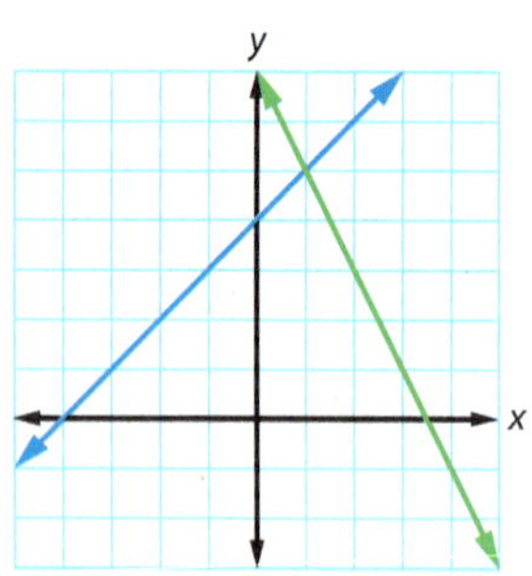

2.

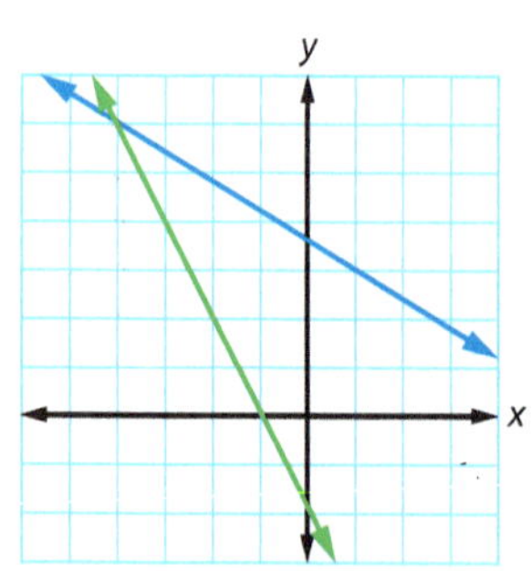

3.

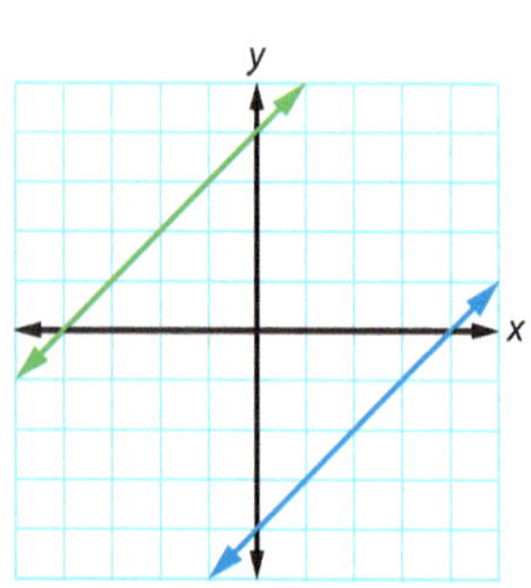

4.

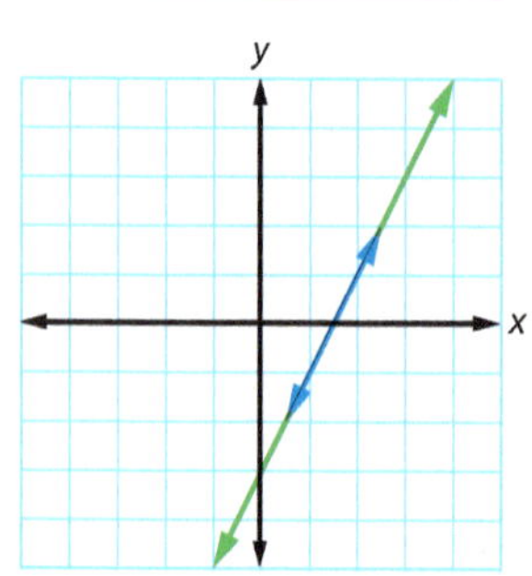

Determine whether (–2, 5) is a solution to each system.

5. $x + y = 3$
$2x + y = 1$

6. $x + 2y = 8$
$3x + y = 1$

7. $x - y = 7$
$2x + 3y = 4$

8. $3x + 2y = 4$
$5x + 3y = 5$

For each system, determine whether the given ordered pair is a solution.

9. $(2, -5)$
$x + 3y = 13$
$6x + y = 7$

10. $(7, 1)$
$2x - 5y = 9$
$x + y = 8$

11. $(-3, -2)$
$x + 2y = 1$
$3x - 2y = 5$

12. $(-1, 4)$
$3x - y = -7$
$x - 5y = -21$

13. $(6, -5)$
$5x + 6y = 0$
$2x + y = 7$

14. $(-2, -8)$
$y = -2$
$y = 4x$

Solve each system by graphing.

15. $y = 2x + 5$
$y = x - 1$

16. $y = x + 2$
$y = 4x - 4$

17. $y = -6x + 7$
$y = 1$

18. $y = 3x + 4$
$x = -2$

B. EXERCISES

Solve each system by graphing.

19. $4x + y = 7$
$x - 3y = 18$

20. $x + y = 9$
$-2x + y = -3$

21. $x + 2y = 10$
$9x + 2y = -6$

22. $8x - 4y = -16$
$4x - y = -8$

23. $x - 5y = -30$
$2x - y = 3$

24. $y = -3x + 5$
$3y = -4x + 15$

25. $2y = -x + 10$
$2y = -3x + 14$

26. $-2y = -x + 4$
$2y = -3x - 12$

27. $5x - 3y = -12$
$2x + 3y = -9$

28. $2x - 7y = 21$
$3x + 7y = 14$

29. $x + 2y = 8$
$y = -\frac{x}{2} + 4$

30. $8x - 2y = 6$
$y = 4x + 3$

31. $x = -3$
$x = 4$

32. $x - y = 6$
$-5y = -5x + 30$

33. **MP2** The sum of 2 numbers is −7. The difference of the same 2 numbers is 3.

- **a.** Write 2 equations to represent these conditions.
- **b.** Graph the equations and name the point of intersection.

34. **MP4** Marie's Deli charges a \$45 fee plus \$5 per plate. Just Right Catering charges \$14 per plate.

- **a.** Write a function rule describing the cost of catering for x people using each caterer.
- **b.** Graph the equations to determine the number of people that would result in the same cost. What is that cost?
- **c.** Analyze the graph and determine when Marie's Deli would be cheaper.

35. Miguel sells jackets for \$30 each. He pays \$20 for each jacket and estimates overhead expenses at \$120.

- **a.** Write a function rule, $r(x)$, describing his revenue if he sells x jackets.
- **b.** Write a function rule, $e(x)$, describing his expenses if he buys x jackets and has an overhead of \$120.
- **c.** Graph the equations to determine the number of jackets Miguel must sell to break even (the intersection of the 2 graphs).

36. Picture Perfect Studios charges \$9 per sheet of photos plus an \$18 sitting fee. The Portrait Place charges \$15 per sheet with no sitting fee. For what number of photo sheets will the costs be the same?

37. **Essential Question:** What does the graph of a system of linear equations reveal about its number of solutions?

C. EXERCISES

38. Solve the system by graphing: $9x - 2y = 7$ and $3x + 4y = 14$.

39. **MP4** Mrs. Spencer is considering which rental truck company to use for a move. The cost of renting from Great Haul is \$100 for the first 250 mi and \$0.10/mi thereafter. Mini Massive Movers costs \$80 for trips up to 100 mi and then \$0.15/mi for additional miles. Graph the cost of each truck rental on the same set of axes. Then use the graph to approximate the number of miles at which the truck rental would cost the same. Which company should she use for a 300 mi move?

CUMULATIVE REVIEW

Evaluate each expression.

40. $\left|\frac{7}{26} - \frac{14}{39}\right|$ **[1.1]**

41. $\frac{3}{5} + \frac{12}{35} \cdot \frac{49}{9}$ **[1.4]**

42. $\frac{6^{-2}}{3} \cdot \frac{2^{-1}}{9^{-3}}$ **[1.3]**

Solve each equation for the indicated variable. [2.3]

43. $3x + 4y = 5$ for x

44. $x + 3y = 21$ for y

Solve each inequality.

45. $4(x - 3) \leq 8x$ **[3.2]**

46. $x \leq 4$ and $3x \geq 2x + 4$ **[3.3]**

47. $x < 5$ or $3(x + 2) \geq 2x$ **[3.4]**

48. Which linear equation has a y-intercept above the origin? **[5.4]**

A $y + 5 = 2x$

B $y = 3x - 1$

C $y - 2 = x$

D $x = 8$

E none of these

49. Which r-value(s) indicate a weak correlation between 2 data sets? Select all that apply. **[5.6]**

A -1

B 0.8

C 0.1

D -0.3

E all of these

TECHNOLOGY CORNER (TI-84 PLUS FAMILY)

Solving Systems

You can use your graphing calculator to solve systems of equations. First, you will have to solve the equations for y to enter them into the function editor. After using the [mode] menu to make sure that the calculator is in function mode, press [y=] and enter the equations as shown. Make sure that all stat plots are turned off. If Plot1, Plot2, or Plot3 is highlighted, arrow up and press [enter] to turn off the stat plot.

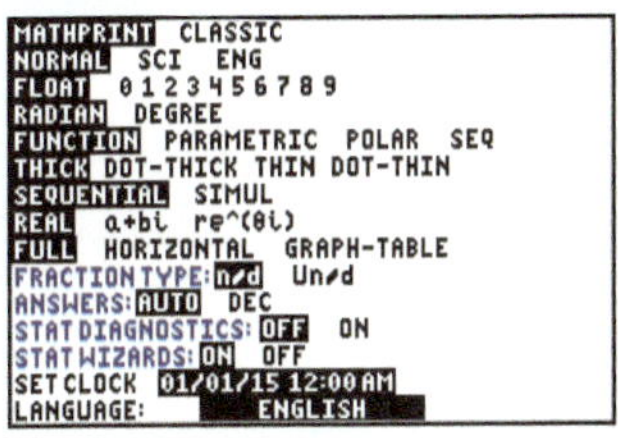

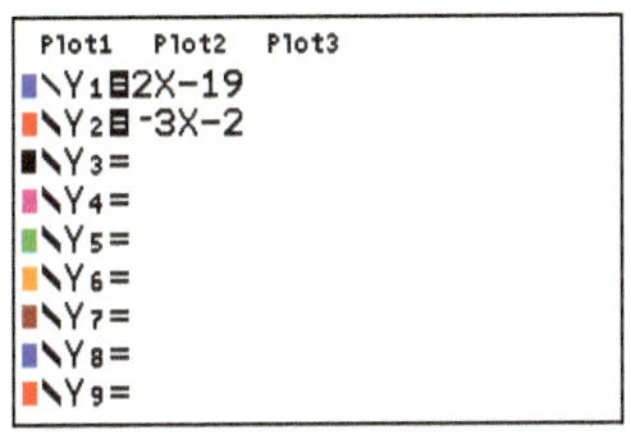

Selecting ZStandard from the [zoom] menu graphs the lines in a window with $-10 \leq x \leq 10$ and $-10 \leq y \leq 10$. The intersection point cannot be seen within this window. Choose Zoom Out from the [zoom] menu. A cursor will be flashing at the point from which the calculator will zoom out. This point can be adjusted if needed. Press [enter] a second time.

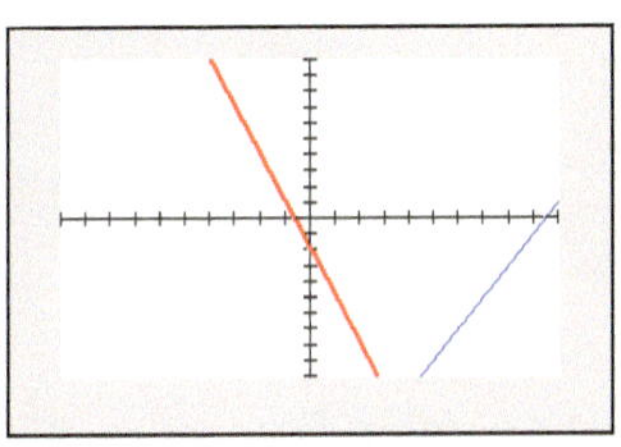

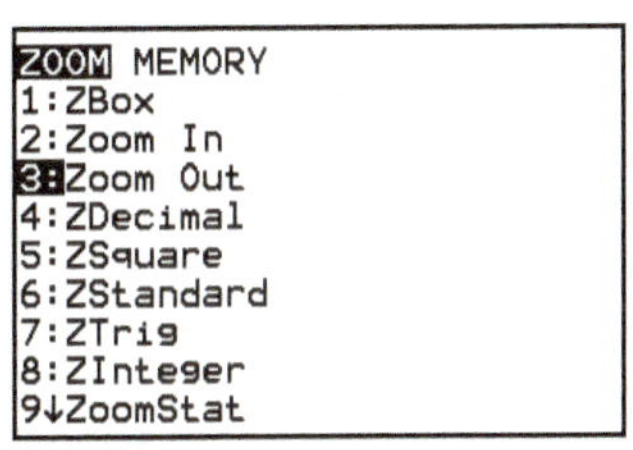

To find the coordinates of the point of intersection, press [2nd] [trace] to access the CALCULATE menu. Choose intersect.

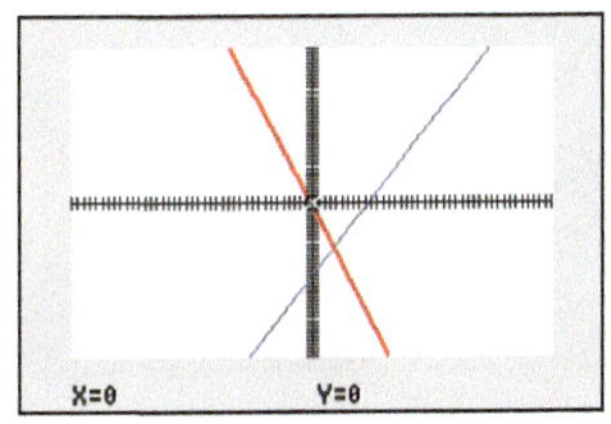

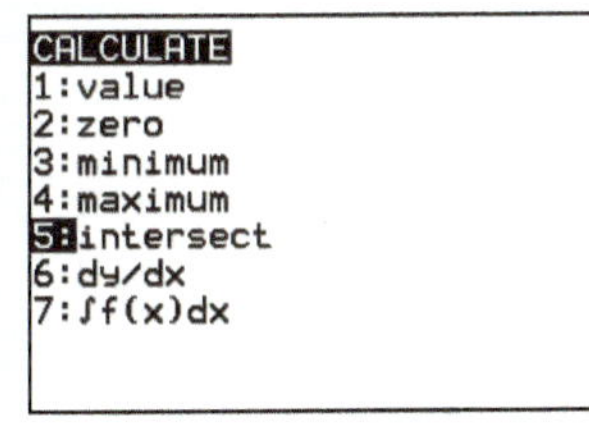

Your calculator will ask for verification of the first curve and the second curve and will give an option to guess the point of intersection. Press [enter] three times. The solution will be given along the bottom of the screen.

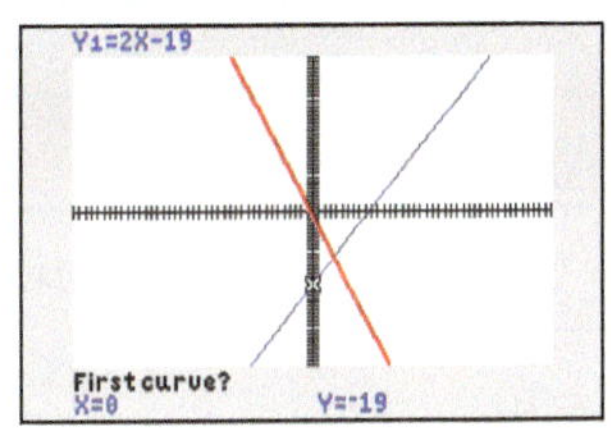

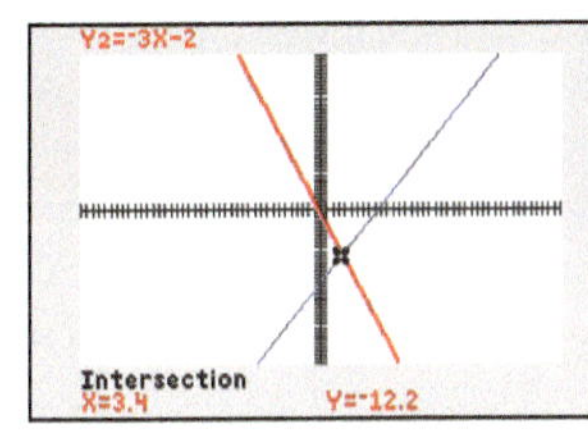

6.2 SOLVING SYSTEMS BY SUBSTITUTION

How does substitution help in solving systems of equations with 2 variables?

After completing this section, I will be able to

- solve systems of linear equations by substitution.
- solve real-world problems by writing and solving a system of linear equations.

The graphing method of solving systems of linear equations works well if the solution involves small integers but becomes awkward when the solution contains large integers or fractions.

The **substitution method** can be used to find a single-variable equation by replacing one of the variables with an equivalent expression.

EXAMPLE 1: Solving by Substitution

Solve the system by substitution.

$x + y = -4$

$x - y = 10$

Answer

$x + y = -4$ $x = -y - 4$	and	$x - y = 10$	**1.** Solve the first equation for x.
		$(-y - 4) - y = 10$	**2.** Substitute the expression for x in the second equation.
		$-2y - 4 = 10$ $-2y = 14$ $y = -7$	**3.** Solve for y.
$x = -(-7) - 4$ $x = 3$	or	$x - (-7) = 10$ $x = 3$	**4.** Substitute -7 for y in either equation and solve for x.
solution: $(3, -7)$			**5.** Write the solution as an ordered pair.
$x + y = -4$ $3 + (-7) = -4$	and	$x - y = 10$ $3 - (-7) = 10$	**6.** The solution checks in both equations.

The graphs of the equations in Example 1 intersect at the point (3, −7).

SKILL CHECK EXERCISES 1, 9

SOLVING A SYSTEM BY SUBSTITUTION

1. Solve either equation for either variable.
2. Substitute the expression into the other equation and solve.
3. Use the value of the known variable to solve for the other variable.
4. Write the solution as an ordered pair.
5. Check the solution by substituting it into each equation.

EXAMPLE 2: Solving by Substitution

Solve the system by substitution.
$2x - 8y = 0$
$3x + 2y = 8$

Answer

$2x - 8y = 0$ $x - 4y = 0$ $x = 4y$	and	$3x + 2y = 8$	**1.** Divide both sides of the first equation by 2 to make it easier to solve for *x*.
		$3(4y) + 2y = 8$	**2.** Substitute the expression for *x* in the second equation.
		$12y + 2y = 8$ $14y = 8$ $y = \frac{8}{14} = \frac{4}{7}$	**3.** Solve for *y*.
$x = 4\left(\frac{4}{7}\right)$ $x = \frac{16}{7}$	or	$3x + 2\left(\frac{4}{7}\right) = 8$ $x = \frac{16}{7}$	**4.** Substitute $\frac{4}{7}$ for *y* in either equation and solve for *x*.
solution: $\left(\frac{16}{7}, \frac{4}{7}\right)$			**5.** Write the solution as an ordered pair.
$x = 4y$ $\frac{16}{7} = 4\left(\frac{4}{7}\right)$	and	$3x + 2y = 8$ $3\left(\frac{16}{7}\right) + 2\left(\frac{4}{7}\right) = 8$ $\frac{48}{7} + \frac{8}{7} = 8$	**6.** The solution checks in both equations.

DOING THE MATH

$$3x + 2\left(\frac{4}{7}\right) = 8$$
$$3x + \frac{8}{7} = \frac{56}{7}$$
$$-\frac{8}{7} \quad -\frac{8}{7}$$
$$\frac{1}{3}(3x) = \frac{1}{3}\left(\frac{48}{7}\right)$$
$$x = \frac{16}{7}$$

SKILL CHECK **EXERCISE 15**

EXAMPLE 3: Solving by Substitution

Solve the system by substitution.
$2x + 3y = 16$
$4x + 5y = 28$

Answer

$2x + 3y = 16$ $2x = -3y + 16$ $x = \frac{-3y + 16}{2}$	and	$4x + 5y = 28$	**1.** Solve the first equation for *x*.
		$4\left(\frac{-3y + 16}{2}\right) + 5y = 28$ $-6y + 32 + 5y = 28$ $-y + 32 = 28$ $-y = -4$ $y = 4$	**2.** Substitute the expression for *x* in the second equation and solve for *y*.
$x = \frac{-3(4) + 16}{2}$ $x = \frac{-12 + 16}{2}$ $x = 2$	or	$4x + 5(4) = 28$ $4x = 8$ $x = 2$	**3.** Substitute 4 for *y* in either equation and solve for *x*.
solution: (2, 4)			**4.** Write the solution as an ordered pair.
$2(2) + 3(4) = 16$	and	$4(2) + 5(4) = 28$	**5.** The solution checks in both equations.

DOING THE MATH

$$4\left(\frac{-3y + 16}{2}\right)$$
$$= \overset{2}{\cancel{4}}\left(\frac{-3y + 16}{\cancel{2}}\right)$$
$$= 2(-3y + 16)$$
$$= -6y + 32$$

SKILL CHECK **EXERCISE 23**

EXAMPLE 4: Solving by Substitution

Solve the system by substitution.

$4x + 3y = 3$

$5x + 2y = 1$

Answer

$$4x + 3y = 3 \quad \text{and} \quad 5x + 2y = 1$$
$$2y = -5x + 1$$
$$y = \frac{-5x + 1}{2}$$

1. Solve the second equation for y since 2 is the smallest coefficient.

$$4x + 3\left(\frac{-5x + 1}{2}\right) = 3$$
$$4x + \frac{-15x + 3}{2} = 3$$
$$2\left(4x + \frac{-15x + 3}{2}\right) = 2(3)$$
$$8x - 15x + 3 = 6$$
$$-7x + 3 = 6$$
$$-7x = 3$$
$$x = -\frac{3}{7}$$

2. Substitute the expression for y in the first equation and solve for x.
   a. Distribute the factor of 3 to remove the parentheses.
   b. Clear the fraction by multiplying both sides of the equation by 2.

$$4\left(-\frac{3}{7}\right) + 3y = 3$$
$$-\frac{12}{7} + 3y = 3$$
$$7\left(-\frac{12}{7} + 3y\right) = 7(3)$$
$$-12 + 21y = 21$$
$$21y = 33$$
$$y = \frac{33}{21} = \frac{11}{7}$$

3. Substitute $-\frac{3}{7}$ for x in either equation and solve for y.

solution: $\left(-\frac{3}{7}, \frac{11}{7}\right)$

4. Write the solution as an ordered pair.

$$4\left(-\frac{3}{7}\right) + 3\left(\frac{11}{7}\right) = 3 \qquad 5\left(-\frac{3}{7}\right) + 2\left(\frac{11}{7}\right) = 1$$

5. The solution checks in both equations.

SKILL CHECK **EXERCISE 27** ☑

AFTERSCHOOLHELP

Solving Systems by Substitution

EXAMPLE 5: Solving a Real-World Problem

Bryan has 84 ft of fencing and wishes to fence a rectangular area so that the length is 6 ft longer than the width. Find the dimensions.

Answer

Let l = the length and w = the width.

1. Assign variables to the unknowns.

$2l + 2w = 84$ and $l = w + 6$

2. Write a system of equations. Use the perimeter formula for the first equation.

$$\begin{aligned} 2(w+6) + 2w &= 84 \\ 2w + 12 + 2w &= 84 \\ 4w + 12 &= 84 \\ 4w &= 72 \\ w &= 18 \end{aligned}$$

3. Substitute the expression for l in the first equation and solve for w.

$l = 18 + 6 = 24$

4. Substitute for w to find l.

$P = 2l + 2w = 2(24) + 2(18) = 48 + 36 = 84$

5. Check using the formula for perimeter.

The dimensions of the rectangular fenced-in area are 18 ft × 24 ft.

SKILL CHECK **EXERCISE 31**

Many of the word problems solved using tables in Chapter 2 can be solved by writing a system of 2 equations with 2 variables.

EXAMPLE 6: Solving a Real-World Problem

Mr. Peterson invested a total of $20,000 between 2 accounts. If the interest rates were 2% and 4.5% and the total interest earned in 2 yr was $1175, how much was invested in each account?

Answer

Let x = the principal for the 2% account and y = the principal for the 4.5% account.

1. Assign variables to the unknowns.

P	***r***	***t***	***I***
x	0.02	2	$0.04x$
y	0.045	2	$0.09y$

2. Make a table using the formula *principal · rate · time = interest.*

Total principal is $20,000.

$x + y = 20{,}000$

Total interest is $1175.

$0.04x + 0.09y = 1175$

3. Write and solve a system of 2 equations.

$y = 20{,}000 - x$

4. Solve the first equation for y.

$$\begin{aligned} 4x + 9y &= 117{,}500 \\ 4x + 9(20{,}000 - x) &= 117{,}500 \\ 4x + 180{,}000 - 9x &= 117{,}500 \\ -5x + 180{,}000 &= 117{,}500 \\ -5x &= -62{,}500 \\ x &= 12{,}500 \end{aligned}$$

5. Clear the second equation of decimals. Substitute the expression for y and solve for x.

$y = 20{,}000 - 12{,}500$

$y = 7500$

6. Substitute 12,500 for x in the first equation and solve for y.

Mr. Peterson invested $12,500 at 2% and $7500 at 4.5%.

7. State the solution to the problem.

$0.04(12{,}500) + 0.09(7500) = 500 + 675 = 1175$

8. **Check:** The total interest earned is $1175.

SKILL CHECK **EXERCISE 35**

VOCABULARY

- substitution method

A. EXERCISES

Choose the best initial step in solving each system; then explain your reasoning. Do not solve the system.

- **A** Solve the first equation for x.
- **B** Solve the first equation for y.
- **C** Solve the second equation for x.
- **D** Solve the second equation for y.

1. $x + 2y = 11$
$3x + 5y = 17$

2. $3x - 2y = 14$
$2x + y = 11$

3. $2x - y = 5$
$x + 3y = 7$

4. $y - x = 19$
$3x + 2y = 7$

5. $2x + 5y = 11$
$3x + 6y = 12$

6. $4x - 2y = 9$
$5x + 6y = 7$

7. $2x + 6y = 8$
$3x - 5y = 13$

8. $2x + 3y = 8$
$5x + 3y = 7$

Solve each system by substitution.

9. $y = 3x$
$4x + y = 49$

10. $y = 3x + 2$
$y = 5x - 6$

11. $x + y = 28$
$x - y = 4$

12. $3c + d = 9$
$c - d = -13$

13. $v + w = 11$
$2v - w = 16$

14. $x + 3y = 18$
$3x - y = -6$

15. $3c - 4d = 31$
$2c + 6d = 12$

16. $2s + 6t = -4$
$9s - 3t = 12$

B. EXERCISES

Solve each system by substitution.

17. $x + 4y = 30$
$2x + 5y = 36$

18. $x + 2y = 9$
$x + y = 28$

19. $y = \frac{1}{2}x$
$x + 3y = 25$

20. $2x - y = -4$
$x + y = -6$

21. $2x - 8y = 7$
$15x + 5y = -10$

22. $4n + 3p = 9$
$3n - 15p = 18$

23. $2x + 5y = 7$
$3x - 5y = 8$

24. $2a + 5b = 7$
$4a - 3b = -25$

25. $3x - 5y = -3$
$3x + 15y = 11$

26. $4x + 7y = 2$
$5x - 3y = -21$

27. $2x + 10y = 9$
$3x - 6y = -4$

28. $3x + 5y = 8$
$2x - 4y = 7$

29. **Essential Question:** How does the substitution method help you solve systems of equations with 2 variables?

Write a system of equations using 2 variables and solve the system by substitution.

30. The sum of 2 numbers is 996. The difference of the larger and twice the smaller is 33. What are the 2 numbers?

31. The perimeter of a garden is 90 ft. Its width is 7 ft less than its length. Find the dimensions.

32. Katie has 2 photo package options. The first one contains sixteen wallet-sized and two 5" × 7" photos and costs $18. The second costs $19 and contains eight wallets and three 5" × 7"s. What is the cost for each size of photo?

33. Haruto needs to make twelve pies on a budget of \$62. If an apple pie costs \$4 to make and a pecan pie costs \$6, how many of each should he plan on making?

34. **MP4** Haruto is making cheesecakes and chocolate pies. Each cheesecake serves 12 people and costs \$10 to make. Each chocolate pie serves eight and costs \$6 to make. He needs to serve 88 people on a budget of \$70. Assuming each guest will eat only 1 serving of dessert, how many of each dessert should he make?

35. Andrew borrowed \$4800 to purchase a car. He borrowed the down payment from his father at 5% interest for 6 months and the rest from the bank at 8% for 3 yr. If the total interest he paid was \$980, how much was the down payment and how much did he borrow from the bank?

36. **MP6** Christy is investing \$6000 in an account that earns 4% interest and \$10,000 in an account that earns 5%. If she plans to keep the 5% investment 1 yr longer than the 4% investment, how long should she leave the money in each account to earn \$5000 in interest?

C. EXERCISES

Solve each system by substitution.

37. $3x + 2y = 3$
$4x - 7y = -\frac{13}{4}$

38. $2x + 5y = 3$
$4x + 3y = -2\frac{2}{5}$

39. $x^2 + y = 5$
$8x^2 - 2y = 0$

40. **MP6** Jessica invested \$4200 in a certificate of deposit that guarantees a 3% return as long as she leaves the money in for at least a full year. One year later she invested \$10,800 in a mutual fund that historically has had an 8% return. How long (to the nearest year) would she have to leave each investment in place for the money to double?

MP5 Use substitution to solve the graphed systems.

41.

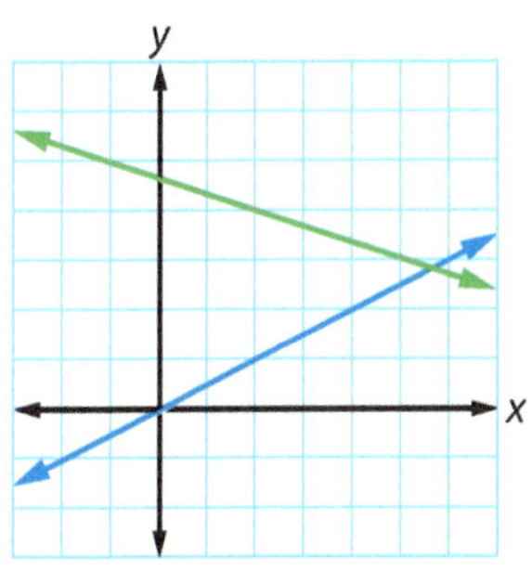

42.

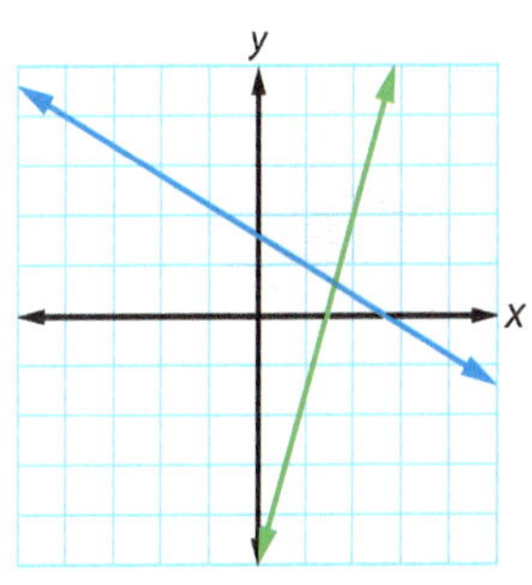

CUMULATIVE REVIEW

Solve each equation or inequality. [3.5–3.6]

43. $|x - 4| = 1$

44. $|x + 8| = 0$

45. $|-5x + 4 + 2x| + 3 = 17$

46. $|x - 4| > 1$

47. $|x + 8| < 0$

48. $|x + 8| > 0$

Write the rule for each function. [4.4]

49. $f = \{(-4, -7), (0, 1), (1, 3), (2, 5)\}$

50. $g = \{(-5, -13), (-2, -4), (0, 2), (3, 11)\}$

51. Which expression correctly represents the percent change from 45 to 60? **[2.5]**

A $\frac{60 - 45}{45} \times 100\%$

B $\frac{45 - 60}{45} \times 100\%$

C $\frac{60 - 45}{60} \times 100\%$

D $\frac{45 - 60}{60} \times 100\%$

E none of these

52. Which of the following are arithmetic sequences? **[4.5]**

I. 2, 4, 7, 11, 16, . . .

II. −5, 3, 11, 19, 27, . . .

III. −3, −9, −15, −21, −27, . . .

A I only

B II only

C III only

D II and III

E I, II, and III

MIND OVER MATH

Sudoku is a game of logic in which the digits 1–9 are placed in each row, each column, and each smaller 3 × 3 grid without repetition. Copy the partially filled grid and complete this puzzle. General hints and more difficult puzzles are readily available on the internet.

7				4		9		
	6	5			8			
				5	6	2		
8	9					1		4
		6	4		3			9
1		4				5	2	
	4					8		1
		7				6		
				7	5			

6.3 SOLVING SYSTEMS BY ELIMINATION

Why is subtracting an equation from another equation valid?

After completing this section, I will be able to

- solve systems of linear equations by elimination.
- solve real-world problems by writing and solving a system of linear equations.

The *elimination method* is a third method used to solve systems of equations. By properly applying the properties of equality, one of the variables in the system can be eliminated resulting in a one-step equation.

DEFINITION

The **elimination method** for solving systems of equations involves adding or subtracting multiples of the equations to eliminate one of the variables.

EXAMPLE 1: Solving by Using Addition

Solve the system by elimination.

$4x - 2y = 13$

$8x + 2y = 23$

Answer

$$\begin{aligned} 4x - 2y &= 13 \\ \underline{8x + 2y} &\underline{= 23} \\ 12x &= 36 \\ x &= 3 \end{aligned}$$

1. Since the coefficients of y are opposites, adding the 2 equations will eliminate the y-term.
2. Solve for x.

$$\begin{aligned} 8x + 2y &= 23 \\ 8(3) + 2y &= 23 \\ 24 + 2y &= 23 \\ 2y &= -1 \\ y &= -\tfrac{1}{2} \end{aligned}$$

3. Substitute 3 for x in either original equation and solve for y.

solution: $\left(3, -\frac{1}{2}\right)$

4. Write the solution as an ordered pair.

$4(3) - 2\left(-\frac{1}{2}\right) = 13$ and $8(3) + 2\left(-\frac{1}{2}\right) = 23$

5. The solution checks in both equations.

SKILL CHECK **EXERCISE 5**

The Addition Property of Equality justifies the addition of 2 equations since we are adding the equal expressions from one equation to the sides of the other equation. The system in Example 1 was easy to solve by elimination since the coefficients of y were opposites. When like terms have the same coefficients, the variable can be eliminated by subtracting one equation from the other.

EXAMPLE 2: Solving by Using Subtraction

Solve the system by elimination.

$3x + 2y = 8$

$3x + 5y = 14$

Answer

$$\begin{array}{r} 3x + 2y = 8 \\ -(3x + 5y = 14) \\ \hline -3y = -6 \end{array} \quad \text{or} \quad \begin{array}{r} 3x + 2y = 8 \\ -3x - 5y = -14 \\ \hline -3y = -6 \end{array}$$

1. Because the coefficients of the x-terms are the same, subtract the equations. Note: This is the same as multiplying both sides of the second equation by -1 and then adding the equations.

$$y = 2$$

2. Solve for y.

$$\begin{aligned} 3x + 5y &= 14 \\ 3x + 5(2) &= 14 \\ 3x + 10 &= 14 \\ 3x &= 4 \\ x &= \tfrac{4}{3} \end{aligned}$$

3. Substitute 2 for y in either original equation and solve for x.

solution: $\left(\frac{4}{3}, 2\right)$

4. Write the solution as an ordered pair.

$3\left(\frac{4}{3}\right) + 2(2) = 8$ and $3\left(\frac{4}{3}\right) + 5(2) = 14$

5. The solution checks in both equations.

SKILL CHECK **EXERCISE 9**

The Subtraction Property of Equality allows the equal expressions from one equation to be subtracted from the sides of another equation. Often the coefficients of the like terms do not have the same absolute value. The Multiplication Property of Equality can be applied to one or both equations to obtain the same absolute value for the coefficients of one of the variables.

EXAMPLE 3: Solving by Using Multiples of the Equations.

Solve each system by elimination.

a. $2x + 3y = 5$
$4x - 9y = -10$

b. $2x + 3y = 6$
$3x - 2y = 8$

Answers

a.
$$-2(2x + 3y = 5) \Rightarrow -4x - 6y = -10$$
$$4x - 9y = -10 \Rightarrow \underline{4x - 9y = -10}$$
$$-15y = -20$$

1. To create opposite coefficients, multiply both sides of the first equation by −2. Then add the equations. Note: The first equation could have been multiplied by 3 to eliminate the y-term.

$$y = \frac{-20}{-15} = \frac{4}{3}$$

2. Solve for y.

$$2x + 3y = 5$$
$$2x + 3\left(\frac{4}{3}\right) = 5$$
$$2x + 4 = 5$$
$$2x = 1$$
$$x = \frac{1}{2}$$

3. Substitute $\frac{4}{3}$ for y in either original equation and solve for x.

solution: $\left(\frac{1}{2}, \frac{4}{3}\right)$

4. Write the solution as an ordered pair.

CHECKING THE SOLUTION

$2\left(\frac{1}{2}\right) + 3\left(\frac{4}{3}\right) = 5$

$4\left(\frac{1}{2}\right) - 9\left(\frac{4}{3}\right) = -10$

b.
$$2(2x + 3y = 6) \Rightarrow 4x + 6y = 12$$
$$3(3x - 2y = 8) \Rightarrow \underline{9x - 6y = 24}$$
$$13x = 36$$

1. The coefficients of y have a least common multiple of 6. To create opposite y coefficients, multiply the first equation by 2 and the second by 3.

$$x = \frac{36}{13}$$

2. Solve for x.

$$3(2x + 3y = 6) \Rightarrow 6x + 9y = 18$$
$$2(3x - 2y = 8) \Rightarrow \underline{-(6x - 4y = 16)}$$
$$13y = 2$$

3. Since the x-value is a fraction, use the elimination method a second time to find y. Multiply the first equation by 3 and the second by 2. Then subtract the equations.

$$y = \frac{2}{13}$$

4. Solve for y.

solution: $\left(\frac{36}{13}, \frac{2}{13}\right)$

5. Write the solution as an ordered pair.

CHECKING THE SOLUTION

$2\left(\frac{36}{13}\right) + 3\left(\frac{2}{13}\right) = 6$

$3\left(\frac{36}{13}\right) - 2\left(\frac{2}{13}\right) = 8$

SKILL CHECK **EXERCISE 17** ☑

SOLVING A SYSTEM BY ELIMINATION

1. Multiply both sides of one or both equations so that the coefficients of one of the variables have the same absolute value.
2. Add or subtract the equations to eliminate a variable.
3. Solve the resulting equation.
4. Find the value of the eliminated variable by
   a. substituting the result into an original equation and solving or
   b. eliminating the other variable.
5. Write the solution as an ordered pair.
6. Check the solution by substituting it into each equation.

EXAMPLE 4: Solving a Real-World Problem

The aquarium offers a family discount for 5 or more people. If the Moore family paid \$82 for 2 adults and 3 children and the Smythe family paid \$130 for 3 adults and 5 children, what was the discounted price for children and adults?

AfterSchoolHelp

Solving Systems by Elimination

Answer

Let a = the discounted adult ticket price and c = the discounted child ticket price.

1. Assign variables to the unknowns.

the Moores $2a + 3c = 82$

the Smythes $3a + 5c = 130$

2. Write and solve a system of 2 equations.

$3(2a + 3c = 82) \Rightarrow 6a + 9c = 246$

$-2(3a + 5c = 130) \Rightarrow -6a - 10c = -260$

$-c = -14$

$c = 14$

a. The coefficients of a have a least common multiple of 6. To create opposite a coefficients, multiply the first equation by 3 and the second by -2.

$a = -\frac{3}{2}(14) + 41$

$a = -21 + 41$

$a = 20$

b. Substitute 14 for c in the first equation and solve for a.

ISOLATING THE VARIABLE

$2a + 3c = 82$

$2a = -3c + 82$

$a = -\frac{3}{2}c + 41$

The discounted adult tickets were \$20 and the discounted child tickets were \$14.

3. State the solution to the problem.

$2(20) + 3(14) = 82$ and $3(20) + 5(14) = 130$

4. The solution checks in both equations.

SKILL CHECK **EXERCISE 33**

VOCABULARY

- elimination method

A. EXERCISES

Determine the multipliers needed to a) eliminate the x-terms and b) eliminate the y-terms. Sometimes a multiplier is not needed; if so, state "just add" or "just subtract." Do not solve.

Example:
$2x + 3y = 3$
$6x - 3y = 1$
Answer: a) 1st equation by -3; b) just add

1. $7x + y = 35$
$x + y = 5$

2. $x + 4y = 6$
$9x - 4y = -66$

3. $2x - 7y = 26$
$x - 7y = 18$

4. $2x + 6y = 5$
$2x - 3y = -4$

Solve each system by elimination.

5. $x + y = 5$
   $x - y = -3$

6. $2x + y = 9$
   $-2x - 3y = 13$

7. $x - y = -2$
   $6x + y = -19$

8. $x + 2y = 13$
   $x + 6y = 45$

9. $2f + 9g = 30$
   $2f + g = -2$

10. $2j + 3k = 19$
    $2j + 7k = 47$

B. EXERCISES

Determine the multipliers needed to a) eliminate the *x*-terms and b) eliminate the *y*-terms. Do not solve.

11. $2x + 3y = 10$
    $4x - 5y = -24$

12. $x + y = 14$
    $2x - 3y = -17$

13. $6x - 11y = -26$
    $x + 3y = -14$

14. $9x + 4y = 10$
    $3x - 10y = 26$

Solve each system by elimination.

15. $2y + x = 9$
    $3x - 2y = 7$

16. $x - 3y = 10$
    $3y + x = 8$

17. $2x + y = 15$
    $x + 2y = 18$

18. $2b + 3a = 18$
    $5a - 3b = 11$

19. $5x - 4y = 7$
    $7x - 3y = 2$

20. $6x - 5y = -38$
    $4x - 17y = 2$

21. $9x + 7y = 12$
    $7x + 3y = 24$

22. $2x + 5y = -45$
    $19x - 8y = -39$

23. $5r + s = 17$
    $3r + s = 12$

24. $x + 2y = 9$
    $3x + y = 16$

25. $2x + 4y = 5$
    $3x - 2y = 7$

26. $5x - 4y = 3$
    $2x + 3y = 17$

27. $3x - 2y = 5$
    $4x + 3y = 8$

28. $\frac{x}{2} + y = 9$
    $x + y = 14$

29. **MP3** Describe and correct Lana's error in solving the system.

$2x - 3y = -5$
$6x + 15y = 57$

Step	Lana's Work
1.	$5(2x - 3y = -5) \Rightarrow 10x - 15y = -5$ $6x + 15y = 57 \Rightarrow 6x + 15y = 57$ $16x = 52$
2.	$x = \frac{52}{16} = \frac{13}{4}$
3.	$6\left(\frac{13}{4}\right) + 15y = 57$ $\frac{39}{2} + 15y = 57$ $15y = \frac{75}{2}$ $y = \frac{5}{2}$
4.	solution: $\left(\frac{13}{4}, \frac{5}{2}\right)$

30. **MP3** Describe and correct Tobias's error in solving the system.

$5x - 2y = -11$
$5x + 3y = 34$

Step	Tobias's Work
1.	$5x - 2y = -11$ $-(5x + 3y = 34)$ $y = 23$
2.	$5x + 3(23) = 34$ $5x + 69 = 34$ $5x = -35$ $x = -7$
3.	solution: $(-7, 23)$

Write a system of equations and solve by elimination.

31. The sum of 2 numbers is 621, and their difference is 109. Find the 2 numbers.

32. The difference of 2 numbers is 62, and the sum of twice the larger and three times the smaller is 304. Find the 2 numbers.

33. The senior class sold 245 tickets to their class play. The adult tickets sold for \$4 each and the student tickets sold for \$2.50 each. The total income from the ticket sales was \$725. How many of each type of ticket were sold?

34. MP4 The exam will have true/false and short answer questions. The short answer questions are worth 4 points each, and the true/false questions are worth 2 points each. There will be 50 questions worth a total of 150 points. How many of each type of question are on the exam?

35. Westside Christian School's athletic director ordered practice balls for the baseball and softball teams. He cannot remember how many of each he ordered. He knows that he ordered a total of 43 balls, costing the athletic department \$165. If baseballs cost \$3 each and softballs cost \$5 each, how many of each type of ball did he order?

36. **Essential Question:** Why is it valid to add an equation to or subtract an equation from another?

C. EXERCISES

Solve each system by elimination.

37. $3x - y = \frac{23}{5}$
$15x + 25y = 90$

38. $\frac{1}{2}x + \frac{1}{5}y = 1$
$\frac{3}{2}x + \frac{1}{5}y = -7$

Write a system of equations and solve the system by elimination.

39. Four less than five times a number is twelve more than another number. Three times the first number less twice the second number is eight. Find the 2 numbers.

40. MP1 Jackie was mailing 2 oz and 3 oz letters. The total weight of the letters was 88 oz. A 2 oz letter cost \$0.61 to mail, and a 3 oz letter cost \$0.78 to mail. She spent \$24.95 to mail the letters. How many of each type did she mail?

CUMULATIVE REVIEW

Simplify each expression. [1.6]

41. $(2x + 3y + 5) + (-2x + 4y + 9)$

42. $(3x - 4y + 9) - (2x - 4y + 2)$

43. $\frac{1}{2}(2x^2 + 4x + 2)$

44. $\frac{2x^2 + 4x + 2}{2}$

45. $\frac{24x + 51}{3}$

46. $6\left(\frac{x + 7}{3}\right)$

Find the slope of the line passing through the given points. [5.2]

47. $(-8, 2)$ and $(1, 20)$

48. $(-3, -16)$ and $(18, -4)$

49. What is the length of each side of an equilateral triangle that has a perimeter of $2\frac{5}{8}$ ft? **[1.2]**

A $\frac{7}{8}$ ft
B $\frac{8}{63}$ ft
C $1\frac{1}{7}$ ft
D $7\frac{7}{8}$ ft
E none of these

50. Which of the absolute value functions has a vertex to the left of the y-axis? **[4.7]**

A $f(x) = |x| + 1$
B $g(x) = |x + 1|$
C $h(x) = |x| - 1$
D $j(x) = |x - 1|$
E none of these

6.4 SPECIAL SYSTEMS

What are the possible relationships between 2 lines?

After completing this section, I will be able to

- solve a given linear system by using an appropriate method.
- classify linear systems as consistent independent, consistent dependent, or inconsistent.
- relate the slopes and y-intercepts of lines in each type of linear system.
- explain 2 assumptions necessary for this classification of linear systems.

The 3 methods covered in this chapter for solving systems—graphing, substitution, and elimination—will yield the solution for a system of linear equations. Using the best method for each problem will save time and help avoid errors.

METHOD	DISADVANTAGE	ADVANTAGE
substitution	can have challenging arithmetic	is easiest when a single variable is isolated
elimination	may involve rearranging and multiplying equations	works well when a coefficient is not a factor of the other coefficients
graphing	may not produce precise answers when the solution does not consist of integers	may be helpful for analyzing real-world problems

EXAMPLE 1: Using Substitution to Solve a System

Solve the system.

$x + y = -4$
$x = -3y$

Answer

$x + y = -4$ and $x = -3y$

1. Since the second equation is already solved for x, use substitution.

$-3y + y = -4$
$-2y = -4$
$y = 2$

2. Substitute the expression for x in the first equation and solve for y.

$x = -3(2)$
$x = -6$

3. Substitute 2 for y in either equation to solve for x.

solution: $(-6, 2)$

4. Write the solution as an ordered pair.

SKILL CHECK ***EXERCISE 15***

EXAMPLE 2: Using Elimination to Solve a System

Solve the system.

$3x + 4y = 3$

$2x - 7y = 31$

Answer

$$2(3x + 4y = 3) \Rightarrow 6x + 8y = 6$$

$$-3(2x - 7y = 31) \Rightarrow -6x + 21y = -93$$

$$29y = -87$$

$$y = -3$$

1. Since solving for any of the variables would result in a fraction, use elimination.

$$3x + 4y = 3$$

$$3x + 4(-3) = 3$$

$$3x - 12 = 3$$

$$3x = 15$$

$$x = 5$$

2. Substitute −3 for *y* in either equation and solve for *x*.

solution: (5, −3)

3. Mentally check the solution in both equations.

SKILL CHECK **EXERCISE 21**

EXAMPLE 3: Using Graphing to Solve a System

Darren estimates he can burn 5 calories per minute walking and 8 calories per minute riding his bike. How much time should Darren spend doing each exercise if he wants to burn 270 calories in 45 min?

Answer

Let w = time spent walking and r = time spent riding.

1. Define the variables.

$5w + 8r = 270$

$w + r = 45$

2. Write the equations.

$5w + 8r = 270$ and $w + r = 45$

$8r = -5w + 270$ $\quad$ $r = -w + 45$

$r = -0.625w + 33.75$

3. Since a visual representation is desirable, and if technology is available, use graphing. Rewrite the equations in function form and use your graphing utility to graph the system.

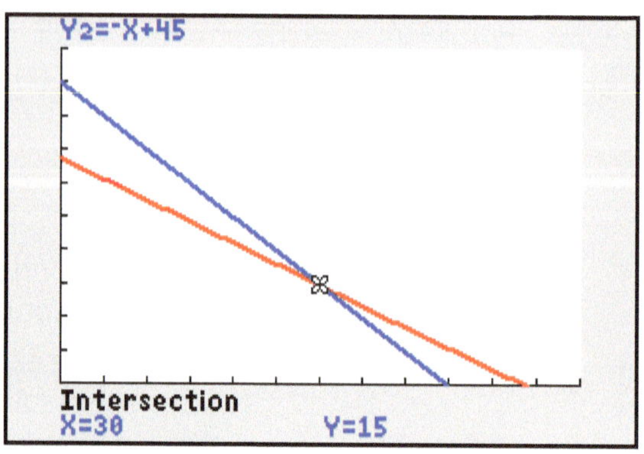

solution: (30, 15)

4. Verify the answer by substituting it in both equations.

SKILL CHECK **EXERCISE 13**

Systems of linear equations can be classified according to the number of their solutions. Systems with no solution are called inconsistent because there is no ordered pair that makes both equations true. Linear systems are **inconsistent** when the graphs are parallel lines. **Consistent** systems have 1 or more ordered-pair solutions that satisfy all equations in the system. If a consistent system has only 1 solution, it is **independent**. Systems whose graphs are intersecting lines are consistent independent systems. A consistent system with an infinite number of ordered-pair solutions is called **dependent**. When the graphs of the equations coincide, the system is a consistent dependent system.

EXAMPLE 4: Identifying an Inconsistent System

Solve the system.

$x + y = 7$

$x + y = 5$

Answer

$$\begin{aligned} x + y &= 7 \\ -(x + y &= 5) \\ \hline 0 &= 2 \end{aligned}$$

1. Subtracting the second equation from the first produces a false statement.

system type: inconsistent

solution: ∅

2. When a false statement results, the system is inconsistent and there are no solutions.

SKILL CHECK **EXERCISE 19**

Solving the system by graphing illustrates why there is no solution.

$x + y = 7$ and $x + y = 5$

$y = -x + 7$ $\quad$ $y = -x + 5$

1. Write each equation in slope-intercept form.

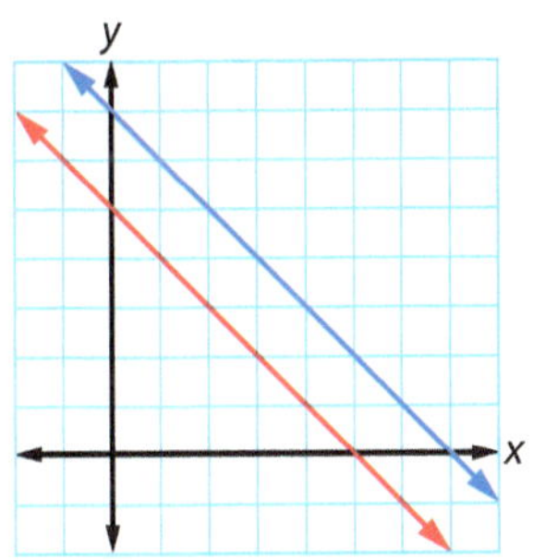

2. Graph both lines on the same coordinate plane.
Equation 1: $m = -1$ and $b = 7$
Equation 2: $m = -1$ and $b = 5$

AFTERSCHOOLHELP

Special Systems

By analyzing the slope-intercept forms of the equations in Example 4, you can see that the lines have the same slope and different y-intercepts. Therefore, the lines are parallel, and there is no point of intersection that satisfies both equations.

If the resulting statement is a true statement, such as $0 = 0$, then the equations represent coinciding lines, and the solution consists of all the points on the line. This type of solution occurs in a system that is consistent dependent.

EXAMPLE 5: Identifying a Consistent System

Solve the system.

$x = -y + 2$

$3x + 3y = 6$

Answer

$x = -y + 2$ and $3x + 3y = 6$	**1.** Since x is already isolated in the first equation, use substitution.
$3(-y + 2) + 3y = 6$	**2.** Substitute the expression for x in the second equation.
$-3y + 6 + 3y = 6$ $6 = 6$	**3.** Solve. The result is a true statement.
system type: consistent dependent	**4.** When a true statement results, the system is consistent dependent, and the graphs of the equations coincide.
solution: $y = -x + 2$	**5.** Use the slope-intercept form of the line to state the solution.

When both equations are written in slope-intercept form, the equations are identical: $y = -x + 2$.

SKILL CHECK **EXERCISE 23**

TYPES OF SYSTEMS

System	consistent independent	consistent dependent	inconsistent
Number of Solutions	1 ordered pair	infinite number (all the points on a line)	no solution
Graph	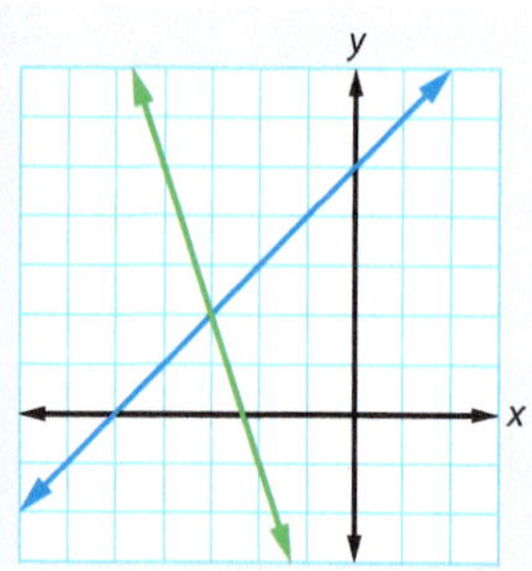	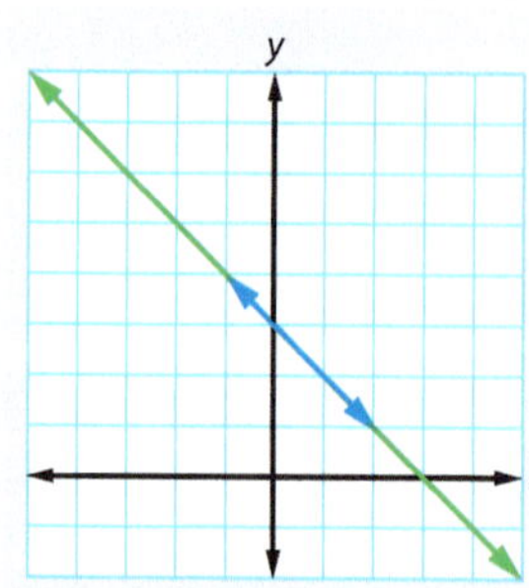	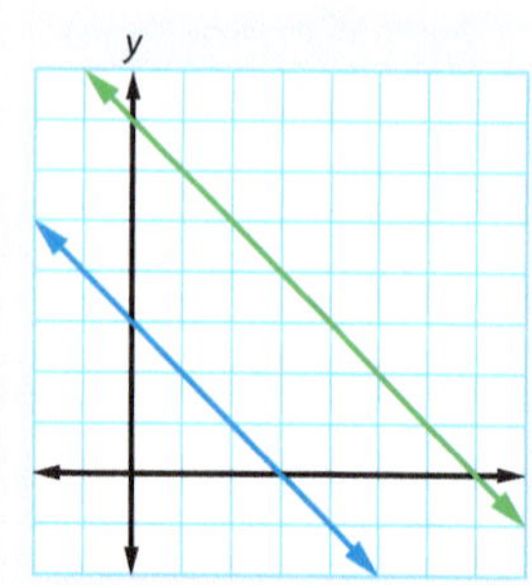
Slope-Intercept Comparison	$m_1 \neq m_2$	$m_1 = m_2$ $b_1 = b_2$	$m_1 = m_2$ $b_1 \neq b_2$
Results from Substitution or Elimination	an ordered pair (x, y)	a true statement $(0 = 0)$	a false statement $(0 = -2)$

Notice that this classification system assumes that lines are straight and coplanar. Lines in some geometries are not straight, and non-coplanar lines do not have to be parallel to be inconsistent. While mathematical reasoning is very efficient at determining consistency with assumptions, we must look to Scripture for truth.

EXAMPLE 6: Classifying Systems

Which pair(s) of equations form a consistent independent system?
Which pair(s) form an inconsistent system?
Equation 1: $x + y = 9$
Equation 2: $x + y = 4$
Equation 3: $x - 2y = 6$

Answer

Equation 1	Equation 2	Equation 3
$x + y = 9$	$x + y = 4$	$x - 2y = 6$
$y = -x + 9$	$y = -x + 4$	$-2y = -x + 6$
		$y = \frac{1}{2}x - 3$

1. Write each equation in slope-intercept form.

$m_1 = -1$	$m_2 = -1$	$m_3 = \frac{1}{2}$
$b_1 = 9$	$b_2 = 4$	$b_3 = -3$

2. Determine m and b for each line.

consistent independent systems:
Equations 1 and 3
Equations 2 and 3

3. Equations that have different slopes intersect at 1 point.

inconsistent system:
Equations 1 and 2

4. Equations that have the same slope are parallel lines.

SKILL CHECK **EXERCISE 27**

VOCABULARY

- consistent
- dependent
- inconsistent
- independent

A. EXERCISES

Choose the best method for solving each system.

A graphing **B** substitution **C** elimination

1. $x + 3y = 7$
$5x + 11y = 3$

2. $2x - 5y = 3$
$4x + 7y = 9$

3. $5x - 3y = 3$
$2x + 7y = 6$

4. $y = -4$
$2x + 5y = 13$

5. Abby is comparing the costs of 2 dressmakers for her bridesmaids' dresses. Given the differing costs per dress and consultation fee for 2 different seamstresses, compare the 2 choices and decide which seamstress would be the better choice according to how she increases or decreases her wedding party.

Choose all that apply.

6. a system with 2 parallel lines
7. a system with 1 ordered-pair solution
8. a system with 2 coinciding lines
9. a system with no solution
10. a system with 2 intersecting lines
11. a system with an infinite number of solutions (all the points on a line)

A consistent
B dependent
C inconsistent
D independent

Solve each system by graphing.

12. $x + y = -2$
$x - y = 4$

13. $2x - 3y = 3$
$y = 2x + 3$

Solve each system by substitution.

14. $y = -2x + 4$
$y = x + 1$

15. $h = -4g + 7$
$g - 3h = 18$

Solve each system by elimination.

16. $x + 2y = 8$
$-x + 5y = 6$

17. $4x - 3y = 8$
$2x + 3y = 4$

B. EXERCISES

Solve each system. Then classify it as consistent independent, consistent dependent, or inconsistent. If the solution is an entire line, write its equation in slope-intercept form.

18. $8x - 3y = -12$
$y - 4 = -8$

19. $x + y = -4$
$x + y = 8$

20. $2x + 2y = 14$
$x + y = 7$

21. $a - b = -2$
$4a - b = 4$

22. $c = -2d + 10$
$9c + 2d = -6$

23. $4x + 3y = 7$
$4x = 7 - 3y$

24. $x + y = 18$
$2x + 2y = 10$

25. $2x - y = 11$
$6x - 3y = -24$

26. $x + y = 10$
$2y = 20 - 2x$

27. $2x + 4y = 10$
$3x - y = 18$

28. $4r + 3s = 9$
$r - 5s = 6$

29. $5x - y = -42$
$-44 + y = 5x$

30. $15x - 24y = 7$
$32y + 9 = 20x$

31. $10x - 12y = 14$
$35 + 30y = 25x$

32. $2(x + y) - 5(x + 1) = 3$
$-3x + 2y = 8$

33. **Essential Question:** What are the possible relationships between 2 lines?

34. What is true about the slopes and y-intercepts of 2 linear equations that form an inconsistent system?

35. What is true about the slopes and y-intercepts of 2 linear equations that form a consistent dependent system?

36. **MP8** What is true about the slopes and y-intercepts of 2 linear equations that form a consistent independent system?

37. What 2 assumptions are required to classify systems as inconsistent, consistent dependent, and consistent independent?

Write and solve a system of equations for each word problem.

38. An isosceles triangle (2 congruent sides) has a perimeter of 48 cm. The noncongruent side is 3 cm more than half of one of the other sides. Find the length of each side of the triangle.

39. Jessie and Kayla decided to purchase books and DVDs for the youth library at their church. Jessie purchased 8 books and 4 DVDs, and Kayla purchased 5 books and 6 DVDs. Jessie's total cost was $80 and Kayla's was $92. If all the books sold for the same amount and all the DVDs had the same price, how much was the cost per item?

40. MP4 Charles wants to rent a car for the weekend. After researching the costs, he narrows his choice to the following 2 options: 1) Easy Rental: $50 per day + $0.15 per mile driven, and 2) Clean Rental: $40 per day + $0.20 per mile driven. Write a system of equations for the costs of renting a car from each company. Graph the system and determine when the companies will cost the same amount. When would it be better to use Easy Rental?

C. EXERCISES

41. MP5 Explain why graphing is not a good choice for solving this system.

$9x - 2y = 7$
$3x + 4y = 14$

42. Explain why substitution is not a good choice for solving this system.

$9x - 2y = 7$
$3x + 4y = 14$

43. Jackson is trying to make a decision between 2 colleges. College A's tuition for a semester is a $450 fee and $325 per credit hour. College B's tuition is $350 per credit hour with no additional fees. At how many credit hours would both college tuition costs be the same? When would College A be less expensive?

CUMULATIVE REVIEW

44. If y varies directly with x, use the given values of x and y to state the constant of variation and write the equation describing the variation. **[4.6]**

a. $x = 9, y = 18$

b. $x = 7, y = 4$

45. What do the graphs of the 2 equations in exercise 44 have in common? **[4.6]**

The weight of the atmosphere at sea level is 14.7 pounds per square inch (psi). This is also referred to as 1 atmosphere (atm). Underwater pressure increases by 14.7 psi for every 33 ft of depth. Absolute pressure at ocean depths includes the atmospheric pressure at sea level.

46. State the rate of change in pressure per foot of ocean depth in psi/ft (to the nearest thousandth). **[5.2]**

47. Write a function rule modeling the absolute pressure, $p(d)$, in pounds per square inch at a given ocean depth, d. **[5.3]**

48. The Mariana Trench of the western Pacific contains the deepest spot in Earth's oceans, Challenger Deep, which is 35,840 ft below sea level. Find the absolute pressure at this depth (to the nearest whole number). **[5.3]**

49. Convert the pressure found in exercise 48 to atmospheres. **[2.4]**

50. The crush level of a submarine (the depth where the outside pressure will crush the submarine's steel hull) has been estimated to be 2400 ft. Determine the pressure that is estimated to crush the submarine (to the nearest whole number). **[5.3]**

51. Convert the pressure found in the previous exercise to atmospheres. **[2.4]**

In 2012 the Deepsea Challenger became the second manned vessel to reach the bottom of the Challenger Deep, 52 yr after the Trieste.

52. First Nibs sells 2 types of chocolate for \$7.50/lb and \$11.00/lb. If the company sold a total of 35 lb of chocolate for \$316.75, which equation(s) can be used to determine how much of each chocolate was sold? Choose all that apply. **[2.8]**

A $7.5x + 11(35 - x) = 316.75$

B $75(35 - x) + 110x = 3167.5$

C $750x + 1100(3500 - 100x) = 31{,}675$

D $750(35 - x) + 1100x = 31{,}675$

E none of these

53. Which line(s) pass through the point $(-5, 3)$? Choose all that apply. **[5.4]**

A $y = \frac{1}{2}x + \frac{11}{2}$

B $y - 3 = \frac{1}{2}(x - 5)$

C $x - 2y = 11$

D $3x + 6y = 3$

E none of these

MATH IN HISTORY
KARL FRIEDRICH GAUSS (1777–1855)

German mathematician and physicist Karl Gauss is viewed by math historians as one of the three greatest mathematicians of all time. While still a teenager, he solved three previously unsolved math problems and became the first to construct a regular heptadecagon (seventeen-sided polygon) using only a compass and a straightedge.

He also developed the method of least squares, a crucial advancement in statistics. On a scatter plot, the vertical distances from each point to a curve are measured. All the measurements are then squared to eliminate the negative values from points below the curve. The sum of these squares measures how well the curve fits the data. Gauss used this method to relocate a newly discovered asteroid, Ceres, after it had disappeared behind the sun. This method is used today to determine the equation that best fits a set of data (this is how your calculator derived such lines to answer questions in the previous chapter). From such an equation, industrial and financial decisions are made to help us manage the earth's resources and improve people's lives.

Another major discovery was his proof of the law of quadratic reciprocity, which was later included in his 1801 number theory masterpiece, *Arithmetical Investigations*. That treatise developed modular arithmetic, which today is used to prevent credit card fraud and to track books in large libraries, among many other beneficial uses.

Gauss also gave the first full and correct proof of the Fundamental Theorem of Arithmetic, which states that every natural number has a unique prime factorization. His doctoral thesis discussed the Fundamental Theorem of Algebra, which states that every algebraic equation has at least one solution. His broad range of interests enabled Gauss to make important contributions to all existing branches of mathematics.

6.5 MOTION PROBLEMS

How can a system of equations be used to solve motion problems?

After completing this section, I will be able to

- solve real-world motion problems by using systems of linear equations.

The motion problems covered in Section 2.7 were solved using 1 variable. Some motion problems can be solved only by using a system of equations with more than 1 variable.

EXAMPLE 1: Finding Rates

Two cars start toward each other at the same time from towns 612 mi apart and pass each other in 6 hr. One travels 2 mi/hr faster than the other. How fast is each car traveling?

Answer

Let r_f = the rate of the faster car and r_s = the rate of the slower car.

1. Use meaningful subscripted variables to represent unknowns.

	r	t	d
Faster Car	r_f	6	$6r_f$
Slower Car	r_s	6	$6r_s$

2. Make a drawing and a table. Fill in the first 2 columns and then use the formula $rt = d$ to fill in the third column.

$r_f = r_s + 2$ and $6r_f + 6r_s = 612$

3. Write the system.
Equation 1: The faster car is 2 mi/hr faster than the slower car.
Equation 2: The total distance is 612 mi.

$$6(r_s + 2) + 6r_s = 612$$
$$6r_s + 12 + 6r_s = 612$$
$$12r_s + 12 = 612$$
$$12r_s = 600$$
$$r_s = 50$$

4. Solve the system using substitution since the first equation is already solved for r_f.

$$r_f = 50 + 2$$
$$r_f = 52$$

5. Substitute 50 for r_s in the first equation and solve for r_f.

The slower car travels at 50 mi/hr and the faster car travels at 52 mi/hr.

6. Interpret the solution.

SKILL CHECK **EXERCISE 23**

EXAMPLE 2: Finding Time

Logan rode her bike 70 mi in 5 hr. She averaged 12 mi/hr at the beginning of the trip and 20 mi/hr for the remainder. How far did Logan travel at 20 mi/hr?

Answer

Let t_1 = the time traveling at 12 mi/hr and t_2 = the time traveling at 20 mi/hr.

1. Use meaningful subscripted variables to represent unknowns.

	r	*t*	*d*
First Leg	12	t_1	$12t_1$
Second Leg	20	t_2	$20t_2$

2. Make a table and fill it in with the known information.

$12t_1 + 20t_2 = 70$ and $t_1 + t_2 = 5$

3. Write the system.
Equation 1: The total distance is 70 mi.
Equation 2: The total time is 5 hr.

$$\begin{aligned} 12t_1 + 20t_2 &= 70 &\Rightarrow\quad 12t_1 + 20t_2 &= 70 \\ -12(t_1 + t_2) &= -12(5) &\Rightarrow\quad -12t_1 - 12t_2 &= -60 \\ & & 8t_2 &= 10 \\ & & t_2 &= 1.25 \text{ hr} \end{aligned}$$

4. Multiply the second equation by −12 and eliminate the t_1 term by adding the equations.

$20(1.25) = 25$

5. Substitute 1.25 for t_2 to find the distance she traveled on the second leg.

Logan traveled 25 mi at 20 mi/hr.

6. Interpret the solution.

SKILL CHECK **EXERCISE 19**

Problems in which the rates are affected by wind or water currents can be solved using 2 variables: the rate at which an object, such as a boat or plane, travels in still air or still water, and the rate at which a wind or water current moves. Traveling into a headwind or against a current decreases your rate of travel. Moving downstream or with a tailwind increases your speed.

No Wind or Current	Headwind 25 mi/hr	Tailwind 25 mi/hr	Downstream 3 mi/hr Current	Upstream 3 mi/hr Current
r	$r-25$	$r+25$	$r+3$	$r-3$

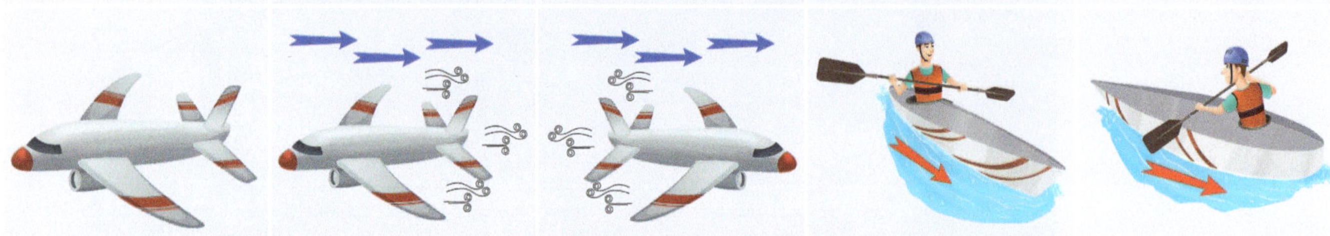

EXAMPLE 3: Finding Rate of a Current

When his Sunday school class went canoeing, Brent paddled 12 mi down the river in 3 hr. After paddling back upstream for 2 hr, he had traveled 4 mi. How fast does Brent paddle in still water? What is the rate of the current?

Answer

Let b = Brent's paddling rate and c = the rate of the current.

1. Define variables for the unknowns.

	r	t	d
Downstream	$b+c$	3	12
Upstream	$b-c$	2	4

2. Make a table and fill it in with the known information.

$3(b+c)=12$ and $2(b-c)=4$

3. Use $rt=d$ to produce the 2 equations for the system.

$$\frac{\cancel{3}(b+c)}{\cancel{3}}=\frac{12}{3} \Rightarrow b+c=4$$

$$\frac{\cancel{2}(b-c)}{\cancel{2}}=\frac{4}{2} \Rightarrow b-c=2$$

$$2b=6$$

$$b=3$$

4. Both equations can be simplified before solving the system. Eliminate the c terms by adding the equations.

$b+c=4$

$3+c=4$

$c=1$

5. Substitute 3 for b in the first simplified equation and solve for c.

Brent's paddling rate is 3 mi/hr.
The rate of the current is 1 mi/hr.

6. Interpret the solution.

SKILL CHECK **EXERCISES 3, 9**

The term *knot* is the unit of speed referring to nautical miles per hour (1 knot ≈ 1.151 mi/hr). The unit is often used to measure speeds in aviation and sea navigation.

EXAMPLE 4: Finding Wind Speed

When flying the 1900 nautical miles to California, James flew nonstop for 4 hr. Flying home by the same route and under the same weather conditions, the flight took 3 hr 30 min. What rate would the airplane have flown with no wind? What was the wind speed?

Answer

AFTERSCHOOLHELP

Motion Problems

Let p = the rate of the airplane with no wind and w = the wind speed.

1. Define variables for the unknowns.

	r	t	d
Flight to CA	$p - w$	4	$4(p - w)$
Flight Home	$p + w$	3.5	$3.5(p + w)$

2. Make a table and fill it in. Fill in the third column by multiplying across.

$4(p - w) = 1900$ and $3.5(p + w) = 1900$

3. Write the system. Since the distance for each leg of the trip is 1900 nautical miles, set each distance equal to 1900.

$$\frac{4(p-w)}{4} = \frac{1900}{4} \qquad 3.5p + 3.5w = 1900$$
$$p - w = 475 \qquad 35p + 35w = 19{,}000$$

4. Solve. Since 1900 is divisible by 4, simplify the first equation. Remove the parentheses and clear the decimals in the second equation.

$$35(p - w) = 35(475) \Rightarrow 35p - 35w = 16{,}625$$
$$35p + 35w = 19{,}000 \Rightarrow 35p + 35w = 19{,}000$$
$$70p = 35{,}625$$
$$p \approx 509$$

5. Multiply the first equation by 35 and eliminate the w terms. The answer is rounded to the nearest knot.

$$p - w = 475$$
$$509 - w \approx 475$$
$$-w \approx -34$$
$$w \approx 34$$

6. Substitute 509 for p in the first simplified equation and solve for w. Since the airplane's rate was approximated, the wind speed will be as well.

The airplane's rate of speed is about 509 knots. The wind speed is about 34 knots.

7. Interpret the solution.

SKILL CHECK **EXERCISES 1, 7**

A. EXERCISES

Write an algebraic expression that models each rate.

1. a bicyclist riding against a wind of w mi/hr if his speed on a calm day is 12 mi/hr
2. a blimp that travels r mi/hr in still air when a 10 mi/hr tailwind is blowing
3. a canoe going downstream in a current of c mi/hr if the canoe averages 4 mi/hr in still water
4. a boat with a speed of z mi/hr in still water travels upstream on Pine Creek against a current of 2 mi/hr
5. a plane that flies 500 knots in still air flies into a headwind of w knots

Use each table to write a system of equations. Do not solve.

6. Pedro traveled a total of 230 mi in 7 hr.

	r	t	d
Before Lunch	30	t_b	$30t_b$
After Lunch	40	t_a	$40t_a$

a. Define the variables.

b. Write an equation describing the times.

c. Write an equation describing the distances.

7. Marissa went biking on a day when the wind speed was 8 mi/hr. Each leg of the trip was 36 mi.

	r	t	d
With Wind	$b + 8$	1.125	$1.125(b + 8)$
Against Wind	$b - 8$	2.25	$2.25(b - 8)$

a. Define the variable.

b. Write an equation describing the distances.

c. Why would we need only 1 equation to find her normal biking speed?

Make a table for each problem. Do not solve.

8. A blimp travels 2 hr with the wind and 5 hr against it. Let b represent the rate of the blimp with no wind and let w represent the wind speed.

9. Blake rows 6 hr upstream against a current of 3 mi/hr and returns with the same current in 4 hr. Let b represent the rate of the boat in still water.

10. A pilot flies 150 knots in still air. He flies with the wind 4 hr to Albuquerque and then against the wind 9 hr to Miami. Let w represent the wind speed.

Use each table to write and solve a system of equations.

Terry traveled a total of 150 mi on his bicycle. The difference in the rates was 10 mi/hr.

	r	t	d
Going against the Wind	$x - w$	5	$5(x - w)$
Returning with the Wind	$x + w$	2	$2(x + w)$

11. Define the variables.

12. Write an equation relating the rates.

13. Write an equation relating the distances.

14. Solve the system to find the rate of the bicycle with no wind and the wind speed. Round to the nearest tenth.

A plane flew to Omaha and then returned. The speed of the plane with no wind was 100 mi/hr more than the wind speed.

	r	*t*	*d*
With Wind	$x + w$	2	$2(x + w)$
Against Wind	$x - w$	3	$3(x - w)$

15. Define the variables.

16. Write an equation describing the rates.

17. Write an equation describing the distances.

18. Solve to find the rate of the plane with no wind and the wind speed.

B. EXERCISES

Write and solve a system of equations for each word problem.

19. The hiking club went on a hike one Saturday. In the morning they walked at an average speed of 3 mi/hr; in the afternoon they averaged 2 mi/hr for 1 hr more than the number of hours they had walked in the morning. How many hours did they walk in the morning if they walked a total of 22 mi?

20. Felix leaves Centerville at 11:00 AM and travels west at 45 mi/hr. Lane leaves Centerville at 1:00 PM the same day and travels east at 50 mi/hr. At what time will the two be 375 mi apart?

21. Flying with the wind, a plane travels 1200 nautical miles in 4 hr. Flying against the wind, it travels 500 nautical miles in 2 hr. What is the speed of the plane with no wind? What is the wind speed?

22. Mrs. Russell and her son Benjamin live 184 mi apart. If they leave their homes at the same time and head toward each other, they will meet in 2 hr. Benjamin travels 12 mi/hr faster than his mother. How fast does each travel?

23. A southbound train travels for 2 hr and meets a northbound train that has been traveling on a parallel track for 4 hr. The trains started from cities 350 mi apart. What is the speed of each train if the northbound train travels 10 mi/hr slower than the southbound?

24. Find the rate of 2 cars traveling in the same direction if they leave the same place at the same time. One car is one and a half times as fast as the other, and after 4 hr the faster car is 80 mi ahead of the slower car.

25. On a calm day Chuck can bike to Janesville in 3 hr. Yesterday he made it in 2 hr with a 12 mi/hr tailwind. How fast does he bike on a completely calm day?

26. MP4 A plane leaves the Aschaffenburg Airport and travels west at 230 knots. A second plane leaves 3 hr later and travels east at 215 knots. How long has each plane been in the air when the 2 planes are 1135 nautical miles apart?

27. A car leaves Middletown and heads toward Rochester 248 mi away at the same time a car leaves Rochester and heads toward Middletown. If the rate of the first car is 6 mi/hr faster than the rate of the second and if the cars meet in 2 hr, how fast is each traveling?

28. The scouts can raft 15 mi downstream from one camp to another in 3 hr, but it takes 5 hr to paddle back to their camp upstream. What is the scouts' paddling rate in still water? What is the rate of the current in the stream?

29. Two trains leave the station at the same time and travel in opposite directions. The express train travels 14 mi/hr faster than the local train. After 3 hr they are 315 mi apart. What is the rate of each train?

30. **Essential Question:** How can a system of equations be used to solve motion problems?

31. In Section 2.7, single equations were used to solve motion problems. When is a system of equations needed to solve motion problems involving wind or water currents?

C. EXERCISES

Write and solve a system of equations for each word problem.

32. MP2 A train leaves City View at 11:00 AM and heads east at 38 mi/hr. At 2:00 PM another train leaves, traveling east at 53 mi/hr. How long will it take the second train to come within 54 mi of the first train?

33. Mr. Brannon took his Sunday school boys on a camping trip. They biked to the trailhead and then hiked to their camp. They rode at an average speed of 9 mi/hr and walked at a speed of 2 mi/hr in the 6 hr they traveled. How long did they hike if they traveled a total distance of 40 mi?

34. Salmon hatch in freshwater streams and then migrate 70 mi downstream to the ocean in 8 hr. After reaching spawning age, they return upstream to fresh water to spawn and usually die soon after. If the salmon travel 58 mi in 8 hr during the upstream trip, find the rate of the salmons' swimming in still water and the rate of the current.

35. MP7 The Feusse and Sallé families left at noon, traveling to a beach house for vacation. By 2:00 PM they had traveled a combined 186 mi. The Sallés stopped at a rest stop at 3:00 PM, and the Feusses arrived at the rest stop at 3:12 PM. What was each family's average traveling rate?

CUMULATIVE REVIEW

Solve each system by graphing. [6.1]

36. $y = -x + 4$
$y = 3x - 4$

37. $y = 2x - 6$
$y = \frac{4}{5}x$

38. $3x + 4y = -12$
$x - 2y = -4$

39. $2x - 3y = -15$
$6x - 9y = 18$

Solve each system using the method of your choice. Then classify it as consistent independent, consistent dependent, or inconsistent. [6.2–6.4]

40. $5x + 3y = 5$
$2x + y = -2$

41. $3x + 2y = 89$
$3x - 2y = 13$

42. $4x + 2y = 10$
$10x + 5y = 35$

43. $6x - 14y = 10$
$15x = 35y + 25$

44. Which is a correct simplification of $\frac{2^4 \cdot 3^0}{2 \cdot 3^5}$? **[1.3]**

A $2^4 \cdot 3^0$

B $\frac{3^5}{2^3}$

C $\frac{2^3}{3^5}$

D 0

E none of these

45. Which equation could Larry use to determine how much pure antifreeze to add to 6 gal of 40% antifreeze to make a solution that is 50% antifreeze? **[2.8]**

A $2.4 + a = 0.5(6 + a)$

B $6 + 0.4a = 0.5(6 + 0.4a)$

C $24 + 10a = 5(60 + 10a)$

D A and C

E none of these

6.6 MIXTURE PROBLEMS

How can I solve mixture problems by using a system of equations?

After completing this section, I will be able to

- solve real-world mixture problems by using a system of linear equations.

The mixture problems covered in Section 2.8 can also be solved using a system of equations. Some mixture problems combine items with different values such as weight or price. A table containing a variation of the following basic equation can be used to organize the information in the problem.

quantity × cost per item = total cost

EXAMPLE 1: Solving a Mixture Involving Weights

The Nut Shop sells cashews for \$8.15/lb and almonds for \$6.95/lb. The manager wants to make 80 lb of a mix of cashews and almonds that sells for \$7.25/lb. How many pounds of each kind of nut should he mix?

Answer

1. **Read** carefully to assign variables.

Let c = the number of pounds of cashews and a = the number of pounds of almonds.

2. **Plan:** Use a table to organize the information given.

	Number of Pounds	Price per Pound	Total Cost
Cashews	c	8.15	$8.15c$
Almonds	a	6.95	$6.95a$
Mix	80	7.25	580

3. **Solve** a system to answer the question.

   a. Write an equation modeling the prices and a second equation modeling the number of pounds.

$8.15c + 6.95a = 580$ and $c + a = 80$

   b. Multiply the first equation by 100 to clear the decimals. Multiply the second equation by −815 to obtain opposite coefficients.

$$100(8.15c + 6.95a = 580) \Rightarrow 815c + 695a = 58{,}000$$
$$-815(c + a = 80) \Rightarrow \underline{-815c - 815a = -65{,}200}$$
$$-120a = -7200$$
$$a = 60$$

   c. Substitute 60 for a in the second equation and solve for c.

$$c + a = 80$$
$$c + 60 = 80$$
$$c = 20$$

   d. Interpret the solution.

Mix 60 lb of almonds and 20 lb of cashews.

4. **Check** by substituting the solution in the other original equation

$8.15(20) + 6.95(60) = 580$

SKILL CHECK **EXERCISES 1, 13**

SOLVING WORD PROBLEMS WITH 2 VARIABLES

1. **Read** the problem carefully and look for the unknown quantities. Use variables to represent the unknowns.
2. **Plan** by making a table. Fill in as much information as possible.
3. **Solve** a system of equations obtained from information in the table and in the problem.
4. **Check** your answer in the context of the original statement of the problem.

EXAMPLE 2: Solving a Mixture Involving Money

Emmanuel Christian School sent out 150 pieces of mail. Some required \$0.60 postage while others required \$0.84. If the total bill was \$108.48, how many pieces of each type were mailed?

Answer

Let x = the number of 60¢ pieces of mail and y = the number of 84¢ pieces of mail.

1. **Read** carefully to assign variables.

	Pieces of Mail	Price per Item (¢)	Total Price (¢)
\$0.60	x	60	$60x$
\$0.84	y	84	$84y$
Total	150		10,848

2. **Plan:** Use a table to organize the information.

TIP

Using cents instead of dollars simplifies the system to be solved.

3. **Solve** a system to answer the question.

a. Write an equation modeling the number of pieces of mail and a second equation modeling the costs.

$x + y = 150$ and $60x + 84y = 10{,}848$

b. Solve the first equation for x and substitute its value in the second equation.

$x = -y + 150$

$$\begin{aligned} 60(-y + 150) + 84y &= 10{,}848 \\ -60y + 9000 + 84y &= 10{,}848 \\ 24y + 9000 &= 10{,}848 \\ 24y &= 1848 \\ y &= 77 \end{aligned}$$

c. Substitute 77 for y in the first equation and solve for x.

$$\begin{aligned} x + 77 &= 150 \\ x &= 73 \end{aligned}$$

d. Interpret the solution.

The school mailed 73 pieces of mail that cost \$0.60 and 77 that cost \$0.84.

4. **Check** by substituting the solution into the other original equation.

$73(0.60) + 77(0.84) = \$108.48$

SKILL CHECK **EXERCISES 5, 17**

Percentages are used in combining mixtures to obtain a final mixture of a desired strength. When 10 gal of a 5% salt solution is combined with 20 gal of an 8% salt solution, the total amount of salt in the 30 gal can be found by adding the amount of salt from each component.

$$0.05(10) + 0.08(20) = 0.5 + 1.6 = 2.1 \text{ gal of salt}$$

The concentration of the resulting mixture can then be determined.

$$\frac{2.1}{30} = 0.07 = 7\% \text{ salt solution}$$

Consider the difference between adding pure water and adding pure salt to a solution. When adding pure water, you are adding a 0% salt solution. When adding pure salt, you are adding a 100% salt solution.

EXAMPLE 3: Solving a Mixture Involving Percentages

If 3%-fat milk is mixed with fat-free milk to produce 5 gal of 2%-fat milk, how much of each should be used?

Answer

1. **Read** carefully to assign variables.

Let x = the number of gallons of 3%-fat milk and y = the number of gallons of fat-free milk.

2. **Plan:** Use a table to organize the information.

	Number of Gallons	Percent of Fat	Gallons of Fat
3%	x	0.03	0.03x
Fat-free	y	0	0
2%	5	0.02	0.1

3. **Solve** a system to answer the question.

a. Write an equation modeling the number of gallons and a second equation modeling the gallons of fat.

$$x + y = 5 \quad \text{and} \quad 0.03x + 0 = 0.1$$
$$0.03x = 0.1$$

b. Clear the decimals in the second equation and solve for x.

$$3x = 10$$
$$x = \frac{10}{3} \text{ or } 3\frac{1}{3}$$

c. Substitute $\frac{10}{3}$ for x in the first equation and solve for y.

$$\frac{10}{3} + y = 5$$
$$y = \frac{15}{3} - \frac{10}{3}$$
$$y = \frac{5}{3} \text{ or } 1\frac{2}{3}$$

d. Interpret the solution.

$3\frac{1}{3}$ gal of 3%-fat milk should be mixed with $1\frac{2}{3}$ gal of fat-free milk to produce 5 gal of 2%-fat milk.

4. **Check** by substituting the solution into the other original equation.

$$0.03\left(\frac{10}{3}\right) + 0 = 0.1$$

SKILL CHECK ***EXERCISE 19***

AFTERSCHOOLHELP

Mixture Problems

A. EXERCISES

Complete the following tables.

	Quantity	Cost per Item	Total Cost
1.	80 lb of flour	$0.50/lb	
2.	27 apples		$17.55
3.		$7 per bag	$574
4.	16 oz	*d* dollars per oz	
5.	*x* cases	$30/case	

	Retail Price	Discount Rate	Sale Price
6.	20 gal	15% acid	
7.	25 L		7 L of salt
8.		7% acid	10.5 L of acid
9.	*x* gal	12% acid	
10.	200 mL	*p*% salt	

11. What is the percent of salt in pure water?

12. What is the strength of pure acid?

Write and solve a system of equations for each word problem.

13. Kimberly wants to make a mix of nuts and raisins for a party. Nuts cost $0.20/oz and raisins cost $0.12/oz. How many ounces of each should she buy if she wants 48 oz of a mix that costs $0.15/oz?

14. A grocer mixes oranges and bananas to make a 10 lb fruit basket. The oranges cost $0.75/lb and the bananas cost $0.60/lb. How many pounds of each should he use if the basket is to cost $6.90?

15. Caramels cost $3.50/lb and butterscotch candies cost $4.50/lb. A 5 lb bag of mixed candy is to sell for $4.20/lb. How many pounds of each kind of candy should be mixed together?

16. Maria plans to invest a total of $7500 between 2 types of stocks, one yielding 6% interest and the other 15%. How much does she invest in each type of stock if the total interest gained annually is $585?

B. EXERCISES

Write and solve a system of equations for each word problem.

17. Florist Friars advertises roses for $2 a bud and carnations for $0.75 a flower. If Sebastian pays $20.50, excluding tax, for a bouquet of fourteen flowers, how many of each flower are in the bouquet?

18. The gardener plans to seed the front lawn with a new mix of grass seed. He has a mix that is 80% fescue and 15% winter rye. How much fescue and how much of his original mix should he blend together to create 50 lb of an 86% fescue mix?

19. Shawn has a solution that is 30% salt. How many gallons of solution and pure water should he mix to make 40 gal of a diluted solution that is 9% salt?

20. A pharmacist is making an antiseptic of iodine and alcohol. The solution that he now has is 5% iodine. He wants a mixture that is 6% iodine. How much of a 10% iodine solution and how much of his original antiseptic should he mix to make 50 mL of the new antiseptic?

21. Raisin Rich breakfast cereal is 10% protein. A dietician wants to raise the percentage of protein per serving to 20%. The high-protein grain that will be mixed with the cereal is 40% protein. How many grams of each should be mixed together to make a 12 g serving of cereal with the proper percentage of protein?

22. Concrete is made of cement, water, and aggregate (sand and gravel). Aggregate should make up 75% of the total mass of concrete. If you have a concrete mix that is 60% aggregate, how many kilograms of this mix and aggregate should be combined to form 48 kg of the concrete mix with the proper percentage of aggregate?

23. A chemist has two HNO_3 acid and water solutions. He knows that the percentage of acid in solution I is twenty percentage points less than the percentage of acid in solution II. If he mixes 40 mL of solution I with 60 mL of solution II and obtains a 27% acid solution, what percentage of acid is in each solution?

24. Becky makes an antiseptic of alcohol and water. If she buys a solution that is 40% alcohol to make a 12.5% alcohol solution, how much water and alcohol solution should she mix to obtain 128 mL of her homemade antiseptic?

25. Hunter invested a total of $10,000 between 2 accounts. One account yields 10% interest and the other 9.5%. How much did he invest in each account if his total annual interest was $971?

26. Miss Parker invested a certain amount of money at 9.2% interest and another amount, $700 more than the first, at 10.4%. If the total annual interest is $425.60, how much money is in each investment?

Write and solve a system of equations for each word problem. Round to the nearest tenth if necessary.

27. Cool Aunt's sells 2 types of antifreeze. Type A is 40% ethylene glycol, and type B is 52%. An auto mechanic wants 200 gal of antifreeze that protects radiators to −25°F. To give that protection, the solution must be 45% ethylene glycol. How many gallons of each type should be mixed to get the correct strength?

28. A butcher wants to grind round steak and mix it with regular ground beef to make a deluxe ground beef. The round steak sells for $3.89/lb and the ground beef sells for $2.79/lb. How many pounds of each should he mix together to make 50 lb of the deluxe ground beef that will sell for $3.29/lb?

29. An ice cream company is going to mix cream that is 40% fat with milk that is $3\frac{1}{2}$% fat to produce 60 gal of a mixture that is 10% fat. How many gallons of cream and how many gallons of milk should they use?

30. Leigh mixes corn syrup and molasses together to make a special pancake syrup. She can afford to pay $0.35/oz for the mix. If corn syrup costs $0.215/oz and molasses costs $0.52/oz, how many ounces of each should she mix together to make 24 oz of syrup?

31. A pet store owner wants to mix 2 different kinds of seeds together to form a high-grade birdseed. Seed A sells for $0.28/lb, seed B sells for $0.60/lb, and the mixture will sell for $0.50/lb. If he wants to make 85 lb of the seed mixture, how much of each seed should he mix together?

32. Wingo is blending coffee beans that have a floral aroma and acidity with other beans that have a fuller body. The more expensive coffee costs \$22.35/lb and the second coffee costs \$16.96/lb. If he needs 50 lb that cost \$18.75/lb, how much of each type should he use in the mixture?

33. **Essential Question:** How could you solve a mixture problem by using a system of equations?

C. EXERCISES

Write and solve a system of equations for each word problem. Round to the nearest tenth if necessary.

34. MP7 A businessman invests \$20,000 in a dry-cleaning establishment and \$30,000 in a restaurant. What is his expected rate of return (or rate of interest) on each investment if he expects a total of \$4600 annually and the average of the 2 rates is 9%?

35. Mr. Johnson ran for mayor in last week's election. In district 12 he received 85% of the vote, and in district 13 he received 43% of the vote. From the 2 districts he received 10,253 votes. How many voted for Mr. Johnson in each district if a total of 16,460 people voted in the 2 districts?

36. In ordinary ocean water, sodium chloride (the chemical name for ordinary table salt) is 2.8% by mass. The Dead Sea is about 12% sodium chloride. If a chemist takes a sample of water from the Dead Sea and wants to dilute it with ordinary ocean water, how much ocean water and how much Dead Sea water should he combine to form 12 L of a 10% sodium chloride solution?

37. MP4 A farmer can sell 50 bu of beans and 100 bu of corn for \$810, or he can sell 20 bu of beans and 205 bu of corn for \$868.50. What is the selling price of beans and corn per bushel?

CUMULATIVE REVIEW

Simplify, leaving each answer in positive exponential form. [1.3]

38. $2^4 \cdot 2^{-2}$

39. $3^6 \div 3^{-1}$

40. $(3^2)^{-3}$

41. $\frac{2^2 \cdot 3^3 \cdot 5^2}{2^6 \cdot 3^7 \cdot 5^9}$

Determine whether each number is a solution to the inequality. [3.1]

42. $5 - x \ngtr 3$

a. −4

b. 2

c. 6

43. $y - 7 > -5$

a. −3

b. 2

c. 5

Solve and graph the solution on a number line. [3.2]

44. $6y > -10 - 8$

45. $3x - 4 \ngtr x + 2$

46. Which is the explicit formula for the sequence −3, 4, 11, 18, . . .? **[4.5]**

A $a_n = a_{n-1} - 3; a_1 = 7$

B $a_n = a_{n-1} + 7; a_1 = -3$

C $a_n = 7 + (n - 1)(-3)$

D $a_n = -3 + (n - 1)7$

E none of these

47. What is the equation of the line passing through (0, 7) and (−3, 9)? **[5.4]**

A $y - 7 = -\frac{2}{3}x$

B $y - 9 = -\frac{2}{3}(x + 3)$

C $y = -\frac{2}{3}x + 7$

D $2x + 3y = 21$

E all of these

6.7 GRAPHING LINEAR INEQUALITIES

How does the graph of a linear inequality differ from the graph of a linear equation?

After completing this section, I will be able to

- determine whether a point is a solution to a linear inequality.
- graph linear inequalities.
- model real-world situations with linear inequalities.

Linear inequalities are similar to linear equations but contain an inequality sign instead of an equals sign. Remember that inequalities use the signs ≠, <, >, ≥, or ≤ and have an infinite number of solutions.

The graph of $x = 3$ on a number line separates the line into 3 distinct sets of points.

The graph of the linear equation $x = 3$ similarly separates the coordinate plane into 3 distinct sets of points: the points on the line $x = 3$; the region to the left of the line, $x < 3$; and the region to the right of the line, $x > 3$.

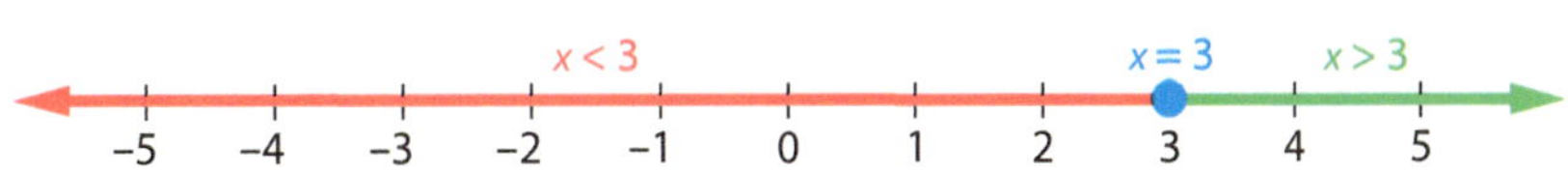

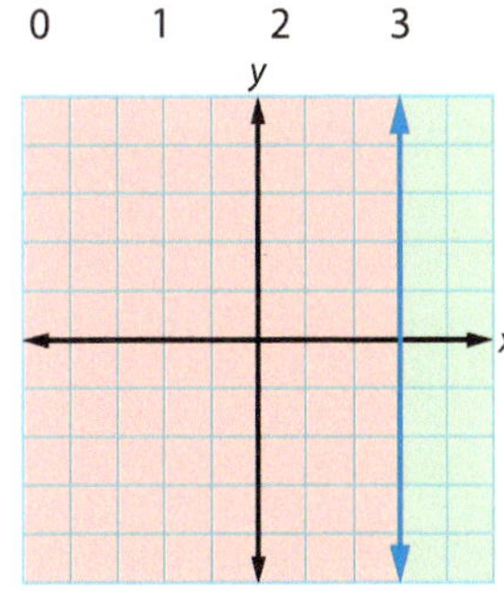

We know that a linear equation such as $2x + y = 3$ includes only points on the line as solutions. The linear equation $2x + y = 3$ will divide the coordinate plane into 3 sets of points: the line itself, which acts as a **boundary line**; the set of points on one side of the line; and the set of points on the other side of the line. The solutions to the linear inequality $2x + y > 3$ will lie on one side of the line called the **half-plane**. Depending upon the type of inequality, the points on the line may or may not be solutions.

EXAMPLE 1: Graphing a Linear Inequality

Graph $2x + y > 3$.

Answer

$2x + y > 3$
$y > -2x + 3$
$m = -2$
y-intercept: $(0, 3)$

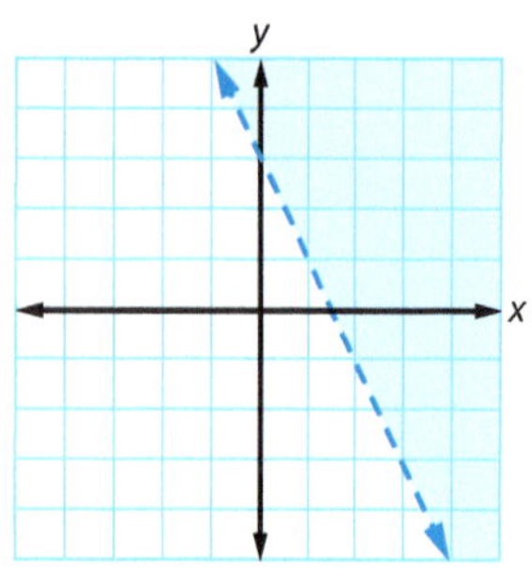

1. Solve the inequality for y to identify the slope and y-intercept of the corresponding linear equation $y = -2x + 3$.
2. Since $y > -2x + 3$ is an inequality, not an equation, the solution set does not include the points on the line. Graph the equation $y = -2x + 3$ with a dashed line to represent the boundary line.
3. Use a convenient point to determine which side of the line contains solutions to the inequality.
   a. The origin, (0, 0), is an easy choice.
   b. Substitute the x- and y-values of (0, 0) into the inequality. Since $2(0) + 0 \not> 3$, the point is not a solution.
   c. Since the test point is not a solution, neither is any other point in that half-plane. The solutions must be located in the half-plane on the other side of the line.
4. The shaded half-plane represents the solution set.

SKILL CHECK **EXERCISE 19**

If the point you choose makes the inequality true, all points in the half-plane on that side of the boundary line will also make the inequality true. In the example above, a point above the line chosen as a test point, such as (3, 4), will provide a true statement: $2(3) + 4 > 3$. This true inequality statement indicates that the test point and all other points in the half-plane above the boundary line are solutions and should be in the shaded region.

GRAPHING AN INEQUALITY

1. Solve for y and identify the slope and y-intercept of the corresponding equality.
2. Graph the corresponding linear equation.
   a. If the inequality is $<$ or $>$, use a dashed line to indicate that points on the line are not solutions.
   b. If the inequality is $\le$ or $\ge$, use a solid line to indicate that points on the line are solutions.
3. Choose a test point that is not on the line and substitute it into the inequality.
   a. If the resulting inequality is true, shade the side containing the test point.
   b. If the inequality is false, shade the other side.

EXAMPLE 2: Graphing a Linear Inequality

Graph $6x - 2y \geq -10$.

Answer

$$
\begin{aligned}
6x - 2y &\geq -10 \\
-2y &\geq -6x - 10 \\
y &\leq \frac{-6x}{-2} + \frac{-10}{-2} \\
y &\leq 3x + 5
\end{aligned}
$$

$m = 3$

y-intercept: (0, 5)

1. Solve the inequality for y and identify the slope and y-intercept. Remember to reverse the inequality sign when multiplying or dividing by a negative number.

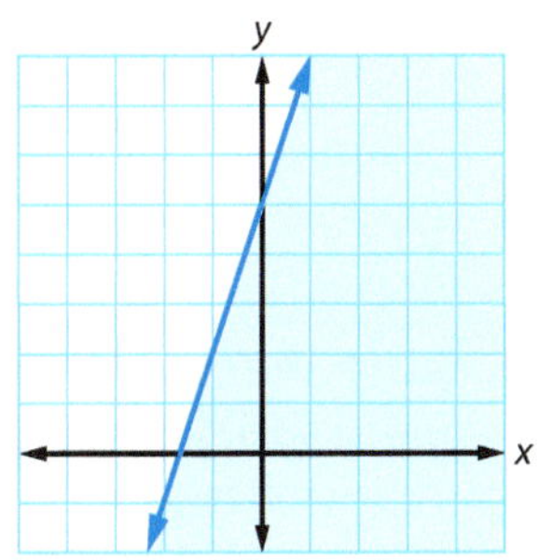

2. Since the inequality sign is ≤, graph the equation $y = 3x + 5$ with a solid line.
3. Use a test point, (0, 0), to determine which side of the line contains solutions to the inequality. Since $6(0) - 2(0) \geq -10$, the point (0, 0) and all points in that half-plane are part of the solution set.
4. Shade all the points on and below the boundary line.

SKILL CHECK **EXERCISE 21** ☑

AFTERSCHOOLHELP

Graphing Linear Inequalities

If the inequality is written in slope-intercept form, it is easy to determine which side of the line should be shaded. On the y-axis, larger values of y are higher and smaller values of y are lower. Therefore, if $y > mx + b$, shade above the line; and if $y < mx + b$, shade below the line. In Example 2, $y \leq 3x + 5$ indicates that the solution is all the points on and below the line.

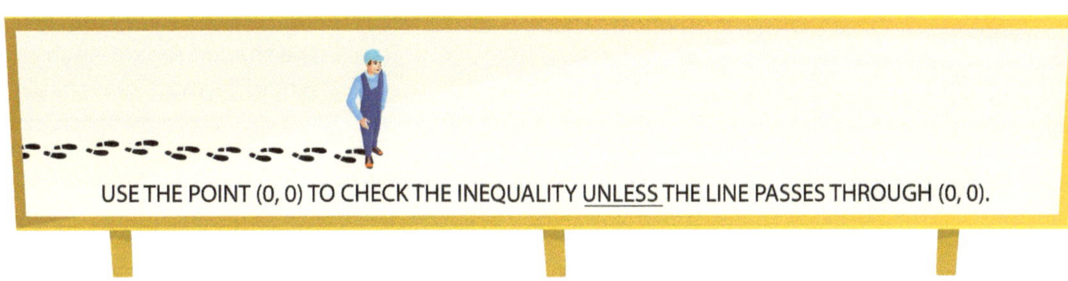

Linear inequalities are often used to model real-world situations in which there are acceptable ranges of values. The domain and range of these applications are often limited to nonnegative numbers.

EXAMPLE 3: Solving a Real-World Problem

Jessica is on a strict weight-loss diet and wants to limit her total caloric intake from carbohydrates and fats to 720 calories each day (she gets the rest of her calories from protein). Each gram of carbohydrates contains 4 calories, and each gram of fats contains 9 calories. Write and graph a linear inequality that models her plan. List several acceptable amounts for daily consumption of carbohydrates and fats.

Answer

Let x = grams of carbohydrates and y = grams of fats.

1. Assign variables to the unknowns.

total calories ≤ 720

2. Write an inequality describing the limited quantity.

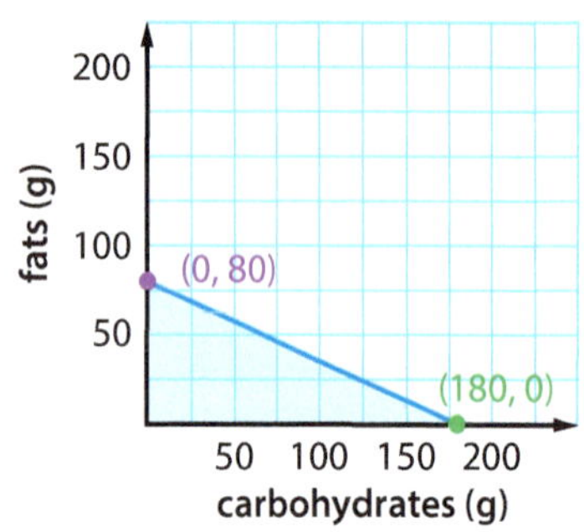

3. Graph the corresponding equation $4x + 9y = 720$ with a solid line using the x-intercept (180, 0) and the y-intercept (0, 80).
4. The point (0, 0) is clearly a solution, so shade below the line to include this point.
5. Note that eating no carbohydrates or fats, 180 g of carbohydrates and no fats, or no carbohydrates and 80 g of fats would be within the plan but would not be very reasonable. Other points within the shaded region, such as (100, 30) or (115, 20) would be more reasonable.

SKILL CHECK **EXERCISE 37** ☑

VOCABULARY

- boundary line
- half-plane

A. EXERCISES

State whether a solid or a dashed line is used in the graph of each inequality.

1. $y \geq -5x + 3$

2. $y < -7x - 9$

3. $2x - 13y > 8$

4. $5x - y \leq 1$

Determine which ordered pairs are solutions to the given inequality.

5. $y < 2x - 9$; $A\ (0, 0)$, $B\ (4, 7)$, $C\ (-2, -13)$, $D\ (3, -5)$

6. $y \geq 3x + 4$; $F\ (0, 0)$, $G\ (-5, -7)$, $H\ (2, -5)$, $J\ (0, 4)$

7. $2x - 5y > 1$; $K\ (0, 0)$, $L\ (3, 1)$, $M\ (-2, -5)$, $N\ (-7, 3)$

8. $7x - y \leq 1$; $P\ (0, 0)$, $Q\ (5, 8)$, $R\ (-1, 0)$, $S\ (-8, -3)$

Determine which side of the line you should shade.

9. $-5x + 8y \geq 7$; (0, 0)

10. $y < 2x + 4$; (0, 0)

11. $y \leq 5x - 6$; (3, 1)

12. $8x - y \geq 4$; (−4, 3)

A the side that includes the given test point

B the side that is opposite the given test point

Match each graph with its inequality.

A

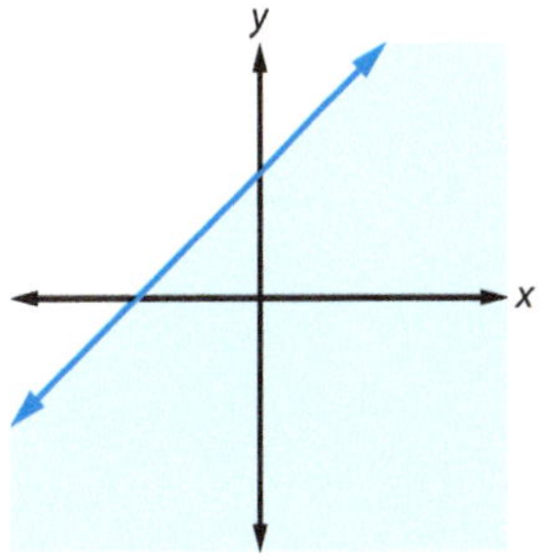

B

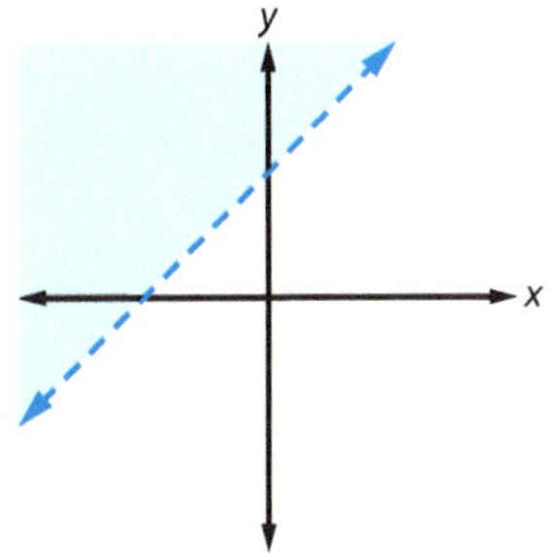

C

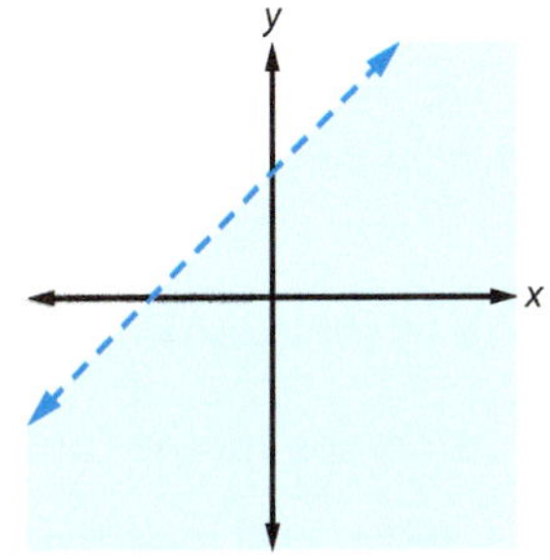

D

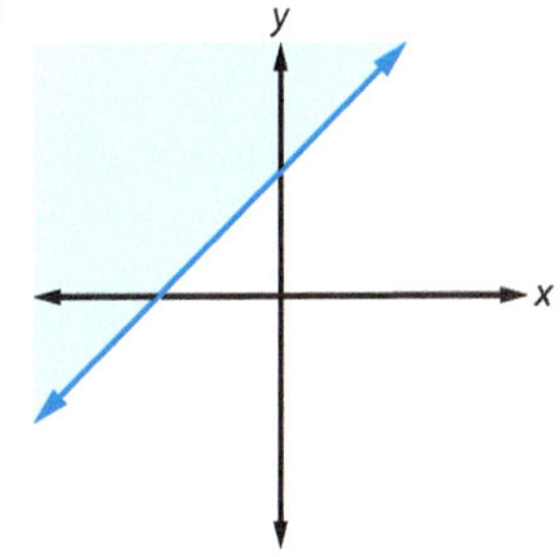

13. $y > x + 2$

14. $y < x + 2$

15. $y \leq x + 2$

16. $y \geq x + 2$

B. EXERCISES

Graph each inequality.

17. $x < -4$

18. $y \geq 5$

19. $y > 3x + 2$

20. $x + y > -2$

21. $2x - y \leq 4$

22. $x - 4y \geq 16$

23. $3y < 7$

24. $-x \leq -2$

Write a linear inequality describing each graph.

25.

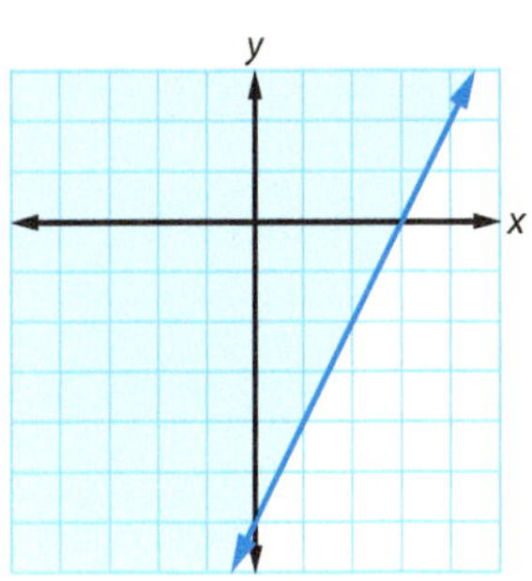

26.

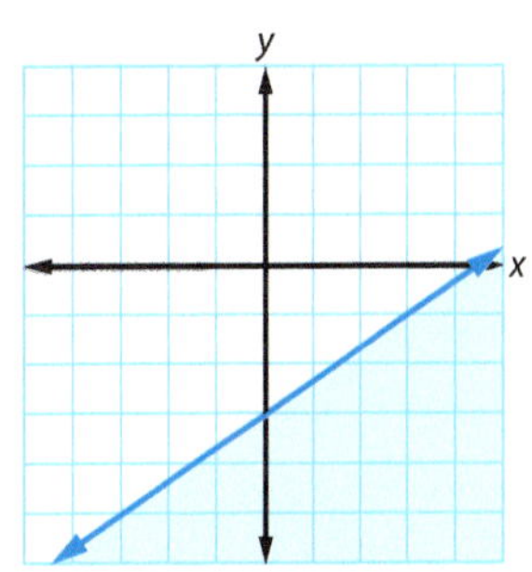

27.

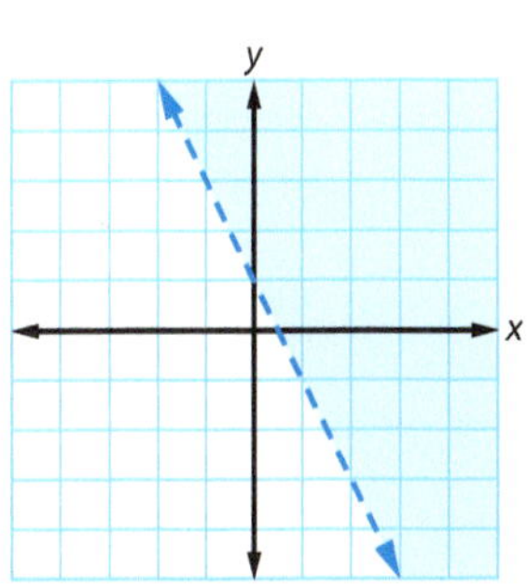

28.

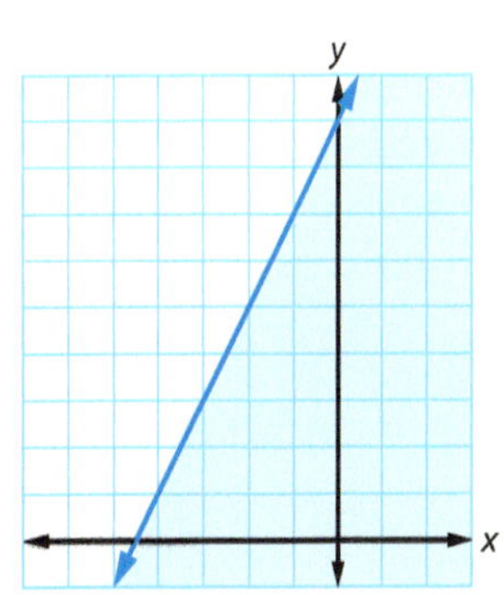

29. Describe how to determine which half-plane to shade to represent the solution set of an inequality.

30. MP3 Describe 3 errors in the following graph of $2x + 5 < y$.

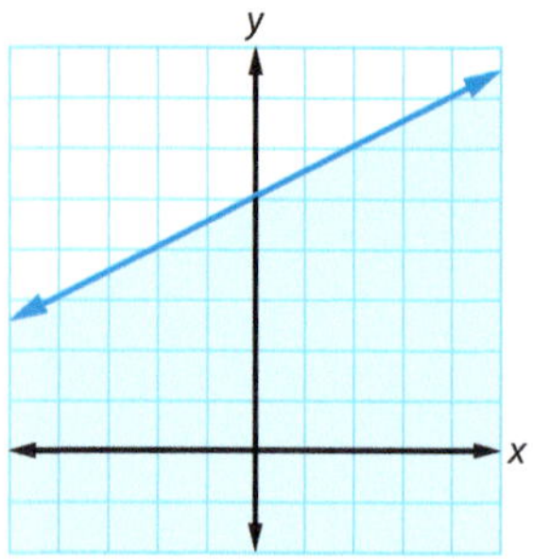

31. MP8 Are linear inequalities relations? Are they functions? Explain.

32. **Essential Question:** How does the graph of a linear inequality differ from the graph of a linear equation?

Translate each statement into a linear inequality; then graph the inequality.

33. A number, y, is at most five times another number, x.

34. The sum of 2 numbers, x and y, is at least five.

Write and graph a linear inequality to model each situation.

35. During this past basketball season, Brittany scored no more than 12 points in a game from free throws (worth 1 point) and 3-point field goals. Provide 1 possible solution from your graph for the number of each type of shot she made. Use x to represent free throws made and y to represent 3-point shots made.

36. Julio can burn 8 calories per minute using a stair-step machine and 10 calories per minute while rowing. He wants to burn at least 100 calories during each workout. Use x to represent minutes rowing and y to represent minutes on the stair-step machine. If he spends 6 min rowing, what is the least number of minutes he needs to spend on the stair-step machine to burn at least 100 calories?

37. MP5 Trey earns a commission of $25 on each computer and $5 on each calculator that he sells. He needs to make at least $325 in commission this week. Use x to represent the number of computers sold and y for the number of calculators sold. If Trey sells 20 calculators, what is the least number of computers he needs to sell to meet his goal?

38. MP4 Amber is making a party mix of mixed nuts and chocolate candies for her sister's wedding reception. She has a budget of $60 and can buy bags of mixed nuts for $4/lb and bags of chocolate candies for $3/lb. Use x to represent the number of pounds of nuts and y for the number of pounds of candy. Provide 1 solution from your graph in which Amber purchases both nuts and candies.

C. EXERCISES

Graph each conjunction or disjunction.

39. $x > 1$ and $x \leq 4$

40. $y \geq 2x + 3$ or $y \leq 2x - 1$

41. MP7 Amber's sister suggested that the party mix described in exercise 38 would be too salty if it exceeded a 3 : 1 ratio of nuts to candy and too sweet if it had less than a 1 : 1 ratio. Write 2 direct variation function rules modeling these ratios and add their graphs to the graph of the inequality from exercise 38. Then state 3 reasonable combinations of nuts and candies for the wedding reception.

42. MP7 Last year's sales records indicate that Trey's calculator-to-computer sales ratio ranged from 2 : 1 to 4 : 1. Write 2 direct variation function rules modeling these ratios and add their graphs to the graph of the inequality from exercise 37. Then state 3 reasonable combinations of calculator and computer sales that would allow Trey to meet his goal.

CUMULATIVE REVIEW

Solve. [2.2]

43. $\frac{3}{8}x - \frac{5}{9} = \frac{4}{3}$

44. $2.5x - 12 = 5.01 - x$

Use proportions to solve. [2.4]

45. If the number of bushels of grain harvested is directly proportional to the number of acres planted and 18 acres produced 414 bu of grain, how many bushels could be produced on 640 acres?

46. If the number of cars sold is directly proportional to the number of salesmen and 3 salesmen sold 11 cars, how many could 5 salesmen have sold? Round to the nearest whole number.

47. If a blueprint's scale is 1 in. : 25 ft, how many feet would be represented by $8\frac{7}{8}$ in. (to the nearest foot)?

48. Two rectangles are similar. If one is 6 cm by 8 cm and the short side of the other is 13.9 cm, what is the length of its other side (to the nearest tenth of a centimeter)?

Solve by graphing. [6.1]

49. Where will the graphs of $y = 6$ and $x + 3y = 12$ intersect?

50. Where will the graphs of $x = -3$ and $x + 2y = 7$ intersect?

51. Which of the following are solutions of the conjunction $y \geq -1$ and $5 > y$? Choose all that apply. **[3.3]**

A -4

B -1

C 0

D 5

E all of these

52. Which of the following are solutions of the disjunction $z < -3$ or $3 \leq z$? Choose all that apply. **[3.4]**

A 0

B 3.1

C -3

D 2.9

E all of these

6.8 SOLVING SYSTEMS OF INEQUALITIES

How can I determine all possible dimensions for fencing in a rectangular area?

After completing this section, I will be able to

- solve systems of inequalities by graphing.
- solve real-world problems by graphing systems of inequalities.

The solution to a *system of inequalities* consists of the ordered pairs that satisfy all the inequalities of the system. This set of points is the intersection of the graphs of the inequalities in the system.

EXAMPLE 1: Solving a System of 2 Inequalities

Solve the system of inequalities.

$3x + y > 4$
$x - y \leq 8$

Answer

$3x + y > 4$ and $x - y \leq 8$
$y > -3x + 4$ $-y \leq -x + 8$
$y \geq x - 8$

1. Write each inequality in slope-intercept form.

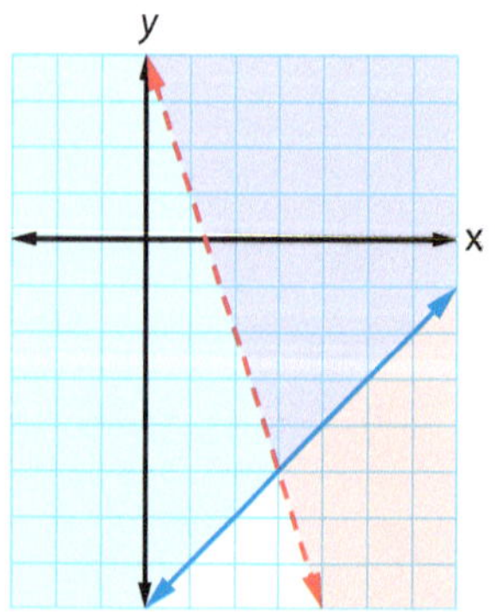

2. Graph each inequality on the same set of axes.

   y-intercept: (0, 4) $m = -3$ Shade above the dashed line.

   y-intercept: (0, –8) $m = 1$ Shade above the solid line.

3. Solution: The purple region indicates the points in the intersection of the 2 graphs.

$3(4) + 0 > 4$ and $4 - 0 \leq 8$

Check: Verify that a point in the region, (4, 0), satisfies both inequalities.

SKILL CHECK **EXERCISE 9**

SOLVING A SYSTEM OF INEQUALITIES

1. Graph all inequalities on the same set of axes.
2. Find the solution set where the shaded regions intersect.
3. Check the result.

EXAMPLE 2: Solving a System of 3 Inequalities

Solve the system of inequalities.

$x > 0$
$y > 0$
$2x - y > 1$

Answer

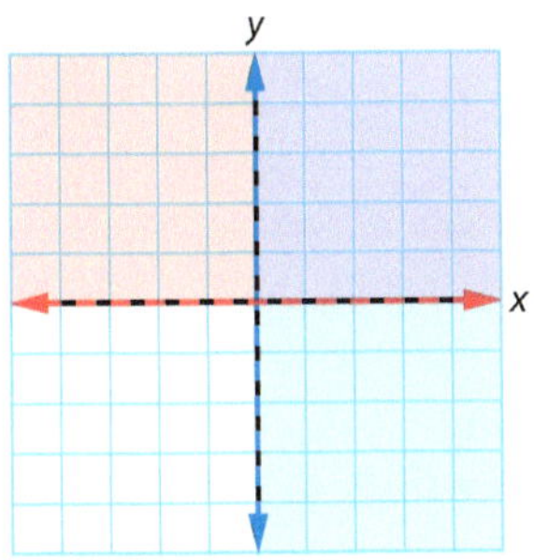

1. Graph the first 2 inequalities.

$x > 0$
Shade to the right of the line $x = 0$ (the y-axis).

$y > 0$
Shade above the line $y = 0$ (the x-axis).

2. Note that the first 2 inequalities limit our solutions to the first quadrant.

$2x - y > 1$
$-y > -2x + 1$
$y < 2x - 1$

3. Graph the third inequality.
y-intercept: $(0, -1)$
$m = 2$
Shade below the dashed line.
4. The solution is all first-quadrant points in the green-shaded area.

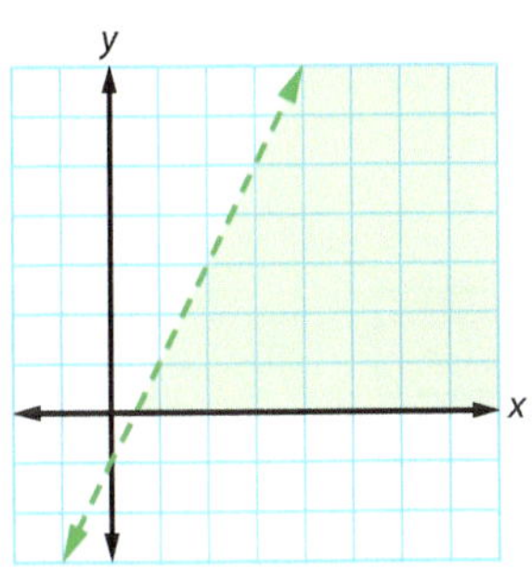

$5 > 0$ $3 > 0$ $2(5) - 3 > 1$
$7 > 1$

Check: Verify that a point in the region, (5, 3), satisfies all 3 inequalities.

SKILL CHECK **EXERCISE 25**

The avalanche pictured is on a nearly vertical slope. The graph shows that such avalanches are rare and indicates the relative risks for different slopes. We can write a system modeling the high-risk area.

Avalanche Risk for Mountain Slope

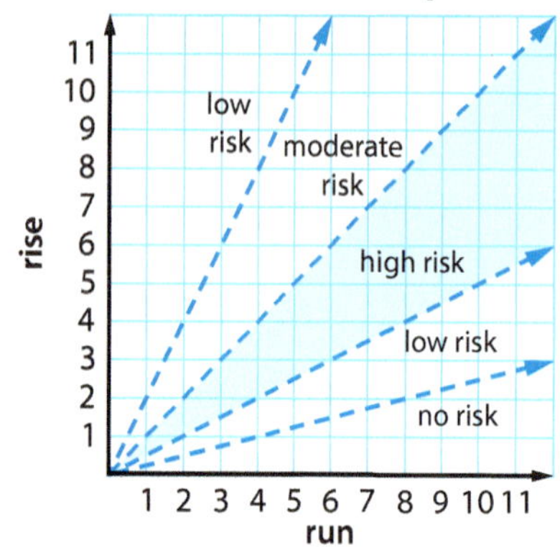

Annapurna, which is avalanche prone, is considered the most dangerous mountain to climb in the world.

AfterSchoolHelp

Solving Systems of Inequalities

The lower boundary's y-intercept is the origin, and its slope is $\frac{1}{2}$. Since the shaded area is above this boundary line, the inequality is $y > \frac{1}{2}x$. The y-intercept of the upper boundary is also (0, 0), but its slope is 1. The shaded area is under this boundary, so the inequality is $y < x$. The high-risk area can be modeled by the following system.

$$y > \frac{1}{2}x$$
$$y < x$$

EXAMPLE 3: Writing Inequalities

Write the system of inequalities represented by the graph.

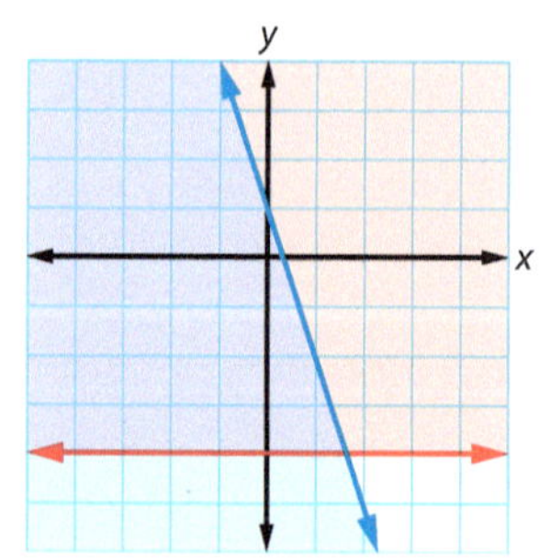

Answer

y-intercept: (0, −4) $m = 0$ $y = -4$	y-intercept: (0, 1) $m = -3$ $y = -3x + 1$	**1.** Find the equations of the boundary lines.
$y \geq -4$	$y \leq -3x + 1$	**2.** Write the inequalities. The solution includes points above the solid red line and points below the solid blue line.

SKILL CHECK ***EXERCISES 15, 31***

Recall that points on dashed boundary lines are not included in the solution set (Examples 1–2), but points located on a solid boundary line are included in the solution set (Example 3).

To find solutions for real-world problems involving multiple constraints (such as maximum cost and minimum production), model each constraint with an inequality and then graph the inequalities. All points within the solution set meet each constraint.

EXAMPLE 4: Solving a Real-World Problem

Ben wants to spend less than \$60 for the materials to make pens and pencils on his wood lathe. Each pencil kit costs \$5, and each pen kit costs \$3. He wants to make at least 15 total pens and pencils to give away as Christmas gifts.

a. Write and graph a system of inequalities that describes the given information.

b. Use the graph to determine whether Ben can make 8 pens and 8 pencils.

Answer

a. Let x = the number of pens and y = the number of pencils.

Inequality 1 and Inequality 2

$3x + 5y < 60$ $\quad$ $x + y \geq 15$

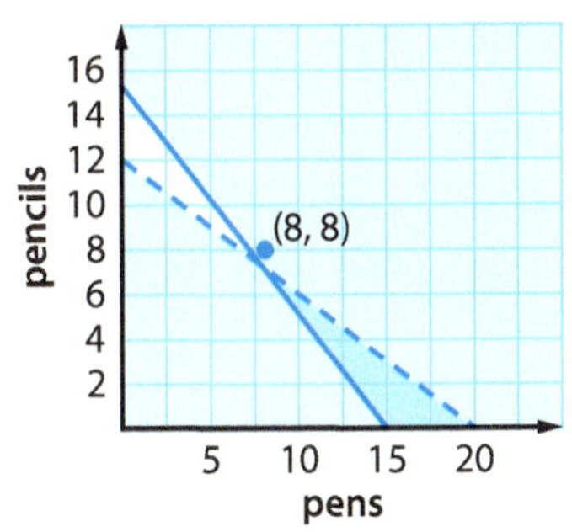

1. Define the variables.
2. Write the system. Inequality 1: The total cost is less than \$60. Inequality 2: The total number of pens and pencils is at least 15.
3. Graph each inequality on the same set of axes.
4. Graph the point that represents 8 pens and 8 pencils.

b. Since the point is outside the solution, Ben cannot make 8 pens and 8 pencils and meet the given parameters.

SKILL CHECK EXERCISE 33

A. EXERCISES

Without graphing, identify which quadrant each system describes.

1. $x > 0$, $y > 0$

2. $x < 0$, $y < 0$

3. $x < 0$, $y > 0$

4. $x > 0$, $y < 0$

For each system, determine whether the given ordered pair is a solution.

5. $(2, 3)$
$2x + 3y \geq 5$
$x + y \leq 7$

6. $(3, 0)$
$x + 3y < 15$
$x + 3y > 6$

7. $(4, 2)$
$y \leq -x + 6$
$x < 4$

8. $(-4, -6)$
$2x + 3y < -9$
$x - 2y \geq -4$

Graph each system of inequalities. List 1 point in your solution set.

9. $y \leq x - 2$
$y \geq 3x - 5$

10. $y > \frac{1}{2}x - 1$
$y < -3x - 3$

11. $y \geq -2x + 6$
$y < \frac{1}{3}x - 5$

12. $x \geq 0$
$y < x + 5$

13. $y > 4$
$y \leq -2x + 3$

14. $x \leq 5$
$y > -\frac{1}{2}x + 6$

B. EXERCISES

Write the system represented by each graph.

15.

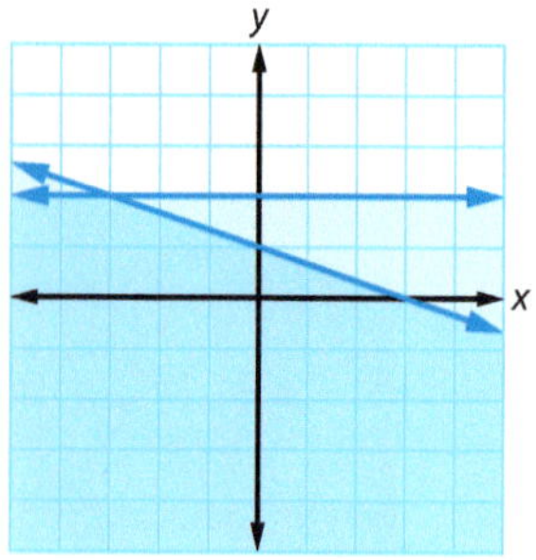

16.

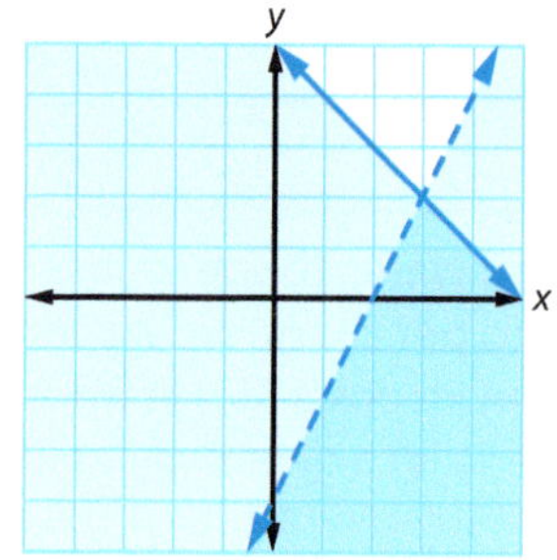

17.

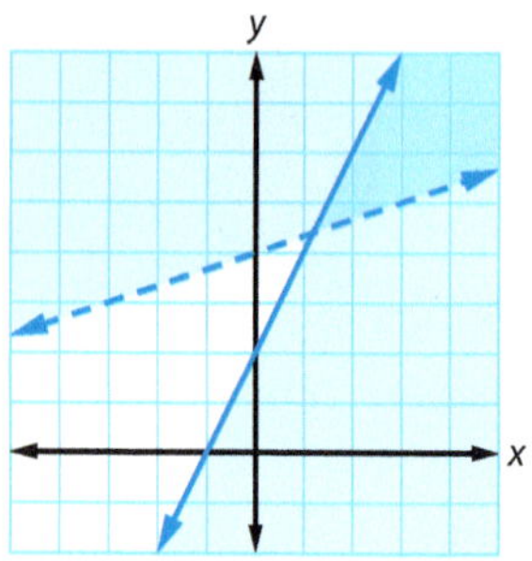

18.

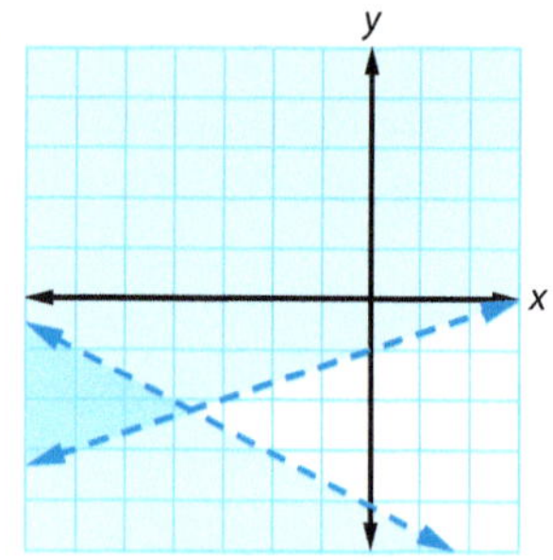

Graph each system of inequalities.

19. $x + y \geq 2$
$x - y < 4$

20. $5x + y > 2$
$2x + y \leq 1$

21. $3x - 2y \geq 4$
$x + y \leq 3$

22. $2x - y < 5$
$3x + y \geq 4$

23. $x + 5y < 15$
$3x + 2y \leq 8$

24. $x + 2y \geq -4$
$x + 2y < 6$

25. $x \geq 0$
$y \geq 0$
$5x + 6y \leq 30$

26. $x \geq 0$
$y \leq 2$
$x + y > 1$

27. $x \geq 0$
$y \geq 0$
$3x + y \leq 8$
$x + 2y \leq 6$

28. $x \geq 0$
$y \geq 0$
$2x + y \leq 4$
$x + 4y \leq 8$

Write the system represented by each graph.

29.

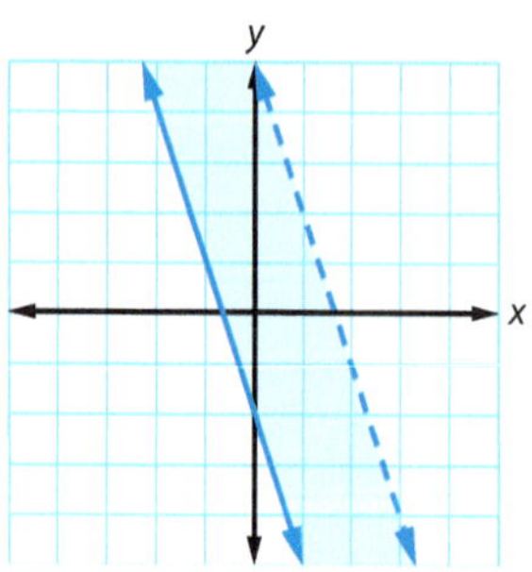

30.

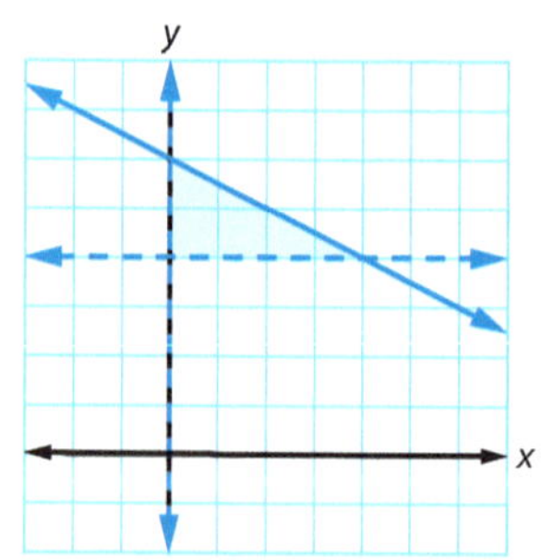

31.

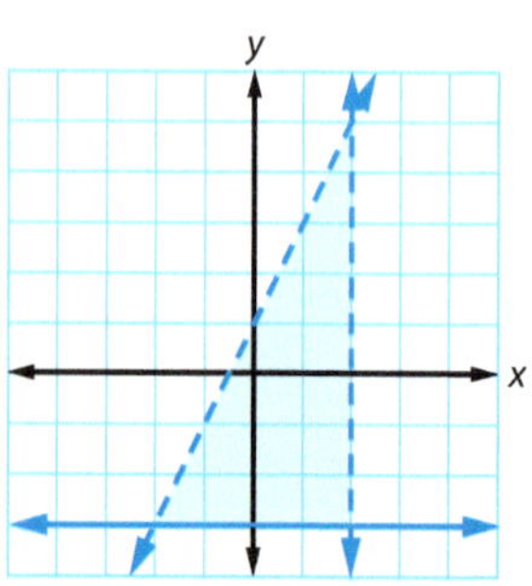

32.

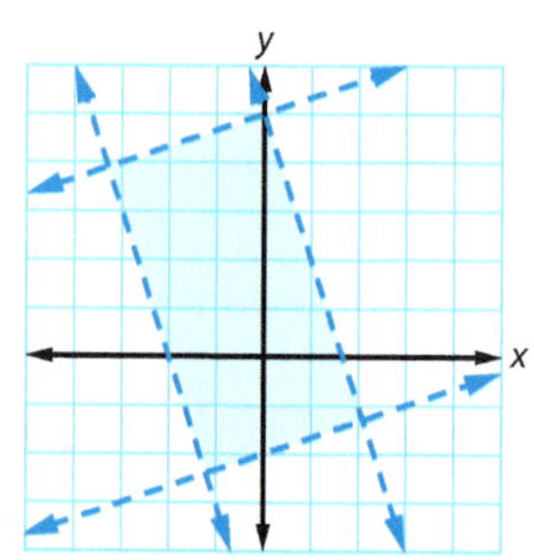

Write a system of inequalities and solve to find possible solutions.

33. Thomas is making desserts for a banquet. A chocolate silk pie costs \$12 to make and serves 8. A small lemon chiffon cake costs \$7 and serves 12. He must serve at least 72 people and needs to keep his costs under \$70.

a. Let x represent the number of chocolate silk pies, and let y represent the number of lemon chiffon cakes. Write one inequality describing the cost, another describing the number of servings, and then two more assuming that the number of pies and cakes cannot be negative.

b. Graph the system of inequalities.

c. Assuming that Thomas makes at least two of each dessert, complete the first 2 columns of the table by listing his options.

d. Complete the table by determining the number of servings and the cost for each option.

Chocolate Silk	Lemon Chiffon	Number of Servings	Cost

34. Phillip is investing at most \$16,000 between 2 different accounts. The first investment is averaging a 5% return, and the second yields 8%. He wishes to make at least \$5500 in a 5 yr time span.

a. Complete the table using the $I = Prt$ formula.

b. Write a system of 4 inequalities, three describing the principal and one describing the interest.

c. Graph the system of inequalities. (*Hint*: Use a scale of 1000.)

d. List three of Phillip's options. Choose answers that are multiples of \$2000.

P	r	t	I

35. **Essential Question:** Tristan is building a rectangular dog pen. The width, x, must be more than 10 ft, and the length, y, must be more than 12 ft; but the perimeter must not be more than 60 ft. Write and graph a system of inequalities that models these constraints and give one possible option.

36. **MP8 Essential Question:** How can you model all possible solutions to a real-world problem that has to meet multiple constraints?

C. EXERCISES

37. MP2 Explain why the system $4x + 5 < y$ and $4x - y > -5$ has no solutions.

38. MP4 Elizabeth makes doll clothes to sell online. She can make a dress for a 12 in. doll in 2.5 hr and a dress for a 19 in. doll in 2 hr. The materials for a smaller dress cost \$2, and the materials for a larger dress cost \$3. If she wishes to spend no more than 80 hr sewing and must spend less than \$100, write and graph a system of inequalities that models her options. If she sells the dresses in sets of 5, which option is closest to both limits of cost and time?

Write a system which produces the graphed solution set.

39. The graph is a polygon with vertices at $(-5, 7)$, $(-5, 1)$, $(-1, 4)$, and $(-1, 2)$.

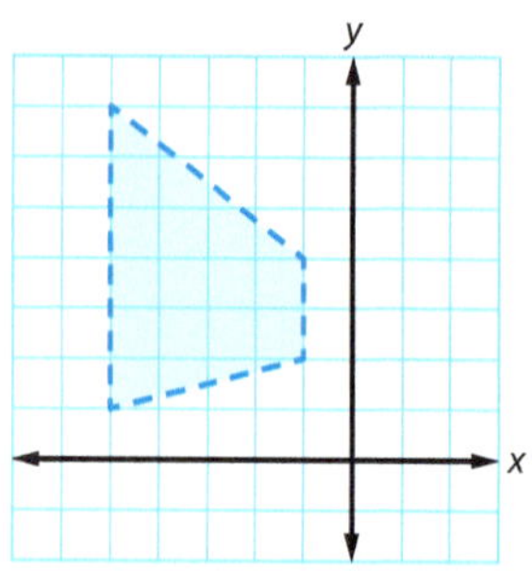

40. MP1 The graphed polygon has vertices $(1, -1)$, $(1, 2)$, $(4, 5)$, $(4, 2)$, $(7, 2)$ and $(7, -1)$.

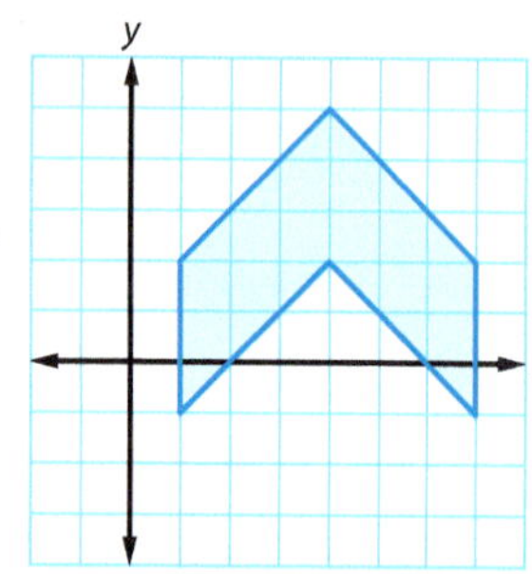

CUMULATIVE REVIEW

Solve each system by substitution. [6.2]

41. $y = x + 8$
$y = 2x - 3$

42. $x + 4y = 10$
$3x - y = -9$

43. $3x + y = 7$
$2x + 3y = -14$

Solve each system by elimination. [6.3]

44. $x + y = 3$
$x - y = 9$

45. $3x + 4y = -8$
$6x + 5y = -1$

46. $2x - 7y = 6$
$3x - 2y = -8$

Write and solve a system of equations for each word problem. [6.5–6.6]

47. Enzo biked 40 km with the wind in 1 hr. On the return trip, it took him twice as long to ride the same distance against the wind. How fast would he travel by bike on a calm day?

48. Aissatou has a solution that is 25% acid. How much pure water and solution must be mixed to create 20 L of solution that is 15% acid?

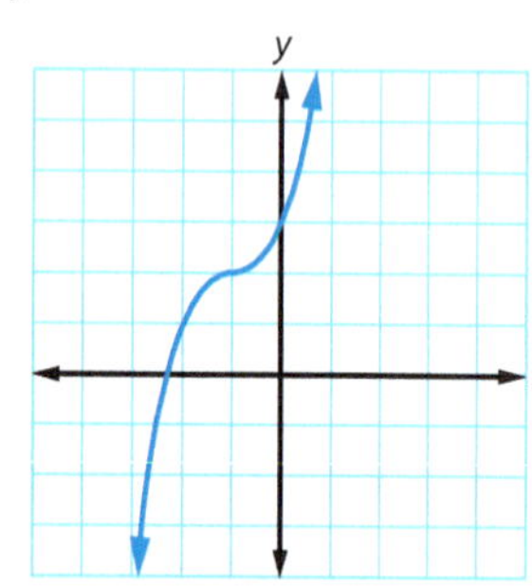

49. Which system is considered "consistent independent"? **[6.4]**

A 2 parallel lines

B 2 lines intersecting at 1 point

C 2 coinciding lines

D 2 inequalities with no intersection

E all of these

50. Is the above graphed relation a function? **[4.2]**

A Yes, it passes the vertical line test.

B Yes, it passes the horizontal line test.

C No, it fails the vertical line test.

D No, it fails the horizontal line test.

E not enough information given

APPLICATION PROBLEMS
HISTORIAS DE LA BIBLIA

In 2012 Comforting Mercies Ministries had a burden to provide Bible story CDs to the Spanish-speaking communities around the globe. Over the years hundreds of thousands of CDs have been provided to children in schools, given to the public at large community events, and distributed door to door. With CDs emphasizing the gospel as well as church information provided on the CD jackets, many lives have been impacted for eternity.

Much planning was necessary to successfully complete the mission. The ability to solve the linear systems studied in this chapter enabled planners to make wise decisions during this process. As you will see, mathematics played a crucial role in fulfilling this vision.

Comforting Mercies focused on areas in which the need was the greatest and found missionaries eager to support the Bible story CD project. With half of its population under the age of 30, Mexico is the largest Spanish-speaking country. Consider the populations of the two largest Spanish-speaking countries.

Year	Mexico	Colombia
1950	26,280,000	11,590,000
1960	34,999,000	15,950,000
1970	51,910,000	21,330,000
1980	68,870,000	26,890,000
1990	83,400,000	33,200,000
2000	99,530,000	39,770,000
2010	111,000,000	46,000,000
2020	125,998,000	50,930,000

1. Make a new table with each population rounded to the nearest million.
2. Using 2 different colors, make a scatter plot of each data set on the same set of axes.
3. Using the scatter plots and data from exercises 1–2, draw and write the equations for the trend lines that model the populations for Mexico and Colombia.
4. If the trends continue, will the population of Colombia catch up with that of Mexico? Explain.

Record Holders will produce the Bible story CDs for \$0.18 each with a one-time stamping fee of \$50. Digital Dzign will produce the CDs for \$0.20 each and waive the stamping fee.

5. Write a function rule describing the cost of producing x CDs using each company.
6. Use substitution to determine the number of CDs at which the cost is the same for both companies.
7. If the ministry needs 10,000 CDs, which company should be used? How much will the CDs cost?

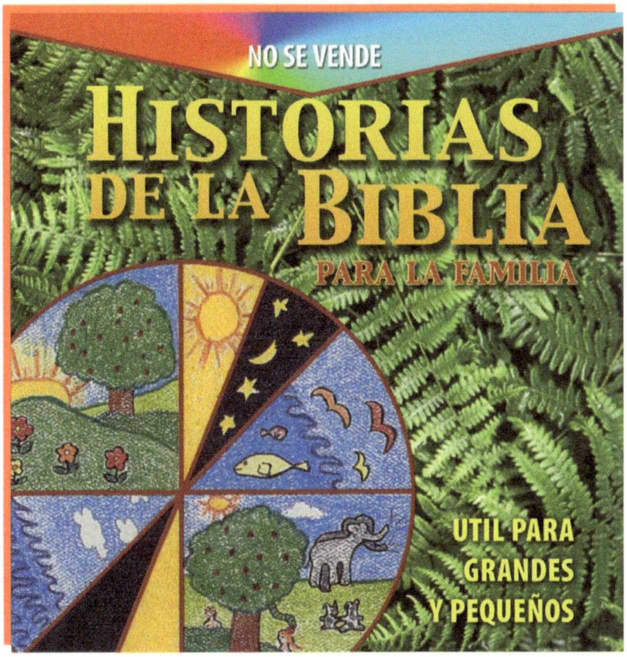

Special printers are needed to print the heavy jackets that hold the CDs.

8. Express Press, an online company, charges $0.049 per jacket for printing, while the Copy Lobby, a local company, charges $0.055 per jacket.

   a. Determine each company's printing fees for 10,000 CD jackets.

   b. Express Press also charges a shipping and handling fee of $1.15/lb. Determine the total cost of having them provide the CD jackets if each jacket weighs 0.4 oz.

9. The youth group from a local church stuffs the CDs into jackets and boxes them for transport to the mission field. In January the group stuffed 5000 CDs in an hour. In February, with an additional 10 people, they did another 7000 in the same amount of time. Assuming that the young people worked at the same rate, how many CDs were stuffed per person per hour?

10. How many young people would be needed to stuff the 10,000 newly produced CDs during the 1 hr youth meeting?

Operation Salvation was a project to distribute 500,000 CDs in Mexico during the country's two-hundredth anniversary celebration. Money was raised for the project through church and individual donations. Christians in the United States were encouraged to sponsor a box of CDs, and Mexican churches donated toward the cost of the jackets.

11. $70,000 was needed for the first 400,000 CDs, including the shipping cost of $4300. Write and solve an equation to find the cost of producing each CD (to the nearest tenth of a cent).

12. Find the cost per CD after shipping is added.

The producer needs to ship 200,000 CDs from California to Texas. The maximum weight for each skid is 2000 lb. Each box contains 500 CDs and weighs about 16.5 lb with packaging.

13. Write and solve an inequality that describes the number of boxes that can be placed on each skid. What is the maximum number of boxes per skid?

14. How many skids will be needed and how many boxes should be on each skid if the same number of boxes are to be packed on each skid?

15. Each skid itself weighs 50 lb and must be considered in the shipping cost in addition to the weight of the potential freight. Determine the total weight of a fully loaded skid.

16. If each CD weighs 0.5 oz, what percent of the total shipping weight is due to the packaging of the CDs and shipping materials (to the nearest tenth of a percent)?

If the CDs are carried onto a plane, the size of each suitcase and its weight are important. A typical carry-on suitcase weighs 6 lb when empty and has interior dimensions of 12″ × 12″ × 8.5″. Each CD in its jacket weighs 0.9 oz and takes up 1.5 in.^{3}

17. Write an inequality that models the number of CDs that can be placed in a suitcase without exceeding a 25 lb weight limit for a carry-on.

18. Write a second inequality that models the number of CDs that will fit in the suitcase.

19. Write a simplified compound inequality that models the number of CDs that will fit in the suitcase. State the largest number of CDs that should be packed in it.

6 CHAPTER REVIEW

Key Concepts

Solving Systems by Graphing [6.1]

- Graph both equations on the same coordinate plane.
- The ordered pair representing the intersection of the lines is the system's solution.
- Real-world problems require an interpretation of the ordered pair.

Solving Systems by Substitution [6.2]

- Solve either equation for either variable.
- Substitute the expression into the other equation and solve.
- Substitute the result into an equation to find the other variable.
- Check by substituting both values into the other equation or by graphing the system.

Solving Systems by Elimination [6.3]

- Multiply the equation(s) to get a pair of opposite or identical terms.
- Add or subtract the equations to eliminate one of the variables.
- Solve the resulting equation to find the value of the remaining variable.
- Substitute the value into an original equation and solve for the other variable.

Special Systems [6.4]

TYPES OF SYSTEMS			
System	consistent independent	consistent dependent	inconsistent
Number of Solutions	1 ordered pair	infinite number (all the points on a line)	no solution
Graph	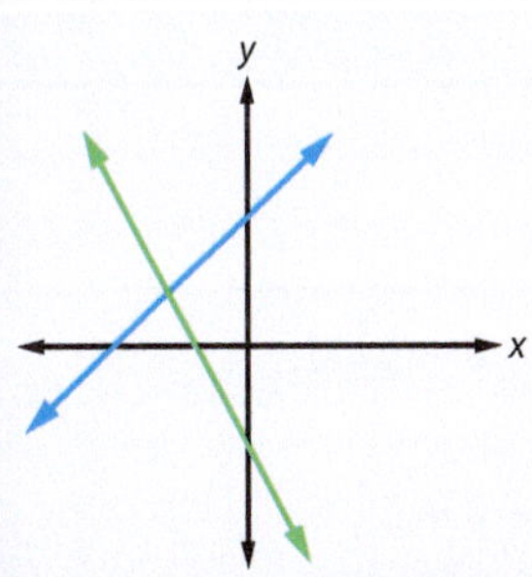	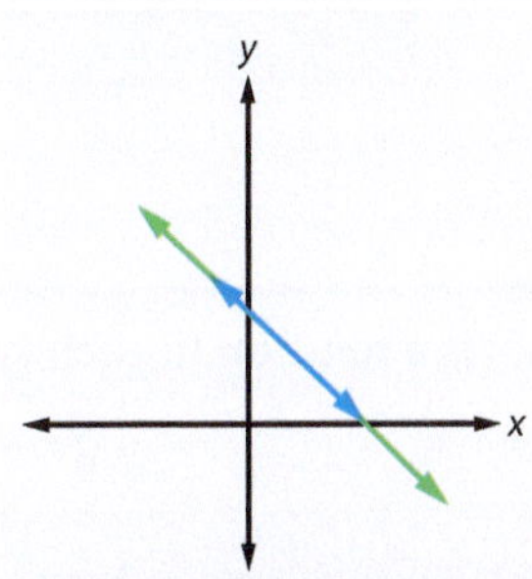	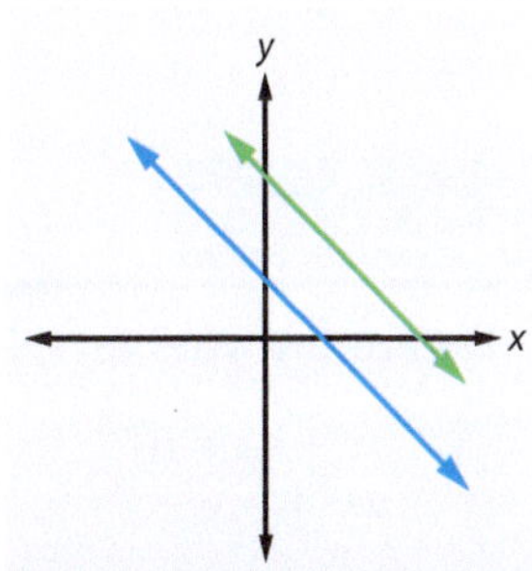
Slope-Intercept Comparison	$m_1 \neq m_2$	$m_1 = m_2$ $b_1 = b_2$	$m_1 = m_2$ $b_1 \neq b_2$
Results from Substitution or Elimination	an ordered pair (x, y)	a true statement $(0 = 0)$	a false statement $(0 = -2)$

Motion & Mixture Problems [6.5–6.6]

- Solving word problems with 2 variables:
  - **Read** the problem carefully and look for the unknown quantities. Use variables to represent the unknowns.
  - **Plan** by making a table. Fill in as much information as possible.
  - **Solve** a system of equations obtained from information in the table and in the problem.
  - **Check** your answer in the context of the original statement of the problem.

Graphing Linear Inequalities [6.7]

- Solve for y and identify the slope and y-intercept of the corresponding equality.
- Graph the corresponding linear equation.
  - If the inequality is $<$ or $>$, use a dashed line to indicate that points on the line are not solutions.
  - If the inequality is $\leq$ or $\geq$, use a solid line to indicate that points on the line are solutions.
- Choose a test point that is not on the line and substitute it into the inequality.
  - If the resulting inequality is true, shade the side containing the test point.
  - If the inequality is false, shade the other side.

Solving Systems of Inequalities [6.8]

- Graph all inequalities on the same set of axes.
- Find the solution set where the shaded regions intersect.
- Check the result.

Biblical Worldview Shaping

Reasoning with Algebra

How does human reasoning differ from Scriptural truth?

Human reasoning, even valid reasoning, does not necessarily lead to truth since the initial assumptions may be false. Scriptural truth is God's inerrant revelation.

EXERCISES

Determine whether (–2, 3) is a solution to each system. [6.1]

1. $3x + 4y = 6$
$y = 1 - x$

2. $5x - 2y = 7$
$x + y = 1$

3. $x + y < 2$
$x + 2y \geq 4$

Solve each system by graphing. [6.1]

4. $x + y = 2$
$2x - y = -5$

5. $2x + y = 1$
$x + 3y = -12$

6. $x + 3y = 5$
$y = 2$

Solve each system by substitution. [6.2]

7. $x + 2y = 4$
$2x + 3y = 8$

8. $10x + 2y = -28$
$4x + 3y = -20$

9. $5x + y = 8$
$-10x - 2y = -3$

Solve each system by elimination. [6.3]

10. $x - 2y = -11$
$3x + 2y = 23$

11. $5x - 3y = -25$
$6x + 2y = -2$

12. $2x + 3y = 9$
$4x - y = 2$

Using 2 variables, write and solve a system of equations for each word problem. [6.2–6.3]

13. The sum of 2 numbers is −23, and their difference is 5. Find the 2 numbers.

14. Eleven is the sum of twice a number and three times a second number. The second number is equal to the difference of two and the first number. Find the 2 numbers.

15. The cost of renting the ice-skating rink is \$200 plus \$10 per skater. The cost of renting the roller-skating rink is \$300 plus \$7 per skater. For what number of people would the cost be almost the same?

16. Mr. Arnold invested \$1550 between 2 accounts. One paid 7.5% interest and the other 8.25%. How much did he invest in each account if his first year's return on his investment was \$121.20?

Solve each system. [6.4]

17. $6x + y = 7$
$y = 1$

18. $8x - 3y = -12$
$y - 4 = -8$

19. $x - y = -2$
$4x - y = 4$

20. $y = 2x - 3$
$y = \frac{1}{3}x + 7$

21. $x - 2y = -42$
$3x - y = -31$

22. $2x - 7y = 3$
$-4x + 14y = -6$

23. $2x + 5y = 1$
$x - 5y = 3$

24. $0.5x - 2.1y = 7.2$
$1.4x + 0.3y = 7.8$

25. $x + 3y = 11$
$x + 3y = 7$

26. $3x + y = -2$
$5x - 3y = 12$

27. Describe the relationship of 2 lines that have the same slope but different y-intercepts. **[6.4]**

28. Describe the relationship of 2 lines that have different slopes. **[6.4]**

29. Describe the relationship of 2 lines that have different slopes but the same y-intercept. **[6.4]**

30. Describe the relationship of 2 lines that have the same slope and the same y-intercept. **[6.4]**

Determine whether each system has 1 solution, no solutions, or an infinite number of solutions. Then classify the system as consistent independent, consistent dependent, or inconsistent. [6.4]

31. $x + y = 5$
$x + y = 2$

32. $8x - 3y = -12$
$y - 4 = -8$

33. $x + y = 10$
$2y = 20 - 2x$

34. $3x + y = 6$
$\frac{2}{3}y = 5 - 2x$

35. $3x + 5y = 1$
$4x - 3y = 11$

36. $3x + y = 4$
$2y = 8 - 6x$

Using 2 variables, write and solve a system of equations for each word problem. [6.5–6.6]

37. Two cars leave from the same location but travel in opposite directions. The first car travels an average of 50 mi/hr and leaves 15 min before the second car, which travels an average of 55 mi/hr. How long will each car travel before they are 200 mi apart?

38. Two fishing boats leave Sandy Cove at the same time traveling in the same direction. One boat is traveling three times as fast as the other boat. After 5 hr the faster boat is 80 mi ahead of the slower boat. What is the speed of each boat?

39. A plane travels 800 mi to Chicago in 4 hr against a headwind. The return trip with a tailwind of the same speed takes 2.5 hr. Find the speed of the plane in still air and the average wind speed.

40. Tessa wants to mix chocolates that cost $4/lb with nuts that cost $6/lb to make 6 lb of a mixture that costs $31. How many pounds of each should she include in the mix?

41. Adrien needs 5 gal of a 23% salt solution. He has a 35% salt solution. How much pure water and how much 35% solution should he mix to get the desired solution? Round to the nearest tenth.

42. Sugar Tooth Candy Company needs 300 gal of a 32% sucrose solution for a certain kind of candy. The company has a solution that is 60% sucrose and a solution that is 25% sucrose. How many gallons of each should the company mix together to obtain the desired solution?

43. The graph of $y > mx + b$ is drawn by graphing $y = mx + b$ with a ________ (solid / dashed) line and shading the region ________ (above / below) the line. **[6.7]**

44. Explain why a linear inequality cannot be a function. **[6.7]**

Graph each inequality. [6.7]

45. $y > -2x + 6$

46. $5x - 3y \geq 10$

Graph each system of inequalities. [6.8]

47. $2x + y < 6$
$3x - y \geq 4$

48. $3x + 5y \geq 12$
$2x - 3y < -6$

Write the system represented by each graph. [6.8]

49. 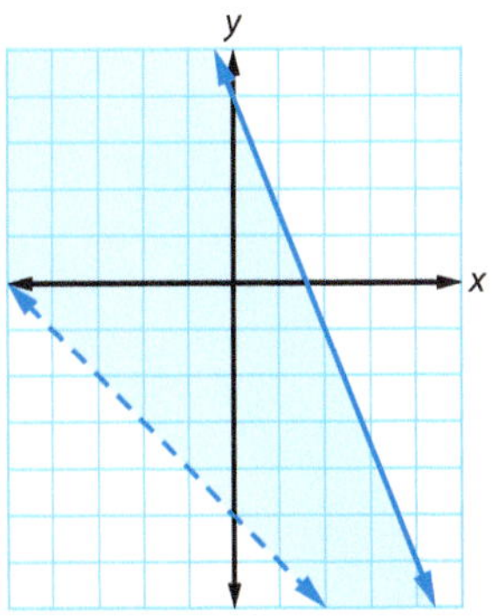

50. 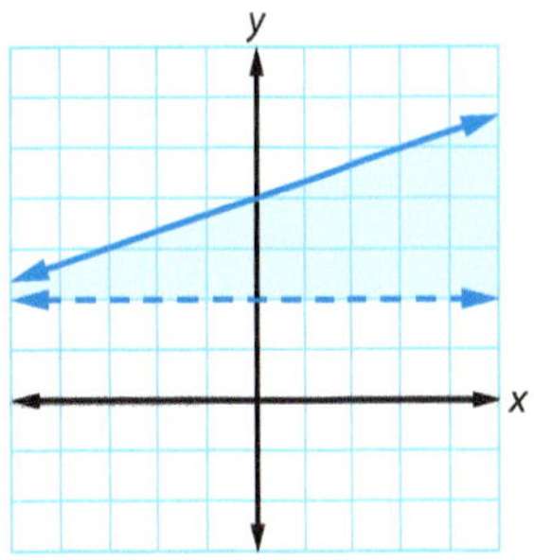

Write the system of inequalities and solve to graph and find solutions. [6.8]

51. Remington is hunting quail and mourning doves. During the current small game season, the law allows him to catch no more than 15 mourning doves per day and limits quail to 12 per day. He is allowed to hunt at most 20 small game birds total per day. Graph the system of inequalities and use it to determine the highest number of quail, x, he can hunt if he has caught 11 mourning doves the same day.

52. Kyla charges \$20/hr for babysitting and \$30/hr for tutoring services. She hopes to make at least \$300 this week while working at most 15 hr. If x represents hours babysitting, use your graph to determine a solution.

53. **BWS Essential Question:** Explain how valid reasoning can lead to false conclusions.

CHAPTER 07
EXPONENTS

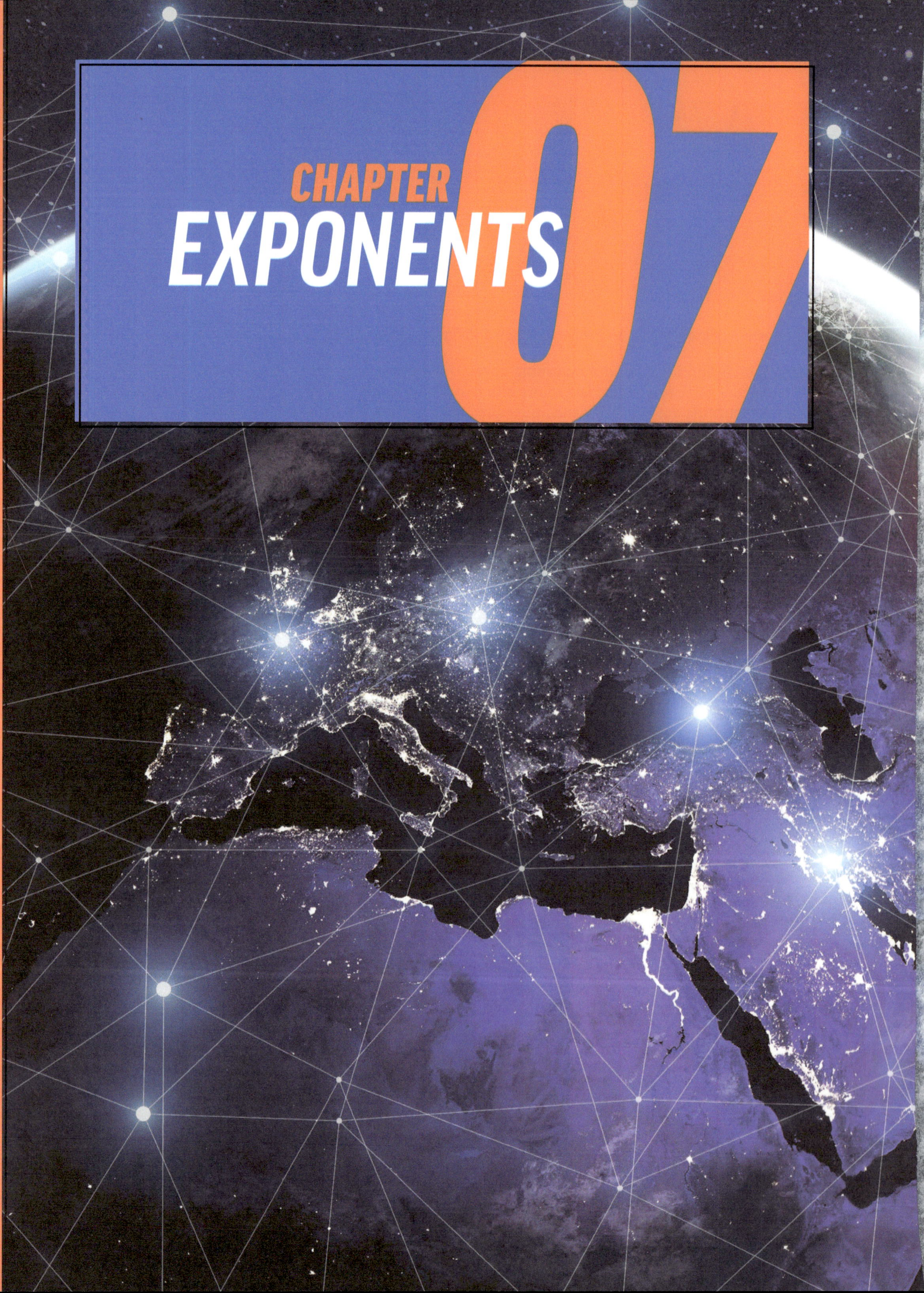

CHAPTER OVERVIEW

7.1 PROPERTIES OF EXPONENTS

What does a quantity with a negative exponent represent?

After completing this section, I will be able to

- apply product, power, and quotient properties of exponents to simplify algebraic expressions.
- simplify powers containing a 0 or negative exponent.
- simplify algebraic expressions containing integral exponents and multiple terms.

If m^2 and m represent the magnifications of each lens in a typical compound microscope, then the total magnification of the microscope can be represented by the expression $m^2 \cdot m$.

The principles used to simplify numerical expressions also apply to algebraic expressions. Review the following properties of exponents covered in Chapter 1.

PROPERTIES OF EXPONENTS			
Product Property	$x^a \cdot x^b = x^{a+b}$	$x^5 \cdot x^2 = x^{5+2} = x^7$	To multiply with like bases, add the exponents.
Power Property	$(x^a)^b = x^{ab}$	$(x^5)^2 = x^{5(2)} = x^{10}$	To raise a power to a power, multiply the exponents.
Quotient Property	$\frac{x^a}{x^b} = x^{a-b}$ for $x \neq 0$	$\frac{x^5}{x^2} = x^{5-2} = x^3$	To divide with like bases, subtract the exponents.

Often multiple properties are needed to simplify an algebraic expression.

EXAMPLE 1: Using Properties of Exponents

Simplify.

a. $y^4(y^3)^2$ **b.** $3x^2 \cdot 5x$ **c.** $\frac{24x^2y^6z^7}{-8xy^2z^5}$

Answers

a. $y^4(y^3)^2 = y^4y^6$ — **1.** Apply the Power Property of Exponents.

$= y^{4+6} = y^{10}$ — **2.** Apply the Product Property of Exponents.

b. $3x^2 \cdot 5x$
$= (3 \cdot 5)(x^2 \cdot x)$
$= 15x^3$

1. Apply the Commutative and Associative Properties of Multiplication.
2. Apply the Product Property of Exponents.

c. $\dfrac{24x^2y^6z^7}{-8xy^2z^5} = -3x^{2-1}y^{6-2}z^{7-5} = -3xy^4z^2$

SKILL CHECK **EXERCISES 3, 7**

The Commutative and Associative Properties can be used to show that the power of a product is equal to the product of the powers. In general, this can be stated as follows.

PROPERTIES OF EXPONENTS		
Power of a Product Property	$(abc)^n = a^nb^nc^n$	$(2a^2c^4)^3 = 2^3(a^2)^3(c^4)^3 = 8a^6c^{12}$

EXAMPLE 2: Simplifying the Power of a Product

Simplify $(-2m^4n)^3\left(\frac{1}{6}mn^2\right)^2$.

Answer

$= (-2)^3(m^4)^3n^3\left(\frac{1}{6}\right)^2m^2(n^2)^2$

$= -8m^{12}n^3\left(\frac{1}{36}\right)m^2n^4$

$= -8\left(\frac{1}{36}\right)m^{12+2}n^{3+4}$

$= -\frac{2}{9}m^{14}n^7$

1. The power of a product is the product of the powers.
2. Apply the Power Property of Exponents.
3. Apply the Product Property of Exponents.
4. Simplify.

SKILL CHECK **EXERCISE 11**

Recall that negative exponents were defined in Chapter 1 as indicating a quotient.

$$x^{-n} = \frac{1}{x^n}$$

This definition is consistent with the Quotient and Product Properties of Exponents.

$$a^2a^{-6} = a^{2+(-6)} = a^{-4} \text{ and } a^2a^{-6} = \frac{a^2}{a^6} = \frac{\cancel{a}\cancel{a}}{\cancel{a}\cancel{a}aaaa} = \frac{1}{a^4},$$
$$\text{so } a^{-4} = \frac{1}{a^4}.$$

$$\frac{b^3}{b^5} = b^{3-5} = b^{-2} \text{ and } \frac{b^3}{b^5} = \frac{\cancel{b}\cancel{b}\cancel{b}}{\cancel{b}\cancel{b}\cancel{b}bb} = \frac{1}{b^2}, \text{ so } b^{-2} = \frac{1}{b^2}.$$

Notice the result of applying this definition to a negative exponent in the denominator.

$$\frac{1}{x^{-n}} = \frac{1}{\frac{1}{x^n}} = 1 \cdot \frac{x^n}{1} = x^n, \text{ so } \frac{1}{x^{-n}} = x^n.$$

A quantity with a negative exponent is equivalent to the reciprocal of that quantity with a positive exponent. This definition implies that

$$a^{-2} = \frac{1}{a^2} \text{ and that } \frac{1}{5^{-2}} = 5^2 = 25.$$

The meaning of a power with an exponent of 0 is also defined to be consistent with the Quotient Property.

$$x^0 = 1 \text{ since } \frac{x^3}{x^3} = x^{3-3} = x^0 \text{ and } \frac{x^3}{x^3} = 1.$$

EXAMPLE 3: Using Zero & Negative Exponents

Write an equivalent expression in positive exponential form.

a. $2x^{-4}y^2$ **b.** $\left(-\frac{1}{3}x^2\right)^{-3}$ **c.** $\frac{34x^{-3}y^2z^{-5}}{-18x^{-6}y^2z^{-2}}$

Answers

a. $2x^{-4}y^2 = \frac{2}{x^4}(y^2) = \frac{2y^2}{x^4}$

Use the definition of a negative exponent to write the expression using only positive exponents.

b. $\left(-\frac{1}{3}x^2\right)^{-3} = \left(-\frac{1}{3}\right)^{-3}x^{2(-3)}$

$= (-3)^3x^{-6} = -27\left(\frac{1}{x^6}\right) = -\frac{27}{x^6}$

Use the definition of a negative exponent to write the expression using only positive exponents.

c. $\frac{34x^{-3}y^2z^{-5}}{-18x^{-6}y^2z^{-2}}$

$= \frac{34}{-18}x^{-3-(-6)}y^{2-2}z^{-5-(-2)}$

1. Apply the Quotient Property of Exponents.

$= -\frac{17}{9}x^3y^0z^{-3}$

$= -\frac{17x^3(1)}{9z^3} = -\frac{17x^3}{9z^3}$

2. Simplify and write the expression using only positive exponents.

SKILL CHECK **EXERCISE 13**

AFTERSCHOOLHELP

Properties of Exponents

By viewing a power as repeated multiplication, we can show that the power of a quotient is equal to the quotient of the powers: $\left(\frac{a}{b}\right)^n = \frac{a^n}{b^n}$.

PROPERTIES OF EXPONENTS		
Power of a Quotient Property	$\left(\frac{a}{b}\right)^n = \frac{a^n}{b^n}$	$\left(\frac{2}{b^3}\right)^4 = \frac{2^4}{(b^3)^4} = \frac{16}{b^{12}}$

EXAMPLE 4: Simplifying the Power of a Quotient

Simplify $\left(\frac{y}{x^2}\right)^3(x^3y^{-4})^2$, leaving all variables in the numerator.

Answer

$\left(\frac{y}{x^2}\right)^3(x^3y^{-4})^2 = \frac{y^3}{x^6}(x^6y^{-8})$

1. Apply the Power of a Quotient Property and the Power Property.

$= x^{6-6}y^{3+(-8)} = x^0y^{-5}$

2. Apply the Quotient Property and the Product Property.

$= y^{-5}$

3. Simplify, leaving the variable in the numerator.

SKILL CHECK **EXERCISE 25**

An expression containing negative exponents in the denominator is not considered to be simplified. An expression may be considered simplified if all variables are listed in the numerator. However, it is generally preferable to use only positive exponents in a simplified expression.

EXAMPLE 5: Simplifying an Expression

Simplify.

a. $\dfrac{-2x^{-2}(-3xy^{-2})}{12y^3}$ **b.** $\dfrac{8x^2z^{-3}}{y}\left(\dfrac{yz^{-4}}{2x^5}\right)^{-2}$

Answers

a. $\dfrac{-2x^{-2}(-3xy^{-2})}{12y^3}=\dfrac{-2(-3)}{12}x^{-2+1}y^{-2-3}=\dfrac{1}{2}x^{-1}y^{-5}$

$=\dfrac{1}{2xy^5}$

1. Apply the Product and Quotient Properties of Exponents.
2. Rewrite the expression using only positive exponents.

b. $\dfrac{8x^2z^{-3}}{y}\left(\dfrac{yz^{-4}}{2x^5}\right)^{-2}=\dfrac{8x^2z^{-3}}{y}\cdot\dfrac{y^{-2}z^8}{2^{-2}x^{-10}}$

$=8(4)x^{2-(-10)}y^{-2-1}z^{-3+8}$

$=32x^{12}y^{-3}z^5=\dfrac{32x^{12}z^5}{y^3}$

1. Apply the Power of a Quotient Property.
2. Apply the Product and Quotient Properties of Exponents.
3. Rewrite the expression using only positive exponents.

or

$\dfrac{8x^2z^{-3}}{y}\left(\dfrac{yz^{-4}}{2x^5}\right)^{-2}=\dfrac{8x^2z^{-3}}{y}\cdot\dfrac{y^{-2}z^8}{2^{-2}x^{-10}}$

$=\dfrac{8x^2}{yz^3}\cdot\dfrac{2^2x^{10}z^8}{y^2}$

$=\dfrac{32x^{12}z^5}{y^3}$

1. Apply the Power of a Quotient Property.
2. Use the definition of a negative exponent to write the expression using only positive exponents.
3. Apply the Product and Quotient Properties of Exponents.

SKILL CHECK **EXERCISE 21**

A. EXERCISES

1. The equation $(ab)^n = a^nb^n$ means that the power of a product is the same as the _______ of the _______.
2. The equation $\left(\frac{a}{b}\right)^n = \frac{a^n}{b^n}$ means that the power of a quotient is the same as the _______ of the _______.

Simplify, leaving each answer in positive exponential form.

3. x^3x^5
4. $x^2y^3x^5y^2$
5. $3w^3\cdot 4w^4$
6. $\dfrac{18x^7}{2x^2}$
7. $\dfrac{-12a^3b^6}{16a^2b}$
8. $\dfrac{a^4a^2}{a^5a}$
9. $(a^4)^3$
10. $c^5(c^2)^4$
11. $(-2b^2cd^5)^2$
12. $\dfrac{1}{z^{-2}}$
13. $\dfrac{x^{-1}}{y^{-1}}$
14. $\left(\dfrac{x^2}{y^5}\right)^3$

B. EXERCISES

Simplify, leaving all variables in the numerator.

15. $4abc(-3a^3bc)$

16. $\dfrac{-6x^5y^2z^3}{7x^2y^2z^2}$

17. $16x^6y^{-4} \div 2xy^5$

18. $\dfrac{a^{-1}b^{-2}c^{-3}}{a^3b^2c}$

Simplify, leaving each answer in positive exponential form.

19. $(5x^3)^3(2x)^2$

20. $(6x^3y^7)^2\left(-\frac{1}{2}x^2y^3\right)^3$

21. $\dfrac{38x^2y^3z^{-1}}{2x^{-2}y^{-3}}$

22. $\left(\dfrac{x^2y^3}{y}\right)^{-2}$

23. $\dfrac{(abc^2)^2}{-3bc}$

24. $\left(\dfrac{x}{y}\right)^3\left(\dfrac{x}{7y}\right)^{-2}$

25. $\left(\dfrac{3x}{y^4}\right)^{-2}\left(\dfrac{x^2}{y}\right)^{-1}$

26. $\left(\frac{1}{3}x^3y^2z\right)^2(2x^3y^2z)^3(18x)$

27. **Essential Question:** What is the equivalent of a number raised to a negative power?

Write a simplified algebraic expression for each quantity.

28. area

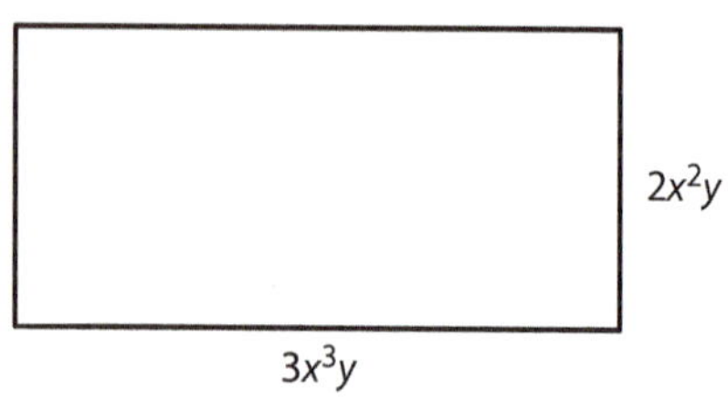

29. volume

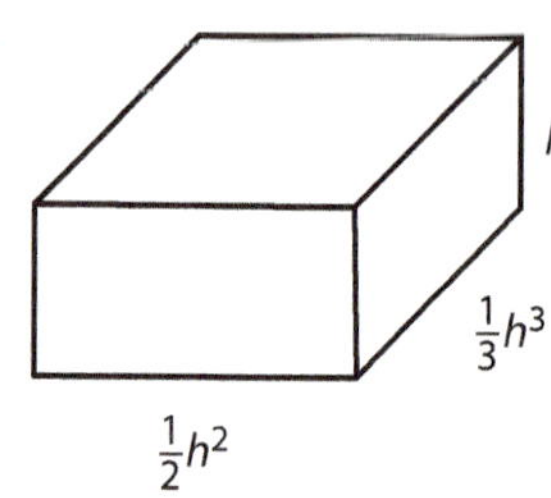

30. height

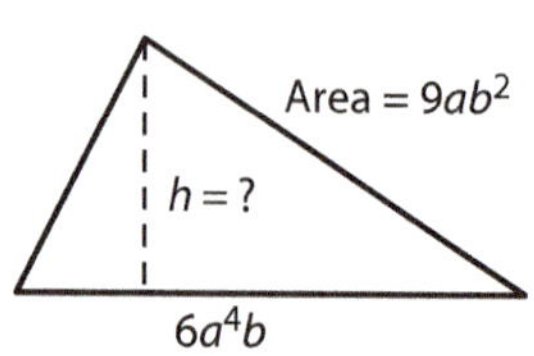

31. width

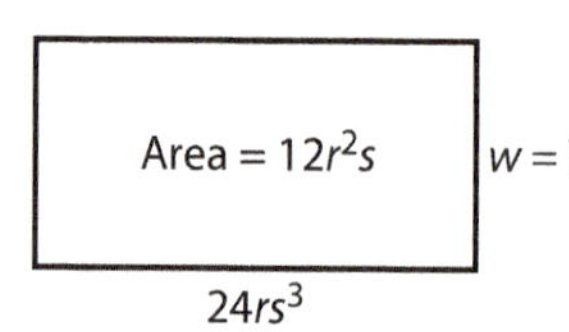

32. MP2 the ratio of the volume of a sphere ($V = \frac{4}{3}\pi r^3$) to the volume of a circular cylinder containing that sphere ($V = \pi r^2 H$ with $H = 2r$)

33. the volume of a cube with an edge of length $3a^3b$

Simplify.

34. $x^n \cdot x^3$

35. $\dfrac{a^{z+2}}{a^z}$

36. MP7 $\dfrac{(b^3)^x}{b^{-5}}$

37. MP7 $(y^z)^5 \cdot y^3$

C. EXERCISES

Simplify each expression.

38. $\frac{1}{2}rs^{-3}(2r^{-2}s - 6r^{-1}s^3)$

39. $m^{-6}n^2(3m^2n^{-4} + 4m^9n^{-2})$

40. $2h^2g^4h^5g - 7hg^2h^6g^3 - 4g^2h^3h^4g^3$

41. $2wzw^2z + w^2z^3 - 2wz^3w + 3wz^2w^2$

42. MP3 Use the Commutative and Associative Properties to show that $(abc)^3 = a^3b^3c^3$.

43. MP3 Show that $\left(\frac{a}{b}\right)^3 = \frac{a^3}{b^3}$ (i.e., the power of a quotient is the quotient of the powers).

CUMULATIVE REVIEW

44. Define b^x, where x is a natural number. **[1.3]**

Simplify each expression using the Distributive Property. [1.6]

45. $5x^3 + 7x^2 - 4x^3$

46. $(6m^4 - m^2) - (5m^4 - 2m^2)$

State the domain and range of each relation and whether it is a function. [4.1]

47. $\{(-2, 8), (4, 9), (8, 9), (9, 4)\}$

48. $\{(-19.2, -3.5), (6.4, 18.2), (1.7, -18)\}$

Rebekah and Justine know that the time it takes to travel the 300 mi to their destination is determined by the rate at which they travel. [4.1, 4.6]

49. Define the independent variable and the dependent variable in this relation.

50. Is the relation a direct variation, an inverse variation, or neither?

51. Write a function rule modeling the relation between the 2 variables.

52. Which has a value of 0? **[1.3, 1.6, 5.2]**

- **A** the coefficient of x in "$x + 2$"
- **B** the value of y if $5 = 5^y$
- **C** 3^0
- **D** the slope of a horizontal line
- **E** none of these

53. Which system has a line as a solution set? **[6.8]**

- **A** $y \le -\frac{2}{5}x + 4$
  $y \ge -\frac{2}{5}x + 4$
- **B** $y < 4x - 1$
  $y > 4x - 1$
- **C** $y \le \frac{1}{2}x + 8$
  $y \ge -2x + 8$
- **D** $x \ge 3$
  $y \ge 3$
- **E** none of these

MIND OVER MATH

ACROSS

1. 36 plus the number of days in 9 regular years
5. cups in 82 gal
9. digits > 5
10. White House street address number
11. 2^9
13. date the Eagle landed on the moon
17. chapters in the New Testament
18. $\sqrt{450{,}000}$ to the nearest integer
19. date Lincoln was shot
23. Fahrenheit temperature at which water boils
25. square inches in a square yard
28. feet in $\frac{3}{4}$ mi
31. sum of all primes < 100
32. pounds in a ton

DOWN

1. degrees in a circle
2. normal body temperature (Celsius)
3. days in February 2025
4. middle year of the twentieth century
5. 37^2 minus the minutes in 4 hr
6. inches in a yard
7. base of the decimal system
8. America's age in 1976
12. 31 in the binary system
13. a multiple of 191
14. CCI
15. the number of the Beast
16. weeks in $18\frac{3}{4}$ yr
20. 1020 subtracted from feet in a mile
21. UCDB on the telephone
22. product of 2 primes < 40 whose difference is 34
24. centimeters in a meter
26. years in 2 decades
27. 8 dozen
29. degrees in a right angle
30. seconds in a minute

7.2 SCIENTIFIC NOTATION

How does my worldview affect how I view math?

After completing this section, I will be able to

- convert numbers between standard and scientific notation.
- apply properties of exponents to evaluate expressions with numbers in scientific notation.
- solve real-world problems using scientific notation.
- explain why it is tempting to view math as amoral.

Very large or very small numbers are cumbersome to read, and they can make calculations difficult. Scientific notation provides a more concise and readable form. Computations can be readily done in scientific notation as you study God's amazing universe at both microscopic and galactic levels. Remember that even though math may seem to be a morally neutral tool, our ability to apply mathematics is part of God's good creation and therefore is not amoral. A biblical worldview demands that Scripture be the ultimate standard for what is true.

DEFINITION

A number is written in **scientific notation** when it is expressed as a product in the form $a \times 10^n$, where $1 \le a < 10$.

EXAMPLE 1: Converting to Scientific Notation

Express each number in scientific notation.

a. The estimated number of atoms in a human body is 7,000,000,000,000,000,000,000,000,000.

b. The mass of an electron is estimated at 0.0000000000000000000000000000009109 kg.

Answers

a. 7,000,000,000,000,000,000,000,000,000 $= 7 \times 10^{27}$ atoms

Move the decimal point 27 places to the left (dividing by 10^{27}) and multiply the result by 10^{27}.

b. 0.0000000000000000000000000000009109 $= 9.109 \times 10^{-31}$ kg

Move the decimal point 31 places to the right (multiplying by 10^{31}) and multiply the result by 10^{-31}.

SKILL CHECK EXERCISES 1, 3

To convert a number written in scientific notation into standard notation, complete the indicated multiplication or division by moving the decimal the indicated number of decimal places to the right or left.

EXAMPLE 2: Converting to Standard Form

Express each number in standard notation.

a. Some scientists estimate there are 2×10^9 stars in the Milky Way Galaxy.

b. The radius of a hydrogen atom is 2.5×10^{-11} meters.

Answers

a. $2 \times 10^9 = 2{,}000{,}000{,}000$	Multiply by 10^9 by moving the decimal point 9 places to the right.
b. $2.5 \times 10^{-11} = 0.000000000025$	Divide by 10^{11} (the equivalent of multiplying by 10^{-11}) by moving the decimal point 11 places to the left.

SKILL CHECK **EXERCISE 9**

The Commutative and Associative Properties of Multiplication and the properties of exponents are used in calculating the products, powers, and quotients of numbers in scientific notation.

EXAMPLE 3: Evaluating Products, Powers & Quotients in Scientific Notation

Simplify.

a. $(7.5 \times 10^{-5})(2 \times 10^3)$

b. $(1.2 \times 10^{-4})^2$

c. $(2.4 \times 10^7) \div (6 \times 10^2)$

Answers

a. $(7.5 \times 10^{-5})(2 \times 10^3)$

$= (7.5 \times 2)(10^{-5} \times 10^3)$

1. Reorder and regroup.

$= 15 \times 10^{-2}$

2. Add exponents to multiply the powers of 10.

$= (1.5 \times 10^1) \times 10^{-2}$

$= 1.5 \times 10^{-1}$

3. Express the result in scientific notation.

b. $(1.2 \times 10^{-4})^2$

$= 1.2^2 \times (10^{-4})^2$

1. The power of a product is the product of the powers.

$= 1.44 \times 10^{-8}$

2. Multiply exponents to find the power of a power. Notice that the result is already in scientific notation.

c. $\dfrac{2.4\times10^{7}}{6\times10^{2}}=\dfrac{2.4}{6}\times\dfrac{10^{7}}{10^{2}}$

1. Write the division problem as a product of fractions.

$=0.4\times10^{5}$

2. Subtract exponents to divide the powers of 10.

$=(4\times10^{-1})\times10^{5}$

$=4\times10^{4}$

3. Express the result in scientific notation.

SKILL CHECK **EXERCISE 15**

Before adding or subtracting numbers in scientific notation, make sure that both numbers are expressed using the same power of 10. It is usually easier to rename the number with the lower power of 10 to match the higher power of 10.

EXAMPLE 4: Evaluating Sums & Differences in Scientific Notation

Simplify, expressing each answer in scientific notation.

a. $(2.3\times10^{6})+(5\times10^{6})$

b. $(5.4\times10^{5})-(4.7\times10^{4})$

Answers

a. $(2.3\times10^{6})+(5\times10^{6})$

$=(2.3+5)\times10^{6}$

$=7.3\times10^{6}$

The Distributive Property implies that $ac+bc=(a+b)c$.

b. $(5.4\times10^{5})-(4.7\times10^{4})$

$=(5.4\times10^{5})-[(4.7\times10^{-1})\times(10^{4}\times10^{1})]$

$=(5.4\times10^{5})-(0.47\times10^{5})$

1. Multiplying 10^{4} by 10^{1} to get 10^{5} requires the decimal factor to be multiplied by 10^{-1}.

$=(5.4-0.47)\times10^{5}$

$=4.93\times10^{5}$

2. The Distributive Property can be used when the powers of 10 are the same.

SKILL CHECK **EXERCISE 21**

AFTERSCHOOLHELP

Scientific Notation

The prefixes in the metric system are another way to shorten the notation used to represent a number. Knowing the prefixes and values in this chart will simplify calculations involving very large or very small numbers.

PREFIX	10^{n}	VALUE
tera (T)	10^{12}	trillion
giga (G)	10^{9}	billion
mega (M)	10^{6}	million
kilo (K or k)	10^{3}	thousand
milli (m)	10^{-3}	thousandth
micro (μ)	10^{-6}	millionth
nano (n)	10^{-9}	billionth

EXAMPLE 5: Applying Scientific Notation

The mass of Earth is about 6×10^{24} kg, the mass of Mars is about 6×10^{23} kg, and the mass of Jupiter is about 2×10^{27} kg.

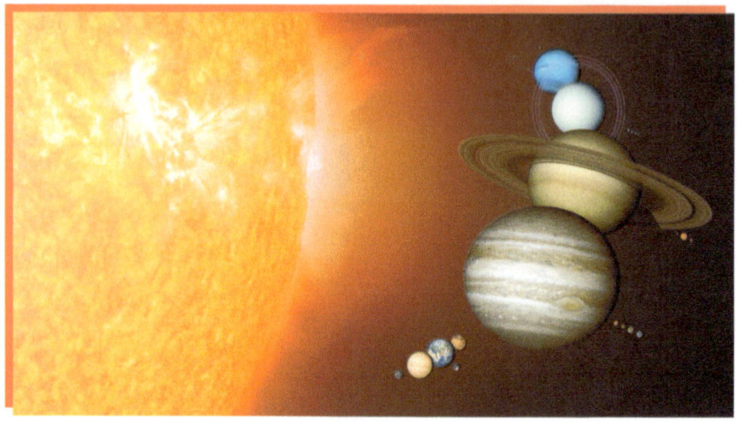

a. What is the difference in the masses of Earth and Mars in grams?

b. Approximately how many times more massive is Jupiter than Earth?

Answers

a. $6 \times 10^{23} = (6 \times 10^{-1}) \times (10^{23} \times 10^{1}) = 0.6 \times 10^{24}$

$(6 \times 10^{24}) - (0.6 \times 10^{24}) = 5.4 \times 10^{24}$ kg

1. Rename the number with the lower power of 10 before subtracting.

$5.4 \times 10^{24}\text{ kg} = (5.4 \times 10^{24}) \times 10^{3}\text{ g} = 5.4 \times 10^{27}\text{ g}$

2. Multiply by the value corresponding to the metric prefix.

b. $\dfrac{2 \times 10^{27}\ \cancel{kg}}{6 \times 10^{24}\ \cancel{kg}}$

1. Since the units are the same, divide the numbers.

$= \frac{1}{3} \times \frac{10^{27}}{10^{24}} = 0.\overline{3} \times 10^{3}$

2. Simplify.

$\approx (3 \times 10^{-1}) \times 10^{3}$

$= 3 \times 10^{2}$

Jupiter is about 300 times more massive than Earth.

SKILL CHECK **EXERCISE 31**

VOCABULARY

- scientific notation

A. EXERCISES

Express each number in scientific notation.

1. 390,000,000

2. 273,000

3. 0.000052

4. 0.0032

5. 13.6×10^{7}

6. 1067×10^{-8}

7. 235.6×10^{-4}

8. 135.8×10^{8}

Express each number in standard notation.

9. 2.54×10^{8}

10. 3.4×10^{7}

11. 1.53×10^{-4}

12. 2.89×10^{-6}

B. EXERCISES

Simplify, expressing each answer in scientific notation.

13. $(2 \times 10^2)(3 \times 10^5)$

14. $(6 \times 10^7)(4 \times 10^7)$

15. $(3 \times 10^5)^4$

16. $(9 \times 10^8) \div (3 \times 10^3)$

17. $(4 \times 10^{11}) \div (8 \times 10^4)$

18. $(2 \times 10^2)^{-3}$

19. $(4.3 \times 10^3) + (5.2 \times 10^3)$

20. $(9.5 \times 10^6) - (7 \times 10^5)$

21. $(2.8 \times 10^9) - (3 \times 10^7)$

22. $(1.2 \times 10^8) + (4.7 \times 10^6)$

23. $(3.5 \times 10^9)(8.6 \times 10^{-7})$

24. $(1.25 \times 10^{-9}) \div (5 \times 10^{-4})$

25. $(7.5 \times 10^{-15})^2$

26. $(4 \times 10^{13}) + (5.5 \times 10^{15})$

27. $(3.4 \times 10^7) - (6.7 \times 10^6)$

28. $(8.5 \times 10^6) - (5 \times 10^5)$

Express each answer in scientific notation. Round to 2 digits after the decimal point and refer to the table of prefixes as needed.

29. The diameter of the smallest capillary in the mature human body is about 0.008 mm. The diameter of the largest artery, the aorta, is about 25,000 μm. Using scientific notation, express each of these measurements in meters.

30. The speed of light is approximately 300 million meters per second. Earth is 1.5×10^8 km from the sun. How many seconds does it take light from the sun to reach Earth?

31. Saturn's average distance from the sun is 887 million miles, and Earth's average distance from the sun is 93 million miles. How many miles closer to the sun is Earth?

32. A pinhead has a diameter of about 1.5 mm, and a red blood cell has a diameter of about 7.5 μm. Approximately how many red blood cells will fit across the diameter of a pinhead?

33. **MP1** The human body contains about 30 trillion red blood cells with a diameter of about 7.5 μm each. If these cells were lined up end-to-end, determine the length of the line in meters and in kilometers.

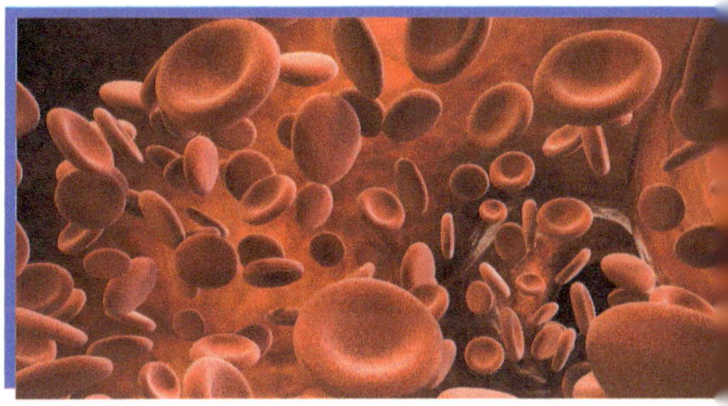

Red blood cells carry oxygen from the lungs to the rest of the body and then carry CO_2 back to the lungs for exhalation.

34. **MP4** An internet search engine returned 150,000,000 hits from its database for the keyword "math" in 0.18 sec.

- **a.** What was the average time per hit?
- **b.** Express this time in nanoseconds.
- **c.** How much faster is a search engine which averages 2.1×10^{-10} sec/hit?

35. **Essential Question:** How does my worldview affect how I view math?

C. EXERCISES

Simplify.

36. $\dfrac{(3 \times 10^7)(7 \times 10^3)}{3 \times 10^5}$

37. $\dfrac{(5 \times 10^6)(9 \times 10^2)}{3 \times 10^{-2}}$

38. $\dfrac{(4 \times 10^5)^2}{(8 \times 10^{-3})(20 \times 10^{12})}$

39. $\dfrac{(2.4 \times 10^{-2})(25 \times 10^{12})}{(2.5 \times 10^2)(1.2 \times 10^3)}$

40. $\dfrac{(4 \times 10^8)(4.5 \times 10^{11})}{(1.5 \times 10^8)(2 \times 10^6)}$

41. $\dfrac{(6 \times 10^{10})(3 \times 10^7)}{(2 \times 10^9) - (2 \times 10^8)}$

Engineering notation is a variation of scientific notation in which a number is written as a product in the form $a \times 10^n$, where $1 \le a < 1000$ and n is a multiple of 3. Engineering notation allows the power of 10 to be replaced with a label.

Example: The number 12,500,000 would be written as 12.5×10^6 and read as 12.5 million.

MP6 Express each number in engineering notation; then state the number with the appropriate label in place of the power of 10.

42. 3.14×10^5

43. 189,000,000,000

44. 5.385×10^{-5}

45. 0.0001

CUMULATIVE REVIEW

Write a function rule for each translation of $f(x) = |x|$. [4.7]

46. 3 units left, 4 units up

47. 11 units up, 2 units right

48. 1.76 units down, 5.5 units right

49. $\frac{3}{2}$ units left, $\frac{5}{9}$ units down

Write the equation of each line in standard form. [5.5]

50. a line parallel to $5x + 2y = -3$ and passing through $(-4, 5)$

51. a line perpendicular to $x + 4y = 7$ and passing through $(-1, 6)$

Solve each system by graphing. [6.1]

52. $y = -x + 7$
$y = \frac{2}{3}x + 2$

53. $y = -\frac{1}{2}x - 1$
$y = x - 4$

54. Which system is inconsistent? **[6.4]**

- **A** $y = 3x + 4$
  $y = 3x - 2$
- **B** $y = -\frac{2}{5}x + 1$
  $y = \frac{5}{2}x + 1$
- **C** $y = 5x - 8$
  $y = \frac{1}{3}x + 3$
- **D** $3x - 4y = -11$
  $6x - 8y = -22$
- **E** none of these

55. When simplified, which value of m will result in a positive power of y in the denominator of $(xy^2)^3(x^4y^m)^2$? **[7.1]**

- **A** -3
- **B** -1
- **C** 0
- **D** 4
- **E** none of these

TECHNOLOGY CORNER (TI-84 PLUS FAMILY)

Graphing Exponential Functions

The graphing calculator is especially helpful for comparing the graphs of several different functions. Use [y=] and the [^] key to enter the illustrated exponential functions.

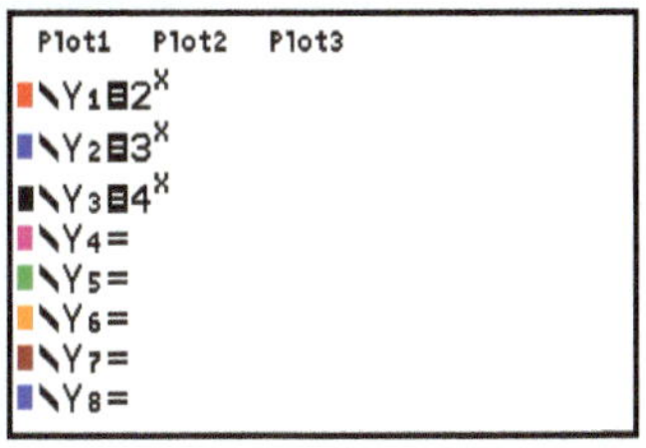

Select [graph] and use [window] to adjust the portion of the graph that is illustrated.

Select [trace] and use [▲] and [▼] to move between the graphs of different functions. Use [◄] and [►] to reveal the coordinates of the points on a graph.

These graphs illustrate that the greater the value of b in functions of the form $y = b^x$, the more rapidly the function increases in the first quadrant.

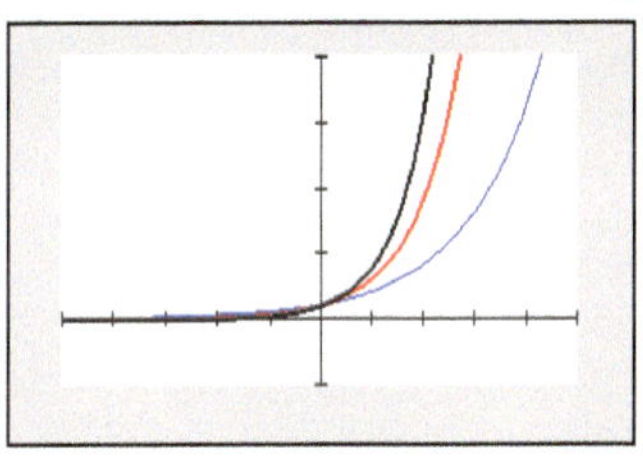

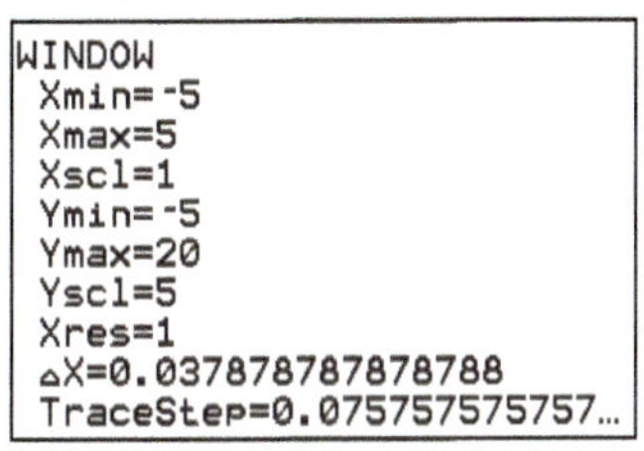

Enter functions Y_4, Y_5, and Y_6 for comparison with the previous functions.

From these graphs it can be seen that the graph of $y = \left(\frac{1}{b}\right)^x$ is a mirror image or reflection of the graph of $y = b^x$, with the y-axis acting as the "mirror."

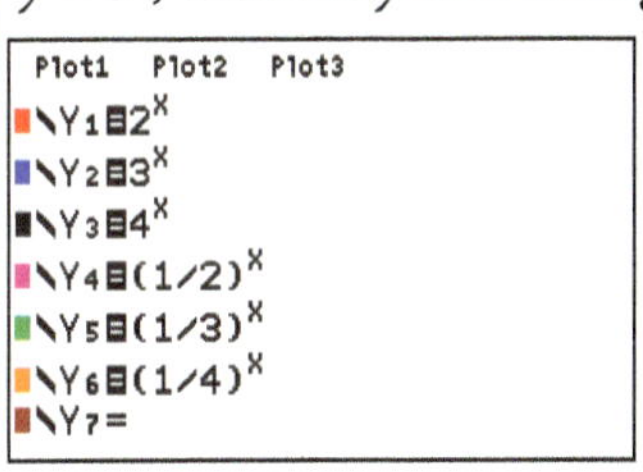

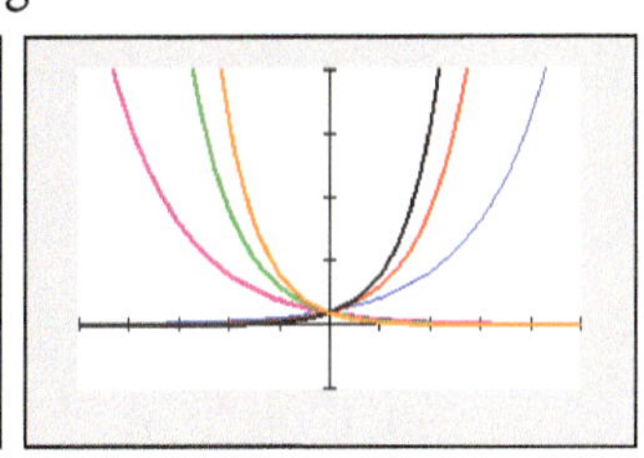

You can also explore what happens when the original graphs are modified to the form $y = b^{-x}$.

Use the [ins] function ([2nd] [del]) to insert a negative sign ([(-)]) before the x in the first 3 functions.

When the functions are graphed, it appears as though there are only 3 functions graphed. Using the [trace] function will help you see that Y_2 and Y_5 have the same graph. Which other functions have the same graph?

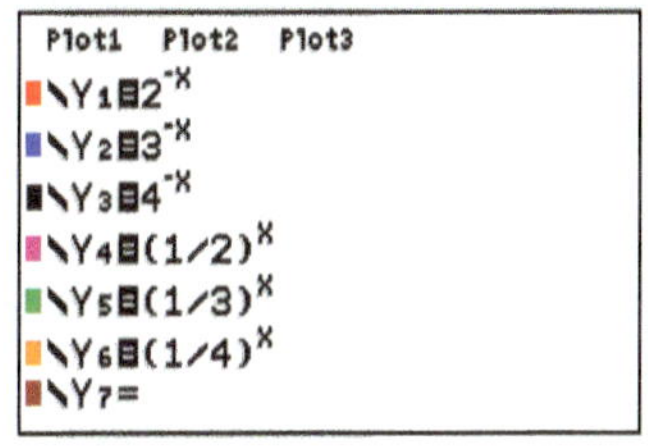

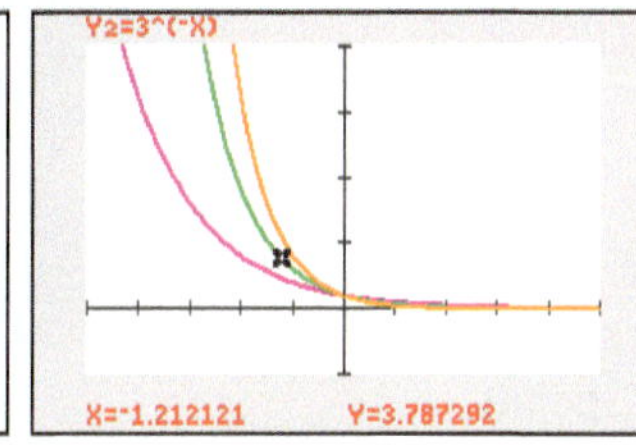

This concept can also be seen by turning off the graphs of the other functions. Moving the cursor to the highlighted = in a function and selecting [enter] removes the highlighting and turns the graph of that function off. You can turn it back on by selecting [enter] again.

This exploration leads to the conjecture that $y = b^{-x}$ and $y = \left(\frac{1}{b}\right)^x$ are identical functions.

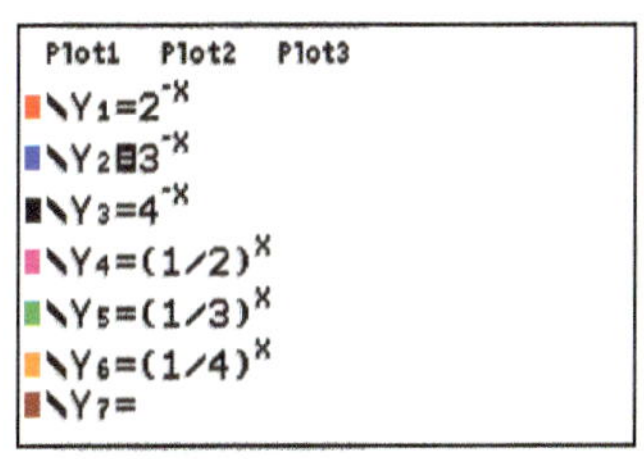

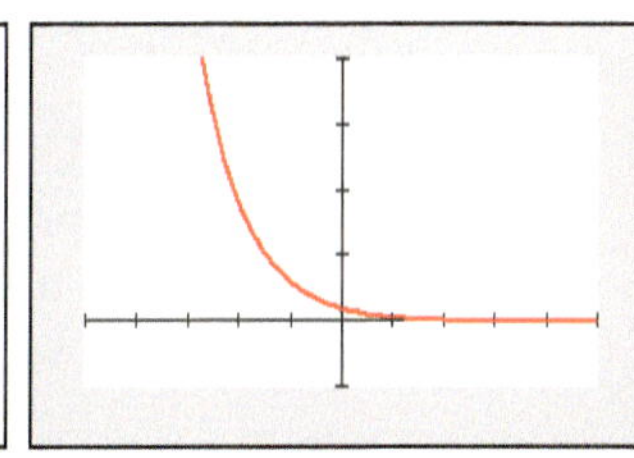

7.3 POWER FUNCTIONS

How are parent functions useful in graphing?

After completing this section, I will be able to

- identify the shape of the parent power functions $y = x^2$ and $y = x^3$ by plotting points.
- describe translations of power functions.
- graph translations of power functions by identifying the vertex or point of inflection.
- write basic power function rules from descriptions or graphs.

The parabolic shape of suspension cables helps lengthy suspension bridges bear the weight of hundreds of vehicles.

Previously, you have graphed linear functions and absolute value functions by plotting a sufficient number of points. By using this same method, the general shape of power functions can be determined.

Recall that the parent function is the simplest form of a family of functions. Consider the parent functions $y = x^2$ and $y = x^3$. These are called *power functions.*

DEFINITION

A **power function** is a function of the form $y = ax^n$, where the base x is a variable and the exponent n and coefficient a are constants.

Using the domain of $D = \{-3, -2, -1, 0, 1, 2, 3\}$ we can plot enough points to determine the general shape of the graph of the parent function $f(x) = x^2$.

Make a table of ordered pairs.

x	$f(x) = x^2$	$f(x)$	$(x, f(x))$
−3	$f(-3) = (-3)^2 = 9$	9	(−3, 9)
−2	$f(-2) = (-2)^2 = 4$	4	(−2, 4)
−1	$f(-1) = (-1)^2 = 1$	1	(−1, 1)
0	$f(0) = 0^2 = 0$	0	(0, 0)
1	$f(1) = 1^2 = 1$	1	(1, 1)
2	$f(2) = 2^2 = 4$	4	(2, 4)
3	$f(3) = 3^2 = 9$	9	(3, 9)

Plot the points and connect them with a smooth curve.

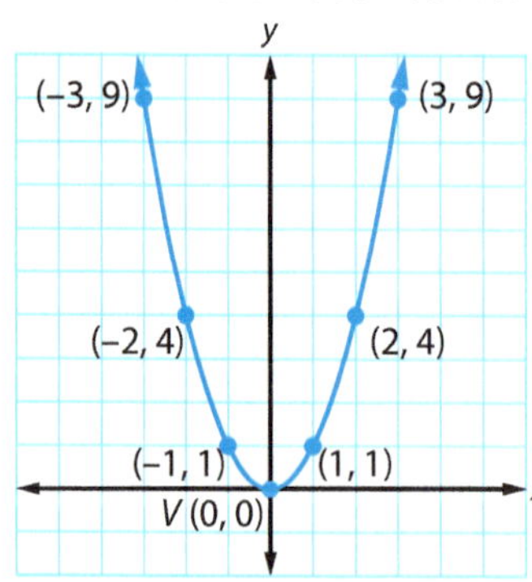

Notice that the graph of a power function, unlike that of a linear function, is curved. The curved graph of $y = x^2$ is called a **parabola**. A parabola has a turning point called the **vertex**. The vertex, V, on the graph displayed is at the point (0, 0). This turning point represents the minimum value of the function $y = x^2$.

EXAMPLE 1: Graphing a Vertical Translation

Graph the functions $y = x^2$, $y = x^2 + 2$, and $y = x^2 - 3$ on the same coordinate plane. Compare the graphs.

Answer

x	−3	−2	−1	0	1	2	3
x^2	9	4	1	0	1	4	9
$x^2 + 2$	11	6	3	2	3	6	11
$x^2 - 3$	6	1	−2	−3	−2	1	6

1. Make a table of ordered pairs for each function.

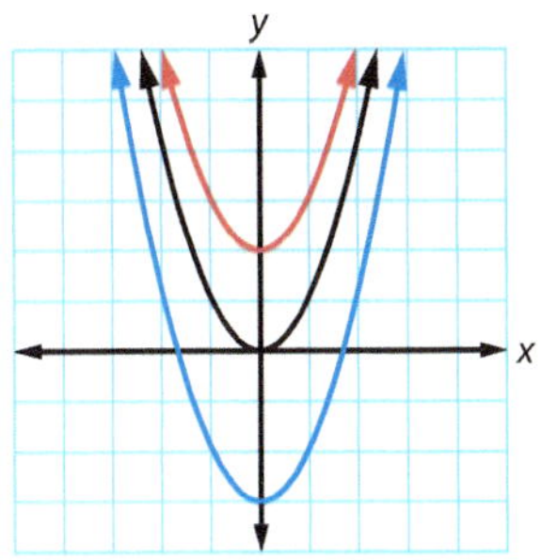

2. Use the ordered pairs to determine the location of each graph.

$y = x^2 + 2$ The graph shifts 2 units up when 2 is
vertex: (0, 2) added to the parent function.
$y = x^2 - 3$ The graph shifts 3 units down when 3 is
vertex: (0, −3) subtracted from the parent function.

3. Note that these 2 graphs are vertical translations of the parent function $y = x^2$.

SKILL CHECK **EXERCISE 19**

Recall that a translation of the graph of a function is a sliding of the graph without any rotating or resizing. A translation of a function results when a number is added to or subtracted from either the input or the output of the function. Adding to or subtracting from the output of a function results in a vertical translation. Given $y = x^2 + k$, the graph will shift k units vertically.

EXAMPLE 2: Graphing a Horizontal Translation

Graph the functions $y = x^2$, $y = (x - 2)^2$, and $y = (x + 3)^2$ on the same coordinate plane. Compare the graphs.

Answer

x	−4	−3	−2	−1	0	1	2	3	4
x^2	16	9	4	1	0	1	4	9	16
$(x - 2)^2$	36	25	16	9	4	1	0	1	4
$(x + 3)^2$	1	0	1	4	9	16	25	36	49

1. Make a table of ordered pairs for each function.

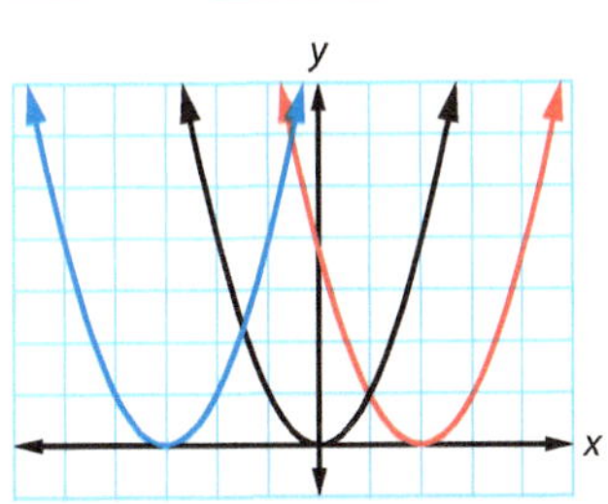

2. Use the ordered pairs to determine the location of each graph.

$y = (x - 2)^2$ The graph shifts 2 units right when 2 is
vertex: (2, 0) subtracted within the squared expression.
$y = (x + 3)^2$ The graph shifts 3 units left when 3 is
vertex: (−3, 0) added within the squared expression.

3. Note that these 2 graphs are horizontal translations of the parent function $y = x^2$.

SKILL CHECK **EXERCISE 20**

Adding to or subtracting from x, the input, results in a horizontal translation. Given $y = (x - h)^2$, the graph will shift h units horizontally.

FUNCTION COMPARISONS

Absolute Value Function

$f(x) = |x|$
vertex: (0, 0)

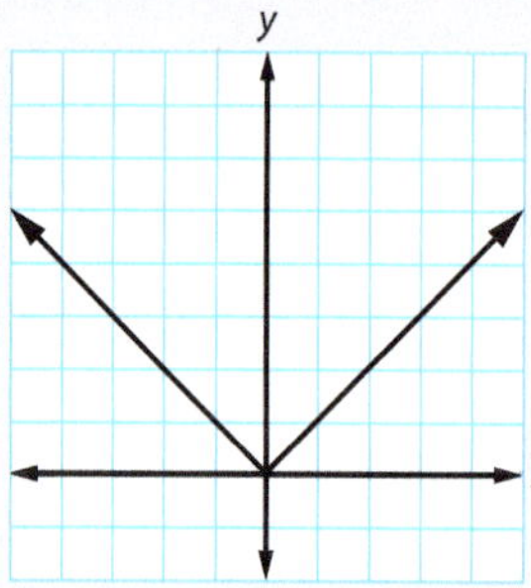

$f(x) = |x| + k$
vertex: $(0, k)$

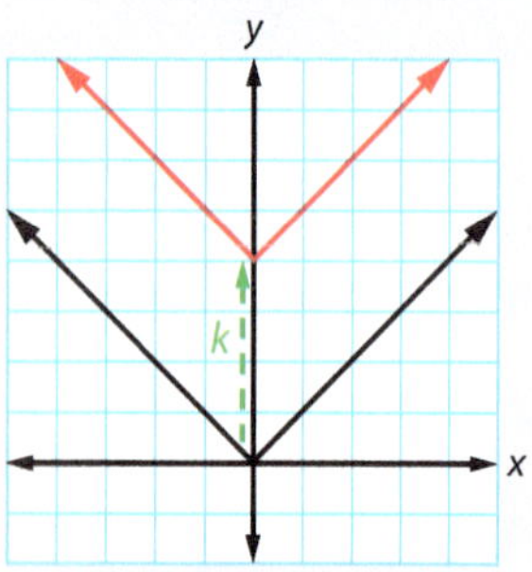

$f(x) = |x| - k$
vertex: $(0, -k)$

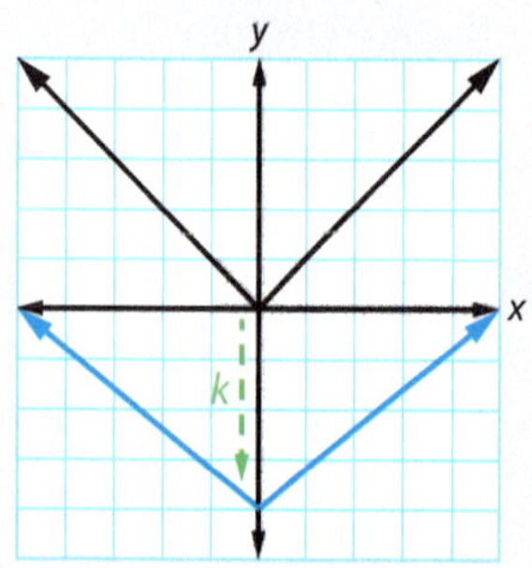

$f(x) = |x + h|$
$= |x - (-h)|$
vertex: $(-h, 0)$

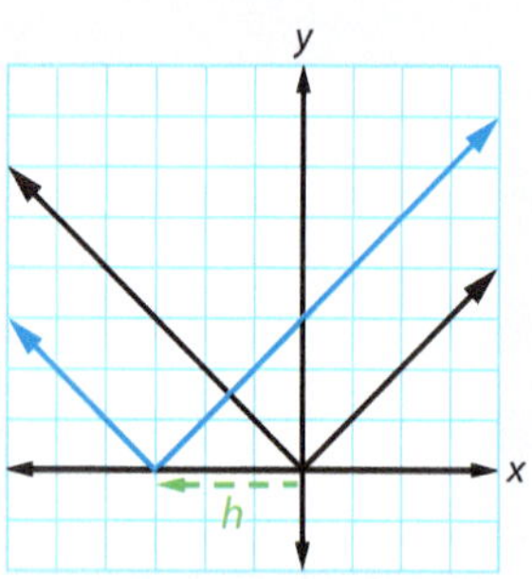

$f(x) = |x - h|$
vertex: $(h, 0)$

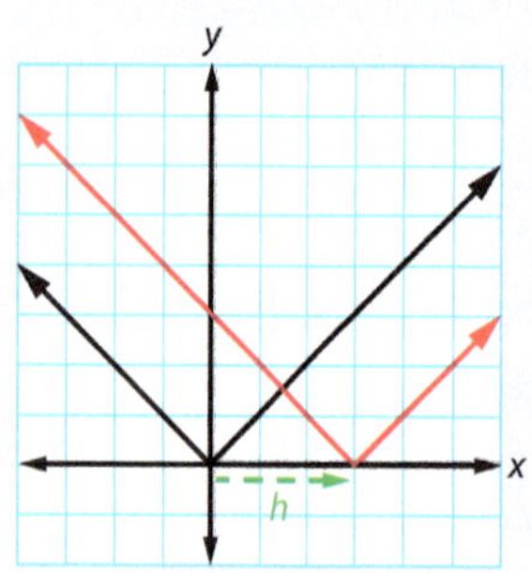

Power Function

$f(x) = x^2$
vertex: (0, 0)

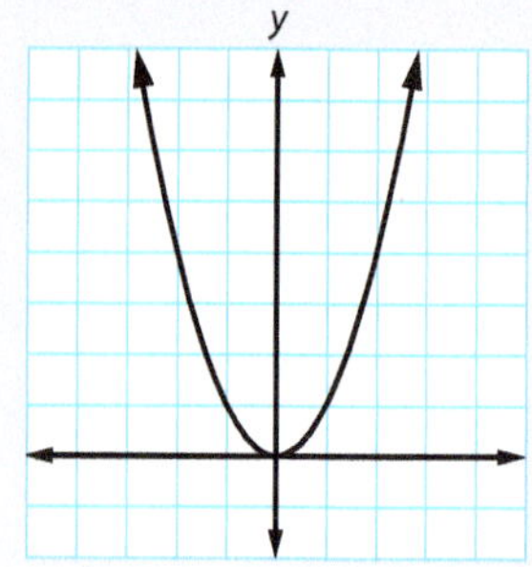

$f(x) = x^2 + k$
vertex: $(0, k)$

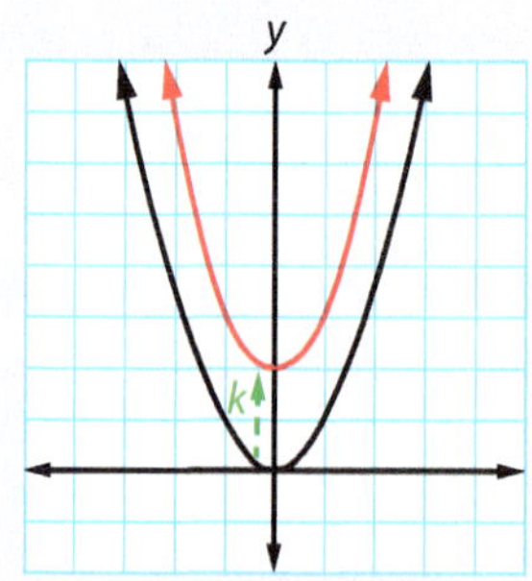

$f(x) = x^2 - k$
vertex: $(0, -k)$

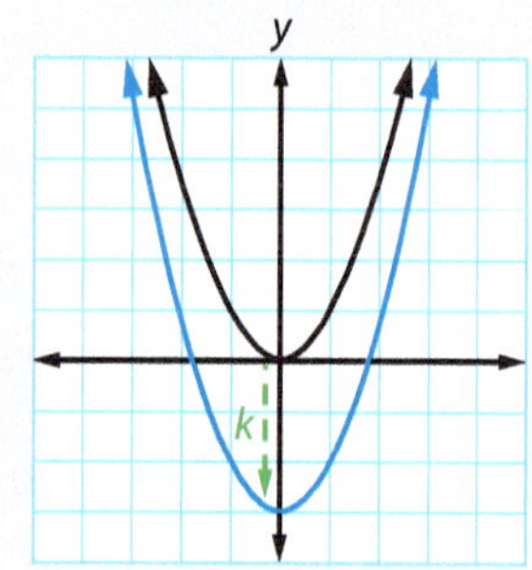

$f(x) = (x + h)^2$
$= (x - (-h))^2$
vertex: $(-h, 0)$

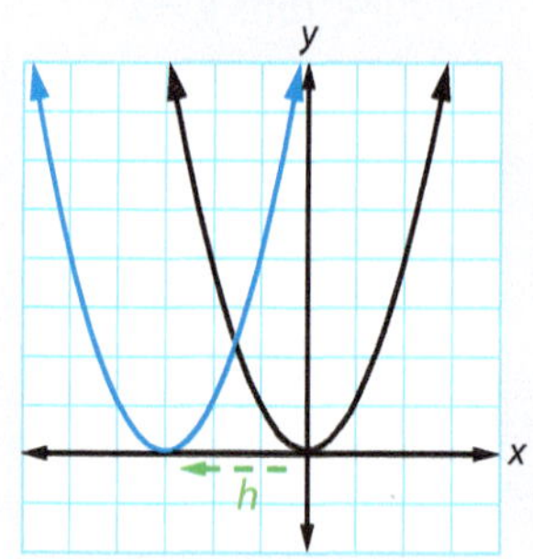

$f(x) = (x - h)^2$
vertex: $(h, 0)$

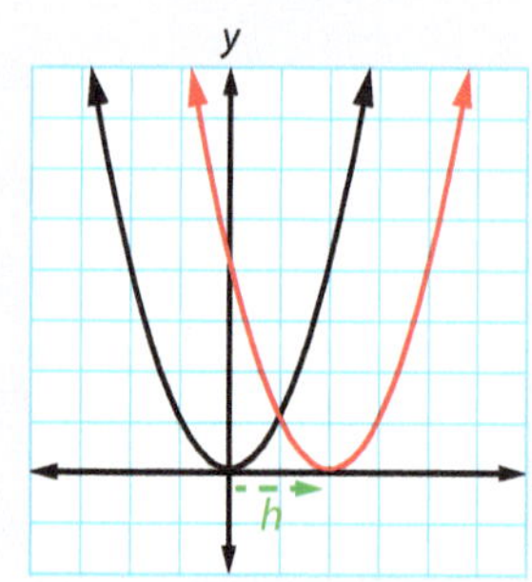

Knowing the general shape of a parent function allows other similar functions to be graphed more easily. Adding to or subtracting from both the input and the output of the parent function will result in a horizontal and vertical shift. Given $y = (x - h)^2 + k$, the graph of $y = x^2$ will shift h units horizontally and k units vertically, and the vertex (turning point) of the function will be located at (h, k). This form is called the **vertex form** of a parabola. Locating the vertex makes it easier both to write the function in vertex form and to sketch the graph.

EXAMPLE 3: Finding the Function Rule

Write the function rule that describes the graph.

a.

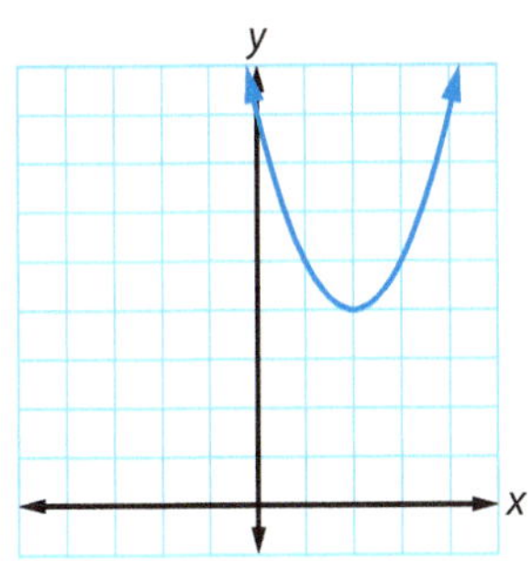

b.

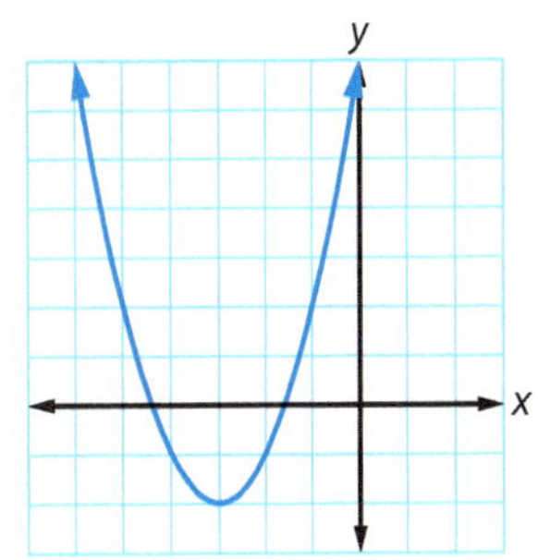

Answers

a. $y = (x - h)^2 + k$	**1.** Begin by writing the general vertex form of a parabola.
$(h, k) = (2, 4)$	**2.** Locate the vertex.
$y = (x - 2)^2 + 4$	**3.** Substitute the values of the vertex into the form.
b. $y = (x - h)^2 + k$	**1.** Begin by writing the general vertex form of a parabola.
$(h, k) = (-3, -2)$	**2.** Locate the vertex.
$y = (x - (-3))^2 + (-2)$ $y = (x + 3)^2 - 2$	**3.** Substitute the values of the vertex into the form and simplify.

SKILL CHECK **EXERCISE 33**

Now consider the power function $y = x^3$. The graph of $y = x^3$ can be drawn using a table of ordered pairs as illustrated.

x	−2	−1	0	1	2
x^3	−8	−1	0	1	8

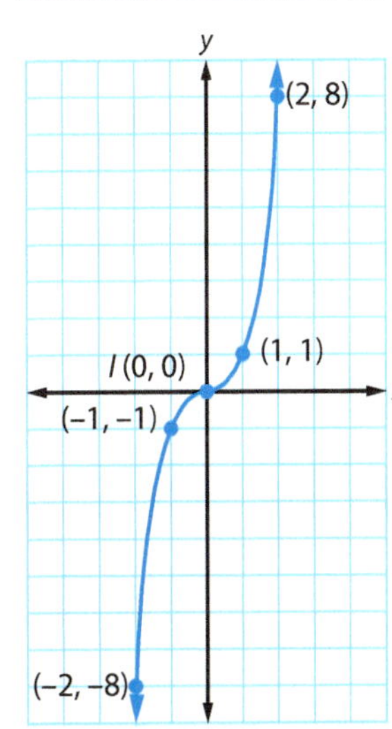

The **point of inflection** of a function is similar to the vertex of a parabola, except that the graph continues in the same basic direction instead of curving back. In the graph of $y = x^3$, the point of inflection, I, is at $(0, 0)$. In general, the point of inflection on the graph of $y = (x - h)^3 + k$ is located at (h, k).

EXAMPLE 4: Describing Translations of the $y = x^3$ Function

AFTERSCHOOLHELP

Power Functions

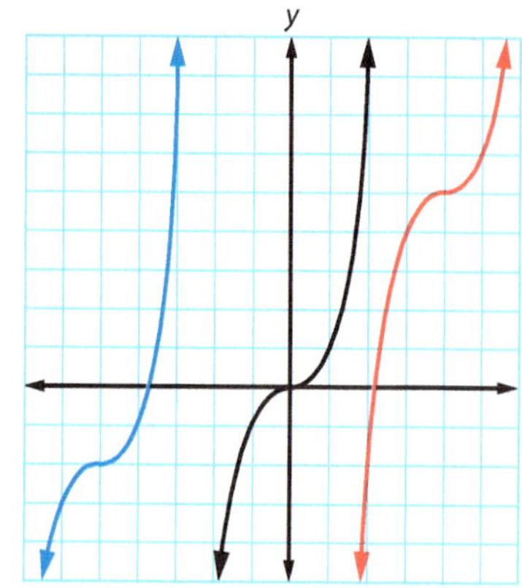

Study the graphs of the following functions and complete the table describing the translations in reference to the parent function $y = x^3$, shown in black.

$y = (x - 4)^3 + 5$

$y = (x + 5)^3 - 2$

Answer

Translated Function	Change from Parent Function $y = x^3$	Type of Translation	Translation of Graph	Point of Inflection (*h*, *k*)
$y = (x - 4)^3 + 5$	−4 from input	horizontal	4 units right	(4, 5)
	+5 to output	vertical	5 units up	
$y = (x + 5)^3 - 2$	+5 to input	horizontal	5 units left	(−5, −2)
	−2 from output	vertical	2 units down	

SKILL CHECK **EXERCISES 6, 11**

EXAMPLE 5: Graphing Translations of Power Functions

Identify the vertex or point of inflection and graph the function.

a. $y = (x - 1)^2 - 3$ **b.** $y = (x + 2)^3 + 1$

Answers

a. $y = (x - h)^2 + k$

$y = (x - 1)^2 + (-3)$

vertex: $(h, k) = (1, -3)$

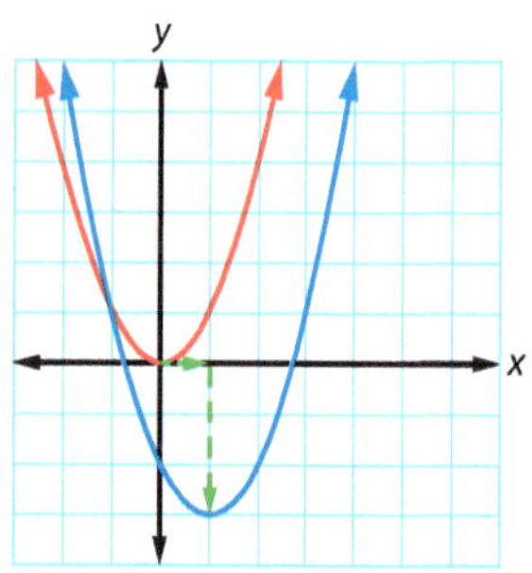

1. Compare to the general vertex form of the function to identify the vertex.

2. Graph the function as a translation of $y = x^2$ with the vertex at (1, −3).

b. $y = (x - h)^3 + k$

$y = (x - (-2))^3 + 1$

vertex: $(-2, 1)$

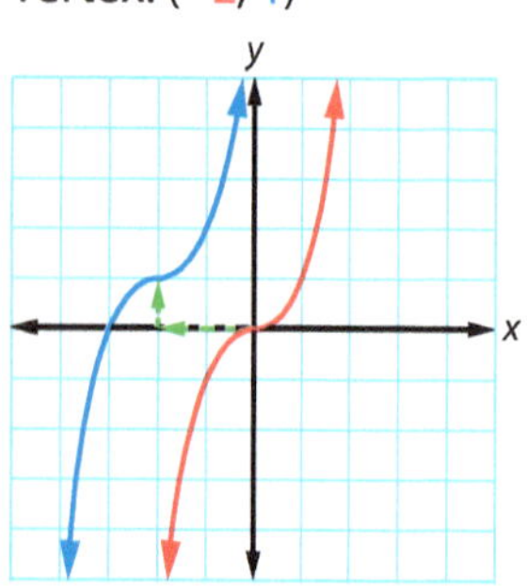

1. Compare to the parent power function.

2. Graph the function as a translation of $y = x^3$ with the point of inflection at (−2, 1).

SKILL CHECK **EXERCISE 21**

VOCABULARY

- parabola
- point of inflection
- power function
- vertex
- vertex form

A. EXERCISES

Fill in the blank.

1. The turning point of a parabola is called the ______.
2. The turning point of a parabola is easily discerned when functions are written in ______.

Describe the translation(s) of the graph of $y = x^2$ or $y = x^3$ that produce the graph of each function.

3. $y = x^2 + 5$
4. $y = (x - 4)^3$
5. $y = (x + 7)^2$
6. $y = x^2 - 12$
7. $y = (x - 5)^2 + 6$
8. $y = (x + 1)^3 - 9$
9. $y = (x - h)^2 + k$
10. $y = (x + h)^2 - k$

State the coordinates of the vertex or point of inflection of each function.

11. $f(x) = (x - 3)^2 + 1$
12. $f(x) = (x - 15)^2 - 9$
13. $f(x) = (x + 8)^3 + 6$
14. $f(x) = (x + 3)^3 - 1$
15. $f(x) = x^3 + 19$
16. $f(x) = (x + 12)^2$
17. $f(x) = (x - 7)^3$
18. $f(x) = x^2 - 5$
19. $f(x) = (x - h)^2 - k$
20. $f(x) = (x + h)^2 + k$

B. EXERCISES

State the coordinates of the vertex or point of inflection. Then create a table of at least 4 more ordered pairs and graph each function.

21. $f(x) = x^2 + 3$
22. $f(x) = (x - 1)^2$
23. $f(x) = (x + 2)^3$
24. $f(x) = x^3 - 2$
25. $f(x) = (x - 4)^2 + 1$
26. $f(x) = (x + 1)^2 - 5$

Write a function rule for each translation of the parent function $y = x^2$.

27. 10 units right
28. 5 units down
29. 3 units left and 12 units down
30. 7 units right and 2 units up
31. 2 units right and 1 unit up
32. 6 units left and 3 units down

Write the equation for each graph.

33.

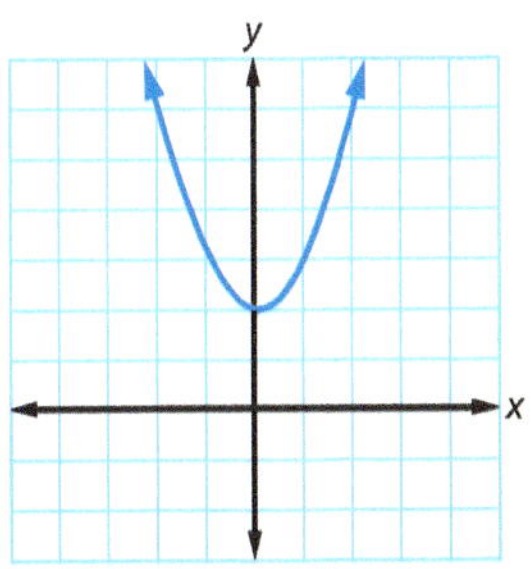

34.

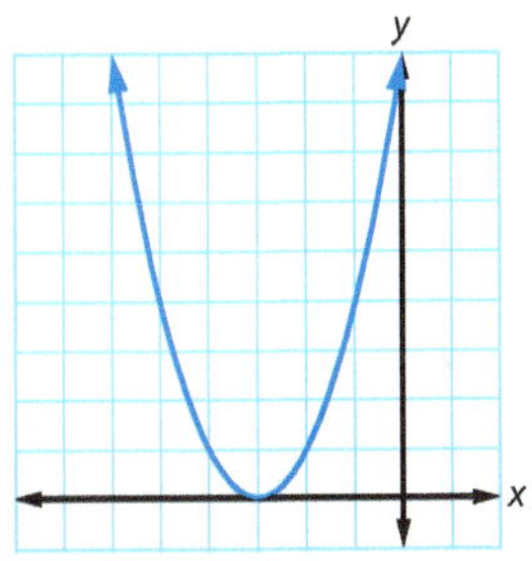

35.

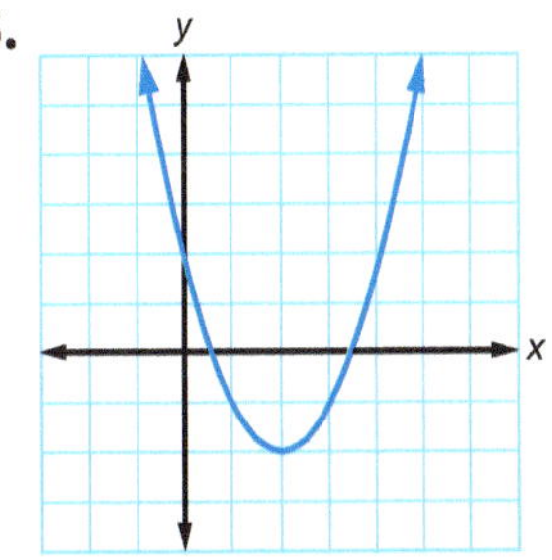

36.

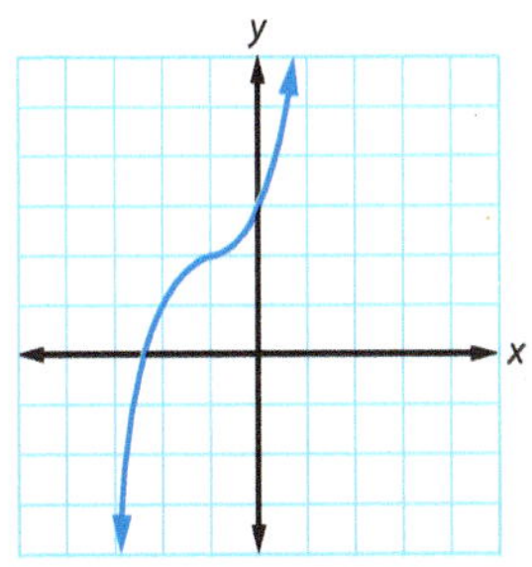

37. **Essential Question:** How are parent functions useful in graphing?

C. EXERCISES

Create a table of at least 5 ordered pairs and graph each function.

38. $y = 3x^2$

39. $y = \frac{1}{2}x^2$

40. $y = -x^2$

MP8 **Describe the effect on the graph of $y = ax^n$ when each of the following occurs.**

41. as $|a|$ increases

42. as $|a|$ decreases

43. when $a < 0$

CUMULATIVE REVIEW

Use function notation to write the rule for each function. [4.4]

44. $f = \{(-2, 7), (-1, 6), (0, 5), (1, 4), (2, 3)\}$

45. $g = \{(-3, -15), (-1, -7), (1, 1), (3, 9), (5, 17)\}$

46.

x	−6	−2	2	6	10
h(x)	4	6	8	10	12

Describe the linear correlation of each data set as (a) strong positive, (b) weak positive, (c) little or none, (d) weak negative, or (e) strong negative. [5.6]

47.

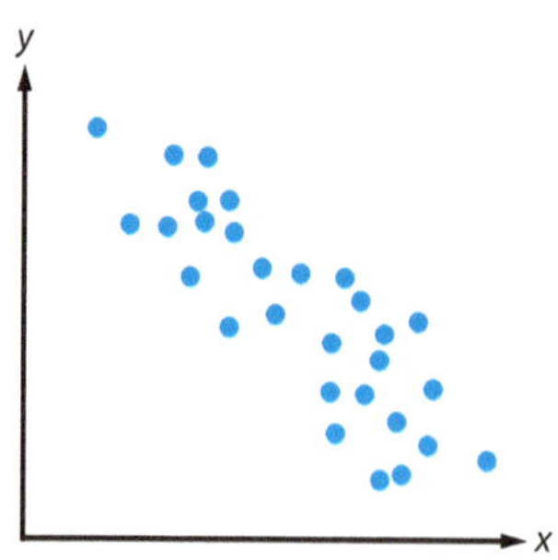

48.

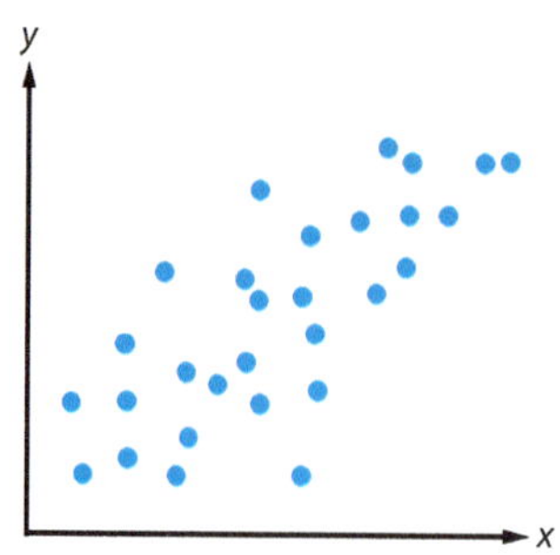

49.

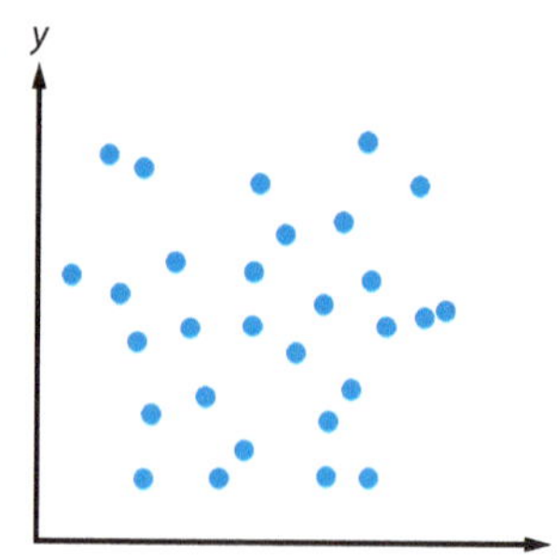

Graph each inequality on its own coordinate plane. [6.7]

50. $y \leq -2x + 3$

51. $y > \frac{3}{4}x$

52. Which equation has 2 solutions? **[3.5]**

A $5|2x - 1| = -10$

B $-|x + 1| + 3 = 2$

C $|x| + 8 = 8$

D $6 + |4x| = 2$

E none of these

53. What is the value of m if y is inversely proportional to x? **[5.6]**

x	2	4	6
y	16	8	m

A 48

B 4

C $\frac{3}{16}$

D 0

E none of these

7.4 EXPONENTIAL FUNCTIONS

How does a biblical worldview affect how a person uses math?

After completing this section, I will be able to

- define exponential functions.
- graph exponential functions by plotting ordered pairs.
- solve real-world problems using exponential functions.
- explain how math can be used in an unethical way.

While linear functions can be used to model some situations, exponential functions are powerful tools capable of representing many scientific and financial phenomena. With this power comes the potential for abuse. A biblical worldview requires that these tools be used in a way that honors God and does not harm others.

Gordon Moore, co-founder of Intel, predicted that the number of transistors per square inch on integrated circuits would double every 18 months.

In a linear function, the change in the dependent variable is constant during each unit of time.

$y = 3x$	x	1	2	3	4
	y	3	6	9	12

+3 +3 +3

In an exponential function, the dependent variable increases by the same factor during each unit of time.

$y = 3^x$	x	1	2	3	4
	y	3	9	27	81

×3 ×3 ×3

DEFINITION

An **exponential function** is a function of the form $f(x) = ab^x$, where $b > 0$ and $b \neq 1$.

EXAMPLE 1: Graphing Exponential Functions of the Form $y = b^x$

Create a table of ordered pairs and graph each function.

a. $y = 2^x$ **b.** $y = \left(\frac{1}{2}\right)^x$

Answers

a.

x	2^x	Ordered Pair
−3	$2^{-3} = \frac{1}{2^3} = \frac{1}{8}$	$\left(-3, \frac{1}{8}\right)$
−2	$2^{-2} = \frac{1}{2^2} = \frac{1}{4}$	$\left(-2, \frac{1}{4}\right)$
−1	$2^{-1} = \frac{1}{2}$	$\left(-1, \frac{1}{2}\right)$
0	$2^0 = 1$	(0, 1)
1	$2^1 = 2$	(1, 2)
2	$2^2 = 4$	(2, 4)
3	$2^3 = 8$	(3, 8)

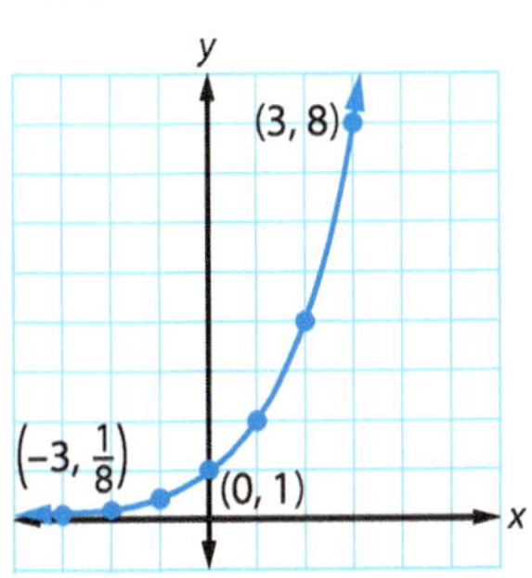

b.

x	$\left(\frac{1}{2}\right)^x$	Ordered Pair
−3	$\left(\frac{1}{2}\right)^{-3} = 2^3 = 8$	(−3, 8)
−2	$\left(\frac{1}{2}\right)^{-2} = 2^2 = 4$	(−2, 4)
−1	$\left(\frac{1}{2}\right)^{-1} = 2^1 = 2$	(−1, 2)
0	$\left(\frac{1}{2}\right)^0 = 1$	(0, 1)
1	$\left(\frac{1}{2}\right)^1 = \frac{1}{2}$	$\left(1, \frac{1}{2}\right)$
2	$\left(\frac{1}{2}\right)^2 = \frac{1}{4}$	$\left(2, \frac{1}{4}\right)$
3	$\left(\frac{1}{2}\right)^3 = \frac{1}{8}$	$\left(3, \frac{1}{8}\right)$

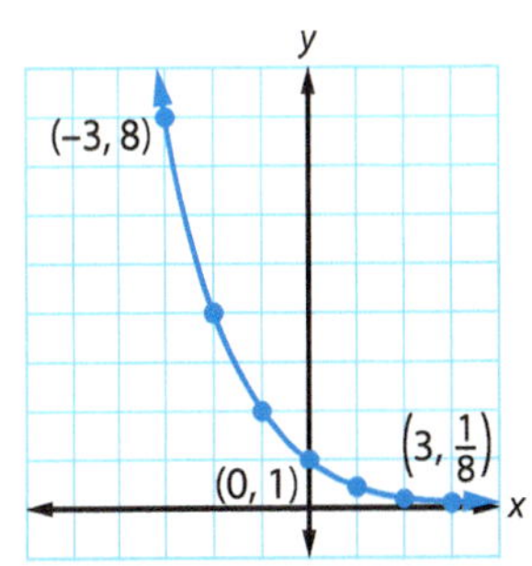

SKILL CHECK ***EXERCISE 7***

Both graphs have a y-intercept of (0, 1) and approach the x-axis but never intersect it. We call the line $y = 0$ (the x-axis) in these graphs an **asymptote,** a line that a graph approaches but never intersects. The function $y = 2^x$ continually increases from left to right, while $y = \left(\frac{1}{2}\right)^x$ decreases.

EXAMPLE 2: Graphing Exponential Functions of the Form $y = ab^x$

Graph $y = 3 \cdot 2^x$ on the same coordinate plane as $y = 2^x$.

Answer

x	$3 \cdot 2^x$	Ordered Pair
−3	$3 \cdot 2^{-3} = 3\left(\frac{1}{8}\right) = \frac{3}{8}$	$\left(-3, \frac{3}{8}\right)$
−2	$3 \cdot 2^{-2} = 3\left(\frac{1}{4}\right) = \frac{3}{4}$	$\left(-2, \frac{3}{4}\right)$
−1	$3 \cdot 2^{-1} = 3\left(\frac{1}{2}\right) = \frac{3}{2}$	$\left(-1, \frac{3}{2}\right)$
0	$3 \cdot 2^0 = 3(1) = 3$	(0, 3)
1	$3 \cdot 2^1 = 3(2) = 6$	(1, 6)
2	$3 \cdot 2^2 = 3(4) = 12$	(2, 12)
3	$3 \cdot 2^3 = 3(8) = 24$	(3, 24)

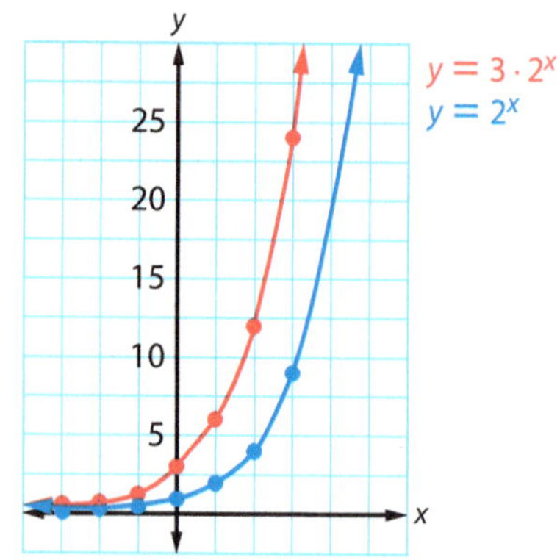

KEYWORD SEARCH

exponential function applet

SKILL CHECK EXERCISE 11

How does the curve of $y = a \cdot 2^x$ differ from that of $y = 2^x$? For every value of x, the corresponding value of y will be a times its previous value. Since every graph of $y = bx$ crosses the y-axis at (0, 1), the curve $y = ab^x$ will cross the axis at (0, a).

EXAMPLE 3: Applying Exponential Functions

Photocopiers and graphics programs allow the user to enlarge or reduce an image using a percentage of the original, p. The equation $y = p^x$ can be used to find the effect of successive enlargements or reductions. What percent of the original is the image produced by each of the following? Round to the nearest tenth of a percent.

a. 3 consecutive reductions set at 75% **b.** 4 consecutive enlargements set at 150%

Answers

a.	$p = 0.75; y = 0.75^x$	**1.** Model the problem with an exponential function.
	$y = 0.75^3 \approx 0.422$	**2.** Evaluate the function.
	about 42.2% of the original	**3.** Convert the decimal to a percent.
b.	$p = 1.5; y = 1.5^x$	**1.** Model the problem with an exponential function.
	$y = 1.5^4 \approx 5.063$	**2.** Evaluate the function.
	about 506.3% of the original	**3.** Convert the decimal to a percent.

SKILL CHECK EXERCISE 19

AFTERSCHOOLHELP

Exponential Functions

Most investments involve compound interest, where interest is regularly credited to the account so that it can begin to earn interest along with the principal. Compound interest can be modeled with an exponential function. While the exponential growth of a long-term investment is impressive, it is important to understand that over the same period the value of money will be decreasing due to inflation. Failing to point out the effect of inflation is one way a financial planner might use math in an unethical way.

EXAMPLE 4: Applying Exponential Functions

The total amount of a \$10,000 investment account earning 5% annual interest compounded yearly can be modeled by the equation $y = 10{,}000(1.05)^n$, where n is the number of years the amount is invested. Assuming there are no withdrawals from the account, find the balance after the following numbers of years.

a. 1 yr **b.** 2 yr **c.** 5 yr **d.** 10 yr
e. 25 yr **f.** 50 yr

Answers

a. $y = 10{,}000(1.05)^1 = \$10{,}500.00$

b. $y = 10{,}000(1.05)^2 = 10{,}000(1.1025) = \$11{,}025.00$

c. $y = 10{,}000(1.05)^5 \approx \$12{,}762.82$

Use a calculator to evaluate the more complicated expressions.

d. $y = 10{,}000(1.05)^{10} \approx \$16{,}288.95$

e. $y = 10{,}000(1.05)^{25} \approx \$33{,}863.55$

f. $y = 10{,}000(1.05)^{50} \approx \$114{,}674.00$

VOCABULARY
- asymptote
- exponential function

SKILL CHECK **EXERCISE 23**

A. EXERCISES

1. Match each function with its description.

a. $y = 2^x$ I linear
b. $y = 2x$ II exponential
c. $y = x^2$ III power

2. A horizontal line that an exponential function approaches but never touches is an ______.

Evaluate.

3. $y = \left(\frac{1}{2}\right)^x$ when $x = 5$

4. $y = 3^x$ when $x = -4$

5. $y = 3 \cdot 4^x$ when $x = 3$

6. $y = -3 \cdot 2^x$ when $x = -3$

Complete each table and then graph each exponential function.

7. $y = 3^x$

x	−2	−1	0	1	2
y					

8. $y = \left(\frac{1}{3}\right)^x$

x	−2	−1	0	1	2
y					

9. $y = \left(\frac{2}{3}\right)^x$

x	−3	−2	−1	0	1	2	3
y							

10. $y = \left(\frac{3}{2}\right)^x$

x	−3	−2	−1	0	1	2	3
y							

11. $y = 2 \cdot 3^x$

x	−2	−1	0	1	2
y					

12. $y = 2 \cdot \left(\frac{1}{3}\right)^x$

x	−2	−1	0	1	2
y					

13. $y = 3 \cdot \left(\frac{2}{3}\right)^x$

x	−3	−2	−1	0	1	2	3
y							

14. $y = 4 \cdot \left(\frac{3}{2}\right)^x$

x	−3	−2	−1	0	1	2	3
y							

15. Which of the following is the x-intercept of the graph of $y = ab^x$?

A (1, 0)

B (0, 1)

C $(a, 0)$

D $(0, a)$

E There is none.

16. Which of the following is the y-intercept of the graph of $y = ab^x$?

A (1, 0)

B (0, 1)

C $(a, 0)$

D $(0, a)$

E There is none.

17. Examine your graphs from exercises 7–14 and make a conjecture about all exponential functions $y = ab^x$ where $a > 0$ and $0 < b < 1$.

18. Examine your graphs from exercises 7–14 and make a conjecture about all exponential functions $y = ab^x$ where $a > 0$ and $b > 1$.

B. EXERCISES

Write an exponential function modeling consecutive enlargements or reductions by the given percentage. Then find the single setting that would accomplish the same result as the given number of consecutive enlargements or reductions. Round to the nearest whole percent.

19. 125%; 2

20. 60%; 3

21. 75%; 4

22. 125%; 3

23. The balance of an account earning 2% annual interest with an initial investment of $1000 can be modeled by the equation $y = 1000(1.02)^x$, where x is the number of years the money is invested. Assuming there are no withdrawals from the account, find the balance after the following numbers of years.

a. 1 yr

b. 2 yr

c. 10 yr

d. 25 yr

e. 50 yr

24. Under ideal conditions, the population of bacteria in a petri dish doubles every hour. If the equation $y = 500 \cdot 2^h$ models the population of bacteria in the dish after h hours, how many bacteria are in the petri dish at each of the following times?

a. initially (at $h = 0$ hr)

b. after 3 hr

c. after 12 hr

d. after 1 day

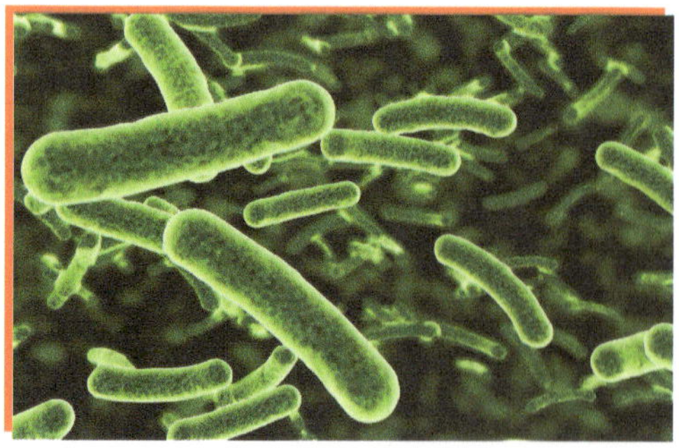

On Earth there are more bacteria than all other life forms combined.

Complete each table and then graph each exponential function.

25. $y = -3^x$

x	−2	−1	0	1	2
y					

26. $y = -\left(\frac{1}{3}\right)^x$

x	−2	−1	0	1	2
y					

27. $y = 3^{-x}$

x	−2	−1	0	1	2
y					

28. $y = \left(\frac{1}{3}\right)^{-x}$

x	−2	−1	0	1	2
y					

29. Compare the graphs of $y = 3^x$ and $y = -3^x$ (exercises 7 and 25) or the graphs of $y = \left(\frac{1}{3}\right)^x$ and $y = -\left(\frac{1}{3}\right)^x$ (exercises 8 and 26) to make a conjecture about the graphs of $y = b^x$ and $y = -b^x$.

30. Compare the graphs of $y = 3^x$ and $y = 3^{-x}$ (exercises 7 and 27) or the graphs of $y = \left(\frac{1}{3}\right)^x$ and $y = \left(\frac{1}{3}\right)^{-x}$ (exercises 8 and 28) to make a conjecture about the graphs of $y = b^x$ and $y = b^{-x}$.

31. Complete the table and then graph the function $y = 2^x + 3$.

x	−2	−1	0	1	2
y					

32. Compare the graphs of $y = 2^x$ and $y = 2^x + 3$ (Example 1*a* and exercise 31) to make a conjecture about the graphs of $y = b^x$ and $y = b^x + c$.

Match each equation to its graph.

33. $y = 2^x$

34. $y = \left(\frac{1}{2}\right)^x$

35. $y = -2^x$

36. $y = 2^x - 3$

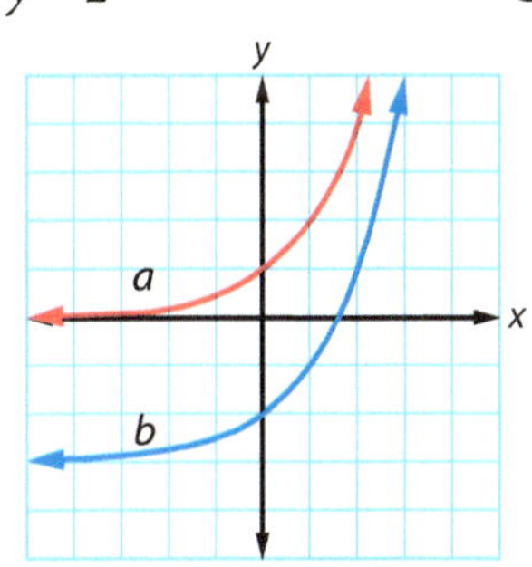

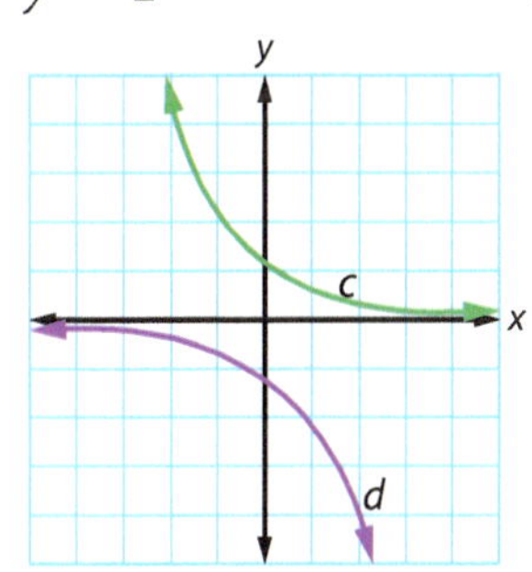

KEYWORD SEARCH

reflection graph applet

37. Essential Question: How does a biblical worldview affect how a person uses math?

C. EXERCISES

Your new employer proposes to pay you 2 pennies for your first day of work, 4 pennies for the second, 8 for the third, and so on until the thirtieth day of work.

38. **MP4** When you model your daily pay, *p*, with the exponential function $p = ab^d$, what are the values of *a* and *b*?

39. Use a calculator to determine your pay on the thirtieth day.

Answer the following exercises regarding the general exponential function $y = ab^x$.

40. The definition of the exponential function $y = ab^x$ states that $b \neq 1$. Write the simplified function rule and describe the graph of the function if $b = 1$.

41. **MP2** If $a > 0$, will the function $y = ab^x$ always increase as x increases? Explain.

42. **MP3** The graphs of $y = 2^x$ and $y = \left(\frac{1}{2}\right)^{-x}$ are identical. Show algebraically that these functions are equal.

CUMULATIVE REVIEW

Simplify. [7.1]

43. $(r^2)^{-2}s^4s^{-1}r^5$

44. $27m^{-3}n(-3mn^2)^{-2}$

45. $\frac{a^{-4}abc^{-2}}{c^{-1}c^{-2}b^3}$

46. $\frac{z^{-2}z^{-1}y^{-2}}{x^{-2}xy^3z^{-4}}$

Simplify, expressing each answer in scientific notation. [7.2]

47. $(6 \times 10^{14}) \div (8 \times 10^3)$

48. $(2.2 \times 10^{-8}) - (6 \times 10^{-9})$

49. To reach the stomach, food travels down the esophagus, which is about 15,000 μm wide. Express the width of the esophagus in meters using scientific notation.

50. The Holy of Holies in Solomon's temple was a cube with each edge approximately 9.1×10^2 cm in length (1 Kings 6:19–20). Express the volume (in cm^3) of this space in scientific notation rounded to the nearest hundredth.

51. Which of the following parabolas has a vertex in quadrant I? **[7.3]**

A $t(x) = (x - 1)^2 + 5$

B $u(x) = (x + 1)^2 + 5$

C $v(x) = (x - 1)^2 - 5$

D $w(x) = (x + 1)^2 - 5$

E all of these

52. Which function has a point of inflection $(-3, 4)$? **[7.3]**

A $y = (x - 3)^3 + 4$

B $y = (x + 3)^3 + 4$

C $y = (x - 4)^3 + 3$

D $y = (x + 4)^3 - 3$

E none of these

CAREERS IN MATH
COMPUTER PROGRAMMER/SOFTWARE ENGINEER

Most of us experience firsthand the impact that computers have on our daily lives. Modern coffee makers are programmed to regulate brewing time and temperature in order to extract the exact amount of flavor from a bean without creating a bitter aftertaste. Smart refrigerators connect to other devices, allowing you to monitor the temperature, see what groceries you need when you are at the store, or even be informed when items in your fridge are out of date. The vehicles we drive are loaded with computer programs, including autopilot features, to provide a more efficient and safer ride. Homeowners of smart homes have the ability to shut windows, turn lights on or off, regulate temperature, and monitor the grounds via an advanced security system. Cooking, cleaning, and even cutting the grass can all be accomplished with the help of computer technology. While computers play an ever-increasing role in our lives, none of these advances would be possible without programmers.

A computer programmer writes code, the logical sequence of instructions for a computer, called a program. A college degree in computer science is recommended since programmers need to be familiar with multiple programming languages and techniques. Programmers write, debug, and maintain code for a wide range of applications, including education, gaming, finances, robotic control, and data management. Mathematical and analytical skills learned in high-school math provide a good foundation for the pursuit of a programming career.

A software engineer's main responsibility is overseeing the development of computer programs. A software engineer (SE) may or may not write the actual code for the software. The SE is involved in defining and assessing needs and then ensuring the writing and implementation of a program that best meets those needs. He works closely with the programmers to ensure the success of their efforts, including testing the software and identifying how bugs might affect it. Most software engineering jobs require a degree in computer science or computer information systems.

Advanced technologies continue to create growing demands for programmers and software engineers in every business and ministry. These occupations help provide effective management of our God-given resources.

7.5 EXPONENTIAL GROWTH & DECAY

What is the difference between simple and compound interest?

After completing this section, I will be able to

- classify exponential functions as exhibiting exponential growth or decay.
- describe characteristics of an exponential growth or decay function.
- write functions modeling real-world exponential growth and decay.
- solve compound interest problems.

KEYWORD SEARCH

growth decay factor

Wildlife management often involves restocking lakes and ponds to maintain desired fish populations.

Exponential growth and decay occur when the value of a function has the same percent increase or decrease during each unit of time.

Using the method described in the previous section, a table of values or a graph can be used to model an exponential growth or decay function.

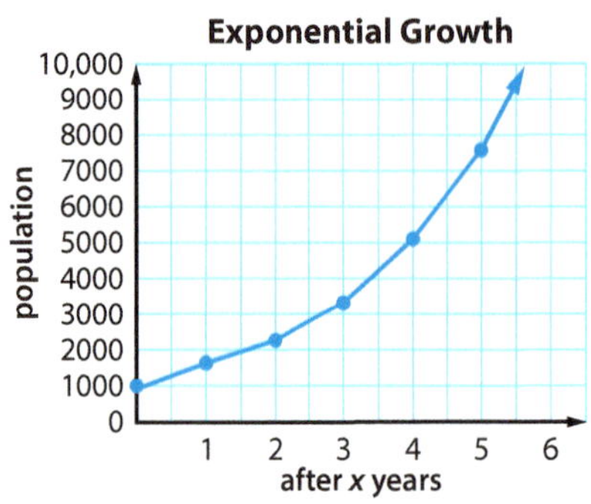

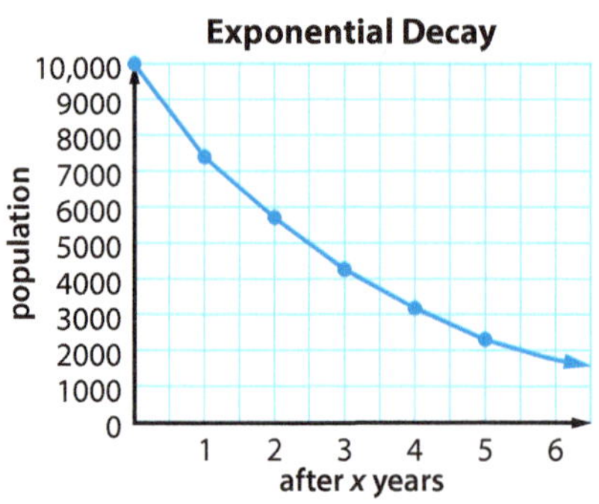

DEFINITIONS

Given an initial amount, A, and the **growth rate**, r (where $r > 0$), the exponential function $f(t) = A(1 + r)^t$ will model **exponential growth**.

The quantity $(1 + r)$ is called the **growth factor**.

If $-1 < r < 0$ the growth rate can be better described as a **decay rate**, and the exponential function $f(t) = A(1 + r)^t$ can be rewritten as $f(t) = A(1 - r)^t$, modeling **exponential decay**.

In an exponential decay function, the quantity $(1 - r)$ is called the **decay factor**.

growth or decay factor

$f(t) = A(1 \pm r)t$

growth or decay rate

EXAMPLE 1: Identifying Characteristics of an Exponential Function

For each exponential function, identify the initial amount, the growth or decay rate, and the growth or decay factor ($b = 1 \pm r$). Then state whether the function models exponential growth or exponential decay.

a. $f(t) = 500(1 - 0.75)^t$

b. $y = 100(1 + 2.5)^x$

c. $f(d) = 2^d$

d. $y = 10{,}000\left(\frac{1}{2}\right)^x$

Answers

	Initial Amount	Rate (r)	Factor (b)	Type	
a.	500	−0.75 or −75%	0.25	decay	Notice that $r < 0$.
b.	100	2.5 or 250%	3.5	growth	Notice that $r > 0$.
c.	1	1 or 100%	2	growth	Rewrite as $f(d) = 1(1 + 1)^d$.
d.	10,000	$-\frac{1}{2}$ or −50%	$\frac{1}{2}$	decay	Rewrite as $y = 10{,}000\left(1 - \frac{1}{2}\right)^x$.

SKILL CHECK ***EXERCISE 1***

EXAMPLE 2: Applying Exponential Growth

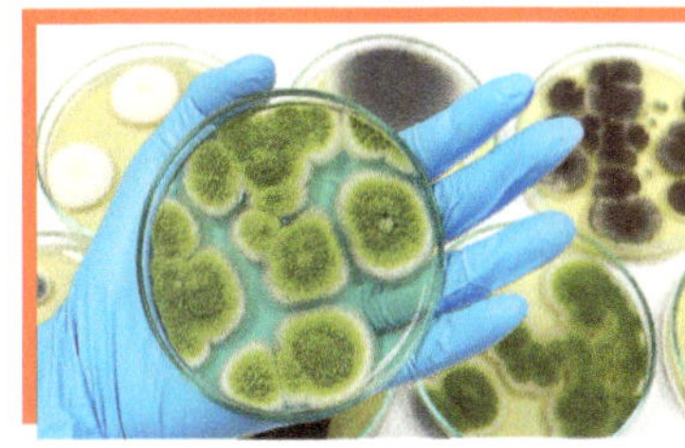

A 5 g mold culture increases in size by 200% every day. Write an exponential function modeling this growth and find the size of the culture after 1 week.

Answer

$f(d) = 5(1 + 2)^d$ or $f(d) = 5(3)^d$

$f(7) = 5(3)^7 = 5(2187) = 10{,}935$ g

1. The rate of change is $r = 200\% = 2$, and $A = 5$.
2. Let $d = 7$ and evaluate.

SKILL CHECK ***EXERCISE 33***

EXAMPLE 3: Applying Exponential Decay

In a single elimination tournament, half of the teams are eliminated in each round.

a. Write an exponential function representing the number of teams left after each round if the tournament begins with 64 teams.

b. Determine the number of teams left after 3 rounds.

c. Graph the function to find how many rounds it takes to determine a champion.

Answers

a. $f(n) = 64\left(1 - \frac{1}{2}\right)^n$ or $f(n) = 64\left(\frac{1}{2}\right)^n$ — The rate of change is $r = -\frac{1}{2}$, and $A = 64$.

b. $f(3) = 64\left(\frac{1}{2}\right)^3 = 64\left(\frac{1}{8}\right) = 8$ teams — Let $n = 3$ and evaluate.

c.

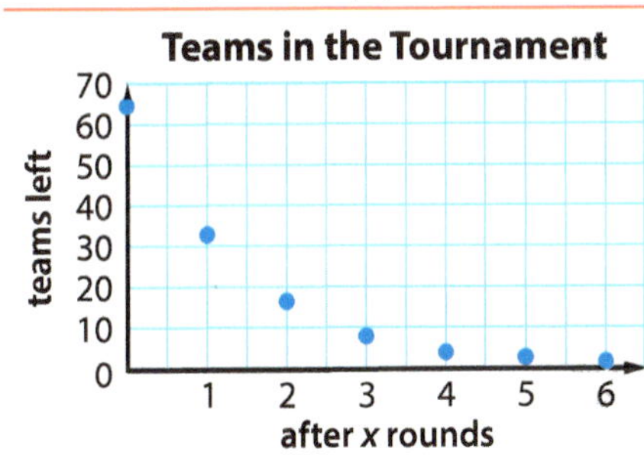

After 6 rounds, only the champion is left.

n	$f(n)$
0	$f(0) = 64\left(\frac{1}{2}\right)^0 = 64(1) = 64$
1	$f(1) = 64\left(\frac{1}{2}\right)^1 = 64\left(\frac{1}{2}\right) = 32$
2	$f(2) = 64\left(\frac{1}{2}\right)^2 = 64\left(\frac{1}{4}\right) = 16$
3	$f(3) = 64\left(\frac{1}{2}\right)^3 = 64\left(\frac{1}{8}\right) = 8$
4	$f(4) = 64\left(\frac{1}{2}\right)^4 = 64\left(\frac{1}{16}\right) = 4$
5	$f(5) = 64\left(\frac{1}{2}\right)^5 = 64\left(\frac{1}{32}\right) = 2$
6	$f(6) = 64\left(\frac{1}{2}\right)^6 = 64\left(\frac{1}{64}\right) = 1$

SKILL CHECK ***EXERCISE 11***

Many items decrease in value by the same percent each year. This **depreciation** can be modeled by an exponential decay function.

EXAMPLE 4: Calculating Depreciation

An automobile that sold for \$30,000 depreciates about 12% each year.

a. Write an exponential function modeling the value of the car.

b. Determine the value (to the nearest hundred dollars) of the car after 5, 10, and 15 yr.

c. Find the loss of value for the car during its first 5 yr, its second 5 yr, and its third 5 yr.

Answers

a. $f(t) = 30{,}000(1 - 0.12)^t = 30{,}000(0.88)^t$	The initial amount, A, is \$30,000. The growth rate is −12% or −0.12, and the growth factor is 0.88.
b. $f(5) = 30{,}000(0.88)^5 \approx \$15{,}800$ $f(10) = 30{,}000(0.88)^{10} \approx \$8400$ $f(15) = 30{,}000(0.88)^{15} \approx \$4400$	Evaluate the function at 5, 10, and 15 yr.
c. $30{,}000 - 15{,}800 = \$14{,}200$ $15{,}800 - 8400 = \$7400$ $8400 - 4400 = \$4000$	Find the difference of the values at the beginning and end of each time period.

SKILL CHECK **EXERCISE 31**

Compound interest is an important example of exponential growth. Simple interest is based on the initial principal only (see Section 2.6). Compound interest is based on the principal plus previously accumulated interest, resulting in faster financial growth.

If interest is compounded *annually*, it is credited to the account at the end of the year. Other typical compounding periods are *semiannually* (twice a year), *quarterly* (4 times a year), *monthly* (12 times a year), and *daily* (365 times a year).

Interest rate per compounding period is determined by dividing the annual percentage rate (APR) by the number of compounding periods in a year. An investment earning 6% APR compounded monthly will grow at a rate of $\frac{0.06}{12} = 0.005 = 0.5\%$ each month. Its growth would be modeled by the function

$$f(m) = A(1 + 0.005)^m = A(1.005)^m,$$

where m is the number of months the money is invested.

When determining the value of an account, the number of compounding periods must be used instead of the number of years. To find the value after 10 yr of a \$10,000 investment at 6% compounded monthly, use 10(12) = 120 months.

$$f(120) = 10{,}000(1.005)^{120} \approx \$18{,}193.97$$

AFTERSCHOOLHELP

Exponential Growth & Decay

COMPOUND INTEREST FORMULA

Compound interest is a special case of exponential growth.

$A = P(1 + r)^x$, where	A = the account balance after t years, P = the principal amount, x = the number of periods, and r = the periodic interest rate.
If there are n compounding periods in each of t years, then	$r = \frac{APR}{n}$ and $x = nt$.

EXAMPLE 5: Solving a Compound Interest Problem

Mr. Mendoza invests \$6000 in a certificate of deposit (CD) earning 4% interest compounded quarterly.

a. Write an exponential function modeling the value of his investment.

b. What is the account balance after 10 yr?

c. Create a table of ordered pairs and graph the function.

Answers

a. $f(q) = 6000(1 + 0.01)^q = 6000(1.01)^q$

The initial amount, A, is \$6000. The quarterly interest rate, r, is $\frac{0.04}{4} = 0.01$.

b. $f(40) = 6000(1.01)^{40} \approx \$8933.18$

Four interest payments per year for 10 yr is 40 compounding periods.

c.

Value of Investment

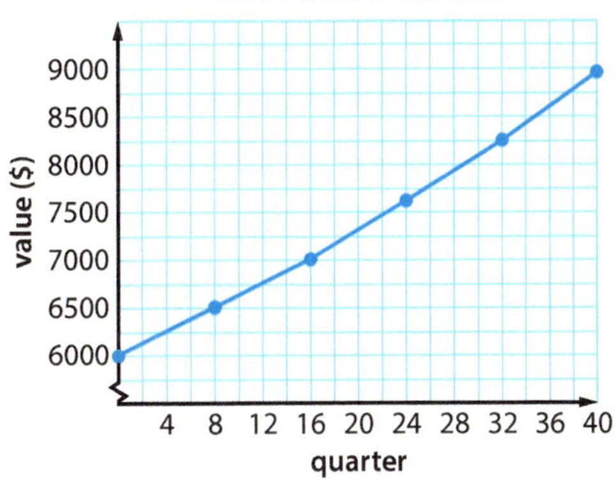

q	$f(q)$
0	$6000(1.01)^0 = \$6000$
8	$6000(1.01)^8 \approx \$6497$
16	$6000(1.01)^{16} \approx \$7035$
24	$6000(1.01)^{24} \approx \$7618$
32	$6000(1.01)^{32} \approx \$8250$
40	$6000(1.01)^{40} \approx \$8933$

SKILL CHECK **EXERCISES 25, 27**

VOCABULARY

- decay factor
- decay rate
- depreciation
- exponential decay
- exponential growth
- growth factor
- growth rate

A. EXERCISES

For each exponential function, identify the initial amount, the growth rate (as a percent), and the growth factor. Then state whether the function models exponential growth or exponential decay.

1. $y = 35(1 + 0.05)^x$

2. $f(t) = 16\left(1 - \frac{1}{5}\right)^t$

3. $y = \left(\frac{7}{10}\right)^x$

4. $f(p) = 10{,}000(1.1)^p$

Write each described exponential function in the form $f(t) = ab^t$. Then state whether the function models exponential growth or exponential decay.

5. initial amount: 10; growth rate: 150%

6. initial amount: 0.5; growth rate: −30%

7. initial amount: 81; growth rate: $\frac{1}{2}$

8. initial amount: 125; growth factor: 0.2
9. initial amount: 75; growth factor: 0.5
10. initial amount: 60,000; growth factor: $\frac{3}{2}$

Write an exponential function modeling each description.

11. This year's annual sales of $800,000 are projected to decrease by 10% each year.
12. The school committee plans for the student body of 200 to grow by 3% each year.
13. A culture containing 1500 bacteria doubles in size each day.
14. A used car purchased for $4000 retains only three-fourths of its value each year.

$5000 is invested at 6% APR for 10 yr. Determine the number of compounding periods and the rate per compounding period when interest is compounded as follows.

15. annually
16. semiannually
17. quarterly
18. monthly
19. daily

B. EXERCISES

Write an exponential expression for the value of each investment. Do not evaluate.

20. $5000 compounded annually at 4% APR for 10 yr
21. $5000 compounded semiannually at 4% APR for 10 yr
22. $2000 compounded quarterly at 3% APR for 20 yr
23. $37,855 compounded monthly at 4% APR for 30 yr
24. $500 compounded daily at 6% APR for 10 yr

Write an exponential function and use it to find the amount in each account.

25. $10,000 compounded quarterly at 6% APR for 12 yr
26. $4500 compounded quarterly at 4.4% APR for 8 yr
27. $2000 compounded semiannually at 6% APR for 20 yr
28. $2000 compounded monthly at 6% APR for 20 yr
29. $5000 compounded annually at 4.5% APR for 6.5 yr
30. The population of Greenwood has increased by about 2% each year since 2020, when it had 25,000 residents.
    a. Write an exponential function, $f(x)$ (where x is the number of years after 2020), modeling the population of Greenwood.
    b. Use the function to approximate the population in 2030, 2040, 2050, and 2060.
    c. Graph the function, showing Greenwood's population through the year 2070.

31. A patient takes 500 mg of a medication, and the amount of medication in the patient's body decreases by 10% each hour.

   a. Write an exponential function, $f(x)$, modeling the amount of medication in the patient's body.

   b. How much medication, to the nearest milligram, is in the patient's body after 4 hr?

   c. Graph the function, showing the amount of medication in the patient's body for the first 8 hr.

32. A ball dropped from a height of 10 m bounces to a height that is $\frac{2}{5}$ of its previous height.

   a. Write an exponential function, $f(x)$, modeling the height of each bounce.

   b. How high is the 5th bounce of the ball?

   c. Graph the function, showing the first 5 bounces of the ball.

33. A culture of 500 bacteria doubles every 20 min.

   a. Write an exponential function, $f(p)$ (where p is the number of 20 min periods), modeling the number of bacteria.

   b. Determine the number of bacteria after 2 hr.

   c. Determine the approximate number of bacteria after 1 day.

34. MP4 A piece of paper is 0.01 cm thick, and its thickness doubles each time it is folded in half.

   a. Write an exponential function, $f(n)$, modeling the thickness of the paper after n folds.

   b. How thick would the paper be after 10 folds?

   c. Graph the function, showing the thickness of the paper for the first 10 folds.

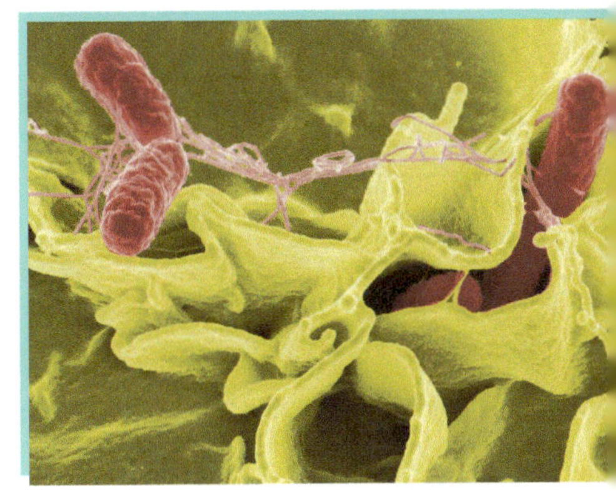

The rod-shaped *Salmonella* bacteria, which cause illnesses such as food poisoning and typhoid fever, have a diameter between 0.7 and 1.5 μm.

35. A pair of squirrels is introduced to an island, and their population doubles every 3 months.

   a. Write an exponential function, $f(m)$, modeling the number of squirrels on the island after m months.

   b. Write an exponential function, $f(n)$, modeling the number of squirrels on the island after n years.

   c. Use either of your functions to determine the number of squirrels on the island after 3 yr. Then use the other function to check your work.

36. **Essential Question:** Jessica invested $1000 into an account that earns 6% APR for 3 yr.

   a. Determine the balance at the end of 3 yr if the account earns simple interest.

   b. Determine the balance at the end of 3 yr for accounts with interest compounded annually, quarterly, and monthly.

   c. Into which of the accounts above would it be wisest to invest money?

   d. In general, assuming 2 accounts have the same APR but different compounding schedules, which will grow money faster?

C. EXERCISES

Exponential functions model growth if $r > 0$ or decay if $-1 < r < 0$.

37. Describe the value of the growth factor when the function models exponential growth.

38. MP2 Describe the value of the growth factor when the function models exponential decay.

39. What values are not possible for the growth factor? Explain.

40. MP1 An investment of $10,000 earns 6% APR for 10 yr. To the nearest dollar, how much more is earned if the investment is compounded as follows?

- **a.** semiannually instead of annually
- **b.** quarterly instead of semiannually
- **c.** monthly instead of quarterly
- **d.** monthly instead of annually

CUMULATIVE REVIEW

Determine whether the given ordered pair is a solution of the system $2x - 3y > 12$ and $y \le -\frac{2}{3}x + 4$. [6.8]

41. (2, 1)

42. (6, −5)

Write the system represented by each graph. [6.8]

43.

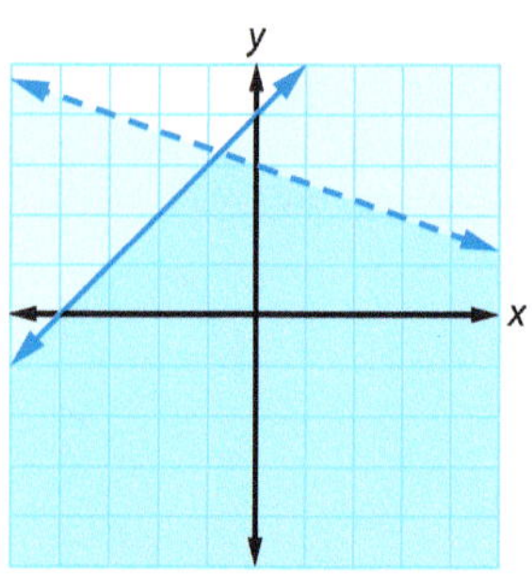

44.

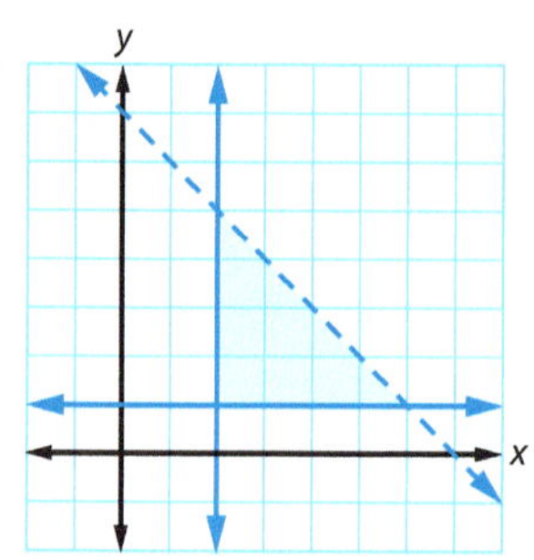

Solve each system. Then classify it as consistent independent, consistent dependent, or inconsistent. If the solution is an entire line, write its equation in slope-intercept form. [6.4]

45. $y = -2x + 3$
$4x + 2y = 6$

46. $5x + 2y = 8$
$x - 2y = 4$

47. $2x - 3y = 6$
$-4x + 6y = 12$

48. $15x - 5y = 10$
$y = 3x - 2$

49. Which of the following sequences below is arithmetic? **[4.5]**

- A 1, 5, 25, 125, 625, . . .
- B 6, −12, 18, −24, 30, . . .
- C 3, 5, 9, 15, 23, . . .
- D $1, \frac{1}{2}, \frac{1}{4}, \frac{1}{8}, \frac{1}{16}, \ldots$
- E none of these

50. Find the explicit formula for the arithmetic sequence: −18, −13, −8, −3, 2, ... **[4.5]**

- A $a_n = a_{n-1} - 18; a_1 = 5$
- B $a_n = a_{n-1} - 5; a_1 = -18$
- C $a_n = 5 + (n - 1)(-18)$
- D $a_n = -18 + (n - 1)(-5)$
- E none of these

7.6 GEOMETRIC SEQUENCES

How is a geometric sequence useful in the real world?

After completing this section, I will be able to

- define a geometric sequence.
- find missing terms in a geometric sequence by determining the sequence's pattern.
- describe geometric sequences with multiple representations.
- solve real-world problems involving geometric sequences.

Recall from Section 4.5 that an arithmetic sequence is a sequence with a common difference between any 2 consecutive terms. The pattern formed in an arithmetic sequence is obtained by adding the common difference, d.

A *geometric sequence* is a sequence with a *common ratio* between any 2 consecutive terms. The pattern formed in a geometric sequence is obtained by multiplying by the common ratio, r.

God determines the sex of each precious newborn.

DEFINITION

A **geometric sequence** is a sequence in which the ratio between any 2 consecutive terms is a constant called the **common ratio**.

EXAMPLE 1: Identifying a Geometric Sequence

Determine whether each sequence is geometric. If so, state the common ratio and the next 3 terms.

a. 1, 3, 9, 18, . . . **b.** 2, 4, 8, 16, . . .

Answers

a. 1 3 9 18 (×3 ×3 ×2)

It is not a geometric sequence since there is not a common ratio.

b. 2 4 8 16 (×2 ×2 ×2)

32, 64, 128

1. It is a geometric sequence with a common ratio of 2.
2. Multiply by 2 to get the next 3 terms.

SKILL CHECK **EXERCISE 7**

A geometric sequence can be expressed as a set of ordered pairs using its position in the sequence, n, and its value, a_n. The relation formed from Example 1b is {(1, 2), (2, 4), (3, 8), (4, 16), . . .} and its graph is displayed to the right. Graphs of geometric sequences with positive common ratios will lie on the graph of an exponential function. Unlike exponential functions, the domain of geometric sequences is restricted to $\mathbb{N} = \{1, 2, 3, \ldots\}$.

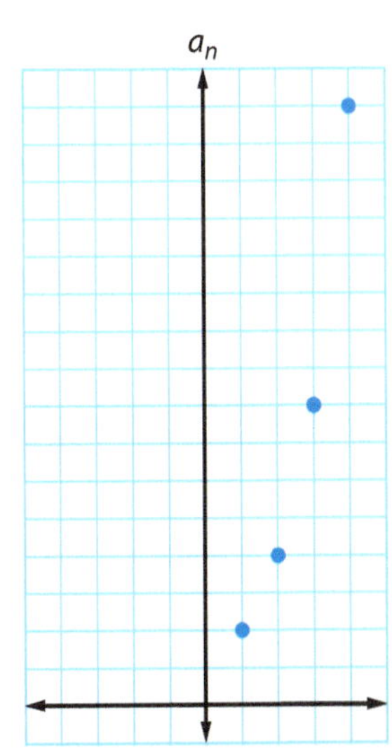

Recall that a recursive formula can be used to generate successive terms in a patterned sequence. Successive terms in a geometric sequence can be generated by multiplying the previous term by the common ratio, r.

RECURSIVE FORMULA FOR GEOMETRIC SEQUENCES

$$a_n = ra_{n-1}$$

Recall that a_n denotes the nth term and a_{n-1} denotes the previous term.

EXAMPLE 2: Displaying Geometric Sequences

Use a table, a graph, and a list to represent the first 4 terms in a geometric sequence where $a_1 = \frac{1}{3}$ and $r = -3$.

Answer

n	$-3a_{n-1}$	a_n	Ordered Pair
1	given	$\frac{1}{3}$	$\left(1, \frac{1}{3}\right)$
2	$-3 \cdot \frac{1}{3}$	−1	(2, −1)
3	$-3 \cdot (-1)$	3	(3, 3)
4	$-3 \cdot 3$	−9	(4, −9)

1. Make a table displaying each term's position and its corresponding value.

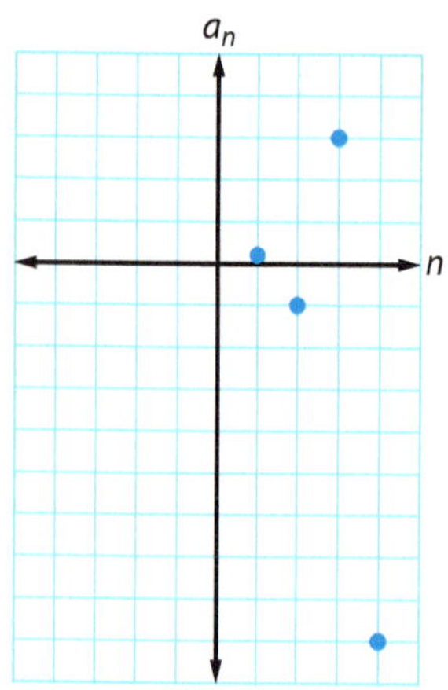

2. Graph the points.

$\frac{1}{3}, -1, 3, -9$

3. List the first 4 terms in the sequence.

SKILL CHECK **EXERCISE 17**

As with arithmetic sequences, the recursive formula works well for finding the first few terms of a sequence, but a more generalized formula is often helpful when finding a specific term in the sequence.

The **explicit formula for a geometric sequence** or *formula for the nth term* can be derived using the following pattern.

$$
\begin{aligned}
& a_1 \\
& a_2 = a_1 \cdot r \\
& a_3 = a_2 \cdot r \quad = (a_1 \cdot r)r \\
& \qquad\qquad\quad = a_1 \cdot r^2 \\
& a_4 = a_3 \cdot r \quad = (a_1 \cdot r^2)r \\
& \qquad\qquad\quad = a_1 \cdot r^3 \\
& a_5 = a_4 \cdot r \quad = (a_1 \cdot r^3)r \\
& \qquad\qquad\quad = a_1 \cdot r^4 \\
& a_n = a_{n-1} \cdot r = (a_1 \cdot r^{n-2})r \\
& \qquad\qquad\quad = a_1 \cdot r^{n-1}
\end{aligned}
$$

EXPLICIT FORMULA FOR A GEOMETRIC SEQUENCE

$$a_n = a_1 r^{n-1}$$

EXAMPLE 3: Finding the *n*th Term of a Geometric Sequence

Write an explicit formula for the *n*th term of the following sequence and find the 8th term.

a. 16, 8, 4, . . . **b.** −3, 15, −75, . . .

Answers

a. $r = \frac{8}{16} = \frac{1}{2}$

1. Determine the common ratio.

$a_n = 16\left(\frac{1}{2}\right)^{n-1}$

2. Substitute a_1 and r into the explicit formula.

$a_8 = 16\left(\frac{1}{2}\right)^{8-1} = 16\left(\frac{1}{2^7}\right)$
$= 16\left(\frac{1}{128}\right) = \frac{1}{8}$

3. Substitute 8 for n and simplify.

b. $r = \frac{15}{-3} = -5$

1. Determine the common ratio.

$a_n = -3(-5)^{n-1}$

2. Substitute a_1 and r into the explicit formula.

$a_8 = -3(-5)^{8-1}$
$= -3(-5)^7 = 234{,}375$

3. Substitute 8 for n and simplify.

SKILL CHECK **EXERCISE 25** ☑

Geometric sequences are useful for modeling real-world situations in which something is increasing or decreasing by a common factor at regular intervals. Identifying a term and the common factor can enable us to make estimates or predictions.

EXAMPLE 4: Applying Sequences to the Real World

AFTERSCHOOLHELP

Geometric Sequences

Suppose a ball dropped from a height of 40 ft rebounds to 80% of its previous height on each bounce.

a. List the height of the ball after each of the first 3 bounces.

b. Write an explicit formula for the *n*th term of the sequence.

c. Find the height of the ball after the 10th bounce.

Answers

a. $a_1 = 0.8(40) = 32$
$a_2 = 0.8(32) = 25.6$
$a_3 = 0.8(25.6) = 20.48$

Use the recursive formula to generate the first 3 terms of the sequence.

The height of the ball after each of the first 3 bounces is 32 ft, 25.6 ft, and 20.48 ft.

b. $a_n = 32(0.8)^{n-1}$

Substitute a_1 and r into the explicit formula.

c. $a_{10} = 32(0.8)^{10-1}$
$= 32(0.8)^9 \approx 4.29$

Evaluate the formula for the 10th term.

The height of the ball will be approximately 4.3 ft after the 10th bounce.

SKILL CHECK **EXERCISE 33** ☑

EXAMPLE 5: Applying Sequences to the Real World

The cellular growth in the human fetal forebrain from week 13 to week 20 roughly matches an exponential growth modeled by the function $f(n) = (3 \times 10^9) \cdot 1.23^{n-1}$.

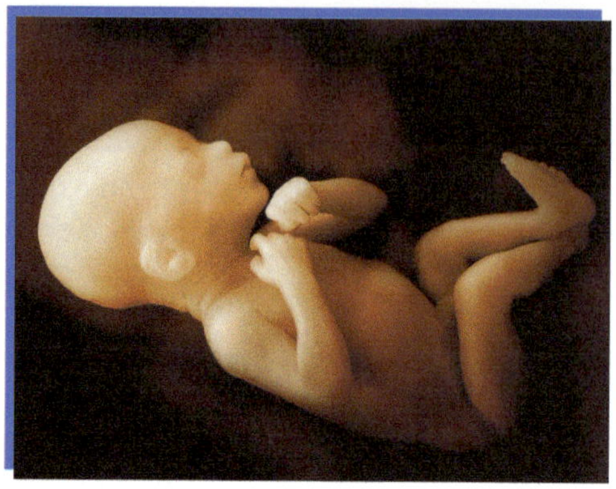

a. What does 3×10^9 represent?

b. What is the approximate number of cells in the fetal forebrain at week 20?

Answers

a. 3×10^9 represents the number of cells in the fetal forebrain at week 13.

b. $f(8) = (3 \times 10^9) \cdot 1.23^{8-1}$ Week 20 would be the 8th term since week 13 is $f(1)$.

$= (3 \times 10^9) \cdot 1.23^7$

$= (3 \cdot 1.23^7) \times 10^9$

$\approx 12.8 \times 10^9$

At week 20, the fetal forebrain has approximately 1.28×10^{10} cells.

SKILL CHECK **EXERCISE 35** ☑

VOCABULARY

- common ratio
- geometric sequence

A. EXERCISES

1. True or false: Any patterned sequence is either arithmetic or geometric.

2. A geometric sequence has a common ______.

Find the common ratio for each sequence.

3. 8, 32, 128, 512, . . .

4. 6, −60, 600, −6000, . . .

5.

n	1	2	3	4
a_n	125	25	5	1

6.

n	1	2	3	4
a_n	9	6	4	$\frac{8}{3}$

Determine whether each sequence is arithmetic, geometric, or neither. Explain.

7. 4, −20, 100, −500, . . .

8. −1, 2, 6, 11, . . .

9.

n	1	2	3	4
a_n	48	24	8	2

10.

n	1	2	3	4
a_n	−70	−56	−42	−28

11.

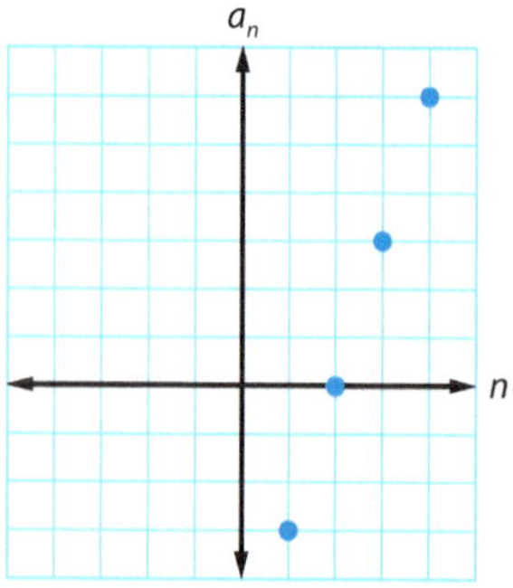

12.

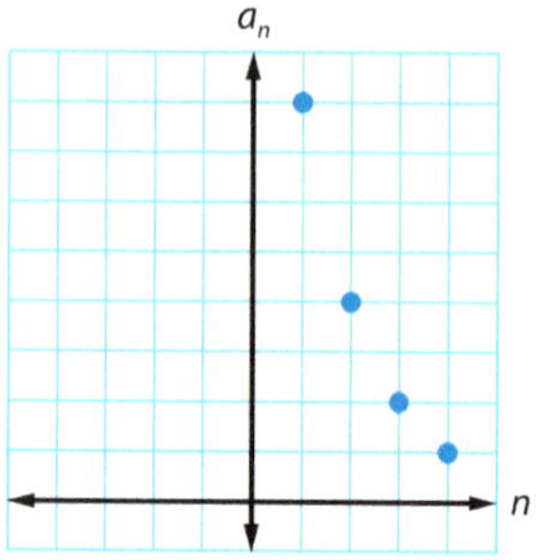

Generate the next 4 terms of each sequence using the given formulas.

13. $a_1 = 10; a_n = 5a_{n-1}$

14. $a_1 = \frac{1}{4}; a_n = -2a_{n-1}$

15. $a_1 = 0.8; a_n = 0.8 \cdot 10^{n-1}$

16. $a_1 = 6; a_n = 6 \cdot 3^{n-1}$

B. EXERCISES

Graph the first 4 terms of each geometric sequence described.

17. first term: 20; common ratio: $\frac{1}{2}$

18. $a_1 = \frac{1}{8}; r = 4$

19. $a_1 = 4; r = -1$

20. $a_n = -16\left(\frac{1}{4}\right)^{n-1}$

Write the specified formula for each sequence.

21. 3, −6, 12, −24, 48, . . . ; recursive

22. −7, −21, −63, −189, −567, . . . ; recursive

23. $2, 1, \frac{1}{2}, \frac{1}{4}, \frac{1}{8}, \ldots$; explicit

24. $\frac{1}{25}, -\frac{1}{5}, 1, -5, 25, \ldots$; explicit

Write the explicit formula and find the specified term for each geometric sequence.

25. −15, −30, −60, −120, . . . ; a_{14}

26. $\frac{1}{2401}, \frac{1}{343}, \frac{1}{49}, \frac{1}{7}, \ldots; a_9$

27. 500, −300, 180, −108, . . . ; a_7

28. 15.625, 6.25, 2.5, 1, . . . ; a_{10}

29. 256, 384, 576, 864, . . . ; a_9

30. 0.19, −0.57, 1.71, −5.13, . . . ; a_{13}

Describe and correct the error made in each statement.

31. Penny determined that the common ratio of the sequence −3, −9, −27, −81, . . . is −3.

32. Ali used the explicit formula, $a_8 = -2(4)^8 = -131{,}072$, to find the 8th term in a sequence with a common ratio of 4 and initial value of −2.

Round each answer to the nearest hundredth.

33. A basketball dropped from an initial height of 10 ft reaches 60% of its previous height.

- **a.** List the height the ball reaches after each of the first 3 bounces.
- **b.** Write a recursive formula modeling the ball's maximum height after the nth bounce.
- **c.** After which bounce will the ball achieve a maximum height of 0.7776 ft?

34. A pendulum swings a length of 400 cm on its first swing, 340 cm on its second swing, and 289 cm on its third swing. Write an explicit formula of the geometric sequence modeling the length of each pendulum swing and find the distance the pendulum travels on the tenth swing.

35. In 2020 Solar Systems developed a microprocessor featuring 2.3×10^{10} transistors. Each year, they want to increase the number of transistors according to the model $f(n) = (2.3 \times 10^{10}) \cdot 1.41^{n-1}$.

- **a.** Explain the meaning of 2.3×10^{10} and 1.41 in the context of the problem.
- **b.** How many transistors does the company expect to pack into each microprocessor in 2028?

36. **MP4** Lyle is reducing the size of a photograph. Reducing at a preset setting, the length of the photograph is 8 in. after the first reduction, 7.6 in. after the second reduction, and 7.22 in. after the third reduction.

a. By what percentage is the length reduced each time?

b. Write an explicit formula modeling the length of the photograph after the nth reduction. Find the length after 12 reductions.

37. **Essential Question:** How are geometric sequences useful in the real world?

C. EXERCISES

38. For what value(s) of r will the graph of a geometric sequence lie upon

a. the graph of an exponential growth curve?

b. the graph of an exponential decay curve?

c. neither an exponential growth nor decay curve?

39. **MP3** Describe and correct the error Jacob made below in writing the recursive formula for the sequence 2, 4, 8, 32, 256, . . .
"$\frac{8}{4} = 2$ and $\frac{4}{2} = 2$, so the sequence is geometric. $r = 2$ and $a_n = 2a_{n-1}$, $a_1 = 2$."

40. **MP7** The 18th term of a geometric sequence is 1,033,121,304. The 20th term is 2,324,522,934. What is the value of the 1st term?

41. **MP3** Show that $\frac{(a_{n-1})(a_{n-2})}{a_{n-3}}$ is equivalent to the explicit formula for a geometric sequence.

CUMULATIVE REVIEW

Write the equation of each line in slope-intercept form. [5.4]

42. a line with a y-intercept of $(0, -3)$ and a slope of $\frac{1}{2}$

43. a line passing through $(-5, -2)$ and $(-2, 4)$

Write and solve a system of equations for each word problem. [6.5–6.6]

44. On Saturday Tami pedaled to the state park at 21 mi/hr. After lunch she rode back at 14 mi/hr. If her total riding time was 5 hr, how far was it to the park?

45. A speedboat goes 15 mi down the river in 30 min and takes 45 min to return. Determine the rate of the boat in still water and the rate of the current.

46. How many milliliters of distilled water and concentrated (18%) acetic acid should be mixed to make 1 L of diluted (4%) acetic acid? Round to the nearest milliliter.

47. Mrs. Fields wants to split her \$10,000 investment between a mutual fund that has averaged a 6% annual return and a more conservative bond fund that has averaged a 2% annual return. How much should she invest in each fund if she wants to earn \$500 in interest during the next year?

Simplify each expression. [7.1]

48. $(12x^{-4}y^3)^2\left(-\frac{1}{4}x^3y^{-1}\right)^3$

49. $\left(\frac{3x}{y^4}\right)^2\left(\frac{y}{x^2}\right)^{-3}$

50. Which of the expressions below cannot be simplified using the Distributive Property? **[1.6]**

A $-4(x + 5) - y + 3x^2$

B $y^3 + 5x + 2x^2 - 7x - y^3$

C $a - (2b - 3c) + \frac{8}{5}a$

D $\frac{1}{2}(xy) - 4x^2 + y^2 + 5y^3$

E none of these

51. Which expression equals $m - n$? **[1.6]**

A $-3n + 2m + 4n - m$

B $7n - 11m - 8n - (-10m)$

C $5m - \frac{8m + 2n}{2}$

D $n - (2n + m)$

E all of these

APPLICATION PROBLEMS
THE INTERNET

The internet is an extremely powerful tool, providing valuable information in countless areas. A simple search can provide billions of results in a fraction of a second. The power of the internet is in its vastness. This vastness provides both true and false information, both godly and ungodly content. God has given us dominion, resources, and liberties, not so that we would use them as opportunities to sin, but so that we would serve one another in love and serve God with fear (Gal. 5:13; Heb. 12:28).

How vast is the internet? The amount of information can be expressed in zettabytes or 10^{21} bytes, where each byte, made up of 8 bits (0s and 1s), may represent a single character. A kilobyte is 1024 or 2^{10} bytes. A megabyte is 1024 kilobytes; it is therefore 1024(1024) bytes or $2^{10}(2^{10}) = 2^{20}$ bytes. These names are derived from the base 10 approximations of the exact number of bytes. For example, the actual 2^{10} or 1024 bytes in 1 KB can be estimated as 1000 or 10^3 bytes.

1. The current projection for annual global Internet Protocol (IP) traffic in 2022 is approaching 5 zettabytes. To gain more appreciation for the size of a zettabyte, follow the patterns to complete the table below.

	Exact Number of Bytes		Approximate Number of Bytes	
	Power of 1024	Power of 2	In Words	Power of 10
kilobyte (KB)	1024^1	2^{10}	1 thousand	10^3
megabyte (MB)	1024^2	2^{20}	1 million	10^6
gigabyte (GB)	1024^3			
terabyte (TB)				
petabyte (PB)				
exabyte (EB)				
zettabyte (ZB)				

Bandwidth is the amount of data that can be transferred from one point to another in a given amount of time. These rates of transfer are usually expressed in bits per second. Older dial-up modems typically transferred data at a rate of 56 Kbps. Broadband technology transmits data simultaneously over multiple frequencies (a "broad band"). A T3 network connection (also known as DS3) can transfer data at a rate up to 45 Mbps, equivalent to using more than 600 ordinary telephone lines. Fiber optic connections such as OC-3 have a transfer rate of 155 Mbps, and newer technologies allow data transfer rates up to 2000 Mbps (equivalent to more than 26,000 telephone lines). What used to take several minutes to download can now be accomplished in a fraction of a second.

A text file containing the entire King James Version of the Bible (with a brief introduction but no images or extra commentary) is about 2000 pages long and contains 4,720,600 characters, including chapter and verse numbers. This file is 4.8 MB (about 38.4 megabits) in size.

2. Assuming 1 megabit = 1000 kilobits, how long would it take (to the nearest tenth of a second) to download the entire 38.4 megabit KJV Bible file with an older 56 Kbps (kilobits per second) modem connection?

3. How long would it take (to the nearest hundredth of a second) to transfer the entire 38.4 megabit KJV Bible file across a 45 Mbps T3 connection?

4. How long would it take (to the nearest hundredth of a second) to transfer the entire 38.4 megabit KJV Bible file across a 155 Mbps OC-3 connection?

5. How long would it take (to the nearest hundredth of a second) to transfer the entire 38.4 megabit KJV Bible file across a 2000 Mbps connection?

Actual download times are almost always longer because the stated maximum theoretical rates are seldom achieved, even under ideal conditions.

Bandwidth is not the only thing growing; data storage on the internet is growing by zettabytes per year. Massive amounts of text, image, and video files are added to cyberspace each day. It has been estimated that the amount of digital data on the internet will grow to more than 175 zettabytes by the year 2025.

Use the conventional approximations for data measurements found in exercise 1 to complete the following exercises.

6. Express the number of bytes in 1 zettabyte in standard notation.

7. Express the number of bytes in 1 zettabyte in scientific notation.

8. Express the number of bytes in 175 zettabytes in scientific notation.

9. The forecasted population of the world in 2025 is estimated to be 8,184,000,000.
   a. Using scientific notation, calculate the approximate number of bytes of digital data on the internet per person in the world in the year 2025.
   b. Determine the number of smartphones with 64 GB of storage needed to store the amount of data from part a.

In 2020 about 90% of the 3.3×10^8 people in the United States used the internet, compared to about 59% of the world's population of 7.8×10^9. At that time, approximately 3.1×10^{11} emails were sent each day, with more than 47% of those emails being classified as spam or unwanted.

Express each answer in both scientific and standard notation.

10. How many internet users were there in the United States in 2020?

11. How many internet users were there in the world in 2020?

12. What was the average number of daily emails per internet user in the world (to the nearest whole number)?

13. A micro-blogging service sent 540 million micro-messages in just 5 days. What was the average number of messages sent per second through this service?

14. Explain one way believers can use this powerful tool to share their faith in Christ.

7 CHAPTER REVIEW

Key Concepts

Properties of Exponents [7.1]

PROPERTIES OF EXPONENTS			
Product Property	$x^a \cdot x^b = x^{a+b}$	$x^5 \cdot x^2 = x^{5+2} = x^7$	To multiply with like bases, add the exponents.
Power Property	$(x^a)^b = x^{ab}$	$(x^5)^2 = x^{5(2)} = x^{10}$	To raise a power to a power, multiply the exponents.
Quotient Property	$\frac{x^a}{x^b} = x^{a-b}$ for $x \neq 0$	$\frac{x^5}{x^2} = x^{5-2} = x^3$	To divide with like bases, subtract the exponents.
Power of a Product Property	$(abc)^n = a^n b^n c^n$	$(2a^2c^4)^3 = 2^3(a^2)^3(c^4)^3 = 8a^6c^{12}$	To find the power of a product, raise each factor to the power.
Power of a Quotient Property	$\left(\frac{a}{b}\right)^n = \frac{a^n}{b^n}$	$\left(\frac{2}{b^3}\right)^4 = \frac{2^4}{(b^3)^4} = \frac{16}{b^{12}}$	To find the power of a quotient, raise each part of the quotient to the power.

Scientific Notation [7.2]

- A positive number is in scientific notation when written in the form $N \times 10^P$ where $1 \leq N < 10$.
- Converting from scientific notation to standard form involves moving the decimal place in N.
  - If $P \geq 1$, move the decimal point P places right.
  - If $P \leq -1$, move the decimal point $|P|$ places left.
- Add the exponents when multiplying and subtract the exponents when dividing.
  - $(M \times 10^x) \times (N \times 10^y) = MN \times 10^{x+y}$
  - $(M \times 10^x) \div (N \times 10^y) = \frac{M}{N} \times 10^{x-y}$
- Convert to the same power of 10 when adding or subtracting.
  - $(M \times 10^x) + (N \times 10^x) = (M + N) \times 10^x$
  - $(M \times 10^x) - (N \times 10^x) = (M - N) \times 10^x$
- Apply the Power of a Product Property when scientific notation is raised to a power.
  - $(M \times 10^x)^y = M^y \times 10^{xy}$

Power Functions [7.3]

- Graph functions of the form $y = (x - h)^2 + k$ or $y = (x - h)^3 + k$ by making adjustments to their parent graphs: $y = x^2$ or $y = x^3$.

POWER FUNCTIONS ($y = x^2$)

$f(x) = x^2$
vertex: (0, 0)

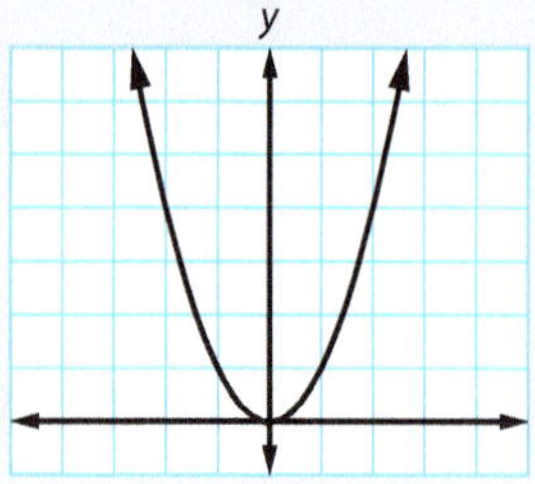

$f(x) = x^2 + k$
vertex: (0, k)

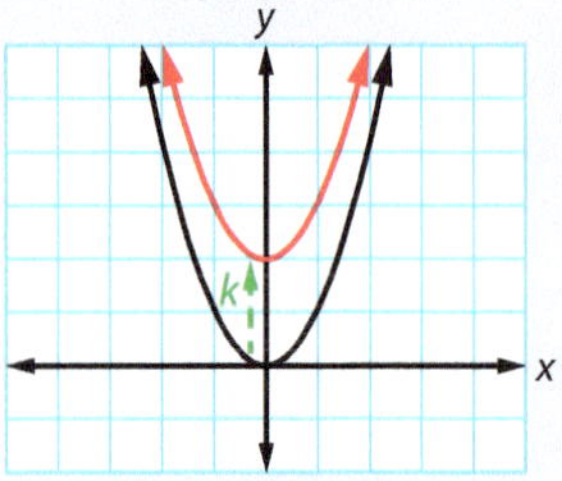

$f(x) = x^2 - k$
vertex: (0, $-k$)

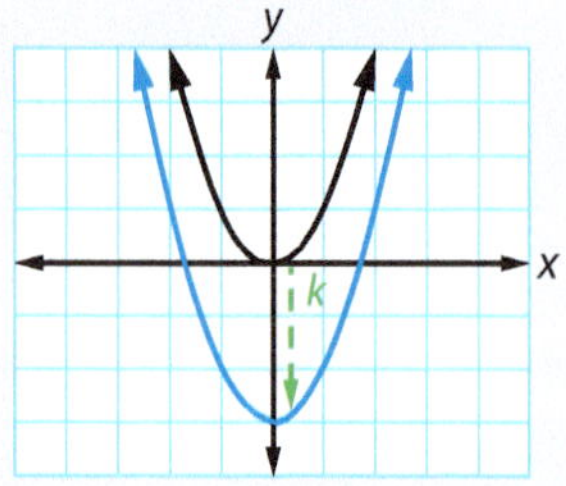

$f(x) = (x + h)^2$
$= (x - (-h))^2$
vertex: ($-h$, 0)

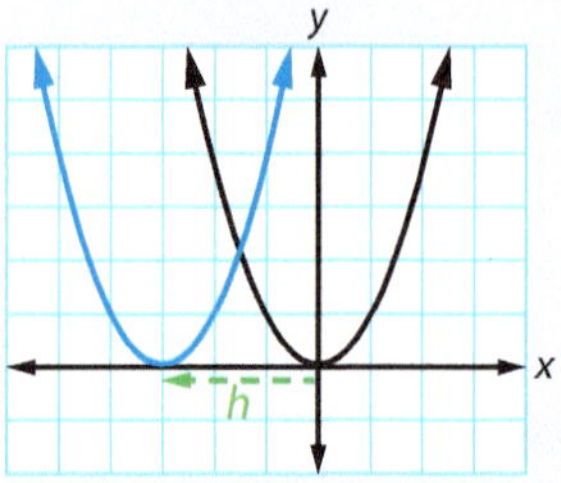

$f(x) = (x - h)^2$
vertex: (h, 0)

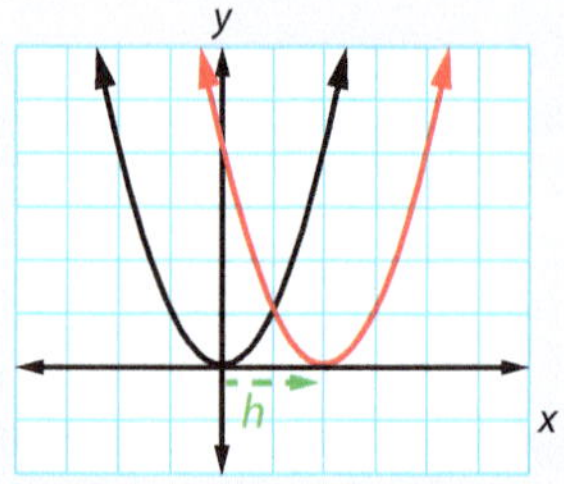

$f(x) = (x - h)^2 + k$
vertex: (h, k)

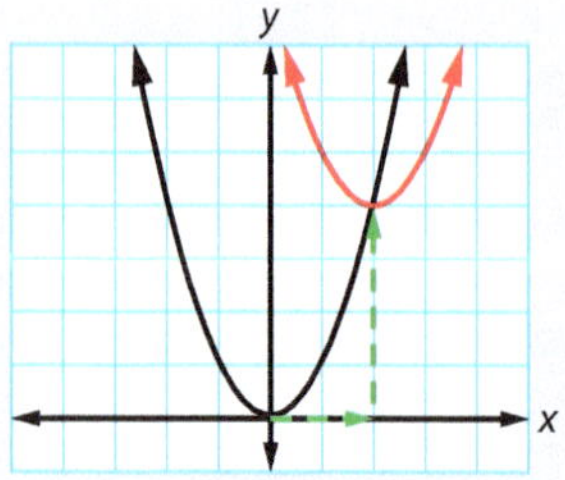

POWER FUNCTIONS ($y = x^3$)

$f(x) = x^3$
point of inflection: (0, 0)

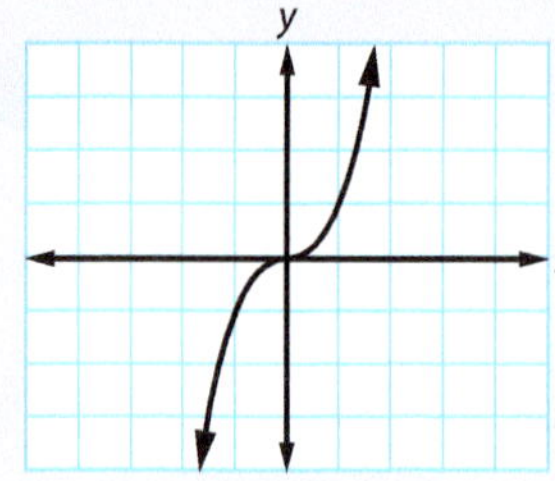

$f(x) = x^3 + k$
point of inflection: (0, k)

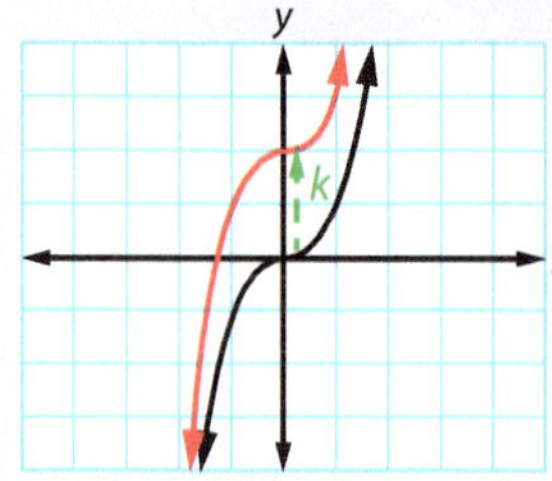

$f(x) = x^3 - k$
point of inflection: (0, $-k$)

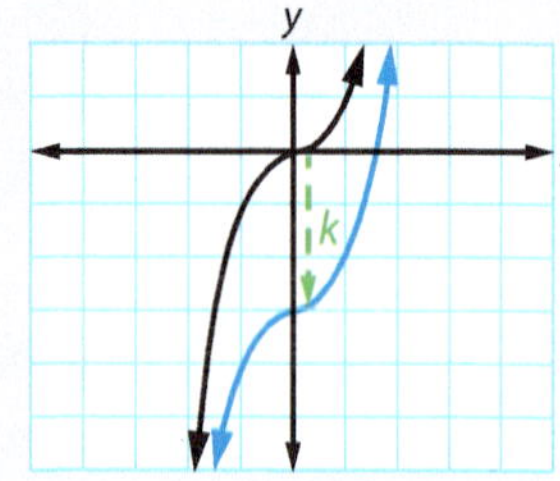

$f(x) = (x + h)^3$
$= (x - (-h))^3$
point of inflection: ($-h$, 0)

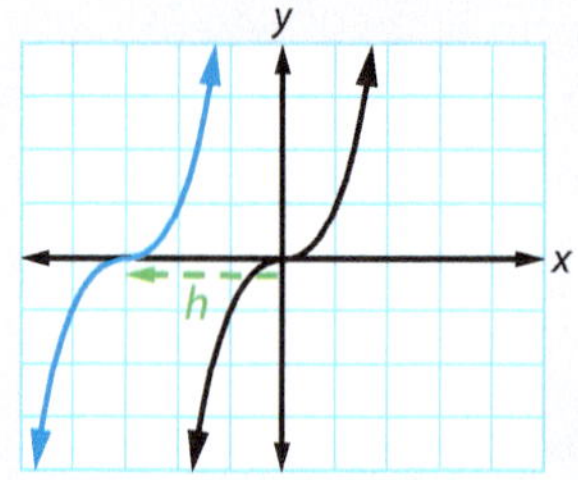

$f(x) = (x - h)^3$
point of inflection: (h, 0)

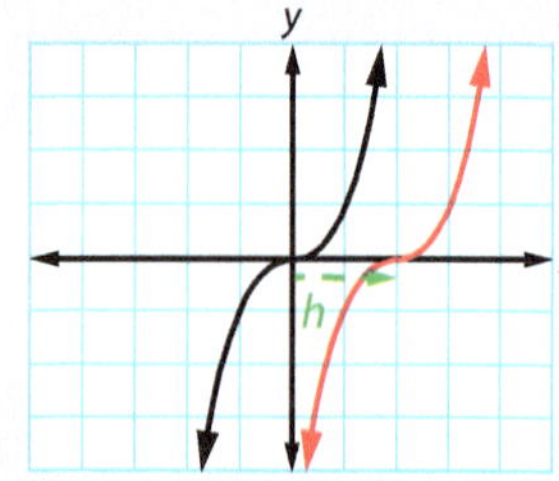

$f(x) = (x - h)^3 + k$
point of inflection: (h, k)

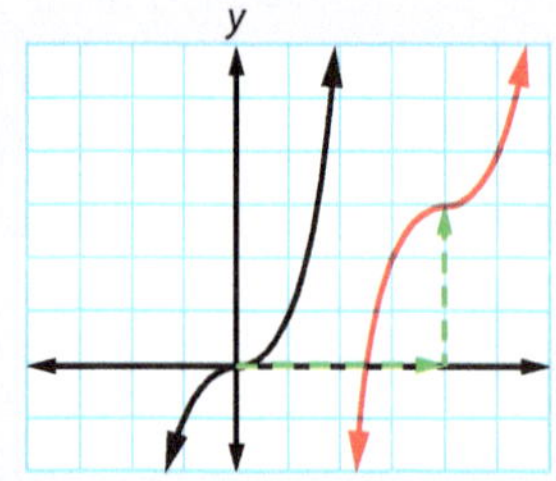

Exponential Functions [7.4]

- An exponential function is a function of the form $f(x) = ab^x$, where $b > 0$ and $b \neq 1$.
- To graph an exponential function, plot a sufficient number of points (typically 7) and connect with a curved line.

Exponential Growth & Decay [7.5]

	EXPONENTIAL FUNCTIONS	
	Exponential Growth	Exponential Decay
function	$f(t) = A(1 + r)^t$	$f(t) = A(1 - r)^t$
A	initial amount	
r	growth rate	decay rate
$(1 \pm r)$	growth factor $(1 + r)$	decay factor $(1 - r)$
t	time	

- A common application for exponential growth is the compound interest formula:

COMPOUND INTEREST FORMULA	
$A = P(1 + r)^x$, where	A = the account balance after t years, P = the principal amount, x = the number of periods, and r = the periodic interest rate.
If there are n compounding periods in each of t years, then	$r = \frac{APR}{n}$ and $x = nt$.

- APR refers to annual percentage rate.

Geometric Sequences [7.6]

- A geometric sequence is a sequence with a common ratio between any 2 consecutive terms.
- The nth term of the sequence is denoted a_n (a_1 is the first term, a_2 is the second, and so on).

GEOMETRIC SEQUENCE FORMULAS	
Recursive	Explicit
$a_n = ra_{n-1}$, where a_1 is given	$a_n = a_1r^{n-1}$

Biblical Worldview Shaping

Algebra and Ethics

What does morality have to do with mathematics?

Mathematics is not amoral; it is a good and worthwhile endeavor that has been corrupted by the Fall. To view mathematics as amoral is to make it an idol that can judge between right and wrong. Using math in unethical ways violates God's command to love our neighbor.

EXERCISES

Simplify. [7.1]

1. $(6m^2n^5)(-3mn^7)$
2. $(3g^2h)^3$
3. $\frac{8m^4n^2p^5}{2m^2np}$
4. $\frac{y^2}{x^{-3}y^2}$
5. $\frac{(y^{2a})^3}{y^{4a}}$
6. $2^5 \cdot 2^{x-5}$
7. Simplify, leaving all variables in the numerator. **[7.1]**
   a. $\frac{-3m^{-3}r^7s^{-4}}{mr^2s^2}$
   b. $\frac{-3x^2y^{-3}}{x^{-1}y}\left(-\frac{y}{2x^3}\right)^2$
8. Simplify, leaving each answer in positive exponential form. **[7.1]**
   a. $\frac{a^{-2}b^2}{a^{-1}b^{-2}}$
   b. $5x^{-3}y^2z^4(-3x^{-3}yz^{-5})$
   c. $\frac{x^{-5}y^3}{z^{-2}}\left(\frac{xz^3}{y}\right)^2$
   d. $(-2a^3b^{-5})^{-2}$
9. Express each number in scientific notation. **[7.2]**
   a. 0.000319
   b. 0.00032×10^8

Simplify, expressing each answer in scientific notation. [7.2]

10. $(9 \times 10^8) \div (3 \times 10^3)$
11. $(6 \times 10^{-2})(5 \times 10^3)$
12. $(5 \times 10^4)^3$
13. $(9 \times 10^3) - (7 \times 10^2)$
14. $(4 \times 10^{-3}) + (5 \times 10^{-2})$

Express each answer in scientific notation. [7.2]

15. An angstrom, the unit used to express the size of atoms, is one ten-millionth of a millimeter. Write this in meters using scientific notation.
16. Light travels at 300 million meters per second. How many kilometers does it travel in 1 hr?

Describe the translation(s) of the graph of $y = x^2$ or $y = x^3$ that produce the graph of each function. Then create a table of at least 4 more ordered pairs and graph each function. [7.3]

17. $y = x^3 - 2$
18. $y = (x - 3)^2 + 2$

State the coordinates of the vertex or point of inflection for each function. [7.3]

19. $f(x) = (x - 1)^2$
20. $f(x) = x^3 + 7$
21. $f(x) = (x + 3)^2 + 5$
22. $f(x) = (x - 2)^3 - 4$

Write a function rule describing each graph or translated function. [7.3]

23. $y = x^2$; 5 units left and 3 units up
24. $y = x^3$; 3 units right and 2 units down
25.
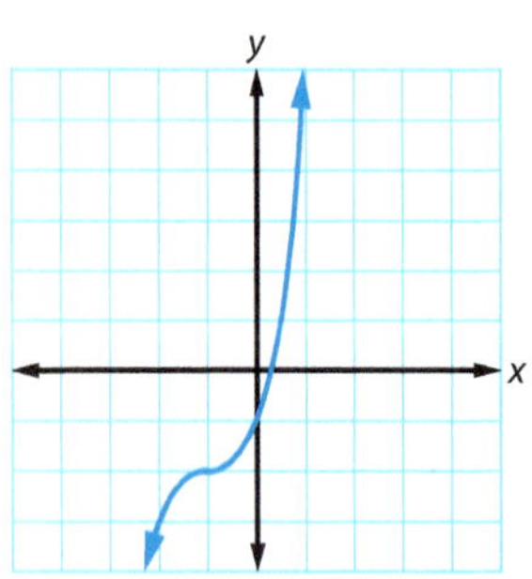

26.
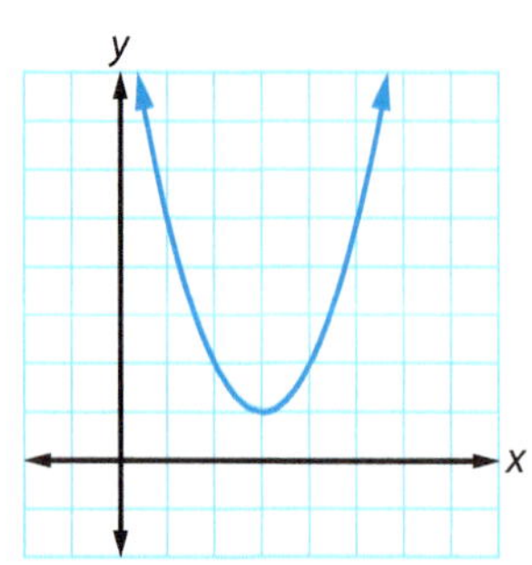

27. Label each function as a power function or an exponential function. **[7.4]**
    a. $y = -2 \cdot 3^x$
    b. $y = -2 \cdot x^3$

Create a table of at least 5 ordered pairs and graph each function. [7.4]

28. $y = 3^x$
29. $y = 2(1.5)^x$
30. $y = \left(\frac{3}{5}\right)^x$
31. $y = 4\left(\frac{1}{2}\right)^x$
32. What is the y-intercept of the graph of functions of the form $y = b^x$ where $b > 0$? **[7.4]**
33. Under what conditions is the graph of the exponential function $y = ab^x$ decreasing? **[7.4]**

Write an exponential function modeling consecutive enlargements or reductions by the given percentage. Then find the single setting that would accomplish the same result as the given number of consecutive enlargements or reductions. [7.4]

34. 80%; 3

35. 120%; 2

36. Write an exponential function modeling the growth in Mr. Leed's salary if he projects his base pay of $50,000 will increase by 5% each year. What will his salary be 10 yr from now (to the nearest dollar)? **[7.4]**

For each exponential function, identify the initial amount, the growth rate (as a percent), and the growth factor. Then state whether the function models exponential growth or exponential decay. [7.5]

37. $y = 1.5^x$

38. $f(t) = 10{,}000\left(1 - \frac{2}{3}\right)^t$

$5000 is invested at 3% APR for 10 yr. Determine the number of compounding periods and the rate per compounding period when interest is compounded as follows. Then determine the ending balance. [7.5]

39. quarterly

40. monthly

41. An initial population of 500 mosquitoes can triple each day. **[7.5]**

- **a.** State the growth rate (as a percent) and the growth factor for the mosquito population.
- **b.** Write an exponential function, $f(d)$, modeling the number of mosquitoes.
- **c.** Determine the number of mosquitoes after 1 week.

42. It takes about 1 week for half of the uranium-237 atoms present in a sample to undergo beta decay. A laboratory determines that a sample contains 16 g of U-237. **[7.5]**

- **a.** Write an exponential function, $f(w)$, modeling the amount of U-237 in the sample.
- **b.** Approximately how much U-237 is left in the sample after 5 weeks?
- **c.** Graph the function, showing the amount of U-237 in the sample during the first 5 weeks.

43. Explain the difference between an arithmetic sequence and a geometric sequence. **[7.6]**

Represent each geometric sequence using the method described. [7.6]

44. 48, 24, 12; graph the first 6 terms

45. $a_n = -2a_{n-1}$; $a_1 = 5$; list the first 6 terms

46. $a_n = 16\left(\frac{3}{2}\right)^{n-1}$; list the first 6 ordered pairs

Write the recursive formula and the explicit formula for the *n*th term of the following sequences and find the 10th term. [7.6]

47. −2, −6, −18

48. 48, 12, 3

49. A ball dropped from a height of 60 ft bounces to 60% of its previous height.

- **a.** List the maximum height of the ball after each of the first 3 bounces.
- **b.** Write an explicit formula for the *n*th term of the sequence.
- **c.** Find the height of the ball after the 10th bounce.

50. **BWS Essential Question:** What is the biblical standard for the ethical use of mathematics?

CHAPTER 08

POLYNOMIAL OPERATIONS

CHAPTER OVERVIEW

8.1 CLASSIFYING & EVALUATING POLYNOMIALS

How are polynomials classified?

After completing this section, I will be able to

- define a polynomial.
- classify a polynomial by its number of terms and its degree.
- evaluate a polynomial using given values for the variables.

The energy of this wave at Waimea Bay, Hawaii, is modeled by the monomial $\frac{1}{8}\rho gH^2$, where ρ is the water density, g is the acceleration due to gravity, and H is the height of the wave.

A *polynomial* is a special type of algebraic expression that can be written as a sum of terms without any variable divisors. Polynomials have useful applications in many fields, such as physics, chemistry, engineering, business, and finance, and can also be used to model God's complex creation.

RECALL

A term is a constant, a variable, or the product of a constant and 1 or more variables.

DEFINITION

A **polynomial** is an expression that can be written as a sum of terms consisting of constants, variables with nonnegative integral exponents, or products of a constant and 1 or more such variables.

Polynomials can be classified by their number of terms.

DEFINITIONS

A **monomial** is a polynomial with exactly 1 term.
A **binomial** is a polynomial with exactly 2 terms.
A **trinomial** is a polynomial with exactly 3 terms.

EXAMPLES OF POLYNOMIALS				NOT POLYNOMIALS
Monomials	Binomials	Trinomials	Other Polynomials	
$4y^3$	$a - 14$	$2a^2b^5 + 8c^4 - 2a^5$	$x^4 + \frac{1}{2}x^3 - x^2 + 9$	x^3y^{-2}
$8ab^2c$	$2x^2 + 3x$	$5x^2 + 2x - 6$	$x^3y + 5xy^2 + 3xy - 7y^5$	3^x
$\frac{1}{2}x$	$5x^2 + y^3$	$3y - 10z + \frac{1}{\pi}$	$a^4b - a^3b^2 + a^2b^3 - ab^4$	$\sqrt{x}$

Polynomials can also be classified by their degree.

DEFINITION

The **degree of a term** is the sum of the exponents of its variables.

EXAMPLE 1: Finding the Degree of a Term

State the degree of each term.

a. $4x^2y$ **b.** $3x$ **c.** $5x^3y^2z^4$ **d.** 7

Answers

Find the sum of the variable exponents.

a.	$4x^2y$	degree: $2 + 1 = 3$
b.	$3x$	degree: 1
c.	$5x^3y^2z^4$	degree: $3 + 2 + 4 = 9$
d.	7	degree: 0

SKILL CHECK **EXERCISE 1**

The **degree of a polynomial** is the highest degree of any term in the polynomial.

AFTERSCHOOLHELP

Classifying & Evaluating Polynomials

EXAMPLE 2: Finding the Degree of a Polynomial

State the degree of the polynomial $3x^3y + 2x - 4xy^2$.

Answer

$3x^3y + 2x - 4xy^2$

first term	$3x^3y^1$	$3 + 1 = 4$
second term	$2x^1$	1
third term	$-4x^1y^2$	$1 + 2 = 3$

The degree of the polynomial is 4.

1. Find the degree of each term.

2. The degree of the polynomial is equal to the highest degree of any of its individual terms.

SKILL CHECK **EXERCISE 5**

It is customary to write polynomials in a **standard form**, where the terms are arranged in order of descending degrees of one of the variables. The polynomial

$$4x^2 + 3 + x + 6x^3$$

should be written with terms arranged in descending degrees of x:

$$6x^3 + 4x^2 + x + 3.$$

This allows polynomials with 1 variable to be quickly classified according to their degree.

POLYNOMIAL	CLASSIFICATION		
	Degree	# of Terms	Name
$6x^3 + 4x^2 + x + 3$	3	4	polynomial
$5x^2 - 2x - 3$	2	3	trinomial
$-2x^4$	4	1	monomial
$8x + 1$	1	2	binomial
-8	0	1	monomial

Polynomials are evaluated by substituting the given values and following the order of operations.

EXAMPLE 3: Evaluating a Polynomial

Evaluate $2x^2 - 5x + 6$ when $x = 3$.

Answer

$2(3)^2 - 5(3) + 6$
$= 2(9) - 5(3) + 6$
$= 18 - 15 + 6 = 9$

1. Substitute 3 for x in the trinomial.
2. Simplify, following the order of operations.

SKILL CHECK **EXERCISE 21**

VOCABULARY

- binomial
- degree of a polynomial
- degree of a term
- monomial
- polynomial
- standard form
- trinomial

EXAMPLE 4: Evaluating a Polynomial in 2 Variables

Evaluate $5x^3y + 2x - 3y + 5$ when $x = -2$ and $y = 1$.

Answer

$5(-2)^3(1) + 2(-2) - 3(1) + 5$
$= 5(-8)(1) + 2(-2) - 3(1) + 5$
$= -40 - 4 - 3 + 5 = -42$

1. Substitute -2 for x and 1 for y.
2. Evaluate.

SKILL CHECK **EXERCISE 29**

A. EXERCISES

State the degree of each term.

1. $4x^5y^2$
2. $-5x$
3. 1492
4. $5x^3y$

Classify each expression as a monomial, binomial, trinomial, other polynomial, or not a polynomial. Then state the degree of each polynomial.

5. $4x^3y^5 + 2xy$
6. $-5xy^3z^2$
7. $-4x^2y + 5xy + y - 8$
8. $3x^3y + 2x^2 - 5$
9. $\frac{4y^3}{x^2} + y^2 - 1$
10. $3x + 4y - 8z + 9$
11. $8x^3 - 2xyz$
12. $4a^{-3} - 1776$
13. $\frac{x^2y^3}{5} + \frac{xy}{2} - \frac{3}{8}$
14. 23

B. EXERCISES

Write each polynomial in standard form. Then classify it by its degree and its number of terms.

15. $3x + 5 - 2x^2$
16. $9 + 4x$
17. $\frac{1}{3}x^2 - 6x^5 + 12 - 4x$
18. $-12 + 6y + y^3$
19. $2xy - 7x^2y + 6x^4$
20. $-7x^2y^2 + x^4 + 5xy + 3x^3y + y^4$

Evaluate each polynomial for the given value.

21. $x^2 + x + 1$ when $x = 11$

22. $2z^5 + 5z^2$ when $z = 2$

23. $5m - 3$ when $m = 0$

24. $-x^2 - 4x - 7$ when $x = -1$

25. $y^5 - 2y^3 - 7$ when $y = -2$

26. $-2x^4 - x^3 + x^2 - x + 1$ when $x = -4$

Evaluate when $x = 3$, $y = -1$, and $z = 2$.

27. $x^2 + 4z - 8$

28. $5x^2 - 3y^2 - 4z$

29. $2x^3y + x^2y - 5xyz$

30. $5x^3y^2 + 8x^2y$

31. $5x^4y^3z^3$

32. $-3z^3 + 5z^2 - 4z + 7$

33. Essential Question: How are polynomials classified?

C. EXERCISES

Evaluate each polynomial for the given values.

34. $x^2y^3 - y^2 + x^2y$ when $x = \frac{2}{3}$ and $y = \frac{1}{2}$

35. $-a^2b + 2ab + 1$ when $a = 0.5$ and $b = 0.2$

36. $\frac{13}{12}a^3b + a^5b^2$ when $a = 2\pi$ and $b = \frac{3}{\pi^2}$; leave your answer in terms of π.

37. **MP8** A one-variable polynomial of degree n can have at most ________ unlike terms.

38. **MP3** If a polynomial of degree 3 has 2 variables, what is the greatest number of unlike terms the polynomial can have?

CUMULATIVE REVIEW

Match each term to its definition.

39. consistent system **[6.4]**

40. coefficient **[1.6]**

41. conjunction **[3.3]**

42. disjunction **[3.4]**

43. domain **[4.1]**

A the multiplier of the variable in a term

B the set of x-coordinates in a relation

C a system of equations that has at least 1 solution

D a compound statement using "and"

E a compound statement using "or"

Use the exponential function $y = 1.03^x$ for exercises 44–46. [7.5]

44. Does the function model exponential growth or exponential decay?

45. State the growth rate.

46. State the growth factor.

47. Which equation can be used to find 3 consecutive odd integers whose sum is 39? **[2.2]**

A $x + (x + 1) + (x + 3) = 39$

B $x + (x + 2) + (x + 4) = 39$

C $3(2x + 1) = 39$

D A and C

E B and C

48. Solve the equation $3(x - 1) - 14x = 2 - x$. **[2.2]**

A $x = -\frac{1}{2}$

B $x = -\frac{1}{10}$

C $x = \frac{5}{12}$

D $x = -2$

E none of these

8.2 ADDING & SUBTRACTING POLYNOMIALS

How can I add and subtract polynomials?

After completing this section, I will be able to

- add polynomials.
- subtract polynomials.

When adding polynomials, you are simply combining like terms. It may be helpful to align like terms for vertical computation.

A mechanic can calculate the horsepower of an n-cylinder engine with b-inch diameter bores from the formula $hp = \frac{nb^2}{2.5}$.

EXAMPLE 1: Adding Polynomials Vertically

Find each sum.

a. $(3a + 2b - 8c + 8) + (2a - 4b + 6c + 9)$ **b.** $(2x^2 + 3xy + 5y^2) + (8x^2 - 5xy + 3y) + (x^2 + 2y^2)$

Answers

a.
$$\begin{array}{r} 3a + 2b - 8c + 8 \\ \underline{+\,2a - 4b + 6c + 9} \\ 5a - 2b - 2c + 17 \end{array}$$

1. Arrange the polynomials so like terms are aligned vertically.
2. Add the columns.

b.
$$\begin{array}{l} 2x^2 + 3xy + 5y^2 \\ 8x^2 - 5xy \qquad + 3y \\ \underline{+\,x^2 \qquad + 2y^2 \qquad} \\ 11x^2 - 2xy + 7y^2 + 3y \end{array}$$

1. Arrange the polynomials so like terms are aligned vertically.
2. Add the columns.

SKILL CHECK **EXERCISE 3** ☑

KEYWORD SEARCH

add subtract polynomials

Polynomials can be added horizontally by removing parentheses before combining like terms.

AFTERSCHOOLHELP

Adding & Subtracting Polynomials

EXAMPLE 2: Adding Polynomials Horizontally

Find each sum.

a. $(3x^2 + 2x - 5) + (8x^2 - 7x - 10)$ **b.** $(15a^2 - 7ab - 12b + 3c^2) + (8ab - 3b + 12c^2)$

Answers

a. $3x^2 + 2x - 5 + 8x^2 - 7x - 10$

$= 11x^2 - 5x - 15$

1. Remove the parentheses.
2. Combine like terms.

b. $15a^2 - 7ab - 12b + 3c^2 + 8ab - 3b + 12c^2$

$= 15a^2 + ab - 15b + 15c^2$

1. Remove the parentheses.
2. Combine like terms.

SKILL CHECK **EXERCISE 9** ☑

Recall that subtraction of real numbers is defined as addition of the opposite. Similarly, subtraction of polynomials can be performed by adding the opposite of each term.

EXAMPLE 3: Subtracting Polynomials Horizontally

Simplify $\left(3x^2+\frac{2}{3}x-5\right)-\left(4x^2-\frac{4}{3}x+2\right)$.

Answer

$$\left(3x^2+\frac{2}{3}x-5\right)+(-1)\left(4x^2-\frac{4}{3}x+2\right)$$
$$=3x^2+\frac{2}{3}x-5-4x^2+\frac{4}{3}x-2$$
$$=-x^2+2x-7$$

1. Subtract by adding the opposite of each term in the polynomial being subtracted.
2. Combine like terms.

SKILL CHECK **EXERCISE 13**

DOING THE MATH

$\frac{2}{3}+\frac{4}{3}=\frac{6}{3}=2$

Polynomials can also be subtracted vertically by adding the opposite. Be sure to write the opposite of each term and align like terms before adding.

EXAMPLE 4: Subtracting Polynomials Vertically

Find each difference.

a.
$$\begin{array}{r} \left(\frac{5}{4}a-7b+6c+10\right) \\ \underline{-\left(\frac{3}{4}a+4b-8c-\ \ 9\right)} \end{array}$$

b. $(4x^2+3.5xy-y^2)-(4y^2-8.2xy+5)$

Answers

a.
$$\begin{array}{r} \frac{5}{4}a-\ \ 7b+\ \ 6c+10 \\ \underline{+\left(-\frac{3}{4}a-\ \ 4b+\ \ 8c+\ \ 9\right)} \\ \frac{1}{2}a-11b+14c+19 \end{array}$$

1. Change the problem to an addition problem by writing the additive inverse of each term in the second polynomial.
2. Add the columns.

DOING THE MATH

$\frac{5}{4}-\frac{3}{4}=\frac{2}{4}=\frac{1}{2}$

b.
$$\begin{array}{l} (4x^2+\ \ 3.5xy-\ \ y^2) \\ \underline{\ \ -(\ \ -8.2xy+4y^2+5)} \\ 4x^2+\ \ 3.5xy-\ \ y^2 \\ \underline{\ \ +(\ \ +8.2xy-4y^2-5)} \\ 4x^2+11.7xy-5y^2-5 \end{array}$$

1. Align like terms.
2. Change the problem to an addition problem by writing the opposite of each term in the second polynomial.
3. Combine like terms.

SKILL CHECK **EXERCISE 7**

Notice the result of adding or subtracting a polynomial will always be a polynomial.

A. EXERCISES

Add.

1.
$$\begin{array}{r} 3m+\ \ 4 \\ 7m+\ \ 2 \\ \underline{+9m-12} \end{array}$$

2.
$$\begin{array}{r} 5k^2-3k+2 \\ \underline{+8k^2+9k-8} \end{array}$$

3.
$$\begin{array}{r} 4m^2+2m\qquad \\ 8m-9 \\ \underline{+3m^2\qquad\quad +4} \end{array}$$

4.
$$\begin{array}{r} 3ay+4y-\ \ 5z \\ 3y+10z \\ \underline{+7ay-5y\qquad} \end{array}$$

Subtract.

5. $$\begin{array}{r} (5a + 3) \\ \underline{-\,(2a - 5)} \end{array}$$

6. $$\begin{array}{r} (3a^2 + 2a - 4) \\ \underline{-\,(a^2 \qquad - 6)} \end{array}$$

7. $$\begin{array}{r} (x + 2y - 3) \\ \underline{-\,(4x - 8y + 3)} \end{array}$$

8. $$\begin{array}{r} (x^2 + 4xy - 10y^2) \\ \underline{-\,(4x^2 - 8xy + 3y^2)} \end{array}$$

Find each sum.

9. $(5x + y) + (4x + 2y)$
10. $(2x^2 - 5x + 3) + (2x^2 + 7x + 5)$
11. $(3x^3 + 4x - y) + (9x^2 - 5x + 7) + (2x^3 + 3x + 8)$
12. $(4a + 9b - 10d) + (5a - 3b + 6d) + (18a - 2c)$

Find each difference.

13. $(6x + y - 7) - (5x - y + 7)$
14. $(3x^2y - x + 5) - (3x^2y - x + 5)$
15. $(a^3 + 4b^2 - 3c) - (2a^3 - 12c)$
16. $(x^5 - x^4 + 8x - 7) - (x^4 + 2x - 2)$
17. $(y^2 + 3y + 8) - (9y^2 - 4y - 3)$
18. $(5x + 2y + 9) - (7x - 3y - 7)$

B. EXERCISES

Simplify each expression.

19. $(3x^2 + 5) - (12x^2 - 7x) + (3x - 2)$
20. $(4x - 2y + 8) + (3x + 9y - 4) - (2x - y + 4)$
21. $\left(\frac{1}{2}x + \frac{1}{5}y\right) + \left(\frac{3}{2}x + \frac{8}{7}\right) + \left(\frac{1}{3}x - \frac{4}{5}y\right)$
22. $(-4x^3 + 3x^2 - 5x + 10) + (-6x^2 + 3x + 9) + (5x^2 - 10)$
23. $(a^3b + a^2b^3 - 4ab) - (12a^2b^3 + 9)$
24. $(3x^2 - 10xy + 7y^2) - (2x^2 + 8xy + 8y^2) + (-9x^2 + 3y^2)$
25. $(12a^2 - 23ab - 6b^2) + (4a^2 - b^2) - (3a^2 + 2ab + b^2) - (a^2 + ab + b^2)$
26. Subtract $2x + 9$ from $3x - 7$.
27. Subtract $4x - 7y - 9$ from 38.
28. **MP3** Isaiah concluded the difference of $3v^4 - 4v^2 + v$ and $v^4 + v - 9$ was $2v^4 - 4v^2 + 2v - 9$. Describe and correct his error.

Write a polynomial expression for the perimeter of each figure.

29. 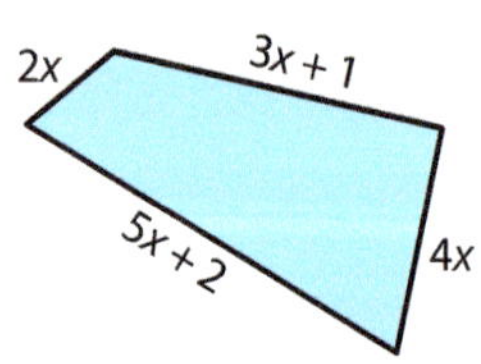

30. 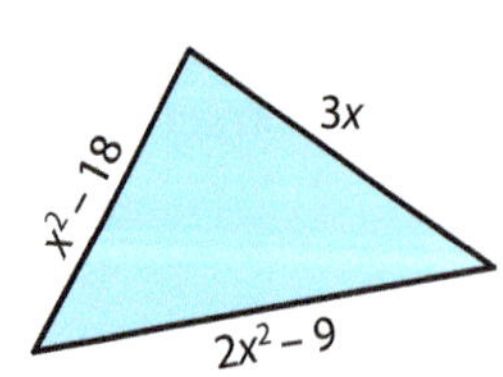

31. 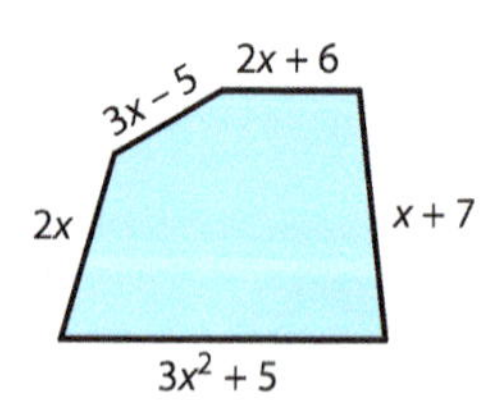

32. The length of a rectangle is $5n + 3$, and its width is $2n - 6$. Write a polynomial expression for its perimeter.
33. **MP4** The perimeter of a rectangle is $6x + 6$, and its width is $x + 2$. Write a polynomial expression for its length.
34. Owen is x years old, Suzette is 5 yr older than Owen, and Jane is 3 yr younger than Owen. Write a polynomial expression representing their combined ages.
35. **Essential Question:** How can I add and subtract polynomials?

C. EXERCISES

Find each missing expression.

36. $(-3x^2 + 2x - 5) - (x^3 - 2x^2 + 4x + 7) +$ ______ $= 2x^3 + 4x^2 - 8x - 10$

37. $\left(\frac{1}{2}x^2 + \frac{1}{3}xy + \frac{1}{4}y^2\right) - \left(\frac{1}{2}x^2 + \frac{3}{2}xy\right) +$ ______ $= \frac{1}{6}x^2 - \frac{7}{6}xy + \frac{3}{8}y^2$

38. Determine which has the greater perimeter: a square with a side equal to $3x + 6$, or a rectangle with a length of $4x + 8$ and a width of $2x + 3$ (where x is a positive number).

39. **MP2** During the championship game, Tony made n points; Neal scored 4 more points than Tony; Cedric had twice as many points as Neal; Rob scored 2 more points than Cedric; and Moe had half as many points as Rob. What was the combined score of these 5 players?

40. **MP3** Write a polynomial expression for the sum of any 2 consecutive numbers and use the expression to explain why this sum must always be odd.

CUMULATIVE REVIEW

Use an example to illustrate each statement. [1.1]

41. Subtraction is not commutative.

42. Subtraction is not associative.

State each property.

43. Addition Property of Equality **[2.1]**

44. Addition Property of Inequality **[3.1]**

Simplify, leaving each answer in positive exponential form. [7.1]

45. $(3x^2)(5x^3)$

46. $(-4a^2b)(3ab)^2$

47. $\left(\frac{x^4}{y^7}\right)^2\left(\frac{x^2}{y^5}\right)^{-3}$

48. $\frac{x^5y^7z^9}{y^4z^{11}}$

49. What is the y-intercept of the graph of $f(x) = -3(1.2)^x$? **[7.4]**

- **A** $(0, 1.2)$
- **B** $(1.2, 0)$
- **C** $(0, -3)$
- **D** $(-3, 0)$
- **E** $f(x)$ has no y-intercept.

50. Which exponential function produces the values in the table below? **[7.4]**

x	−1	0	1	2
y	108	36	12	4

- **A** $y = 12(3)^{x-1}$
- **B** $y = 12\left(\frac{1}{3}\right)^x$
- **C** $y = 36\left(\frac{1}{3}\right)^x$
- **D** $y = 36(3)^x$
- **E** none of these

TECHNOLOGY CORNER (TI-84 PLUS FAMILY)

Checking Polynomial Products

The graphing calculator can be used to check the results of simplifying a polynomial expression. Use [y=] to enter the polynomial expression $(x^2 + 3x + 2)(x - 1)$ as Y_1. Graph the function using the Zstandard option from the [zoom] menu.

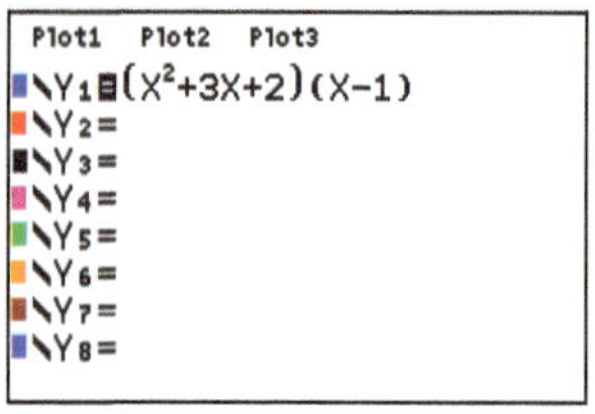

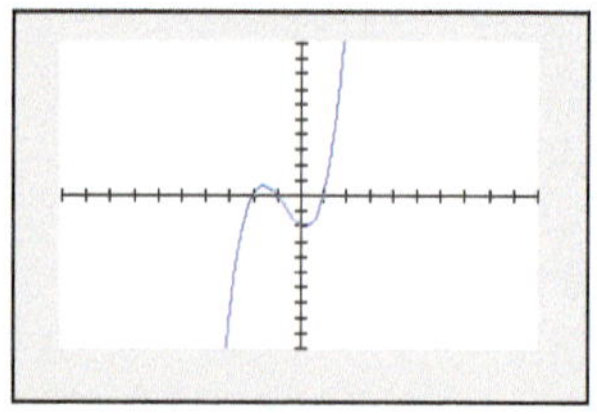

Then use [y=] to enter a possible answer, $x^3 + 2x^2 - x + 2$, as Y_2. Move the cursor onto the \ icon to the left of Y_2, press [enter], and use the arrow keys to select the ⊸ icon. Select OK.

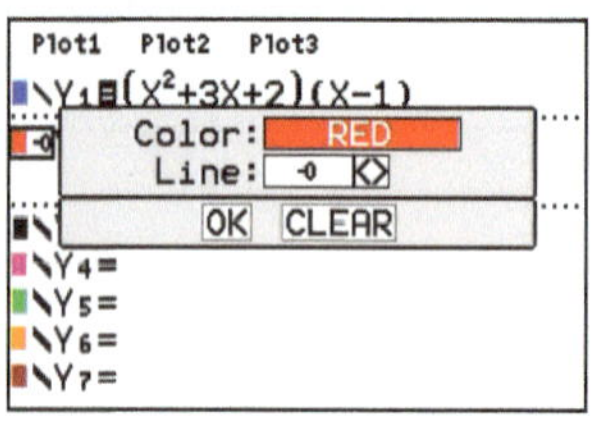

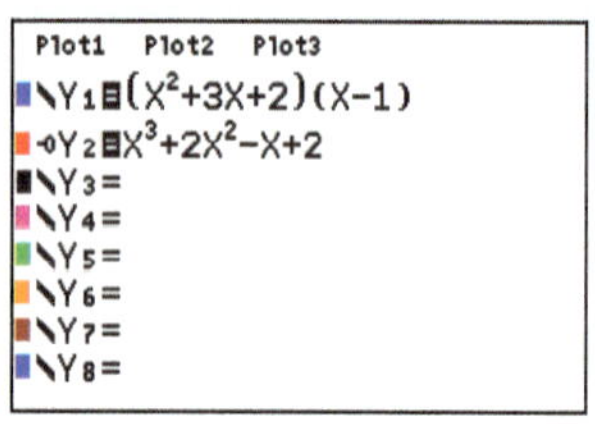

Be sure SEQUENTIAL is selected in the [mode] menu and select [graph] to draw the graphs of Y_1 and Y_2. If the polynomials are equal, the graphs will coincide. Notice the effect of selecting the ⊸ icon before Y_2. These expressions cannot be equal since their graphs do not coincide.

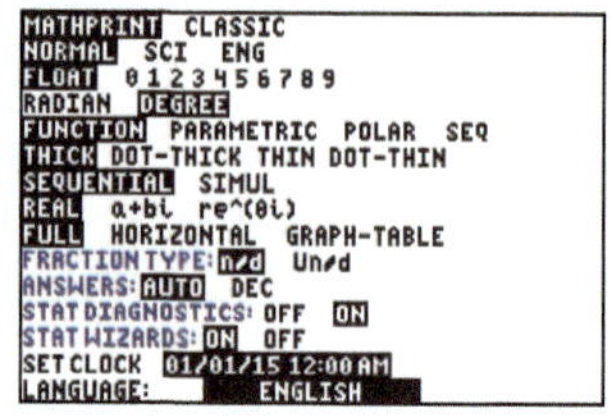

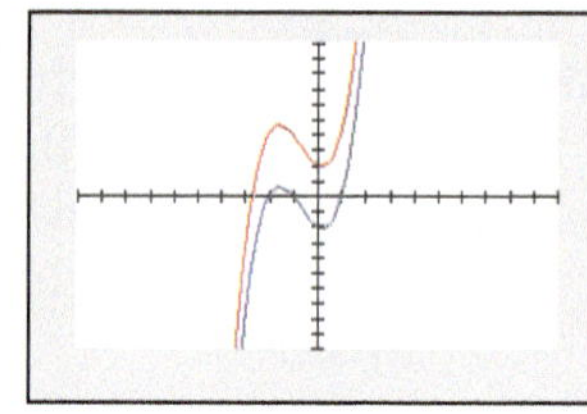

Re-evaluate the product and edit Y_2 to match the new answer, $x^3 + 2x^2 - x - 2$. It appears that the graphs coincide and that the expressions are equal.

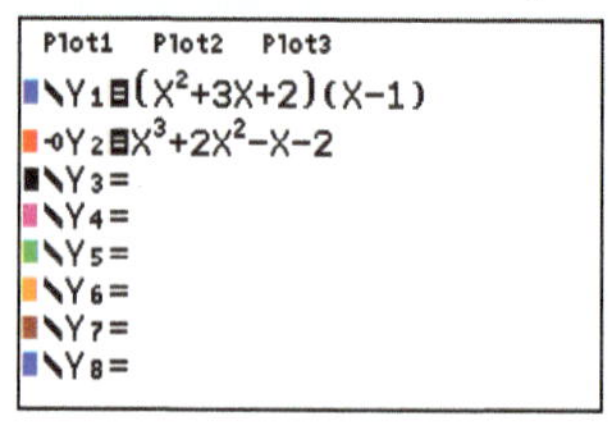

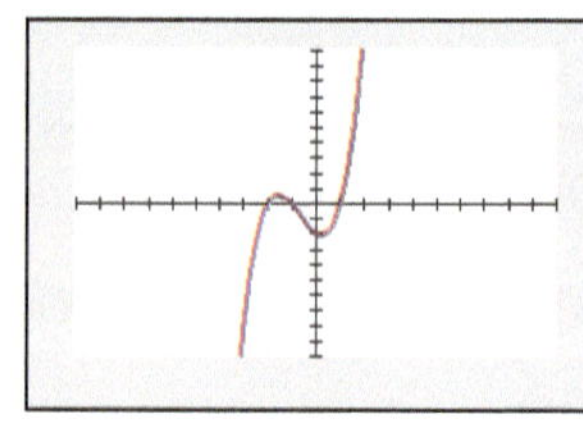

8.3 MULTIPLYING POLYNOMIALS

How can I verify that the Distributive Property works?

After completing this section, I will be able to

- model polynomial multiplication with algebra tiles.
- multiply a polynomial by a monomial.
- multiply any 2 polynomials.
- explain how to check results obtained by using the Distributive Property.

Death Valley had the highest ever recorded temperature: 56.7°C. This can be written in °F by using the formula $\frac{9}{5}C + 32$.

The Distributive Property is used in multiplying a number by a sum. It can also be used in finding the product of a monomial and a polynomial.

Finding the area of the rectangle illustrates this process.

$$2x(x + 2) = 2x^2 + 4x$$

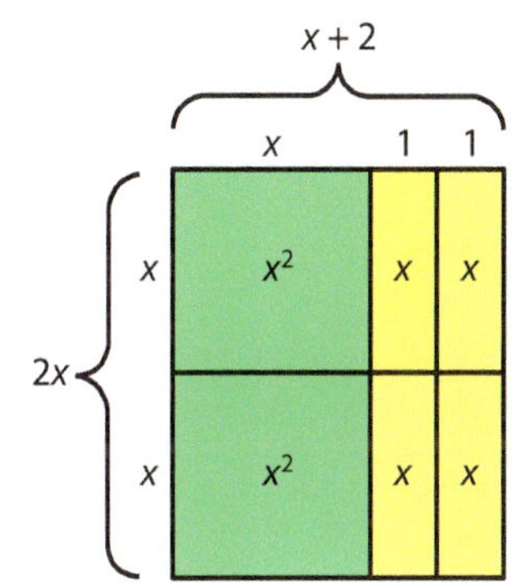

To multiply a monomial by a polynomial, use the Distributive Property to multiply each term of the polynomial by the monomial. This multiplication can be arranged vertically, similar to the format used for the multiplication of whole numbers.

$$\begin{array}{r} 512 \\ \times\ \ 4 \\ \hline 2048 \end{array} \qquad \begin{array}{r} x^2 - 3x + 8 \\ \times \qquad\quad 4x \\ \hline 4x^3 - 12x^2 + 32x \end{array}$$

EXAMPLE 1: Distributing a Monomial by Using 2 Methods

Multiply $3x(x^2 - 7x + 8)$.

Answer

horizontal method

$3x(x^2 - 7x + 8)$

$= 3x(x^2) + 3x(-7x) + 3x(8)$

$= 3x^3 - 21x^2 + 24x$

vertical method

$$\begin{array}{r} x^2 - 7x + 8 \\ \times \qquad\quad 3x \\ \hline 3x^3 - 21x^2 + 24x \end{array}$$

SKILL CHECK ***EXERCISE 5***

EXAMPLE 2: Distributing a Multivariable Monomial

Multiply $2xy^2(3x - 8xy + 5y - 3z)$.

Answer

$$2xy^2(3x - 8xy + 5y - 3z)$$

$$= 2xy^2(3x) + 2xy^2(-8xy) + 2xy^2(5y) + 2xy^2(-3z)$$

$$= 6x^2y^2 \quad -16x^2y^3 \quad +10xy^3 \quad -6xy^2z$$

$$= 6x^2y^2 - 16x^2y^3 + 10xy^3 - 6xy^2z$$

SKILL CHECK **EXERCISE 21**

When 2 binomials are multiplied, the Distributive Property requires each term in the first binomial to be distributed to each term in the second binomial.

Finding the area of the rectangle illustrates this process.

$$(2x + 1)(x + 2)$$
$$= 2x(x + 2) + 1(x + 2)$$
$$= 2x^2 + 4x + x + 2$$
$$= 2x^2 + 5x + 2$$

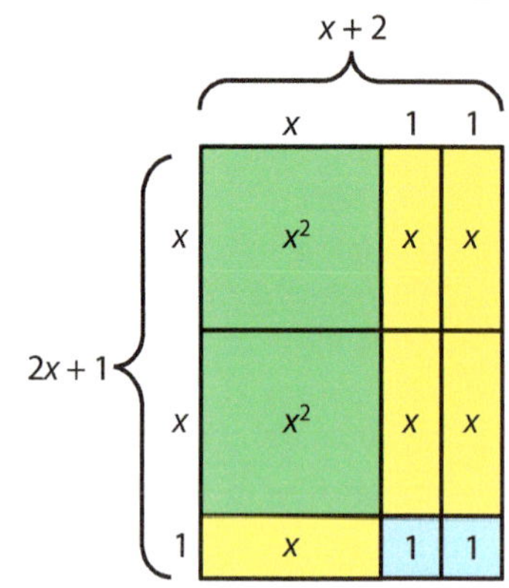

EXAMPLE 3: Multiplying 2 Binomials

Multiply $(3x - 7)(2x - 5)$.

Answer

$$(3x - 7)(2x - 5)$$

$$= 3x(2x - 5) - 7(2x - 5)$$

$$= 6x^2 - 15x - 14x + 35$$

$$= 6x^2 - 29x + 35$$

SKILL CHECK **EXERCISE 15**

When both polynomials have 2 or more terms, using the vertical multiplication format helps to align the like terms that will be added. Notice how each term in the lower binomial is multiplied by each term in the upper trinomial.

$$\begin{array}{r} 512 \\ \times\ 24 \\ \hline 2048 \\ 1024\ \ \\ \hline 12{,}288 \end{array} \qquad \begin{array}{r} x^2 - 3x + 8 \\ \times \qquad 4x - 7 \\ \hline -7x^2 + 21x - 56 \\ 4x^3 - 12x^2 + 32x \qquad\quad \\ \hline 4x^3 - 19x^2 + 53x - 56 \end{array}$$

KEYWORD SEARCH

multiply polynomials

MULTIPLYING POLYNOMIALS

1. Arrange each polynomial in descending powers of the same variable.
2. Multiply each term of the multiplier by each term of the multiplicand.
3. Combine like terms to obtain the final product.

EXAMPLE 4: Multiplying Polynomials Vertically

DOING THE MATH

$$-\tfrac{4}{3}x + 9x = -1\tfrac{1}{3}x + 9x$$
$$= 7\tfrac{2}{3}x = \tfrac{23}{3}x$$

Express mixed numeral coefficients as improper fractions.

Multiply $\left(x - \frac{1}{3}\right)(3x^2 + 9 + 4x)$.

Answer

$$\begin{array}{r} 3x^2 + \ \ 4x + 9 \\ \times \qquad\quad x - \frac{1}{3} \\ \hline -1x^2 - \ \frac{4}{3}x - 3 \\ 3x^3 + 4x^2 + \ \ 9x \qquad \\ \hline 3x^3 + 3x^2 + \frac{23}{3}x - 3 \end{array}$$

1. Arrange the terms in descending powers of x.
2. Multiply each term in the lower binomial by each term in the upper trinomial, aligning like terms in the same column.
3. Combine like terms.

SKILL CHECK **EXERCISE 7**

By examining both the distributive process for the general case below and the diagram, you can see why each term in the first factor is multiplied by each term in the second factor. The resulting product is a polynomial representing the total area in the diagram.

$$(a + b)(c + d + e) = a(c + d + e) + b(c + d + e)$$
$$= ac + ad + ae + bc + bd + be$$

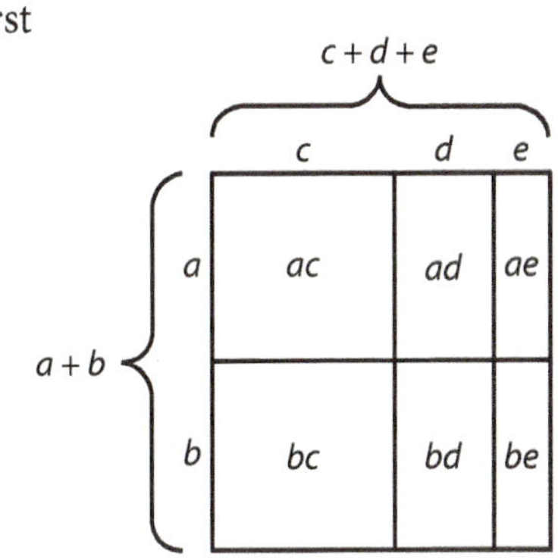

EXAMPLE 5: Multiplying Polynomials Horizontally

AFTERSCHOOLHELP

Multiplying Polynomials

Multiply $\left(2a + \frac{3}{4}b\right)\left(7a + \frac{4}{5}b - \frac{5}{6}c\right)$.

Answer

$$\left(2a + \tfrac{3}{4}b\right)\left(7a + \tfrac{4}{5}b - \tfrac{5}{6}c\right)$$
$$= 2a\left(7a + \tfrac{4}{5}b - \tfrac{5}{6}c\right) + \tfrac{3}{4}b\left(7a + \tfrac{4}{5}b - \tfrac{5}{6}c\right)$$
$$= 14a^2 + \tfrac{8}{5}ab - \tfrac{5}{3}ac + \tfrac{21}{4}ab + \tfrac{3}{5}b^2 - \tfrac{5}{8}bc$$
$$= 14a^2 + \tfrac{137}{20}ab - \tfrac{5}{3}ac + \tfrac{3}{5}b^2 - \tfrac{5}{8}bc$$

SKILL CHECK **EXERCISE 29**

Notice that multiplying polynomials will always yield a polynomial.

A. EXERCISES

Write the polynomial multiplication modeled by each diagram and simplify.

1.

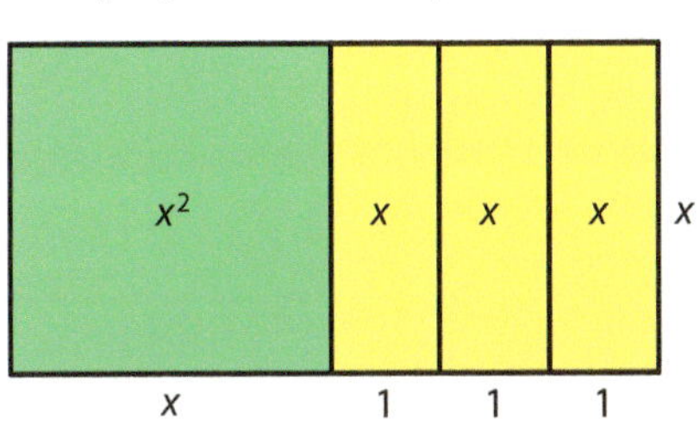

2.

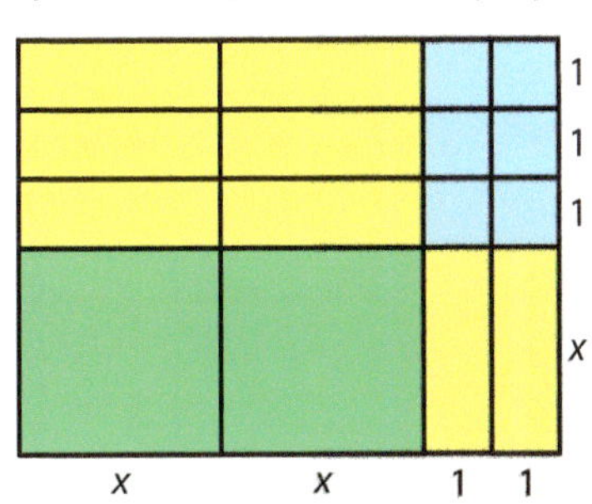

3.

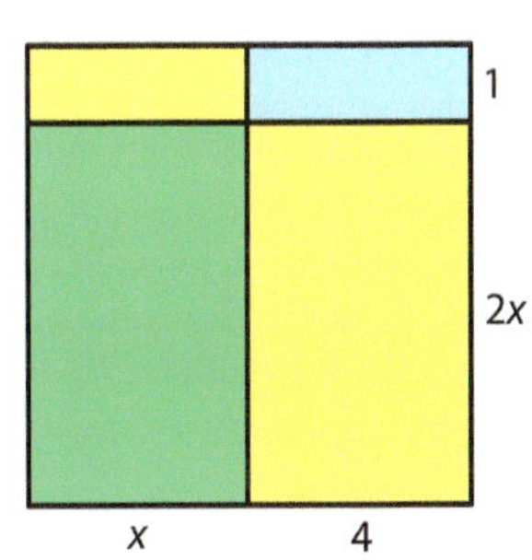

Find each product.

4. $$\begin{array}{r} 2x^2 + 3x - 8 \\ \underline{\times \qquad\qquad 7x} \end{array}$$

5. $$\begin{array}{r} 5a + 10b \\ \underline{\times \qquad -4b} \end{array}$$

6. $$\begin{array}{r} 3x + 4 \\ \underline{\times\ 2x + 1} \end{array}$$

7. $$\begin{array}{r} 3x - 6 \\ \underline{\times\ 3x + 5} \end{array}$$

8. $$\begin{array}{r} 3a^2 - 2a - 4 \\ \underline{\times \qquad\quad a - 7} \end{array}$$

9. $$\begin{array}{r} 6x^2 + 3xy - 9y^2 \\ \underline{\times \qquad\qquad 2x + y} \end{array}$$

Find each product.

10. $3a(k + 2m - p)$

11. $-7k(8k^2 + 9k + 6)$

12. $-2k(k^3 + 6k^2 + 3k - 9)$

13. $3a^2(4a^4 + 5a - 18)$

14. $(x + 7)(x + 5)$

15. $(y - 4)(y - 3)$

16. $(b + 2)(b - 1)$

17. $(c - 3)(c + 8)$

B. EXERCISES

Find each product.

18. $6m(4m^2 - 3mn - 10)$

19. $-8x^3y(3x^2 + 7xy + 10)$

20. $\frac{1}{5}a^2(25a^3 + 15a^2 - 50)$

21. $16x\left(\frac{1}{4}x^2 + \frac{1}{8}xy + \frac{1}{2}\right)$

22. $(2a + 4b)(3a - 9)$

23. $(4b^2 - 8)(-2b^2 + 6b)$

24. $(2x^2 + 5)(3x^2 - 9)$

25. $(-3x^2 + 4y)(4xy + 3y^2)$

26. $(2x^2 - x + 4)(3x - 7)$

27. $(2x^2 - 5x + 8)(3x + 6)$

28. $(x + y)(x^2 + xy + y^2)$

29. $(x - 3)(x^2 + 3x + 9)$

30. $(x^2 - 3x - 9)(x^2 + 4x - 6)$

31. $(2x^2 - 5x + 7)(3x^2 - 7x - 4)$

32. $(a + b)(7a + 3b - 9c + d)$

33. **MP7** $(x - 7)(x - 5)(x - 3)$

34. **Essential Question:** Given: $(3x + 2)(2x - 5)$.

a. Show that $(3x + 2)(2x - 5) = 6x^2 - 11x - 10$.

b. Evaluate $(3x + 2)(2x - 5)$ for $x = -4, 0, 2$.

c. Evaluate $6x^2 - 11x - 10$ for $x = -4, 0, 2$.

d. How do I know the Distributive Property works?

Write a polynomial expression for the area of each figure.

35.

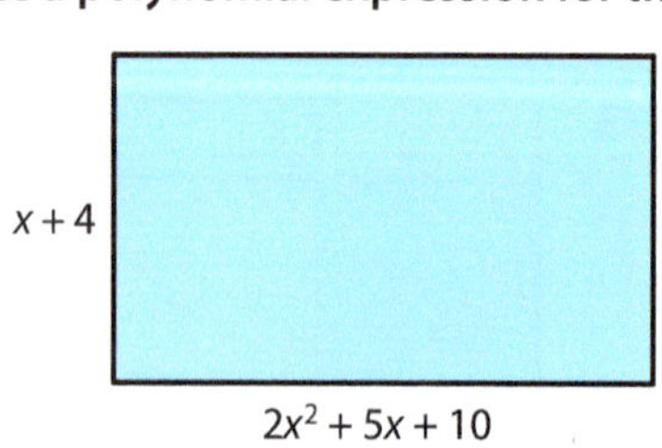

36.

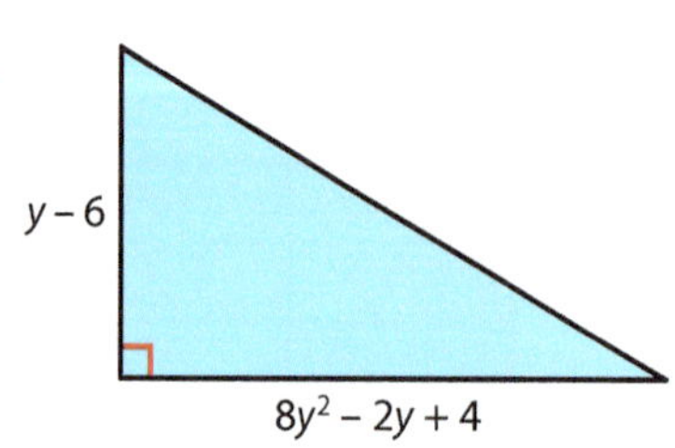

37. **MP1**

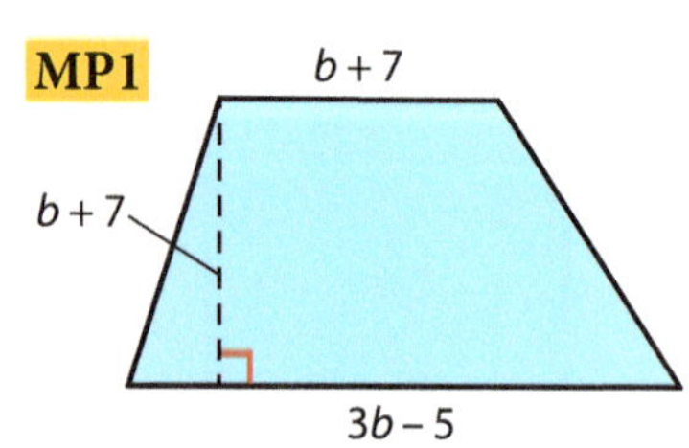

C. EXERCISES

Simplify each expression.

38. $-3x(x + 4) - (x + 5)(x - 5)$

39. $(2x + 3)(x - 7) - (3x - 1)(x - 5)$

40. **MP2** Write a polynomial expression for the product of 3 consecutive integers.

41. Draw a diagram illustrating the multiplication of $(x + 3)(2x + 1)$. State the product and explain how this answer is illustrated by your diagram.

42. Explain how the Distributive Property is used to multiply $(x^2 + 2x + 1)(x^2 - 2x - 1)$ and state the product.

43. **MP1** Write a polynomial expression for the product of the sum and the difference of 2 numbers x and y.

CUMULATIVE REVIEW

Solve each equation. [3.5]

44. $|2x + 3| = 4$

45. $|3x - 7| = 5$

Solve each system. [6.2–6.3]

46. $2x - y = 7$
$3x + 2y = 0$

47. $5x + 4y = 0$
$y = -2x + 3$

Simplify, expressing each answer in scientific notation. [7.2]

48. $(4.4 \times 10^{-5}) + (3.8 \times 10^{-4})$

49. $(5.1 \times 10^{9}) \times (3 \times 10^{-4})$

Express each answer in scientific notation. [7.2]

50. Earth's average distance from the sun is 1.5×10^8 mi, while Neptune's is 4.5×10^9 mi. Neptune's average distance from the sun is how many times that of Earth?

51. Neptune's farthest distance from the sun, its aphelion, is 4.55 billion miles; and its closest distance from the sun, its perihelion, is 4.45 billion miles. What is the difference of these distances in miles?

52. Which inequality has $(-1, 4)$ as a solution but not $(2, 14)$? **[6.7]**

A $y \le -x$

B $y < 3x + 8$

C $y \ge 2$

D $y > 3 - x$

E none of these

53. Find the 13th term of the sequence 2, −6, 18, −54, ... **[7.6]**

A −3,188,646

B 1,062,882

C −24,576

D 12,288

E none of these

8.4 MULTIPLYING BINOMIALS BY USING FOIL

How can I model the area needed to put a pool in my yard?

After completing this section, I will be able to

- find the product of 2 binomials by using the FOIL method.
- create polynomial models of real-world problems involving binomials.

As described in Section 8.3, multiplying 2 binomials produces 4 products. These products can be organized in a pattern that facilitates mental computation.

Notice that the process of applying the Distributive Property can be easily remembered using the acronym FOIL.

First, Outer, Inner, Last

$$(a + b)(c + d)$$

$$= ac + ad + bc + bd$$

F O I L

EXAMPLE 1: Using the FOIL Method

Multiply $(x + 2)(x + 6)$.

Answer

$(x + 2)(x + 6)$

$= x^2 + 6x + 2x + 12$

$= x^2 + 8x + 12$

1. Find the sum of the products of the first, outer, inner, and last terms.
2. Combine like terms.

SKILL CHECK **EXERCISE 1**

When the binomials are written in descending degree, the products of the inner and outer terms are often like terms that need to be combined.

EXAMPLE 2: Using the FOIL Method

Multiply $(y + 6)(4y - 9)$.

Answer

$(y + 6)(4y - 9)$

$= 4y^2 - 9y + 24y - 54$

$= 4y^2 + 15y - 54$

1. Find the sum of the products of the first, outer, inner, and last terms.
2. Combine like terms.

SKILL CHECK **EXERCISE 17**

AFTERSCHOOLHELP

Multiplying Binomials by Using FOIL

KEYWORD SEARCH
FOIL method

Recall that exponents represent repeated multiplication. Therefore, $(a + b)^2 = (a + b)(a + b)$. Applying the FOIL method produces $a^2 + 2ab + b^2$.

EXAMPLE 3: Squaring a Binomial

Multiply $(2x - 5)^2$.

Answer

$(2x - 5)^2 = (2x - 5)(2x - 5)$ 1. Square the quantity by multiplying it by itself.

$= 4x^2 - 10x - 10x + 25$ 2. Use FOIL.

$= 4x^2 - 20x + 25$ 3. Combine like terms.

SKILL CHECK **EXERCISE 21**

The FOIL pattern works well with higher degree binomials.

EXAMPLE 4: Using the FOIL Method

Find each product.

a. $(3a^2 - 5)(a^2 + 3)$ **b.** $(2x + 3)(4x^2 - 3)$

Answers

a. $(3a^2 - 5)(a^2 + 3) = 3a^4 + 9a^2 - 5a^2 - 15$ 1. Use FOIL.

$= 3a^4 + 4a^2 - 15$ 2. Combine like terms.

b. $(2x + 3)(4x^2 - 3) = 8x^3 - 6x + 12x^2 - 9$ 1. Use FOIL.

$8x^3 + 12x^2 - 6x - 9$ 2. Write the answer in standard form.

SKILL CHECK **EXERCISE 29**

Binomials and their products can be used to model real-world applications.

EXAMPLE 5: Modeling a Real-World Situation

A rectangular swimming pool has sides that are in a ratio of 3 : 2. If the pool is surrounded by a 5 ft wide concrete walkway, write a polynomial expression for the total area of the pool and walkway.

Answer

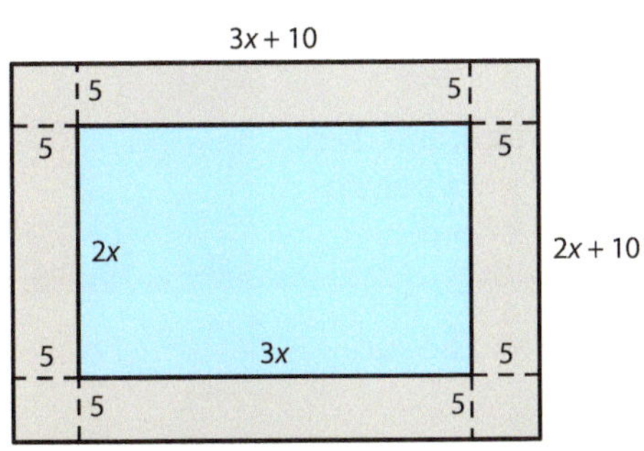

1. Draw a diagram to model the problem. Since the length and width of the pool are in a 3 : 2 ratio, represent them with 2x and 3x. Add 10 to these distances.

$A = (2x + 10)(3x + 10)$ 2. Recall that *area* = *length* × *width*.

$= 6x^2 + 20x + 30x + 100$ 3. Use FOIL and combine like terms.

$= 6x^2 + 50x + 100$

SKILL CHECK **EXERCISE 37**

A. EXERCISES

Multiply using the FOIL method.

1. $(x + 5)(x + 2)$
2. $(y - 7)(y + 9)$
3. $(b - 3)(b - 4)$
4. $(x - 2)(x + 1)$
5. $(v + 1)(v - 5)$
6. $(n - 4)(n - 4)$
7. $(t - 6)(t - 7)$
8. $(d + 9)(d - 8)$
9. $(y - 5)(y + 5)$
10. $(g + 3)(g - 3)$
11. $(a - 7)(a + 7)$
12. $(x - 8)(x - 8)$
13. $(c - 1)(c - 1)$
14. $(y + 9)(y + 9)$
15. $(x + 3)(x + 3)$

B. EXERCISES

Multiply using the FOIL method.

16. $(3x + 2)(x + 4)$
17. $(5z - 6)(3z + 2)$
18. $(2c - 9)(4c - 11)$
19. $(a + 6)^2$
20. $(3f - 2)^2$
21. $(-7g - 10)^2$
22. $(11x + 12)^2$
23. $(5z + 6)(2z - 7)$
24. $(c + 4)(2c - 9)$
25. $(3h + 4)(3h - 4)$
26. $(2x - 13)(2x + 13)$
27. $(2v + 4)(3v - 6)$
28. $(5w + 6)(w - 4)$
29. $(3x^2 + 6)(x^2 - 1)$
30. $(z^3 + 4)(2z^3 - 11)$
31. $(2b^3 + 3b)(3b^3 + 4b)$
32. $(a^2 - a)(a^3 + 1)$
33. $(4x^3 + 2)(5x^3 - 6x^2)$
34. Write a polynomial expression for the area of a rectangle with a length of $4x + 3$ and a width of $2x - 4$.
35. Write a polynomial expression for the area of a triangle if its base is $2x + 12$ and its height is $x + 7$.
36. MP4 Marco's Pollo, a chicken farm, is looking to increase the size of their existing chicken coops. Each coop is 6 ft by 10 ft, and the extension of the longer side of each coop will be twice the length of the extension of the shorter side. Write a polynomial expression for the area of an expanded coop.
37. The common widescreen aspect ratio is 16 : 9. If a television has a 1 in. frame on each side and across the top and a 3 in. frame across the bottom, write a polynomial expression for the total area of the front of the television.

38. The length of a soccer field is 40 yd longer than its width, and there is a 5 yd wide grass border around the entire field. Write a polynomial expression for the total area of the field and grass border.
39. **Essential Question:** The Maloney family wants to put a pool in their backyard. It will have sides in a ratio of 5 : 3 and be surrounded by a 6 ft wide concrete walkway.
    a. Write a polynomial expression modeling the total area of the pool and walkway.
    b. The Maloneys have 800 ft^2 of space to put in the pool and walkway. What is the largest whole number value of x for which this will fit in their yard? How much free space will be left over?
    c. Using this value of x, what will be the dimensions of the pool itself?

C. EXERCISES

Simplify each expression.

40. $(8x + 3)(3x - 2) + (x - 4)(x + 4)$

41. $(-2x - 5)(x - 12) - (3x + 2)(-x + 7)$

42. MP4 Mirza is making an open-top box by cutting squares with side length x out of the 4 corners of a 9 in. × 12 in. piece of cardboard and folding up the sides. Write a polynomial expression for the area of the base of this box.

Write a polynomial expression for each shaded area.

43.

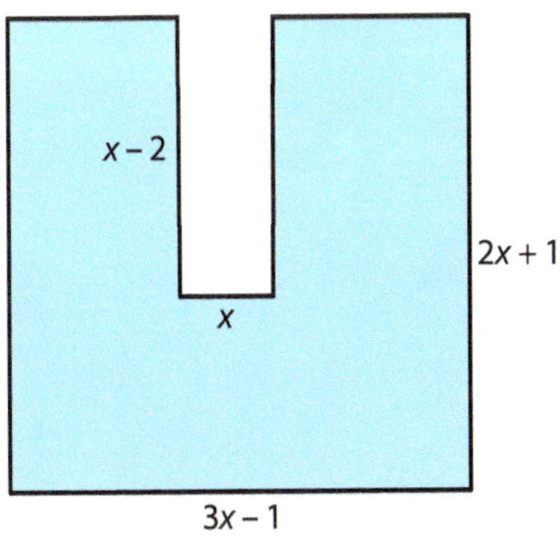

44. MP6

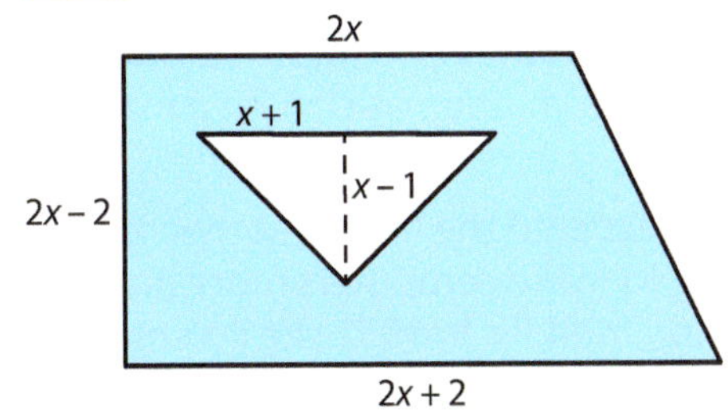

CUMULATIVE REVIEW

Classify each expression as a monomial, binomial, trinomial, or other polynomial. Then state the degree of each polynomial. [8.1]

45. $-6a^2b^3c + 17abc$

46. $3x^5yz^4 - xyz^5 + 5xy + z^6$

Simplify each expression. [8.2]

47. $(8a - 9b + 5) - (-9a + 8b + 4)$

48. $(10c - 3cd) - (4d - 2cd) + (cd + c)$

49. $(22f^2g + 9fg^2 - 7) + (fg^2 - 6fg) + (3f^2g - 5fg + 14)$

Find each product. [8.3]

50. $\begin{array}{r} 5x^3 - 4x - 11 \\ \times \qquad\quad 8x \\ \hline \end{array}$

51. $\begin{array}{r} 4x - 7 \\ \times\ 4x + 6 \\ \hline \end{array}$

52. $\begin{array}{r} 3a^2 + 8ab - 6b^2 \\ \times \qquad 2a - 5b \\ \hline \end{array}$

53. Which line is not parallel to the others? **[5.5]**

A $y = \frac{2}{3}x - 1$

B $y + 6 = \frac{2}{3}(x - 4)$

C $2x - 3y = 15$

D $\frac{2}{3}x + y = 5$

E All are parallel.

54. Which line is perpendicular to $y = -3x - 13$? **[5.5]**

A $y = 3x + 13$

B $y = -\frac{1}{3}x - \frac{1}{13}$

C $2x + 6y = -78$

D $3x - 9y = 54$

E none of these

MATH IN HISTORY
GRACE HOPPER (1906–92)

Grace Hopper was a distinguished American naval officer, mathematician, and computer scientist. After studying math and physics at Vassar College, she attended Yale University, earning her PhD in 1934. She taught mathematics at Vassar from 1931 to 1943. During World War II she joined the Naval Reserve and was assigned to the Bureau of Ordnance Computation Project at Harvard University, where she soon began programming the Harvard Mark computers used in the war effort. After the war, she remained in the Naval Reserve, but she also began working as a senior mathematician for a major computer corporation.

Hopper was a leader in the field of software development, seeking to make computers more programmer and application friendly. In 1952 she invented the first computer compiler, which is an intermediate program that converts a programmer's instructions into the computer's machine language. Her compiler work for the UNIVAC computer led to the development of the popular computer business language COBOL in 1959.

She was recalled to active duty in 1967 and served until 1986, when she retired as Rear Admiral Hopper at the age of 80, the oldest active-duty officer in the United States. At her retirement celebration, she was awarded the Defense Distinguished Service Medal. Hopper received 37 honorary degrees between 1972 and 1987 and was buried with full naval honors at Arlington National Cemetery.

Grace Hopper is a shining example of a fruitful life lived in the service and ideals of her country. She diligently developed her God-given talents to become a successful student, a distinguished educator, a business innovator, and a national hero. Her life story of diligence and service should be a challenge to us to develop our God-given abilities into a life of service and fruitfulness.

8.5 SPECIAL PRODUCTS

Why do special products produce consistent results?

After completing this section, I will be able to

- use patterns to find the square of a sum or difference.
- use patterns to find the product of conjugates.
- explain why we can define properties that are consistent and reliable.

The expression $(40 - x)^2$ represents the dimensions of each pictured corner lot due to a roadway expansion plan.

Recognizing the patterns in special binomial products can save you time.

The first special product is the *square of a sum* which can be illustrated by the area of the square displayed to the right.

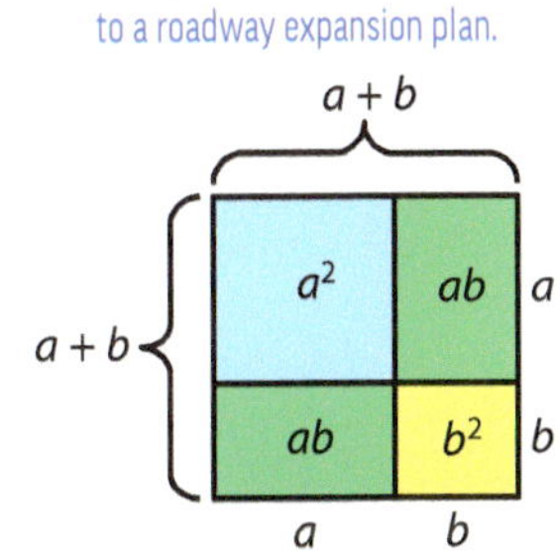

$(a + b)^2 = a^2 + 2ab + b^2$

EXAMPLE 1: Finding the Square of a Sum

Simplify $(a + b)^2$.

Answer

$(a + b)^2 = (a + b)(a + b)$	**1.** Apply the definition of a power.
$= a^2 + ab + ab + b^2$	**2.** Use FOIL.
$= a^2 + 2ab + b^2$	**3.** Combine like terms.

SKILL CHECK **EXERCISE 5**

Notice that the first and last terms of the trinomial are the squares of the first and last terms of the binomial. The trinomial's middle term is twice the product of the 2 terms. This pattern can be used to quickly square a binomial sum.

$$(a + b)^2 = a^2 + 2ab + b^2$$

EXAMPLE 2: Using the Square of a Sum Pattern

Use the square of a sum pattern to expand the following expressions.

a. $(x + 2)^2$ **b.** $(3x + 4)^2$ **c.** $\left(4x + \frac{1}{2}\right)^2$

Answers

a. $(x + 2)^2 = x^2 + 2(x)(2) + 2^2$ **1.** Use $(a + b)^2 = a^2 + 2ab + b^2$.

$= x^2 + 4x + 4$ **2.** Simplify.

b. $(3x + 4)^2 = (3x)^2 + 2(3x)(4) + 4^2$ **1.** $a = 3x$ and $b = 4$.

$= 9x^2 + 24x + 16$ **2.** Simplify.

c. $\left(4x + \frac{1}{2}\right)^2 = (4x)^2 + 2(4x)\left(\frac{1}{2}\right) + \left(\frac{1}{2}\right)^2$ **1.** $a = 4x$ and $b = \frac{1}{2}$.

$= 16x^2 + 4x + \frac{1}{4}$ **2.** Simplify.

SKILL CHECK **EXERCISE 7** ☑

A similar pattern can be seen in the *square of a difference.*

$$(a - b)^2 = (a - b)(a - b)$$
$$= a^2 - ab - ab + b^2$$
$$= a^2 - 2ab + b^2$$

The 2 equal products of the inner and the outer terms are negative, so the middle term of the trinomial is $-2ab$. The last term of the trinomial is positive since it is the product of 2 factors of $-b$. This pattern can be used to quickly find the square of a difference.

$$(a - b)^2 = a^2 - 2ab + b^2$$

EXAMPLE 3: Using the Square of a Difference Pattern

Use the square of a difference pattern to expand the following expressions.

a. $(40 - x)^2$ **b.** $(2x - 5y)^2$ **c.** $(1.5x - 8y^2)^2$

Answers

a. $(40 - x)^2 = 40^2 - 2(40)(x) + x^2$ **1.** Use $(a - b)^2 = a^2 - 2ab + b^2$.

$= 1600 - 80x + x^2$ **2.** Simplify.

b. $(2x - 5y)^2 = (2x)^2 - 2(2x)(5y) + (5y)^2$ **1.** $a = 2x$ and $b = 5y$.

$= 4x^2 - 20xy + 25y^2$ **2.** Simplify.

c. $(1.5x - 8y^2)^2 = (1.5x)^2 - 2(1.5x)(8y^2) + (8y^2)^2$ **1.** $a = 1.5x$ and $b = 8y^2$.

$= 2.25x^2 - 24xy^2 + 64y^4$ **2.** Simplify.

SKILL CHECK **EXERCISE 9** ☑

Now consider the product of 2 binomials, one indicating the sum of 2 terms, $a + b$, and the other indicating the difference of the same terms, $a - b$. Two such binomial expressions are called **conjugates**.

EXAMPLE 4: Finding the Product of Conjugates

Multiply $(a + b)(a - b)$.

Answer

$(a + b)(a - b) = a^2 - ab + ab - b^2$	**1.** Use FOIL.
$= a^2 - b^2$	**2.** The sum of the middle terms equals 0.

SKILL CHECK **EXERCISE 11**

A *product of conjugates* is the difference of 2 squares.

$$(a + b)(a - b) = a^2 - b^2$$

EXAMPLE 5: Using the Product of Conjugates

AFTERSCHOOLHELP

Special Products

Use the product of conjugates pattern to expand the following expressions.

a. $(x - 4)(x + 4)$ **b.** $(3x + 7y)(3x - 7y)$

Answers

a.	$(x - 4)(x + 4) = x^2 - 4^2$	**1.** Use $(a + b)(a - b) = a^2 - b^2$.
	$= x^2 - 16$	**2.** Simplify.
b.	$(3x + 7y)(3x - 7y) = (3x)^2 - (7y)^2$	**1.** $a = 3x$ and $b = 7y$.
	$= 9x^2 - 49y^2$	**2.** Simplify.

SKILL CHECK **EXERCISE 13**

With sufficient practice, you should be able to use the patterns to quickly compute the following products mentally.

SPECIAL PRODUCTS OF BINOMIALS	
square of a sum	$(a + b)^2 = a^2 + 2ab + b^2$
square of a difference	$(a - b)^2 = a^2 - 2ab + b^2$
product of conjugates	$(a + b)(a - b) = a^2 - b^2$

The consistency in mathematical patterns is a reflection of the foundational order and consistency of creation. These patterns can also be used to quickly compute some numerical products.

EXAMPLE 6: Applying Special Patterns

Compute the following expressions using special product patterns.

a. 87×93 **b.** 72^2

Answers

a.	$87 \times 93 = (90 - 3)(90 + 3)$	**1.** Write as the product of conjugates.
	$= 90^2 - 3^2$	**2.** Use $(a + b)(a - b) = a^2 - b^2$.
	$= 8100 - 9 = 8091$	**3.** Simplify.
b.	$72^2 = (70 + 2)^2$	**1.** Write as the square of a sum.
	$= 70^2 + 2(70)(2) + 2^2$	**2.** Use $(a + b)^2 = a^2 + 2ab + b^2$.
	$= 4900 + 280 + 4 = 5184$	**3.** Simplify.

SKILL CHECK **EXERCISES 17, 19**

VOCABULARY

- conjugate

A. EXERCISES

Fill in the blanks in the following equations.

1. $(a + b)^2 = a^2 + \_\_\_\_\_\_ + b^2$

2. $(a - b)^2 = a^2 \_\_\_\_\_\_ 2ab \_\_\_\_\_\_ b^2$

3. $(a + b)(a - b) = a^2 \_\_\_\_\_\_ b^2$

Find each square.

4. $(x + 5)^2$

5. $(y - 4)^2$

6. $(6x - 1)^2$

7. $(2a + 3)^2$

8. $(a + 3b)^2$

9. $(x - 9y)^2$

Find each product.

10. $(x - 1)(x + 1)$

11. $(y + 8)(y - 8)$

12. $(3c + 1)(3c - 1)$

13. $(5z - 3)(5z + 3)$

14. $(3x + 2y)(3x - 2y)$

15. $(2d + 11)(2d - 11)$

B. EXERCISES

Express each of the following as the product of conjugates and then multiply mentally.

16. $15 \cdot 25$

17. $27 \cdot 33$

18. $36 \cdot 44$

Express each of the following as the square of a sum or a difference and then multiply.

19. 29^2

20. 63^2

21. 52^2

Simplify each expression.

22. $(5x - 3)^2$

23. $(4x - 3y)(4x + 3y)$

24. $(6y + 3)^2$

25. $(20x - 9y)(20x + 9y)$

26. $(y^3 - 8)(y^3 + 8)$

27. $\left(12a + \frac{1}{3}\right)^2$

28. $(1.5x - 10)^2$

29. $(0.8x - 11y)^2$

30. $\left(\frac{3}{4}xy - \frac{2}{3}\right)\left(\frac{3}{4}xy + \frac{2}{3}\right)$

31. $\left(\frac{10}{3}x^2 - \frac{3}{2}y\right)^2$

32. $(a + 3)(a - 3) - (a + 2)^2$

33. $(b - 2)^2 - (b + 2)(b - 2)$

Describe and correct each error in Agatha's work.

34. MP3 $(x + 2)^2 = x^2 + 4$

35. MP3 $(x - 1)^2 = x^2 - 2x - 1$

36. **Essential Question:** Explain why we know special products will produce reliable results.

C. EXERCISES

37. $(y - 6)(y + 2)(y + 6)$

38. $(2z - 3)(2z + 3)(z + 3)$

39. Show that the square of a difference pattern can be derived from the square of a sum pattern in which $(a - b)^2 = [a + (-b)]^2$.

40. MP8 Determine the pattern for $(a + b)^3$.

a. Multiply $(a + b)(a + b)^2$.

b. Use the pattern developed in part a to expand $(a + 2)^3$.

c. Use the pattern developed in part a to expand $(2x + 5y)^3$.

d. Use the pattern developed in part a and $[a + (-b)]^3$ to expand $(a - b)^3$.

41. MP3 Show that $(a - b)^2 = a^2 - 2ab + b^2$ by using the diagram.

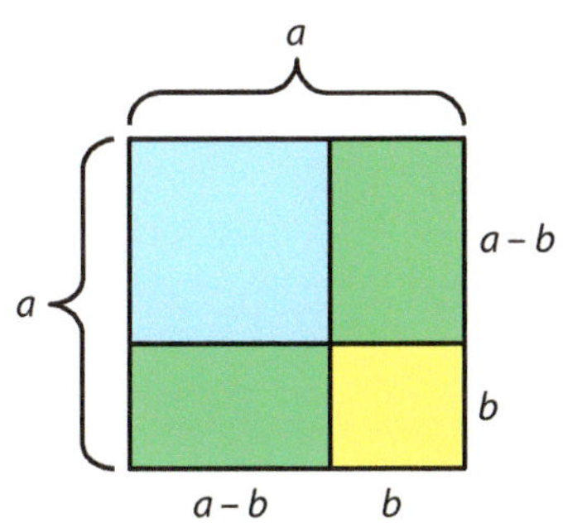

42. MP6 Calculate the shaded area in the diagram in 2 different ways to show that $(a - b)(a + b) = a^2 - b^2$.

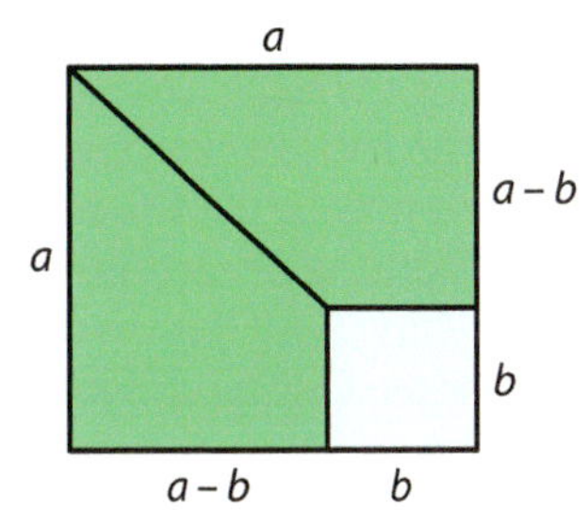

CUMULATIVE REVIEW

43. What is the multiplicative identity? **[1.2]**

44. What is the multiplicative inverse of $-\frac{3}{7}$? **[1.2]**

Solve. [2.2]

45. $4(2x - 1) = 21$

46. $3(x + 7) - (2x - 4) = 0$

47. $\frac{4(y + 1)}{5} + 6 = 27$

48. $\frac{-4(2y - 1)}{15} = -\frac{4}{3}$

Use the exponential function $y = 3\left(\frac{1}{2}\right)^x$ for exercises 49–50. [7.4]

49. State the graph's y-intercept.

50. Is the graph increasing or decreasing?

51. If y is directly proportional to x and $y = 20$ when $x = 10$, find y when $x = 50$. **[4.6]**

A 25

B 100

C 4

D $\frac{1}{2}$

E none of these

52. If y varies inversely with x and $y = 6$ when $x = 2$, find y when $x = 24$. **[4.6]**

A 72

B 8

C $\frac{1}{2}$

D $\frac{1}{3}$

E none of these

8.6 DIVIDING POLYNOMIALS

How can I divide a polynomial by a binomial?

After completing this section, I will be able to

- divide a polynomial by a monomial.
- divide a polynomial by a binomial.

Steel expands with an increase in temperature according to the formula $e = \frac{st}{1.1 \cdot 10^5}$, where s is the length of steel and t is the temperature change (in °F).

The manipulation of polynomials improves your math skills, providing the steps necessary for expanding your skills in new realms of mathematics, science, and technology. In science, a ratio of polynomials may be used to represent the relative change between 2 chemicals during a reaction. In software engineering, cyclic redundancy checking (CRC) is used to ensure the accurate transfer of data over the internet. CRC is also known as "polynomial code checksum" because of the polynomial equations used in its error-checking algorithms.

When a polynomial is multiplied by a monomial, the monomial is distributed to each term. The division of a polynomial by a monomial is similar.

$$\frac{a+b+c}{d} = (a+b+c)\cdot\frac{1}{d} = \frac{a}{d}+\frac{b}{d}+\frac{c}{d}$$

To divide a polynomial by a monomial, divide each term of the polynomial by the monomial.

EXAMPLE 1: Dividing by a Monomial

Divide.

a. $(12x^2 + 20x) \div 4x$ **b.** $\frac{8x^3 + 16x^2 - 4x}{-4x}$

Answers

a. $\frac{12x^2}{4x} + \frac{20x}{4x}$

1. Divide each term of the polynomial by the monomial.

$= 3x + 5$

2. Simplify each term.

$4x(3x+5) = 12x^2 + 20x$

3. Check by multiplying the quotient by the divisor to obtain the dividend.

b. $\frac{8x^3}{-4x} + \frac{16x^2}{-4x} + \frac{-4x}{-4x}$

1. Distribute the division to each term.

$= -2x^2 - 4x + 1$

2. Simplify each term.

$-4x(-2x^2 - 4x + 1) = 8x^3 + 16x^2 - 4x$

3. Check using multiplication.

SKILL CHECK **EXERCISES 3, 7**

LONG DIVISION REVIEW

$$\begin{array}{r} 51 \\ 32\overline{)1635} \\ \underline{160} \\ 35 \\ \underline{32} \\ 3 \end{array}$$

$1635 \div 32 = 51\frac{3}{32}$

Dividing by a monomial is fairly simple and, with practice, can often be done mentally. When the divisor is a polynomial, we can adopt the long division algorithm often used for dividing whole numbers.

EXAMPLE 2: Dividing by a Binomial

Divide $(x^2 + 6x - 9)$ by $(x + 2)$.

Answer

$$x + 2\overline{)x^2 + 6x - 9}$$

1. Write the problem using the long division format. Be sure that the terms of the dividend and the divisor are in order of descending degree.

$$\begin{array}{r} x \\ x + 2\overline{)x^2 + 6x - 9} \\ \underline{-(x^2 + 2x)} \\ 4x - 9 \end{array}$$

2. a. Divide the first term of the dividend by the first term of the divisor and write the result in the quotient.
   b. Multiply the entire divisor by x, aligning the product under the dividend.
   c. Subtract and bring down the next term to obtain a new dividend.

$$\begin{array}{r} x + 4 \\ x + 2\overline{)x^2 + 6x - 9} \\ \underline{-(x^2 + 2x)} \\ 4x - 9 \\ \underline{-(4x + 8)} \\ -17 \end{array}$$

3. a. Divide the first term of the new dividend by the first term of the divisor and write the result in the quotient.
   b. Multiply the entire divisor by 4, aligning the product under the dividend.
   c. Subtract to obtain a new dividend. Notice that there is no next term to bring down, making −17 the remainder.

$$\begin{array}{r} x + 4 - \frac{17}{x+2} \\ x + 2\overline{)x^2 + 6x - 9} \end{array}$$

4. Write the final quotient, placing the remainder over the divisor.

SKILL CHECK **EXERCISE 21**

Just as the division of whole numbers can be checked by multiplying the divisor by the quotient, the division of polynomials can be checked by using the same process.

CHECKING DIVISION	
Whole Numbers	**Polynomials**
$32\left(51 + \frac{3}{32}\right)$	$(x + 2)\left(x + 4 - \frac{17}{x+2}\right)$
$= 32(51) + \cancel{32}\left(\frac{3}{\cancel{32}}\right)$	$= (x + 2)(x + 4) + \cancel{(x+2)}\left(\frac{-17}{\cancel{x+2}}\right)$
$= 1632 + 3$	$= x^2 + 6x + 8 - 17$
$= 1635$	$= x^2 + 6x - 9$

Notice that the quotient of $x + 4 - \frac{17}{x+2}$ does not meet the definition of a polynomial. The division of polynomials will not always yield a polynomial result.

DIVIDING POLYNOMIALS

1. Arrange the terms in both the dividend and the divisor in descending degree.
2. Complete the divide, multiply, and subtract procedure.
   a. Divide the first term of the dividend by the first term of the divisor and write the result in the quotient.
   b. Multiply the result by the entire divisor and place the product under the dividend.
   c. Subtract this product from the dividend and bring down the next term to obtain a new dividend.
3. Repeat step 2 until the degree of the new dividend is less than the degree of the divisor. Place any remainder over the divisor and add the resulting rational expression to the quotient.
4. Check your results by multiplying your answer by the divisor.

AFTERSCHOOLHELP

Dividing Polynomials

EXAMPLE 3: Dividing by a Binomial

Divide

a. $\dfrac{2x^4 - x^3 + 5x^2 + 9x - 6}{x - 1}$ **b.** $(x^3 - 8) \div (x - 2)$ **c.** $(3a^3 + a - 5a^2 + 2) \div (1 + 3a)$

Answers

a. Follow the steps listed in the table.

$$
\begin{array}{r}
2x^3 + x^2 + 6x + 15 + \frac{9}{x-1} \\
x - 1\,\overline{)\,2x^4 - x^3 + 5x^2 + 9x - 6} \\
\underline{-(2x^4 - 2x^3)} \qquad\qquad\qquad \\
x^3 + 5x^2 \qquad\qquad \\
\underline{-(x^3 - x^2)} \qquad\qquad \\
6x^2 + 9x \qquad \\
\underline{-(6x^2 - 6x)} \qquad \\
15x - 6 \\
\underline{-(15x - 15)} \\
9
\end{array}
$$

The quotient is $2x^3 + x^2 + 6x + 15 + \frac{9}{x - 1}$.

$(x - 1)\left(2x^3 + x^2 + 6x + 15 + \frac{9}{x - 1}\right) = (x - 1)(2x^3 + x^2 + 6x + 15) + 9$
$= 2x^4 - x^3 + 5x^2 + 9x - 6$

b. $(x^3 - 8) \div (x - 2)$

Since there are no x^2 and x terms in the dividend, you will need to show placeholders for them in the long division. Insert 0s as coefficients of the missing terms.

$$
\begin{array}{r}
x^2 + 2x + 4 \\
x - 2\,\overline{)\,x^3 + 0x^2 + 0x - 8} \\
\underline{-(x^3 - 2x^2)} \qquad\qquad \\
2x^2 + 0x \qquad \\
\underline{-(2x^2 - 4x)} \qquad \\
4x - 8 \\
\underline{-(4x - 8)} \\
0
\end{array}
$$

The quotient is $x^2 + 2x + 4$.

$(x - 2)(x^2 + 2x + 4) = x^3 + 2x^2 + 4x - 2x^2 - 4x - 8 = x^3 - 8$

c. $(3a^3 + a - 5a^2 + 2) \div (1 + 3a)$

First arrange each polynomial in descending powers of a; then divide.

$$
\begin{array}{r}
a^2 - 2a + 1 + \frac{1}{3a+1} \\
3a + 1\,\overline{)\,3a^3 - 5a^2 + a + 2} \\
\underline{-(3a^3 + a^2)} \qquad\qquad \\
-6a^2 + a \qquad \\
\underline{-(-6a^2 - 2a)} \qquad \\
3a + 2 \\
\underline{-(3a + 1)} \\
1
\end{array}
$$

The quotient is $a^2 - 2a + 1 + \frac{1}{3a + 1}$.

$(3a + 1)\left(a^2 - 2a + 1 + \frac{1}{3a + 1}\right) = (3a + 1)(a^2 - 2a + 1) + 1 = 3a^3 - 5a^2 + a + 2$

KEYWORD SEARCH

polynomial long division

SKILL CHECK **EXERCISES 33, 35**

A. EXERCISES

Find each quotient.

1. $(5x^2 + 25x) \div 5x$

2. $(x^3 + x^2 - x) \div x$

3. $(8x^3 - 2x) \div 2x$

4. $(abc + bcd) \div bc$

5. $\dfrac{3x^2 + 15}{3}$

6. $\dfrac{25ab^3 - 35a}{5a}$

7. $\dfrac{7a^2b - 42ab^2}{7ab}$

8. $\dfrac{16x^5y + 8xy}{8xy}$

9. $(24a^2b - 12ab + 9b^2) \div 3b$

10. $(81x^3 - 27x^2y + 9xy^3) \div 3x$

11. $(15x^3 - 3x^2 + 27x) \div (-2x)$

12. $(-4a^3b^2 + 32ab^3 - 18b^5) \div (-3b^2)$

13. $\dfrac{3n^4 + 5n^3 - n^2}{n^2}$

14. $\dfrac{34a^3b^2 - 10a^2b^3 + 2ab^5}{ab}$

15. $\dfrac{x^4 + 9x^3 - 5x}{3x}$

16. $\dfrac{7a^5b^2c^2 - 4a^3bc^4 + 3bc^8}{-2bc^2}$

B. EXERCISES

Find each quotient.

17. $(x^2 + x - 6) \div (x + 3)$

18. $(x^2 - 7x + 12) \div (x - 4)$

19. $(3x^2 - 26x - 9) \div (x - 9)$

20. $(2x^2 + 10x + 12) \div (x + 2)$

21. $(a^2 + 4a + 12) \div (a - 6)$

22. $(5y^2 - 2y + 9) \div (y + 1)$

23. $(24x^2 + 13x - 9) \div (x + 1)$

24. $\dfrac{x^2 + 4x - 3}{x - 1}$

25. $\dfrac{6y^2 - 7y - 5}{2y + 1}$

26. $\dfrac{3x^2 + 11x - 15}{3x - 4}$

Describe and correct the error in each student's work.

27. Shawn says the quotient is $a^2 + a$.

$$\begin{array}{r}
a^2 + \quad a + \quad 7 - \;\; 7 \\
a - 5\overline{)a^3 - 4a^2 + 2a - 42} \\
\underline{-(a^3 - 5a^2)} \qquad\qquad \\
a^2 + 2a \qquad\quad \\
\underline{-(a^2 - 5a)} \qquad\quad \\
7a - 42 \\
\underline{-(7a - 35)} \\
-7
\end{array}$$

28. Lisa says the quotient is $y^2 + 5y + \dfrac{y + 3}{y + 4}$.

$$\begin{array}{r}
y^2 + 5y \qquad\qquad \\
y + 4\overline{)y^3 + 9y^2 + 21y + 3} \\
\underline{-(y^3 + 4y^2)} \qquad\qquad \\
5y^2 + 21y \quad \\
\underline{-(5y^2 + 20y)} \quad \\
y + 3
\end{array}$$

29. The remainder of a polynomial quotient will always be a ______ degree than the divisor.

30. **Essential Question:** How can I divide a polynomial by a binomial?

Divide. Be careful to use placeholders and descending order of powers.

31. $(2 + x^2 - x) \div (3 + x)$

32. $(7x^2 + 8 - 3x) \div (4 + x)$

33. $(3x - 2x^2 + x^3) \div (x - 4)$

34. $(x^2 - 25) \div (x - 5)$

35. $(x^3 - 27) \div (x - 3)$

36. $(x^4 - 16) \div (x - 2)$

37. $\dfrac{x^4 - 8x^2 + 16}{x^2 + 4}$

38. $\dfrac{3x^3 - 5x^2 + 4x + 2}{x^2 - 2x + 2}$

Write a polynomial expression for each of the following.

39. the height of the parallelogram

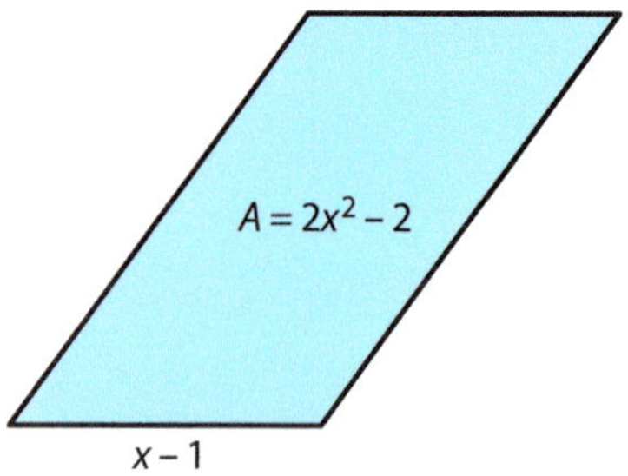

40. the width of the rectangle

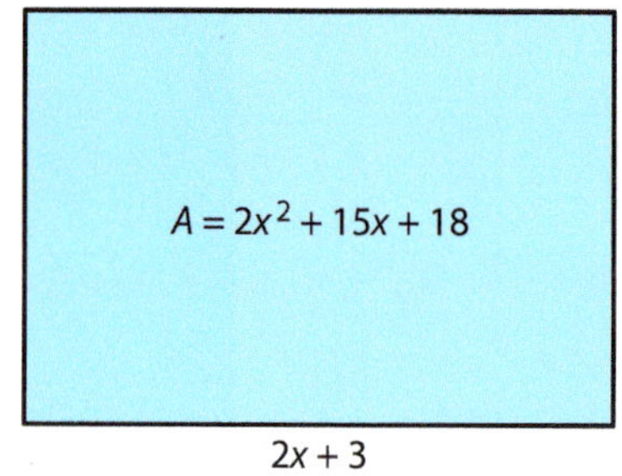

C. EXERCISES

Find each quotient.

41. $(x^5 - 2x^2 + 6) \div (x^2 + 4)$

42. $\left(-6m^3 - 2m^2 + \frac{7}{2}m + \frac{3}{2}\right) \div (2m + 1)$

Write a polynomial expression for each of the following.

43. **MP2** the height of the triangle

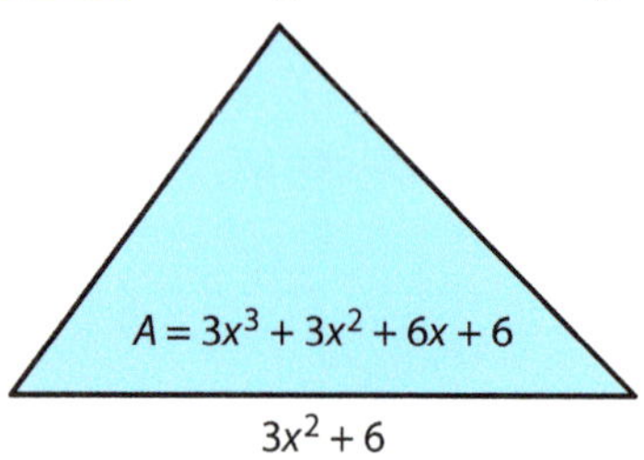

44. **MP2** the height of the trapezoid

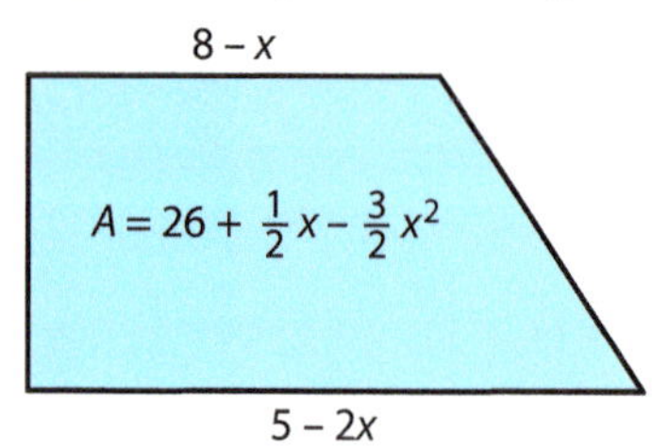

45. **MP6** For what value of m will there be a remainder of 0 when $x^3 - mx^2 + mx - 4$ is divided by $x + 1$?

CUMULATIVE REVIEW

Evaluate. [1.3]

46. $\frac{1}{2^{-1}}\left(\frac{3}{2}\right)^{-1}(-2)^{-2}$

47. $4^{-2}(-1^2)\left(\frac{1}{4}\right)^{-3}$

Write the equation of each line in slope-intercept form. [5.4–5.5]

48. a line passing through (4, 1) and (2, 6)

49. a line perpendicular to $y = -\frac{1}{2}x + 4$ and passing through (−3, −6)

Choose the correct equation for each graph. **[4.7, 7.3–7.4]**

50.

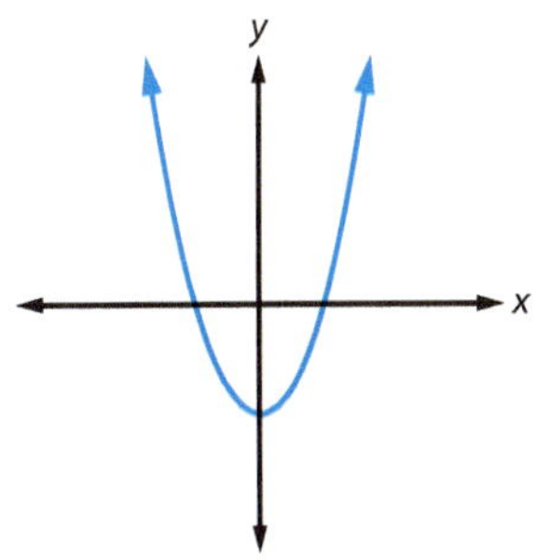

A $y = (x - 3)^2$
B $y = (x + 3)^2$
C $y = x^2 - 3$
D $y = x^2 + 3$

51.

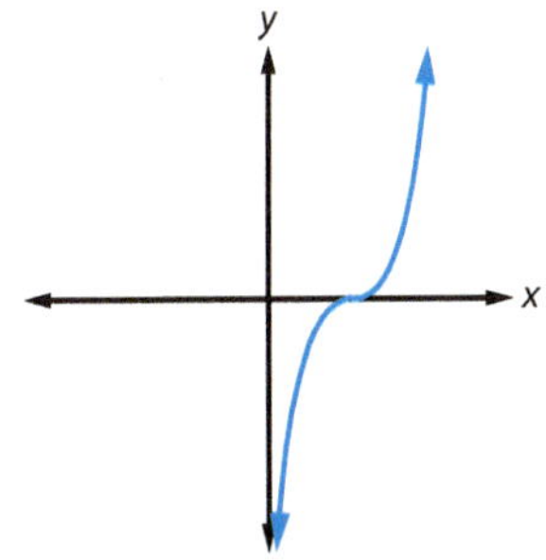

A $y = (x - 2)^3$
B $y = (x + 2)^3$
C $y = x^3 - 2$
D $y = x^3 + 2$

52.

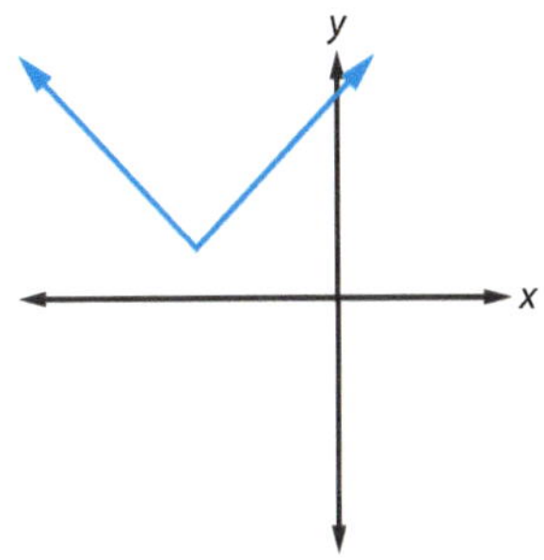

A $y = |x - 3| - 1$
B $y = |x + 3| + 1$
C $y = |x - 3| + 1$
D $y = |x + 3| - 1$

53.

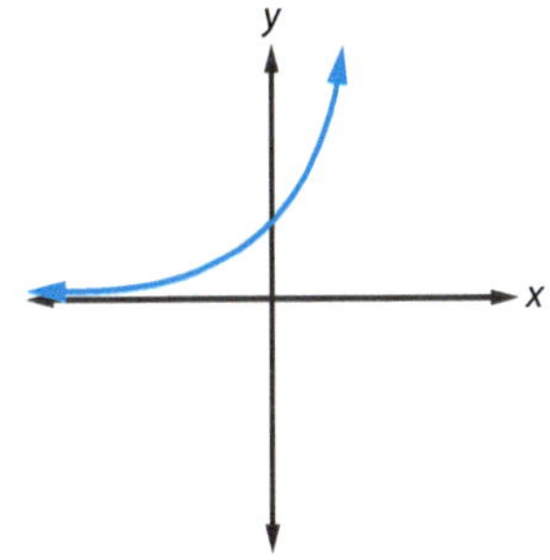

A $y = 2^x$
B $y = \left(\frac{1}{2}\right)^x$
C $y = 3 \cdot 2^x$
D $y = 3 \cdot \left(\frac{1}{2}\right)^x$

54. Which line(s) slope downward from left to right? **[5.2]**

A the line passing through (−1, 5) and (2, 2)
B the line with $m = -\frac{1}{4}$
C the line passing through (10, 0) and (1, 9)
D A and B
E A, B, and C

55. Which of the following products will be a binomial? **[8.5]**

A $(3a + b)^2$
B $(5 - 4c)(5 + 4c)$
C $(8d - 5g)^2$
D B and C
E A, B, and C

MIND OVER MATH

Find the digit that will properly fill in each box in the following division problems.

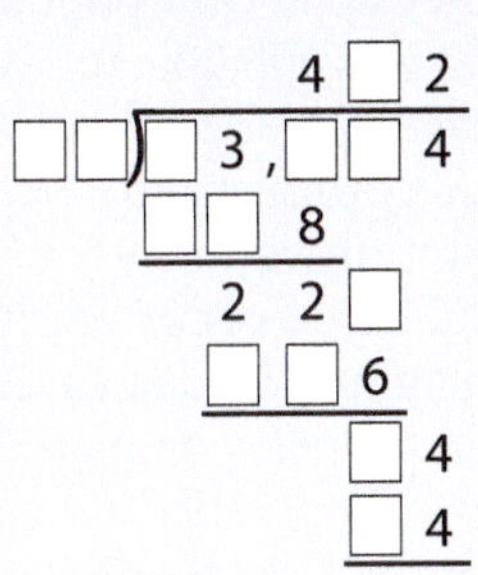

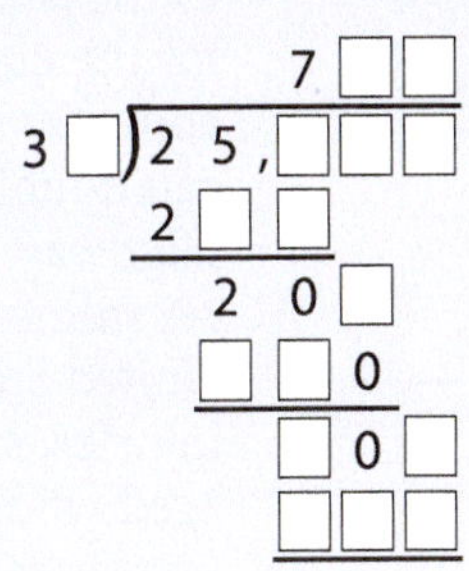

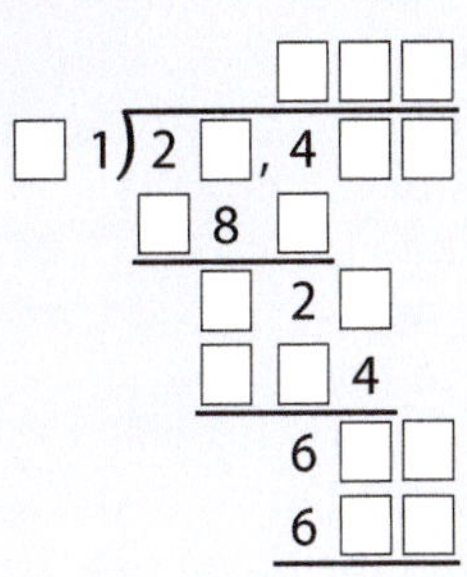

APPLICATION PROBLEMS
MODELING POPULATION GROWTH

Population modeling offers a mathematical method of predicting changes, with the primary objective of managing resources and meeting basic needs. Management of wildlife populations focuses on population control, while the modeling of human populations concentrates on resource management.

Population growth can be modeled by the following exponential function:

$P(t) = A(1 + r)^t$, where A = the initial population,
r = the average periodic growth rate, and
t = the number of periods (most often in years).

A spreadsheet is a powerful mathematical tool that can be used to efficiently solve complex mathematical problems. Though the answers to the following problems can be found using a calculator, you will benefit from learning how to use a spreadsheet to quickly model population growth. While most spreadsheets follow the same basic principles, formula formats may vary slightly between software products.

1. In 2010 the population of Wilmington, NC, was 106,476. Use the exponential growth function to estimate Wilmington's 2018 population, assuming the city's annual growth rate was approximately 0.83%. Enter the following formula into cell C1 of a spreadsheet: =106476*(1.0083)^8.

Create a spreadsheet modeling the growth of Wilmington's population.

	A	B
1	0	
2	1	
3		
4		

Enter 0 in cell A1 and enter 1 in cell A2. Highlight both cells as shown and place the cursor on the little square in the lower right-hand corner of the highlighted area. This is called the *fill handle*. Drag the fill handle down to row 41. If this procedure (referred to as *replicating* the formula) is done correctly, the cells will be filled with sequential numbers to 40. Next, enter the year 2010 in cell B1 and 2011 in cell B2. Highlight these 2 cells and drag to replicate the years to 2050. *Hint*: Double-clicking the fill handle will automatically fill down to the end of an adjacent column of data.

Change the formula in cell C1 to the following: =106476*(1.0083)^A1. Then highlight cell C1 and double-click or drag the fill handle down to row 41. Notice how the variable A1 is automatically changed as the formula is replicated down the column. The cell references are copied relative to each cell's position on the spreadsheet. The result represents a yearly estimate of Wilmington's population given a consistent growth rate of 0.83% over a period of 40 yr.

2. Predict Wilmington's 2050 population, assuming an annual growth rate of 0.83%.

3. Comparing Wilmington's 2010 and 2050 populations, name 3 areas of major concern for the city developers.

In the early 1900s the white-tailed deer population for the entire United States was 500,000; today it is about 30 million. Unmanaged, deer populations can double every 2–3 yr. According to the National Highway Traffic Safety Administration, about 1.5 million deer-vehicle collisions occur yearly in the United States, resulting in about 150 deaths, 10,000 injuries, and $1.1 billion in property damage. Unmanaged wildlife can lead to overpopulation or extinction. Unrestrained population growth leaves deer prone to starvation during harsh winters, and malnutrition lowers their resistance to the spread of disease. Population modeling can help us make wise choices that benefit both wildlife and people.

Create a spreadsheet modeling the growth of the deer population in the state of Wisconsin.

Enter 1,500,000 in cell A1 to represent an initial deer population. Enter a growth rate of 0.04 in cell A2. Enter 0 and 1 in cells B1 and B2 respectively. Replicate the values in column B to 20 by highlighting the 2 values and dragging the fill handle.

Now enter the growth function in cell C1 as =$A$1*(1+$A$2)^B1.

	A	B	C	D
1	1500000	0	=$A$1*(1+$A$2)^B1	
2	0.04	1		
3		2		
4		3		

The $ signs signify an absolute reference. When you replicate this formula, the "A" and "1" in the formula will not change. Without the $ signs, these elements would change relative to the cell position. Now replicate the formula in C1 down the column to show 20 yr of population growth.

Use your spreadsheet to answer the following questions.

4. If Wisconsin's deer population is 1.5 million and the managed population growth is 4%, what is the estimated population in 10 yr?
5. If the deer population of 1.5 million goes unmanaged and the growth rate is 29%, what is the estimated population in 10 yr? (*Hint*: Change the growth rate in your spreadsheet.)
6. What is the predicted unmanaged deer population in 20 yr if the growth rate is 23%?

The US population was estimated to be 329,484,000 in July 2020. The US population growth rate has slowed considerably in recent decades, from more than 2% in 1950 to less than 0.4% in 2020. The consistent downward trend in the population growth rate since 2013 may soon approach 0.

Create a spreadsheet modeling the population growth of the United States.

Enter the July 2020 US population in cell A1 and a growth rate of 0.003 in cell A2. Enter 0 and 1 in cells B1 and B2 respectively. Enter 2020 and 2021 in cells C1 and C2 respectively. Replicate the values in columns B and C to 40 yr and 2060 by highlighting these 4 cells and dragging the fill handle.

Enter the following growth formula in cell D1: =$A$1*(1+$A$2)^B1.

	A	B	C	D	E
1	329484000	0	2020	=$A$1*(1+$A$2)^B1	
2	0.003	1	2021		
3		2	2022		
4		3	2023		
5		4	2024		

Recall that the $ signs signify absolute references, so those elements do not change when a formula is replicated. Replicate the formula in cell D1 down the column to model the predicted population until 2060.

Use your spreadsheet to answer the following questions.

7. What is the predicted US population for 2060 with a growth rate of 0.3%?
8. What would the 2060 US population be if the annual growth rate were 1.1%? (*Hint*: Change the growth rate in your spreadsheet.)
9. Assuming an annual 0.3% growth rate, predict the first year in which the US population would exceed 500 million people. Be careful to replicate the data in all related columns.
10. Name 2 major factors that would cause the prediction in exercise 9 to be significantly inaccurate.
11. The world's population grew from about 6.034 billion to 6.114 billion from 1999 to 2000, and from about 7.673 billion to 7.753 billion from 2019 to 2020. Find the percent increase in the world's population during each period (to the nearest tenth of a percent).
12. The world's population was estimated to be about 7.960 billion in 2022 and increasing with an annual growth rate of about 1.1%. Use these estimates to predict the world's population in 2040 (to the nearest million).

8 CHAPTER REVIEW

Key Concepts

Classifying & Evaluating Polynomials [8.1]

- Polynomials can be classified by their degree and the number of terms.
  - The degree of a term is the sum of the exponents of its variables.
  - The degree of a polynomial is determined by the highest degree of any term in the polynomial.
  - Polynomials can be more specifically named depending on the number of terms (monomial: 1 term; binomial: 2 terms; trinomial: 3 terms).
- To evaluate polynomials, substitute values for the variable(s) and simplify the resulting expression by using the order of operations.

Adding & Subtracting Polynomials [8.2]

- Addition of polynomials is the same as combining like terms.
- Subtraction can be written as adding the opposite and then the like terms can be combined.

Multiplying Polynomials [8.3]

MULTIPLYING POLYNOMIALS

1. Arrange each polynomial in descending powers of the same variable.
2. Multiply each term of the multiplier by each term of the multiplicand.
3. Combine like terms to obtain the final product.

Multiplying Binomials by Using FOIL [8.4]

- FOIL is an acronym that helps find the product of binomials.

First, Outer

$(a + b)(c + d)$

Inner, Last

$= ac + ad + bc + bd$

F O I L

Special Products [8.5]

- square of a sum: $(a + b)^2 = a^2 + 2ab + b^2$
- square of a difference: $(a - b)^2 = a^2 - 2ab + b^2$
- product of conjugates: $(a + b)(a - b) = a^2 - b^2$

Dividing Polynomials [8.6]

DIVIDING POLYNOMIALS

1. Arrange the terms in both the dividend and the divisor in descending degree.
2. Complete the divide, multiply, and subtract procedure.
   a. Divide the first term of the dividend by the first term of the divisor and write the result in the quotient.
   b. Multiply the result by the entire divisor and place the product under the dividend.
   c. Subtract this product from the dividend and bring down the next term to obtain a new dividend.
3. Repeat step 2 until the degree of the new dividend is less than the degree of the divisor. Place any remainder over the divisor and add the resulting rational expression to the quotient.
4. Check your results by multiplying your answer by the divisor.

Biblical Worldview Shaping

Foundations of Algebra

How has God interpreted the world for us?

God has designed the world in a particular way. Even if we are not aware of or dispute this fact, we are subject to the limitations and freedoms inherent in the way God has created the world. We find this interpretation in both special and general revelation.

EXERCISES

Classify each expression as a monomial, binomial, trinomial, other polynomial, or not a polynomial. Then state the degree of each polynomial. [8.1]

1. $x - 5$
2. 7
3. $3x^2y - 6xy + 4y$
4. $4x^{-3} + 3x^{-2} - x^{-1} + 5$

Evaluate when $x = 2$, $y = -1$, and $z = 0$. [8.1]

5. $4x^3 + 2x - 5z$
6. $8y^4 + 7y^2 + 3y + 9$
7. $3x^2 - 5xz$
8. $6y^2 - 3y - 7x$

Simplify each expression. [8.2–8.4]

9. $(2x^2 + 3x - 5) + (7x^2 - 9)$
10. $(r^2 + 2t) + (r^2 + 6rt - 9t)$
11. $(5a^2 - ab - 3b^2) - (6a^2 - 7ab + 10b^2)$
12. $(15a^2 - 7ab - 12b + 3c^2) - (8ab - 3b + 12c^2)$
13. $(x^2 + y^2) - (x + y^2) + (12x^2 + y)$
14. $\left(\frac{1}{2}x^2 + \frac{1}{3}x\right) + \left(\frac{2}{3}x - \frac{1}{2}\right) - \left(\frac{1}{3}x^2 - \frac{1}{2}\right)$
15. $3x(x^2 - 5x + 4)$
16. $7xy^2(3x - 5xy^4)$
17. $-3a^4bc^2(2a^3b + 7abc^5)$
18. $(x^2 + 3x + 5)(x + 2)$

19. $(x + 2)(x^2 - 2x + 4)$

20. $(2b^2 - b)(b - 6)$

21. $(x + 4)(x - 6)$

22. $(a + 5)(a + 7)$

23. $(x - 9)(x - 11)$

24. $(a + 3b)(a - 7b)$

25. $(3x - 7)(2x + 5)$

26. $(x^3 - 2x)(2x^3 - 3x)$

27. Noah wants to mount a picture with dimensions in a ratio of 4 : 3. He uses a custom frame that is 2 in. wide on 3 sides and 3 in. wide on the top. Write a polynomial expression for the total area of the picture and frame. **[8.4]**

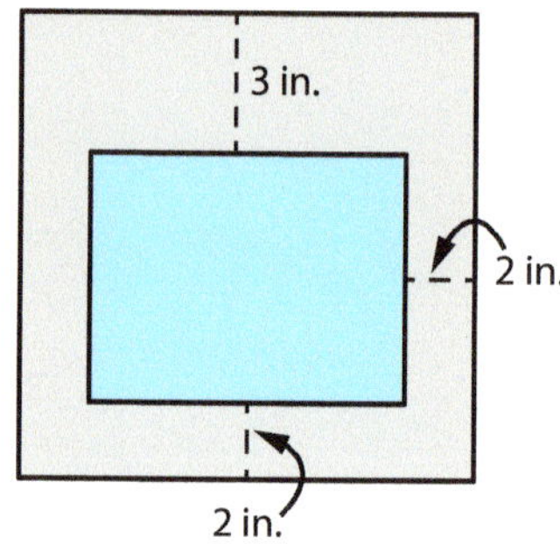

28. Priya, an interior designer, painted a circle with a radius of r inches on the wall of a facility. In another space she wants to paint a circle with a radius of 5 in. more than twice that of the first circle. Write a polynomial expression modeling the area of the larger circle. **[8.4]**

Simplify each expression. [8.5–8.6]

29. $(y - 6)(y + 6)$

30. $(x + 3)^2$

31. $(x - 2y)^2$

32. $(2y - 3)^2$

33. $(4a - 5b)^2$

34. $(7xy + 3z^3)(7xy - 3z^3)$

35. $(12c^5 + 15d^4)^2$

36. $(x + 3)(x - 3)^2$

37. $\frac{25x^2 - 35x}{5x}$

38. $\frac{21x^3y^4 + 7x^2y}{7xy}$

39. $(8x^2y^3z + 12xy^5) \div 2xy^2$

40. $(15x^3 + 27x^2 - 12x) \div 3x$

41. $(6x^2 + 21x + 9) \div (x + 3)$

42. $(x^2 + x + 1) \div (x - 1)$

43. $\frac{x^2 + 5}{x - 2}$

44. $\frac{8x^3 + 27}{3 + 2x}$

Write a polynomial expression for each of the following. [8.2–8.5]

45. the width of the rectangle

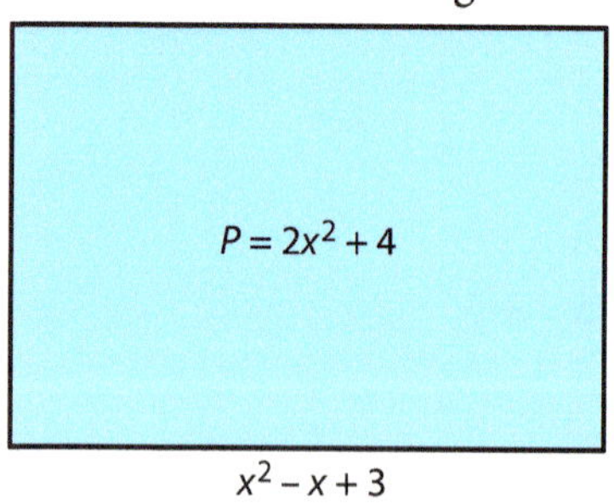

46. the perimeter of the triangle

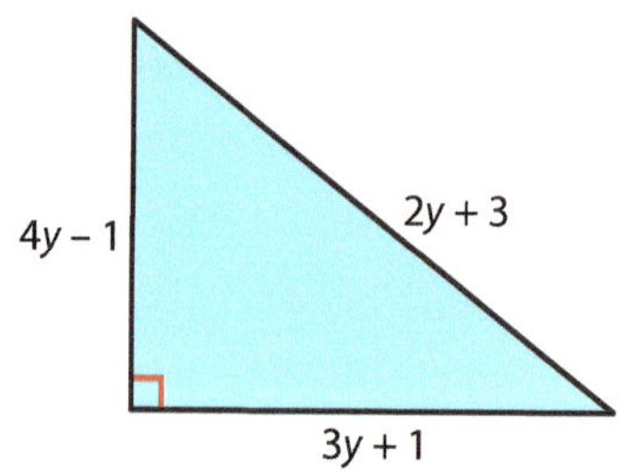

47. the area of the rectangle

48. the area of the triangle

49. the area of the shaded region

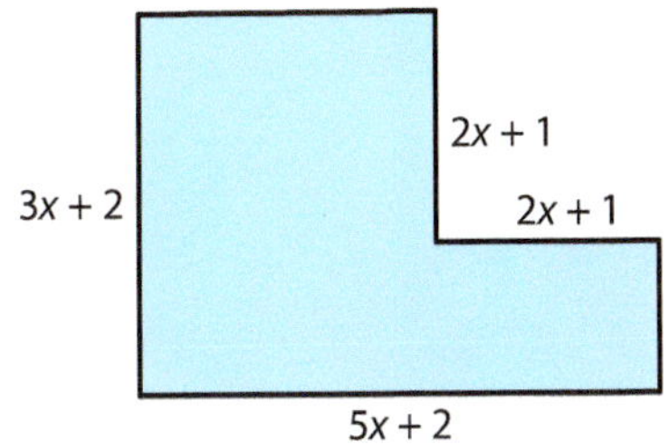

50. the area of the shaded region

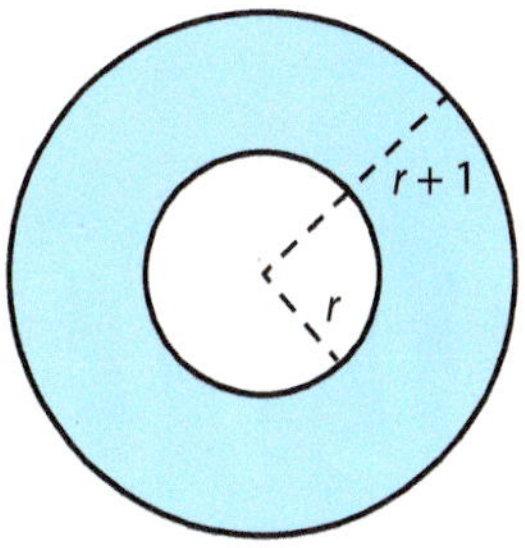

51. Corners with side length x are cut from an 8.5 in. × 11 in. piece of cardboard, and the resulting flaps are folded up and fastened to make a box. Write a polynomial expression in standard form modeling each of the following.

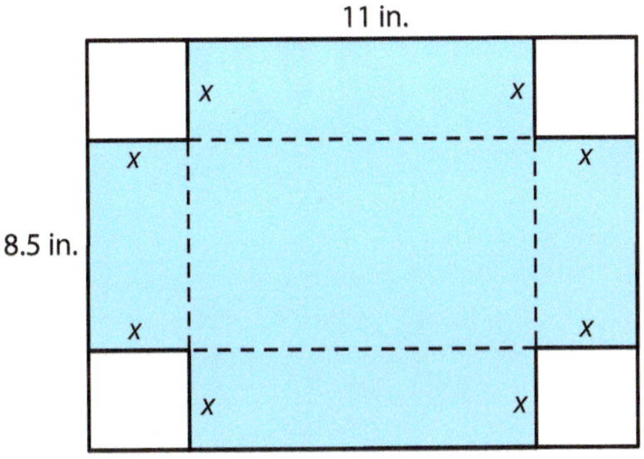

a. the length of the box

b. the width of the box

c. the area of the bottom of the resulting box

d. the volume of the resulting box

52. **BWS Essential Question:** In algebra, to say "God has interpreted the world for us" means which of the following? Choose all that apply.

A Scripture is a textbook for algebra.

B God has designed the world in a particular way.

C Scripture does not apply to algebra.

D There are freedoms and limitations in algebra based on God's design.

CHAPTER 09
FACTORING POLYNOMIALS

CHAPTER OVERVIEW

9.1 FACTORING BY USING THE DISTRIBUTIVE PROPERTY

How is the Distributive Property used in factoring?

After completing this section, I will be able to

- identify the GCF of the terms of a polynomial.
- factor a common monomial or binomial from a polynomial.
- explain how the Distributive Property justifies factoring out a monomial from a polynomial.

A *factor* is one of 2 or more quantities multiplied to form a product. When you learned to find the greatest common factor (GCF) of 2 numbers, you factored each number into its prime factors. Factoring is the process of writing an algebraic or numerical expression as a product.

EXAMPLE 1: Finding the GCF

Find the GCF of the given monomials.

a. 108 and 72 **b.** $15x$ and $42x^2y$

Answers

a. $108 = 2 \cdot 2 \cdot 3 \cdot 3 \cdot 3$
$72 = 2 \cdot 2 \cdot 2 \cdot 3 \cdot 3$

1. Factor each number into its prime factors.

$\text{GCF} = 2 \cdot 2 \cdot 3 \cdot 3 = 36$

2. Find the GCF by finding the product of the shared factors.

b. $15x = 3 \cdot 5 \cdot x$
$42x^2y = 2 \cdot 3 \cdot 7 \cdot x \cdot x \cdot y$

1. Factor each term into its prime factors.

$\text{GCF} = 3 \cdot x = 3x$

2. Find the GCF by finding the product of the shared factors.

SKILL CHECK **EXERCISES 1, 5** ☑

In this section you will use a common monomial factor to factor a polynomial by applying the Distributive Property in reverse.

DEFINITIONS

The **greatest common factor** of a polynomial is the product of all factors shared by every term of the polynomial.

Two numbers or terms are considered **relatively prime** if their greatest common factor is 1.

EXAMPLE 2: Factoring Out a Monomial

Factor each of the polynomials.

a. $8x^2 - 16$ **b.** $42a^3 + 14a$

Answers

a. $8x^2 = 2 \cdot 2 \cdot 2 \cdot x \cdot x$ $16 = 2 \cdot 2 \cdot 2 \cdot 2$	**1.** Factor each term into its prime factors. Both terms contain 3 factors of 2.
$\text{GCF} = 2 \cdot 2 \cdot 2 = 8$	**2.** Find the GCF.
$8x^2 - 16 = 8(x^2) - 8(2)$ $= 8(x^2 - 2)$	**3.** Factor the GCF, 8, from each term of the polynomial.
$8(x^2 - 2) = 8x^2 - 16$	**4.** Check using the Distributive Property.
b. $42a^3 = 2 \cdot 3 \cdot 7 \cdot a \cdot a \cdot a$ $14a = 2 \cdot 7 \cdot a$ $\text{GCF} = 2 \cdot 7 \cdot a = 14a$	**1.** Factor each term and determine the GCF. Both terms contain 1 factor each of 2, 7, and a.
$42a^3 + 14a$ $= 14a(3a^2) + 14a(1)$	**2.** Use the Commutative and Associative Properties to factor the GCF from each term.
$= 14a(3a^2 + 1)$	**3.** Apply the Distributive Property in reverse to obtain 2 factors.
$14a(3a^2 + 1) = 42a^3 + 14a$	**4.** Check using the Distributive Property.

SKILL CHECK **EXERCISES 7, 11**

Notice that we "proved" the answer was correct by appealing to the Distributive Property. That means our "faith" is in the property. The problems worked out in Example 2 are evidence that the property holds, but if they appeared not to be true, we would recheck our work to find our error, instead of rejecting the property. A similar relation is true between the patterns we see through math and science and our faith in Scripture. Patterns can provide evidence that supports Scripture, but Scripture is the ultimate standard.

FACTORING COMMON MONOMIALS FROM A POLYNOMIAL

1. Find the GCF of all terms in the polynomial.
2. Divide the GCF out of each term (applying the Distributive Property in reverse).
3. Write the factorization as the GCF times the sum of the quotients.
4. Check your solution by multiplying the 2 factors.

Not all polynomials can be factored using these methods. If no common factor other than 1 is found, then the polynomial is considered *prime*.

USING PRIME FACTORS

$36x^2y$
$= 2 \cdot 2 \cdot 3 \cdot 3 \cdot x \cdot x \cdot y$
$9xy^3$
$= 3 \cdot 3 \cdot x \cdot y \cdot y \cdot y$
$27xy$
$= 3 \cdot 3 \cdot 3 \cdot x \cdot y$
$\text{GCF} = 3 \cdot 3 \cdot x \cdot y$

EXAMPLE 3: Factoring Out a Monomial

Factor each polynomial, if possible.

a. $36x^2y + 9xy^3 - 27xy$ **b.** $51x^3 + 26y$

Answers

a. $2^2 \cdot 3^2x^2y + 3^2xy^3 - 3^3xy$

$\text{GCF} = 3^2xy = 9xy$

1. The GCF uses the lowest power of all common factors.

$9xy(2^2x) + 9xy(y^2) - 9xy(3)$

2. Factor the GCF from each term.

$= 9xy(4x + y^2 - 3)$

3. Write the factorization as the GCF times the sum of the quotients.

$9xy(4x + y^2 - 3)$
$= 36x^2y + 9xy^3 - 27xy$

4. Check using the Distributive Property.

b. $51x^3 + 26y$

$3 \cdot 17x^3 + 2 \cdot 13y$

1. After factoring each term, you can see that the only common factor is 1 and that the terms are relatively prime.

prime

2. When the terms are relatively prime, the polynomial is prime and will not factor.

SKILL CHECK **EXERCISES 19, 21**

AFTERSCHOOLHELP

Factoring by Using the Distributive Property

A polynomial can contain a common binomial factor, as in

$$3(x + y) - y(x + y).$$

You can factor out the common binomial by considering $(x + y)$ as a single quantity and applying the Distributive Property:

$$3(x + y) - y(x + y) = (x + y)(3 - y).$$

This is just like factoring the expression

$$3z - yz = z(3 - y),$$

where z represents the binomial.

EXAMPLE 4: Factoring Out a Binomial

Factor $x(x + 1) + 2(x + 1)$.

Answer

$x(x + 1) + 2(x + 1)$

1. The binomial $x + 1$ is a factor of each term.

$(x + 1)(x + 2)$

2. Apply the Distributive Property in reverse to write the sum as a product of polynomials.

SKILL CHECK **EXERCISE 27**

VOCABULARY

- greatest common factor
- relatively prime

A. EXERCISES

Find the GCF of each pair of monomials. If the GCF is 1, write *relatively prime*.

1. 45 and 75

2. 33 and 51

3. $24x$ and $32xy$

4. $7xz^2$ and $15y$

5. x^2y^3 and x^3y^2

6. $4ab^2$ and $10a^2b$

Factor each expression. If the polynomial will not factor, write *prime*.

7. $3x + 3y$ **8.** $ax + ay$ **9.** $8x - 32y$

10. $-50x + 10$ **11.** $a^2 - a$ **12.** $x^2y + wz$

13. $7x^2 - 49xy + 28y^2$ **14.** $6a^2 + 48a - 24$ **15.** $5x + 7y - 11z$

B. EXERCISES

Factor each expression. If the polynomial will not factor, write *prime*.

16. $300x^2 - 100x$ **17.** $6a^2b + 3ab^2$ **18.** $10a^2bc^3 + 5ab^2c$

19. $91a^3b^2 + 13a^2b$ **20.** $-4\pi r + 2\pi x$ **21.** $14x^2y - 15z$

22. $20a^3x + 50ax + 10x$ **23.** $56x^2 + 42xy - 28x$ **24.** $15x^2 - 16y + 19z^3$

25. $7a^5 - 84a^3 + 21a$ **26.** $x^5 + x^3 - x^2 + x$ **27.** $x(x - 2) + 8(x - 2)$

28. $4y(2y - 1) - 3(2y - 1)$ **29.** $5(b + c) + a(b + c)$ **30.** $6x(y - z) - (y - z)$

31. **Essential Question:** Priscilla stated that the factored form of $5a + 7$ was $5(a + 2)$.

- **a.** What property can be applied to check Priscilla's answer?
- **b.** Apply the property mentioned in part a to $5(a + 2)$ to check Priscilla's answer.
- **c.** Which is incorrect: Priscilla's answer or the property mentioned in part a? Explain.

Use factoring to solve.

32. If the area of a rectangle is 42 in.^{2}, list all possible natural number dimensions of the rectangle.

33. Mrs. Shropshire is coordinating the third- and fourth-grade program. She wishes to arrange the children into the fewest number of rows that contain the same number of children. She also does not wish to mix the classes within a row. If there are 60 third graders and 48 fourth graders, how many children should be placed in each row? How many rows will be needed?

34. Crumb Bull Cookie Company is repackaging their cookies into 2 different size boxes that will contain either 52 or 117 cookies each. If each box is to contain rows of the same number of cookies, how many cookies will be in each row?

35. The area of the rectangle to the right is $18x^2 - 6x$, and its width is $6x$. Find the length of the rectangle.

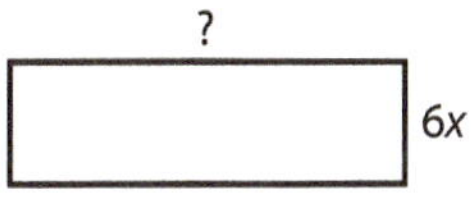

36. MP2 The circumference of a circle is $2x\pi + 2y\pi$. Write the factored form of this expression. What is the radius of the circle?

C. EXERCISES

Factor each expression.

37. $6x^4y^5z^3 - 27x^3y^6z^4 + 18x^2y^4z^5$

38. $34a^7b^8c^2 + 51a^3b^5d^4 - 289a^2b^9c^{10}d$

39. $24xz(x + 1) + 16yz(x + 1) + 40yz(x + 1)$

40. MP7 $3m^7n^8(4m^3 - 5n^2) - 2m^6n^7(20m^4n - 25mn^3) + 11n^5(4m^8 - 5m^5n^2)$

The diagram to the right illustrates an insulated cylinder. Let V represent the volume of the insulated region between the inner and outer cylinders.

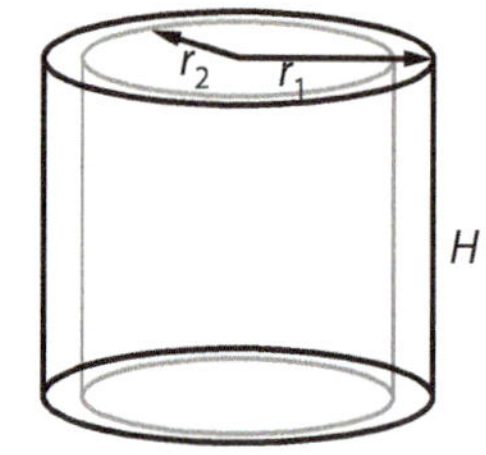

41. MP4 Write an equation modeling the volume of the insulated region. Then factor the right-hand side of the formula.

42. If the height of the cylinder is 12 in. with an outer radius of 5 in. and an inner radius of 4.5 in., find the volume of the region between the 2 cylinders in terms of π.

CUMULATIVE REVIEW

Use the mapping diagram for exercises 43–47. [4.1]

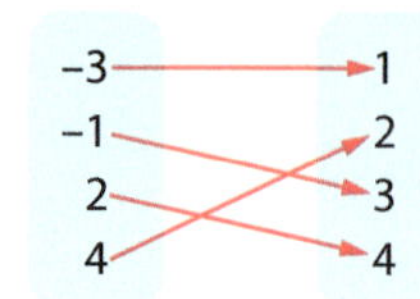

43. State the domain of the relation.

44. State the range of the relation.

45. Is the relation a function?

46. Write the relation as a set of ordered pairs.

47. If $(x, 0)$ was included in the relation, for what value(s) of x would the relation not be a function?

Multiply. [8.4]

48. $(x - 5)(x - 8)$

49. $3(x - 2)(x + 7)$

50. $4y(x + 1)(x - 8)$

51. Which function models a yearly 10% depreciation of a car originally valued at $35,000? **[7.5]**

A $f(x) = 35{,}000(0.9)^x$

B $f(x) = 35{,}000(0.1)^x$

C $f(x) = 35{,}000(-0.1)^x$

D $f(x) = 35{,}000(1.1)^x$

E none of these

52. Find the quotient of $3x^3 - 22x - 15$ and $x - 3$. **[8.6]**

A $3x^2 - 9x + 5$

B $3x^2 + 9x + 5$

C $3x - 13 - \frac{54}{x - 3}$

D $3x - 31 + \frac{78}{x - 3}$

E none of these

MIND OVER MATH

A pentomino is a figure consisting of 5 adjoining congruent squares.

Determine whether each figure can be formed using the pentominoes pictured above without any overlaps or rotations.

1.

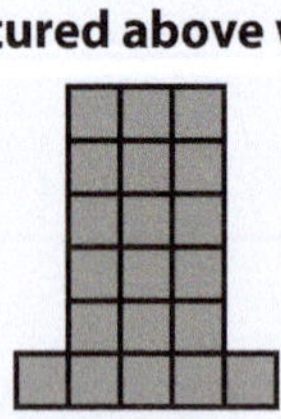

2.

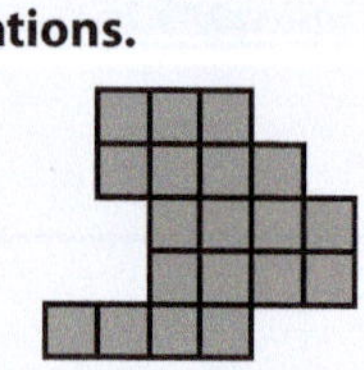

3.

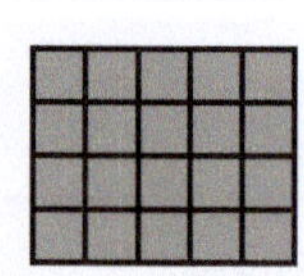

Determine whether each figure can be formed using the pentominoes pictured above if rotations and overlaps are allowed.

4.

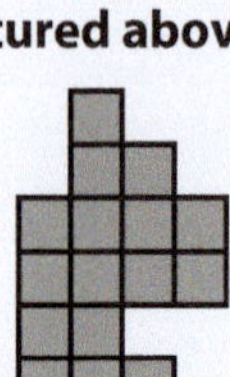

5.

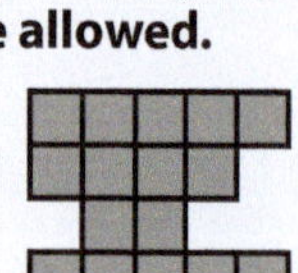

6.

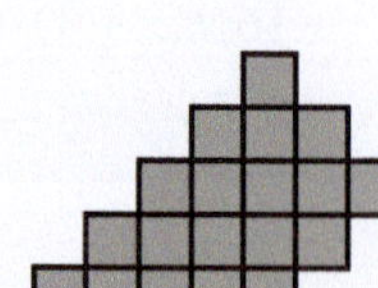

9.2 FACTORING TRINOMIALS OF THE FORM $x^2 + bx + c$

How can I know whether $x^2 + bx + c$ is factorable using integers?

After completing this section, I will be able to

- relate the FOIL method to factoring trinomials.
- factor a trinomial of the form $x^2 + bx + c$.

Recall that the FOIL method can be used to find the product of 2 binomial factors.

$$\begin{aligned}(x + 5)(x + 7) &= x^2 + 7x + 5x + (5 \cdot 7)\\ &= x^2 + (7 + 5)x + 35\\ &= x^2 + 12x + 35\end{aligned}$$

sum of the constants (12) — product of the constants (35)

Notice that the constant term of the trinomial, 35, is the product of the binomial's constants, 5 and 7, and that the coefficient of the middle term, 12, is their sum. To factor a trinomial of the form $x^2 + bx + c$, you must find the 2 binomials that were multiplied to obtain the trinomial. The key is to identify the 2 factors of c whose sum is b. In $x^2 + 12x + 35$, we want to identify 2 factors of 35 whose sum is 12.

EXAMPLE 1: Factoring a Trinomial of the Form $x^2 + bx + c$

Factor $y^2 + 5y + 6$.

Answer

$y^2 + 5y + 6$ $(y \quad)(y \quad)$	**1.** Write the factors of y^2 as the first terms in the binomials.
$6 = 3(2)$ and $3 + 2 = 5$ $(y + 3)(y + 2)$	**2.** Find factors of $+6$ whose sum is $+5$.
$(y + 3)(y + 2)$ $= y^2 + 2y + 3y + 6$ $= y^2 + 5y + 6$	**3.** Multiply the binomials using the FOIL method to check your answer.

SKILL CHECK **EXERCISE 5**

FACTORING TRINOMIALS OF THE FORM $x^2 + bx + c$

1. Write the factors of the first term.
2. Find factors of the constant term, c, whose sum is b. Write these factors as the last terms in the binomials.
3. Check the results by multiplying the binomials.

EXAMPLE 2: Factoring a Trinomial When b Is Negative

Factor $x^2 - 5x + 6$.

Answer

$x^2 - 5x + 6$ $(x \quad)(x \quad)$	**1.** Write the factors of x^2 as the first terms in the binomials.
$6 = -2(-3)$ and $-2 + (-3) = -5$	**2.** Find factors of $+6$ whose sum is -5. A positive product implies that the factors have the same sign.
$(x - 2)(x - 3)$	**3.** Write these factors as the last terms in the binomials.
$(x - 2)(x - 3)$ $= x^2 - 3x - 2x + 6$ $= x^2 - 5x + 6$	**4.** Multiply the binomials using the FOIL method to check your answer.

SKILL CHECK **EXERCISE 7**

DETERMINING THE SIGNS OF THE FACTORS OF c

1. If the sign of c is positive $(x^2 \pm bx + c)$,
   a. the signs are the same and
   b. the factors will have the sign of the trinomial's middle term.
2. If the sign of c is negative $(x^2 \pm bx - c)$,
   a. the signs are different and
   b. the larger factor will have the sign of the trinomial's middle term.

EXAMPLE 3: Factoring a Trinomial When *c* Is Negative

Factor each trinomial.

a. $x^2 + x - 72$ **b.** $x^2 + 3x - 6$

Answers

a. $x^2 + 1x - 72$

$(x \quad)(x \quad)$

$-72 = 9(-8)$ and $9 + (-8) = 1$

$(x + 9)(x - 8)$

$(x + 9)(x - 8)$

$= x^2 - 8x + 9x - 72$

$= x^2 + x - 72$

1. Factor x^2.
2. Find factors of -72 whose sum is $+1$.
3. Check using FOIL.

b. $x^2 + 3x - 6$

$(x \quad)(x \quad)$

$-6 = -1(6)$, but $-1 + 6 \neq 3$.

$-6 = 1(-6)$, but $1 + (-6) \neq 3$.

$-6 = -2(3)$, but $-2 + 3 \neq 3$.

$-6 = 2(-3)$, but $2 + (-3) \neq 3$.

$(x + ?)(x + ?)$

$x^2 + 3x - 6$

1. Factor x^2.
2. Find factors of -6 whose sum is $+3$.
3. Since there are no factors of -6 whose sum is $+3$, the trinomial is prime.

SKILL CHECK **EXERCISES 11, 19**

AFTERSCHOOLHELP

Factoring Trinomials of the Form $x^2 + bx + c$

The first step in factoring any polynomial is to factor out any common monomial factors. When the constant term is a large number, a more organized approach may be required to find the correct pair of factors.

EXAMPLE 4: Applying Multiple Factoring Methods

Write $-64a - 288 + 2a^2$ in descending order, then factor completely.

Answer

$2a^2 - 64a - 288$

$2(a^2 - 32a - 144)$

$2(a \quad)(a \quad)$

Factors	Sum
1(−144)	−143
2(−72)	−70
3(−48)	−45
4(−36)	−32

$2(a + 4)(a - 36)$

$2(a + 4)(a - 36)$

$= 2(a^2 - 36a + 4a - 144)$

$= 2(a^2 - 32a - 144)$

$= 2a^2 - 64a - 288$

1. Rearrange terms in descending order.
2. Factor out the common factor of 2.
3. Factor the first term of the trinomial, a^2.
4. Find factors of -144 whose sum is -32. The signs must be different, with the larger factor being negative. Use an organized list to find the correct factors. Note that the complete set of factors is not listed. Stop when you find the correct sum.
5. To check your answer, multiply the binomials using FOIL and then distribute the 2.

SKILL CHECK **EXERCISE 25**

Some trinomials contain 2 variables. Consider the following product:

$(x + 5y)(x - 4y) = x^2 - 4xy + 5xy - 20y^2$
$= x^2 + xy - 20y^2$.

Notice that the factors of $-20y^2$ are $5y$ and $-4y$. When factoring a trinomial of the form of $x^2 + bxy + cy^2$, be sure to include a factor of y in the second term of each binomial.

EXAMPLE 5: Factoring a Trinomial of the Form $x^2 + bxy + cy^2$

Factor $x^2 + 11xy + 24y^2$.

Answer

$x^2 + 11xy + 24y^2$ $(x \quad y)(x \quad y)$	**1.** Factor the first term, x^2. Factor y^2 from the last term.
$24 = 3(8)$ and $3 + 8 = 11$ $(x + 3y)(x + 8y)$	**2.** Find factors of $+24$ whose sum is $+11$. Both signs must be positive.
$(x + 3y)(x + 8y)$ $= x^2 + 8xy + 3xy + 24y^2$ $= x^2 + 11xy + 24y^2$	**3.** Check using FOIL.

SKILL CHECK **EXERCISE 31**

A. EXERCISES

Match each polynomial with its factorization.

1. $x^2 + 7x + 12$
2. $x^2 - 7x + 12$
3. $x^2 + x - 12$
4. $x^2 - x - 12$

A $(x - 3)(x + 4)$
B $(x - 3)(x - 4)$
C $(x + 3)(x + 4)$
D $(x + 3)(x - 4)$

Factor each trinomial. If the trinomial will not factor, write *prime*.

5. $a^2 + 8a + 15$
6. $x^2 + 15x + 54$
7. $a^2 - 4a + 3$
8. $x^2 + 4x - 32$
9. $b^2 + b - 30$
10. $x^2 - 2x - 3$
11. $a^2 + 2a - 8$
12. $a^2 + 3a - 28$
13. $x^2 + 7x + 4$
14. $a^2 - 4a + 5$
15. $x^2 - 11x + 30$
16. $x^2 - 11x + 10$
17. $y^2 + 7y + 10$
18. $x^2 - 6x + 12$
19. $x^2 - 10x - 40$
20. $a^2 - 13a - 30$

B. EXERCISES

Write each trinomial in descending order and factor completely.

21. $56 + 18x + x^2$
22. $-26y + y^2 + 48$
23. $15x + x^2 - 54$
24. $6b^2 - 42b + 72$
25. $3x^2 - 24x - 27$
26. $ax^2 - 11ax - 12a$
27. $24x + 36 + 3x^2$
28. $4x^2 + 96 - 44x$
29. $6x^3 - 16x^2 + x^4$

Factor each trinomial. If the trinomial will not factor, write *prime*.

30. $x^2 + 20xy + 100y^2$

31. $x^2 + 6xy + 9y^2$

32. $x^2 - 50xy + 96y^2$

33. $x^2 - 18xy + 45y^2$

34. $a^2 - 14ab - 72b^2$

35. $x^2 - 9xy - 24y^2$

36. **Essential Question:** How can I know if $x^2 + bx + c$ is factorable using integers?

37. **MP3** Describe and correct the error Luis made when he factored $x^2 - 10x - 24$ as $(x - 4)(x - 6)$.

38. **MP3** Describe and correct the error Louise made when she concluded that $2x^2 - 18x + 40$ is prime since there are no integral factors of 40 whose sum is -18.

39. The area of a rectangle can be expressed as $x^2 - 18x + 56$. Factor to find binomial expressions representing the length and width. What is the difference between the length and the width?

40. **MP1** The area of the observable portion of the photo is represented by $x^2 + 25x + 60$. Express the area of the entire photo as a polynomial. Then factor to find binomial expressions representing its length and width.

C. EXERCISES

Factor each polynomial.

41. $-ax^3 + 45ax^2 + 144ax$

42. $4a^4b - 12a^3b^2 - 40a^2b^3$

43. $a^6 + 4a^3b^2 - 117b^4$

44. **MP7** Determine all integral values of b for which $x^2 + bx - 6$ is factorable.

CUMULATIVE REVIEW

Use the polynomial $6x^3y^3 - 27x^2y^4z + 8x^2y^3$ for exercises 45–50.

45. State the coefficient of the last term. **[1.6]**

46. Classify the polynomial by its number of terms. **[8.1]**

47. State the degree of the first term. **[8.1]**

48. State the degree of the polynomial. **[8.1]**

49. Divide the polynomial by $9xy^3$. **[8.6]**

50. Factor the polynomial. **[9.1]**

Write an equation and solve. [2.7]

51. Josh leaves his home near Tallahassee and travels west toward Pensacola. Marcus leaves his home near Pensacola at the same time and travels east to meet Josh. If the distance between their homes is 175 mi, Marcus travels at 55 mi/hr, and the trip takes 90 min, find Josh's average speed (to the nearest mile per hour).

52. Arianna and Bethany plan to meet at the church, but Bethany has an extra 10 mi to travel. It took Arianna 12 min to travel 7 mi, and it took Bethany 25 min. What was the average rate for each driver (to the nearest mile per hour)?

53. Which product will be a binomial when simplified? **[8.5]**

A $(2a + 1)^2$

B $(4b - 5)(4b + 5)$

C $(3c - 8)^2$

D $(d - 7)(-7 + d)$

E B and D

54. $(k - 7)^2$ is equivalent to which expression? **[8.5]**

A $2k - 14$

B $k^2 + 49$

C $k^2 - 49$

D $k^2 - 14k - 49$

E $k^2 - 14k + 49$

TECHNOLOGY CORNER (TI-84 PLUS FAMILY)

Graphing Factored Polynomials

In Chapter 7 you graphed *quadratic* functions (containing an x^2 term) and *cubic* functions (containing an x^3 term). Looking at the graphs of these types of functions in factored form can be informative. Set your calculator's graphing window by pressing [zoom] [6]. Then press [y=], [clear] any functions that have been entered previously, and make sure all plots are off.

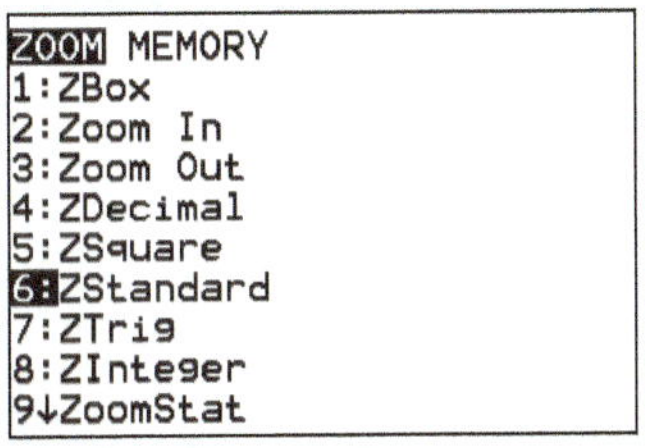

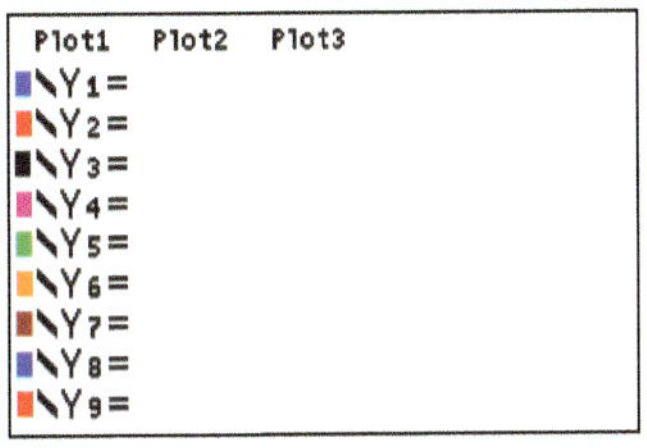

Graph the function $y = (x - 4)(x + 1)$. Where does the parabola cross the x-axis?

Graph the function $y = x(x + 2)(x - 3)$. What are the 3 points where the cubic function crosses the x-axis?

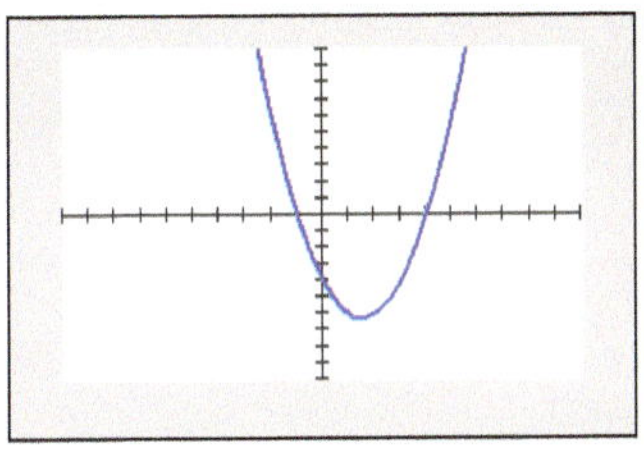

$y = (x - 4)(x + 1)$

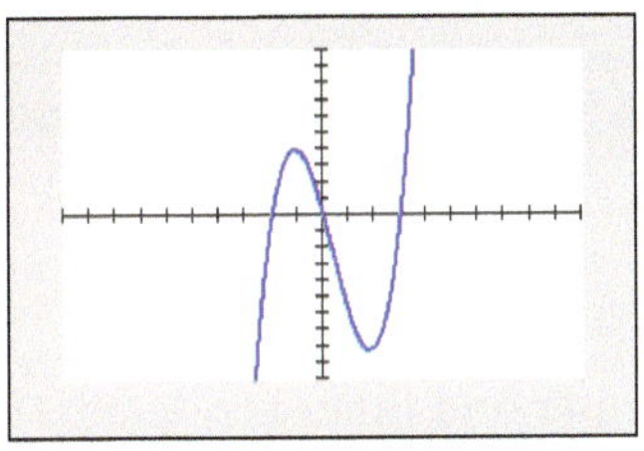

$y = x(x + 2)(x - 3)$

Note that the x-coordinates of the x-intercepts (also called the *zeros* of the function) have the opposite signs as the constant terms in the binomials. This is similar to what we saw in the horizontal translations in Sections 4.7 and 7.3.

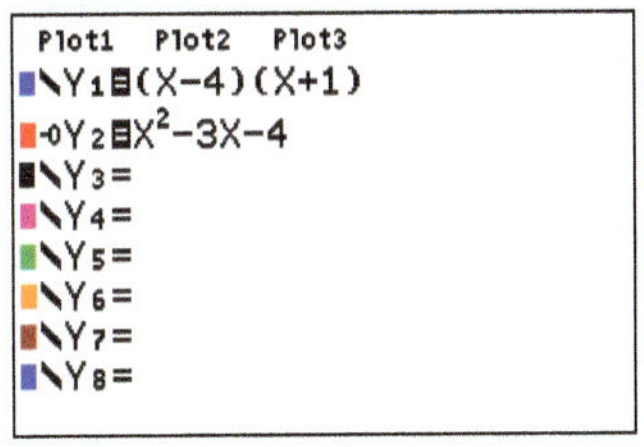

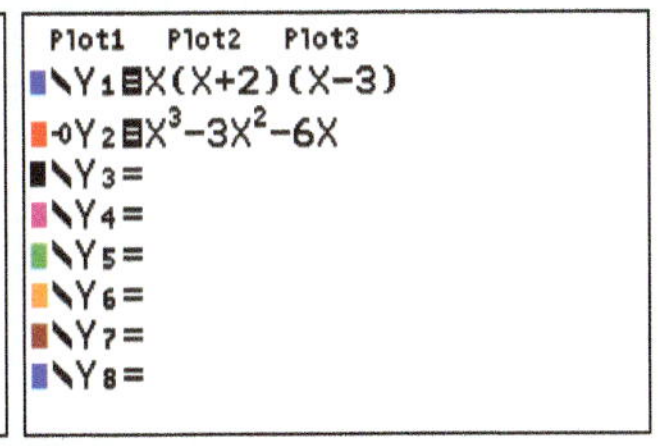

You can also check your factoring by graphing both the standard and factored forms of the expressions as functions. Their graphs should be identical.

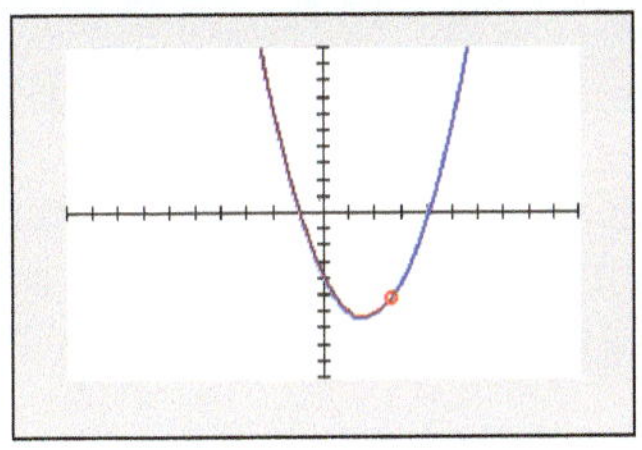

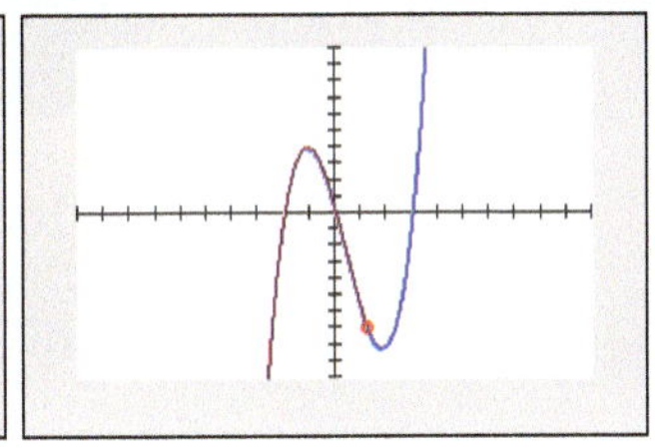

9.3 FACTORING TRINOMIALS OF THE FORM $ax^2 + bx + c$

How do I factor a trinomial of the form $ax^2 + bx + c$, where $a \neq 1$?

After completing this section, I will be able to

- factor a trinomial of the form $ax^2 + bx + c$ or $ax^2 + bxy + cy^2$.

To learn how to factor a trinomial of the form $ax^2 + bx + c$, observe the multiplication of 2 binomials that produce such a trinomial. Multiply the binomials $(3x - 2)(x + 5)$ using the FOIL method.

$$\begin{aligned}(3x-2)(x+5) &= 3x^2 + 15x - 2x - (2 \cdot 5)\\ &= 3x^2 + (15 - 2)x - 10\\ &= 3x^2 + 13x - 10\end{aligned}$$

Notice that the coefficients of the outer and inner products, 15 and −2, have a sum of +13 and that their product, −30, is the product of the leading coefficient and the constant term, 3(−10).

When factoring trinomials of the form $ax^2 + bx + c$, find factors of the product ac whose sum is b. Then rewrite the middle term as a sum of terms using these factors as coefficients.

KEYWORD SEARCH

British factoring method

EXAMPLE 1: Factoring a Trinomial of the Form $ax^2 + bx + c$

Factor $3x^2 + 7x + 2$.

Answer

$3x^2 + 7x + 2$	**1.** Notice that $ac = 6$ and $b = 7$.
$6 = 1(6)$ and $1 + 6 = 7$	**2.** Find factors of $+6$ whose sum is $+7$.
$3x^2 + 1x + 6x + 2$	**3.** Rewrite the middle term, $7x$, as a sum of terms with 1 and 6 as their coefficients.
$(3x^2 + 1x) + (6x + 2)$	**4.** Group the first 2 terms and the last 2 terms and factor the GCF from each grouping.
$x(3x + 1) + 2(3x + 1)$	**a.** Factor x from the first pair of terms and 2 from the second pair.
$(x + 2)(3x + 1)$	**b.** The polynomial now consists of 2 larger terms sharing a common binomial factor of $3x + 1$. Factor it from both terms.

CHECK USING FOIL

$(x + 2)(3x + 1)$
$= 3x^2 + x + 6x + 2$
$= 3x^2 + 7x + 2$

SKILL CHECK EXERCISE 1

FACTORING TRINOMIALS OF THE FORM $ax^2 + bx + c$

1. Find factors of ac whose sum is b.
2. Rewrite the middle term as a sum of terms with these factors of ac as their coefficients.
3. Factor the four-term polynomial by grouping the terms in pairs.
   a. Factor the common monomial from each pair.
   b. Factor the common binomial from each new term.
4. Check your factorization by multiplying the factors.

If the binomials that result from factoring the GCF from each grouping are not identical, an error has been made. Be sure to check your work.

EXAMPLE 2: Factoring a Trinomial When b Is Negative

Factor $2x^2 - 19x + 24$.

Answer

$2x^2 - 19x + 24$	1. Notice that $ac = 48$ and $b = -19$.
$48 = -3(-16)$ and $-3 + (-16) = -19$	2. Find factors of $+48$ whose sum is -19. Both factors must be negative.
$2x^2 - 3x - 16x + 24$	3. Rewrite the middle term, $-19x$, as a sum of terms with -3 and -16 as their coefficients.
$(2x^2 - 3x) + (-16x + 24)$	4. Group the first 2 terms and the last 2 terms and factor the GCF from each grouping.
$x(2x - 3) - 8(2x - 3)$	a. Factor x from the first pair of terms and -8 from the second pair.
$(x - 8)(2x - 3)$	b. Factor the common binomial, $2x - 3$, from both terms.

CHECK USING FOIL

$(x - 8)(2x - 3)$
$= 2x^2 - 3x - 16x + 24$
$= 2x^2 - 19x + 24$

SKILL CHECK **EXERCISE 3**

EXAMPLE 3: Factoring a Trinomial When c Is Negative

Factor $2x^2 + 3xy - 14y^2$.

Answer

$2x^2 + 3xy - 14y^2$	1. Notice that $ac = -28$ and $b = 3$.
$-28 = 7(-4)$ and $7 + (-4) = 3$	2. Find factors of -28 whose sum is $+3$. The signs must be different, with the larger factor being positive.
$2x^2 + 7xy - 4xy - 14y^2$	3. Rewrite the middle term, $3xy$, as a sum of terms with 7 and -4 as their coefficients.
$(2x^2 + 7xy) + (-4xy - 14y^2)$	4. Group the first 2 terms and the last 2 terms and factor the GCF from each grouping.
$x(2x + 7y) - 2y(2x + 7y)$	a. Factor x from the first pair of terms and $-2y$ from the second pair.
$(x - 2y)(2x + 7y)$	b. Factor the common binomial, $2x + 7y$, from both terms.

CHECK USING FOIL

$(x - 2y)(2x + 7y)$
$= 2x^2 + 7xy - 4xy - 14y^2$
$= 2x^2 + 3xy - 14y^2$

SKILL CHECK **EXERCISE 15**

AFTERSCHOOLHELP

Factoring Trinomials of the Form $ax^2 + bx + c$

EXAMPLE 4: Factoring a Trinomial When *a* Is Negative

Factor $-3x^2 + 20 + 7x$.

Answer

$-3x^2 + 7x + 20$

1. Arrange the polynomial in standard form before factoring.

$-1(3x^2 - 7x - 20)$

2. Factor −1 from each term to make $a > 0$. Notice that $ac = -60$ and $b = -7$.

$-60 = -12(5)$ and $-12 + 5 = -7$

3. Find factors of −60 whose sum is −7. The signs must be different, with the larger factor being negative.

$-1(3x^2 - 12x + 5x - 20)$

4. Rewrite the middle term, $-7x$, as a sum of terms with −12 and 5 as their coefficients.

$-1[(3x^2 - 12x) + (5x - 20)]$

5. Group the first 2 terms and the last 2 terms and factor the GCF from each grouping.

$-1[3x(x - 4) + 5(x - 4)]$

   a. Factor $3x$ from the first pair of terms and 5 from the second pair.

$-(3x + 5)(x - 4)$

   b. Factor the common binomial, $x - 4$, from both terms.

SKILL CHECK **EXERCISE 23**

CHECK USING FOIL

$-1(3x + 5)(x - 4)$
$= -1(3x^2 - 12x + 5x - 20)$
$= -1(3x^2 - 7x - 20)$
$= -3x^2 + 7x + 20$

Remember that the first step in factoring a polynomial is to check to see whether you can factor out a common monomial factor before applying other methods.

EXAMPLE 5: Applying Multiple Factoring Methods

Factor $12x^3 - 26x^2 + 10x$.

Answer

$2x(6x^2 - 13x + 5)$

1. Factor the common monomial $2x$ from each term. Then proceed to factor the remaining trinomial. Notice that $ac = 30$ and $b = -13$.

$30 = -10(-3)$ and $-10 + (-3) = -13$

2. Find factors of +30 whose sum is −13.

$2x(6x^2 - 10x - 3x + 5)$

3. Rewrite the middle term, $-13x$, as a sum of terms with −10 and −3 as their coefficients.

$2x[(6x^2 - 10x) + (-3x + 5)]$

4. Group the first 2 terms and the last 2 terms within brackets and factor the GCF from each grouping.

$2x[2x(3x - 5) + (-1)(3x - 5)]$

   a. Factor $2x$ from the first pair of terms and −1 from the second pair.

$2x[(2x - 1)(3x - 5)]$
$= 2x(2x - 1)(3x - 5)$

   b. Factor the common binomial, $3x - 5$, from both terms.

SKILL CHECK **EXERCISE 25**

CHECK USING FOIL

$2x(2x - 1)(3x - 5)$
$= 2x(6x^2 - 10x - 3x + 5)$
$= 2x(6x^2 - 13x + 5)$
$= 12x^3 - 26x^2 + 10x$

EXAMPLE 6: Applying Factoring to the Real World

Factor $3x^2 - 25x - 200$ to find expressions representing the dimensions of the soccer field.

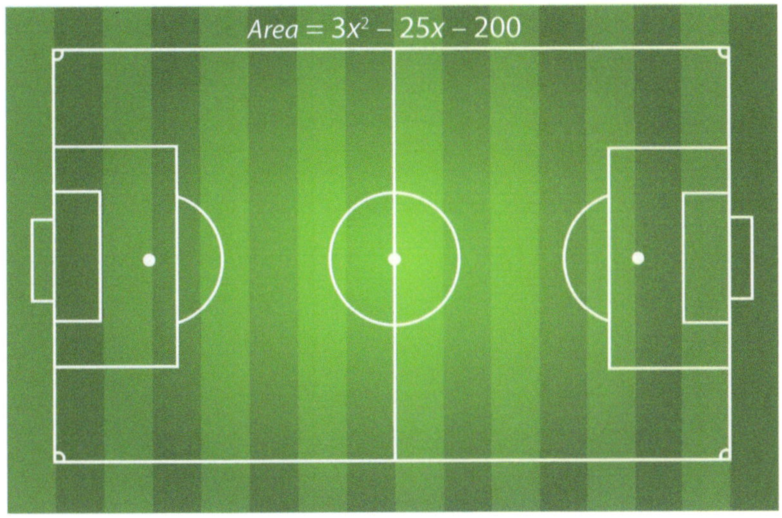

Answer

$3x^2 - 25x - 200$	**1.** Notice that $ac = -600$ and $b = -25$.
$-600 = 15(-40)$ and $15 + (-40) = -25$	**2.** Find factors of -600 whose sum is -25. The signs must be different, with the larger factor being negative.
$3x^2 + 15x - 40x - 200$	**3.** Rewrite the middle term, $-25x$, as a sum of terms with 15 and -40 as their coefficients.
$(3x^2 + 15x) + (-40x - 200)$	**4.** Group the first 2 terms and the last 2 terms and factor the GCF from each grouping.
$3x(x + 5) - 40(x + 5)$	**a.** Factor $3x$ from the first pair of terms and -40 from the second pair.
$(3x - 40)(x + 5)$ The dimensions are $3x - 40$ by $x + 5$.	**b.** Factor the common binomial, $x + 5$, from both terms.

CHECK USING FOIL

$(3x - 40)(x + 5)$
$= 3x^2 + 15x - 40x - 200$
$= 3x^2 - 25x - 200$

SKILL CHECK **EXERCISES 35-37**

A. EXERCISES

Factor each trinomial.

1. $2x^2 + 11x + 12$

2. $4x^2 + 15x + 9$

3. $3x^2 - 14x + 8$

4. $9x^2 - 30x + 25$

5. $3x^2 + 5x - 28$

6. $2x^2 + x - 21$

7. $6x^2 + 5x - 4$

8. $6b^2 - 11b - 10$

9. $15c^2 + 19c + 6$

10. To factor $12x^2 + 17x + 6$, Josh has rewritten it as $(12x^2 + 8x) + (9x + 6)$ and Allen has rewritten it as $(12x^2 + 9x) + (8x + 6)$. Complete their work to show that both ways result in the same answer.

B. EXERCISES

Factor each trinomial.

11. $6a^2 - 5a - 4$

12. $9x^2 + 3x - 2$

13. $5x^2 - 14x - 3$

14. $2x^2 + 13x - 7$

15. $5x^2 + 8xy - 21y^2$

16. $6x^2 - 19xy - 20y^2$

Factor each trinomial. Be sure to factor out any common monomials first.

17. $6x^2 + 3x - 30$

18. $30x^2 - 100x - 250$

19. $16a^2 + 40ab - 24b^2$

20. $-m^2 + mn + 20n^2$

21. $a^2 + 3ab - 28b^2$

22. $x^2 + 3xy - 10y^2$

23. $-x^2 + 7xy + 98y^2$

24. $2x^2 - 8x - 64$

25. $8a^2 - 104a - 112$

26. $-12ab^3 - 9ab^2 + 21ab$

27. $51x^3y - 34x^2y - 85xy$

28. $45x^2 - 3x - 18$

29. $-28x^2 + 26x + 24$

30. $2x^2 - 18xy - 104y^2$

31. $3x^2 + 9xy - 462y^2$

32. **MP3** Describe and correct the error in Dawson's work when he determined that $12x^2 - 8xy - 20y^2$ is prime.

Step	Dawson's Work
1.	$4(3x^2 - 2xy - 5y^2)$
2.	$3(-5) = -15$ and $3 + (-5) = -2$
3.	$4(3x^2 + 3xy - 5xy - 5y^2)$
4.	$4[3x(x + y) - 5y(x - y)]$
5.	cannot factor further

33. **Essential Question:** How can you factor $ax^2 + bx + c$ when $a \neq 1$?

C. EXERCISES

34. **MP2** Perfect Pitch Fútbol Academy uses the polynomial $6x^2 + 85x + 300$ to model the area in square yards of each of its youth soccer fields. Fill in each missing field dimension in the table below.

Age Group	Field Width (yd)	Field Length (yd)
13+		110
U12	55	
U10		65
U8	25	35

MP4 The total area of a garden, including a 2 ft perimeter pathway, can be represented by the trinomial $15x^2 + 32x + 16$.

35. Factor the trinomial.

36. Sketch a diagram of the garden area, including the pathway. Label it with the path's dimensions and with the monomial expressions for the length and width of the interior area of the garden.

37. What is the length-to-width ratio of the garden's plantable area?

MP4 The width of a sidewalk surrounding a swimming pool is represented by x, and the total area of the pool and sidewalk is represented by $4x^2 + 104x + 651$.

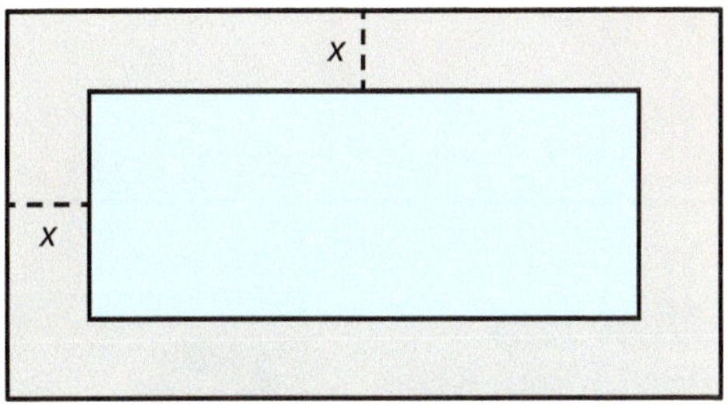

total area = $4x^2 + 104x + 651$

38. Factor to find the dimensions of the pool.

39. Let $x = 5$ ft. Demonstrate that the area found using the binomials from factoring will result in the same area found using the original trinomial.

40. How much does the area change if the sidewalk is widened to 7 ft?

CUMULATIVE REVIEW

41. Solve $5x - 7 = 3x + 11$. **[2.2]**

42. Solve $8 - 7x \geq 15 - (4x + 1)$. **[3.2]**

43. Solve for x and y when $x = 8 - y$ and $3x - 8 = 5y$. **[6.2]**

44. Graph $y < -3x + 5$. **[6.7]**

45. Simplify $(3.5 \times 10^{-11})^2$, expressing the answer in scientific notation. **[7.2]**

46. Simplify $2x(x - 4) + 20x^2 + 3(x^2 - 3) + 8x$. **[8.3]**

Write an equation and solve.

47. A bus leaves Regina for Raleigh averaging 50 mi/hr. Two hours later another bus leaves averaging 55 mi/hr. How long will it take the second bus to catch up with the first bus? **[2.7, 6.5]**

48. Lynda has $5.90 in quarters and dimes. If she has 4 fewer dimes than quarters, how many of each are there? **[2.8]**

49. Which of the following is equivalent to $5 \nleq x$? **[3.1]**

- **A** $x \nleq 5$
- **B** $x \geq 5$
- **C** $x > 5$
- **D** $x < 5$
- **E** none of these

50. Which expression can be used to determine the tenth term of the sequence 6250, 1250, 250, . . . ? **[7.6]**

- **A** $6250\left(\frac{1}{5}\right)^9$
- **B** $6250\left(\frac{1}{5}\right)^{10}$
- **C** $6250(5)^9$
- **D** $6250(5)^{10}$
- **E** none of these

9.4 SPECIAL PATTERNS

What does worldview have to do with seeing patterns in math and science?

After completing this section, I will be able to

- apply the pattern to factor a difference of squares.
- apply the pattern to factor a perfect square trinomial.
- evaluate the claim that my worldview does not affect how I view patterns in math and science.

Integers such as 36 and 121 are called perfect squares because they can be expressed as the square of another integer. Recall that the product of a pair of conjugates is a difference of 2 squares.

$$(x + y)(x - y) = x^2 - y^2$$

Since factoring is the opposite process of multiplication, factoring a *difference of squares binomial* produces a product of conjugates.

$$x^2 - y^2 = (x + y)(x - y)$$

Notice that both terms in the factorization are found by taking the square roots of terms in the original binomial.

The volume of metal used to make a washer is calculated with a binomial: $V = \pi R^2t - \pi r^2t$, where R and r are the inner and outer radii and t is the thickness.

FACTORING THE DIFFERENCE OF SQUARES

$$a^2 - b^2 = (a + b)(a - b)$$

EXAMPLE 1: Factoring the Difference of Squares

Factor each binomial.

a. $x^2 - 64$ **b.** $4a^2 - 49$

Answers

a. $x^2 - 64$

1. Rewrite the binomial as a difference of squares.

$x^2 - 8^2$

$(x + 8)(x - 8)$

2. Apply the difference of squares pattern: $a^2 - b^2 = (a + b)(a - b)$.

b. $4a^2 - 49$

1. Rewrite the binomial as a difference of squares.

$(2a)^2 - 7^2$

$(2a + 7)(2a - 7)$

2. Apply the difference of squares pattern: $a^2 - b^2 = (a + b)(a - b)$.

SKILL CHECK **EXERCISES 1, 5**

When factoring a polynomial, always factor out common factors in the initial step.

EXAMPLE 2: Applying Multiple Factoring Methods

Factor each binomial.

a. $100 - 16t^2$ **b.** $45x^6z^2 - 80y^2$

Answers

a.	$4(25 - 4t^2)$	**1.** Factor out the common factor of 4.
	$4[5^2 - (2t)^2]$	**2.** Rewrite the binomial as a difference of squares.
	$4(5 + 2t)(5 - 2t)$	**3.** Apply the difference of squares pattern.
b.	$5(9x^6z^2 - 16y^2)$	**1.** Factor out the common factor of 5.
	$5[(3x^3z)^2 - (4y)^2]$	**2.** Rewrite the binomial as a difference of squares.
	$5(3x^3z + 4y)(3x^3z - 4y)$	**3.** Apply the difference of squares pattern.

SKILL CHECK **EXERCISE 23**

A difference of squares, such as $x^2 - 25$, could be rewritten as the trinomial $x^2 + 0x - 25$ and factored by finding the factors of -25 whose sum is 0. However, if you write $x^2 + 25$ as the trinomial $x^2 + 0x + 25$ and attempt to find 2 positive factors of 25 whose sum is 0, it quickly becomes apparent that the polynomial is prime. In fact, any binomial that is a sum of squares $(a^2 + b^2)$ cannot be factored using real numbers. Recognizing these patterns provides a much quicker solution.

You have already studied the patterns of 2 perfect squares:

$(x + y)^2 = x^2 + 2xy + y^2$ and
$(x - y)^2 = x^2 - 2xy + y^2$.

While these *perfect square trinomials* can be factored using the previous techniques, recognizing their patterns will enable you to factor quickly. Notice that the first and last terms of each trinomial are positive and are perfect squares. The middle term is twice the product of the square roots of the first and last terms.

FACTORING PERFECT SQUARE TRINOMIALS

$$a^2 + 2ab + b^2 = (a + b)^2$$
$$a^2 - 2ab + b^2 = (a - b)^2$$

EXAMPLE 3: Factoring Perfect Square Trinomials

Factor each trinomial.

a. $x^2 + 10x + 25$ **b.** $16x^2 - 56xy + 49y^2$

Answers

a.	$x^2 + 2(x)(5) + 5^2$	**1.** The first and last terms are perfect squares. The middle term is twice the product of their square roots.
	$(x + 5)^2$	**2.** Apply the square of a sum pattern: $a^2 + 2ab + b^2 = (a + b)^2$.
b.	$16x^2 - 56xy + 49y^2$ $(4x)^2 - 2(4x)(7y) + (7y)^2$	**1.** The first and last terms are perfect squares. The middle term is the opposite of twice the product of their square roots.
	$(4x - 7y)^2$	**2.** Apply the square of a difference pattern: $a^2 - 2ab + b^2 = (a - b)^2$.

SKILL CHECK **EXERCISES 17, 19**

AFTERSCHOOLHELP

Special Patterns

Often perfect square trinomials will be evident when a common monomial is factored out.

EXAMPLE 4: Applying Multiple Factoring Methods

Factor each trinomial.

a. $12x^2 - 36x + 27$ **b.** $x^6 - 8x^3z + 16z^2$

Answers

a. $12x^2 - 36x + 27$	**1.** Factor out the common factor of 3.
$3(4x^2 - 12x + 9)$	
$3[(2x)^2 - 2(2x)(3) + 3^2]$	**2.** The first and last terms are perfect squares. The middle term is the opposite of twice the product of their square roots.
$3(2x - 3)^2$	**3.** The negative middle term indicates that the square of a difference pattern should be applied: $a^2 - 2ab + b^2 = (a - b)^2$.
b. $x^6 - 8x^3z + 16z^2$	**1.** The first and last terms are perfect squares. The middle term is the opposite of twice the product of their square roots.
$(x^3)^2 - 2(x^3)(4z) + (4z)^2$	
$(x^3 - 4z)^2$	**2.** Apply the square of a difference pattern.

SKILL CHECK **EXERCISE 29**

The patterns observed in math reflect the consistency and order in the universe. Christians have a much greater understanding of and appreciation for the consistency demonstrated in math and science because they know that all things were created by Christ (Col. 1:16). Common patterns point to a single Creator of the universe and of life. The consistency and order discovered in the study of mathematics and science reveal God's power, order, and glory, "the invisible things of him from the creation of the world" (Rom. 1:20).

A. EXERCISES

Fill in the missing parts of each factorization.

1. $y^2 - 144 = (y$ ______ $12)(y$ ______ $12)$

2. $x^2 - y^2 = (x$ ______ $y)(x$ ______ $y)$

3. $49x^2 - 4 = (7x -$ ______$)(7x +$ ______$)$

4. $9x^2 - 100 = (3x -$ ______$)(3x +$ ______$)$

5. $36a^2 - 49 = ($______ $- 7)(6a +$ ______$)$

6. $25x^2 - 49 = (5x -$ ______$)($______ $+ 7)$

Identify whether the trinomial is a perfect square by writing *yes* or *no*. Justify your answer.

7. $x^2 + 6x + 9$

8. $x^2 + 14x + 49$

9. $x^2y - 8xy + 16y^2$

10. $a^2b^2 - 12abc + 36c^2$

11. $4x^2 - 10x + 25$

12. $a^2 - 4ab - 4b^2$

13. Cassie says that $x^2 + 16x + 8$ is in the form of a perfect square trinomial since $8(x)$ doubled is $16x$. Explain her error.

Factor each expression.

14. $a^2 - 225b^2$

15. $a^2x^2 - y^2$

16. $x^2 - 6x + 9$

17. $x^2 - 16x + 64$

18. $m^2 + 24m + 144$

19. $16a^2 + 8ax + x^2$

20. $a^2 - 10ab + 25b^2$

21. $a^2 - 14a + 49$

B. EXERCISES

Factor. Be sure to factor any common monomials first. If the polynomial will not factor, write *prime*.

22. $2a^2 - 242$

23. $3x^2 - 147$

24. $9x^2 + 12x + 4$

25. $36x^2 + 12x + 1$

26. $a^2x^4 - 16b^2$

27. $a^4 - a^2b^2$

28. $x^8 + 26x^5 + 169$

29. $9x^8 + 30x^4z + 25z^2$

30. $4 - 16x^2$

31. $49x^5 - 9x^7$

32. $x^4 + 0.6x^2y + 0.09y^2$

33. $xt^4 - 20xt^2 + 100$

34. $63x^6 - 175y^{10}$

35. $147a^{10} - 507b^8$

36. **MP3** Jackson initially factored $64x^2 - 100y^2$ as $(8x - 10y)(8x + 10y)$. He noticed that each binomial factor had a GCF of 2 and wrote his final answer as $2(4x - 5y)(4x + 5y)$. Describe and correct his error.

37. **Essential Question:** How do we view patterns in math and science from a biblical perspective?

C. EXERCISES

Factor each expression.

38. **MP1** $x^6 - x^2$

39. $27x^5y - 72x^3y^2 + 48xy^3$

40. **MP7** $169a^2 - (b - 5)^2$

41. **MP8** Use the pattern found in part a to factor in parts b and c.

- **a.** Simplify the product of $(a + b)(a^2 - ab + b^2)$.
- **b.** Factor $v^3 + 64$.
- **c.** Factor $27x^3 + 8$.

CUMULATIVE REVIEW

Factor each polynomial. [9.1–9.3]

42. $2ax + 4a$

43. $-40xy^2 + 5xy + 20x$

44. $x^2 + 8x - 9$

45. $y^2 - 11y + 30$

46. $a^2 + 5ab - 36b^2$

47. $8x^2 + 14x + 3$

48. $25x^2 - 20xy + 4y^2$

49. $6r^2 + 29rp - 5p^2$

50. Factor $3x^2 + 3x - 18$ completely. **[9.2]**

- **A** $(x - 3)(x + 6)$
- **B** $3(x - 3)(x + 6)$
- **C** $3(x^2 + x - 6)$
- **D** $3(x - 2)(x + 3)$
- **E** prime

51. For which value of k will $x^2 + kx + 16$ be a prime polynomial? **[9.2]**

- **A** -10
- **B** 17
- **C** 8
- **D** -6
- **E** none of these

CAREERS IN MATH
ENGINEER

Engineers apply math and science to solve technical problems. They design, build, test, and evaluate products. There are many different engineering fields, but all require an in-depth knowledge of mathematics, including algebra, trigonometry, calculus, and physics. The following examples illustrate the creativity and diversity of engineering.

Civil engineers design and build infrastructure such as roads and bridges. They calculate the direction and amount of stress on bridge girders to determine how much load a bridge will hold. The best and most economical route for a road, its curvature, and the banking on a curve must also be determined.

Electrical engineers design and build power grids, transformers, and delivery systems, as well as electric motors. They must understand the concepts of resistance, voltage, and impedance, and how to determine the size and type of wiring that should be used.

Aeronautical engineers design and build aircraft and the complex systems within them. These engineers must design to specifications for performance and safety.

Mechanical engineers design, build, and test machines that produce, transmit, and use power, such as commercial air conditioning systems.

Chemical engineers convert raw materials into useful forms and develop new materials.

All these fields deal with creating products that glorify our Lord through calling attention to His creation. The finished products then help us fulfill the Creation Mandate as stewards of the earth.

9.5 FACTORING COMPLETELY

How do I know when a polynomial is completely factored?

After completing this section, I will be able to

- describe a completely factored polynomial.
- completely factor binomials and trinomials.
- factor polynomials with 4 unlike terms by grouping.

In later chapters you will use the tools learned in this chapter to solve equations that require factoring completely.

A polynomial is factored completely when it cannot be factored further over the set of integers.

EXAMPLE 1: Factoring a Binomial Completely

Factor $5ax^5 - 5axy^4$.

Answer

$5ax(x^4 - y^4)$	**1.** Factor out the common monomial $5ax$.
$5ax(x^2 + y^2)(x^2 - y^2)$	**2.** Factor the difference of squares.
$5ax(x^2 + y^2)(x - y)(x + y)$	**3.** Factor the resulting difference of squares. Remember that a sum of squares, such as $x^2 + y^2$, cannot be factored using real numbers.
$5ax(x^2 + y^2)(x - y)(x + y)$ $= 5ax^5 - 5axy^4$	**4.** Multiplying the factors produces the original expression.

SKILL CHECK **EXERCISE 5**

EXAMPLE 2: Factoring a Trinomial Completely

Factor $x^4 - 9x^2 + 8$.

Answer

$x^4 - 9x^2 + 8$	**1.** There are no common monomials and no special patterns.
$(x^2 - 1)(x^2 - 8)$	**2.** Factor by finding the factors of +8 whose sum is −9.
$(x - 1)(x + 1)(x^2 - 8)$	**3.** Factor the first binomial again since it is a difference of squares.

SKILL CHECK **EXERCISE 17**

In some expressions you can factor common factors from more than 1 grouping of the terms to leave a common binomial.

EXAMPLE 3: Factoring by Grouping

Factor each polynomial.

a. $2xy - 7x + 10y - 35$ **b.** $2y^3 + 5y^2 - 8y - 20$ **c.** $a^3 - 5a^2 - 4a + 20$

Answers

a. $2xy - 7x + 10y - 35$ $(2xy - 7x) + (10y - 35)$	**1.** Group the first 2 and the last 2 terms since there are no common monomials and no special patterns.
$x(2y - 7) + 5(2y - 7)$	**2.** Factor out a common monomial from each binomial.
$(x + 5)(2y - 7)$	**3.** Factor out the common binomial, $2y - 7$, from both terms.
b. $2y^3 + 5y^2 - 8y - 20$ $(2y^3 + 5y^2) + (-8y - 20)$	**1.** Group the first 2 and the last 2 terms since there are no common monomials and no special patterns.
$y^2(2y + 5) - 4(2y + 5)$	**2.** Factor out a common monomial from each binomial.
$(y^2 - 4)(2y + 5)$	**3.** Factor out the common binomial, $2y + 5$, from both terms.
$(y - 2)(y + 2)(2y + 5)$	**4.** Factor $y^2 - 4$.
c. $a^3 - 5a^2 - 4a + 20$ $(a^3 - 5a^2) + (-4a + 20)$	**1.** Group the first 2 and the last 2 terms since there are no common monomials and no special patterns.
$a^2(a - 5) - 4(a - 5)$	**2.** Factor out a common monomial from each binomial.
$(a^2 - 4)(a - 5)$	**3.** Factor out the common binomial, $a - 5$, from both terms.
$(a - 2)(a + 2)(a - 5)$	**4.** Factor $a^2 - 4$.

AFTERSCHOOLHELP

Factoring Completely

SKILL CHECK **EXERCISES 13, 31**

STRATEGIES FOR FACTORING COMPLETELY OVER THE SET OF INTEGERS

1. Factor out common monomials.
2. Consider the type of polynomial and specific factoring strategies.

Binomials	Trinomials	Polynomials
Difference of Squares $a^2 - b^2 = (a - b)(a + b)$	**a. Perfect Square Trinomial** $a^2 \pm 2ab + b^2 = (a \pm b)^2$ **b.** $\boldsymbol{x^2 + bx + c}$ **or** $\boldsymbol{x^2 + bxy + cy^2}$ Find factors of c that sum to b. **c.** $\boldsymbol{ax^2 + bx + c}$ **or** $\boldsymbol{ax^2 + bxy + cy^2}$ Rewrite the middle term by using factors of ac whose sum is b as coefficients, and factor by grouping.	Factor by grouping.

3. Continue factoring until each factor is no longer factorable.
4. Check.

A. EXERCISES

1. Match each polynomial below with the step needed for it to be factored completely.

   a. $(x^2 + 25)(x^2 - 12x + 36)$
   b. $(xy + bz)(xz - yz - z)$
   c. $(a + b)^2(3a - 12b + 20)$
   d. $(x + y)(9x^4 - 16y^4)$

   I Factor a common monomial.
   II Factor the perfect square trinomial.
   III Factor the difference of squares.
   IV factored completely

Factor each polynomial completely.

2. $5a^2 + 10a + 5$
3. $x^3 - 2x^2 + x$
4. $200 - 2a^2$
5. $8a^8 - 2a^6$
6. $3x^2 - 18x + 24$
7. $3a^5 - 4a^3$
8. $5x^2 - 70x + 245$
9. $m^{10} + 2m^5 + 1$

Factor each polynomial by grouping.

10. $x(2x - y) + y(2x - y)$
11. $4(b + c) - m(b + c)$
12. $(at - aw) + (bt - bw)$
13. $(xt - 3xw) + (2t - 6w)$
14. $a(m + n) + b(m + n) - 2(m + n)$
15. $m(2a + b) - n(2a + b) + 2n(2a + b)$

B. EXERCISES

Factor each polynomial completely.

16. $x^4 - 16$
17. $3x^3 - 33x^2 + 84x$
18. $8a^2 - 48ab^2 + 72b^4$
19. $a^4x^4 - 81$
20. $18x^2 - 3x - 36$
21. $30y^2 - 35y - 100$
22. $40w^2 - 62w + 24$
23. $70x^2 - 32x - 24$
24. $wy + wz + xy + xz$
25. $xy + 5x + 3y + 15$
26. $xy - 6x + 7y - 42$
27. $x^2 - 2x - 3xy + 6y$
28. $6x^2 + xy^2 + 24x + 4y^2$
29. $(2ax^2 + bx^2) - (2ay^2 + by^2)$
30. $y^3 - 3y^2 - 4y + 12$
31. $a^3 + 2a^2 - 9a - 18$
32. $5y^3 - 15y^2 - 5y + 15$
33. $3x^3 - 6x^2 - 48x + 96$
34. **Essential Question:** How can I determine when a polynomial is completely factored?

C. EXERCISES

MP5 Given 1 factor of the polynomial, factor the polynomial completely.

35. $(x - 2)$ is a factor of $x^3 - 3x^2 - 10x + 24$.
36. $(x + 2)$ is a factor of $x^3 + x^2 - 8x - 12$.
37. $(y + 1)$ is a factor of $2y^3 - 9y^2 + 4y + 15$.
38. $(y + 4)$ is a factor of $3xy^2 + 18xy + 24x - y^2 - 6y - 8$.

MP7 Factor each expression completely.

39. $x^2 + 2xy + y^2 - 9$
40. $x^2 + 2xy + y^2 + 2x + 2y + 1$

CUMULATIVE REVIEW

Write and solve a system of equations for each word problem. [6.5–6.6]

41. Geoff flew his twin-engine plane from Cincinnati to St. Louis. He made the 350 mi flight against a sustained wind in 2 hr. If flying conditions were the same for the return trip, which took 10 min less, what was the average speed of the plane with no wind and what was the wind speed (to the nearest mile per hour)?

42. Cameron and his friend took a canoe trip 12 mi up a lazy river in 3 hr and returned in 2 hr 24 min. What was the rate of the current, and what would be the boys' paddling rate in still water?

43. Vanessa needs a 25% saline solution. If she has a 40% solution on hand, how much water and how much 40% solution should she mix to obtain 3 gal of a 25% saline solution?

44. A metalsmith is working to create an alloy of copper and nickel. If he currently has an alloy that is 20% copper, how much of the 20% alloy and how much copper should he use to produce 10 lb of an alloy that is 60% copper?

Write a function rule for each described power function. [7.3]

45. translation of $y = x^3$ six units left and five units down

46. translation of $y = x^2$ three units up and ten units right

47. parabola with vertex at $(-11, 8)$

48. cubic function with point of inflection at $(9, 1)$

49. Which of these expressions when evaluated is not equal to an integer? **[1.3]**

A $\sqrt{4^3}$

B $\sqrt[3]{8^2}$

C $\sqrt[3]{(-1)^4}$

D $\sqrt{2^3}$

E all of these

50. Which polynomial has the highest degree? **[8.1]**

A $y^{11} + 4$

B $u + v + w + x + y$

C $a^3b^4 - cd + a^3c^4d^5$

D $1 - 4a^4b^3c^2d$

E All have the same degree.

APPLICATION PROBLEMS
GLACIERS

Mont Blanc, the tallest mountain in the Alps, is located on the French-Italian border and is home to several glaciers. At its base lies the French town of Chamonix, the location of the first Olympic winter games in 1924. The area remains a popular location for mountain climbers and skiers.

Tourists note the beauty of the mountain, but the locals know all too much about the potential dangers of the glaciers. In 1892 a glacial lake hidden beneath a thick layer of ice broke through Mont Blanc's glacier ice. Approximately 200,000 cubic meters of water cascaded down the mountain, flooding the village of Saint-Gervais-les-Bains and killing 175 people.

Since then, the community has worked hard to monitor the ice movement of the glaciers and has searched for ways to determine whether hidden pools of water are developing under the glaciers. A large tunnel system through the mountain has been used to facilitate this monitoring, but recently satellite imagery has proved a more valuable tool.

The local power plants use the meltwater from the glaciers to produce electricity in hydroelectric power plants. Harnessing God-given natural resources and using them to benefit people is an example of how we can fulfill God's command in Genesis 1 to subdue the earth.

Probing is a method used to measure snow and ice depth. Modern technology, such as satellite imagery, allows for more efficient measurements.

As climates around the world go through cyclical weather patterns, glaciers can expand or shrink. Since glaciers are often a significant source of both water and electrical power, concern about glacial retreat is not unjustified. The hydroelectric plant at the base of Mont Blanc had to reposition its water intake because of glacial retreat.

1. If the glacier Mer de Glace has lost 8.3% of its length during the last 130 yr and the glacier is currently 11 km long, how long was the glacier 130 yr ago? How much length has been lost?
2. Suppose the midsection of the glacier has lost 150 m, or 27%, of its thickness.
   a. What was the glacier's thickness (to the nearest tenth of a meter) when first measured?
   b. Convert this original measurement to the nearest foot.
   c. What is the glacier's current thickness in meters and in feet?

Interestingly, Mont Blanc has been called "the mountain that will not shrink." Due to less temperature variation as altitude increases, glacier thickness above the equilibrium line (snow line) has not shown significant change over the last 100 yr.

As the climate goes through different cycles, Christians understand that God controls everything about the climate, including the ice (Job 38:29–30). Is the earth warmer now than it was in the 1970s and 1980s? The answer is yes, but does this necessarily mean that it will continue to grow warmer? The answer to this question is known only by the Creator, who controls the weather.

KEYWORD SEARCH

Mont Blanc shrink

Some glaciers are obviously shrinking, and some are disappearing. The yearly glacier net mass balance (the difference between the amount of accumulated snow and the amount that melts or evaporates) of Washington's Southern Cascade Glacier was studied over a span of 50 yr. A sample of the data is given.

year	net mass balance
18	−7.8
20	−5.8
22	−6.0
24	−5.0
26	−6.5
28	−9.0

year	net mass balance
30	−9.9
32	−10.3
34	−12.0
36	−15.8
38	−16.8
40	−19.8

3. Make a scatter plot of the data. (The Chapter 5 Technology Corner shows how to do this on a graphing calculator.)

4. Since the points do not appear to be in a line, use a graphing calculator (or spreadsheet) to complete a quadratic regression on the data.

```
EDIT CALC TESTS
1:1-Var Stats
2:2-Var Stats
3:Med-Med
4:LinReg(ax+b)
5:QuadReg
6:CubicReg
7:QuartReg
8:LinReg(a+bx)
9↓LnReg
```

a. Write the function rule that models the data set.

b. Graph the function on your scatter plot.

c. Use the function rule to estimate the net mass balance in year 60.

5. Including data from the entire 50 yr study presents a different picture. Create a scatter plot by entering the data set below.

year	net mass balance
1	0.0
4	0.2
6	−3.0
8	−3.0
10	−4.0
12	−4.2
14	−5.3
16	−6.0
18	−7.8
20	−5.8
22	−6.0
24	−5.0
26	−6.5

year	net mass balance
28	−9.0
30	−9.9
32	−10.3
34	−12.0
36	−15.8
38	−16.8
40	−19.8
42	−21.8
44	−22.1
46	−23.3
48	−22.0
50	−23.0

6. The points from the expanded data set appear to be on a line.

a. Determine the equation of the line of best fit that models the data set.

b. Use the line of best fit to estimate the net mass balance in year 60.

c. Why might your estimates from exercises 4c and 6b differ?

The Les Bois hydroelectric power plant monitors the ice movement of the Mont Blanc glaciers using a system of tunnels that run under the glaciers. An instrument attached to a bicycle wheel was used to determine that the ice on the bottom layer of one of the glaciers moves down the mountain at an average rate of 1 in./hr.

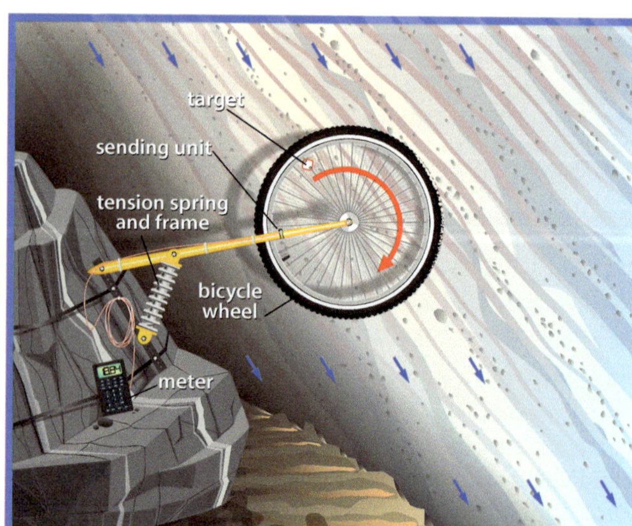

7. Assuming that the ice continues to flow at the current average, how long would it take a portion of ice to move from the glacier's head down the length of the 6.8 mi glacier?

8. Eight stakes were evenly spaced along a horizontal line across the top of a glacier, and their movement (in meters) was recorded. Use a graphing calculator or a spreadsheet to make a scatter plot of the data.

stake	movement (m)
1	230
2	235
3	250
4	265

stake	movement (m)
5	270
6	245
7	230
8	210

9. Use a graphing calculator (or spreadsheet) to complete a quadratic regression on the data.

   a. State the equation of a parabola that models the data set.

   b. Graph the function on your scatter plot.

In 2010 scientists used satellite imagery to locate a hidden glacial lake estimated to contain 65,000 m^3 of water. The lake was found to be at an altitude of approximately 3200 m, making it a challenge for the engineers hired to drain it.

10. Use unit analysis to convert the following measurements.

   a. altitude: 3200 m to feet

   b. volume: 65,000 m^3 to cubic feet

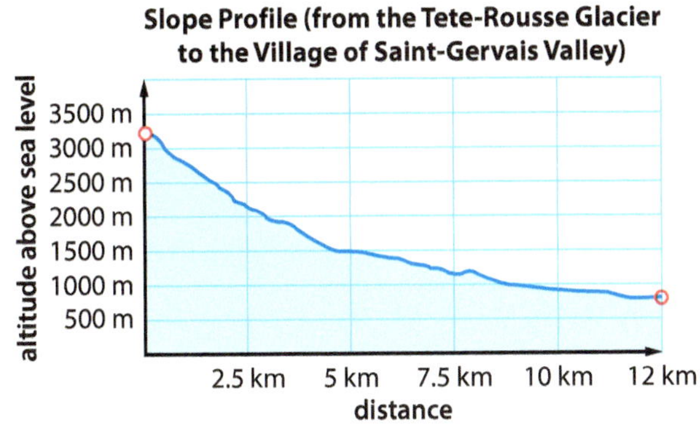

11. Given the elevation profile for the path from the glacier (the red circle on the left) to the village of Saint-Gervais-les-Bains (the red circle on the right), estimate each of the following slopes as a decimal (to the nearest hundredth) and then convert to a percent grade.

   a. the average slope for the first 2.5 km (measured horizontally)

   b. the average slope from the glacier to the village

12. It has been estimated that it would take 15 min for the water from the lake to reach the village of Saint-Gervais-les-Bains, which is about 12 km from the glacier.

   a. Determine the speed of the water in kilometers per minute.

   b. What is the rate in miles per hour (to the nearest whole number)?

9 CHAPTER REVIEW

Key Concepts

STRATEGIES FOR FACTORING COMPLETELY OVER THE SET OF INTEGERS

1. Factor out common monomials. [9.1]
2. Consider the type of polynomial and specific factoring strategies.

Binomials	Trinomials	Polynomials
Difference of Squares $a^2 - b^2 = (a - b)(a + b)$ [9.4]	a. **Perfect Square Trinomial** $a^2 \pm 2ab + b^2 = (a \pm b)^2$ [9.4] b. $\mathbf{x^2 + bx + c}$ **or** $\mathbf{x^2 + bxy + cy^2}$ Find factors of c that sum to b. [9.2] c. $\mathbf{ax^2 + bx + c}$ **or** $\mathbf{ax^2 + bxy + cy^2}$ Rewrite the middle term by using factors of ac whose sum is b as coefficients, and factor by grouping. [9.3]	Factor by grouping. [9.5]

3. Continue factoring until each factor is no longer factorable.
4. Check.

If the polynomial cannot be factored, then write *prime*.

Biblical Worldview Shaping

Design through Algebra

Do patterns in nature prove the existence of God?

The patterns and design clearly seen in nature give evidence of a creator, but our faith is based in Scripture, not our own reasoning.

EXERCISES

Factor out the common monomial. [9.1]

1. $3r^2 + 5r$
2. $8x^2y^3 + 6xy^2 - 2x$
3. $4x^3 - 8x^2$
4. $x^4 + 3x^3y + 7x^2$

Factor. [9.2–9.3]

5. $y^2 - 15y + 56$
6. $x^2 + 3xy - 54y^2$
7. $2d^2 + 9d + 7$
8. $9t^2 - 27t + 8$

Rearrange the terms in descending order; then factor. [9.2–9.3]

9. $9x + 14 + x^2$
10. $a^2 - 84 - 5a$
11. $6 + b - 15b^2$
12. $7 + 2y^2 + 15y$

Factor. Begin by factoring out the common monomial. [9.2–9.3]

13. $x^2y + xy - 56y$

14. $7y^2 - 21y + 14$

15. $18x^2 - 100x + 50$

16. $12x^2 - 68x + 80$

Use special patterns to factor. [9.4]

17. $a^2 - 9$

18. $a^2 + 14a + 49$

19. $9x^2 - 30x + 25$

20. $25a^6x^2 - 144y^2$

Factor each polynomial completely. If the polynomial will not factor over the set of integers, write *prime*. [9.5]

21. $2x^2 - 30x + 108$

22. $a^2 - 22a + 121$

23. $y^2 + 17yz + 60z^2$

24. $z^2 - 13z + 48$

25. $9x(2 - x) - 25(2 - x)$

26. $18y^2 - 50$

27. $b^2 + 4$

28. $24x^2 - 96y^2$

29. $5b^2 - 50b + 125$

30. $72ab^2 - 96abc + 32ac^2$

31. $2a^2 + 9a - 18$

32. $15x^2 + xy - 2y^2$

33. $b^2 - 15b - 36$

34. $12x^2 + 78xy + 90y^2$

35. $3a^2 - 10a + 7$

36. $15x^2 + 14y$

37. $12y^2 + 22y - 70$

38. $9r^2 + 9r + 2rs + 2s$

39. $10x^3 + 14x^2 - 15x - 21$

40. $c^4 + 3c^2 + 2$

41. $w^6 + 6w^3z^2 + 9z^4$

42. $(x + y)^2 - 9$

43. $256a^8 - 1$

44. $9x^4 - 288x^2 + 2304$

45. $x^2y + x^2 - xy - 1$

46. Rewrite the formula for the surface area of a cone, $S = \pi rl + \pi r^2$, by factoring the right side of the equation. **[9.1]**

47. The varsity quarterback has been practicing sidearm throws, releasing the ball from a height of 4 ft. Rewrite the formula for the height of the ball over time, $h = 4 + 63t - 16t^2$, by factoring the right side of the equation. **[9.3]**

The total area in square inches of both the picture and its frame is represented by the polynomial $4w^2 + 24w + 35$. [9.3]

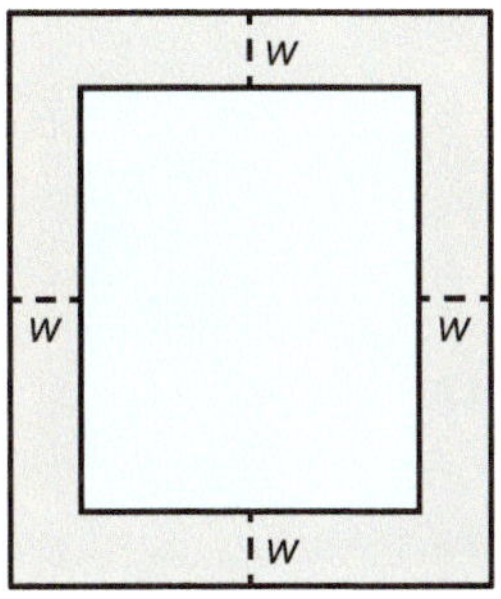

48. Factor the polynomial to find binomial expressions representing the height and width of the outside of the frame.

49. Using the factors found in exercise 48, find the area of the picture alone.

50. Assume that the frame is 2 in. wide.

a. What is the total area of the picture and the frame?

b. What is the area of the frame?

51. **BWS Essential Question:** Which of the following statements is not correct?

A Math and science are useful in recognizing patterns in nature.

B Patterns in nature prove the existence of God.

C Patterns in nature give evidence of a creator.

D How you interpret patterns seen in nature depends on your worldview.

CHAPTER 10
RADICALS

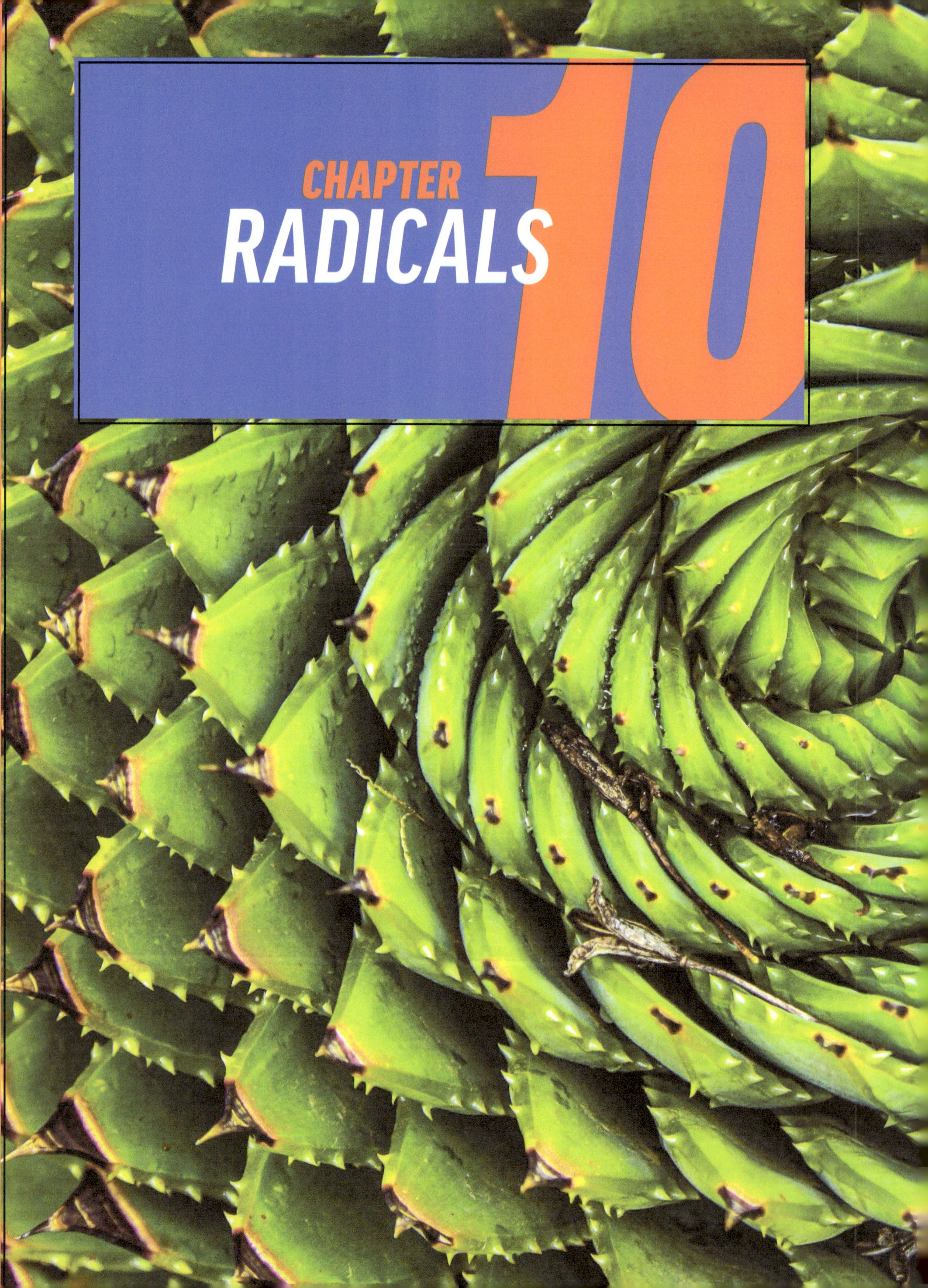

CHAPTER OVERVIEW

10.1 SIMPLIFYING RADICALS

How is an expression with a fractional exponent related to a radical expression?

After completing this section, I will be able to

- convert expressions between radical form and exponential form.
- simplify numerical radicals.
- simplify radicals with variable radicands.

A tsunami wave's speed, V, in miles per hour can be modeled using the formula $V = \frac{15}{22}\sqrt{gd}$.

The definitions of roots and powers allow an equation to be converted from exponential to radical form and vice versa. In general, $b^n = a$ is equivalent to $\sqrt[n]{a} = b$. The radicand is the value obtained when b is raised to the nth power.

Up to this point, the definition of exponents has been limited to integers. Radical expressions are used to give meaning to rational exponents.

Compare the following computations to identify a pattern for converting rational exponents to radical form.

$$\sqrt{4} \cdot \sqrt{4} = 2 \cdot 2 = 4 \text{ and } 4^{\frac{1}{2}} \cdot 4^{\frac{1}{2}} = 4^{\left(\frac{1}{2}+\frac{1}{2}\right)} = 4^1 = 4$$

Both expressions are equivalent to 4, resulting in the following conclusion.

$$\sqrt{4} \cdot \sqrt{4} = 4^{\frac{1}{2}} \cdot 4^{\frac{1}{2}}$$

$$(\sqrt{4})^2 = \left(4^{\frac{1}{2}}\right)^2$$

Therefore, $\sqrt{4} = 4^{\frac{1}{2}}$.

You can convert between the radical form and the exponential form of any expression using the following definition.

EXAMPLE

$8^{\frac{2}{3}} = \sqrt[3]{8^2} = (\sqrt[3]{8})^2$

DEFINITION

Rational exponent: For any real number x and integers a and b, where $b > 1$, the exponential expression $x^{\frac{a}{b}}$ is equivalent to the radical $\sqrt[b]{x^a}$ or $(\sqrt[b]{x})^a$.

EXAMPLE 1: Converting to Exponential Form

Express each in exponential form.

a. $\sqrt[5]{x^3}$ **b.** $\sqrt[4]{7x^3y^2}$

Answers

a. $\sqrt[5]{x^3} = x^{\frac{3}{5}}$ Apply the definition of rational exponent.

b. $\sqrt[4]{7x^3y^2} = 7^{\frac{1}{4}}x^{\frac{3}{4}}y^{\frac{2}{4}}$

1. Apply the definition of rational exponent. Notice that the power of 7 is 1, making its exponent $\frac{1}{4}$.

$= 7^{\frac{1}{4}}x^{\frac{3}{4}}y^{\frac{1}{2}}$

2. Reduce exponents where possible.

SKILL CHECK **EXERCISE 3**

EXAMPLE 2: Converting to Radical Form

Express each in radical form.

a. $5^{\frac{2}{3}}$ **b.** $3^{\frac{1}{2}}x^{\frac{3}{4}}$

Answers

a. $5^{\frac{2}{3}} = \sqrt[3]{5^2} = \sqrt[3]{25}$ — Apply the definition of rational exponent.

b. $3^{\frac{1}{2}}x^{\frac{3}{4}} = 3^{\frac{2}{4}}x^{\frac{3}{4}}$ — **1.** Rename fractional exponents using a common denominator so that their bases can be placed under the same radical.

$= \sqrt[4]{3^2x^3} = \sqrt[4]{9x^3}$ — **2.** Assign the numerators as powers of the factors in the radicand. The common denominator is the index of the radical.

SKILL CHECK **EXERCISE 7**

In mathematics, answers are generally stated in simplest form. When a radical is in simplest form, its radicand contains no perfect power factors. The Product Property of Radicals is used to simplify radicals.

PRODUCT PROPERTY OF RADICALS

For $x \geq 0$ and $y \geq 0$, $\sqrt[n]{xy} = \sqrt[n]{x} \cdot \sqrt[n]{y}$.

EXAMPLE 3: Simplifying Small Radical Numbers

Simplify each radical.

a. $\sqrt{24}$ **b.** $\sqrt{75}$ **c.** $\sqrt[3]{40}$

Answers

a. $\sqrt{24} = \sqrt{2 \cdot 2 \cdot 2 \cdot 3}$ — **1.** Write the radicand as the product of its prime factors.

$= \sqrt{2^2} \cdot \sqrt{6}$ — **2.** Apply the Product Property.

$= 2\sqrt{6}$ — **3.** Simplify.

b. $\sqrt{75} = \sqrt{3 \cdot 5 \cdot 5}$ — **1.** Write the radicand as the product of its prime factors.

$= \sqrt{5^2} \cdot \sqrt{3}$ — **2.** Apply the Product Property.

$= 5\sqrt{3}$ — **3.** Simplify.

c. $\sqrt[3]{40} = \sqrt[3]{2 \cdot 2 \cdot 2 \cdot 5}$ — **1.** Write the radicand as the product of its prime factors.

$= \sqrt[3]{2^3} \cdot \sqrt[3]{5}$ — **2.** Apply the Product Property.

$= 2\sqrt[3]{5}$ — **3.** Simplify.

SKILL CHECK **EXERCISES 9, 13**

The steps outlined below can be used to simplify radical expressions containing numerical or variable factors.

SIMPLIFYING RADICALS

1. Write the radicand as the product of its prime factors.
2. Apply the Product Property to the perfect powers.
3. Simplify the roots of the perfect powers.
4. Multiply the factors inside and outside the radical.

EXAMPLE 4: Simplifying Large Radical Numbers

Simplify each radical.

a. $\sqrt{648}$ **b.** $\sqrt[3]{14{,}580}$

Answers

a. $\sqrt{648} = \sqrt{2 \cdot 2 \cdot 2 \cdot 3 \cdot 3 \cdot 3 \cdot 3}$ — **1.** Write the radicand as the product of its prime factors.

$= \sqrt{2^2} \cdot \sqrt{3^2} \cdot \sqrt{3^2} \cdot \sqrt{2}$ — **2.** Apply the Product Property to the perfect squares.

$= 2 \cdot 3 \cdot 3 \cdot \sqrt{2} = 18\sqrt{2}$ — **3.** Simplify.

b. $\sqrt[3]{14{,}580} = \sqrt[3]{2 \cdot 2 \cdot 3 \cdot 3 \cdot 3 \cdot 3 \cdot 3 \cdot 3 \cdot 5}$ — **1.** Write the radicand as the product of its prime factors.

$= \sqrt[3]{3^3} \cdot \sqrt[3]{3^3} \cdot \sqrt[3]{2 \cdot 2 \cdot 5}$ — **2.** Apply the Product Property to the perfect cubes.

$= 3 \cdot 3 \cdot \sqrt[3]{20} = 9\sqrt[3]{20}$ — **3.** Simplify.

SKILL CHECK **EXERCISE 17**

AFTERSCHOOLHELP

Simplifying Radicals

Care must be taken when the Product Property is used to simplify radicals containing variables. The claim that $\sqrt{x^2} = x$ is true only when x is not negative.

$$\sqrt{2^2} = \sqrt{4} = 2, \text{ but } \sqrt{(-2)^2} = \sqrt{4} = 2 \text{ as well.}$$

If x is negative, then $\sqrt{x^2} = -x$ (the opposite of x), which is a positive number. This can be simplified by using the definition of absolute value to state

$$\sqrt{x^2} = |x|.$$

Consider the following cube roots:

$$\sqrt[3]{2^3} = \sqrt[3]{8} = 2 \text{ and } \sqrt[3]{(-2)^3} = \sqrt[3]{-8} = -2.$$

Notice that in this case, $\sqrt[3]{x^3} = x$ is true for all values of x.

When you are finding even roots, the absolute value is used to ensure a positive result.

DEFINITION

The ***n*th root of x^n** or $\sqrt[n]{x^n} = \begin{cases} |x| & \text{if } n \text{ is even.} \\ x & \text{if } n \text{ is odd.} \end{cases}$

You may be able to save time by factoring the radicand into a product with perfect power factors.

EXAMPLE 5: Simplifying Radicals Containing Variables

Simplify each radical.

a. $\sqrt{32x^3y^2}$ **b.** $\sqrt[3]{16x^2y^4z^6}$

Answers

a. $\sqrt{32x^3y^2}$
$= \sqrt{16 \cdot 2x^2xy^2}$
$= \sqrt{16} \cdot \sqrt{x^2} \cdot \sqrt{y^2} \cdot \sqrt{2x}$
$= 4|x||y|\sqrt{2x}$
$= 4x|y|\sqrt{2x}$

1. Write the radicand as a product with perfect square factors.
2. Apply the Product Property and simplify. Use the absolute value to ensure a positive result.
3. If x were negative, the original expression would not be a real number; therefore, x must be positive and the absolute value is not needed.

b. $\sqrt[3]{16x^2y^4z^6}$
$= \sqrt[3]{8 \cdot 2x^2y^3yz^3z^3}$
$= \sqrt[3]{8} \cdot \sqrt[3]{y^3} \cdot \sqrt[3]{z^3} \cdot \sqrt[3]{z^3} \cdot \sqrt[3]{2x^2y}$
$= 2yzz\sqrt[3]{2x^2y} = 2yz^2\sqrt[3]{2x^2y}$

1. Write the radicand as a product with perfect cube factors.
2. Apply the Product Property and simplify. Notice that the absolute value is not used for cube roots.

SKILL CHECK **EXERCISES 23, 25**

Note that $\sqrt{x^4} = \sqrt{x^2} \cdot \sqrt{x^2} = |x| \cdot |x| = |x|^2 = x^2$. The absolute value is not necessary since x^2 is always positive. In general, the absolute value is necessary only when the simplification of an even root of a variable raised to an even power produces a variable with an odd exponent.

VOCABULARY
- nth root of x^n
- rational exponent

However, a radical with an odd power in the radicand of an even root, such as $\sqrt{x^7}$, is assumed to have a positive base ($x > 0$). Therefore, the radical expression represents a real number and absolute value symbols are not required.

$\sqrt{x^2} = |x|$ $\quad \sqrt{x^3} = x\sqrt{x}$ $\quad \sqrt{x^4} = x^2$

$\sqrt{x^5} = x^2\sqrt{x}$ $\quad \sqrt{x^6} = |x|^3$ $\quad \sqrt{x^7} = x^3\sqrt{x}$

A. EXERCISES

Convert each radical to exponential form.

1. $\sqrt{5}$
2. $\sqrt{3x^3}$
3. $\sqrt[5]{3x^2y^3}$
4. $\sqrt[8]{7^2ab^4}$

Convert each exponential expression to radical form.

5. $3^{\frac{1}{2}}y^{\frac{1}{2}}$
6. $5^{\frac{1}{3}}x^{\frac{2}{3}}$
7. $2^{\frac{4}{9}}y^{\frac{1}{3}}z^{\frac{2}{3}}$
8. $3^{\frac{3}{8}}r^{\frac{1}{4}}s^{\frac{1}{2}}$

Simplify each radical that can be simplified.

9. $\sqrt{12}$
10. $\sqrt{18}$
11. $\sqrt{99}$
12. $\sqrt{15}$
13. $\sqrt[3]{32}$
14. $\sqrt[3]{54}$

B. EXERCISES

Simplify each radical that can be simplified.

15. $\sqrt{200}$

16. $\sqrt{192}$

17. $\sqrt[3]{320}$

18. $\sqrt[3]{160}$

19. $\sqrt[4]{1000}$

20. $\sqrt[7]{512}$

Simplify each radical expression.

21. $\sqrt{x^8}$

22. $\sqrt{y^5}$

23. $\sqrt{y^{10}}$

24. $\sqrt{16x^2}$

25. $\sqrt{18x^2y}$

26. $\sqrt{50x^3y^3}$

27. $\sqrt{9x^4y^2z^6}$

28. $\sqrt{75a^3bc^8}$

29. $\sqrt{\frac{x^6y^6}{100}}$

30. $\sqrt[3]{8x^4y}$

31. $\sqrt[4]{10x^5y^6z^7}$

32. $\sqrt[5]{32a^5b^{15}}$

33. **Essential Question:** How is an expression with a fractional exponent related to a radical expression?

Tsunamis are large ocean waves caused by earthquakes, volcanoes, or landslides on the ocean floor. They can reach speeds as high as 600 mi/hr in deep water and produce waves as tall as 120 ft when they hit the shore. The wave's speed, V, in miles per hour can be modeled using the formula $V = \frac{15}{22}\sqrt{gd}$, where g is the gravitational acceleration of 32 ft/sec^2 and d is the depth of the water in feet.

These before and after satellite photos demonstrate the destructive power of a tsunami.

Use the formula to determine the speed of a wave at the following depths.

34. 242 ft

35. 3872 ft

Skid marks can be used to estimate the speed a car was going before it skidded to a stop. The speed, S, in miles per hour can be modeled using the formula $S = \sqrt{30Df}$, where D is the length of the skid marks in feet and f is the coefficient of friction between the tires and the road.

Use the formula to determine the speed a car was going before it skidded to a stop.

36. skid marks: 40 ft; $f = 0.75$

37. skid marks: 160 ft; $f = 0.75$

38. skid marks: 160 ft; $f = \frac{1}{3}$

C. EXERCISES

Simplify each expression that can be simplified.

39. $625^{\frac{5}{4}}$

40. $\sqrt{9x^2 - 24x + 16}$

41. $\sqrt{16x^2 - 9}$

42. $11xyz\sqrt[4]{2^5 \cdot 3^4 \cdot 5x^{12}y^6z^9}$

43. **MP7** What is the smallest integer value of x for which $\sqrt[4]{12{,}150x}$ is equivalent to an integer?

44. **MP3** Use a counterexample to demonstrate that $\sqrt{x^2} = x$ is not always true. Then explain why $\sqrt{x^2} = |x|$.

CUMULATIVE REVIEW

Simplify each expression. [7.1]

45. $2abc(-5a^2b^3c^5)$

46. $-3x^2y^4z(-2xy^5z)$

47. $(2x^2)^3(2x)^2$

48. $3x^4y^2(2x^5y)^3$

Factor each polynomial. [9.4]

49. $4x^2 - 289$

50. $a^2 + 2ab + b^2$

51. $m^2 + 28m + 196$

52. $12b^8 - 3b^6$

53. A business's profit is the difference of its revenue and costs. Tao Harmonicas models the revenue earned from selling x harmonicas as $65x$ and the cost of producing them as $-\frac{1}{3}x^2 + 250x - 3000$. Which polynomial expression represents the profit? **[8.2]**

A $-\frac{1}{3}x^2 + 185x - 3000$

B $-\frac{1}{3}x^2 + 315x - 3000$

C $\frac{1}{3}x^2 - 185x + 3000$

D $\frac{1}{3}x^2 - 315x + 3000$

E none of these

54. $(5y^2 - 3z)^2$ is equivalent to **[8.5]**

A $10y^2 - 6z$.

B $25y^4 + 9z^2$.

C $25y^4 + 30y^4z^2 - 9z^2$.

D $25y^4 - 30y^2z + 9z^2$.

E none of these

MIND OVER MATH

Solve each equation for the variable. Write the solution in the appropriate squares in the grid.

ACROSS

1. $5x + 35 = 3160$

4. $3y - 105 = 342$

7. $6x^2 - 672 = x^2 + 32(979)$

8. $3x - (8 - x) = 2x(5) - 242$

10. $9x - 16 = 227$

12. $\frac{x}{33} + \frac{x}{9} = 42$

13. $x - 100 = -30$

14. $\frac{x}{28} = 157$

16. $\frac{x}{6} + 27x = 263{,}245$

18. $\frac{x}{2} + 10(x - 3) = 568.5$

19. $x^2 = 144$

20. $13x - 15{,}079 = 52{,}755$

22. $2y - 583 = 4293$

24. $\sqrt{4} - 222 = -5x$

25. $\frac{x}{5} + \frac{x}{25} = 54$

26. $z = \sqrt{625} + 2\sqrt{100}$

27. $2^7 = k + 29$

28. $x^2 - 168 = 3313$

30. $16(2\sqrt[3]{27}) = x - 28$

31. $x = 8(100) + 3$

DOWN

2. $2x = 7 \cdot 8$

3. $x + 599 = 75^2$

4. $10\sqrt{4900} = x - 26^2 - 3$

5. $x - 1 = \sqrt{64} \cdot 6$

6. $25^2 - x = 1$

9. $x^2 = 160{,}000$

11. $x \cdot 56^0 = 73{,}524$

13. $40^3 + 25^3 = x + 19^2 + 30$

15. $3x = 2913$

17. $\frac{1}{4}x = \frac{921}{6}$

20. $24^2 = x + 3^3$

21. $t - 669 = 85^2 + 20^2$

22. $50^2 = g - 58$

23. $\frac{6}{z} = \frac{1}{71} \cdot \frac{1}{2}$

27. $10^2 - 2^3 = e$

29. $2 \cdot 3^2 \cdot 5 = z$

10.2 MULTIPLYING RADICALS

How can I multiply a square root by a cube root?

After completing this section, I will be able to

- multiply numerical radicals.
- multiply radicals with variable radicands.
- explain how to multiply radicals with different indices.

The Product Property of Radicals was used to simplify radicals in Section 10.1. It is also used to multiply radicals. Note that radicals must have the same index before they can be multiplied.

$$\sqrt[n]{x} \cdot \sqrt[n]{y} = \sqrt[n]{xy}$$

EXAMPLE 1: Finding Products of Square Roots

Find each product.

a. $\sqrt{2} \cdot \sqrt{6}$ **b.** $-2\sqrt{45} \cdot 5\sqrt{20}$

Answers

a. $\sqrt{2} \cdot \sqrt{6} = \sqrt{2(2 \cdot 3)}$

1. Apply the Product Property, writing the radicand as the product of its prime factors.

$= \sqrt{2^2 \cdot 3} = 2\sqrt{3}$

2. Simplify the square root of the perfect square.

b. $-2\sqrt{45} \cdot 5\sqrt{20} = -2 \cdot 5\sqrt{45} \cdot \sqrt{20}$

1. Use the Commutative Property to rearrange the factors.

$= -10\sqrt{(5 \cdot 3^2)(2^2 \cdot 5)}$

2. Apply the Product Property, writing the radicand as the product of its prime factors.

$= -10 \cdot 2 \cdot 3 \cdot 5 = -300$

3. Simplify the square root of each perfect square.

SKILL CHECK **EXERCISES 1, 7** ☑

As you become more familiar with factors and perfect squares, you may recognize factors that are perfect squares under the radical. Simplifying each radical before multiplying can save you time and effort.

EXAMPLE 2: Finding Products of Square Roots

Simplify $\sqrt{50} \cdot \sqrt{32}$.

Answer

$\sqrt{25 \cdot 2} \cdot \sqrt{16 \cdot 2}$

1. Apply the Product Property, writing the radicand as a product with perfect square factors.

$= 5\sqrt{2} \cdot 4\sqrt{2}$

2. Simplify the square root of each perfect square.

$= 20 \cdot 2 = 40$

3. Simplify.

SKILL CHECK **EXERCISE 5**

Unless otherwise stated, from here on you can assume that all variables in the radicand represent nonnegative values. Therefore, it will not be necessary to use absolute values to ensure that even roots of variables are positive.

EXAMPLE 3: Finding Products of Square Roots

Simplify $\sqrt{6ab} \cdot \sqrt{2bc}$.

Answer

$\sqrt{6ab} \cdot \sqrt{2bc} = \sqrt{(3 \cdot 2)2ab^2c}$

1. Apply the Product Property, writing the radicand as the product of its prime factors.

$= 2b\sqrt{3ac}$

2. Simplify the square root of each perfect square.

SKILL CHECK **EXERCISE 17**

MULTIPLYING RADICALS

1. Apply the Product Property, writing the radicand as the product of its prime factors.
2. Simplify the root of each perfect power.
3. Simplify the resulting expression.

AFTERSCHOOLHELP

Multiplying Radicals

EXAMPLE 4: Finding Products of Cubed Roots

Simplify each radical expression.

a. $\sqrt[3]{225} \cdot \sqrt[3]{315}$ **b.** $\sqrt[3]{4x^2y^5z} \cdot \sqrt[3]{32xy}$

Answers

a. $\sqrt[3]{225} \cdot \sqrt[3]{315} = \sqrt[3]{(3^2 \cdot 5^2)(3^2 \cdot 5 \cdot 7)}$

$= \sqrt[3]{3^3 \cdot 3 \cdot 5^3 \cdot 7}$

1. Apply the Product Property, writing the radicand as the product of its prime factors.

$= 3 \cdot 5\sqrt[3]{3 \cdot 7}$

2. Simplify the cube root of each perfect cube.

$= 15\sqrt[3]{21}$

3. Simplify.

b. $\sqrt[3]{4x^2y^5z} \cdot \sqrt[3]{32xy} = \sqrt[3]{2^2 \cdot 2^5x^3y^6z}$

1. Apply the Product Property, writing the radicand as the product of its prime factors.

$= \sqrt[3]{2^6 \cdot 2x^3y^6z}$

$= 2^{\frac{6}{3}}x^{\frac{3}{3}}y^{\frac{6}{3}}\sqrt[3]{2z}$

$= 2^2xy^2\sqrt[3]{2z}$

2. Simplify the cube root of each perfect cube.
   a. Group factors so the exponents are multiples of 3 (perfect cubes).
   b. Divide the exponents of the perfect cubes by the index, 3. Leave all other factors in the radicand.

$= 4xy^2\sqrt[3]{2z}$

3. Simplify.

SKILL CHECK **EXERCISE 13**

KEYWORD SEARCH

Heron's formula

The area of a triangle is found by multiplying the base length times the height and dividing by 2. But what if the height is not known? Heron's formula, named after the Greek mathematician Heron of Alexandria, can be used to calculate the area of a triangle directly from just the lengths of its 3 sides. The semiperimeter, s, is the sum of the side lengths divided by 2. The formula, proved in a later course, is

$$A = \sqrt{s(s-a)(s-b)(s-c)}, \text{ where } s = \frac{a+b+c}{2}.$$

EXAMPLE 5: Finding Area of a Triangle

The lengths of the sides of a triangle are 6 ft, 8 ft, and 10 ft. Using Heron's formula, find the area of the triangle.

Answer

$s = \frac{6+8+10}{2} = \frac{24}{2} = 12 \text{ ft}$

1. Calculate the triangle's semiperimeter.

$A = \sqrt{12(12-6)(12-8)(12-10)}$

2. Substitute known values into Heron's formula.

$= \sqrt{12(6)(4)(2)}$

$= \sqrt{(2^2 \cdot 3)(3 \cdot 2)(2^2)(2)}$

$= 2 \cdot 3 \cdot 2 \cdot 2 = 24 \text{ ft}^2$

3. Simplify.

SKILL CHECK **EXERCISE 33**

Even though the Product Property of Radicals does not apply to radicals with different indices, these products can be found by reasoning step-by-step using concepts you have previously learned. This deductive reasoning is a very powerful tool. Giving a valid reason for each step ensures that the final answer is correct—meaning that it is consistent with other results obtained within our algebraic system. Example 6 uses deductive reasoning to convert radicals to exponential form.

EXAMPLE 6: Multiplying Radicals with Different Indices

Simplify each radical expression.

a. $\sqrt{2}\cdot\sqrt[3]{2}$ **b.** $\sqrt{3}\cdot\sqrt[3]{2}$

Answers

a. $\sqrt{2}\cdot\sqrt[3]{2}$

$= 2^{\frac{1}{2}}\cdot 2^{\frac{1}{3}}$

1. Convert each radical to rational exponent form.

$= 2^{\frac{1}{2}+\frac{1}{3}} = 2^{\frac{3}{6}+\frac{2}{6}} = 2^{\frac{5}{6}}$

2. Since the bases are the same, add the exponents.

$= \sqrt[6]{2^5} = \sqrt[6]{32}$

3. Convert to radical form.

b. $\sqrt{3}\cdot\sqrt[3]{2}$

$= 3^{\frac{1}{2}}\cdot 2^{\frac{1}{3}}$

1. Convert each radical to rational exponent form.

$= 3^{\frac{3}{6}}\cdot 2^{\frac{2}{6}} = \sqrt[6]{3^3}\cdot\sqrt[6]{2^2}$

2. Find a common denominator and convert to radical form.

$= \sqrt[6]{3^3\cdot 2^2} = \sqrt[6]{27\cdot 4} = \sqrt[6]{108}$

3. Apply the Product Property and simplify.

SKILL CHECK **EXERCISE 25**

A. EXERCISES

Simplify each expression.

1. $\sqrt{5}\cdot\sqrt{3}$

2. $\sqrt{7}\cdot\sqrt{5}$

3. $\sqrt{8}\cdot\sqrt{3}$

4. $\sqrt{18}\cdot\sqrt{3}$

5. $\sqrt{12}\cdot\sqrt{5}$

6. $4\sqrt{15}\cdot 2\sqrt{6}$

7. $-3\sqrt{5}\cdot 4\sqrt{10}$

8. $\sqrt{33}\cdot\sqrt{55}$

9. $\sqrt{28}\cdot\sqrt{35}$

10. $\sqrt{42}\cdot\sqrt{28}$

11. $\sqrt{5}\cdot\sqrt{118}$

12. $\sqrt{54}\cdot\sqrt{48}$

B. EXERCISES

Simplify each expression. Assume that all variables in the radicand represent nonnegative values.

13. $\sqrt[3]{40}\cdot\sqrt[3]{54}$

14. $\sqrt[3]{81}\cdot\sqrt[3]{64}$

15. $\sqrt[4]{32}\cdot\sqrt[4]{64}$

16. $\sqrt{4ab}\cdot\sqrt{5b}$

17. $\sqrt{3x^2y^3}\cdot\sqrt{2xy}$

18. $\sqrt{2w^{15}z^3}\cdot\sqrt{8wz}$

19. $\sqrt{x^2y}\cdot\sqrt{xy^2}\cdot\sqrt{xy}$

20. $4\sqrt{8a^5c^7}\cdot\sqrt{6a^4c}$

21. $2x\sqrt{15x^3y}\cdot\sqrt{21x^4y^3}$

22. $\sqrt[3]{4x^5y^7}\cdot\sqrt[3]{2x^6y}$

23. $\sqrt[3]{25a^4bc^6}\cdot\sqrt[3]{10abc^3}$

24. $\sqrt[4]{32x^3yz^6}\cdot\sqrt[4]{x^2y^2z^2}$

25. $\sqrt[3]{3}\cdot\sqrt{3}$

26. $\sqrt[3]{2}\cdot\sqrt[4]{2}$

27. $\sqrt{3}\cdot\sqrt[3]{9}$

28. The side of a square measures $3\sqrt{5}$ in. What is the area of the square?

29. The width of a rectangle is $2\sqrt{7}$ ft, and its length is $5\sqrt{14}$ ft. Express the area of the rectangle as a simplified radical.

30. The width of a rectangle is $\sqrt[3]{x^4}$ units, and its length is $\sqrt[3]{x^5}$ units. Express the area of the rectangle in simplest terms.

31. The length of an edge of a cube is $\sqrt[3]{3x^5}$ units. Express the volume of the cube in simplest terms.

32. A fabric is patterned with equilateral triangles whose sides measure 6 in. Use Heron's formula to calculate the area of one of the triangles.

33. The lengths of the sides of a triangle are 6 m, 10 m, and 12 m. Use Heron's formula to calculate the area and express the answer in simplest radical form.

34. **Essential Question:** What type of reasoning is used in finding a way to multiply radicals with different indices?

C. EXERCISES

Simplify each expression. State each product in simplest radical form.

35. $\sqrt[3]{3249} \cdot \sqrt[3]{171}$

36. $\sqrt[4]{8000x^{13}y^{12}} \cdot \sqrt[4]{540x^{30}y^{22}}$

37. $\sqrt[4]{2x^3} \cdot \sqrt[5]{2x^3}$

38. $\sqrt[8]{x^7y^3} \cdot \sqrt[6]{xy^5}$

39. **MP8** Use the pattern found in part a to write the expressions in parts b and c in simplest radical form.

a. Write the simplified expression for $\sqrt[m]{a} \cdot \sqrt[n]{a}$.

b. $\sqrt[5]{7} \cdot \sqrt[8]{7}$

c. $\sqrt[9]{65} \cdot \sqrt[7]{65}$

CUMULATIVE REVIEW

Simplify each expression. [7.1, 8.6]

40. $\dfrac{21x^7}{3x^3}$

41. $\dfrac{-2r^3s^5t}{14rs^4t}$

42. $\dfrac{24a^5b^{-3}c}{12a^{-2}b^4c^{-1}}$

43. $\left(\dfrac{2x^{-3}}{x^2}\right)^{-2}$

44. $\dfrac{8x^4y + 16xy}{4xy}$

45. $\dfrac{n^5 + 5n^3 - n^2}{n^2}$

Convert each radical to exponential form. [10.1]

46. $\sqrt[4]{x^2y^3z^4}$

47. $\sqrt[8]{a^7b^4c^2}$

48. Which polynomial is factored completely? **[9.5]**

A $3y(2y - 5) + 4(-2y + 5)$

B $7x(x^2 + 4)(x - 1)$

C $(15z + 6)(2z - 3)$

D $64a^2 - 25$

E B and C

49. Which step is not included in factoring $3x^3 + x^2 - 48x - 16$ completely? **[9.5]**

A Factor a monomial.

B Factor a binomial.

C Factor a perfect square trinomial.

D Factor a difference of squares.

E All the above steps are needed.

TECHNOLOGY CORNER (TI-84 PLUS FAMILY)

Roots & Rational Exponents

Decimal approximations of roots can be found easily on your calculator. The $\sqrt{\ }$ function ([2nd] [x^2]) automatically keeps everything entered underneath the root symbol, making calculations such as $\sqrt{9+16}$ quite simple. To evaluate expressions such as $\sqrt{9}+16$, use [▶] to exit the radicand before entering the rest of the expression.

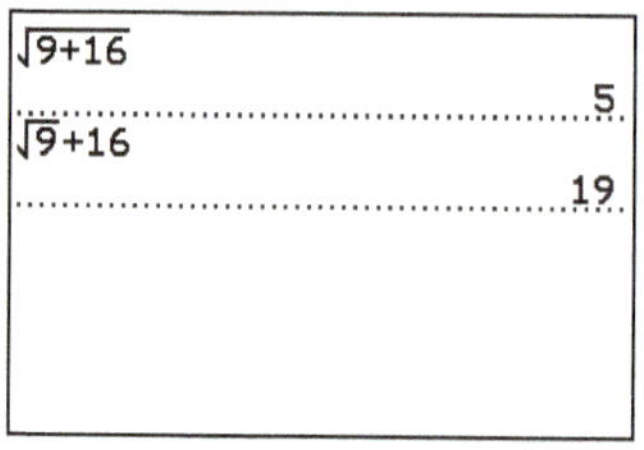

The [math] menu allows you to enter cube roots and other roots.

To evaluate $\sqrt[3]{-8}$, enter [math] [4] [(-)] 8 [enter].

To evaluate $\sqrt[4]{81}$, enter [4] [math] [5] 81 [enter].

(Note that you can also enter an index, such as 4, after pressing [math] [5].)

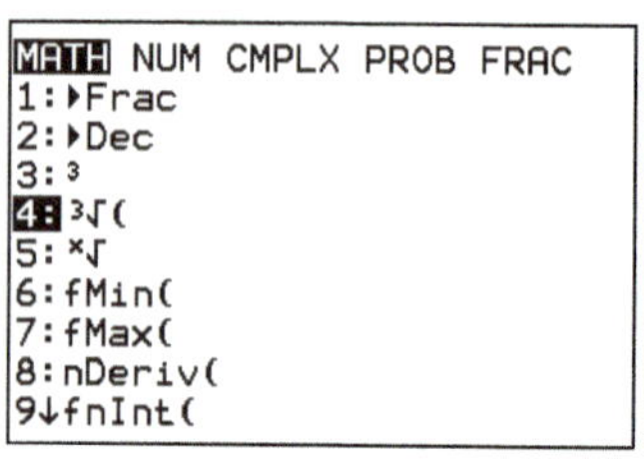

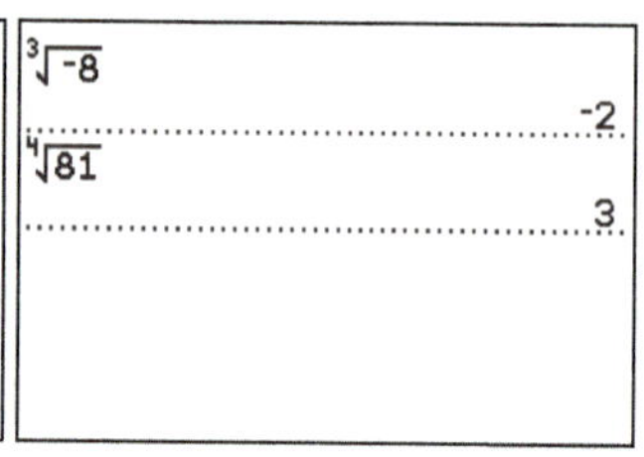

Rational exponents can be entered using the [^] key followed by the [FRACTION] option ([alpha] [y=] [1] or [alpha] [X,T,θ,n]).

To evaluate $4^{\frac{3}{2}}$, enter 4 [^] [FRACTION] 3 [▼] 2.

Evaluate $(-8)^{\frac{1}{3}}$ using [(] [(-)] 8 [)] [^] [FRACTION] 1 [▼] 3.

The reciprocal key, [x^{-1}], can be used as an alternative for entering fractional exponents with a numerator of 1: [(-)] 8 [^] 3 [x^{-1}].

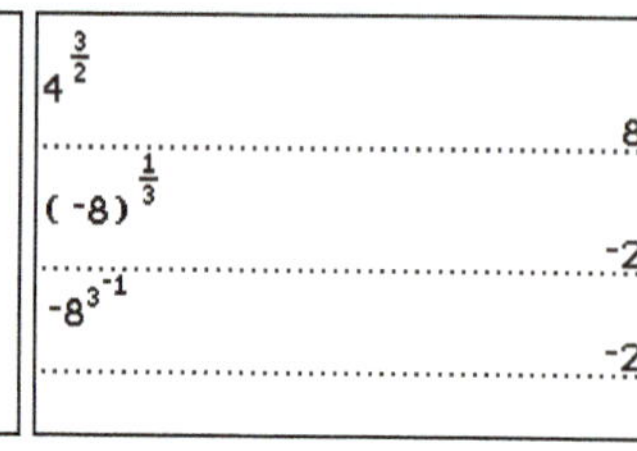

10.3 DIVIDING RADICALS

How can I make a radical in a denominator rational?

After completing this section, I will be able to

- divide numerical radicals.
- divide radicals with variable radicands.
- rationalize denominators containing a radical.

The body surface area of an adult is used to determine the dosage regimen of some medical treatments.

The Product Property of Radicals, $\sqrt[n]{xy} = \sqrt[n]{x} \cdot \sqrt[n]{y}$, is useful when multiplying and simplifying radicals. A similar property applies to division.

QUOTIENT PROPERTY OF RADICALS

For $x \geq 0$ and $y > 0$, $\sqrt[n]{\frac{x}{y}} = \frac{\sqrt[n]{x}}{\sqrt[n]{y}}$.

The numerical example below demonstrates how a square root can be found with and without the use of the Quotient Property.

$$\sqrt{\frac{16}{25}} = \sqrt{\left(\frac{4}{5}\right)^2} = \frac{4}{5} \text{ and } \frac{\sqrt{16}}{\sqrt{25}} = \frac{4}{5}, \text{ so } \sqrt{\frac{16}{25}} = \frac{\sqrt{16}}{\sqrt{25}}.$$

EXAMPLE 1: Simplifying Quotients of Square Roots

Simplify each expression.

a. $\frac{\sqrt{40}}{\sqrt{5}}$ **b.** $\frac{\sqrt{15}}{\sqrt{12n^2}}$

Answers

a. $\frac{\sqrt{40}}{\sqrt{5}} = \sqrt{\frac{40}{5}}$ — **1.** Apply the Quotient Property of Radicals.

$= \sqrt{8}$ — **2.** Simplify the radicand.

$= 2\sqrt{2}$ — **3.** Simplify the radical.

b. $\frac{\sqrt{15}}{\sqrt{12n^2}} = \sqrt{\frac{15}{12n^2}} = \sqrt{\frac{5}{4n^2}}$ — **1.** Apply the Quotient Property and reduce the radicand.

$= \frac{\sqrt{5}}{\sqrt{4n^2}} = \frac{\sqrt{5}}{2n}$ — **2.** Apply the Quotient Property to simplify the denominator.

SKILL CHECK **EXERCISES 1, 7**

Fractions with radicals in the denominator $\left(\frac{\sqrt{5}}{\sqrt{7}}\right)$ or radicals with fractions in the radicand $\left(\sqrt{\frac{5}{7}}\right)$ are not considered to be simplified.

These expressions cannot be simplified using the Quotient Property, as $\frac{\sqrt{40}}{\sqrt{5}}$ and $\frac{\sqrt{15}}{\sqrt{12n^2}}$ were in the previous example.

When a radical expression cannot be simplified using the Quotient Property, a process called **rationalizing the denominator** can be used to remove a radical from the denominator. During this process, the radical is multiplied by an expression equal to 1, changing the denominator from an irrational radical to an integer.

EXAMPLE 2: Rationalizing the Denominator

Simplify each expression.

a. $\frac{\sqrt{5}}{\sqrt{7}}$ **b.** $\frac{\sqrt{3a^3b}}{\sqrt{6ab^3}}$

Answers

AfterSchoolHelp

Dividing Radicals

a. $\frac{\sqrt{5}}{\sqrt{7}} \cdot \frac{\sqrt{7}}{\sqrt{7}}$

1. Multiply by 1 in the form of $\frac{\sqrt{7}}{\sqrt{7}}$.

$= \frac{\sqrt{35}}{\sqrt{7} \cdot \sqrt{7}} = \frac{\sqrt{35}}{7}$

2. Simplify the denominator.

b. $\frac{\sqrt{3a^3b}}{\sqrt{6ab^3}} = \sqrt{\frac{3a^3b}{6ab^3}}$

1. Apply the Quotient Property and reduce the radicand.

$= \sqrt{\frac{3a^2}{6b^2}} = \frac{a}{b}\sqrt{\frac{1}{2}}$

$= \frac{a\sqrt{1}}{b\sqrt{2}} \cdot \frac{\sqrt{2}}{\sqrt{2}} = \frac{a\sqrt{2}}{2b}$

2. Apply the Quotient Property again to rewrite the fraction; then multiply by 1 in the form of $\frac{\sqrt{2}}{\sqrt{2}}$.

SKILL CHECK **EXERCISES 9, 23**

Expressions with roots other than square roots are rationalized using similar methods. However, it will not help to change the radicand in the denominator into a perfect square. If you are taking a cube root, the radicand must be a perfect cube. For a fourth root, the radicand must be a perfect fourth power, and so on.

RATIONALIZING THE DENOMINATOR

1. Write the radicals in fractional form, $\frac{\sqrt[n]{a}}{\sqrt[n]{b}}$.
2. Simplify both the numerator and the denominator as much as possible.
3. If a radical remains in the denominator, multiply by the form of 1 that makes the radicand in the denominator a perfect power.
4. Simplify both the numerator and the denominator, and reduce the fraction if needed.

EXAMPLE 3: Simplifying Quotients of Cube Roots

Simplify each expression.

a. $\frac{10\sqrt[3]{2}}{\sqrt[3]{25}}$ **b.** $\frac{9\sqrt[3]{5a}}{25\sqrt[3]{15ab}}$ **c.** $\frac{\sqrt[3]{a^2bc^2}}{\sqrt[3]{27ab^3c^3}}$

Answers

a. $\dfrac{10\sqrt[3]{2}}{\sqrt[3]{25}} \cdot \dfrac{\sqrt[3]{5}}{\sqrt[3]{5}} = \dfrac{10\sqrt[3]{10}}{\sqrt[3]{5^3}}$

$= \dfrac{10\sqrt[3]{10}}{5} = 2\sqrt[3]{10}$

1. Since the radical is a cubed root, factors of 5 are needed to get a perfect cube.
2. Simplify.

b. $\dfrac{9\sqrt[3]{5a}}{25\sqrt[3]{15ab}} = \dfrac{9\cancel{\sqrt[3]{5}}\,\cancel{\sqrt[3]{a}}}{25\cancel{\sqrt[3]{5}}\,\sqrt[3]{3}\,\cancel{\sqrt[3]{a}}\,\sqrt[3]{b}} = \dfrac{9}{25\sqrt[3]{3b}}$

$= \dfrac{9}{25\sqrt[3]{3b}} \cdot \dfrac{\sqrt[3]{3^2b^2}}{\sqrt[3]{3^2b^2}} = \dfrac{\overset{3}{\cancel{9}}\sqrt[3]{9b^2}}{25 \cdot \cancel{3} \cdot b} = \dfrac{3\sqrt[3]{9b^2}}{25b}$

1. Reduce the fraction by canceling the common radical factors.
2. Since the radical is a cubed root, you need 3 factors of $3b$ to get a perfect cube.

VOCABULARY

- rationalizing the denominator

c. $\dfrac{\sqrt[3]{a^2bc^2}}{\sqrt[3]{27ab^3c^3}} = \sqrt[3]{\dfrac{a^2bc^2}{27ab^3c^3}} = \dfrac{1}{3}\sqrt[3]{\dfrac{a}{b^2c}}$

$= \dfrac{1}{3}\left(\dfrac{\sqrt[3]{a}}{\sqrt[3]{b^2c}}\right) \cdot \dfrac{\sqrt[3]{bc^2}}{\sqrt[3]{bc^2}} = \dfrac{\sqrt[3]{abc^2}}{3bc}$

1. Apply the Quotient Property and reduce the radicand.
2. Apply the Quotient Property again and rationalize the denominator.

SKILL CHECK **EXERCISES 21, 27**

A. EXERCISES

Simplify each radical expression.

1. $\dfrac{\sqrt{4}}{\sqrt{25}}$ **2.** $\dfrac{\sqrt{80}}{\sqrt{20}}$ **3.** $\dfrac{\sqrt{5}}{\sqrt{4}}$ **4.** $\dfrac{\sqrt{15}}{\sqrt{5}}$

5. $\dfrac{\sqrt{15}}{\sqrt{20}}$ **6.** $\dfrac{6\sqrt{14}}{\sqrt{18}}$ **7.** $\dfrac{\sqrt{30}}{3\sqrt{250}}$ **8.** $\dfrac{2}{\sqrt{3}}$

9. $\dfrac{3}{\sqrt{13}}$ **10.** $\dfrac{3}{\sqrt{5}}$ **11.** $\dfrac{1}{\sqrt{6}}$ **12.** $\dfrac{\sqrt{3}}{\sqrt{2}}$

13. $\dfrac{\sqrt{14}}{\sqrt{21}}$ **14.** $\dfrac{\sqrt{114}}{\sqrt{32}}$ **15.** $\dfrac{3\sqrt{5}}{\sqrt{15}}$ **16.** $\dfrac{3\sqrt{4}}{\sqrt{32}}$

B. EXERCISES

Simplify each radical expression.

17. $\sqrt[3]{\dfrac{1}{2}}$ **18.** $\sqrt[3]{\dfrac{1}{9}}$ **19.** $\dfrac{\sqrt[3]{9}}{\sqrt[3]{36}}$ **20.** $\dfrac{6\sqrt[3]{7}}{\sqrt[3]{4}}$

21. $\dfrac{15\sqrt[3]{5}}{4\sqrt[3]{25}}$ **22.** $\dfrac{\sqrt[3]{5}}{\sqrt[3]{12}}$ **23.** $\dfrac{\sqrt{5xy}}{\sqrt{10x^2y^3}}$ **24.** $\dfrac{\sqrt{r^3s^2t}}{\sqrt{rst}}$

25. $\dfrac{15\left(\sqrt{2tx^2z^3} \cdot \sqrt{7txz^2}\right)}{\sqrt{10t^3x^2}}$ **26.** $\dfrac{\sqrt{15x^4y^2}}{\left(\sqrt{xyz} \cdot \sqrt{3y^2}\right)}$ **27.** $\dfrac{\sqrt[3]{15x^2y^4}}{\sqrt[3]{5x^3y}}$ **28.** $\dfrac{30\sqrt[3]{9a^3b^4c}}{\sqrt[3]{27ab^5c}}$

29. **MP3** Did Emilia simplify $\dfrac{3}{\sqrt{6}}$ correctly? If not, what is the correct answer?

$\dfrac{3}{\sqrt{6}} = \dfrac{1}{\sqrt{2}} = \dfrac{1}{\sqrt{2}} \cdot \dfrac{\sqrt{2}}{\sqrt{2}} = \dfrac{\sqrt{2}}{2}$

30. **MP3** Describe and correct any error Niklas made in finding the quotient $8 \div \sqrt[3]{4}$.

$8 \div \sqrt[3]{4} = \dfrac{8}{\sqrt[3]{4}} \cdot \dfrac{\sqrt[3]{4}}{\sqrt[3]{4}} = \dfrac{8\sqrt[3]{4}}{4} = 2\sqrt[3]{4}$

31. **Essential Question:** How do you rationalize the denominator of a fraction?

32. The area of a rectangle is $36\sqrt{5}$ in.^{2}, and its length is $4\sqrt{15}$ in. Express the width of the rectangle as a simplified radical.

33. The area of a triangle is $15\sqrt{2}$ cm², and its base is $2\sqrt{5}$ cm. Express the height of the triangle as a simplified radical.

Doctors calculate the body surface area (BSA) of an adult to determine the proper dosage of treatments such as chemotherapy. The Mosteller formula, BSA $= \sqrt{\frac{HW}{3600}}$, gives the BSA in square meters, where H is the person's height in centimeters and W is the person's mass in kilograms.

Determine the BSA of each of the following individuals, expressing each answer as both a simplified radical and a decimal approximation to the nearest tenth.

34. 180 cm; 60 kg

35. 175 cm; 72 kg

C. EXERCISES

Simplify each expression.

36. $\left(\sqrt{18x^2y^4} \cdot \sqrt{9xy^5}\right) \div \left(\sqrt{15x^4y^2} \cdot \sqrt{6xy^9}\right)$

37. $\left(2\sqrt{8a^3b^2} \cdot 3\sqrt{4ab^3}\right) \div \left(\sqrt{24a^5b} \cdot \sqrt{6a^3b^9}\right)$

38. $\sqrt[4]{4096} \div \sqrt[3]{512}$

39. MP7 $\sqrt[4]{56x^3y^4} \div \sqrt{18x^3y^2}$

40. MP8 We recognize $\frac{1}{\sqrt{a}}$ to be the multiplicative inverse of $\sqrt{a}$ (if $a \neq 0$), since $\sqrt{a}\left(\frac{1}{\sqrt{a}}\right) = 1$. However, this expression is usually written with a rationalized denominator: $\frac{\sqrt{a}}{a}$. Write the multiplicative inverse of each of the following expressions with a rationalized denominator.

a. $\sqrt[3]{a}$ **b.** $\sqrt[4]{a}$ **c.** $\sqrt[n]{a}$

CUMULATIVE REVIEW

Graph each inequality. [6.7]

41. $y > -x + 3$

42. $2x + 3y \leq 12$

Simplify. [8.2]

43. $7x^2 - 4xy - 5y - 3x^2 + 6y$

44. $(a^3 + 4b^2 - 3c) + (2a^3 - 10c)$

45. $(2x^3y + x^2y^3 - 4xy) - (12x^2y^3 + 15)$

Factor. [9.2–9.4]

46. $3x^2 + 16x + 5$

47. $2b^2 - 28b + 98$

48. $9x^4 + 6x^2y^2 + y^4$

49. Which of the following has a rational product? **[10.2]**

A $\sqrt{(-4)^2} \cdot \sqrt{4}$

B $\sqrt{5} \cdot \sqrt{20}$

C $\sqrt[3]{18} \cdot \sqrt[3]{12}$

D A and B

E all of these

50. In which product will the middle term have a negative coefficient? **[8.4]**

A $(x + 4)(x + 9)$

B $(x - 5)(x + 5)$

C $(x + 11)(x - 10)$

D $(x - 3)(x + 2)$

E all of these

10.4 ADDING & SUBTRACTING RADICALS

When does $\sqrt{x} + \sqrt{y} = \sqrt{x + y}$?

After completing this section, I will be able to

- define like radicals.
- add and subtract numerical radicals.
- add and subtract radicals with variable radicands.

Like radicals are radicals that have the same radicand and the same index, and they can be combined by applying the Distributive Property. The Distributive Property is used to add 2 expressions, such as $2x + 3x$.

$$2x + 3x = (2 + 3)x = 5x$$

Similarly, like radicals are added or subtracted using the Distributive Property.

$$2\sqrt{5} + 3\sqrt{5} = (2 + 3)\sqrt{5} = 5\sqrt{5}$$

The time, t, required to free-fall d feet is $t = \frac{\sqrt{d}}{4}$.

CONNECT

like variables:
same letter
same exponent

like radicals:
same index
same radicand

ADDING & SUBTRACTING RADICALS

1. Simplify each radical.
2. Combine like radicals by adding or subtracting their rational factors.

EXAMPLE 1: Adding & Subtracting Square Roots

Simplify each expression.

a. $4\sqrt{3} + 2\sqrt{3}$ **b.** $\sqrt{45} + \sqrt{18} - \sqrt{5}$

Answers

a. $4\sqrt{3} + 2\sqrt{3}$
$= (4 + 2)\sqrt{3} = 6\sqrt{3}$

Since each radical is already simplified, combine the like radicals by adding their rational factors.

b. $\sqrt{45} + \sqrt{18} - \sqrt{5}$
$= \sqrt{3^2 \cdot 5} + \sqrt{3^2 \cdot 2} - \sqrt{5}$
$= 3\sqrt{5} + 3\sqrt{2} - \sqrt{5}$
$= 2\sqrt{5} + 3\sqrt{2}$

1. Simplify each radical that can be simplified.
2. Combine like radicals.

SKILL CHECK **EXERCISES 5, 11**

EXAMPLE 2: Adding & Subtracting Cube Roots

Simplify each expression.

a. $-8\sqrt[3]{5}+9\sqrt[3]{40}-2\sqrt[3]{135}$ **b.** $5\sqrt[3]{16}-3\sqrt{8}+6\sqrt[3]{54}$

Answers

a. $-8\sqrt[3]{5}+9\sqrt[3]{40}-2\sqrt[3]{135}$

$=-8\sqrt[3]{5}+9\sqrt[3]{2^3\cdot 5}-2\sqrt[3]{3^3\cdot 5}$

$=-8\sqrt[3]{5}+18\sqrt[3]{5}-6\sqrt[3]{5}$

$=4\sqrt[3]{5}$

1. Simplify each radical.
2. Combine like radicals.

b. $5\sqrt[3]{16}-3\sqrt{8}+6\sqrt[3]{54}$

$=5\sqrt[3]{2^3\cdot 2}-3\sqrt{2^2\cdot 2}+6\sqrt[3]{2\cdot 3^3}$

$=10\sqrt[3]{2}-6\sqrt{2}+18\sqrt[3]{2}$

$=28\sqrt[3]{2}-6\sqrt{2}$

1. Simplify each radical.
2. Combine like radicals.

SKILL CHECK **EXERCISES 13, 23**

AFTERSCHOOLHELP

Adding & Subtracting Radicals

EXAMPLE 3: Adding & Subtracting Radicals Containing Variables

Simplify each expression.

a. $2x\sqrt{50x}-5\sqrt{8x^3}$ **b.** $\sqrt{2x^2}+\sqrt{75x^4}-2x^2\sqrt{3}-x\sqrt{18}$

Answers

a. $2x\sqrt{5^2\cdot 2x}-5\sqrt{2^2\cdot 2x^2x}$

$=2x\cdot 5\sqrt{2x}-5\cdot 2x\sqrt{2x}$

$=10x\sqrt{2x}-10x\sqrt{2x}$

$=0$

1. Simplify each radical.
2. Combine like radicals.

b. $\sqrt{2x^2}+\sqrt{3\cdot 25x^4}-2x^2\sqrt{3}-x\sqrt{2\cdot 9}$

$=x\sqrt{2}+5x^2\sqrt{3}-2x^2\sqrt{3}-3x\sqrt{2}$

$=-2x\sqrt{2}+3x^2\sqrt{3}$

1. Simplify each radical. You may recognize factors that are perfect squares.
2. Combine like radicals.

SKILL CHECK **EXERCISES 27, 29**

VOCABULARY

- like radicals

A. EXERCISES

1. Like radicals have the same ______ and the same ______.
2. Are $5\sqrt{10}$ and $-4\sqrt[3]{10}$ like radicals? Explain.

Simplify.

3. $\sqrt{3} + 5\sqrt{3}$
4. $2\sqrt{6} + 3\sqrt{6}$
5. $\sqrt{3} + 4\sqrt{2} + \sqrt{3}$
6. $2\sqrt{5} - 6\sqrt{5} + 3\sqrt{7}$
7. $\sqrt{8} + 3\sqrt{2}$
8. $\sqrt{20} + 4\sqrt{5}$
9. $5\sqrt{12} - 6\sqrt{3}$
10. $\sqrt{8} + 3\sqrt{2} - 4\sqrt{2}$
11. $-3\sqrt{3} + \sqrt{18} + \sqrt{27}$
12. $\sqrt[3]{54} + 5\sqrt[3]{2}$
13. $\sqrt[3]{32} - \sqrt[3]{108}$
14. $\sqrt[3]{24} - \sqrt[3]{375}$

Complete the following table.

	x	y	$\sqrt{x}$	$\sqrt{y}$	$\sqrt{x} + \sqrt{y}$	$\sqrt{x + y}$
15.	25	144				
16.	27	12				
17.	45	75				
18.	50	0				

19. **Essential Question:** Based on the results of exercises 15–18, what conclusion can you draw about the expressions $\sqrt{x} + \sqrt{y}$ and $\sqrt{x + y}$?

B. EXERCISES

Simplify each expression.

20. $\sqrt{5} + \sqrt{20} - \sqrt{72}$
21. $4\sqrt{2} - 6\sqrt{72} + 8\sqrt{32}$
22. $\sqrt[3]{16} + \sqrt[3]{18}$
23. $\sqrt[3]{40} - \sqrt[3]{875}$
24. $\sqrt[3]{16} + \sqrt[3]{81} + \sqrt[3]{250}$
25. $3\sqrt[3]{108} - 2\sqrt[3]{32} + 4\sqrt[3]{64}$
26. $4\sqrt{x} + 3\sqrt{y} - 2\sqrt{x}$
27. $3\sqrt{ab} + 4\sqrt{a} - 5\sqrt{b} + 5\sqrt{b}$
28. $\sqrt{x^2y} - x\sqrt{y}$
29. $\sqrt{32a^3} + \sqrt{2a^3} - 3\sqrt{6}$
30. $\sqrt{\frac{x^2}{16}} + \sqrt{\frac{x^2}{9}}$
31. $\sqrt{\frac{x^2}{16} + \frac{x^2}{9}}$
32. $\sqrt[3]{2} + \sqrt{18} + \sqrt[3]{250}$
33. $\sqrt[3]{3a^3} + \frac{\sqrt{300}}{10} - a\sqrt[3]{81}$
34. The width of a rectangle is $2\sqrt{3}$ units, and its length is three times longer than its width. Express the perimeter of the rectangle as a simplified radical.
35. Express the perimeter of the polygon to the right as a simplified radical expression.

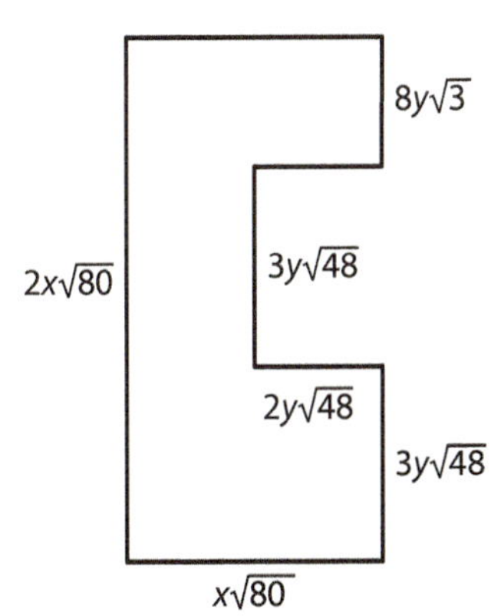

C. EXERCISES

36. Use Heron's formula (Section 10.2) to calculate the area of a triangle whose sides are $4\sqrt{6}$ cm, $3\sqrt{6}$ cm, and $6\sqrt{6}$ cm.

37. Express the area of the rectangle to the right as a simplified radical expression.

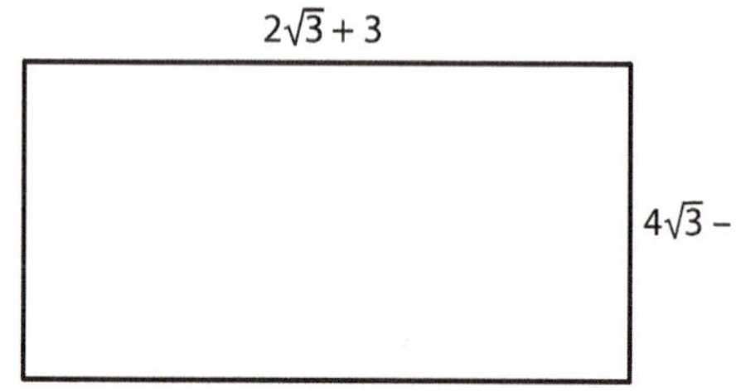

Simplify each expression.

38. $2\sqrt[3]{16x^6y^4} - xy\sqrt[3]{2x^3y}$

39. $2y\sqrt[3]{72x^5y^2} - 3x\sqrt[3]{243x^2y^5}$

40. $\sqrt{x^3 - 8x^2 + 16x} + \sqrt{64x^3} + \sqrt{x^2 + 4x + 4}$

Simplify by converting to exponential form with a common denominator. Express each answer as a simplified radical.

41. $\sqrt{a^3b} + a\sqrt[6]{a^3b^3}$

42. $5y\sqrt[3]{x^5y} - 3\sqrt[6]{x^{10}y^8}$

CUMULATIVE REVIEW

Solve each equation. [2.2]

43. $3x - 18 = x + 10$

44. $5(x - 8) = 2x - 5x$

Find each quotient. [8.6]

45. $(x^2 - 2x - 35) \div (x + 5)$

46. $(6x^2 + 11x - 31) \div (3x - 5)$

Factor each polynomial. [9.4–9.5]

47. $x^2 - 32x + 256$

48. $4x^2 - 16y^2$

49. $x^3 - 4x^2 - 9x + 36$

50. $x^3 + 5x^2 - 16x - 80$

51. Which radical(s) are simplified? **[10.1]**

A $-3\sqrt{30}$

B $\sqrt[3]{36}$

C $5\sqrt{6}$

D $2\sqrt[3]{50}$

E all of these

52. Which quotient always requires rationalizing the denominator to simplify? **[10.3]**

A dividing a rational number by a rational number

B dividing an irrational number by an irrational number

C dividing an irrational number by a rational number

D dividing a rational number by an irrational number

E B and D

MATH IN HISTORY
THE BERNOULLI FAMILY

The Bernoulli family settled in Switzerland after fleeing religious persecution from Catholics in Belgium. This leading family of European mathematicians produced eight mathematicians within three generations.

While complying with his parents' desire that he obtain degrees in philosophy and theology, Jakob (1654–1705) independently studied mathematics and astronomy. He was appointed professor of mathematics at the University of Basel in 1687. Jakob's work in differential and integral calculus is applied today in electrical engineering, and his law of large numbers is widely used in statistics and insurance studies. It is applied in risk aversion, by insurance companies in particular, since they must pay off on some policies earlier than expected and on others much later than expected or not at all.

Unfortunately, the family is remembered for its internal conflicts as well as its many contributions to the field of mathematics. After failing as an apprentice in his father's business, Johann (1667–1748) entered the University of Basel to study medicine and was tutored by his older brother Jakob in mathematics. Johann quickly became Jakob's equal as they studied Leibniz's new calculus together. Johann's boasts led to unfair public criticism by Jakob, and their collaboration eventually turned to bitter rivalry. While Johann served as the chair of mathematics at Groningen University, he tutored Guillaume de l'Hôpital in Leibniz's methods of calculus. In 1696 l'Hôpital published the first differential calculus book without acknowledging that the work was based on Bernoulli's lectures. Johann did not return to Basel until eight years later, when he accepted his brother's long-coveted position after his death. His accomplishments were in optical phenomena, analytical trigonometry, and exponential calculus.

Johann's nephew, all three of his sons, and two of his grandsons became famous mathematicians and scientists. Johann competed with his son Daniel (1700–1782), falsely dating his work on hydraulics in 1732, prior to Daniel's work entitled *Hydrodynamica* (completed in 1734).

Ecclesiastes 4:4 tells us that a person will never find satisfaction in his work if it is motivated by envy. While the members of this family contributed many advances in calculus, probability, differential equations, and science, how much more could have been done if they had cooperated with each other?

10.5 THE PYTHAGOREAN THEOREM

When is deductive reasoning not useful?

After completing this section, I will be able to

- apply the Pythagorean Theorem and its converse to solve problems involving right triangles.
- determine the length and midpoint of segments in the coordinate plane.
- evaluate the claim that the Pythagorean Theorem proves that Euclidean geometry is true.

Surveyors can measure distances across bodies of water by using special instruments and applying the Pythagorean Theorem.

The Pythagorean Theorem is one of the most recognized and popular theorems ever discovered. It is deeply rooted in a rich history and has blossomed with a variety of applications that are widely used today.

DEFINITION

A **theorem** is a statement that has been proved to be true.

THE PYTHAGOREAN THEOREM

If the legs of a right triangle have lengths a and b and the hypotenuse has length c, then $a^2 + b^2 = c^2$.

RECALL

The **hypotenuse** is always across from the right angle and is the longest side of the right triangle.

NOTE

When finding length, the value must be positive, so we use the principal square root.

EXAMPLE 1: Finding the Missing Length of a Right Triangle

Find each missing length using right triangle ABC shown.

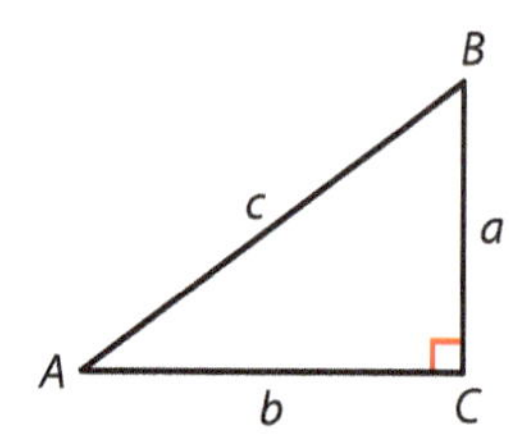

a. Find c if $a = 12$ and $b = 9$.

b. Find b if $a = 10$ and $c = 20$.

Answers

a. $a = 12$, $b = 9$, and c = length of the hypotenuse

1. Identify the known lengths of the sides and assign variables, making sure that the hypotenuse is c.

$$a^2 + b^2 = c^2$$
$$12^2 + 9^2 = c^2$$

2. Substitute values into the Pythagorean Theorem.

$$144 + 81 = c^2$$
$$c^2 = 225$$

3. Solve for c^2.

$$c = 15 \text{ units}$$

4. Take the square root of each side.

b. $a = 10$, $c = 20$ (length of the hypotenuse), and b = unknown leg length

$$a^2 + b^2 = c^2$$
$$10^2 + b^2 = 20^2$$
$$b^2 = 300$$
$$b = \sqrt{300} = \sqrt{100 \cdot 3}$$
$$= 10\sqrt{3} \text{ units}$$

1. Identify the known lengths and assign variables.
2. Substitute values into the Pythagorean Theorem and solve for b^2.
3. Take the square root of each side and simplify.

SKILL CHECK **EXERCISES 1, 5**

The converse of the Pythagorean Theorem can be used to determine whether a triangle is a right triangle.

KEYWORD SEARCH

Pythagorean theorem proofs

CONVERSE OF THE PYTHAGOREAN THEOREM

If a triangle has side lengths a, b, and c such that $a^2 + b^2 = c^2$, then the triangle is a right triangle.

EXAMPLE 2: Determining a Right Triangle

Determine whether each triangle with given side lengths is a right triangle.

a. $4\sqrt{2}$, 7, and 9 **b.** $3\sqrt{6}$, $3\sqrt{5}$, and 10

Answers

a.

$$a^2 + b^2 = c^2$$
$$(4\sqrt{2})^2 + 7^2 \stackrel{?}{=} 9^2$$
$$32 + 49 \stackrel{?}{=} 81$$
$$81 = 81$$

1. Substitute the 2 shorter lengths for a and b and the longest length for c.
2. Evaluate each side.

The triangle is a right triangle since $a^2 + b^2 = c^2$.

b.

$$a^2 + b^2 = c^2$$
$$(3\sqrt{6})^2 + (3\sqrt{5})^2 \stackrel{?}{=} 10^2$$
$$54 + 45 \stackrel{?}{=} 100$$
$$99 \neq 100$$

1. Substitute the 2 shorter lengths for a and b and the longest length for c.
2. Evaluate each side.

The triangle is not a right triangle since $a^2 + b^2 \neq c^2$.

SKILL CHECK **EXERCISE 11**

Applying the Pythagorean Theorem to points on the coordinate plane allows you to find the distance between any 2 graphed points.

EXAMPLE 3: Finding Distance between Points

Use the Pythagorean Theorem to find the distance between P_1 (−3, 1) and P_2 (5, 4).

Answer

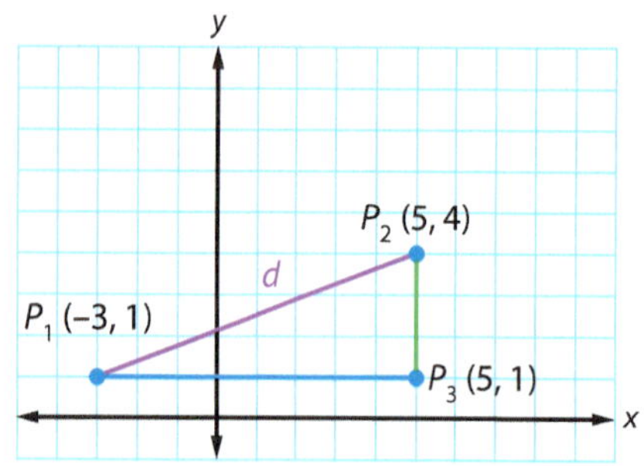

1. Graph the points and draw the segment connecting them to represent the unknown distance, d.
2. Form a right triangle by drawing a vertical line from P_2 and a horizontal line from P_1. These lines form a right angle at their intersection, P_3 (5, 1).

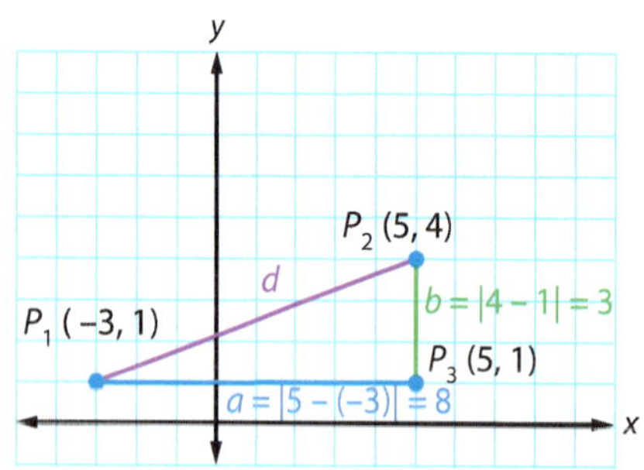

3. Determine the lengths of the legs.
   a. Since P_3 and P_1 have the same y-coordinate, a is the distance between their x-coordinates.
   b. Since P_3 and P_2 have the same x-coordinate, b is the distance between their y-coordinates.

$$8^2 + 3^2 = d^2$$
$$64 + 9 = d^2$$
$$d^2 = 73$$
$$d = \sqrt{73}$$

4. Substitute into the Pythagorean Theorem and solve for d. Note that $\sqrt{73}$ cannot be simplified.

The distance between P_1 and P_2 is $\sqrt{73} \approx 8.5$ units.

SKILL CHECK **EXERCISE 15** ☑

While the procedure from Example 3 could be used to find the distance between any 2 points on the Cartesian coordinate plane, a formula can be developed by applying this procedure to the generic points P_1 (x_1, y_1) and P_2 (x_2, y_2).

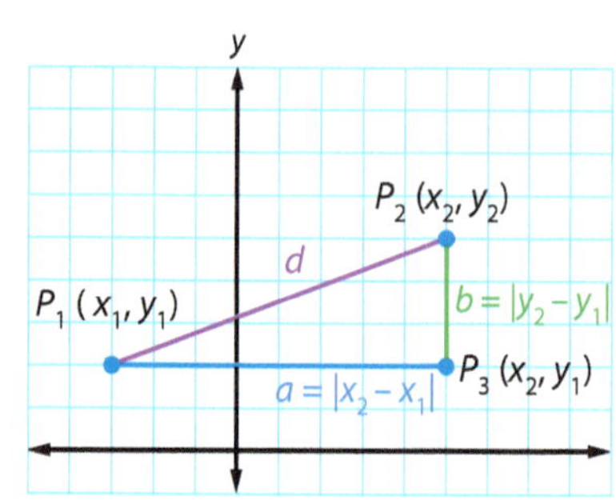

The length of the horizontal leg is the distance between the x-coordinates,

$$a = |x_2 - x_1|;$$

and the length of the vertical leg is the distance between the y-coordinates,

$$b = |y_2 - y_1|.$$

Using the Pythagorean Theorem produces

$$d^2 = |x_2 - x_1|^2 + |y_2 - y_1|^2.$$

Taking the square root of both sides and dropping the absolute value signs (since squaring always yields a positive result) produces the following formula.

AFTERSCHOOLHELP

The Pythagorean Theorem

DISTANCE FORMULA

The distance, d, between any 2 points $P_1\ (x_1, y_1)$ and $P_2\ (x_2, y_2)$ is given by the formula $d = \sqrt{(x_2 - x_1)^2 + (y_2 - y_1)^2}$.

Applying the distance formula allows you to find the distance between any 2 points without graphing.

EXAMPLE 4: Using the Distance Formula

Use the distance formula to find the distance between the given points.

a. (2, 1) and (5, 1) **b.** (−2, −9) and (6, −5)

Answers

a. Let $(x_1, y_1) = (2, 1)$ and $(x_2, y_2) = (5, 1)$.

$$d = \sqrt{(x_2 - x_1)^2 + (y_2 - y_1)^2}$$
$$= \sqrt{(5-2)^2 + (1-1)^2}$$
$$= \sqrt{3^2 + 0^2} = \sqrt{9} = 3$$

The distance is 3 units.

1. Assign variables to the given points.
2. Substitute into the distance formula and simplify.

b. Let $(x_1, y_1) = (-2, -9)$ and $(x_2, y_2) = (6, -5)$.

$$d = \sqrt{(x_2 - x_1)^2 + (y_2 - y_1)^2}$$
$$= \sqrt{[6-(-2)]^2 + [-5-(-9)]^2}$$
$$= \sqrt{8^2 + 4^2} = \sqrt{64 + 16} = \sqrt{80} = 4\sqrt{5}$$

The distance is $4\sqrt{5}$, or almost 9 units.

1. Assign variables to the given points.
2. Substitute into the distance formula and simplify.

SKILL CHECK **EXERCISE 17** ☑

The distance formula extends the formula for the length of a segment on a number line, $l = |b - a|$, to any 2 points in the coordinate plane. Likewise, the formula for the midpoint of a segment on a number line, $m = \frac{a+b}{2}$, can be extended to any segment in the coordinate plane.

MIDPOINT FORMULA

The midpoint of a segment with endpoints $P_1\ (x_1, y_1)$ and $P_2\ (x_2, y_2)$ is $M\left(\frac{x_1 + x_2}{2}, \frac{y_1 + y_2}{2}\right)$.

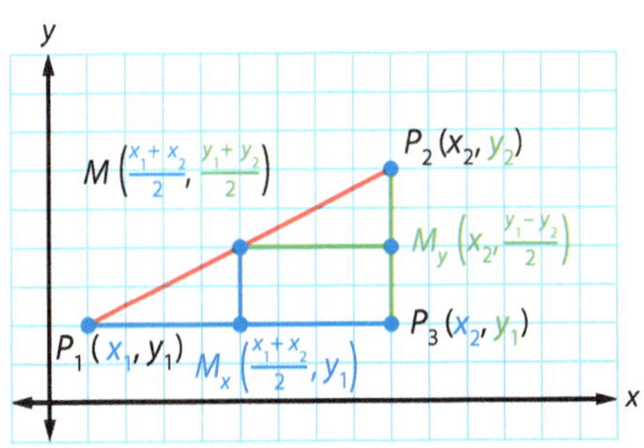

The midpoint formula can also be applied to find an endpoint of a segment when the midpoint and other endpoint are given.

EXAMPLE 5: Applying the Midpoint Formula

Find the missing point, given a line segment with endpoints P_1 and P_2 and midpoint M.

a. $P_1\,(-3, 1)$ and $P_2\,(5, 4)$; find M.

b. $P_1\,(2\sqrt{3}, \sqrt{5})$ and $M\,(-\sqrt{3}, 3\sqrt{5})$; find P_2.

Answers

a. Let $(x_1, y_1) = (-3, 1)$ and $(x_2, y_2) = (5, 4)$.

1. Assign variables to the given points.

$$\left(\frac{x_1 + x_2}{2}, \frac{y_1 + y_2}{2}\right) = \left(\frac{-3 + 5}{2}, \frac{1 + 4}{2}\right) = \left(\frac{2}{2}, \frac{5}{2}\right)$$

2. Substitute into the midpoint formula and simplify.

The midpoint is $\left(1, \frac{5}{2}\right)$.

b. Let $(x_1, y_1) = (2\sqrt{3}, \sqrt{5})$ and $\left(\frac{x_1 + x_2}{2}, \frac{y_1 + y_2}{2}\right) = (-\sqrt{3}, 3\sqrt{5})$.

1. Assign variables to the given points.

$$\left(\frac{x_1 + x_2}{2}, \frac{y_1 + y_2}{2}\right) = \left(\frac{2\sqrt{3} + x_2}{2}, \frac{\sqrt{5} + y_2}{2}\right) = (-\sqrt{3}, 3\sqrt{5})$$

2. Substitute into the midpoint formula.

$$\frac{2\sqrt{3} + x_2}{2} = -\sqrt{3} \qquad \frac{\sqrt{5} + y_2}{2} = 3\sqrt{5}$$

$$2\sqrt{3} + x_2 = -2\sqrt{3} \qquad \sqrt{5} + y_2 = 6\sqrt{5}$$

$$x_2 = -4\sqrt{3} \qquad y_2 = 5\sqrt{5}$$

3. Solve for x_2 and y_2.

The endpoint P_2 is $(-4\sqrt{3}, 5\sqrt{5})$.

SKILL CHECK **EXERCISES 13, 27**

VOCABULARY

- hypotenuse
- theorem

The Pythagorean Theorem can be proved from the assumptions of Euclidean geometry by using deductive reasoning. But that means only that it is consistent with those assumptions, not that the assumptions are necessarily true. The Pythagorean Theorem, for example, does not prove that planes are flat (an assumption of Euclidean geometry) even though that assumption is necessary for the theorem to hold. If we assume planes are spherical, then a different set of theorems would be deduced, and the Pythagorean Theorem would not hold. As useful as deductive reasoning is in mathematics, it does have limitations.

A. EXERCISES

Find the length of the unknown side of each right triangle.

	a	b	c
1.	3	4	
2.		12	13
3.	4	5	
4.	3	3	
5.	2		8
6.		6	12
7.		7	9
8.	2		5

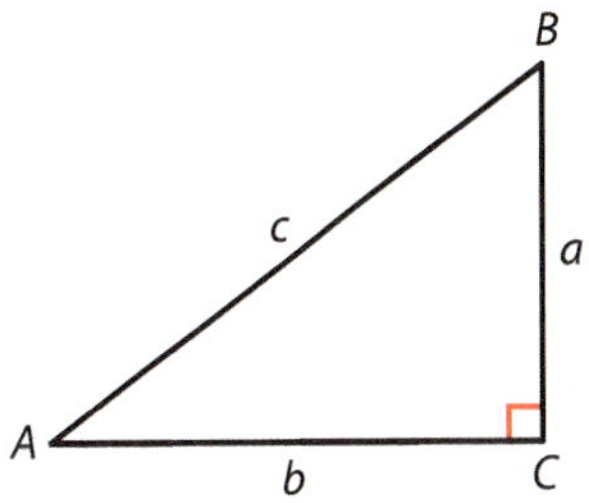

Determine whether a triangle with the given side lengths is a right triangle.

9. 10, 8, and 6

10. 7, 14, and 17

11. 4, $\sqrt{3}$, and 1

12. $4\sqrt{2}$, $3\sqrt{2}$, and $5\sqrt{2}$

Find the midpoint of a segment with the given endpoints.

13. (2, 4) and (6, 6)

14. (9, −12) and (7, −4)

B. EXERCISES

Find the distance between the given pairs of points. Express each answer as a simplified radical.

15. (3, 4) and (0, 0)

16. (3, 2) and (9, 10)

17. (−3, −1) and (2, −13)

18. (1, 8) and (4, −1)

19. $(2\sqrt{7}, -3)$ and $(\sqrt{7}, 5)$

20. $(5\sqrt{3}, \sqrt{24})$ and $(-\sqrt{12}, \sqrt{54})$

Find the length of the third side of each right triangle in which *c* is the hypotenuse.

21. $a = 3$; $b = \sqrt{7}$

22. $b = \sqrt{51}$; $c = 10$

23. $a = \sqrt{11}$; $b = \sqrt{13}$

Find the missing endpoint of a segment with the given endpoint and midpoint.

24. $P_1\,(-6, 8)$ and $M\,(1, 2)$

25. $P_1\,(-5, 0)$ and $M\,(-7.5, -7)$

26. $P_1\,(2\sqrt{3}, -6)$ and $M\,(3\sqrt{3}, 1)$

27. $P_1\,(5\sqrt{2}, 3\sqrt{5})$ and $M\,(3\sqrt{2}, -2\sqrt{5})$

Find the perimeter of each figure.

28.

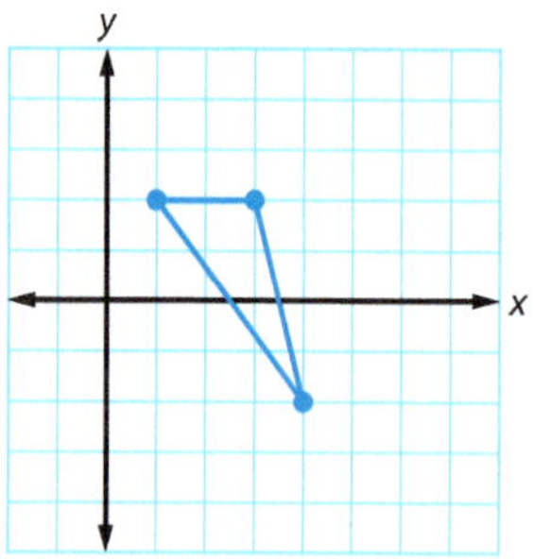

29.

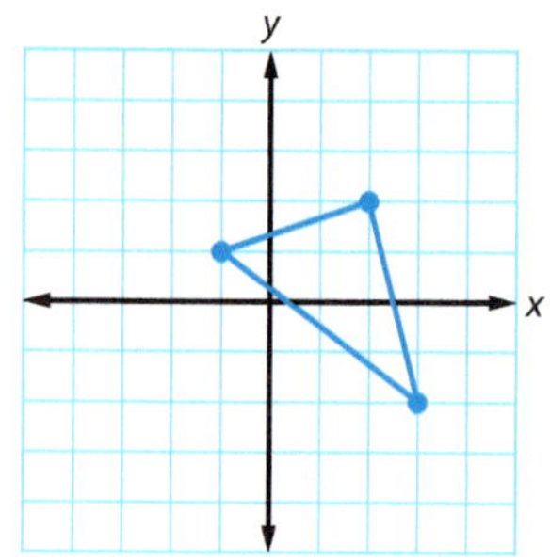

30.

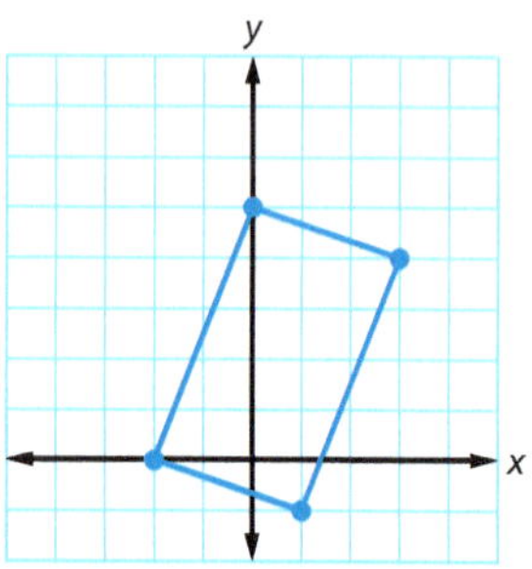

31.

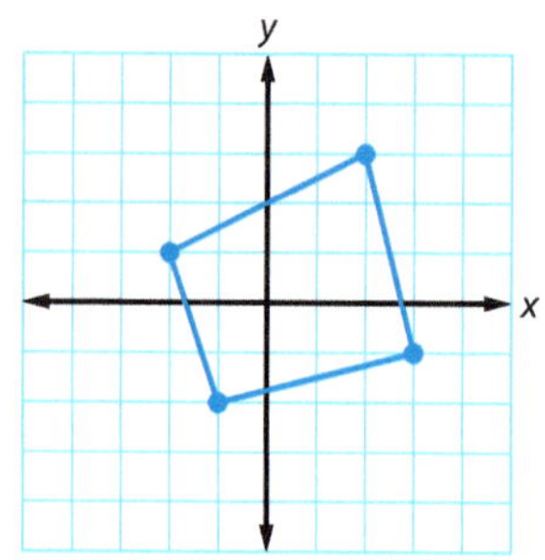

32. **Essential Question:** Why does the Pythagorean Theorem not prove that Euclidean geometry is true?

33. A surveyor is measuring between property stakes *A* and *B* across a pond. He knows the distances from *C* to each of the 2 property stakes and that angle *C* is a right angle. What is the distance from *A* to *B*?

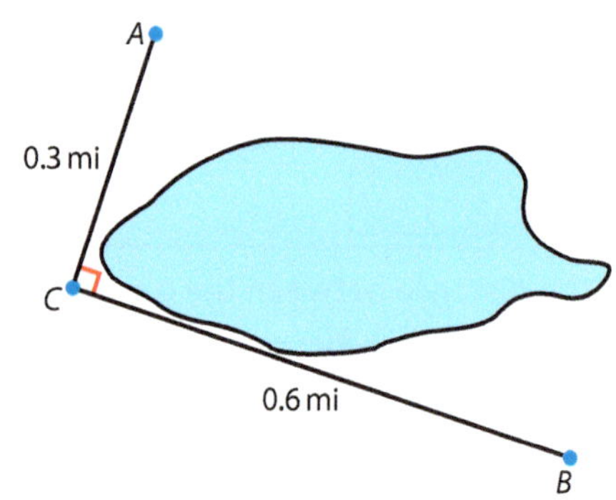

34. The bases are each 90 ft apart on a baseball diamond. Find the distance a player has to throw the ball to make an out at first base if he is standing at third base.

35. A destroyer is located at point (4, 9) and an aircraft carrier is at point (15, 3), where the numbers represent distances in nautical miles. What is the distance (to the nearest tenth of a nautical mile) between the 2 vessels?

36. The M. T. Container Manufacturing Company is planning to make a new box 3 in. high and 10 in. long. If the M. T. box measures 14 in. diagonally across the bottom, what is the width of the box?

37. A construction worker needs to pour a concrete slab for a garage. His carpenter's square (a right-angle tool) is much too small to ensure perfect right angles, so he depends on the Pythagorean Theorem. He measures along the concrete forms as illustrated in the diagram and adjusts the forms to create a perfect right angle. What length of *c* will create a right angle?

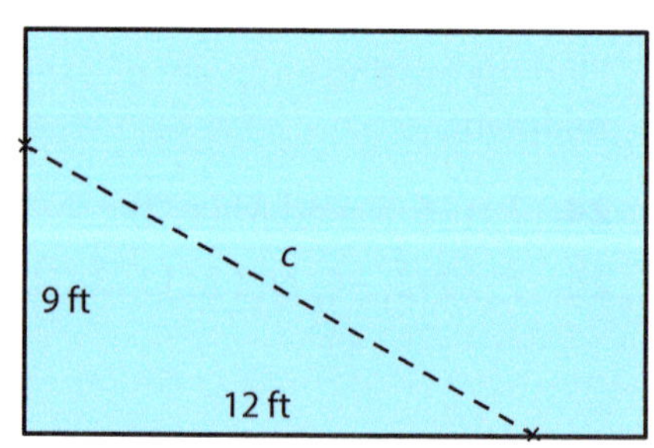

C. EXERCISES

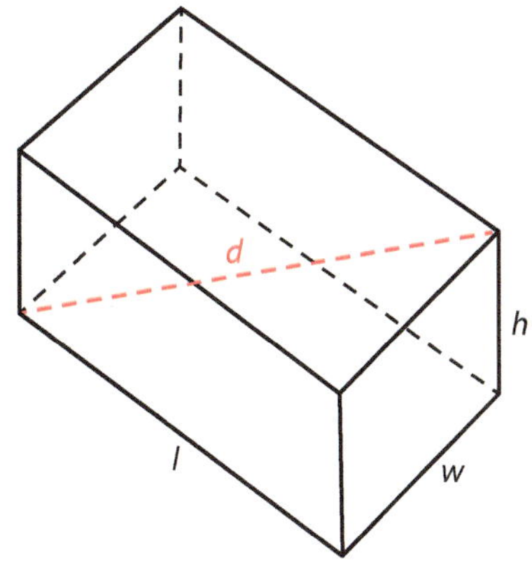

38. The dimensions of a rectangular box are $l = 15$ in., $w = 12$ in., and $h = 9$ in. Use the Pythagorean Theorem to find the following.

a. the length of a diagonal of the bottom of the box

b. the length of a diagonal of the box, d

39. **MP4** The dimensions of a rectangular box are l, w, and h. Use the Pythagorean Theorem to derive expressions for the following.

a. the length of a diagonal of the bottom of the box

b. the length of a diagonal of the box, d

40. **MP6** Mr. Ferguson is going to construct a ramp to his barn. The barn loft is 48 in. above ground level, and the base of the ramp needs to be 20 ft from the barn.

a. How long (in feet and inches) must the ramp be (to the nearest inch)?

b. What is the slope of the ramp? Express your answer as both a fraction and a percent.

A Pythagorean triple consists of 3 positive integers a, b, and c, such that $a^2 + b^2 = c^2$. Commonly used Pythagorean triples include (3, 4, 5) and (6, 8, 10). If m and n are positive integers with $m > n$, then $(m^2 - n^2, 2mn, m^2 + n^2)$ is a Pythagorean triple.

KEYWORD SEARCH
Pythagorean triple

41. Given $m = 12$ and $n = 10$, find the related Pythagorean triple and verify that it is a Pythagorean triple.

42. **MP8** Show algebraically that $(m^2 - n^2, 2mn, m^2 + n^2)$ is a Pythagorean triple.

CUMULATIVE REVIEW

Simplify each radical. [10.1]

43. $\sqrt{50x^6}$

44. $\sqrt{63y^3}$

Simplify each expression. [10.2–10.4]

45. $\sqrt{10} \cdot \sqrt{18}$

46. $\sqrt[3]{50} \cdot \sqrt[3]{15}$

47. $\dfrac{15\sqrt{x}}{\sqrt{3y}}$

48. $\dfrac{8}{\sqrt[3]{5}}$

49. $6\sqrt[3]{2x^4} - x\sqrt[3]{16x}$

50. $-3\sqrt{25z} + \sqrt{256z}$

51. Which polynomial is the conjugate of $2x - 9$? **[8.5]**

A $2x + 9$

B $2x - 9$

C $9 - 2x$

D $9x - 2$

E none of these

52. What is the coefficient of x after multiplying $(2x + a)(3x + 5)$? **[8.4]**

A 6

B $5a$

C $10 + 3a$

D $30a$

E none of these

10.6 MULTIPLYING & DIVIDING RADICAL EXPRESSIONS

How can I use multiplication to divide by sums and differences containing radicals?

After completing this section, I will be able to

- multiply sums and differences involving radicals.
- divide sums and differences involving radicals.

You have learned how to multiply and divide polynomials and how to multiply and divide radicals. By combining these skills, you will be able to multiply and divide more complicated radical expressions. A **radical expression** is an algebraic expression containing at least 1 radical. Even though radical expressions such as $\sqrt{3} + 4$ or $\sqrt{6} - \sqrt{3}$ are not polynomials, the same principles apply when multiplying radical expressions with more than 1 term.

Mechanical engineers can use the formula $K_f = \frac{K_t - 1}{1 + \sqrt{\frac{a}{r}}}$ to identify the "fatigue stress factor," which affects the integrity of objects subjected to repeated stress, such as the adjustment pin of a leg press machine.

EXAMPLE 1: Applying the Distributive Property

Simplify each radical expression.

a. $\sqrt{5}(\sqrt{3} - 2)$ **b.** $2\sqrt{6}(\sqrt{15} - 3\sqrt{10})$

Answers

a. $\sqrt{5}(\sqrt{3} - 2) = \sqrt{15} - 2\sqrt{5}$ Apply the Distributive Property.

b. $2\sqrt{6}(\sqrt{15} - 3\sqrt{10})$

$= 2\sqrt{(2 \cdot 3)(3 \cdot 5)} - 6\sqrt{(2 \cdot 3)(2 \cdot 5)}$ **1.** Apply the Distributive Property, writing each radicand as the product of its prime factors.

$= 2 \cdot 3\sqrt{2 \cdot 5} - 6 \cdot 2\sqrt{3 \cdot 5}$ **2.** Simplify the square root of each perfect square.

$= 6\sqrt{10} - 12\sqrt{15}$ **3.** Simplify.

SKILL CHECK ***EXERCISE 1***

EXAMPLE 2: Multiplying Radical Expressions

Simplify each expression.

a. $(\sqrt{7} + \sqrt{2})(\sqrt{3} - 6)$ **b.** $(\sqrt{5} - 7)(\sqrt{5} + 7)$ **c.** $(\sqrt{3} - \sqrt{2})^2$

Answers

a. $\sqrt{7} \cdot \sqrt{3} - 6\sqrt{7} + \sqrt{2} \cdot \sqrt{3} - 6\sqrt{2}$ **1.** Multiply using the FOIL method.

$= \sqrt{21} - 6\sqrt{7} + \sqrt{6} - 6\sqrt{2}$ **2.** Note that all terms are simplified and there are no like terms to combine.

b. $(\sqrt{5})^2 - 7^2$

$= 5 - 49 = -44$

1. Recognize a product of conjugates: $(a + b)(a - b) = a^2 - b^2$.
2. Simplify.

c. $(\sqrt{3})^2 - 2(\sqrt{3})(\sqrt{2}) + (\sqrt{2})^2$

$= 3 - 2\sqrt{6} + 2 = 5 - 2\sqrt{6}$

1. Recognize the square of a difference: $(a - b)^2 = a^2 - 2ab + b^2$.
2. Simplify.

SKILL CHECK **EXERCISES 5, 11**

MULTIPLYING RADICAL EXPRESSIONS

1. Follow the same rules that you follow for multiplying polynomials.
2. After multiplying, simplify radicals and combine like radicals.

EXAMPLE 3: Multiplying Multiple Radical Expressions

Simplify $(\sqrt{3} - 2)(\sqrt{3} + 2)(\sqrt{6} - 4)$.

Answer

$[(\sqrt{3})^2 - 2^2](\sqrt{6} - 4)$
$= (3 - 4)(\sqrt{6} - 4)$
$= -1(\sqrt{6} - 4)$

1. Apply the product of conjugates pattern to multiply the first 2 radical expressions.

$= -\sqrt{6} + 4 = 4 - \sqrt{6}$

2. Simplify the product. Use the Commutative Property of Addition to rewrite the answer.

SKILL CHECK **EXERCISE 21**

RECALL

The conjugate of $a + b$ is $a - b$, and their product is $a^2 - b^2$.

Recall that expressions with radicals in the denominator are not simplified. Expressions such as $\frac{4}{\sqrt{5} - 2}$ cannot be simplified by multiplying by the radical $(\sqrt{5})$ as was done in Section 10.3. Instead, multiplying the numerator and denominator by the conjugate of $\sqrt{5} - 2$ produces an equivalent expression without radicals in the denominator.

EXAMPLE 4: Dividing Radical Expressions

AFTERSCHOOLHELP

Multiplying & Dividing Radical Expressions

Simplify $\frac{4 + \sqrt{3}}{5 + \sqrt{3}}$.

Answer

$\frac{4 + \sqrt{3}}{5 + \sqrt{3}} \cdot \frac{5 - \sqrt{3}}{5 - \sqrt{3}}$

1. Multiply the numerator and the denominator by the conjugate of the denominator.

$= \frac{20 - 4\sqrt{3} + 5\sqrt{3} - 3}{25 - 3}$

2. Multiply the numerator using the FOIL method. The product of conjugates in the denominator equals the difference of their squares.

$= \frac{17 + \sqrt{3}}{22}$

3. Simplify.

SKILL CHECK **EXERCISES 17, 19**

DIVIDING RADICAL EXPRESSIONS

1. Multiply the numerator and the denominator by the conjugate of the denominator.
2. Simplify.

EXAMPLE 5: Dividing Radical Expressions

Simplify each expression.

a. $\frac{(\sqrt{m} - \sqrt{n})}{(\sqrt{m} + \sqrt{n})}$ **b.** $\frac{3\sqrt{2} + 4}{2\sqrt{3} + 5}$

Answers

a. $\frac{\sqrt{m} - \sqrt{n}}{\sqrt{m} + \sqrt{n}} \cdot \frac{\sqrt{m} - \sqrt{n}}{\sqrt{m} - \sqrt{n}}$

1. Multiply the numerator and the denominator by the conjugate of the denominator.

$= \frac{(\sqrt{m})^2 - 2(\sqrt{m})(\sqrt{n}) + (\sqrt{n})^2}{(\sqrt{m})^2 - (\sqrt{n})^2}$

2. Apply the square of a difference in the numerator. The product of conjugates in the denominator equals the difference of their squares.

$= \frac{m - 2\sqrt{mn} + n}{m - n}$

3. Simplify.

b. $\frac{3\sqrt{2} + 4}{2\sqrt{3} + 5} \cdot \frac{2\sqrt{3} - 5}{2\sqrt{3} - 5}$

1. Multiply the numerator and the denominator by the conjugate of the denominator.

$= \frac{6\sqrt{6} - 15\sqrt{2} + 8\sqrt{3} - 20}{4(3) - 25}$

2. Multiply the numerator using the FOIL method. The product of conjugates in the denominator equals the difference of their squares.

$= \frac{6\sqrt{6} - 15\sqrt{2} + 8\sqrt{3} - 20}{-13}$

3. Simplify.

SKILL CHECK **EXERCISE 33**

VOCABULARY

- radical expression

A. EXERCISES

Simplify each expression.

1. $\sqrt{3}(\sqrt{2} - 1)$

2. $\sqrt{5}(5 + \sqrt{3})$

3. $\sqrt{7}(\sqrt{21} + \sqrt{2})$

4. $\sqrt{6}(\sqrt{18} - \sqrt{2})$

5. $(2 - \sqrt{3})(2 + \sqrt{3})$

6. $(\sqrt{3} - \sqrt{11})(\sqrt{3} + \sqrt{11})$

7. $(\sqrt{7} - \sqrt{10})^2$

8. $(\sqrt{5} - \sqrt{3})^2$

9. $(\sqrt{7} + 8)^2$

10. $(7 + \sqrt{3})(4 - \sqrt{11})$

11. $(3 - \sqrt{10})(4 + \sqrt{20})$

12. $(\sqrt{12} - \sqrt{3})(\sqrt{5} + \sqrt{2})$

State the conjugate of each radical expression.

13. $-3 + \sqrt{5}$

14. $\sqrt{7} - \sqrt{11}$

Simplify each expression.

15. $\frac{1}{2 + \sqrt{3}}$

16. $\frac{2 + \sqrt{6}}{3 - \sqrt{6}}$

17. $\frac{\sqrt{11}}{\sqrt{3} + \sqrt{2}}$

18. $\frac{9}{3 + \sqrt{2}}$

19. $\frac{1}{\sqrt{2} - 1}$

20. $\frac{7\sqrt{3}}{2\sqrt{4} + 1}$

B. EXERCISES

Simplify each expression.

21. $(\sqrt{5} - 3)(\sqrt{2} + 1)(\sqrt{5} + 3)$

22. $(3 - \sqrt{2})^2(\sqrt{5} + 3)$

23. $\sqrt{5}(\sqrt{35} + 5)(\sqrt{15} + 2)$

24. $(x\sqrt{3} - 5)(x\sqrt{6} - 3)$

25. $\sqrt{x}(\sqrt{3} - 5)(\sqrt{x^3} + 2)$

26. $\sqrt{xy^2}(\sqrt{x} - \sqrt{y})(\sqrt{x} + \sqrt{y})$

27. $\sqrt{5}(\sqrt{3} - \sqrt{2})(\sqrt{5} - 6)$

28. $\sqrt{x}(x + \sqrt{8})^2$

29. $3\sqrt{5}(\sqrt{7} - \sqrt{3})(\sqrt{7} + \sqrt{3})$

30. $\sqrt{7}(\sqrt{3} + \sqrt{21})^2$

31. $\frac{2}{\sqrt{3} - 1}$

32. $\frac{\sqrt{5} - 3}{\sqrt{5} + 3}$

33. $\frac{5}{\sqrt{2} - \sqrt{6}}$

34. $\frac{\sqrt{5} + 1}{\sqrt{7} - 2}$

35. $\frac{16}{\sqrt{7} + \sqrt{3}}$

36. $\frac{\sqrt{8}}{3\sqrt{2} - \sqrt{3}}$

37. Express the area of the rectangle as a simplified radical expression.

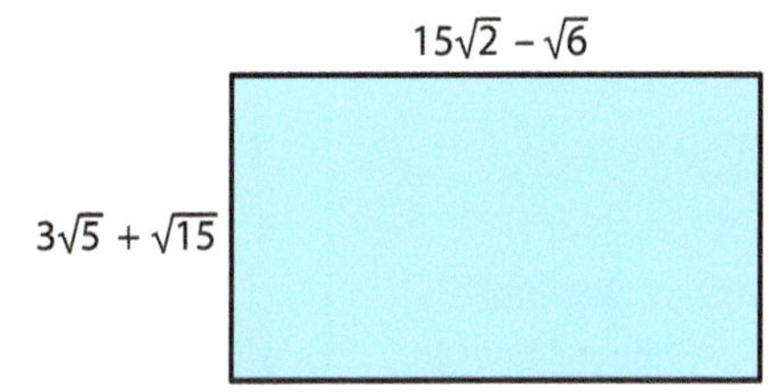

38. **Essential Question:** How can you use multiplication to divide by sums and differences containing radicals?

C. EXERCISES

MP1 Simplify each expression.

39. $\frac{\sqrt{2} + \sqrt{5}}{\sqrt{3} - \sqrt{10}}$

40. $\frac{3\sqrt{2} - 8\sqrt{7}}{4\sqrt{3} + \sqrt{5}}$

41. $(\sqrt{x} - 4)^2(\sqrt{x} - 7)^2$

42. Find the length of the hypotenuse of a right triangle with legs of length $2 + \sqrt{15}$ units and $2\sqrt{3} - \sqrt{5}$ units.

43. **MP2** Solve for x.

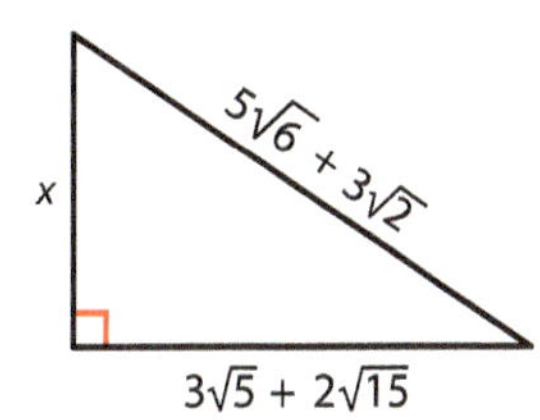

CUMULATIVE REVIEW

Translate each word phrase into an algebraic expression. [1.5]

44. three less than twice the square root of a number

45. twice the square root of three less than a number

Express each number in scientific notation. [7.2]

46. 0.000015

47. $15{,}000 \times 10^{-2}$

Find each product. [8.4–8.5]

48. $(5x - 3)(3x + 2)$

49. $(4x - 7)^2$

Factor each expression. [9.3]

50. $x^2 + 4xy - 45y^2$

51. $2a^2 - 18ab - 44b^2$

52. Which exponential function has the growth rate with the highest absolute value? **[7.5]**

A $a(x) = 500(1.21)^x$

B $b(x) = 300(0.75)^x$

C $c(x) = 1.35(1.005)^x$

D $d(x) = -700(0.88)^x$

E none of these

53. Which binomial is prime? **[9.1]**

A $12a^2b + 39cd^2$

B $10a^4b^3z^2 - 7xyz$

C $57m^3n + 15n^2pt$

D $9pqh - 52rgk$

E All are factorable.

10.7 RADICAL EQUATIONS

When is solving radical equations used in the real world?

After completing this section, I will be able to

- define a radical equation.
- solve radical equations.
- identify extraneous solutions to radical equations.

Radical equations are used during accident investigations.

Many real-world situations can be modeled using radical equations. Equations such as $\sqrt{22.5x} = 50$ or $3 + \sqrt{x} = 24$ are called **radical equations** because they contain a variable in the radicand. Like all equations, radical equations are solved by applying inverse operations.

EXAMPLE 1: Solving Radical Equations

Solve each equation.

a. $\sqrt{x} = 10$ **b.** $\sqrt{2x+5} = 8$

Answers

a.

$$(\sqrt{x})^2 = 10^2$$
$$x = 100$$

1. Square both sides of the equation to eliminate the square root.

$$\sqrt{100} = 10$$

2. Check your solution in the original equation.

b.

$$\sqrt{2x+5} = 8$$
$$(\sqrt{2x+5})^2 = 8^2$$

1. Square both sides.

$$2x + 5 = 64$$
$$2x = 59$$
$$x = \frac{59}{2}$$

2. Solve.

$$\sqrt{2\left(\frac{59}{2}\right) + 5} = 8$$
$$\sqrt{64} = 8$$

3. Check your solution in the original equation.

SKILL CHECK **EXERCISES 1, 9**

When solving any equation, it is always best practice to substitute an answer in the original equation to verify a solution and check your work. When solving a radical equation, it is possible to do the work correctly but get an invalid result.

How is an invalid answer possible? Consider the following.

$-2 = 2$	a false statement
$(-2)^2 = 2^2$	square both sides
$4 = 4$	a true statement

Squaring both sides of the equation did not result in both sides being multiplied by the same number, so a false statement was changed into a true statement. When squaring both sides of an equation, it is imperative that the solution be checked in the original equation to determine its validity. If a solution does not make the original equation true, then it is invalid and is called an **extraneous solution.**

SOLVING RADICAL EQUATIONS

1. Isolate the radical on one side of the equation.
2. Square both sides of the equation.
3. Solve the resulting equation.
4. Check any solutions in the original equation.

EXAMPLE 2: Solving Radical Equations

Solve each equation.

a. $8 + \sqrt{x-3} = 4$ **b.** $10 - 2\sqrt{x+1} = 4$

Answers

a. $8 + \sqrt{x-3} = 4$

$\sqrt{x-3} = -4$

1. Isolate the radical.

$\left(\sqrt{x-3}\right)^2 = (-4)^2$

2. Square both sides.

$x - 3 = 16$

$x = 19$

3. Solve.

$8 + \sqrt{19-3} = 4$

$8 + 4 = 4$; false, ∴ no solution

4. Check your solution in the original equation. Since $12 \neq 4$, there is no solution and 19 is extraneous.

b. $10 - 2\sqrt{x+1} = 4$

$-2\sqrt{x+1} = -6$

$\sqrt{x+1} = 3$

1. Isolate the radical.

$\left(\sqrt{x+1}\right)^2 = 3^2$

2. Square both sides.

$x + 1 = 9$

$x = 8$

3. Solve.

$10 - 2\sqrt{8+1} = 4$

$10 - 2\sqrt{9} = 4$

$10 - 2(3) = 4$

$10 - 6 = 4$

4. Check your solution in the original equation.

SKILL CHECK **EXERCISES 11, 13**

AfterSchoolHelp

Radical Equations

When skid marks are present, accident investigators can measure the length of the skid marks to determine the speed of the car involved (see Section 10.1 exercises 36–38). The formula $S = \sqrt{30Df}$ is used to determine the distance required for a car to skid to a stop. S is the car's speed in miles per hour, D is the length of the skid marks in feet, and f is the coefficient of friction between the tires and the road.

EXAMPLE 3: Solving a Real-World Problem

Use the formula $S = \sqrt{30Df}$ and a coefficient of friction of 0.75 to determine the stopping distance for a car traveling 50 mi/hr.

Answer

$\sqrt{30D(0.75)} = 50$

$(\sqrt{22.5D})^2 = 50^2$

1. Since the radical is already isolated, square both sides.

$22.5D = 2500$

$D = \frac{2500}{22.5} = 111.\overline{1}$

2. Solve.

$\sqrt{30(111.\overline{1})(0.75)}$

$= \sqrt{2500} = 50$

3. Check your solution in the original equation.

It will take about 111 ft for the car to stop.

SKILL CHECK **EXERCISE 23**

VOCABULARY

- extraneous solution
- radical equation

A. EXERCISES

Solve each equation.

1. $\sqrt{x} = 4$

2. $\sqrt{y} = 5$

3. $\sqrt{x + 2} = 8$

4. $\sqrt{y - 5} = 2$

5. $3 + \sqrt{x} = 9$

6. $\sqrt{x} = 3.2$

7. $\sqrt{x} + 5 = 0$

8. $3\sqrt{2r} = 6$

9. $\sqrt{3x + 7} = 2$

10. $3 - \sqrt{x - 5} = 4$

11. $2 - \sqrt{x} = -8$

12. $\sqrt{4a - 3} + 5 = 8$

B. EXERCISES

Solve each equation.

13. $7 + 3\sqrt{x - 3} = -14$

14. $3 + 4\sqrt{x + 3} = 11$

15. $5 + 2\sqrt{y - 3} = 27$

16. $5 + \sqrt{x + 1} = 7$

17. $14 - \sqrt{x - 3} = 10$

18. $8 + 2\sqrt{x} = 18$

19. $12 - 4\sqrt{x - 3} = 7$

20. $5\sqrt{x} - 13 = 2\sqrt{x} + 26$

Write an equation and solve.

21. The square root of twice a number is 12. Find the number.

22. Five more than twice the square root of a number is 17. Find the number.

Use the formula $S = \sqrt{30Df}$ to determine the distance required for a car to skid to a stop.

23. $f = \frac{2}{3}$; $S = 60$ mi/hr

24. $f = 0.8$; $S = 45$ mi/hr

25. **MP2** The surface of Rowley Road has a coefficient of friction of 0.75 when dry. Would a deer standing still in the road be struck if a driver traveling 40 mi/hr applies the brakes from 70 ft away and does not swerve? Justify your answer.

26. Solve the equation $S = \sqrt{30Df}$ for D.

The formula $T = 2\pi\sqrt{\frac{l}{g}}$ relates a pendulum's period (the time it takes to swing back and forth), T, in seconds to its length, l, in centimeters using g, the gravitational acceleration of 981 cm/sec^2.

How long would the pendulum have to be (to the nearest tenth of a centimeter) to make a pendulum with the given period?

27. 1 sec

28. 2 sec

29. Solve the equation $T = 2\pi\sqrt{\frac{l}{g}}$ for l.

30. Using a 20 cm pendulum, Dixon wants to determine the effect of acceleration due to gravity when flying in a plane at an altitude of 11 km. He finds that the pendulum has an average period of 0.9 sec. Determine to the nearest tenth the gravitational acceleration, g, acting on the plane.

Use the formula $V = \frac{15}{22}\sqrt{gd}$ to determine the depth of the water for tsunami waves traveling at the given speeds. V is the wave's speed in miles per hour, g is the gravitational acceleration of 32 ft/sec^2, and d is the depth of the water in feet.

31. 120 mi/hr

32. 180 mi/hr

33. Solve the equation $V = \frac{15}{22}\sqrt{gd}$ for d.

34. Essential Question: Describe some real-world uses for solving radical equations.

C. EXERCISES

Solve each equation.

35. $\sqrt{5x+1} - \sqrt{x+3} = 0$

36. **MP7** $\sqrt{3x-3} + \sqrt{x+1} = 2\sqrt{x+1}$

37. $\sqrt{x} + 2 = \sqrt{9x}$

38. $\sqrt[3]{2x} = 6$

39. $\sqrt[3]{10x+2} = 8$

40. **MP7** $2\sqrt[3]{x+7} = 3\sqrt[3]{x-12}$

CUMULATIVE REVIEW

Write a function rule for each translation of the base function $y = x^2$. [7.3]

41. 3 units left and 2 units up

42. 4 units right and 1 unit down

Find each quotient. [8.6]

43. $\frac{x^3 - x^2 - 18x + 8}{x+4}$

44. $\frac{x^2 - 26x + 169}{x - 13}$

Factor each binomial. [9.4]

45. $x^2 - 9$

46. $x^2 - \frac{1}{9}$

Simplify each expression. [10.4]

47. $\frac{\sqrt{3}}{2} + \frac{\sqrt{3}}{5}$

48. $\frac{4\sqrt{20}}{5} - \frac{5\sqrt{5}}{2}$

49. Evaluate $a^2b - ac^3 + bc$ when $a = -5$, $b = 3$, and $c = -2$. **[8.1]**

A −121

B 29

C 41

D 109

E none of these

50. Two sides of a right triangle measure 6 cm and 8 cm. The third side could measure **[10.5]**

A 10 cm.

B $2\sqrt{7}$ cm.

C 5 cm.

D A and B

E A and C

10.8 SQUARE ROOT FUNCTIONS

How can I quickly graph $y = \sqrt{x-2} + 3$?

After completing this section, I will be able to

- determine the domain of a square root function.
- graph square root functions by plotting coordinate points.
- describe translations and reflections of square root functions.

Radical functions are functions with the independent variable in the radicand. By plotting enough points, the graph of a radical function can be determined. This method can be used to graph cube root and higher root functions, but this section will focus only on square root functions. The square root function $y = \sqrt{x}$ is the parent function for the family of square root functions.

A semiannual interest rate, s, based on an annual rate, r, can be calculated using the function $s(r) = \sqrt{1+r} - 1$.

The square root function has a limited domain, since the square root of a negative number is not defined for real numbers. Therefore, when finding points for graphing a square root function, the value of the radicand must be greater than or equal to 0. This allows us to sketch the graph of the parent function, $y = \sqrt{x}$, by plotting a few key points.

x	$y = \sqrt{x}$
0	0
1	1
4	2
9	3
−4	undefined

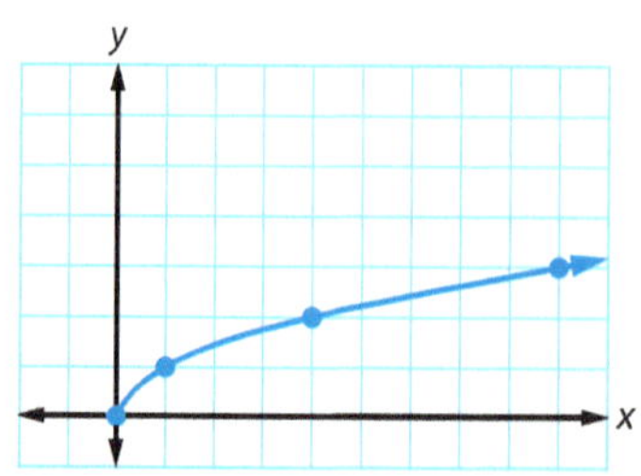

Graph of the Parent Function: $y = \sqrt{x}$

Knowing the domain also helps us determine key points that can be used to graph simple translations of the parent function.

EXAMPLE 1: Graphing $y = \sqrt{x} + k$ & $y = \sqrt{x-h}$

Determine the domain and graph each function.

a. $y = \sqrt{x} + 2$ **b.** $y = \sqrt{x-3}$

Answers

a. $x \geq 0$

1. Determine the domain.

2. Using the domain for x, choose values that make the radicand a perfect square. It is helpful to begin with the value that makes the radicand equal to 0.

x	$y = \sqrt{x} + 2$
0	2
1	3
4	4
9	5

3. Graph the function.

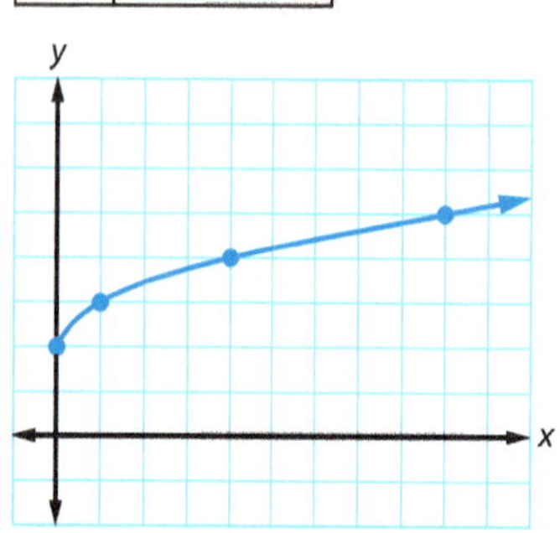

b. $x - 3 \geq 0$

$x \geq 3$

x	$y = \sqrt{x - 3}$
3	0
4	1
7	2
12	3

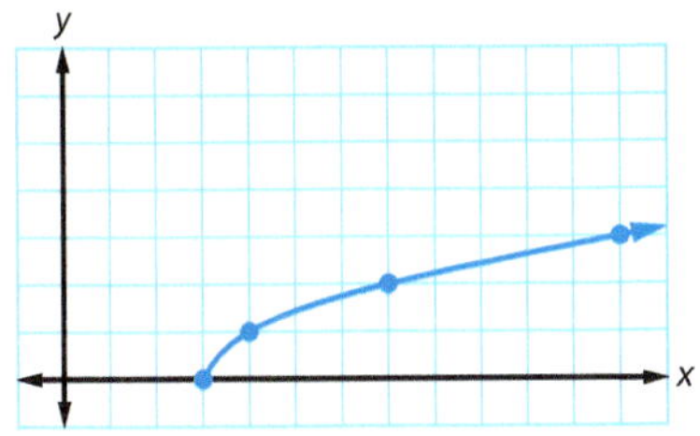

1. Determine the domain.
2. Using the domain for x, choose values that make the radicand a perfect square. It is helpful to begin with the value that makes the radicand equal to 0.
3. Graph the function.

SKILL CHECK **EXERCISES 11, 13**

Translations of $y = \sqrt{x}$ can be easily graphed without finding points by applying the same method used to graph translations of other parent functions ($y = |x|$ and $y = x^2$).

By knowing the general shape of the graph of the parent function, any function of the form $y = \sqrt{x - h} + k$ can be quickly sketched by translating the graph of the parent function h units horizontally and k units vertically.

AFTERSCHOOLHELP

Square Root Functions

EXAMPLE 2: Graphing $y = \sqrt{x - h} + k$

Describe the translation and graph each function without plotting points.

a. $y = \sqrt{x + 4} - 1$ **b.** $y = \sqrt{x - 2} + 3$

Answers

a. $y = \sqrt{x + 4} - 1$

$x + 4 \geq 0$

$x \geq -4$

1. Determine the domain.

The graph of $y = \sqrt{x}$ will be translated −4 units horizontally and −1 unit vertically.

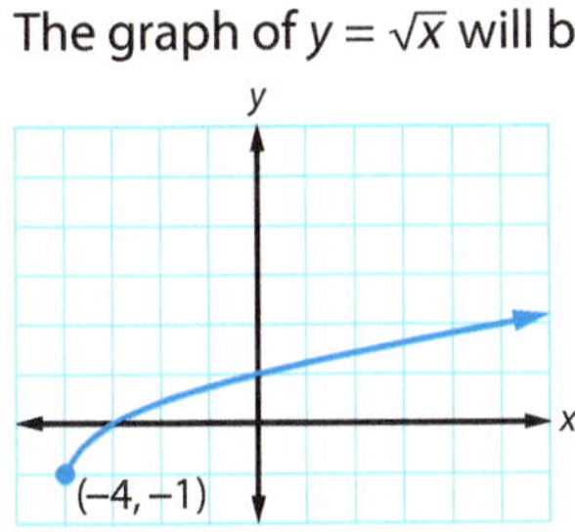

2. Graph the function.

b. $y = \sqrt{x - 2} + 3$

$x - 2 \geq 0$

$x \geq 2$

1. Determine the domain.

The graph of $y = \sqrt{x}$ will be translated +2 units horizontally and +3 units vertically.

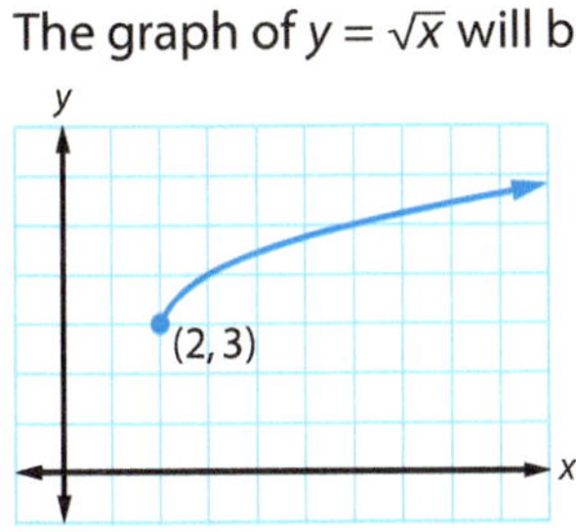

2. Graph the function.

SKILL CHECK **EXERCISE 31**

EXAMPLE 3: Graphing Reflections of $y = \sqrt{x}$

Graph each function and describe its relation to the parent graph.

a. $y = \sqrt{-x}$ **b.** $y = -\sqrt{x}$

Answers

a. $y = \sqrt{-x}$

$-x \geq 0$

$x \leq 0$

1. Determine the domain.

x	$y = \sqrt{-x}$
0	0
−1	1
−4	2
−9	3

2. Using the domain for x, choose values that make the radicand a perfect square.

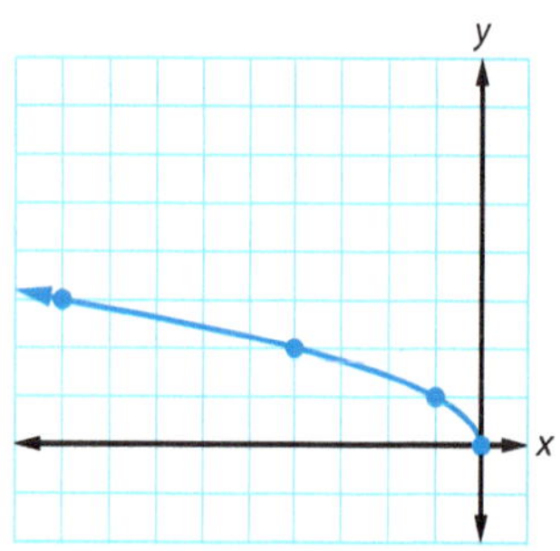

3. Graph the function.

The graph of $y = \sqrt{-x}$ is a reflection of the graph of $y = \sqrt{x}$ across the y-axis.

b. $y = -\sqrt{x}$

$x \geq 0$

1. Determine the domain.

x	$y = -\sqrt{x}$
0	0
1	−1
4	−2
9	−3

2. Using the domain for x, choose values that make the radicand a perfect square.

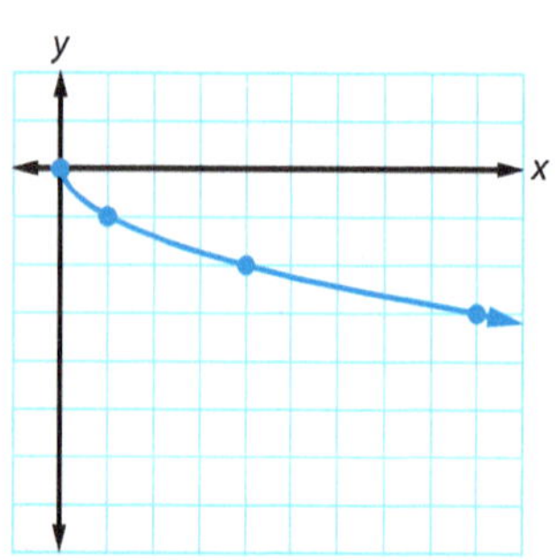

3. Graph the function.

The graph of $y = -\sqrt{x}$ is a reflection of the graph of $y = \sqrt{x}$ across the x-axis.

SKILL CHECK **EXERCISE 15**

Multiplying the radicand or the parent function itself by a factor of −1 will affect the direction of the graph but not the general shape.

EXAMPLE 4: Writing a Function Rule

Describe each transformation of $y = \sqrt{x}$ and write the function rule.

a.

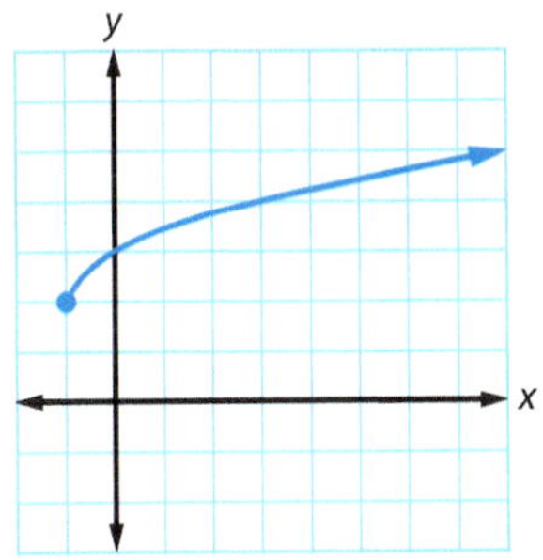

b.

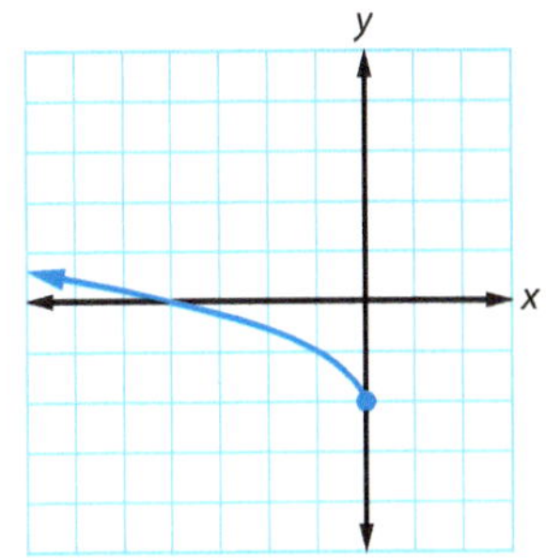

Answers

a. $h = -1; k = 2$

1. $y = \sqrt{x}$ is translated −1 unit horizontally and +2 units vertically.

$y = \sqrt{x - h} + k$

$y = \sqrt{x + 1} + 2$

2. Write the function rule.

b. $y = \sqrt{-x}$

$k = -2$

$y = \sqrt{-x} - 2$

1. $y = \sqrt{x}$ is reflected across the y-axis and then translated −2 units vertically.
2. Write the function rule.

SKILL CHECK **EXERCISES 25, 27**

VOCABULARY

- radical function

A. EXERCISES

Determine the domain of each radical function.

1. $y = \sqrt{x} - 1$

2. $y = \sqrt{x + 3}$

3. $y = \sqrt{x} - 2$

4. $y = \sqrt{x} + 7$

5. $y = \sqrt{x - 5}$

6. $y = \sqrt{x - 3} + 4$

7. $y = \sqrt{x + 2} - 3$

8. $y = \sqrt{-x} + 1$

Make a table of ordered pairs and graph each function.

9. $y = \sqrt{x} - 1$

10. $y = \sqrt{x + 3}$

11. $y = \sqrt{x} - 2$

12. $y = \sqrt{x} + 7$

13. $y = \sqrt{x - 5}$

14. $y = \sqrt{x - 3} + 4$

15. $y = \sqrt{-x} + 1$

16. $y = \sqrt{x + 2} - 3$

B. EXERCISES

Match each function with its graph.

A

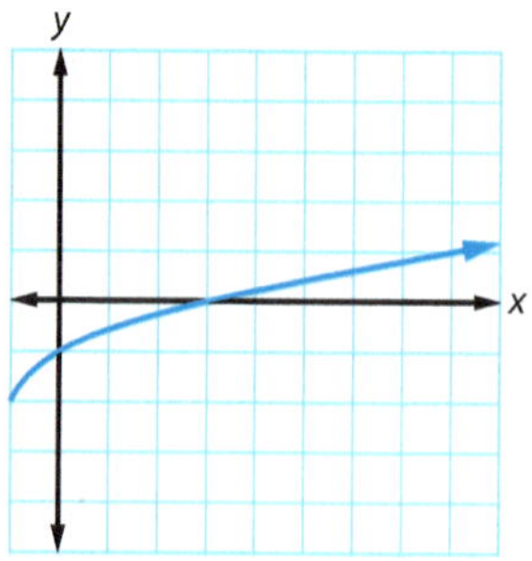

D

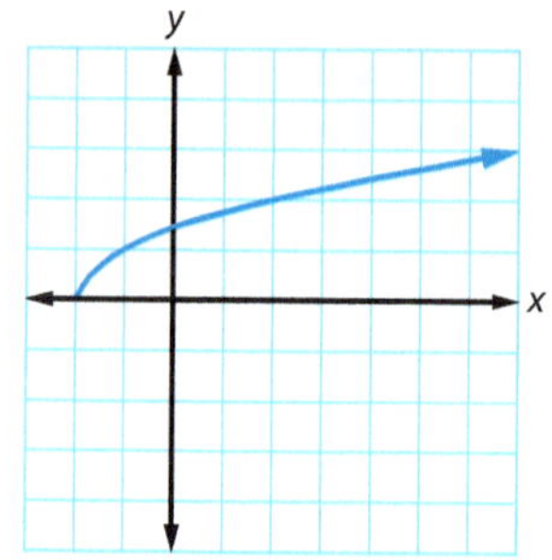

B

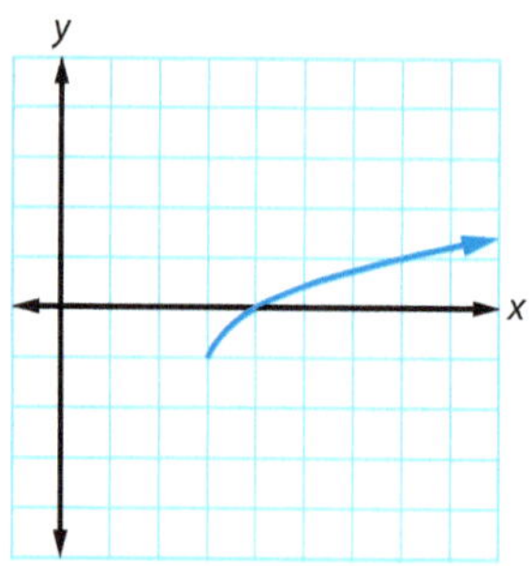

E

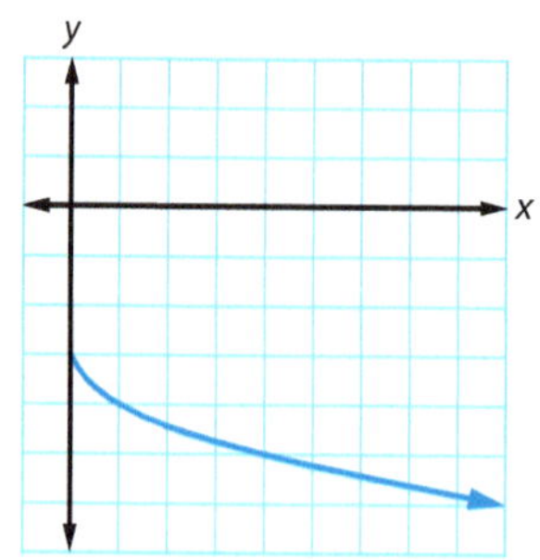

C

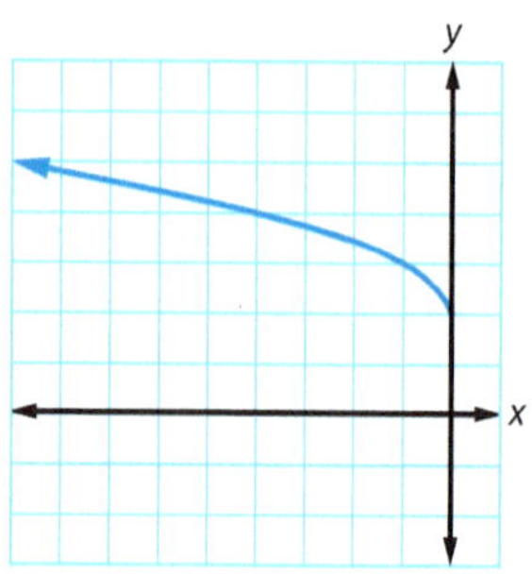

F

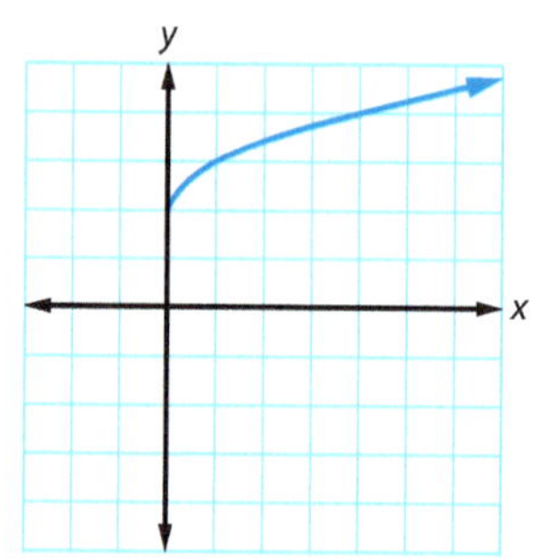

17. $f(x) = \sqrt{x} + 2$

18. $f(x) = \sqrt{x + 2}$

19. $f(x) = \sqrt{x + 1} - 2$

20. $f(x) = \sqrt{x - 3} - 1$

21. $f(x) = \sqrt{-x} + 2$

22. $f(x) = -\sqrt{x} - 3$

Write the function rule for each translation or reflection of the base function $f(x) = \sqrt{x}$.

23. translated 1 unit up

24. translated 3 units left

25. translated 2 units right and 2 units down

26. translated 5 units left and 3 units up

27. reflected across the y-axis and translated 4 units up

28. reflected across the x-axis and translated 3 units down

Sketch each transformation of $y = \sqrt{x}$ without making a table of ordered pairs.

29. $y = \sqrt{x} - 3$

30. $y = \sqrt{x + 2}$

31. $y = \sqrt{x + 3} - 2$

32. $y = \sqrt{x - 2} + 2$

33. $y = -\sqrt{x} - 3$

34. $y = \sqrt{-x} - 4$

35. **Essential Question:** How can a function such as $y = \sqrt{x - 2} + 3$ be quickly graphed?

C. EXERCISES

36. Without graphing, describe the graph of $y = -\sqrt{-x}$.

37. What is the domain of $y = \sqrt[3]{x}$?

38. MP7 Make a table of ordered pairs and graph $y = \sqrt[3]{x}$.

Write the function rule based on the parent graph $y = \sqrt{x}$.

39. reflected across the x-axis, translated 2 units up and 3 units right

40. MP5 endpoint at (4, 1) and intersects (0, 3)

CUMULATIVE REVIEW

Find each missing length. Express each answer as both a simplified radical and estimated to the nearest integer. [10.5]

41.

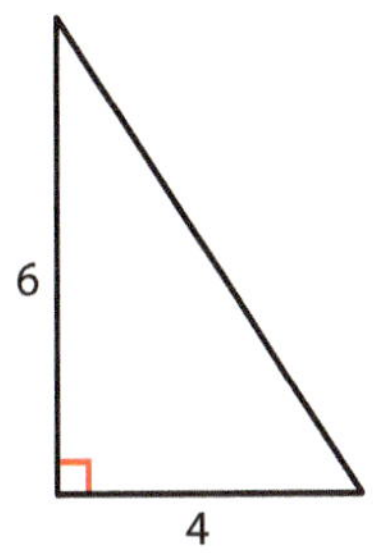

42.

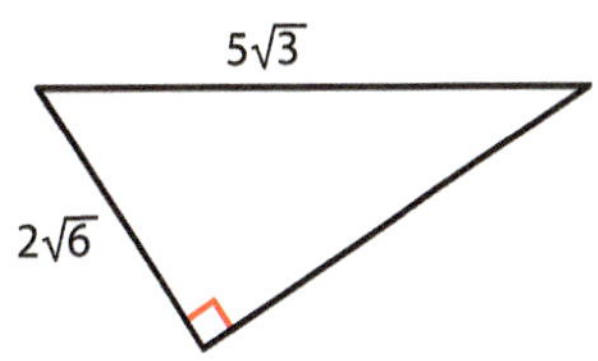

Find the distance between the given pair of points. Then find the midpoint of each segment. [10.5]

43. $(5, -3)$ and $(9, -1)$

44. $(-2, -4)$ and $(7, 8)$

Factor each expression completely. [9.1–9.4]

45. $x^2 - 7x$

46. $x^2 - 2x - 24$

47. $4v^2 - 9w^2$

48. $2x^2y + 30xy + 100y$

49. For which value of a will $3x^2 + ax - 8$ not be factorable over the set of real numbers? **[9.3]**

- A 10
- B -10
- C 14
- D -2
- E none of these

50. Which sequence does not include 2 as an element? **[4.5, 7.6]**

- A 30, 26, 22, . . .
- B $-42, -32, -22, \ldots$
- C $-13, -10, -7, \ldots$
- D $a_1 = \frac{243}{16}; r = \frac{2}{3}$
- E none of these

APPLICATION PROBLEMS
THE GOLDEN RATIO

God's creation reveals some interesting mathematical patterns. The golden ratio, often represented by the lowercase Greek letter phi (ϕ), is a number that is frequently found in nature. You will discover interesting mathematical characteristics of the golden ratio and its relationship to nature, art, graphic design, architecture, and science.

The Fibonacci sequence can be defined recursively as $a_n = a_{n-1} + a_{n-2}$, which implies that each term is the sum of the 2 preceding terms. In the most common Fibonacci sequence, $a_1 = 1$ and $a_2 = 1$, yielding 1, 1, 2, 3, 5, 8, ... Consecutive numbers of this Fibonacci sequence often appear in nature, such as in the spirals of pinecones, pineapples, or sunflower seeds.

The golden ratio sequence can be generated by dividing each Fibonacci term by its previous term. It can be defined recursively as $b_n = \frac{a_{n+1}}{a_n}$, where a_n is a Fibonacci sequence and the first golden ratio sequence term, b_1, is $\frac{a_2}{a_1}$. As n gets larger, the ratio $\frac{a_{n+1}}{a_n}$ gets closer to the golden ratio, ϕ, which is approximately 1.618. The following exercises illustrate how a golden ratio sequence can be used to approximate the value of ϕ.

1. In a spreadsheet, enter a value of 1 in both cell A1 and cell A2. Enter the formula "=A1+A2" in cell A3; then replicate this formula down through cell A26. Enter the formula "=A2/A1" in cell B1; then replicate this formula down through cell B25. To the nearest millionth, what value does the golden ratio sequence approach?
2. Change the values in cell A1 and cell A2 to 2 other numbers. What value does the sequence approach?

The ratio $\frac{a_{n+1}}{a_n}$ uses a Fibonacci sequence to find decimal approximations of the golden ratio. The golden ratio is an irrational number with an exact value of $\phi = \frac{1 + \sqrt{5}}{2}$.

The golden ratio can be defined geometrically using the ratio of the lengths of line segments. If the ratio of the longer part of a segment to its shorter part equals the ratio of the total length of the segment to its longer part, the ratio is the golden ratio.

$$\phi = \frac{a}{b} = \frac{a + b}{a}$$

3. The golden ratio is an irrational number. What is an irrational number?
4. Use a calculator to approximate $\frac{1 + \sqrt{5}}{2}$ to the nearest millionth.
5. A line segment is divided so that $a = 144$ and $b = 89$.
   a. Find the decimal equivalents of $\frac{a}{b}$ and $\frac{a + b}{a}$ to the nearest thousandth.
   b. Is the segment approximately divided into a golden ratio?
6. A line segment is divided so that $a = 3$ and $b = 2$.
   a. Find the decimal equivalents of $\frac{a}{b}$ and $\frac{a + b}{a}$ to the nearest thousandth.
   b. Is the segment approximately divided into a golden ratio?

Mathematicians, artists, scientists, engineers, and even philosophers have long been intrigued by the many interesting definitions and unique characteristics of the golden ratio. By expressing the golden ratio as a continued fraction, we can illustrate one of its unique characteristics.

It can be shown that $\phi = 1 + \cfrac{1}{1 + \cfrac{1}{1 + \cfrac{1}{1 + \cfrac{1}{1 + \dots}}}}$.

It has also been proven that the infinite fraction in the denominator of the second term is equal to ϕ and we can substitute ϕ for this denominator. This can be rearranged to state that

$$\frac{1}{\phi} = \phi - 1.$$

In other words, the reciprocal of the golden ratio can be obtained by simply subtracting 1.

7. Simplify each expression, writing your answer as an improper fraction. Then use a calculator to find each decimal approximation to the nearest thousandth.

a. $1 + \frac{1}{2}$

b. $1 + \cfrac{1}{1 + \cfrac{1}{2}}$

c. $1 + \cfrac{1}{1 + \cfrac{1}{1 + \cfrac{1}{2}}}$

d. $1 + \cfrac{1}{1 + \cfrac{1}{1 + \cfrac{1}{1 + \cfrac{1}{2}}}}$

8. Predict the next 3 improper fractions in the sequence of expressions suggested by the answers to exercise 7. Then find their decimal approximations to the nearest thousandth.

9. Use a calculator to evaluate $\frac{1}{\phi} = \frac{2}{1 + \sqrt{5}}$. Round to the nearest millionth.

10. Use a calculator to evaluate $\frac{1 + \sqrt{5}}{2} - \frac{2}{1 + \sqrt{5}}$.

Another interesting definition of the golden ratio uses an infinite nested radical.

$$\phi = \sqrt{1 + \sqrt{1 + \sqrt{1 + \sqrt{1 + \dots}}}}$$

Squaring both sides produces

$$\phi^2 = 1 + \sqrt{1 + \sqrt{1 + \sqrt{1 + \dots}}}.$$

Since $\phi = \sqrt{1 + \sqrt{1 + \sqrt{1 + \dots}}}$, we can substitute to produce

$$\phi^2 = 1 + \phi.$$

A second unique characteristic of ϕ is that its square can be obtained by simply adding 1.

11. Use a calculator to find the decimal approximation of $\sqrt{1 + \sqrt{1 + \sqrt{1 + \sqrt{1 + \sqrt{1 + \sqrt{1}}}}}}$ to the nearest thousandth.

12. Use a calculator to evaluate $\left(\frac{1 + \sqrt{5}}{2}\right)^2$. Round to the nearest millionth.

13. Use a calculator to evaluate $\left(\frac{1 + \sqrt{5}}{2}\right)^2 - \frac{1 + \sqrt{5}}{2}$.

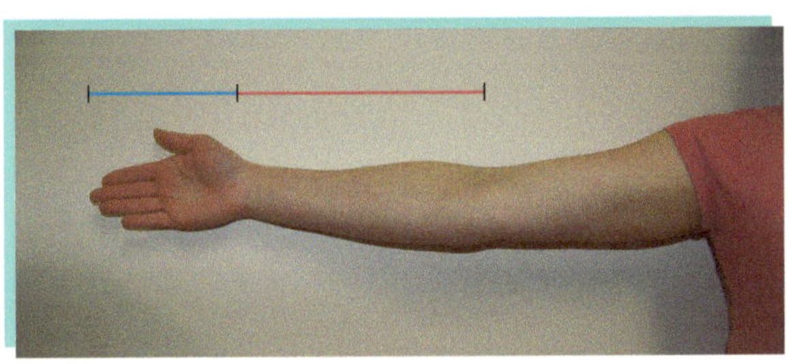

14. Record the following measurements in a table similar to the one below: forearm, hand, total height, and height of navel. Then determine each of your personal ratios (to the nearest thousandth).

15. Using your values as the experimental values and ϕ as the known value, calculate the percent error of each ratio. Record your results in the table.

16. Obtain and record the forearm-to-hand length ratios and total-to-navel height ratios of 4 other people. For each set of 5 measurements, find the average ratio and its percent error when ϕ is used as the known value.

17. Do the results of exercises 14–16 confirm or refute the claim that the golden ratio is found in the design of the human body? Explain your reasoning.

PERCENT ERROR

$$\frac{|\text{experimental value} - \text{known value}|}{\text{known value}} \times 100\%$$

Measurement		Personal Ratio		Percent Error
forearm		$\frac{\text{forearm}}{\text{hand}}$		
hand				
total height		$\frac{\text{total height}}{\text{height of navel}}$		
height of navel				

10 CHAPTER REVIEW

Key Concepts

Simplifying Radicals [10.1]

- For any real number x and integers a and b, where $b > 1$, the exponential expression $x^{\frac{a}{b}}$ is equivalent to the radical $\sqrt[b]{x^a}$ or $(\sqrt[b]{x})^a$.
  - $8^{\frac{2}{3}} = \left(8^{\frac{1}{3}}\right)^2 = (\sqrt[3]{8})^2 = (\sqrt[3]{2^3})^2 = 2^2 = 4$, or
  - $8^{\frac{2}{3}} = (8^2)^{\frac{1}{3}} = \sqrt[3]{8^2} = \sqrt[3]{64} = 4$
- Radicals can be simplified by applying the Product Property of Radicals.

PRODUCT PROPERTY OF RADICALS
For $x \geq 0$ and $y \geq 0$, $\sqrt[n]{xy} = \sqrt[n]{x} \cdot \sqrt[n]{y}$.

SIMPLIFYING RADICALS	$\sqrt{360}$
1. Write the radicand as the product of its prime factors.	$= \sqrt{2 \cdot 2 \cdot 2 \cdot 3 \cdot 3 \cdot 5}$
2. Apply the Product Property to the perfect powers.	$= \sqrt{2^2} \cdot \sqrt{3^2} \cdot \sqrt{2 \cdot 5}$
3. Simplify the roots of the perfect powers.	$= 2 \cdot 3 \cdot \sqrt{2 \cdot 5}$
4. Multiply the factors inside and outside the radical.	$= 6\sqrt{10}$

- $\sqrt[n]{x^n} = \begin{cases} |x| & \text{if } n \text{ is even.} \\ x & \text{if } n \text{ is odd.} \end{cases}$
  - $\sqrt{x^2} = |x|$
  - $\sqrt[3]{x^3} = x$. Absolute value symbols are not needed since the result can be negative.

Multiplying Radicals [10.2]

- The Product Property of Radicals indicates that radicals must have the same index to be multiplied together.

MULTIPLYING RADICALS	$\sqrt{10} \cdot \sqrt{6}$
1. Apply the Product Property, writing the radicand as the product of its prime factors.	$= \sqrt{2 \cdot 5 \cdot 2 \cdot 3} = \sqrt{2^2 \cdot 5 \cdot 3}$
2. Simplify the root of each perfect power.	$= \sqrt{2^2} \cdot \sqrt{5 \cdot 3} = 2 \cdot \sqrt{5 \cdot 3}$
3. Simplify the resulting expression.	$= 2\sqrt{15}$

Dividing Radicals [10.3]

- Some radicals containing fractions can be simplified using the Quotient Property of Radicals.

QUOTIENT PROPERTY OF RADICALS	$\sqrt{\frac{30}{40}}$
For $x \geq 0$ and $y > 0$, $\sqrt[n]{\frac{x}{y}} = \frac{\sqrt[n]{x}}{\sqrt[n]{y}}$.	$= \sqrt{\frac{3}{4}} = \frac{\sqrt{3}}{\sqrt{4}} = \frac{\sqrt{3}}{2}$

- When a radical expression cannot be simplified using the Quotient Property, a process called rationalizing the denominator can be used to remove a radical from the denominator.

RATIONALIZING THE DENOMINATOR	$\sqrt{\frac{40}{30}}$
1. Write the radicals in fractional form, $\frac{\sqrt[n]{a}}{\sqrt[n]{b}}$.	$=\sqrt{\frac{4}{3}}=\frac{\sqrt{4}}{\sqrt{3}}$
2. Simplify both the numerator and the denominator as much as possible.	$=\frac{2}{\sqrt{3}}$
3. If a radical remains in the denominator, multiply by the form of 1 that makes the radicand in the denominator a perfect power.	$=\frac{2}{\sqrt{3}}\cdot\frac{\sqrt{3}}{\sqrt{3}}$
4. Simplify both the numerator and the denominator and reduce the fraction if needed.	$=\frac{2\sqrt{3}}{3}$

Adding & Subtracting Radicals [10.4]

- In order to add or subtract radicals, both the indices and the radicands must match.

ADDING & SUBTRACTING RADICALS

1. Simplify each radical.
2. Combine like radicals by adding or subtracting their rational factors.

The Pythagorean Theorem [10.5]

THE PYTHAGOREAN THEOREM

If the legs of a right triangle have lengths a and b and the hypotenuse has length c, then $a^2 + b^2 = c^2$.

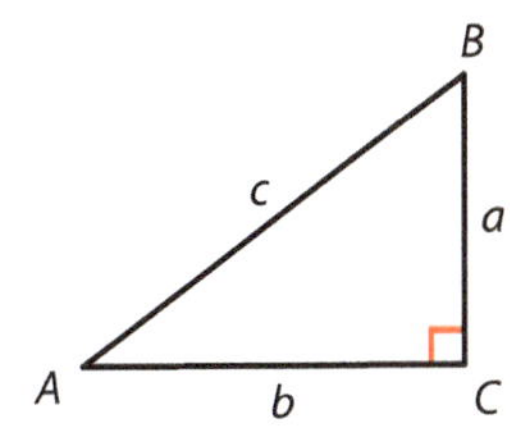

- The converse of the Pythagorean Theorem can be used to determine whether a triangle is a right triangle.

CONVERSE OF THE PYTHAGOREAN THEOREM

If a triangle has side lengths a, b, and c such that $a^2 + b^2 = c^2$, then the triangle is a right triangle.

- Formulas based on right triangles can be used to find the distance and midpoint between 2 points.

DISTANCE FORMULA

The distance, d, between any 2 points $P_1\ (x_1, y_1)$ and $P_2\ (x_2, y_2)$ is given by the formula $d=\sqrt{(x_2-x_1)^2+(y_2-y_1)^2}$.

MIDPOINT FORMULA

The midpoint of a segment with endpoints $P_1\ (x_1, y_1)$ and $P_2\ (x_2, y_2)$ is $M\left(\frac{x_1+x_2}{2}, \frac{y_1+y_2}{2}\right)$.

Multiplying & Dividing Radical Expressions [10.6]

MULTIPLYING RADICAL EXPRESSIONS	$(\sqrt{2}-3)(\sqrt{2}+5)$
1. Follow the same rules that you follow for multiplying polynomials (FOIL).	$=\sqrt{4}-3\sqrt{2}+5\sqrt{2}-15$
2. After multiplying, simplify radicals and combine like radicals.	$=2-3\sqrt{2}+5\sqrt{2}-15$ $=2\sqrt{2}-13$

DIVIDING RADICAL EXPRESSIONS	$\dfrac{\sqrt{2}-3}{\sqrt{2}+5}$
1. Multiply the numerator and the denominator by the conjugate of the denominator.	$=\dfrac{\sqrt{2}-3}{\sqrt{2}+5}\cdot\dfrac{\sqrt{2}-5}{\sqrt{2}-5}=\dfrac{\sqrt{4}-3\sqrt{2}-5\sqrt{2}+15}{\sqrt{4}-25}$
2. Simplify.	$=\dfrac{2-3\sqrt{2}-5\sqrt{2}+15}{2-25}=\dfrac{17-8\sqrt{2}}{-23}$ or $\dfrac{-17+8\sqrt{2}}{23}$

Radical Equations [10.7]

- Radical equations are equations whose unknown is in the radicand of a radical expression.

SOLVING RADICAL EQUATIONS
1. Isolate the radical on one side of the equation.
2. Square both sides of the equation.
3. Solve the resulting equation.
4. Check any solutions in the original equation.

- Extraneous solutions are solutions that do not make the original equation true.

Square Root Functions [10.8]

- Radical functions are functions with the independent variable in the radicand.
- Graph functions of the form $y=\sqrt{x-h}+k$ by translating the parent function: $y=\sqrt{x}$.
- Graph reflections of the parent function $y=\sqrt{x}$ in the forms of $y=\sqrt{-x}$ and $y=-\sqrt{x}$.

GRAPHING RADICAL FUNCTIONS

$y = \sqrt{x}$
endpoint: (0, 0)

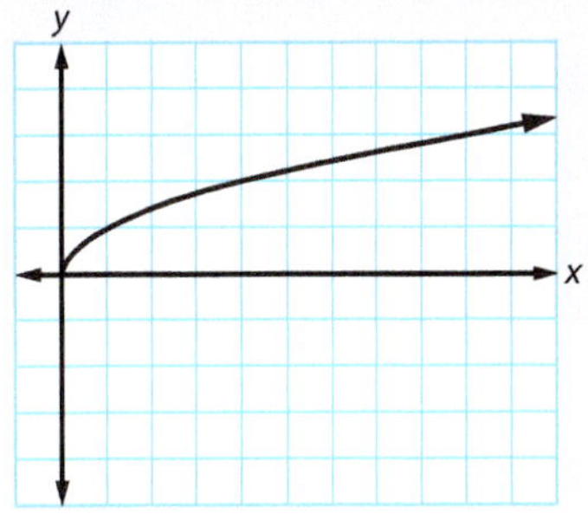

$y = \sqrt{x + h}$
endpoint: (−h, 0)

$y = \sqrt{x - h}$
endpoint: (h, 0)

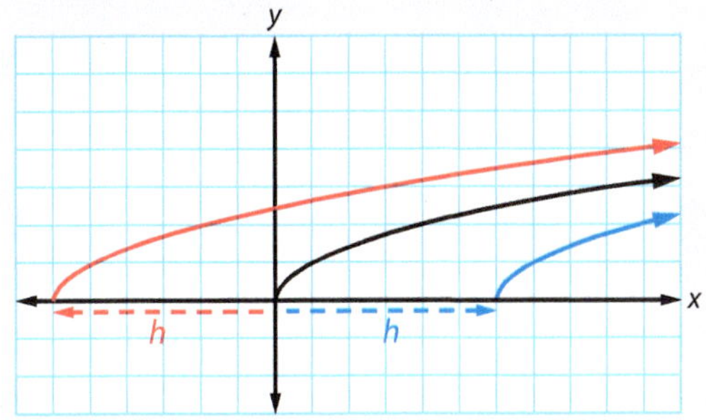

$y = \sqrt{x} + k$
endpoint: (0, k)

$y = \sqrt{x} - k$
endpoint: (0, −k)

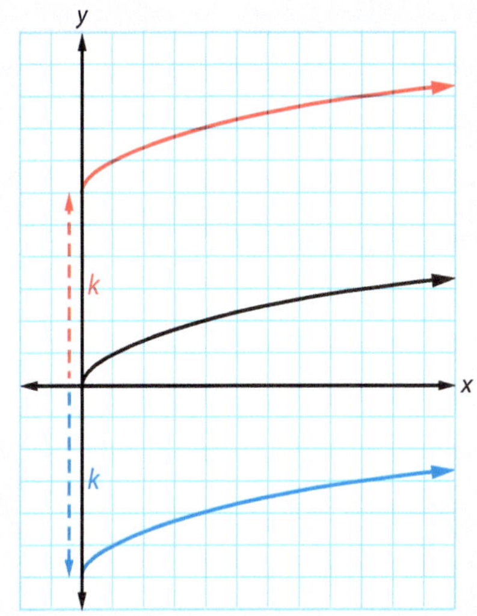

$y = \sqrt{x - h} + k$
endpoint: (h, k)

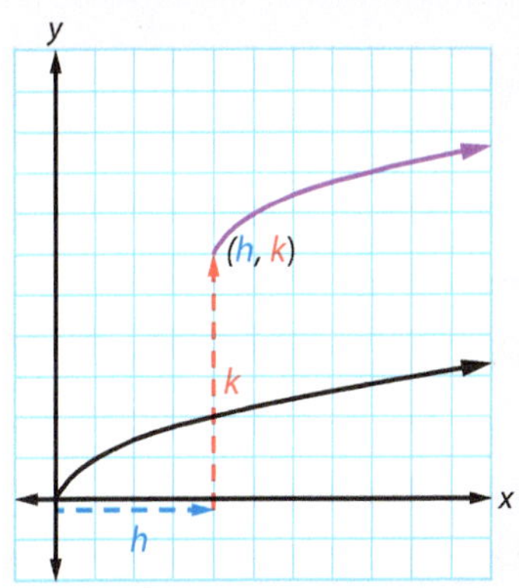

$y = \sqrt{-x}$
endpoint: (0, 0)

reflection across the y-axis

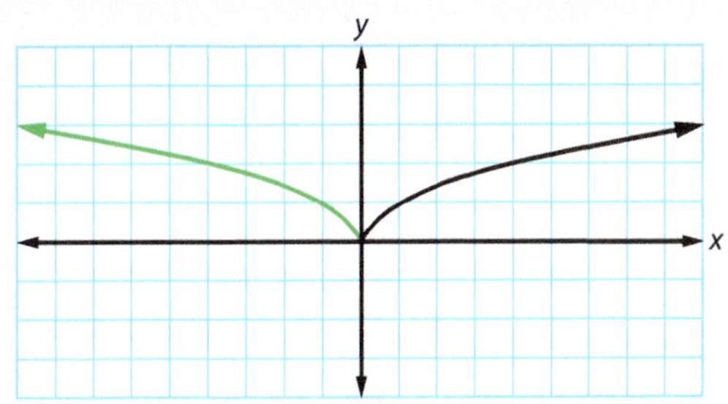

$y = -\sqrt{x}$
endpoint: (0, 0)

reflection across the x-axis

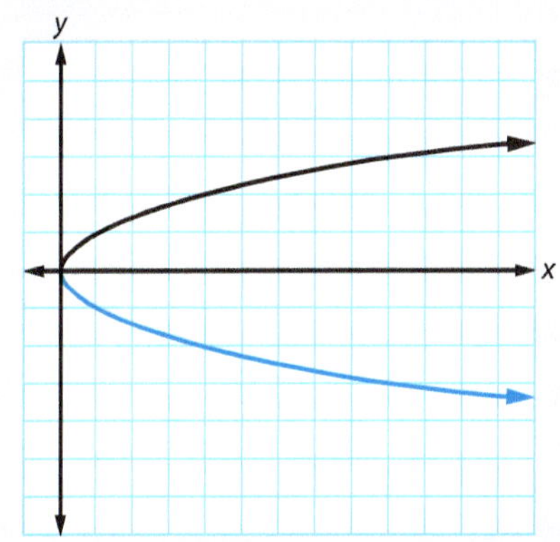

Biblical Worldview Shaping

Reasoning in Algebra

What is the ultimate standard for judging any claim or idea?

Since our faith is in God and His inspired word, Scripture is the ultimate standard by which we test any claim or idea.

EXERCISES

Convert each radical to exponential form. [10.1]

1. $\sqrt{2x^3}$

2. $\sqrt[4]{3y^2}$

Convert each exponential expression to simplified radical form. [10.1]

3. $3^{\frac{2}{3}}a^{\frac{5}{3}}$

4. $(16xy)^{\frac{1}{3}}z^{\frac{4}{3}}$

Simplify. If the expression is not a real number, state "not real." Do not assume that variables in the radicand are nonnegative. [10.1]

5. $\sqrt[3]{-8}$

6. $\sqrt{-25}$

7. $\sqrt{84}$

8. $\sqrt[5]{1728}$

9. $\sqrt{12a^3b^4}$

10. $\sqrt[3]{81x^2y^4}$

Simplify each expression. Assume that all variables in the radicand represent nonnegative values. [10.2–10.4, 10.6]

11. $\sqrt[3]{21} \cdot \sqrt[3]{9}$

12. $\sqrt{8} \cdot \sqrt{18}$

13. $\sqrt{20ab} \cdot \sqrt{63a^2b^3}$

14. $\sqrt[3]{9a^5b^2} \cdot \sqrt[3]{3ab^4}$

15. $\sqrt[3]{5} \cdot \sqrt{5}$

16. $\frac{5}{\sqrt{6}}$

17. $\frac{\sqrt[3]{3}}{\sqrt[3]{27}}$

18. $\frac{\sqrt{7x}}{\sqrt{27x^2y}}$

19. $\sqrt{2x^{12}y^3} \div \sqrt{8x^4y^9}$

20. $\sqrt{6} - \sqrt[3]{24} + 5\sqrt{6}$

21. $\sqrt{18} + \sqrt{12} + \sqrt{8}$

22. $\sqrt[3]{7} + \sqrt[3]{56} - \sqrt[3]{189}$

23. $a\sqrt{32a} + \sqrt{8a^3} - \sqrt{18a^5}$

24. $\sqrt{35}(\sqrt{7} + \sqrt{5})$

25. $\sqrt{10g}\left(\sqrt{2g} - 3\sqrt{15}\right)$

26. $(\sqrt{2} + \sqrt{3})(\sqrt{2} - \sqrt{3})$

27. $(5\sqrt{2} - \sqrt{3})^2$

28. $(\sqrt{2} - \sqrt{7})(\sqrt{2} + \sqrt{7})(\sqrt{3} - 2)$

29. $\frac{6}{3 + \sqrt{2}}$

30. $\frac{30}{3\sqrt{2} + 2\sqrt{3}}$

31. $\frac{3\sqrt{3}}{\sqrt{3} - 3\sqrt{2}}$

The length of each edge of a cube is $2\sqrt[3]{5}$ cm. [10.2]

32. What is the area of a side of the cube?

33. What is the volume of the cube?

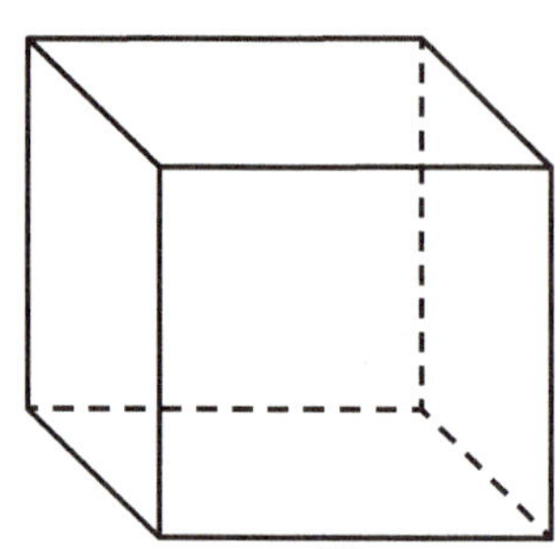

$2\sqrt[3]{5}$ cm

Find each missing length, expressing each answer as a simplified radical. [10.5]

34.

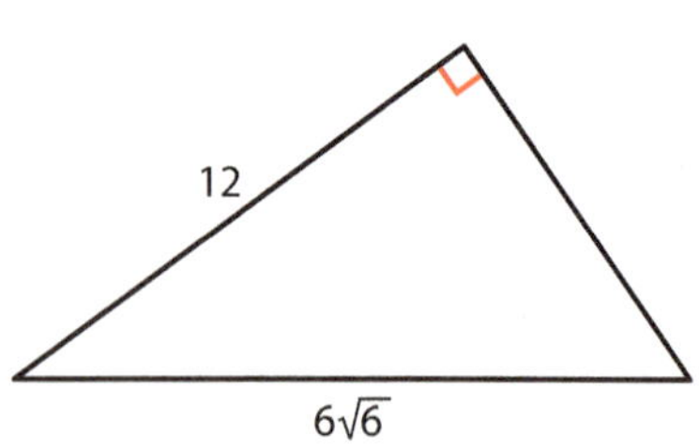

35.

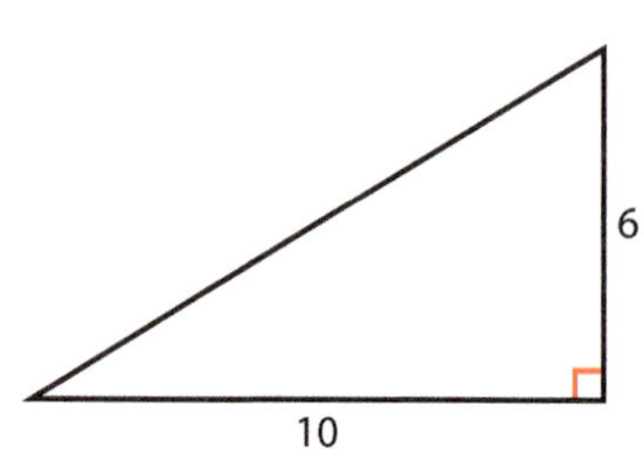

Find the distance between the given pairs of points. Express each answer as a simplified radical. [10.5]

36. (1, 7) and (−3, 9)

37. (4, −5) and (−5, 4)

Find the missing point given a line segment with endpoints P_1 and P_2 and midpoint M. [10.5]

38. $P_1\,(1, 7\sqrt{2})$ and $P_2\,(-3, 9\sqrt{2})$

39. $P_1\,(-3\sqrt{5}, 2\sqrt{10})$ and $M\,(-2\sqrt{5}, -2\sqrt{10})$

Solve and check your solution. [10.7]

40. $\sqrt{x} = 16$

41. $\sqrt{x+5} = 25$

42. $\sqrt{x-2} + 8 = 5$

43. $10 - 3\sqrt{x+7} = 4$

Use the formula $T = 2\pi\sqrt{\frac{l}{g}}$ to answer exercises 44–45. T is a pendulum's period in seconds, l is its length in feet, and g is the gravitational acceleration of 32 ft/sec². [10.7]

44. On March 31, 1851, Léon Foucault used a 220 ft long pendulum hung in the Panthéon of Paris to demonstrate the rotation of the earth to Napoleon Bonaparte. How long did it take for the pendulum to swing back and forth? Express your answer as both a simplified radical expression and a decimal approximation to the nearest tenth of a second.

45. If the Foucault pendulum in Chicago's Museum of Science and Industry takes 9 sec to swing back and forth, how long is this pendulum (to the nearest tenth of a foot)?

46. The aspect ratio of many widescreen televisions is 16 : 9. If the advertised 42 in. screen of a television is its diagonal measurement, determine the width and the height of the screen (to the nearest tenth of an inch).

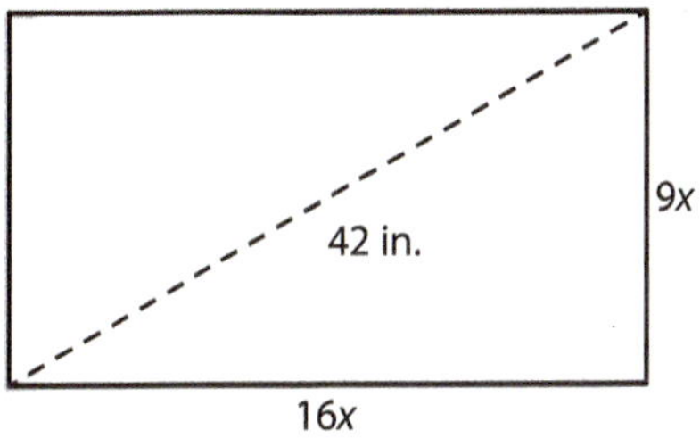

Determine the domain and make a table of ordered pairs; then graph each function. [10.8]

47. $y = \sqrt{x} - 3$

48. $y = \sqrt{x-2}$

49. $y = \sqrt{-x} + 3$

50. Write the function rule for the given graph. [10.8]

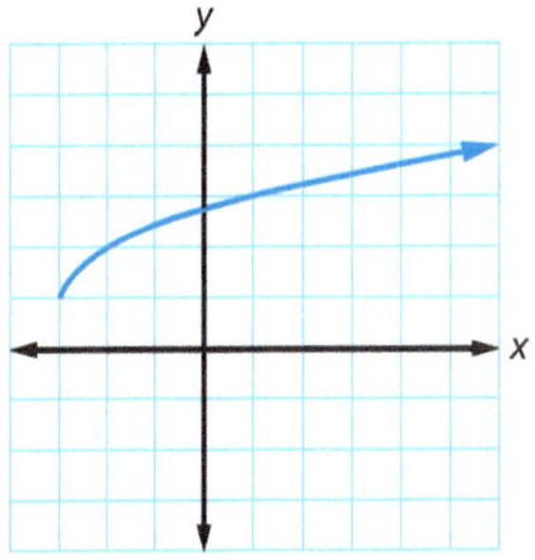

51. Write the function rule for the translation of $y = \sqrt{x}$ three units down and five units left. [10.8]

52. **BWS Essential Question:** Finding the remnants of Noah's ark would

A prove that the Bible is true.

B be consistent with the biblical account.

C prove the world is flat.

D validate the Pythagorean Theorem.

CHAPTER 11

QUADRATIC EQUATIONS & FUNCTIONS

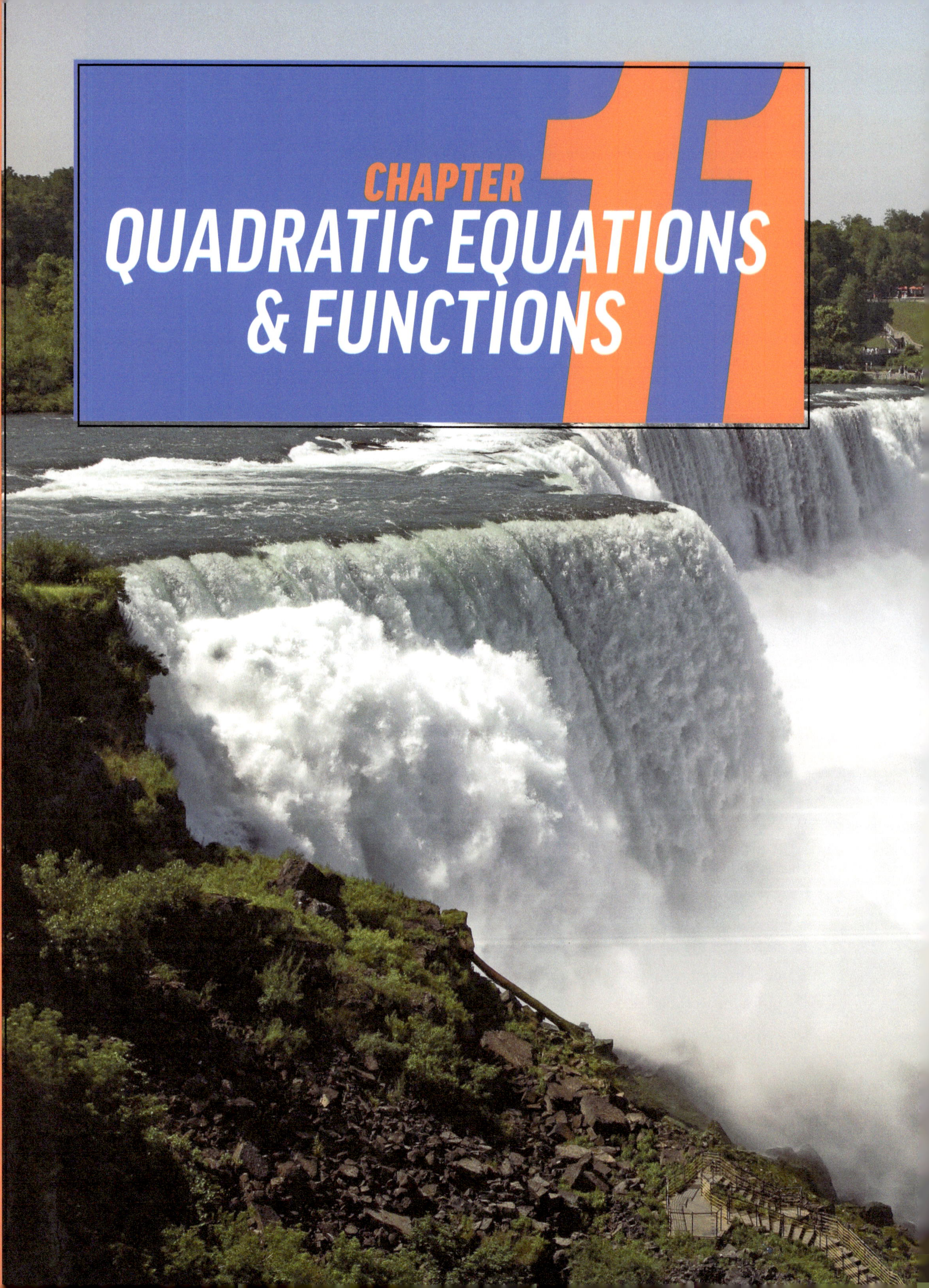

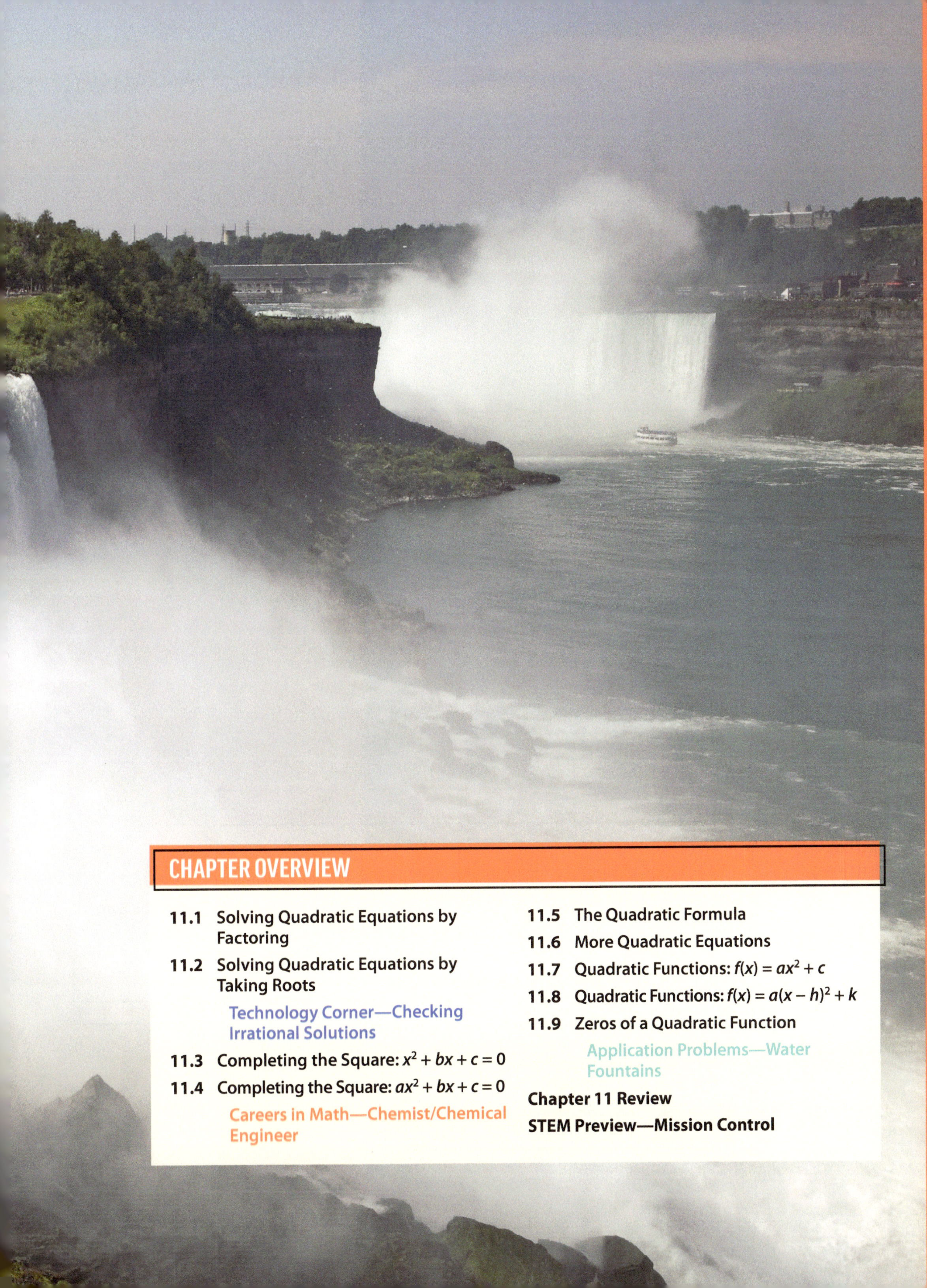

CHAPTER OVERVIEW

11.1 SOLVING QUADRATIC EQUATIONS BY FACTORING

What do I know about multiple factors if their product is 0?

After completing this section, I will be able to

- solve quadratic equations.
- solve real-world problems involving quadratic equations.

You graphed quadratic functions in Section 7.3 and factored quadratic expressions in Chapter 9. In this section, you will be solving quadratic equations. **Quadratic equations** are equations of the second degree. It is common to write quadratic equations in descending order or *standard form.*

DEFINITION

The **standard form** of a quadratic equation is $ax^2 + bx + c = 0$, where a, b, and c are real numbers and $a \neq 0$.

In the equation $xy = 0$, either $x = 0$, $y = 0$, or both equal 0. This illustrates the Zero Product Property and is used in solving quadratic equations.

ZERO PRODUCT PROPERTY

If $pq = 0$, then $p = 0$ or $q = 0$.

EXAMPLE 1: Solving by Applying the Product Property

WRITING SOLUTIONS

Solutions can be written separately: $x = 5, x = -2$; as a list: $x = 5, -2$; or as a set: $\{5, -2\}$.

Solve $(x - 5)(x + 2) = 0$.

Answer

$x - 5 = 0$	or	$x + 2 = 0$	**1.** Set each factor equal to 0 using the Zero Product Property.
$x = 5$		$x = -2$	**2.** Solve each linear equation.

There are 2 solutions: $x = 5, -2$.

SKILL CHECK **EXERCISE 1**

Many quadratic equations will need to be factored before you can apply the Zero Product Property. Remember to look for common factors as your first step in factoring.

EXAMPLE 2: Solving by Factoring

Solve each equation.

a. $8x^2 - 16x = 0$ **b.** $x^2 = -8x - 12$

Answers

a.

$8x^2 - 16x = 0$ $8x(x - 2) = 0$	**1.** Factor $8x$ from each term.
$8x = 0$ or $x - 2 = 0$	**2.** Apply the Zero Product Property.
$x = 0$ $\qquad x = 2$	**3.** Solve each equation.
$8(0)^2 - 16(0) = 0$ $8(2)^2 - 16(2) = 32 - 32 = 0$	**4.** Check each solution by substituting into the original equation.

b.

$x^2 = -8x - 12$ $x^2 + 8x + 12 = 0$	**1.** Write the equation in standard form.
$(x + 2)(x + 6) = 0$	**2.** Factor the trinomial.
$x + 2 = 0$ or $x + 6 = 0$	**3.** Apply the Zero Product Property.
$x = -2$ $\qquad x = -6$	**4.** Solve each equation.
$(-2)^2 + 8(-2) + 12 = 4 - 16 + 12 = 0$ $(-6)^2 + 8(-6) + 12 = 36 - 48 + 12 = 0$	**5.** Check each solution by substituting into the original equation.

SKILL CHECK **EXERCISES 13, 17**

AFTERSCHOOLHELP

Solving Quadratic Equations by Factoring

SOLVING QUADRATIC EQUATIONS BY FACTORING

1. Write the equation in standard form: $ax^2 + bx + c = 0$.
2. Factor the polynomial side of the equation.
3. Set each factor that contains a variable equal to 0 using the Zero Product Property.
4. Solve the resulting equations.
5. Check your solutions in the original equation.

EXAMPLE 3: Solving by Factoring

Solve each equation.

a. $2x^2 + 8x - 42 = 0$ **b.** $3x^2 = 147$

Answers

a.

$2x^2 + 8x - 42 = 0$ $2(x^2 + 4x - 21) = 0$ $2(x + 7)(x - 3) = 0$	**1.** Factor the trinomial.
$2 \neq 0$ or $x + 7 = 0$ or $x - 3 = 0$	**2.** Apply the Zero Product Property.
$x = -7$ $\qquad x = 3$	**3.** Solve each equation.
$2(-7)^2 + 8(-7) - 42 = 98 - 56 - 42 = 0$ $2(3)^2 + 8(3) - 42 = 18 + 24 - 42 = 0$	**4.** Check each solution by substituting into the original equation.

b.

$3x^2 = 147$ $3x^2 - 147 = 0$	**1.** Write the equation in standard form.
$3(x^2 - 49) = 0$	**2.** Factor 3 from each term.
$3(x - 7)(x + 7) = 0$	**3.** Factor the binomial.
$3 \neq 0$ or $x - 7 = 0$ or $x + 7 = 0$	**4.** Apply the Zero Product Property.
$x = 7$ $\qquad x = -7$	**5.** Solve each equation.

SKILL CHECK **EXERCISE 27**

Quadratic equations will have at most 2 real-number solutions. The Zero Product Property can be applied to equations with higher degrees after they have been factored. A third-degree equation such as $4x^3 = 14x^2 + 8x$ can have up to 3 real-number solutions since it is a third-degree polynomial.

EXAMPLE 4: Solving by Factoring

Find the solution set for the equation $4x^3 = 14x^2 + 8x$.

Answer

$$4x^3 = 14x^2 + 8x$$

1. Set the equation equal to 0.

$$4x^3 - 14x^2 - 8x = 0$$

$$2x(2x^2 - 7x - 4) = 0$$

2. Factor $2x$ from each term.

$$2x(2x^2 - 8x + 1x - 4) = 0$$

$$2x[2x(x - 4) + 1(x - 4)] = 0$$

$$2x(2x + 1)(x - 4) = 0$$

3. Factor the trinomial using the technique you learned in Section 9.3.

$$2x = 0 \quad \text{or} \quad 2x + 1 = 0 \quad \text{or} \quad x - 4 = 0$$

4. Apply the Zero Product Property.

$$x = 0 \qquad 2x = -1 \qquad x = 4$$

$$x = -\frac{1}{2}$$

5. Solve each equation.

$$4(0)^3 - 14(0)^2 - 8(0) = 0$$

$$4\left(-\frac{1}{2}\right)^3 - 14\left(-\frac{1}{2}\right)^2 - 8\left(-\frac{1}{2}\right) = -\frac{1}{2} - \frac{7}{2} + 4 = 0$$

$$4(4)^3 - 14(4)^2 - 8(4) = 256 - 224 - 32 = 0$$

6. Substitute each solution into the original equation.

SKILL CHECK **EXERCISE 31**

Dividing both sides of an equation by a variable expression causes a solution to be lost. In the previous example, dividing both sides of the original equation by x would eliminate the solution of 0.

EXAMPLE 5: Solving a Real-World Problem

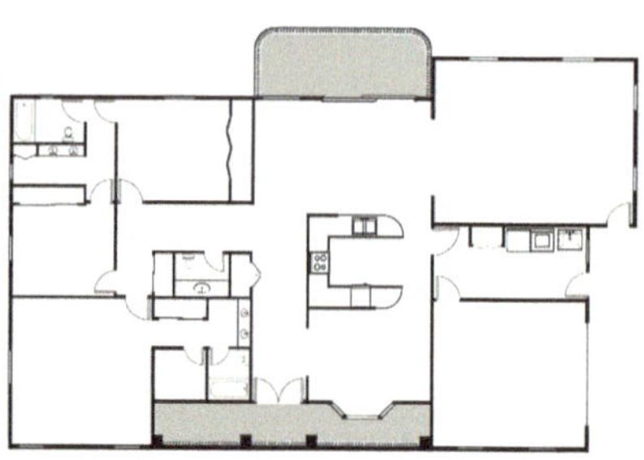

An architect wishes to design a room for a home addition in which the length is 5 ft longer than the width. If the room needs to be 204 ft², find the dimensions.

Answer

Let w = the width and $w + 5$ = the length.

1. Assign variable expressions to the unknowns.

$$(w + 5)w = 204$$

2. Use the formula $A = lw$ to write an equation modeling the problem.

$$w^2 + 5w = 204$$

3. Solve the resulting quadratic equation.

$$w^2 + 5w - 204 = 0$$

   a. Write the equation in standard form.

$$(w + 17)(w - 12) = 0$$

   b. Factor the trinomial.

$$w + 17 = 0 \quad \text{or} \quad w - 12 = 0$$

   c. Apply the Zero Product Property.

$$w = -17 \qquad w = 12$$

   d. Solve each equation.

dimensions: 12 ft × 17 ft

4. Interpret the solutions to your equation. Since −17 is not realistic for length, the width must be 12 ft and the length is 12 + 5 = 17 ft.

Area = 12 ft × 17 ft = 204 ft²

5. Check your solution in the original problem.

SKILL CHECK **EXERCISE 37**

VOCABULARY
- quadratic equations
- standard form

A. EXERCISES

Solve.

1. $(x - 2)(x + 4) = 0$
2. $(y + 5)(y + 2) = 0$
3. $2z(z + 7) = 0$
4. $3x(x + 12) = 0$
5. $8x(x + 1)(x - 10) = 0$
6. $-5a(a - 1)(3a + 2) = 0$
7. $(4b + 1)(3b + 2) = 0$
8. $(2x - 11)(4x - 5) = 0$

Solve by factoring.

9. $y^2 + 6y + 9 = 0$
10. $a^2 - 8a + 16 = 0$
11. $x^2 - 9 = 0$
12. $y^2 - 225 = 0$
13. $a^2 + 5a - 36 = 0$
14. $x^2 - x - 56 = 0$

B. EXERCISES

Solve by factoring.

15. $x^2 - x = 0$
16. $x^2 + 3x = 0$
17. $c^2 = 7c$
18. $d^2 = 5d$
19. $x^2 - 10x = -16$
20. $m^2 - 2m = 35$
21. $p^2 = 2p + 24$
22. $x^2 - 12 = -4x$
23. $y^2 = 169$
24. $16y^2 = 9$
25. $4x^2 = 9$
26. $2a^2 + 100 = -30a$
27. $3x^2 - 216 = 63x$
28. $8x^2 + 10x = -3$
29. $10t^2 = -7t + 12$
30. $14x^2 + 12 = 29x$
31. $5r^3 + 5r^2 = 10r$
32. $4s^3 = 16s$
33. **MP3** Maggie's work to solve $x^2 - 2x = 3$ is shown below. Her answer was $x = 3, 5$. Describe and correct her errors.

$$x^2 - 2x = 3$$
$$x(x - 2) = 3$$
$$x = 3 \quad \text{or} \quad x - 2 = 3$$
$$x = 5$$

34. **Essential Question:** What do I know about multiple factors if their product is 0?

Write a quadratic equation and solve.

35. Two consecutive integers have a product of 240. Find the integers.
36. Two consecutive integers have a product that is 14 less than ten times the smaller integer. Find the integers.
37. **MP4** A gardener currently has a vegetable garden 12 ft wide and 20 ft long. He wishes to increase the area to 560 ft^2 by adding the same amount to the garden's length and width. How many feet should be added? What are the new dimensions?
38. The difference of 2 numbers is 6 and the difference of their squares is 60. Find the numbers.

C. EXERCISES

Write a quadratic equation and solve.

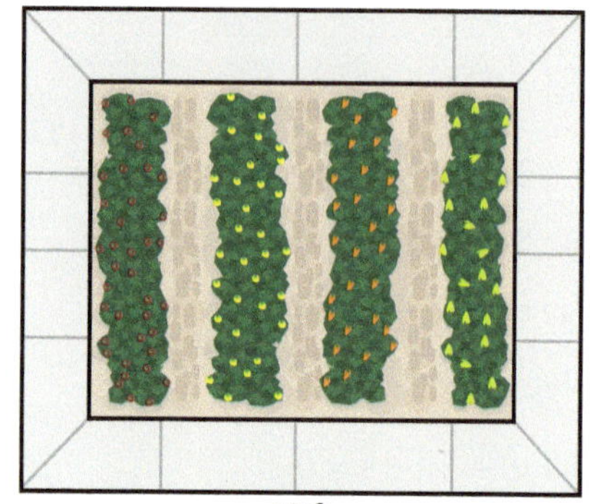

39. Javier said, "If five times my age in 8 yr is subtracted from the square of my present age, the result is 86." Find Javier's present age.

40. MP4 Brady wants to lay a sidewalk within his flower garden, which is 52 ft long and 44 ft wide. How wide should the sidewalk be if he desires to keep 1748 ft^2 for the garden area?

41. Solve $15x^3 = 17x^2 + 42x$.

42. MP7 Use the fact that $x - 2$ is a factor of $x^3 + 9x^2 + 2x - 48$ to solve the equation $x^3 + 9x^2 + 2x - 48 = 0$.

CUMULATIVE REVIEW

Find all positive values for *k* such that the trinomial will factor over the set of integers. [9.2]

43. $x^2 + kx + 21$

44. $x^2 - kx + 12$

45. $x^2 + kx - 12$

Simplify each radical expression. [10.1, 10.6]

46. $\sqrt{108}$

47. $\sqrt{12^2 - 4(-3)(-1)}$

48. $\sqrt{22}(4\sqrt{20} - 3\sqrt{55})$

Find each missing length of a right triangle in which *c* is the hypotenuse. [10.5]

49. $a = 21$ and $c = 29$

50. $a = 8$ and $b = 4\sqrt{3}$

51. Which square root equation has no real-number solutions? **[10.7]**

A $\sqrt{x - 8} - 6 = -3$

B $-2\sqrt{1 - x} = -14$

C $\sqrt{2x + 3} = 0$

D $3\sqrt{5x} + 4 = 19$

E none of these

52. Which transformations of the parent function $f(x) = \sqrt{x}$ produce the function $g(x) = -\sqrt{x} - 5$? **[10.8]**

A reflected across *x*-axis, translated 5 units down

B reflected across *y*-axis, translated 5 units down

C reflected across *y*-axis, translated 5 units left

D reflected across *x*-axis, translated 5 units right

E none of these

MIND OVER MATH

How quickly can you find each product? Would it be faster with a calculator? Why or why not?

1. $\frac{583}{12} \cdot 8 \cdot 134 \cdot \frac{52}{77} \cdot \frac{6}{901} \cdot 596 \cdot 0$

2. $\frac{7}{19} \cdot \frac{19}{5} \cdot \frac{5}{41} \cdot \frac{41}{37} \cdot \frac{37}{8} \cdot \frac{8}{3} \cdot \frac{3}{7}$

3. $\frac{19}{23} \cdot \frac{14}{15} \cdot \frac{37}{40} \cdot \frac{15}{19} \cdot \frac{23}{27} \cdot \frac{11}{14} \cdot \frac{27}{37}$

11.2 SOLVING QUADRATIC EQUATIONS BY TAKING ROOTS

If I drop an object, how long will it take to hit the ground?

After completing this section, I will be able to

- describe solutions of the equation $x^2 = c$.
- solve quadratic equations of the form $ax^2 - c = 0$ by taking roots.
- solve quadratic equations of the form $(x + b)^2 = c$ by taking roots.
- solve real-world problems involving quadratics.

Equations of the form $x^2 - c = 0$ in which c is a perfect square can be solved using the factoring method from the previous section.

$$x^2 - 64 = 0$$
$$(x - 8)(x + 8) = 0$$
$$x - 8 = 0 \quad \text{or} \quad x + 8 = 0$$
$$x = 8 \quad \text{or} \quad x = -8$$

Another method of solving equations of this form is to isolate the squared term and apply the Square Root Property.

It is said that Galileo dropped balls from the Leaning Tower of Pisa and showed that the falling of objects can be described by a quadratic relationship.

SQUARE ROOT PROPERTY

1. If $x^2 = c$ and $c \geq 0$, then $x = \pm\sqrt{c}$.
   a. If $c > 0$, then x has 2 real solutions.
   b. If $c = 0$, then x has 1 real solution.
2. If $x^2 = c$ and $c < 0$, then x has no real solutions.

EXAMPLE 1: Solving Equations of the Form $ax^2 - c = 0$

Solve each equation.

a. $x^2 - 64 = 0$ **b.** $2x^2 - 24 = 0$

Answers

a. $x^2 - 64 = 0$ — 1. Isolate the squared variable.

$x^2 = 64$

$\sqrt{x^2} = \sqrt{64}$ — 2. Take the square root of each side.

$|x| = 8$ — 3. Recall that $\sqrt{x^2} = |x|$.

$x = \pm 8$ — 4. Solve.

$8^2 - 64 = 0$ — 5. Substitute each solution into the original equation.

$(-8)^2 - 64 = 0$

b. $2x^2 - 24 = 0$ — 1. Isolate the squared variable.

$2x^2 = 24$ — a. Isolate the term containing x^2.

$x^2 = 12$ — b. Isolate x^2 by dividing both sides by 2.

$x = \pm\sqrt{12}$ — 2. Apply the Square Root Property.

$x = \pm 2\sqrt{3}$ — 3. Simplify.

$2(2\sqrt{3})^2 - 24 = 2(12) - 24 = 0$ — 4. Substitute each solution into the original equation.

$2(-2\sqrt{3})^2 - 24 = 2(12) - 24 = 0$

SKILL CHECK **EXERCISES 5, 11** ☑

SOLVING EQUATIONS OF THE FORM $ax^2 - c = 0$

1. Isolate the squared expression.
2. Apply the Square Root Property.
3. Simplify the radical and solve each equation.

The same procedure can be applied to equations containing a polynomial that is squared.

EXAMPLE 2: Solving Equations of the Form $(x + b)^2 = c$

AFTERSCHOOLHELP

Solving Quadratic Equations by Taking Roots

Solve $(x + 2)^2 = 36$.

Answer

$(x + 2)^2 = 36$	**1.** The left side is already a perfect square.
$x + 2 = \pm 6$	**2.** Apply the Square Root Property.
$x = -2 \pm 6$ $x = 4, -8$	**3.** Solve by adding −2 to both sides. Note that -2 ± 6 means $-2 + 6$ or $-2 - 6$.
$(4 + 2)^2 = 36$ and $(-8 + 2)^2 = 36$	**4.** Both answers are solutions.

SKILL CHECK **EXERCISE 17**

EXAMPLE 3: Describing Solutions of Quadratic Equations

Describe the solution for each equation and solve.

a. $(x - 5)^2 = 21$ **b.** $(x + 2)^2 + 8 = 8$ **c.** $3(x^2 + 5) + 2 = -4$

Answers

a. $(x - 5)^2 = 21$ Since $c > 0$, then x will have 2 real solutions.	**1.** The left side is already a perfect square.
$x - 5 = \pm\sqrt{21}$	**2.** Apply the Square Root Property.
$x = 5 \pm \sqrt{21}$	**3.** Solve and state the solutions as simplified radical expressions.
$x \approx 5 \pm 4.58 = 0.42, 9.58$	**4.** Use a calculator to find the decimal approximations.
b. $(x + 2)^2 + 8 = 8$ $(x + 2)^2 = 0$ Since $c = 0$, then x will have 1 real solution.	**1.** Isolate the squared term.
$x + 2 = \sqrt{0} = 0$	**2.** Apply the Square Root Property.
$x = -2$	**3.** Solve.
c. $3(x^2 + 5) + 2 = -4$ $3x^2 + 15 = -6$ $3x^2 = -21$ $x^2 = -7$ Since $-7 < 0$, then x will have no real solutions.	**1.** Isolate x^2.
$x = \pm\sqrt{-7}$	**2.** Apply the Square Root Property.
There are no real-number solutions for x.	**3.** No real number can be squared to produce −7.

SKILL CHECK **EXERCISES 25, 27**

CHECK

$(0.42 - 5)^2 = (-4.58)^2$
$= 20.9764 \approx 21$

$(9.58 - 5)^2 = 4.58^2$
$= 20.9764 \approx 21$

Assuming negligible air resistance, the height of a dropped object can be modeled by the function $h(t) = -16t^2 + h_i$, where t is the time in seconds and h_i is the initial height in feet.

EXAMPLE 4: Solving a Real-World Problem

After hitting his golf ball into the water, Mark Wetzel had to play his next shot from the drop area. If he dropped the ball from a height of 5 ft, how long did it take for the ball to reach the ground?

Answer

$h(t) = -16t^2 + h_i$

$0 = -16t^2 + 5$

1. Substitute the initial height of 5 ft and find t when the ball is at a height of 0 ft.

$16t^2 = 5$

$t^2 = \frac{5}{16}$

2. Isolate t^2.

$t = \pm\sqrt{\frac{5}{16}} = \pm\frac{\sqrt{5}}{4}$

3. Apply the Square Root Property and simplify.

$t \approx 0.56$ sec

4. Calculate the decimal approximation of the answer. Only the positive answer is a reasonable solution.

SKILL CHECK **EXERCISE 35**

A. EXERCISES

Solve using the Square Root Property.

1. $x^2 = 25$ **2.** $x^2 = 121$ **3.** $x^2 = 6$

4. $x^2 = 15$ **5.** $x^2 - 100 = 0$ **6.** $x^2 - 16 = 0$

7. $x^2 - 5 = 31$ **8.** $x^2 + 3 = 147$ **9.** $x^2 + 5 = 0$

10. $3x^2 = 15$ **11.** $3x^2 = 51$ **12.** $-4x^2 = 64$

B. EXERCISES

Solve each equation by taking roots.

13. $3x^2 = 81$ **14.** $3x^2 = 72$ **15.** $4x^2 = 50$

16. $6x^2 = 40$ **17.** $(x + 4)^2 = 25$ **18.** $(x - 6)^2 = 13$

19. $(x + 2)^2 = 12$ **20.** $(4x + 1)^2 = 9$ **21.** $(9x - 2)^2 = 16$

22. $4(x + 1)^2 = 9$ **23.** $9(x - 2)^2 = 16$ **24.** $2x^2 - 17 = x^2 + 3$

Describe the solution for each equation and solve.

25. $(x + 3)^2 = 121$ **26.** $3x^2 - 10 = x^2 - 24$

27. $(x - 5)^2 + 17 = 0$ **28.** $4(x + 2)^2 = 0$

29. $4x^2 - 16 = 3x^2 - 16$ **30.** $5x^2 + x - 2 = x^2 + x + 7$

Write a quadratic equation and solve. Use a calculator and the decimal approximation of 3.14 for π.

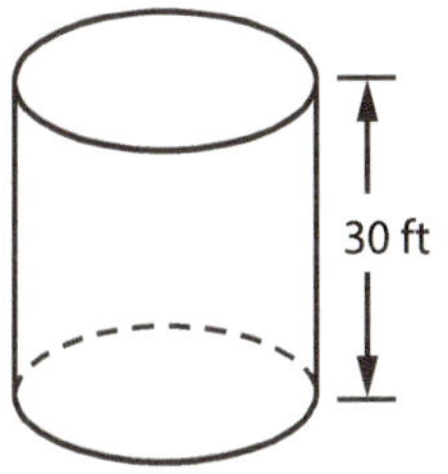

31. A circular sprinkler is advertised to have coverage of about 450 ft². Find its range (radius) to the nearest foot.

32. The volume of the circular cylinder shown is 45,600 ft³. Find its radius to the nearest foot.

33. **Essential Question:** Let t be the amount of time in seconds after dropping an object from an initial height of h_i (in feet). The height of the dropped object after t sec is modeled by $h(t) = -16t^2 + h_i$.

   a. When determining how long it will take the object to hit the ground, what value should be substituted for $h(t)$?

   b. Using the value from part a, solve the equation $h(t) = -16t^2 + h_i$ for t.

   c. Using the equation from part b, how long (to the nearest second) will it take for an object dropped from a plane flying at 36,000 ft above the earth to reach the ground?

Write a quadratic equation and solve. Assume that air resistance is negligible.

34. Dave dropped an object partway up his rock climb. If the object took exactly 2 sec to hit the ground, how high had he climbed?

35. A construction worker dropped a bolt from the twelfth floor, a height of 120 ft. How long (to the nearest tenth of a second) did it take to reach the ground?

36. **MP2** How many seconds will it take for an object to fall 200 ft? 400 ft? Explain why the time does not double.

37. Each layer of the Leaning Tower of Pisa is known as a cornice. From the side of the tower closer to the ground, Galileo dropped the same ball from the base of the third and fifth cornices. He found it took 7 and 8.5 sec, respectively, for the ball to reach the ground. What is the difference in vertical height between the bases of the third and fifth cornices?

Galileo Galilei (1564–1642) of Pisa, Italy, is best known for his work in astronomy but was also a student of mathematics and physics.

C. EXERCISES

Solve each equation by taking roots.

38. $2(5x - 3)^2 = 7$

39. **MP7** $4x^2 - 20x + 25 = 32$

40. **MP4** An architect plans to center a square building that will be 150 ft on each side within a 40,000 ft^2 square lot. How far will the resulting courtyard extend in each direction from the building?

41. **MP8** Another common formula that models objects in free fall is $d = \frac{1}{2}gt^2$, where d is the distance fallen, g is the gravitational acceleration, and t is the time.

   a. Use the distance formula $d = |b - a|$ to write an expression for the distance an object falls from an initial height of h_i to its current height, h_c.

   b. Substitute the expression for distance into the equation $d = \frac{1}{2}gt^2$ and solve for h_c. Be sure to consider the relative magnitudes of h_c and h_i when interpreting the absolute value.

   c. The gravitational acceleration near the surface of the earth is approximately 32 ft/sec^2. Substitute this value into the derived formula for h_c to verify the function rule for free fall used in Example 6.

   d. Convert 32 ft/sec^2 to meters per second squared (to the nearest tenth of a meter). Then state the function rule for the height of an object in free fall for t seconds with an initial height of h_i meters.

CUMULATIVE REVIEW

Evaluate each expression. [1.2]

42. $\frac{6}{7} \div 2$

43. $\frac{5}{6} \div 2$

Translate each phrase into a numerical expression. Then evaluate your expression. [1.4]

44. half of $\frac{4}{5}$, squared

45. half of $\frac{3}{8}$, squared

Factor and solve. [11.1]

46. $x^2 - 14x - 51 = 0$

47. $9x^2 - 64 = 0$

48. $x^4 - 13x^3 + 36x^2 = 0$

49. $4x^3 + 4x^2 + x = 0$

50. Which polynomial is correctly paired with its factored form? **[9.4]**

A $w^2 - 9w + 6; (w - 3)^2$

B $4x^2 + 49; (2x + 7)^2$

C $y^2 + 16y - 64; (y - 8)^2$

D $z^2 - 10z + 25; (z - 5)^2$

E none of these

51. Which expression is equivalent to $\frac{20}{\sqrt[3]{2}}$? **[10.3]**

A $\sqrt[3]{10}$

B $10\sqrt[3]{2}$

C $10\sqrt[3]{4}$

D $5\sqrt[3]{4}$

E none of these

TECHNOLOGY CORNER (TI-84 PLUS FAMILY)

Checking Irrational Solutions

Your graphing calculator can be used to find decimal equivalents of answers in radical form and to check your answers in the original equation. The [ans] function ([2nd] [(-)]) may be used to insert the previous answer into the expression to be calculated. As shown on the right, you can also store answers as alphabetical characters. Use arrow keys to highlight recent entries and press [enter] to find and edit a similar expression without reentering the entire expression. Note that you can also access and select entries in your history by repeatedly selecting the [entry] function ([2nd] [enter]).

The answers to quadratic equations often come in conjugate pairs, such as $\frac{3 \pm \sqrt{5}}{7}$, the solutions to $49x^2 - 42x + 4 = 0$. To find the decimal equivalent of $\frac{3+\sqrt{5}}{7}$, press [(] 3 [+] [2nd] [x^2] 5 [▸] [)] [÷] 7 [enter]. Alternatively, press [alpha] [y=] [1] to enter the expression as a fraction.

To find the decimal equivalent of the second solution, $\frac{3-\sqrt{5}}{7}$, press [entry] and use [◂] to place the cursor over the [+] sign. Change it to [−] and press [enter]. (The previous entry can also be chosen by highlighting it using [▴] and pressing [enter].)

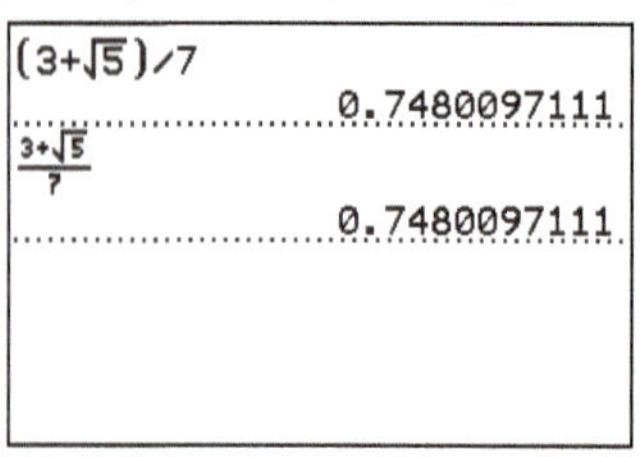

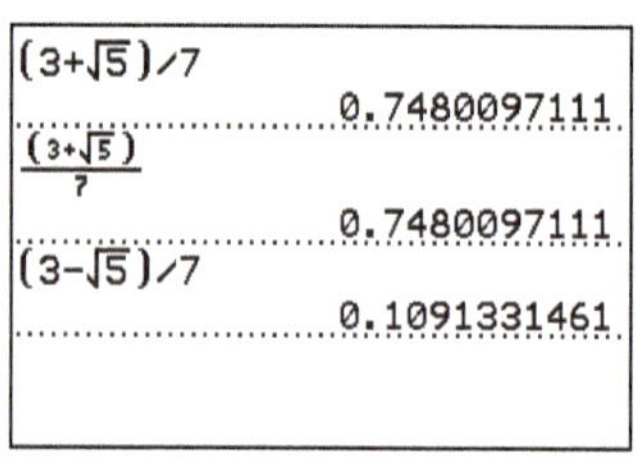

You can check solutions by storing them as alphabetical characters. Enter the first solution and store it as "A" by pressing [sto→] [alpha] [math]. Store the second solution as "B" by pressing [sto→] [alpha] [apps].

Enter $49x^2 - 42x + 4$ into the calculator using A ([alpha] [math]) and B ([alpha] [apps]) in place of x. Use [entry] to type the expression with B more easily. Since each expression evaluates to 0, both solutions are correct.

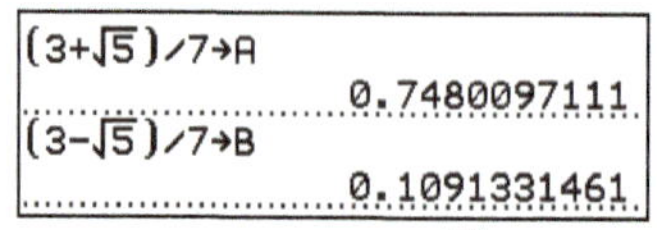

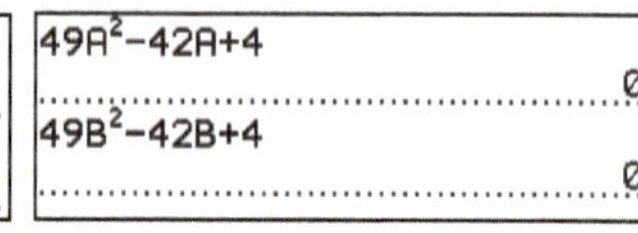

Test whether $\frac{3+\sqrt{6}}{7}$ is a solution by storing it as C ([sto→] [alpha] [prgm]), pressing [entry] twice, and replacing B with C.

You can also type $\frac{3+\sqrt{6}}{7}$ and press [enter] to display the decimal solution. Press [entry] twice to display the quadratic expression and use [ans] as the variable. Since the expression does not evaluate to 0, this is not a solution.

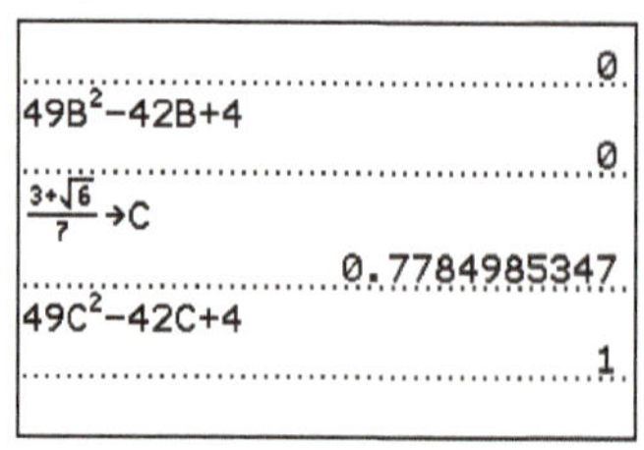

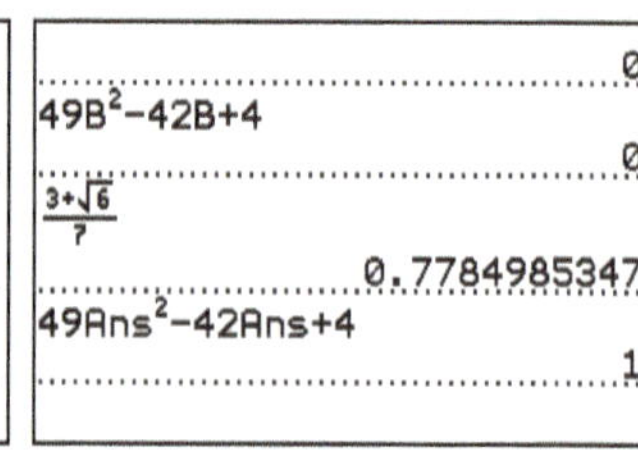

11.3 COMPLETING THE SQUARE: $x^2 + bx + c = 0$

How can I use factoring patterns to solve quadratic equations?

After completing this section, I will be able to

- complete the square for expressions of the form $x^2 + bx$.
- solve quadratic equations of the form $x^2 + bx + c = 0$ by completing the square.

Some quadratic equations, such as $a^2 + 6a + 2 = 0$, cannot be solved by factoring. The method discussed in Section 11.2 cannot be applied since isolating the squared expression, a^2, leaves a variable in the other side of the equation, $-6a - 2$.

However, by adding the correct number to both sides of the equation, you can create an equivalent equation that will allow you to use the method described in Section 11.2 to solve the equation.

Determining exactly which number should be added to both sides of the equation $a^2 + 6a + 2 = 0$ is simpler when the constant term is moved to the right side of the equation: $a^2 + 6a = -2$. Looking at a geometric representation of $a^2 + 6a$ can help you determine which number should be added to both sides in order to solve the equation.

1. Create a geometric representation for $a^2 + 6a$.

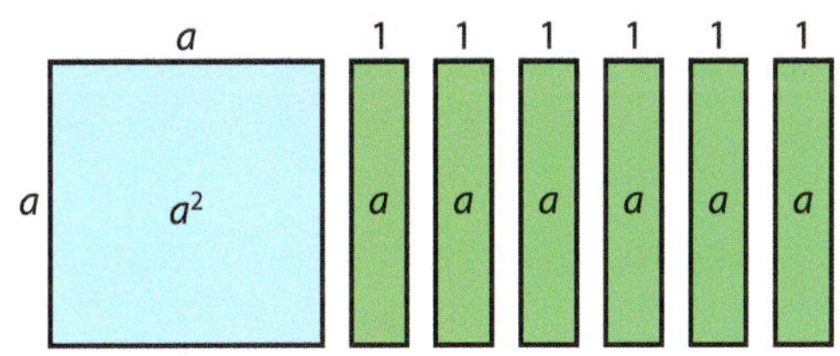

2. Arrange the tiles to form the upper and left edges of a square.

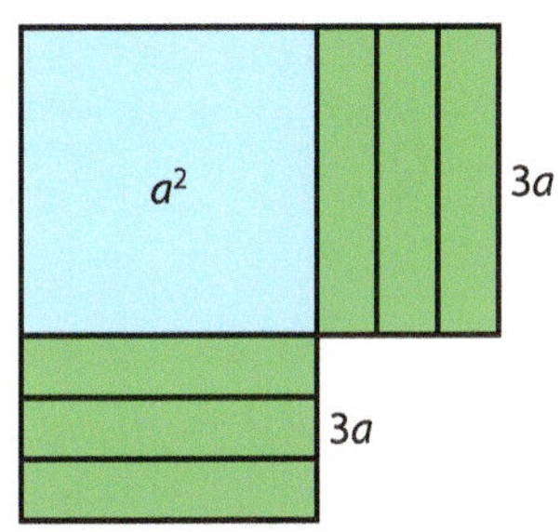

3. Complete the square. Note that half of the 6 strips were used on each side of the square, resulting in the addition of 9 unit squares to complete the square's shape.

When the square of one-half of the coefficient of the a term, $\left(\frac{6}{2}\right)^2 = 3^2 = 9$, is added to both sides of $a^2 + 6a = -2$, the left side can be factored as a perfect square trinomial. This allows you to solve the equation by taking roots.

$$a^2 + 6a + 9 = -2 + 9$$
$$(a + 3)^2 = 7$$
$$a + 3 = \pm\sqrt{7}$$
$$a = -3 \pm \sqrt{7}$$

The process of adding a number to the expression $x^2 + bx$ to produce a perfect square trinomial is called *completing the square.*

COMPLETING THE SQUARE

To complete the square of an expression in the form of $x^2 + bx$, add the square of half the coefficient of x: $x^2 + bx + \left(\frac{b}{2}\right)^2$.

EXAMPLE 1: Completing the Square

For each expression, complete the square and factor the resulting trinomial.

a. $x^2 - 16x$ **b.** $x^2 + \frac{3}{5}x$

Answers

a. $\left(\frac{-16}{2}\right)^2 = (-8)^2 = 64$ — **1.** Find the square of half the coefficient of x.

$x^2 - 16x + 64$ — **2.** Add the result to the binomial.

$(x - 8)^2$ — **3.** Factor the perfect square trinomial.

b. $\left(\frac{1}{2} \cdot \frac{3}{5}\right)^2 = \left(\frac{3}{10}\right)^2 = \frac{9}{100}$ — **1.** Find the square of half the coefficient of x.

$x^2 + \frac{3}{5}x + \frac{9}{100}$ — **2.** Add the result to the binomial.

$\left(x + \frac{3}{10}\right)^2$ — **3.** Factor the perfect square trinomial.

SKILL CHECK **EXERCISES 3, 11**

You can use this method to solve any quadratic equation of the form $x^2 + bx + c = 0$.

EXAMPLE 2: Solving Quadratic Equations

Solve each equation.

a. $x^2 + 8x = 9$ **b.** $a^2 - 3 = 7a$

Answers

a. $x^2 + 8x + 16 = 9 + 16$

1. Complete the square by adding $\left(\frac{8}{2}\right)^2 = 16$ to both sides.

$(x + 4)^2 = 25$

2. Factor the left side of the equation and simplify the right side.

$x + 4 = \pm 5$

3. Apply the Square Root Property.

$x = -4 \pm 5$

$x = 1, -9$

4. Solve.

$1^2 + 8(1) = 1 + 8 = 9$

$(-9)^2 + 8(-9) = 81 - 72 = 9$

5. Check.

b. $a^2 - 3 = 7a$

$a^2 - 7a = 3$

1. Write the equation in $x^2 + bx = c$ form by adding 3 to and subtracting $7a$ from both sides.

$a^2 - 7a + \left(\frac{7}{2}\right)^2 = 3 + \frac{49}{4}$

2. Complete the square by adding $\left(\frac{-7}{2}\right)^2 = \left(\frac{7}{2}\right)^2 = \frac{49}{4}$ to both sides.

$\left(a - \frac{7}{2}\right)^2 = \frac{61}{4}$

3. Factor the left side of the equation and simplify the right side.

$a - \frac{7}{2} = \pm\frac{\sqrt{61}}{2}$

4. Apply the Square Root Property.

$a = \frac{7 \pm \sqrt{61}}{2} = \frac{7 + \sqrt{61}}{2}, \frac{7 - \sqrt{61}}{2}$

$\approx 7.41, -0.41$

5. Solve.

SKILL CHECK **EXERCISES 19, 27**

Example 2b illustrates that the sign of b can be ignored when $\frac{b}{2}$ is squared. Leaving the expression as $\left(\frac{b}{2}\right)^2$ may make the factoring of the perfect square trinomial easier.

SOLVING QUADRATIC EQUATIONS BY COMPLETING THE SQUARE

1. Write the equation in the form $x^2 + bx = c$.
2. Complete the square by adding $\left(\frac{b}{2}\right)^2$ to both sides of the equation.
3. Factor the trinomial.
4. Apply the Square Root Property: If $x^2 = c$, then $x = \pm\sqrt{c}$.
5. Solve the resulting equations.

EXAMPLE 3: Solving a Real-World Problem

Cassia increased the size of her square garden to 550 ft^2 by adding 4 ft to its width and 7 ft to its length. Find the dimensions of the original garden.

Answer

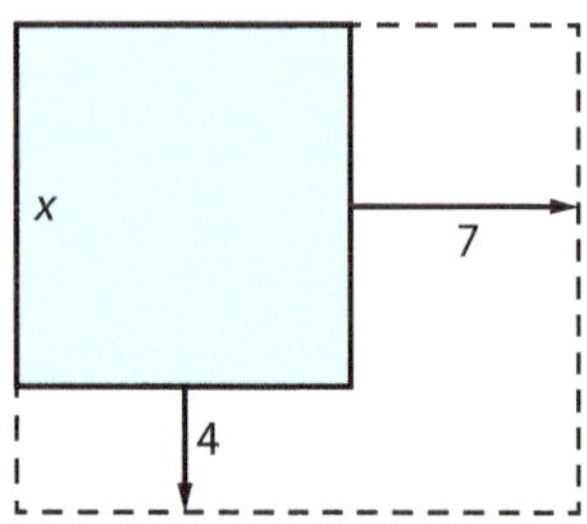

1. Draw a diagram of the problem and assign variable expressions to the unknowns.

Let x = the original length and width,
$x + 4$ = the new width, and
$x + 7$ = the new length.

$$(x + 7)(x + 4) = 550$$

2. Use the area formula to write an equation for the new area.

$$x^2 + 11x + 28 = 550$$
$$x^2 + 11x = 522$$

3. Write the equation in $x^2 + bx = c$ form.

$$x^2 + 11x + \left(\frac{11}{2}\right)^2 = 522 + \frac{121}{4}$$

4. Complete the square by adding $\left(\frac{11}{2}\right)^2 = \frac{121}{4}$ to both sides.

$$\left(x + \frac{11}{2}\right)^2 = \frac{2088}{4} + \frac{121}{4} = \frac{2209}{4}$$

5. Factor the left side and simplify the right side.

$$x + \frac{11}{2} = \pm\sqrt{\frac{2209}{4}} = \pm\frac{47}{2}$$

6. Apply the Square Root Property.

$$x = \frac{-11 \pm 47}{2}$$

7. Solve. Note that a negative answer is not reasonable.

$$x = \frac{-11 + 47}{2} = 18 \text{ or } x = \frac{-11 - 47}{2} = -29$$

The original garden was 18 ft on each side.

SKILL CHECK **EXERCISE 37**

AFTERSCHOOLHELP

Completing the Square: $x^2 + bx + c = 0$

While the integral answers in Example 3 indicate that the equation could have been solved by factoring, finding the correct factors of 522 could be challenging.

A. EXERCISES

What number must be added to complete each square?

1. $x^2 + 8x$

2. $x^2 - 12x$

3. $x^2 + 16x$

4. $x^2 - 10x$

5. $x^2 - 2x$

6. $x^2 + 52x$

7. $x^2 + 3x$

8. $x^2 + 5x$

9. $x^2 - 11x$

10. $x^2 - 9x$

11. $x^2 + \frac{3}{2}x$

12. $x^2 + \frac{5}{6}x$

13. $x^2 - \frac{10}{11}x$

14. $x^2 - \frac{4}{5}x$

Solve each equation by completing the square.

15. $a^2 + 2a = 15$

16. $x^2 + 6x = -5$

17. $x^2 + 4x = 5$

18. $y^2 - 6y = 7$

B. EXERCISES

Solve each equation by completing the square. State irrational solutions both as simplified radical expressions and as decimal approximations rounded to the nearest hundredth.

19. $x^2 - 10x - 24 = 0$

20. $a^2 + 6a - 55 = 0$

21. $x^2 - 2x - 99 = 0$

22. $m^2 - 14m + 40 = 0$

23. $m^2 - \frac{4}{3}m = 0$

24. $x^2 + \frac{5}{6}x = 0$

25. $x^2 + x = 2$

26. $a^2 - 7a = -10$

27. $x^2 + 20x + 15 = 0$

28. $x^2 - 12x - 5 = 0$

29. $x^2 - 18x + 9 = 0$

30. $x^2 - 8x - 2 = 0$

31. $x^2 = 8x - 4$

32. $x^2 = 4 + 12x$

33. $4x = x^2 - 3$

34. $10x = x^2 + 7$

35. **Essential Question:** What must be added to both sides of $x^2 + bx = c$ to complete the square? Explain which special factoring pattern this produces and how this helps us to solve the quadratic equation.

36. **MP1** Find 2 integers such that one is 6 more than the other and their product is 216.

37. Increasing the width of a square by 2 in. and the length by 7 in. causes the area of the resulting rectangle to be 204 in.^{2} What was the size of the original square?

38. **MP4** Roger is putting down flooring in a recently expanded room. The length is now 2.5 ft longer than the width and the area of the room is 174 ft^2. What are the dimensions of the room?

C. EXERCISES

Solve each equation by completing the square.

39. $x^2 + 3x + 1 = 0$

40. $3m^2 + 15m - 36 = 0$

41. **MP7** $(2t - 5)(2t + 2) = (t - 3)(3t + 2)$

42. **MP2** The line segment to the right is divided into 2 parts such that the ratio of the longer part, x, to the shorter part, 2, equals the ratio of the total length of the segment to its longer part. Determine the missing length.

CUMULATIVE REVIEW

Simplify each expression. [8.3, 8.6]

43. $\frac{4x^2 + 12x - 20}{4}$

44. $\frac{3x^2 + 5x - 15}{3}$

45. $\frac{\frac{1}{2}x^2 - 3x + 1}{\frac{1}{2}}$

46. $2\left(\frac{1}{2}x^2 - 3x + 1\right)$

Solve each equation. [10.7]

47. $\sqrt{x} = 9$

48. $\sqrt{x} + 10 = 3$

49. $2\sqrt{x} = 5$

50. $-3\sqrt{x - 4} + 9 = 0$

51. Which line is an asymptote of the graph of $y = ab^x$? **[7.4]**

A $x = 0$

B $x = b$

C $y = 0$

D $y = a$

E none of these

52. What is the simplified form of $\sqrt[3]{50a^2} \cdot \sqrt[3]{12ab}$? **[10.2]**

A $10a\sqrt{6ab}$

B $2a\sqrt[3]{75b}$

C $10a\sqrt[3]{6ab}$

D $40a\sqrt[3]{3b}$

E none of these

11.4 COMPLETING THE SQUARE: $ax^2 + bx + c = 0$

How can I calculate how long it will take for a projectile to hit the ground?

After completing this section, I will be able to

- solve quadratic equations of the form $ax^2 + bx + c = 0$ by completing the square.
- solve real-world problems involving quadratic equations by completing the square.

The method for completing the square used in Section 11.3 works only when the leading coefficient is 1. Given $ax^2 + bx + c = 0$, where $a \neq 1$, you will need to first divide both sides of the equation by a, the leading coefficient.

The height of a baseball in motion can be modeled by $h(t) = -16t^2 + v_i t + h_i$.

EXAMPLE 1: Completing the Square

Solve $4x^2 + 16x - 8 = 0$.

Answer

$\frac{4x^2 + 16x - 8}{4} = \frac{0}{4}$ $x^2 + 4x - 2 = 0$	**1.** Divide both sides of the equation by the leading coefficient, 4.
$x^2 + 4x = 2$	**2.** Write the equation in $x^2 + bx = c$ form.
$x^2 + 4x + 4 = 2 + 4$	**3.** Complete the square by adding $\left(\frac{4}{2}\right)^2 = 4$ to both sides.
$(x + 2)^2 = 6$	**4.** Factor the trinomial and simplify the right side.
$x + 2 = \pm\sqrt{6}$	**5.** Apply the Square Root Property.
$x = -2 \pm \sqrt{6}$	**6.** Solve.

SKILL CHECK **EXERCISES 1, 7**

EXAMPLE 2: Completing the Square

Solve each quadratic.

a. $3x^2 - 2x - 1 = 0$ **b.** $2x^2 = 10x + 3$

Answers

a.

$$\frac{3x^2 - 2x - 1}{3} = \frac{0}{3}$$

1. Divide both sides of the equation by the leading coefficient, 3.

$$x^2 - \frac{2}{3}x - \frac{1}{3} = 0$$

$$x^2 - \frac{2}{3}x = \frac{1}{3}$$

2. Write the equation in $x^2 + bx = c$ form.

$$x^2 - \frac{2}{3}x + \left(\frac{1}{3}\right)^2 = \frac{1}{3} + \frac{1}{9}$$

3. Complete the square by adding $\left(\frac{1}{2} \cdot \frac{2}{3}\right)^2 = \left(\frac{1}{3}\right)^2 = \frac{1}{9}$ to both sides.

$$\left(x - \frac{1}{3}\right)^2 = \frac{4}{9}$$

4. Factor the trinomial and simplify the right side.

$$x - \frac{1}{3} = \pm\frac{2}{3}$$

5. Apply the Square Root Property.

$$x = 1, -\frac{1}{3}$$

6. Solve.

DOING THE MATH

$$x - \frac{1}{3} = \pm\frac{2}{3}$$
$$+\frac{1}{3} \quad +\frac{1}{3}$$
$$x = \frac{2}{3} + \frac{1}{3} = 1$$
$$\text{or} = -\frac{2}{3} + \frac{1}{3} = -\frac{1}{3}$$

b.

$$2x^2 - 10x = 3$$

$$\frac{2x^2 - 10x}{2} = \frac{3}{2}$$

$$x^2 - 5x = \frac{3}{2}$$

1. Write the equation in $x^2 + bx = c$ form by subtracting $10x$ from both sides and then dividing both sides by 2.

$$x^2 - 5x + \left(\frac{5}{2}\right)^2 = \frac{3}{2} + \frac{25}{4}$$

2. Complete the square by adding $\left(\frac{5}{2}\right)^2 = \frac{25}{4}$ to both sides.

$$\left(x - \frac{5}{2}\right)^2 = \frac{31}{4}$$

3. Factor the trinomial and simplify the right side.

$$x - \frac{5}{2} = \pm\frac{\sqrt{31}}{2}$$

4. Apply the Square Root Property.

$$x = \frac{5}{2} \pm \frac{\sqrt{31}}{2} \text{ or } \frac{5 \pm \sqrt{31}}{2}$$

5. Solve.

SKILL CHECK **EXERCISES 13, 17**

AFTERSCHOOLHELP

Completing the Square: $ax^2 + bx + c = 0$

In previous sections you have modeled the fall of objects, such as a dropped golf ball, whose initial vertical velocity is 0. Projectiles (such as a thrown ball or a diver jumping off a board) have an initial velocity greater than 0 when propelled by force rather than being dropped. If a projectile has an initial vertical velocity v_i (measured in ft/sec) and h_i is the initial height, then the function that models the height at time t is $h(t) = -16t^2 + v_i t + h_i$.

EXAMPLE 3: Solving a Real-World Problem

A player pops a baseball straight up from an initial height of 3 ft with an initial velocity of 88 ft/sec. How long after the batter hits the ball is the ball 15 ft high?

Answer

$h(t) = -16t^2 + v_it + h_i$ $15 = -16t^2 + 88t + 3$	**1.** Substitute the given information into the formula.
$16t^2 - 88t = -12$ $\frac{16t^2 - 88t}{16} = -\frac{12}{16}$ $t^2 - \frac{11}{2}t = -\frac{3}{4}$	**2.** Write the equation in $ax^2 + bx = c$ form. Then divide both sides by 16 so that $a = 1$.
$t^2 - \frac{11}{2}t + \left(\frac{11}{4}\right)^2 = -\frac{3}{4} + \frac{121}{16}$	**3.** Complete the square by adding $\left(\frac{1}{2} \cdot \frac{11}{2}\right)^2 = \left(\frac{11}{4}\right)^2 = \frac{121}{16}$ to both sides.
$\left(t - \frac{11}{4}\right)^2 = \frac{109}{16}$	**4.** Factor the trinomial and simplify the right side.
$t - \frac{11}{4} = \pm\frac{\sqrt{109}}{4}$	**5.** Apply the Square Root Property.
$t = \frac{11 \pm \sqrt{109}}{4}$	**6.** Solve.
$t \approx 5.36, 0.14$ sec	**7.** The baseball is 15 ft high after 0.14 sec and after 5.36 sec.

KEYWORD SEARCH

completing the square

SKILL CHECK **EXERCISE 33**

A. EXERCISES

Solve by completing the square.

1. $2x^2 + 12x = 4$

2. $3y^2 - 12y = -3$

3. $3a^2 - 24a = 0$

4. $5b^2 + 60b = -55$

5. $2y^2 + 16y = 6$

6. $6x^2 + 24x = -12$

7. $2s^2 + 12s + 28 = 0$

8. $3r^2 + 72r = -24$

9. $2x^2 - 16x = 8$

10. $5y^2 - 40y + 100 = 0$

B. EXERCISES

Solve by completing the square.

11. $2y^2 + y = 3$

12. $2x^2 - x = 10$

13. $4c^2 - 8c - 5 = 0$

14. $9d^2 + 36d - 13 = 0$

15. $2x^2 - x = 1$

16. $3y^2 - y = 2$

17. $2g^2 - g = 2$

18. $2h^2 - 6h = -1$

19. $2y^2 - 3y = -35$

20. $2x^2 - 10x = 3$

21. $3k^2 - 2k - 8 = 0$

22. $3y^2 = -8y - 5$

23. $2x^2 = -3x - 1$

24. $6n^2 = 9n + 27$

25. $4m^2 - 4m = 11$

26. $3x^2 + 18x = 2$

27. $2y^2 + 24y + 1 = 0$

28. $3t^2 + 24t + 1 = 0$

29. Find 2 integers such that the first is 1 less than twice the second and their product is 231.

30. Find 2 integers such that the first is 5 more than twice the second and the sum of their squares is 850.

31. Find the length and width of a one-story rectangular house if its length is 15 ft less than twice its width and its area is 1568 ft^2.

32. **MP4** An architect had originally planned for a 11.5 ft × 20 ft first-floor restroom. He was later allotted an additional 16 ft^2. If both the length and width are extended by the same amount, how much must be added to each dimension?

33. A ball is thrown upward at an initial velocity of 26 ft/sec from a height of 3 ft. Using the function from Example 3, find the approximate length of time (to the nearest tenth of a second) it will take the ball to return to a height of 5 ft.

34. **Essential Question:** Explain what change needs to be made to your equation from exercise 33 to determine how long it will take the ball to hit the ground. Then find the approximate length of time it will take the ball to hit the ground.

C. EXERCISES

35. **MP1** Find 2 integers such that the first is 1 less than twice the second and the difference of their squares is 456.

MP5 The height (in feet) of a football kicked from a team's own 30 yd line can be modeled by the equation $h = -0.02x^2 + 1.2x$, where x is the horizontal distance traveled in yards. Use a calculator to answer the following questions.

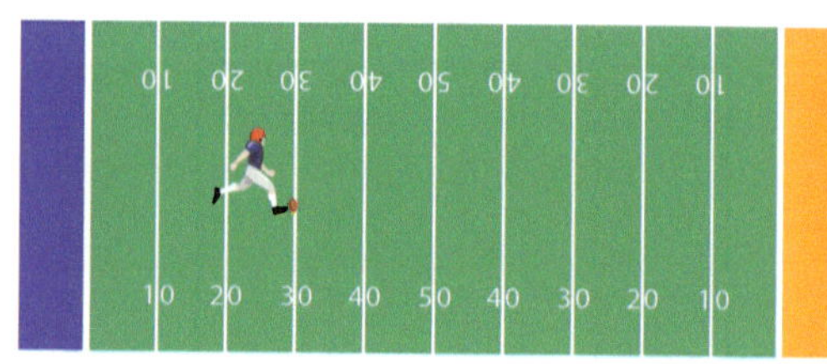

36. On what yard line will the ball hit the ground?

37. Approximately where on the field will the ball return to a height of 4 ft?

38. Will the football reach a height of 20 ft?

Solve each equation by completing the square.

39. $5x^2 + 24x + 1 = 0$

40. $\frac{1}{2}x^2 + 3x = 17$

41. The equation $x^4 + 3x^2 - 2 = 0$ can be solved by completing the square. Let $x^2 = y$ and substitute to get a quadratic equation in y; then solve for y. Approximate y to 2 decimal places. Then find x from y.

CUMULATIVE REVIEW

Simplify each expression. [7.1, 8.5, 10.1, 10.6]

42. $(x^{-3})^{-2}$

43. $\frac{19a^2b^{-3}c}{57a^4b^{-3}c^{-3}}$

44. $(13x^2 + 15z)(13x^2 - 15z)$

45. $(2x + 5)^2 + \frac{1}{(2x + 5)^{-1}}$

46. $\sqrt[3]{378}$

47. $\sqrt[4]{1536}$

48. $\frac{6 + \sqrt{12}}{2}$

49. $\frac{10 + \sqrt{20}}{4}$

50. Which expression is not equivalent to a real number? **[11.2]**

A $\sqrt{13^2 - 4(9)(4)}$

B $\sqrt{21^2 - 4(20)(-5)}$

C $\sqrt{9^2 - 4\left(\frac{3}{4}\right)(27)}$

D $\sqrt{7^2 - 4(-1)(6)}$

E none of these

51. Which term must be added to both sides of $x^2 + bx = c$ to complete the square? **[11.3]**

A $\left(\frac{b}{2}\right)^2$

B $\left(\frac{c}{2}\right)^2$

C $-\left(\frac{c}{2}\right)^2$

D $\left(\frac{bx}{2}\right)^2$

E none of these

CAREERS IN MATH
CHEMIST/CHEMICAL ENGINEER

Chemists can do amazing things with the raw materials found in nature. The microchips in your computer, your cell phone, and even the family car are products of chemical engineering. The silicon in microchips is made from natural sand, or silica. The process involves the removal of oxygen, accomplished by heating silica and carbon to temperatures over 2000°C. Such processes require the scientific and mathematical knowledge and skills of a chemist.

The career of a Christian chemist offers not only the means to make a good living but also many opportunities for demonstrating care and love for others. For instance, chemists work on cures for serious diseases or medicines for those suffering from physical infirmities, as well as the development of new materials for warmth or stronger materials for structural safety. Their efforts help others live more safely and comfortably through the wise use of the natural materials provided in God's creation.

Chemists develop new materials to improve existing products or make new ones. Analytical chemists determine the composition of a substance by identifying its elements and compounds and their relative quantities. Chemical engineers work together with industrial chemists to develop the processes needed to make new products.

God has created an astonishing variety of compounds with many differing chemical properties. Chemists can use their knowledge and skills to harness the potential of God's creation for His glory.

11.5 THE QUADRATIC FORMULA

How are mathematical models helpful and how are they limited?

After completing this section, I will be able to

- solve quadratic equations using the quadratic formula.
- explain the proper use of a mathematical model.

The welder's sparks trace curves called parabolas, which are described by quadratic equations.

Any quadratic equation can be solved by completing the square. By completing the square on the standard form of a quadratic equation, $ax^2 + bx + c = 0$, a formula can be derived that expresses the solutions of any quadratic equation in terms of its coefficients, a, b, and c.

DERIVING THE FORMULA

$ax^2 + bx + c = 0$ $ax^2 + bx = -c$ $x^2 + \frac{b}{a}x = -\frac{c}{a}$	**1.** Write the equation in $x^2 + bx = c$ form. **a.** Subtract c from both sides. **b.** Divide both sides by a.
$x^2 + \frac{b}{a}x + \left(\frac{b}{2a}\right)^2 = -\frac{c}{a} + \frac{b^2}{4a^2}$	**2.** Complete the square by adding $\left(\frac{1}{2} \cdot \frac{b}{a}\right)^2 = \left(\frac{b}{2a}\right)^2 = \frac{b^2}{4a^2}$ to both sides.
$\left(x + \frac{b}{2a}\right)^2 = -\frac{4ac}{4a^2} + \frac{b^2}{4a^2}$ $\left(x + \frac{b}{2a}\right)^2 = \frac{-4ac + b^2}{4a^2}$	**3.** Factor the trinomial and combine the terms on the right side using the common denominator $4a^2$: $-\frac{c}{a} \cdot \frac{4a}{4a} = -\frac{4ac}{4a^2}$.
$x + \frac{b}{2a} = \pm\sqrt{\frac{-4ac + b^2}{4a^2}}$	**4.** Apply the Square Root Property.
$x + \frac{b}{2a} = \frac{\pm\sqrt{b^2 - 4ac}}{2a}$	**5.** Simplify the right side by reordering the terms in the numerator and finding the square root of the denominator.
$x = -\frac{b}{2a} \pm \frac{\sqrt{b^2 - 4ac}}{2a}$ $x = \frac{-b \pm \sqrt{b^2 - 4ac}}{2a}$	**6.** Solve by subtracting $\frac{b}{2a}$ from both sides and combining the numerators over the common denominator.

THE QUADRATIC FORMULA

If $ax^2 + bx + c = 0$, where $a \neq 0$, then $x = \frac{-b \pm \sqrt{b^2 - 4ac}}{2a}$.

EXAMPLE 1: Applying the Quadratic Formula

Use the quadratic formula to solve $x^2 + 5x - 6 = 0$.

Answer

$1x^2 + 5x - 6 = 0$

$a = 1, b = 5, \text{ and } c = -6$

1. Identify the values of a, b, and c from the standard form of the quadratic equation, $ax^2 + bx + c = 0$.

$$x = \frac{-5 \pm \sqrt{5^2 - 4(1)(-6)}}{2(1)}$$

2. Substitute these values into the quadratic formula, $x = \frac{-b \pm \sqrt{b^2 - 4ac}}{2a}$.

$$= \frac{-5 \pm \sqrt{25 - (-24)}}{2} = \frac{-5 \pm \sqrt{49}}{2} = \frac{-5 \pm 7}{2}$$

3. Simplify the radical.

$$x = \frac{-5 + 7}{2} = \frac{2}{2} = 1 \text{ or } x = \frac{-5 - 7}{2} = \frac{-12}{2} = -6$$

4. Simplify the 2 expressions separately.

The solutions are $x = -6, 1$.

SKILL CHECK **EXERCISES 7, 13**

The equation in Example 1 can also be solved by factoring or by completing the square. The advantage of using the quadratic formula is that it can be used to solve any quadratic equation.

KEYWORD SEARCH

quadratic formula calculator

SOLVING QUADRATIC EQUATIONS BY THE QUADRATIC FORMULA

1. Write the equation in standard form: $ax^2 + bx + c = 0$.
2. Identify the values of a, b, and c.
3. Substitute these values into the quadratic formula and simplify.

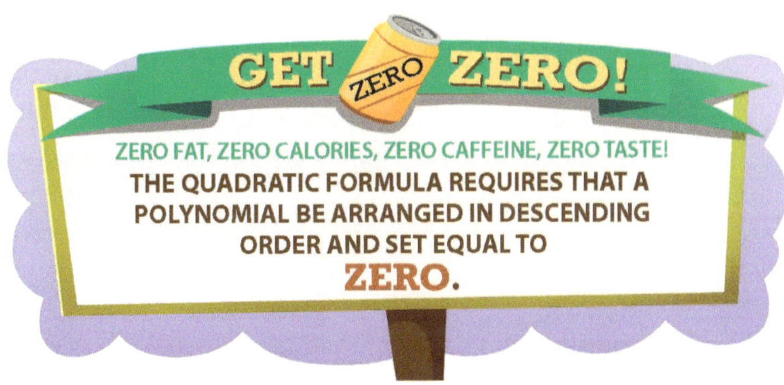

EXAMPLE 2: Applying the Quadratic Formula

Use the quadratic formula to solve $3y^2 - 4y = 8$.

Answer

$3y^2 - 4y - 8 = 0$

1. Write the equation in standard form.

$a = 3, b = -4, \text{ and } c = -8$

2. Identify the values of a, b, and c.

$$y = \frac{-(-4) \pm \sqrt{(-4)^2 - 4(3)(-8)}}{2(3)}$$

3. Substitute these values into the quadratic formula.

$$y = \frac{4 \pm \sqrt{16 + 96}}{6} = \frac{4 \pm \sqrt{112}}{6} = \frac{4 \pm 4\sqrt{7}}{6}$$

4. Simplify the radical.

$$y = \frac{\cancel{2}(2 \pm 2\sqrt{7})}{\cancel{2}(3)} = \frac{2 \pm 2\sqrt{7}}{3}$$

5. Simplify the expression.

SKILL CHECK **EXERCISES 19, 21**

EXAMPLE 3: Solving a Real-World Problem

A box (without a lid) is to be made from a sheet of cardboard 24 in. long and 18 in. wide by cutting squares from each corner and then folding the sides up. To the nearest tenth of an inch, what is the length of each side of the squares that should be cut out of each corner if the bottom of the box is to have an area of 265 in.^{2}?

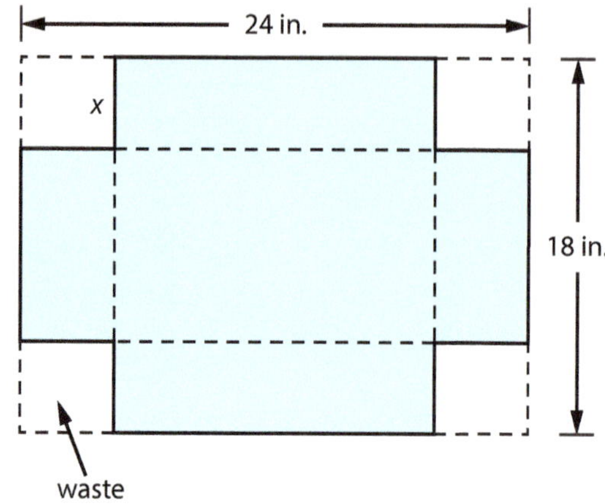

Answer

Let x = the length of each square to be cut out, $24 - 2x$ = the length of the bottom of the box, and $18 - 2x$ = the width of the bottom of the box.

1. Assign expressions to the unknowns.

$$265 = (24 - 2x)(18 - 2x)$$

2. Use the formula $A = lw$ to write an equation modeling the problem.

$$265 = 432 - 84x + 4x^2$$

$$0 = 4x^2 - 84x + 167$$

3. Write the equation in standard form.
   a. Use FOIL to multiply the right side.
   b. Subtract 265 from both sides.

$a = 4$, $b = -84$, and $c = 167$

4. Identify the values of a, b, and c.

$$x = \frac{-(-84) \pm \sqrt{(-84)^2 - 4(4)(167)}}{2(4)}$$

$$= \frac{84 \pm \sqrt{7056 - 2672}}{8} = \frac{84 \pm \sqrt{4384}}{8}$$

5. Substitute these values into the quadratic formula.

$$x = \frac{84 + \sqrt{4384}}{8} \approx 18.8 \text{ in.} \quad \text{or} \quad x = \frac{84 - \sqrt{4384}}{8} \approx 2.2 \text{ in.}$$

6. Calculate the decimal approximations of the possible solutions.

The length of each square is ≈2.2 in.

7. Since 18.8 in. is larger than the width of the cardboard, 2.2 in. is the only reasonable solution.

SKILL CHECK **EXERCISE 35**

AFTERSCHOOLHELP

The Quadratic Formula

Quadratic equations help you model real-world problems such as those described in Example 3. The quadratic formula can be applied to easily find the solutions to these equations, but it cannot interpret the solutions for you. The proper use of a mathematical model requires wisdom, especially in more complex models and in situations that involve moral judgements. Models help you understand and predict, but mathematics alone cannot tell you the right thing to do.

A. EXERCISES

State the values of *a*, *b*, and *c* when the equation is written in standard form. Do not solve.

1. $x^2 + 6x + 5 = 0$

2. $2x^2 - 6x + 1 = 0$

3. $x^2 + 5x = -6$

4. $3y^2 - 5 = 2y$

5. $w^2 - 6 = 0$

6. $z^2 = 7z$

Use the quadratic formula to solve each equation.

7. $x^2 + 6x + 8 = 0$

8. $x^2 + 4x + 3 = 0$

9. $x^2 + 8x + 15 = 0$

10. $x^2 - 9x + 20 = 0$

11. $a^2 - 9 = 0$

12. $b^2 - 9b = -8$

13. $2c^2 - c - 3 = 0$

14. $d^2 + 3d = 28$

15. $y^2 - 6y = -9$

16. $4x^2 - 12x + 9 = 0$

B. EXERCISES

Use the quadratic formula to solve each equation.

17. $2x^2 - 9x + 4 = 0$

18. $3x^2 = 2x + 5$

19. $2y^2 - 5y = -1$

20. $3w^2 + 3 = 7w$

21. $3x^2 = 7$

22. $4x^2 = 5x$

23. $2x^2 - 5x - 10 = 0$

24. $4y^2 - 3y = 8$

25. $7m^2 + 9m = -2$

26. $15x^2 + 22x + 8 = 0$

27. $-3k^2 + 2k + 10 = 0$

28. $4x^2 + 4x + 3 = 0$

29. $8x^2 = 26x + 7$

30. $5r^2 - 7 = 2r$

Use the quadratic formula and a calculator to solve each equation. Round decimal approximations to the nearest hundredth.

31. $3x^2 - 8x + 2 = 0$

32. $3y^2 - 2 = 3y$

33. Find 2 real numbers whose sum is 4 and whose product is 1.

34. MP2 Find 2 integers such that the larger is 2 less than eight times the smaller and the difference of their squares is 884.

35. A box is to be made from a sheet of cardboard 30 in. long by 24 in. wide by cutting squares from each corner and then folding the cardboard up. How tall will the box be if the bottom of the box is to have an area of 310 in.^{2}?

36. MP4 A rectangular piece of cardboard is made into a box by cutting 3 in. squares from each corner and then folding the cardboard up. The length of the cardboard box is 3 in. more than its width. If the volume ($V = lwh$) of the box is 540 in.^{3}, what are the dimensions of the original piece of cardboard?

37. **Essential Question:** How are mathematical models helpful and how are they limited?

C. EXERCISES

Use the quadratic formula and a calculator to solve. Round decimal approximations to the nearest hundredth.

38. $-2.5x^2 + 7.4x - 3.8 = 0$

39. $\frac{2}{3}x^2 - \frac{4}{5}x - \frac{1}{2} = 0$

40. Show how to solve $ry^2 + sy + t = 0$ for y by completing the square. Assume that $r \neq 0$.

MP4 A baseball is thrown from third to first base. The height of the ball is modeled by the equation $h = -0.0034x^2 + 0.41x + 5.5$, where h is the height in feet and x is the horizontal distance in feet.

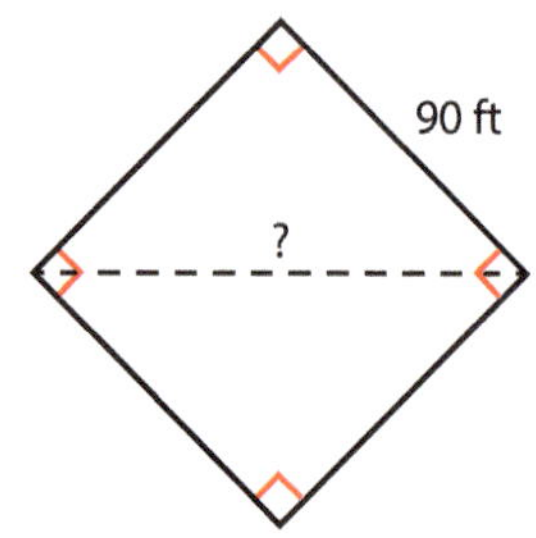

41. How far (to the nearest tenth of a foot) does the ball travel before it returns to a height of 10 ft?

42. Should the first baseman be able to catch the ball? Explain your reasoning.

43. Will the ball reach a height of 17 ft? If so, at what distance(s) from third base?

CUMULATIVE REVIEW

Classify each number as rational, irrational, or not real. [1.1, 10.1]

44. 12

45. $\sqrt{5}$

46. $\frac{\sqrt{169}}{2}$

47. $\frac{-17}{3}$

48. $\sqrt{-17}$

49. $\frac{\sqrt{160}}{4}$

Solve each equation. [10.7]

50. $14 - \sqrt{x + 7} = 9$

51. $6\sqrt{x} - 11 = 2\sqrt{x} + 1$

52. Which quadratic equation cannot be solved by factoring? **[11.1]**

A $x^2 - 4x + 5 = 0$

B $y^3 + 7y^2 = 0$

C $3a^2 - 10a - 8 = 0$

D $c^2 + 96 = -20c$

E All the above are factorable.

53. Which square root function is correctly paired with its domain? **[10.8]**

A $f(x) = \sqrt{x}; x \leq 0$

B $g(x) = \sqrt{x - 2}; x > 2$

C $h(x) = \sqrt{12 - 3x}; x \geq 4$

D $k(x) = 1 - \sqrt{x + 10}; x \leq -9$

E $m(x) = \sqrt{14 - 2x}; x \leq 7$

11.6 MORE QUADRATIC EQUATIONS

What does the discriminant of a quadratic equation indicate?

After completing this section, I will be able to

- describe the solutions of a quadratic equation by evaluating the discriminant.
- solve quadratic equations by using the most efficient method.
- model quadratics by writing equations in standard form using given solutions.

Jørn Utzon designed the "sails" on Australia's Sydney Opera House by using a pair of shells, each represented by a quadratic equation.

When solving a quadratic equation, choose the method that is most appropriate for the problem.

GENERAL METHOD OF SOLVING QUADRATIC EQUATIONS

Write the quadratic equation in the form $ax^2 + bx + c = 0$.

1. If $b = 0$, isolate x^2 and apply the Square Root Property.
2. If the polynomial factors easily, solve the equation by factoring.
3. Otherwise, use the quadratic formula: $x = \dfrac{-b \pm \sqrt{b^2 - 4ac}}{2a}$.

Quadratic equations are usually not solved by completing the square since one of the 3 methods listed above is generally easier. However, knowing how to complete the square is essential to the development of the quadratic formula. This method is also used in graphing quadratic functions and in other applications that will be studied in later courses.

When the quadratic formula is used, the value of the radicand determines the number and type of solutions to the equation. This value, $D = b^2 - 4ac$, is called the **discriminant**. The following table describes the number and type of real-number solutions to $ax^2 + bx + c = 0$ when a, b, and c are rational numbers.

DETERMINING THE NUMBER & TYPE OF SOLUTIONS

$D = b^2 - 4ac$	Solutions	Example
$D > 0$ and a perfect square	2 rational	$x = \frac{1 \pm \sqrt{25}}{4} = \frac{3}{2}, -1$
$D > 0$ but not a perfect square	2 irrational	$x = \frac{1 \pm \sqrt{13}}{4} \approx 1.15, -0.65$
$D = 0$	1 rational	$x = \frac{1 \pm \sqrt{0}}{4} = \frac{1}{4}$
$D < 0$	no real	$x = \frac{1 \pm \sqrt{-9}}{4}$; no real solution

EXAMPLE 1: Describing Solutions

Evaluate the discriminant; then describe the solution(s) of each equation.

a. $x^2 + 5x - 7 = 0$ **b.** $x^2 = 156 - x$ **c.** $3x^2 + 5x = -9$

Answers

a. $x^2 + 5x - 7 = 0$

$a = 1, b = 5$, and $c = -7$

1. Identify the values of a, b, and c when the equation is in standard form, $ax^2 + bx + c = 0$.

$D = b^2 - 4ac$

$= 5^2 - 4(1)(-7) = 25 + 28 = 53$

2. Evaluate the discriminant.

There are 2 irrational solutions.

3. $D > 0$ and is not a perfect square.

b. $x^2 = 156 - x$

$x^2 + x - 156 = 0$

1. Write the equation in standard form.

$D = b^2 - 4ac$

$= 1^2 - 4(1)(-156) = 1 + 624 = 625$

2. Evaluate the discriminant.

There are 2 rational solutions.

3. $D > 0$ and is a perfect square.

c. $3x^2 + 5x = -9$

$3x^2 + 5x + 9 = 0$

1. Write the equation in standard form.

$D = b^2 - 4ac$

$= 5^2 - 4(3)(9) = 25 - 108 = -83$

2. Evaluate the discriminant.

There are no real solutions.

3. $D < 0$

SKILL CHECK **EXERCISES 7, 11**

To find a quadratic equation that has stated values as its solutions, reverse the steps used to solve a quadratic equation by factoring.

EXAMPLE 2: Writing Quadratic Equations

Write a quadratic equation in standard form with the given solutions.

a. $x = 5, -7$ **b.** $x = \frac{2}{3}, -\frac{1}{4}$

Answers

a. $x = 5$ or $x = -7$

$x - 5 = 0$ $\quad x + 7 = 0$

1. Write each solution as a linear equation in the form $ax + b = 0$.

$(x - 5)(x + 7) = 0 \cdot 0$

2. Use the Multiplication Property of Equality to form a quadratic equation.

$x^2 + 7x - 5x - 35 = 0$

$x^2 + 2x - 35 = 0$

3. Use FOIL on the left side to produce an equation in standard form.

b. $x = \frac{2}{3}$ or $x = -\frac{1}{4}$

$3x = 3\left(\frac{2}{3}\right)$ $\quad 4x = 4\left(-\frac{1}{4}\right)$

$3x = 2$ $\quad 4x = -1$

$3x - 2 = 0$ $\quad 4x + 1 = 0$

$(3x - 2)(4x + 1) = 0$

$12x^2 + 3x - 8x - 2 = 0$

$12x^2 - 5x - 2 = 0$

1. Write each solution as a linear equation in the form $ax + b = 0$.
   a. Clear the fractions.
   b. Set each equation equal to 0.
2. Use the Multiplication Property of Equality to form a quadratic equation.
3. Use FOIL on the left side to produce an equation in standard form.

SKILL CHECK **EXERCISE 17**

AFTERSCHOOLHELP

More Quadratic Equations

EXAMPLE 3: Solving a Real-World Problem

A carpenter wants to fit 3 boards together to form a right triangle to use as a brace. The hypotenuse must be 3 ft longer than one leg and 5 ft longer than the other leg. Find the measurements needed for the sides of the triangle.

Answer

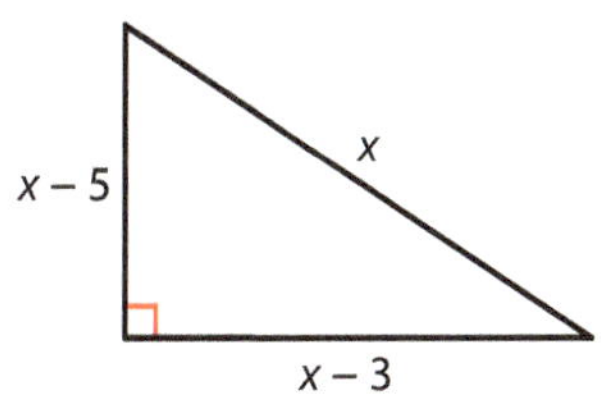

1. Sketch a right triangle. Let x = the hypotenuse, $x - 3$ = the length of one leg, and $x - 5$ = the length of the other leg.

$$a^2 + b^2 = c^2$$
$$(x - 5)^2 + (x - 3)^2 = x^2$$
$$x^2 - 10x + 25 + x^2 - 6x + 9 = x^2$$
$$x^2 - 16x + 34 = 0$$

2. Substitute the expressions into the Pythagorean Theorem.
3. Simplify and write the equation in standard form.

$$D = b^2 - 4ac$$
$$= (-16)^2 - 4(1)(34) = 120$$

4. Evaluate the discriminant. Since $D > 0$ and is not a perfect square, the solutions will be irrational.

$$x = \frac{-(-16) \pm \sqrt{120}}{2(1)} = \frac{16 \pm 2\sqrt{30}}{2} = 8 \pm \sqrt{30}$$

5. Use the quadratic formula to solve, substituting known values into the equation.

$x = 8 + \sqrt{30} \approx 13.48$ or $x = 8 - \sqrt{30} \approx 2.52$

hypotenuse ≈ 13.48 ft

$\text{leg}_1 = x - 5 \approx 13.48 - 5 = 8.48$ ft

$\text{leg}_2 = x - 3 \approx 13.48 - 3 = 10.48$ ft

6. Interpret the results. Using 2.52 ft would give negative lengths for both legs, so 13.48 ft is the only reasonable length for the hypotenuse.

SKILL CHECK **EXERCISE 35**

VOCABULARY

- discriminant

A. EXERCISES

Determine the easiest method for solving. Do not solve.

1. $a^2 - 2a - 8 = 0$
2. $x^2 - 3x - 5 = 0$
3. $x^2 - 7 = 0$
4. $x^2 + 3x = 0$
5. $7x^2 - 15x + 54 = 0$
6. $(x - 21)^2 = 4$

A factoring
B taking roots
C completing the square
D using the quadratic formula

Evaluate the discriminant; then choose the answer describing the solution(s) of each equation.

7. $x^2 + 20x + 51 = 0$
8. $x^2 + 11 = 0$
9. $x^2 - 6x + 9 = 0$
10. $-3x^2 + x = -8$
11. $x^2 + 2x = -9$
12. $1 + 10x = -25x^2$

A 2 rational
B 2 irrational
C 1 rational
D no real

Solve using the method of your choice and describe the solution(s).

13. $a^2 + 3a = -1$
14. $x^2 + 2x + 9 = 0$
15. $2x^2 - 7x + 2 = 0$
16. $4q^2 - 5q - 9 = 0$

B. EXERCISES

Write a quadratic equation in standard form with the given solutions.

17. $x = -8, 4$
18. $x = 0, 5$
19. $x = \frac{1}{2}, \frac{3}{4}$
20. $x = -\frac{1}{2}, \frac{2}{3}$

Solve each equation.

21. $6y^2 = 13y - 2$
22. $3a^2 - 5 = -2a$
23. $9a^2 + 24a = -16$
24. $4r^2 - 5 = r$
25. $x^2 + 8x - 2 = 0$
26. $-t^2 + 4 = 0$
27. $a^2 + 44a + 2 = 0$
28. $x^2 - 4x + 2 = 0$
29. $(3x + 4)(2x - 5) = 4$
30. $(x + 5)(2x - 7) = 9$
31. $(3x - 1)^2 + 6 = 0$
32. $x(5x - 9) + 6 = 0$
33. **Essential Question:** What does the value of the discriminant of a quadratic equation indicate?
34. What are the dimensions of a rectangle whose perimeter is 54 in. and whose area is 180 in.^{2}?
35. The hypotenuse of a right triangle measures 8 in. more than one leg and 3 in. more than the other leg. Find the lengths of the sides of the triangle (to the nearest tenth of an inch).
36. MP4 Sean, the grounds manager of Coutee County Country Club, is purchasing 360° rotating sprinklers with 3 ft more than twice the radius of his rotating sprinklers at home. The club's new sprinklers cover 484π ft^2 of ground. What is the radius Sean's home sprinkler sprays?

C. EXERCISES

Solve each equation. State irrational solutions as decimal approximations rounded to the nearest thousandth.

37. $(y + 4)^2 - y(y + 2) = 9$

38. $(x - 5)^2 + (x - 5) - 2 = 0$

39. $(x - 6)(x + 2) - (x - 6) = 9$

40. $(x - 1)^2 - (3x + 2)(x - 5) = 0$

MP1 **Write a quadratic equation in standard form with the given solutions.**

41. $x = \frac{-3 \pm \sqrt{5}}{3}$

42. $x = \frac{-2 \pm \sqrt{3}}{6}$

CUMULATIVE REVIEW

Describe the translation of the graph of the parent function $y = x^2$ that produces the graph of each function. [7.3]

43. $y = x^2 + 9$

44. $y = x^2 - 4$

45. $y = (x - 6)^2$

46. $y = (x + 7)^2$

Solve each equation. Discard any extraneous solutions. [10.7]

47. $\sqrt{5x + 2} = 4$

48. $-3\sqrt{3x + 1} = -21$

49. Mr. Sutton is taking the scouts camping and has a rope that is to reach from a stake in the ground to the top of a tent pole. If the pole is 8 ft high and the rope is 12 ft long, how far from the pole should he place the stake to make the rope taut? **[10.5]**

50. If one side of a square is increased by 3 in. and the adjacent side is decreased by 1 in., the rectangle formed has an area of 77 in.^{2} Find the length of a side of the square. **[11.3]**

51. Which function models exponential decay? **[7.5]**

- A $f(t) = 20(0.95)^t$
- B $g(t) = -20(0.9)^t$
- C $p(t) = -20(1.15)^t$
- D $y(t) = 20(-1.12)^t$
- E none of these

52. Factor $2x^3 - 5x^2 + 18x - 45$ completely. **[9.5]**

- A $(x + 3)^2(2x - 5)$
- B $(x^2 + 9)(2x - 5)$
- C $(x + 3)(x - 3)(2x - 5)$
- D $2(x + 3)(x - 3)(x - 5)$
- E cannot be factored further

11.7 QUADRATIC FUNCTIONS: $f(x) = ax^2 + c$

How does the value of a in $f(x) = ax^2 + c$ affect the graph of the function?

After completing this section, I will be able to

- graph quadratic functions of the form $f(x) = ax^2 + c$ by plotting coordinate points.
- describe the effect of a and c on the graph of a quadratic function.

Polynomial functions of the second degree are called *quadratic functions.*

DEFINITION

A **quadratic function** is a function that can be written in the form $f(x) = ax^2 + bx + c$, where a, b, and c are real numbers and $a \neq 0$. This form is called the *standard form* of a quadratic function.

In Section 7.3 you were introduced to the parabolic graph of the parent quadratic function $f(x) = x^2$. The graph of any quadratic function is also a parabola.

EXAMPLE 1: Comparing a Vertical Stretch with $f(x) = x^2$

Graph $f(x) = x^2$ and $g(x) = 3x^2$ and compare their graphs.

Answer

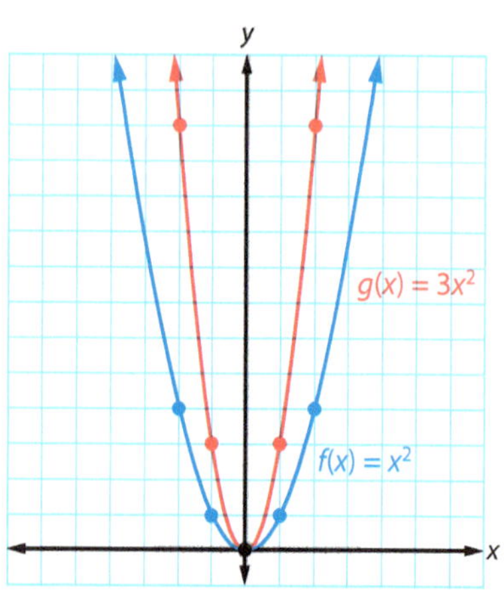

1. Make a table for each function, using at least 5 values for x.

x	$f(x) = x^2$	$g(x) = 3x^2$
0	0	0
1	1	3
−1	1	3
2	4	12
−2	4	12

2. For each function, plot the ordered pairs and draw the parabola.
3. Each point on $g(x)$ is 3 times higher than its corresponding point on $f(x)$, causing the graph to appear "steeper" or "narrower." This is called a vertical stretch.

SKILL CHECK **EXERCISE 13**

GRAPHING QUADRATIC FUNCTIONS

1. Find at least 5 ordered pairs that satisfy the function.
2. Plot the ordered pairs on a Cartesian coordinate plane.
3. Connect the ordered pairs with a smooth curve.

EXAMPLE 2: Comparing Graphs with $f(x) = x^2$

Graph $p(x) = -x^2$ and $q(x) = \frac{1}{2}x^2$ and compare their graphs with the graph of $f(x) = x^2$.

Answer

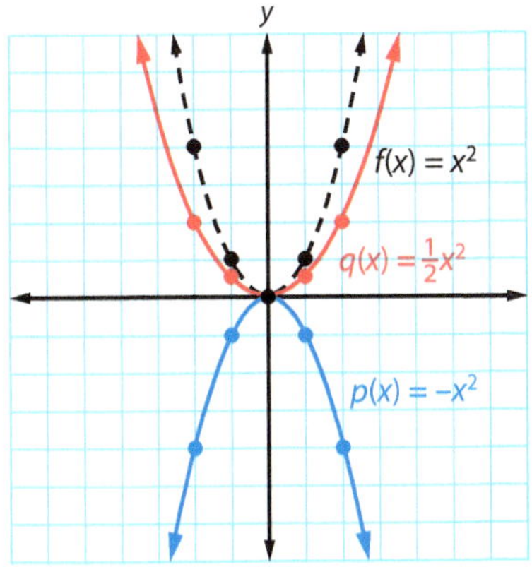

1. Make a table for $p(x)$ and $q(x)$.

x	$p(x) = -x^2$	$q(x) = \frac{1}{2}x^2$
0	0	0
1	−1	$\frac{1}{2}$
−1	−1	$\frac{1}{2}$
2	−4	2
−2	−4	2

2. For each function, plot the ordered pairs and draw the parabola.
3. **a.** Each point on $p(x)$ is located the same distance below the x-axis as its corresponding point on $f(x)$ is above the x-axis. Since $p(x)$ has the same shape as $f(x)$, it is said to be the reflection of $f(x)$ across the x-axis.
   b. Each point on $q(x)$ is half the height of its corresponding point on $f(x)$, causing the graph to appear "flatter" or "wider." This is called a vertical shrink.

SKILL CHECK **EXERCISES 21, 23**

Recall that the "turning point" of a parabola is called its *vertex*. Since the vertex is either the lowest or the highest point on the graph, it is called a **minimum** or a **maximum**. The vertices of all the parabolas in Examples 1–2 are at the origin, (0, 0). These parabolas are also symmetric across the y-axis. The vertex falls on this "fold line," called the **line of symmetry**.

THE EFFECT OF a ON THE GRAPH OF $y = ax^2 + bx + c$

If $a > 0$, the graph opens upward.

If $a < 0$, the graph opens downward.

If $|a| > 1$, the graph is a vertical stretch (steeper).

If $|a| < 1$, the graph is a vertical shrink (flatter).

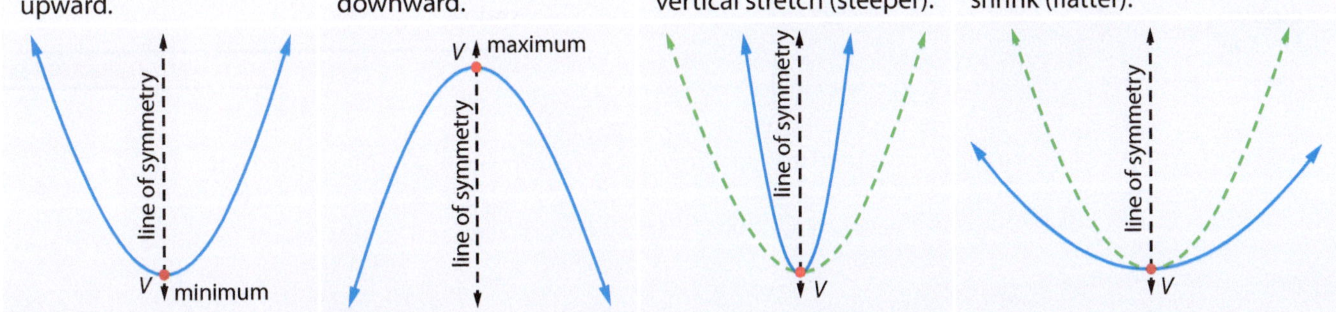

EXAMPLE 3: Graphing a Vertical Translation & Stretch

Graph $s(x) = 3x^2 - 4$ and compare its graph with the graph of $g(x) = 3x^2$.

Answer

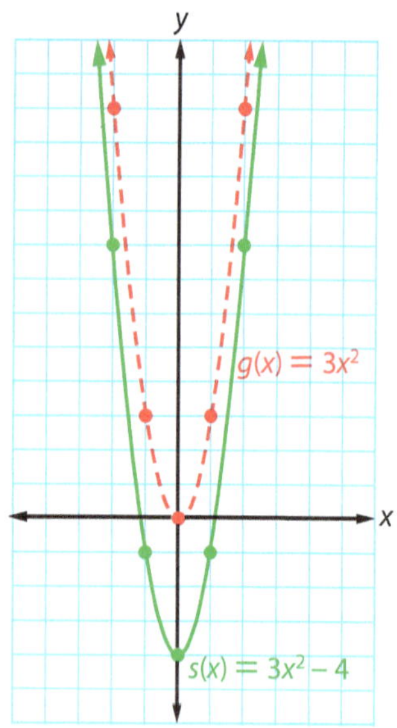

1. Make a table for each function. Note that each value for $s(x)$ is 4 less than the corresponding value for $g(x)$.

x	$g(x) = 3x^2$	$s(x) = 3x^2 - 4$
−2	12	8
−1	3	−1
0	0	−4
1	3	−1
2	12	8

2. Plot the ordered pairs and draw the parabola.
3. The graph of $s(x)$ is the result of translating the graph of $g(x)$ down 4 units.

SKILL CHECK **EXERCISE 19**

AFTERSCHOOLHELP

Quadratic Functions: $f(x) = ax^2 + c$

EXAMPLE 4: Graphing a Vertical Translation & Shrink

Graph $r(x) = -\frac{1}{2}x^2 + 2$ and compare its graph with the graph of $f(x) = -x^2$.

Answer

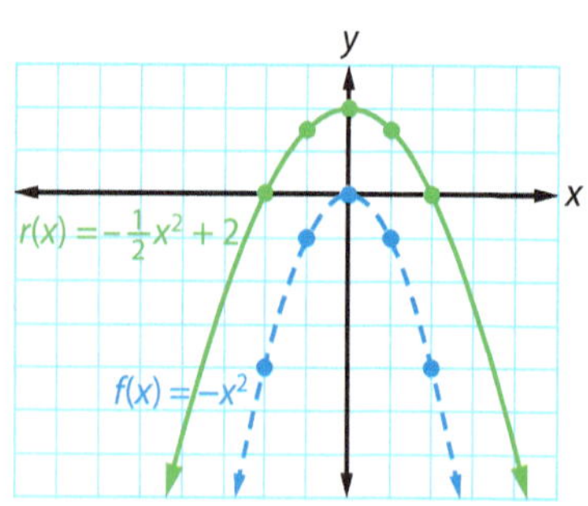

1. Make a table for each function.

x	$f(x) = -x^2$	$r(x) = -\frac{1}{2}x^2 + 2$
−2	−4	0
−1	−1	$\frac{3}{2}$
0	0	2
1	−1	$\frac{3}{2}$
2	−4	0

2. Plot the ordered pairs and draw the parabola.
3. The graph of $r(x)$ is the result of translating the graph of $f(x)$ up 2 units after a vertical shrink.

SKILL CHECK **EXERCISE 27**

VOCABULARY

- line of symmetry
- maximum
- minimum
- quadratic function

A. EXERCISES

For each function, find $f(-1)$, $f(0)$, $f(2)$, and $f(4)$.

1. $f(x) = x^2$
2. $f(x) = -x^2$
3. $f(x) = x^2 - 4$
4. $f(x) = x^2 + 5$
5. $f(x) = -2x^2$
6. $f(x) = 3x^2$
7. $f(x) = 3x^2 - 7$
8. $f(x) = -2x^2 + 8$

Complete the following table.

	Function	Direction of Opening	Vertex		Comparison of Shape to $f(x) = x^2$
			At	Minimum or Maximum	
9.	$f(x) = 4x^2$				
10.	$f(x) = -\frac{1}{3}x^2$				
11.	$f(x) = -x^2 + 4$				
12.	$f(x) = \frac{1}{5}x^2 - 7$				

Graph each quadratic function.

13. $f(x) = 2x^2$
14. $f(x) = x^2 + 2$
15. $f(x) = x^2 - 7$
16. $f(x) = x^2 + 3$
17. $f(x) = -x^2 + 2$
18. $f(x) = -x^2 + 5$
19. $f(x) = 2x^2 - 3$
20. $f(x) = 4x^2 - 5$

B. EXERCISES

Graph each quadratic function.

21. $f(x) = -3x^2$
22. $f(x) = -\frac{1}{3}x^2$
23. $f(x) = \frac{1}{3}x^2$
24. $f(x) = \frac{1}{4}x^2$
25. $f(x) = 2x^2 + 3$
26. $f(x) = -2x^2 + 7$
27. $f(x) = -\frac{1}{2}x^2 + 5$
28. $f(x) = 7x^2 - 10$
29. **Essential Question:** Describe how different values of a in $y = ax^2 + c$ transform the graph of $y = x^2 + c$.
30. **MP4** The path of a projectile recorded in the table below can be modeled by a quadratic function.

horizontal distance (ft), x	0	25	50	75	100				
approximate height (ft), y	1	263	450	563	600	563	450	263	1

   a. Plot the first 5 ordered pairs from the table.
   b. Use the fact that parabolas are symmetric and that the maximum height of the projectile is 600 ft to fill in the missing distances and to finish the graph.
   c. Approximate the horizontal distance traveled by the projectile before it hits the ground.

31. Where is the vertex of any quadratic function of the form $f(x) = ax^2$?
32. What is the x-intercept for graphs of the form $f(x) = ax^2$? What is the y-intercept?

33. MP8 Where is the line of symmetry for any function of the form $f(x) = ax^2 + c$? What is the equation for this line?

34. State the coordinates of the vertex of any quadratic function of the form $f(x) = ax^2 + c$.

35. The graph of $f(x) = \frac{1}{2}x^2 + 5$ is contained in which quadrants of the coordinate plane?

36. The graph of $f(x) = \frac{1}{2}x^2 - 5$ is contained in which quadrants of the coordinate plane?

C. EXERCISES

37. Given $f(x) = 0.2x^2 + 0.3$, do the following.

   a. Find $f(x)$ for $x = 0, \pm1, \pm2$, and ±3.

   b. Graph the function.

   c. State the coordinates of the vertex and whether it is a minimum or maximum point.

   d. State the equation of the line of symmetry.

38. Graph the quadratic function modeling the height, h, of an object dropped from 100 ft at time t (in seconds). *Hint*: Recall that the height of a dropped object is modeled by $h(t) = -16t^2 + h_i$.

Write a function of the form $f(x) = ax^2 + c$ based on the changes to $f(x) = x^2$ as described in each exercise.

39. translated vertically −5 units, a vertical stretch of a, and the graph intersects (1, −2)

40. a vertical shrink of $\frac{1}{3}$, reflected across the x-axis, translated c units vertically, and the graph intersects (6, −5)

41. MP7 the graph intersects both (−4, −9) and (2, 9)

CUMULATIVE REVIEW

State the coordinates of the vertex of each function. [7.3]

42. $f(x) = (x + 6)^2 + 1$

43. $f(x) = (x - 5)^2 - 2$

44. $f(x) = (x - 10)^2$

45. $f(x) = (x + 8)^2 - 15$

Simplify each expression. [10.1, 10.3–10.4]

46. $\frac{15}{\sqrt{3}}$

47. $52^{-\frac{1}{2}}$

48. $2\sqrt{3} + 5\sqrt{12} + 7\sqrt{2} + \sqrt{27}$

49. $-4\sqrt{12} + \sqrt{48} + \sqrt{75}$

50. What steps could be taken to solve $4x^2 - 10x = 5$ by completing the square? **[11.4]**

   A divide both sides by 4; add $\left(-\frac{5}{2}\right)^2$ to both sides

   B divide both sides by 4; add $\left(\frac{5}{4}\right)^2$ to both sides

   C divide both sides by 5; add $(-1)^2$ to both sides

   D add $(-5)^2$ to both sides

   E none of these

51. Which discriminant indicates that a quadratic equation has 2 irrational solutions? **[11.6]**

   A 200

   B −80

   C −169

   D 225

   E none of these

11.8 QUADRATIC FUNCTIONS: $f(x) = a(x - h)^2 + k$

Why is a mathematical model not always the sole consideration in decision-making?

After completing this section, I will be able to

- graph quadratic functions of the form $f(x) = a(x - h)^2 + k$.
- find the vertex of the quadratic function $f(x) = ax^2 + bx + c$.
- graph quadratic functions in standard form.
- defend the claim that mathematical models are useful but limited.

Previously we observed the effects that a and c have on a parabola in the function $f(x) = ax^2 + c$. The effect of adding a middle term, bx, is more easily seen when the equation $f(x) = ax^2 + bx + c$ is written in **vertex form**, $f(x) = a(x - h)^2 + k$.

Writing a quadratic function in vertex form allows you to quickly determine its vertex (h, k) and axis of symmetry. Since the parabola's axis of symmetry is a vertical line through the vertex, its equation is $x = h$.

When $a = 1$, the parabola can be easily graphed by translating the parent function, $f(x) = x^2$, h units horizontally and k units vertically.

EXAMPLE 1: Graphing Quadratics: $f(x) = (x - h)^2 + k$

Graph $p(x) = (x - 2)^2 - 1$.

Answer

$f(x) = (x - h)^2 + k$

$p(x) = (x - 2)^2 + (-1)$

vertex $= (h, k) = (2, -1)$

line of symmetry: $x = 2$

1. Find the vertex and the equation of the parabola's line of symmetry.

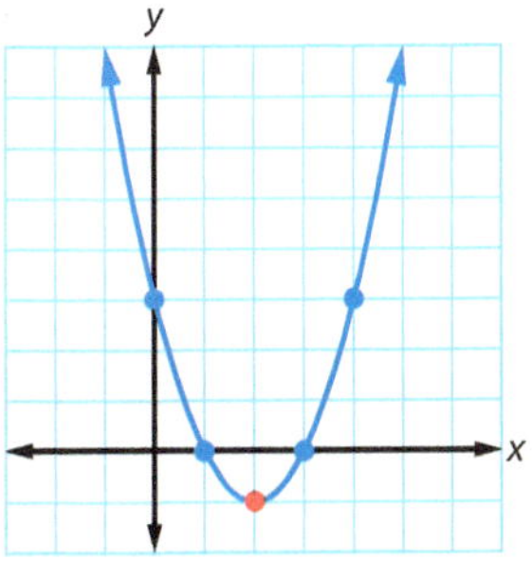

2. Graph the function as a translation of $f(x) = x^2$ with the vertex at $(2, -1)$.

SKILL CHECK **EXERCISE 11**

When $a \neq 1$, making a table of at least 5 values for x becomes necessary to graph the parabola.

EXAMPLE 2: Graphing Quadratics: $f(x) = a(x - h)^2 + k$

Graph $q(x) = -2(x - 3)^2 + 5$.

Answer

$f(x) = a(x - h)^2 + k$

$h = 3$; $k = 5$

vertex: (3, 5)

line of symmetry: $x = 3$

1. Compare the given function with the vertex form to find the vertex and the equation of the parabola's line of symmetry.

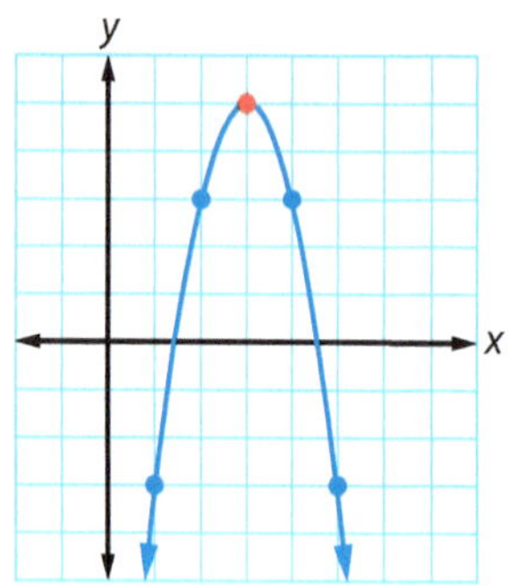

2. Make a table of ordered pairs, including the vertex and several points on either side of it.

x	$-2(x - 3)^2 + 5$	$q(x)$	
1	$-2(1 - 3)^2 + 5 = -2(-2)^2 + 5 = -8 + 5$	−3	
2	$-2(2 - 3)^2 + 5 = -2(-1)^2 + 5 = -2 + 5$	3	
3	$-2(3 - 3)^2 + 5 = -2(0)^2 + 5 = 0 + 5$	5	(vertex)
4	$-2(4 - 3)^2 + 5 = -2(1)^2 + 5 = -2 + 5$	3	
5	$-2(5 - 3)^2 + 5 = -2(2)^2 + 5 = -8 + 5$	−3	

3. Plot the points and draw the parabola.

SKILL CHECK **EXERCISE 19**

AfterSchoolHelp

Quadratic Functions: $f(x) = a(x - h)^2 + k$

The vertex form of a quadratic function conveys much information about the graph of a quadratic function.

Function	$f(x) = a(x - h)^2 + k$	Vertex	Direction of Opening		Comparison of Shape to $f(x) = x^2$	
$f(x) = -x^2$	$f(x) = -1(x - 0)^2 + 0$	(0, 0)	down	$(a < 0)$	same	$(\lvert a \rvert = 1)$
$f(x) = 3x^2 - 4$	$f(x) = 3(x - 0)^2 - 4$	(0, −4)	up	$(a > 0)$	vertical stretch	$(\lvert a \rvert > 1)$
$f(x) = (x + 1)^2$	$f(x) = 1[x - (-1)]^2 + 0$	(−1, 0)	up	$(a > 0)$	same	$(\lvert a \rvert = 1)$
$f(x) = (x - 2)^2 - 1$	$f(x) = 1(x - 2)^2 - 1$	(2, −1)	up	$(a > 0)$	same	$(\lvert a \rvert = 1)$
$f(x) = -\frac{1}{3}(x + 7)^2 + 9$	$f(x) = -\frac{1}{3}[x - (-7)]^2 + 9$	(−7, 9)	down	$(a < 0)$	vertical shrink	$(\lvert a \rvert < 1)$

A quadratic function in standard form, $f(x) = ax^2 + bx + c$, can be changed to the equivalent vertex form, $f(x) = a(x - h)^2 + k$, using a variation of the procedure outlined in Section 11.3 for completing the square.

EXAMPLE 3: Converting to Vertex Form

Write $f(x) = x^2 - 6x + 10$ in vertex form. Identify the vertex and direction of opening.

Answer

$f(x) = (x^2 - 6x) + 10$

1. Group the $ax^2 + bx$ terms.

$f(x) = (x^2 - 6x + 9) + 10 - 9$

2. Complete the square by adding $\left(\frac{6}{2}\right)^2 = 9$ inside the parentheses. Subtract 9 outside the parentheses so the value of the right side of the equation will not change.

$f(x) = (x - 3)^2 + 1$

3. Factor the trinomial inside the parentheses and simplify the constants outside the parentheses.

$a = 1, h = 3, k = 1$

vertex: (3, 1)

The graph opens upward and has the same shape as $y = x^2$.

4. Identify a, h, and k and use them to describe the graph.

a. The vertex is (h, k).

b. Since $a = 1$, the graph opens upward.

SKILL CHECK **EXERCISE 9**

If the function has a leading coefficient other than 1, completing the square is more complicated. By completing the square on $f(x) = ax^2 + bx + c$, we can develop formulas for h and k.

$f(x) = (ax^2 + bx) + c$

$f(x) = a\left(x^2 + \frac{b}{a}x\right) + c$

1. Group the $ax^2 + bx$ terms and then factor out the leading coefficient, a.

$f(x) = a\left(x^2 + \frac{b}{a}x + \frac{b^2}{4a^2}\right) + c - \frac{b^2}{4a}$

2. Complete the square by adding $\left(\frac{1}{2} \cdot \frac{b}{a}\right)^2 = \frac{b^2}{4a^2}$ inside the parentheses. Because of the factor of a outside the parentheses, the value added to the equation is $a\left(\frac{b^2}{4a^2}\right) = \frac{b^2}{4a}$. Subtract this value outside the parentheses.

$f(x) = a\left(x + \frac{b}{2a}\right)^2 + \left(c - \frac{b^2}{4a}\right)$

3. Factor the trinomial.

$h = -\frac{b}{2a}$ and $k = c - \frac{b^2}{4a}$

4. Comparing the resulting function to $f(x) = a(x - h)^2 + k$ provides the general expressions for h and k.

We will use this formula for h, but it is typically easier to find the value for k by substituting the value of h into $f(x)$.

EXAMPLE 4: Using the Line of Symmetry

State the vertex of $f(x) = -2x^2 - 12x - 19$ and then graph the function.

Answer

$f(x) = -2x^2 - 12x - 19$

$h = -\frac{-12}{2(-2)} = -\frac{12}{4} = -3$

$k = f(-3) = -2(-3)^2 - 12(-3) - 19$

$= -18 + 36 - 19 = -1$

vertex: (−3, −1)

1. Find the vertex.

Use $h = -\frac{b}{2a}$.

Use $k = f(h)$.

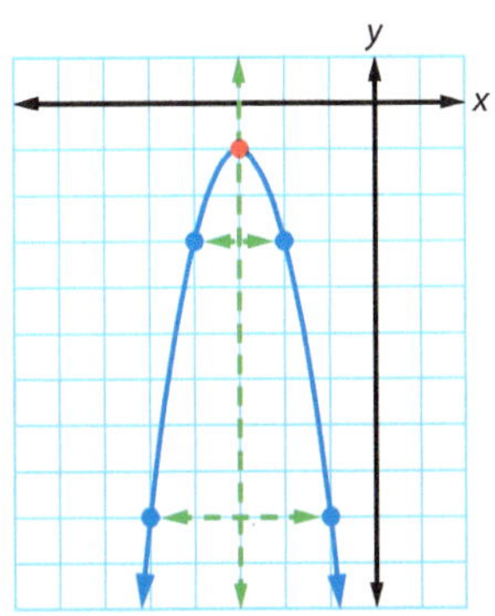

2. Note that $a = -2$, which implies that the graph will open downward and will be stretched vertically by a factor of 2.

3. Make a table of ordered pairs, including 2 points to the right of the vertex.

x	$-2x^2 - 12x - 19$	$f(x)$
−3	vertex	−1
−2	$-2(-2)^2 - 12(-2) - 19 = -8 + 24 - 19$	−3
−1	$-2(-1)^2 - 12(-1) - 19 = -2 + 12 - 19$	−9

4. Graph the 3 points and use the line of symmetry, $x = -3$, to plot the points to the left of the vertex. Then draw the parabola.

SKILL CHECK **EXERCISE 29**

EXAMPLE 5: Solving a Real-World Problem

Sam has 150 ft of fencing to fence off a portion of his backyard for his dog. What dimensions of the fenced portion will maximize the area of the dog pen?

Answer

Let w = the width.

$P = 2l + 2w$

$150 = 2l + 2w$

$2l = 150 - 2w$

length $= 75 - w$

1. Write expressions for the length and width.

$Area = w \cdot l$

$f(w) = w(75 - w)$

$f(w) = -w^2 + 75w$

2. Write the function rule that models the area of the dog pen.

$h = -\frac{b}{2a} = -\frac{75}{2(-1)} = 37.5$

$k = f(37.5) = -(37.5)^2 + 75(37.5) = 1406.25$

The maximum point is (37.5, 1406.25).

3. Find the maximum point on the parabola.

length $= 75 - w = 75 - 37.5 = 37.5$

4. Find the length of the dog pen.

The dimensions necessary to maximize the area of the dog pen are 37.5 ft × 37.5 ft.

SKILL CHECK **EXERCISE 35**

Notice in Example 5 the model provided the exact dimensions necessary to achieve the maximum area. However, Sam might prefer the dog pen to be 25 ft × 50 ft or 15 ft × 60 ft to allow more length for his dog to run. Or maybe a different shape would be necessary based on the contour of the yard. While the model provided the correct dimensions to achieve maximum area, the final decision may require considerations beyond just mathematics.

VOCABULARY

- vertex form

A. EXERCISES

Complete the following table.

	Function	Vertex (*h*, *k*)	Axis of Symmetry	Value of *a*	Direction of Opening	Comparison of Shape to $f(x) = x^2$
1.	$f(x) = (x - 2)^2 + 9$					
2.	$f(x) = -(x + 5)^2 + 1$					
3.	$f(x) = 3(x - 2)^2 - 4$					
4.	$f(x) = \frac{1}{4}x^2 - 6$					
5.	$f(x) = -5(x + 1)^2$					
6.	$f(x) = \frac{2}{3}(x - 8)^2$					
7.	$f(x) = -\frac{7}{4}x^2 + 7$					
8.	$f(x) = -6x^2$					
9.	$f(x) = x^2 - 8x + 9$					
10.	$f(x) = x^2 - 10x - 7$					

Graph each quadratic function.

11. $f(x) = (x - 4)^2 + 2$
12. $f(x) = (x - 2)^2$
13. $f(x) = (x - 1)^2 - 3$
14. $f(x) = -2(x - 2)^2 + 5$
15. $f(x) = (x + 3)^2$
16. $f(x) = (x + 1)^2 - 2$
17. $f(x) = -(x + 4)^2 + 1$
18. $f(x) = (x - 3)^2 - 8$
19. $f(x) = 3(x + 1)^2 + 4$
20. $f(x) = -\frac{1}{2}(x - 2)^2 + 3$

B. EXERCISES

Complete the square to write each quadratic function in vertex form. State the vertex, but do not graph.

21. $f(x) = x^2 + 4x - 12$
22. $f(x) = x^2 - 6x + 8$
23. $f(x) = x^2 + 14x + 40$
24. $f(x) = x^2 - 8x + 15$

For each quadratic function, state the vertex and then graph the function.

25. $f(x) = x^2 + 10x + 24$
26. $f(x) = x^2 - 4x + 5$
27. $f(x) = -x^2 - 8x - 12$
28. $f(x) = -2x^2 + 8x - 7$
29. $f(x) = 3x^2 - 18x + 20$
30. $f(x) = 5x^2 - 10x - 1$

31. **MP2** The profit resulting from manufacturing and selling a product is represented by the function $P(x) = -\frac{1}{4}(x - 200)^2 + 750$, where x is the number of items manufactured.
    - **a.** How much income is gained (or lost) by manufacturing 100 items? 250 items?
    - **b.** How many items should be produced for maximum profit?
    - **c.** What is the maximum profit?

32. The cost equation used by the May Kerr Manufacturing Co. is $C(h) = 2h^2 - 8h + 12$, where h is the number of hours it takes to produce a particular product and $C(h)$ is the cost of production in hundreds of dollars. What is the minimum cost possible?

33. The height of a punted football can be modeled by the function $f(t) = -16t^2 + 66t + 6$, where $f(t)$ is the height in feet and t is the time in seconds. Find the maximum height of the football and the length of time it would take to reach that height. Do not round.

34. **MP4** Julie wants to place 44 ft of picket fencing around a garden to create the largest rectangular area possible.
    - **a.** If x represents the width, write an algebraic expression that represents the length.
    - **b.** Write a function rule, $f(x)$, that models the area of the garden.
    - **c.** What is the maximum point on the parabola?
    - **d.** Find the dimensions of the garden with the maximum area.
    - **e.** What is the maximum area?

35. Pastor Wilbanks wants to fence in a rectangular play area next to the church, using fencing for the remaining 3 sides. The church has 90 ft of fencing budgeted for the project.
    - **a.** If x represents the width, write an algebraic expression that represents the length.
    - **b.** Write a function rule, $f(x)$, that models the area of the play area.
    - **c.** What is the maximum point on the parabola?
    - **d.** Find the dimensions of the maximum area.
    - **e.** What is the maximum area?

36. **Essential Question:** Why is a mathematical model not always the sole consideration in decision-making?

C. EXERCISES

For each quadratic function, state the vertex and then graph the function.

37. $f(x) = \frac{1}{4}x^2 - x - 2$

38. $f(x) = -\frac{1}{2}x^2 - x - 2$

39. **MP4** The altitude of a volcanic bomb (a mass of molten rock spewed from a volcano) can be modeled by the function $h(t) = -4.9t^2 + 80t + 1000$, where $h(t)$ is the altitude of the bomb and t is the time in seconds. Find the maximum height of the projectile (to the nearest tenth of a foot). How long after the eruption (to the nearest second) does the bomb reach its maximum height?

40. **MP7** Determine algebraically the equation of the parabola in vertex form with an axis of symmetry of $x = -5$ that intersects both $(-3, 4)$ and $(-1, 25)$.

CUMULATIVE REVIEW

Solve each system of equations. [6.2]

41. $4x + 3y = 8$
$2x - y = -1$

42. Find 2 numbers such that one is 32 more than twice the other and their difference is 103.

Simplify each expression. [8.3, 8.6]

43. $(x - 4)(x^4 + 4x^3 + 16x^2 + 64x + 256)$

44. $(x^2 + 5x - 7) \div (x + 1)$

Solve each equation. [11.1, 11.5]

45. $x^2 + 8x - 105 = 0$

46. $x^2 = 7x + 144$

47. $2x^2 - 3x + 1 = 0$

48. $18x^3 + 33x^2 - 105x = 0$

49. What is the simplified form of the expression $\frac{5}{6 + \sqrt{11}}$? **[10.6]**

A $\frac{6 - \sqrt{11}}{5}$

B $30 + 5\sqrt{11}$

C $30 - 5\sqrt{11}$

D $\frac{30 + 5\sqrt{11}}{47}$

E none of these

50. Which graph is a vertical stretch of $y = x^2$ with its vertex as a maximum? **[11.7]**

A $f(x) = -2x^2 - 4$

B $f(x) = -\frac{1}{3}x^2 + 5$

C $f(x) = \frac{3}{8}x^2 + 1$

D $f(x) = 5x^2 - 8$

E none of these

11.9 ZEROS OF A QUADRATIC FUNCTION

How can I find the zeros of a function?

After completing this section, I will be able to

- find the y-intercept of a quadratic function.
- find the zeros of a quadratic function.
- graph a quadratic function using its intercepts and vertex.

Some companies use quadratic modeling to determine production goals for maximum profit.

The consistency displayed by many physical phenomena, such as the motion of falling and projected objects, results from God's providential preservation of His creation (Col. 1:16–17; Heb. 1:2–3). In His goodness, He has allowed us to use mathematics to model His creation and to use these models to help those around us. We can see the consistency in His handiwork and marvel at His greatness and faithfulness as demonstrated in His creation (Ps. 19:1). No wonder Paul could write that "the invisible things of him from the creation of the world are clearly seen" (Rom. 1:20).

You have seen how to determine the vertex of a parabola and whether it is a minimum or maximum point. It is also helpful to know where a graph crosses the x- and y-axes. Since the y-intercept always has an x-value of 0, the y-intercept can be found by letting $x = 0$ and solving for $y = f(0)$.

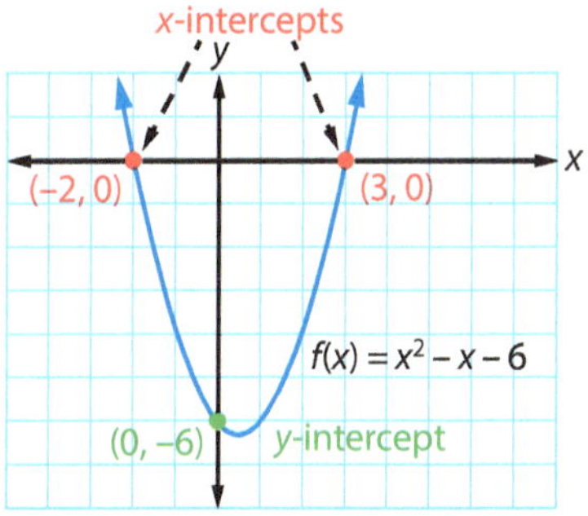

EXAMPLE 1: Finding the y-Intercept

Find the y-intercept of each function algebraically.

a. $g(x) = 2x^2 - 3x + 5$ **b.** $h(x) = (x - 1)^2 + 2$

Answers

a. $g(0) = 2(0)^2 - 3(0) + 5 = 5$

(0, 5)

b. $h(0) = (0 - 1)^2 + 2 = (-1)^2 + 2 = 3$

(0, 3)

1. Evaluate each function when $x = 0$.
2. State the y-intercept.

SKILL CHECK **EXERCISES 1, 7**

The x-intercepts of a graph always have y-values of 0. Therefore, the x-intercepts can be found by letting $y = 0$ or $f(x) = 0$ and solving for x. The x-intercepts of $f(x) = x^2 - x - 6$ are found by setting the equation equal to 0 and solving the resulting quadratic equation, $x^2 - x - 6 = 0$. Since $f(x) = 0$ for these values of x, they are called the *zeros* or *roots* of the function.

DEFINITION

A **zero** (or **root**) **of a function** is a value of x at which the graph of the function crosses the x-axis. It is the x-coordinate of an x-intercept of the graph.

EXAMPLE 2: Finding the Zero(s) of a Quadratic Function

Find the zero(s) of each function algebraically.

a. $r(x) = x^2 - 4x - 5$ **b.** $s(x) = -x^2 + 4x - 4$

Answers

a.	b.	
$x^2 - 4x - 5 = 0$	$-x^2 + 4x - 4 = 0$	**1.** Set each function equal to 0.
$(x - 5)(x + 1) = 0$ $x - 5 = 0$ or $x + 1 = 0$ $x = 5$ $\quad x = -1$	$-1(x^2 - 4x + 4) = 0$ $-1(x - 2)(x - 2) = 0$ $x - 2 = 0$ or $x - 2 = 0$ $x = 2$ $\quad x = 2$	**2.** Solve by factoring.
$x = -1, 5$	$x = 2$	**3.** State the zero(s).

SKILL CHECK **EXERCISE 17**

The zeros of the functions in Example 2 are easily confirmed by their graphs. The single root in the second parabola is often called a **double root** since 2 factors produce the same root. In this case, the parabola's vertex is located at that x-intercept.

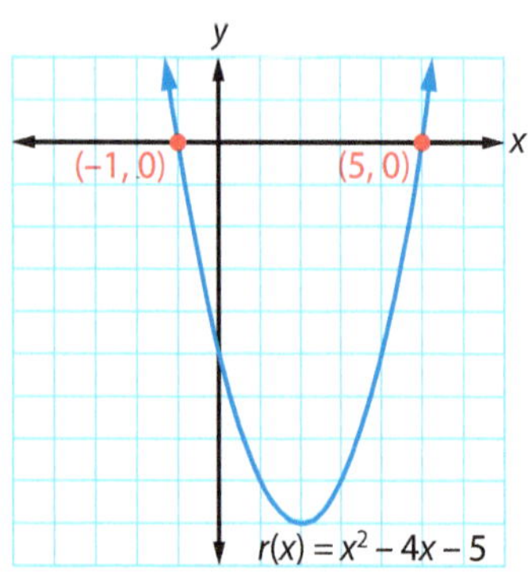

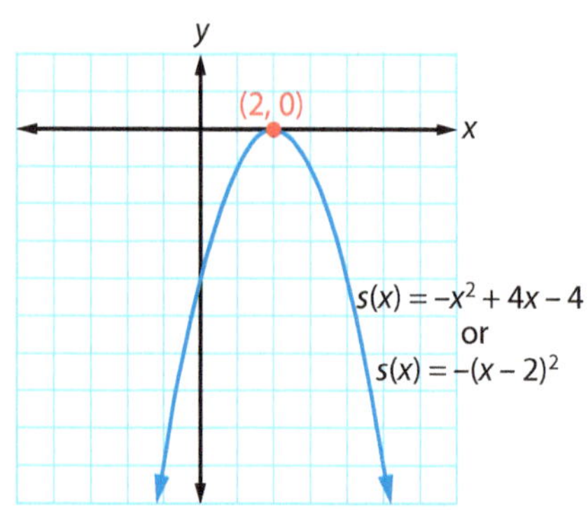

EXAMPLE 3: Finding the Zero(s) of a Quadratic Function

Find the zero(s) of each function (to the nearest hundredth).

a. $p(x) = (x - 1)^2 - 6$ **b.** $q(x) = x^2 - x + 4$

Answers

a. $p(x) = (x - 1)^2 - 6$

$(x - 1)^2 - 6 = 0$

1. Set the function equal to 0.

$(x - 1)^2 = 6$

$x - 1 = \pm\sqrt{6}$

$x = 1 \pm \sqrt{6}$

2. Solve by isolating the squared expression and applying the Square Root Property.

$x \approx 3.45, -1.45$

3. Use a calculator to find the decimal approximations.

b. $q(x) = x^2 - x + 4$

$x^2 - x + 4 = 0$

1. Set the function equal to 0.

$x = \dfrac{-(-1) \pm \sqrt{(-1)^2 - 4(1)(4)}}{2(1)} = \dfrac{1 \pm \sqrt{-15}}{2}$

2. Use the quadratic formula to solve.

There are no real zeros.

3. The discriminant is negative.

SKILL CHECK **EXERCISES 15, 19**

The graph of $p(x) = (x - 1)^2 - 6$ can be used to approximate its roots at -1.5 and 3.5, but an algebraic solution is needed to get exact values. The fact that the graph of $q(x) = x^2 - x + 4$ does not cross the x-axis illustrates why there are no real solutions when the discriminant $(D = b^2 - 4ac)$ is negative.

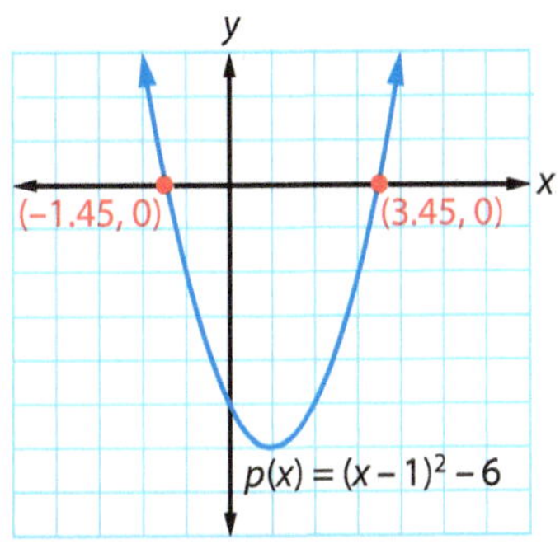

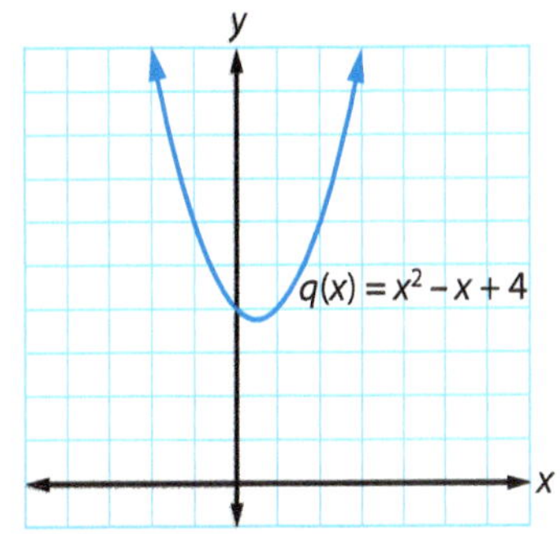

EXAMPLE 4: Using Intercepts to Graph a Quadratic Function

Given the quadratic function $f(x) = 3x^2 + 8x - 3$, find its y-intercept, zeros, and vertex; then graph the function.

Answer

1. Find the y-intercept.

a. Find $f(0)$.

b. State the y-intercept.

$f(x) = 3x^2 + 8x - 3$

$f(0) = 3(0)^2 + 8(0) - 3 = -3$

y-intercept: $(0, -3)$

2. Find the zeros.

a. Set the function equal to 0.

b. Solve by factoring.

$$3x^2 + 8x - 3 = 0$$

$$(3x - 1)(x + 3) = 0$$

$3x - 1 = 0$ or $x + 3 = 0$

$3x = 1$ $\quad x = -3$

$x = \frac{1}{3}$

3. Find the vertex.

a. The x-coordinate of the vertex will be halfway between the 2 zeros.

b. Use $k = f(h)$.

c. State the vertex.

$a = 3$, $b = 8$, and $c = -3$

$$h = \frac{-3 + \frac{1}{3}}{2} = \frac{-\frac{9}{3} + \frac{1}{3}}{2} = -\frac{4}{3} \text{ or } -1\tfrac{1}{3}$$

$$k = f\left(-\tfrac{4}{3}\right) = 3\left(-\tfrac{4}{3}\right)^2 + 8\left(-\tfrac{4}{3}\right) - 3$$

$$= \frac{16}{3} - \frac{32}{3} - \frac{9}{3} = -\frac{25}{3} \text{ or } -8\tfrac{1}{3}$$

$\left(-1\tfrac{1}{3}, -8\tfrac{1}{3}\right)$

4. Graph.

a. Plot the vertex, noting that it is a minimum since $a > 0$.

b. Plot the zeros.

c. Plot the y-intercept and use symmetry to plot a fifth point. Note that the y-intercept is $1\frac{1}{3}$ units to the right of the line of symmetry.

d. Note that the parabola is a vertical stretch compared to the parent function, $y = x^2$, as predicted by $|a| > 1$.

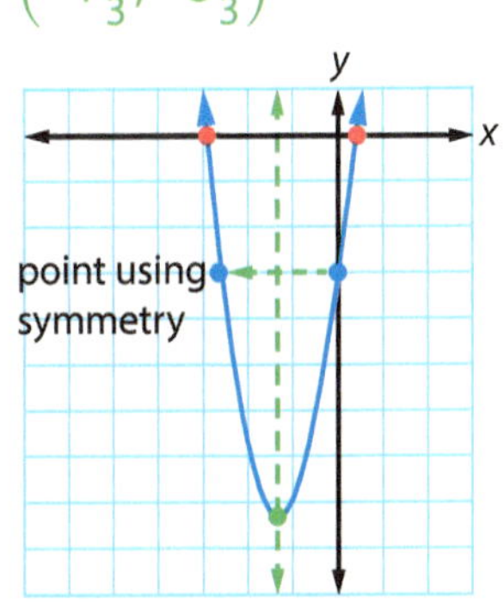

SKILL CHECK **EXERCISE 29**

AFTERSCHOOLHELP

Zeros of a Quadratic Function

VOCABULARY

- double root
- root
- zero

A. EXERCISES

Find the *y*-intercept of each function algebraically.

1. $f(x) = \frac{1}{4}x^2$
2. $f(x) = x^2 + 8$
3. $f(x) = 3x^2 - 5$
4. $f(x) = 4x^2 - 5x$
5. $f(x) = 3x^2 - 2x + 1$
6. $f(x) = \frac{1}{2}x^2 - 2x - 3$
7. $f(x) = (x - 2)^2 + 4$
8. $f(x) = (x + 3)^2 - 5$
9. $f(x) = 2(x - 1)^2 - 7$
10. $f(x) = \frac{1}{2}(x - 4)^2 - 11$

Find the zeros of each function algebraically.

11. $f(x) = x^2 - 2x - 8$
12. $f(x) = x^2 + 6x + 8$
13. $f(x) = (x - 3)^2$
14. $f(x) = -x^2 - 6x - 9$
15. $f(x) = -2(x + 3)^2 - 2$
16. $f(x) = x^2 + 3$
17. $f(x) = 2(x + 1)^2 - 8$
18. $f(x) = (x - 1)^2 - 8$
19. $f(x) = x^2 + 6x - 8$
20. $f(x) = 2x^2 - 5x - 8$

B. EXERCISES

21. What is the *y*-intercept of any function in the form $f(x) = ax^2 + bx + c$?
22. **MP7** Which quadrants contain the graph of any function $f(x) = ax^2 + bx + c$ if $a < 0$ and the discriminant $D < 0$?
23. Which line contains the vertex of any quadratic function with 1 real root?
24. How many zeros are there for any quadratic function whose discriminant is 0?
25. **Essential Question:** How can I find the *x*-intercepts of a function?

For each quadratic function, state the *y*-intercept, zero(s), and vertex; then graph the function.

26. $f(x) = 4x^2$
27. $f(x) = x^2 - 4$
28. $f(x) = x^2 + 3x$
29. $f(x) = x^2 - 6x + 8$
30. $f(x) = x^2 - 5x + 6$
31. $f(x) = (x + 2)^2 - 3$
32. $f(x) = 4x^2 - 12x + 9$
33. $f(x) = x^2 - 4x - 4$
34. $f(x) = -x^2 - 2x - 3$
35. $f(x) = \frac{1}{4}(x - 5)^2 + 3$
36. **MP4** The accountants at Counter Active Appliances model their profits from a particular model of refrigerator with the function $P(x) = -x^2 + 400x - 6000$, where x is the number of refrigerators produced weekly.
    - **a.** How many refrigerators should the company produce weekly to make the maximum profit?
    - **b.** What is the maximum profit?
    - **c.** Use a compound inequality to state the number of refrigerators that need to be produced weekly for the company to make a profit.

37. Dew Process wants to produce water filters for commercial kitchen faucets at a minimum cost to the company. The cost of each filter is modeled by the function $C(x) = x^2 - 140x + 5700$, where x is the number of filters produced.

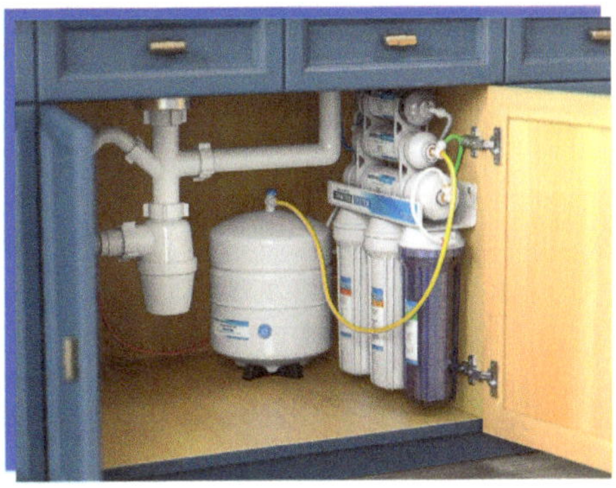

   a. How many filters should the company produce to keep the cost to a minimum?
   b. What is the minimum cost?
   c. When would the company have a cost of $0?

C. EXERCISES

38. MP8 Write a general expression for the y-value of the y-intercept of a graph in the form $f(x) = a(x - h)^2 + k$.

39. MP2 Complete the following steps to determine whether x feet of fencing will enclose more area when arranged as a square or as a circle.
   a. Express the length of a side of a square (s) and of the radius of a circle (r) in terms of x.
   b. Express the area of the square and of the circle in terms of x.
   c. Which area is larger? Justify your answer.

40. Given $f(x) = -1.5x^2 - 7x + 12$, describe its graph as follows.
   a. Find the vertex.
   b. State the equation of the line of symmetry.
   c. Which way does the parabola open?
   d. Is the parabola "steeper" or "flatter" than $y = x^2$?
   e. Find its zeros.

CUMULATIVE REVIEW

Factor completely. [9.5]

41. $2x + 14$

42. $4x^2 - 16$

43. $x^2 + 5x + 6$

44. $5x^2 - 4x - 12$

Solve each equation. [11.1, 11.5]

45. $x^2 + 4x = 0$

46. $x^2 + 7x - 8 = 0$

47. $0 = x^2 - 8x + 4$

48. $3x^2 - 4x - 4 = 0$

49. Which ordered pair is not part of the solution to $y < 2 - x$? **[6.7]**
   A (2, −2)
   B (−1, 2)
   C (0, 2)
   D (−3, 0)
   E none of these

50. Determine the value for b that will result in a vertex of (2, −2) in the function $y = 3x^2 + bx + 10$. **[11.8]**
   A 4
   B −12
   C −2
   D −10
   E none of these

APPLICATION PROBLEMS
WATER FOUNTAINS

What comes to your mind when you see a waterfall or a fountain? Waterfalls such as Niagara Falls in North America and Angel Falls in Venezuela inspire thoughts of the power of God and the beauty of His creation. When Christian mathematicians look at a waterfall and consider the laws of physics in action, they can also see the consistency of God's creation and be reminded of His faithfulness to us.

Man-made fountains are also fascinating to study. The path the water follows as it is projected from the fountainhead can be mathematically modeled by a quadratic equation, whose familiar parabolic form illustrates the trajectory of many projectiles. Sprinklers create variations of fountains that are used to keep our lawns looking nice and, more importantly, to irrigate crops.

Water wheels, invented by the ancient Greeks, are often located close to a waterfall to take advantage of the water's potential energy. Water wheels were used into the twentieth century to operate mills. Similar technology is used in modern hydroelectric power plants.

The distance, d, in feet that water falls to attain vertical velocity, v, in feet per second can be calculated using the formula $d = \frac{v^2}{2g}$, where g is the gravitational acceleration of 32 ft/sec^2.

1. Find the distance water must fall to reach a velocity of 9.8 ft/sec.

2. Solve the equation $d = \frac{v^2}{2g}$ for v to derive the radical equation describing the vertical velocity, v, which water attains after falling a given distance, d. Note that $v \nless 0$ in this context.

3. Water droplets coming off a waterfall fall at the same rate as any other falling object. The function $h(t) = -16t^2 + 65$ models the height of a water droplet at time t.

- **a.** Use the function to determine the height of the waterfall.
- **b.** How long (to the nearest tenth of a second) does it take for a droplet to complete its fall?

4. The path of a water droplet can be modeled by the function $h(x) = ax^2 + 65$, where $h(x)$ is the height of the water and x is its horizontal distance from the top of the waterfall. Substitute each of the following ordered pairs into the equation and solve for a. Then write the function rule that models the path of the droplet.

- **a.** (4, 0)
- **b.** (5.5, 0)
- **c.** (4.75, 0)
- **d.** Model the waterfall by plotting all 3 functions on your graphing calculator. State the size of an appropriate viewing window.

Circular sprinkler systems are used in both agricultural and residential settings. Consider a field whose length is 10 ft longer than its width and whose area is 2376 ft^2.

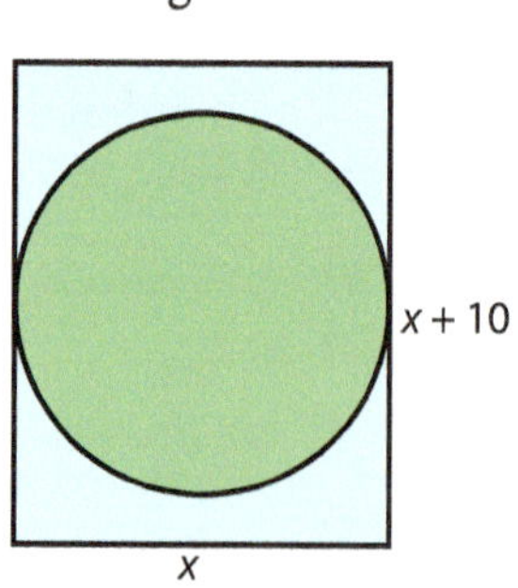

5. Find the dimensions of the field.

6. A single sprinkler head that rotates 360° is positioned in the middle of the field and adjusted to water the entire width of the field.
   a. What is the area that this sprinkler waters (to the nearest square foot)?
   b. What percent of the field is not watered?

7. By placing 2 sprinkler heads in opposite corners to rotate 90° with a radius the same as the width of the field, the farmer notices that practically all the field gets watered.

   a. What is the area that each sprinkler waters (to the nearest square foot)?
   b. Assuming the unwatered areas are negligible, what percent of the field is watered twice?

A landscape designer is planning a fountain and positions 2 fountainheads so that their 2 parabolic streams of water will intersect. These streams can be modeled by the equations $y = -0.14x^2 + 1.7x$ and $y = -0.14x^2 + 3.9x - 22$, where y is the height of the water in feet and x is its distance from the first fountainhead in feet.

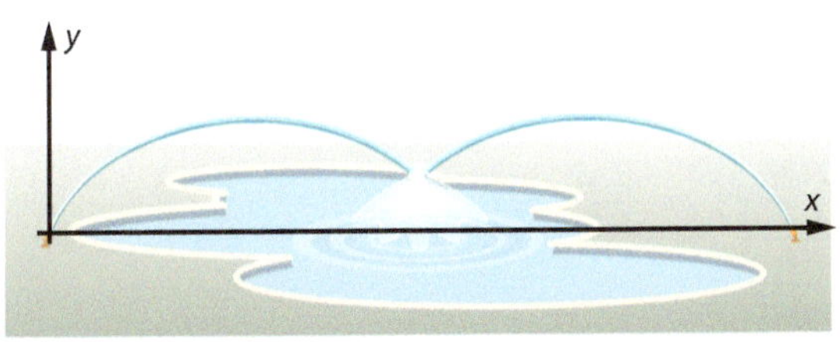

8. Using the elimination method of solving systems of equations (described in Section 6.3) will eliminate the y and x^2 terms. Write the resulting linear equation.

9. Solve for x and then for y.

10. Use these values to describe the location of the intersection of the streams relative to the first fountainhead.

Use your knowledge of symmetry and your graphing calculator to complete a quadratic regression, given the following information. (See the Technology Corner in Chapter 5 but use QuadReg.)

11. Water flowing from a fountain follows a parabolic path. If the water reaches a height of 9 ft and the distance between the fountainhead [use point (0, 0)] and where the water hits the pool is 5 ft, find a quadratic function that models the path of the water.

12. Find the height of the water 1 ft, 2.5 ft, 3 ft, and 3.5 ft away from the fountainhead.

Use the graph below for exercises 13–15.

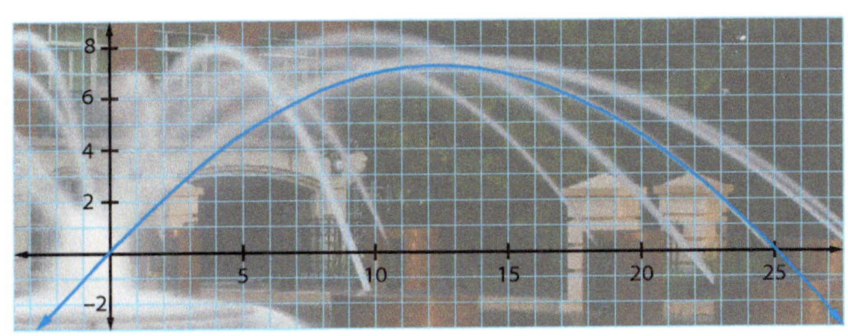

13. Make a table of 6 ordered pairs by estimating the location of the following points on the parabola outlined in the photo above: the vertex, the zeros, and 3 additional points on the parabola.

14. Complete the following steps to write a function that models the parabolic path of the stream of water.
   a. Substitute the values from the vertex into the formula $f(x) = a(x - h)^2 + k$.
   b. Determine the value of a (to the nearest thousandth) by substituting any other point on the parabola into the formula. You may want to use one of the zeros.
   c. Write the function for the parabola.

15. Plot the points from your table and graph your function using a graphing calculator, a spreadsheet, or interactive geometry software. Check that your function's graph models the points from the table.

11 CHAPTER REVIEW

Key Concepts

Solving Quadratic Equations by Factoring [11.1]

SOLVING QUADRATIC EQUATIONS BY FACTORING

1. Write the equation in standard form: $ax^2 + bx + c = 0$.
2. Factor the polynomial side of the equation.
3. Set each factor that contains a variable equal to 0 using the Zero Product Property.
4. Solve the resulting equations.
5. Check your solutions in the original equation.

Solving Quadratic Equations by Taking Roots [11.2]

SOLVING EQUATIONS OF THE FORM $ax^2 - c = 0$

1. Isolate the squared expression.
2. Apply the Square Root Property.
3. Simplify the radical and solve each equation.

- The Square Root Property helps to determine the number and type of solutions.

SQUARE ROOT PROPERTY

1. If $x^2 = c$ and $c \geq 0$, then $x = \pm\sqrt{c}$.
   a. If $c > 0$, then x has 2 real solutions.
   b. If $c = 0$, then x has 1 real solution.
2. If $x^2 = c$ and $c < 0$, then x has no real solutions.

Completing the Square: $ax^2 + bx + c = 0$ [11.3–11.4]

- If $a \neq 1$, then divide each term by a before executing the completing-the-square process.

SOLVING QUADRATIC EQUATIONS BY COMPLETING THE SQUARE

1. Write the equation in the form $x^2 + bx = c$.
2. Complete the square by adding $\left(\frac{b}{2}\right)^2$ to both sides of the equation.
3. Factor the trinomial.
4. Apply the Square Root Property: If $x^2 = c$, then $x = \pm\sqrt{c}$.
5. Solve the resulting equations.

The Quadratic Formula [11.5]

SOLVING QUADRATIC EQUATIONS BY THE QUADRATIC FORMULA
1. Write the equation in standard form: $ax^2 + bx + c = 0$.
2. Identify the values of a, b, and c.
3. Substitute these values into the quadratic formula, $x = \frac{-b \pm \sqrt{b^2 - 4ac}}{2a}$, and simplify.

More Quadratic Equations [11.6]

- The discriminant is used to determine the number and type of solutions.

DETERMINING THE NUMBER & TYPE OF SOLUTIONS		
$D = b^2 - 4ac$	Solutions	Example
$D > 0$ and a perfect square	2 rational	$x = \frac{1 \pm \sqrt{25}}{4} = \frac{3}{2}, -1$
$D > 0$ but not a perfect square	2 irrational	$x = \frac{1 \pm \sqrt{13}}{4} \approx 1.15, -0.65$
$D = 0$	1 rational	$x = \frac{1 \pm \sqrt{0}}{4} = \frac{1}{4}$
$D < 0$	no real	$x = \frac{1 \pm \sqrt{-9}}{4}$; no real solution

Quadratic Functions: $f(x) = ax^2 + c$ [11.7]

GRAPHING QUADRATIC FUNCTIONS
1. Find at least 5 ordered pairs that satisfy the function.
2. Plot the ordered pairs on a Cartesian coordinate plane.
3. Connect the ordered pairs with a smooth curve.

- The vertex will be the minimum or maximum point on a parabola, depending on the value of a.
- The line of symmetry is the line containing the vertex that divides the parabola into 2 identical parts.

THE EFFECT OF a ON THE GRAPH OF $y = ax^2 + bx + c$

If $a > 0$, the graph opens upward.

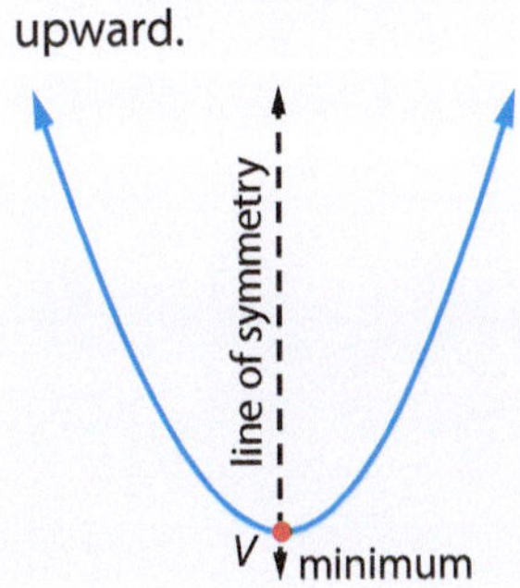

If $a < 0$, the graph opens downward.

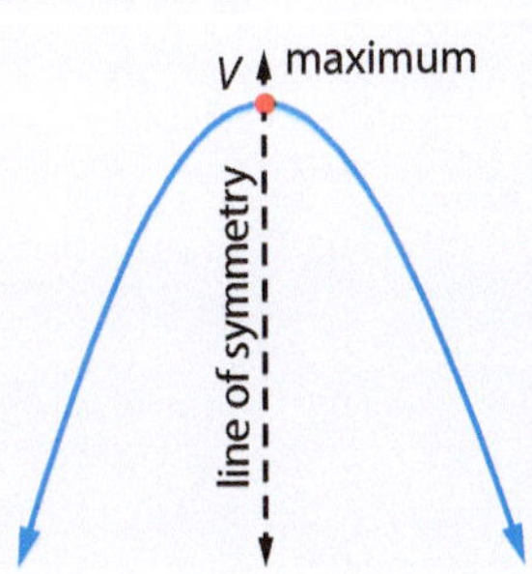

If $|a| > 1$, the graph is a vertical stretch.

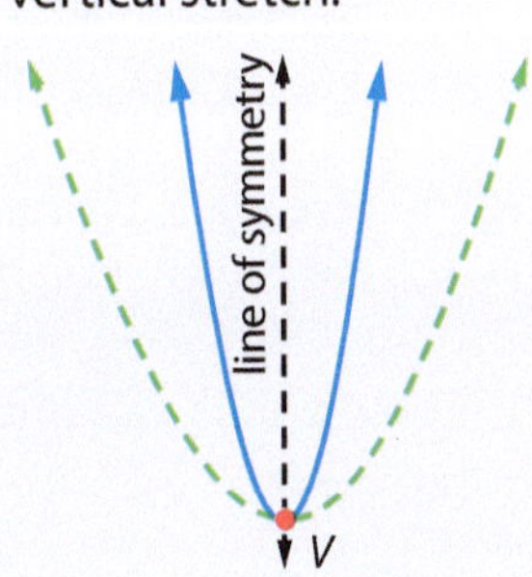

If $|a| < 1$, the graph is a vertical shrink.

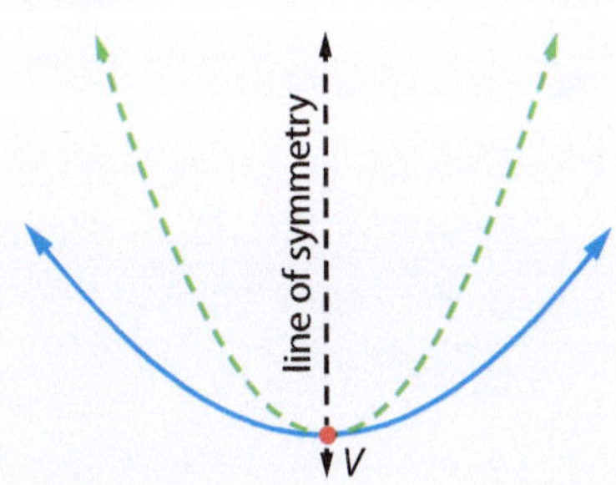

Quadratic Functions: $f(x) = a(x - h)^2 + k$ [11.8]

- Quadratic functions written in vertex form, $f(x) = a(x - h)^2 + k$, reveal key characteristics.
  - vertex: (h, k)
  - axis of symmetry: $x = h$
  - direction of opening: up when $a > 0$; down when $a < 0$
  - comparison to the parent function $f(x) = x^2$: same when $|a| = 1$; stretch when $|a| > 1$; shrink when $|a| < 1$
- Complete the square to convert a quadratic function written in standard form to vertex form.

Zeros of a Quadratic Function [11.9]

- The y-intercept can be found by letting $x = 0$ and solving for $y = f(0)$.
- The zeros (x-intercepts or roots) can be found by letting $f(x) = 0$ and solving for x.
- A double root occurs when solving $f(x) = 0$ yields 1 solution.

Biblical Worldview Shaping

Modeling with Algebra

What is the proper use of a mathematical model?
As long as we understand the assumptions it is based on, a model can help us to make wise decisions, but factors outside the realm of mathematics must also be considered.

EXERCISES

Solve each equation. [11.1]

1. $(x - 2)(x + 3) = 0$
2. $(x + 6)(x - 10) = 0$
3. $2a(4a - 1)(3a + 9) = 0$

Solve each equation by factoring. [11.1]

4. $x^2 + 3x - 10 = 0$
5. $x^2 - 6x - 7 = 0$
6. $8y^3 - 2y = 0$

Solve by taking roots. [11.2]

7. $x^2 = 6$
8. $3y^2 - 7 = 5$
9. $(x - 4)^2 = 11$
10. $25(x - 1)^2 = 12$
11. For each equation, identify whether there are 2, 1, or no real solutions. [11.2]
    a. $(x - 8)^2 = 20$
    b. $(x + 3)^2 + 10 = 9$
    c. $-5(x + 2)^2 = 0$
    d. $5x^2 - 13 = x^2 + 3$

Solve by completing the square. [11.3–11.4]

12. $x^2 + 8x + 7 = 0$

13. $x^2 + 5x = 3$

14. $3t^2 + 18t - 3 = 0$

15. $2x^2 - 20x - 5 = 0$

Use the quadratic formula to solve. [11.5]

16. $x^2 - 6x + 2 = 0$

17. $x^2 - 4x - 21 = 0$

18. $3x^2 + 2x - 5 = 0$

19. $4x^2 - 5 = 2x - 2$

Evaluate the discriminant; then state the number and nature of the root(s). [11.6]

20. $x^2 - 2x + 1 = 0$

21. $x^2 + 4x - 5 = 0$

22. $x^2 + 4x + 5 = 0$

23. $4x^2 + 3x = 3$

Solve each equation. [11.6]

24. $(x - 7)(x + 2) = 0$

25. $a^2 = 16$

26. $x^2 - 9 = 0$

27. $2x^2 + 8x + 9 = 0$

28. $a + 3 = a^2 - 2$

29. $5x^2 = 8x$

30. $x + 7 = x^2 + 10x + 25$

31. $x^2 + 6x + 9 = 0$

Write a quadratic equation in standard form with the given solutions. [11.6]

32. $x = -13, -8$

33. $x = -\frac{3}{5}, \frac{1}{2}$

Write a quadratic equation and solve.

34. Find 2 numbers whose difference is 19 and whose product is −84. **[11.1]**

35. How long does it take for a dropped rock to travel 361 ft to the bottom of a gorge? Use $h(t) = -16t^2 + h_i$. **[11.2]**

36. The area of a rectangular piece of concrete is 120 ft², and its perimeter is 46 ft. What are the dimensions of the rectangle? **[11.6]**

37. Mr. Franks distributes boxes of hot dogs to local grocery stores. The amount of profit he makes depends on the number of boxes of hot dogs he provides to the stores. The profit function is $P(x) = -x^2 + 140x - 3940$, where x is the number of boxes. How many boxes must he provide to make the maximum profit? What is his maximum profit? **[11.8]**

38. Match each quadratic function below with the comparison in shape to the parent function $y = x^2$ and whether its vertex is a maximum or minimum. **[11.7]**

a. $-\frac{1}{8}x^2 - 2$

b. $-3x^2$

c. $5x^2 + 10$

d. $\frac{3}{4}x^2 + 1$

I vertical stretch; maximum

II vertical shrink; maximum

III vertical stretch; minimum

IV vertical shrink; minimum

For each quadratic function, state the vertex, zero(s), and axis of symmetry; then graph the function. [11.7–11.9]

39. $f(x) = x^2 - 6$

40. $f(x) = (x + 6)^2 + 2$

41. $f(x) = 3 + 2x - x^2$

Use the graph of $p(x)$ for exercises 42–45. [11.9]

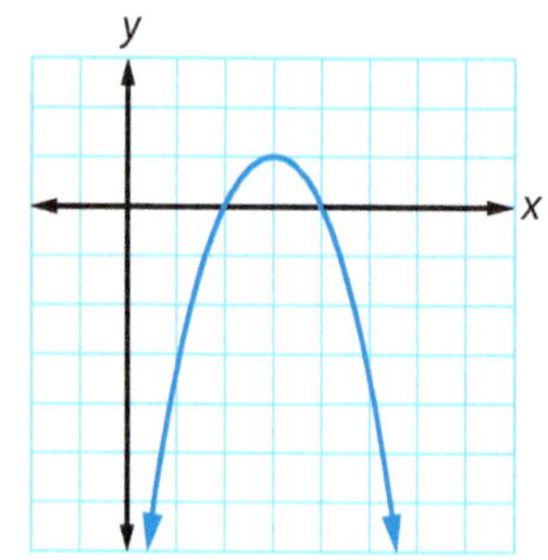

42. Find $p(5)$.

43. Find the zeros.

44. What is the maximum value of the function? Which value of the independent variable produces this maximum value?

45. Find the equation of the line of symmetry.

Given $q(x) = 2x^2 - 8x + 6$, find the following without graphing. [11.7–11.9]

46. Find the y-intercept.

47. Find the zeros.

48. Find the vertex. Is it a minimum or a maximum?

49. Will the graph cross the line $y = 30$? Why or why not?

50. **BWS Essential Question:** Mathematical models

A are all we need to make decisions.

B can accurately predict all future outcomes.

C are useful in making decisions.

D have no limitations.

STEM PROJECT

MISSION CONTROL

In September 1999, after 10 months of space travel, the $125 million Mars climate orbiter, a robotic space probe, was lost as the result of a simple mistake in math. The values for the orbiter's thrusters were given in English units, and NASA entered them as metric units. This resulted in an incorrect trajectory, bringing the orbiter too close to Mars. It either burned up in the Mars atmosphere, or it escaped the pull of Mars and entered an orbit around the sun. In this STEM project your team will build an object launcher and become engineers and mission control, determining the correct trajectories of your launched objects.

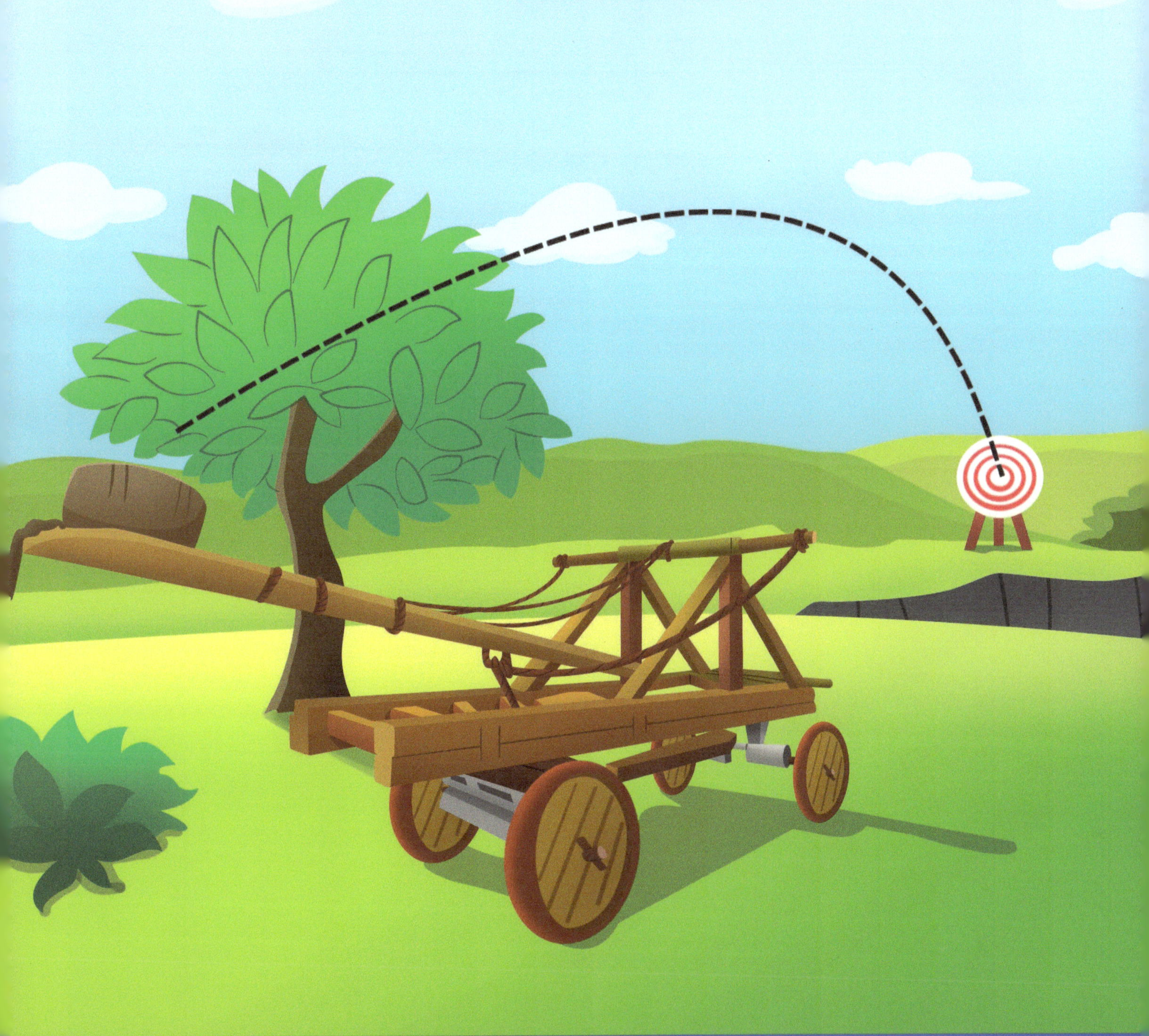

CHAPTER 12
RATIONAL EXPRESSIONS & EQUATIONS

CHAPTER OVERVIEW

12.1 SIMPLIFYING RATIONAL EXPRESSIONS

How can a rational expression be used in medicine?

After completing this section, I will be able to

- determine values for which a rational expression is undefined.
- simplify rational expressions by canceling common factors.
- defend the claim that mathematics helps us see design in creation.

Recall that a rational number is defined as a ratio of integers whose denominator is not 0. The same principles you used to simplify, add, subtract, multiply, and divide rational numbers can be applied to rational expressions.

When a larger tube tapers into a smaller one, the taper per inch, T, is found by the formula $T = \frac{D-d}{L}$, where L is the length of the pipe and D and d are the larger and smaller diameters, respectively.

DEFINITION

A **rational expression** is a ratio of polynomials whose denominator is not 0.

The following are all examples of rational expressions.

1. $\frac{x}{x+2}$
2. $\frac{x-4}{5}$
3. $\frac{x-6}{x-9}$
4. $\frac{2x^2+3x+6}{x^2-8x+7}$
5. $\frac{x+7}{x^2+3x-9}$
6. $\frac{3x^2+4x-5}{x^2+1}$

The result of substituting -2 for x in the first expression is $\frac{-2}{0}$. Therefore, the rational expression is undefined when $x = -2$, and -2 is called an *excluded value*.

DEFINITION

An **excluded value** is a number that makes the rational expression undefined.

To find the excluded values for a rational expression, determine which value(s) of x will make the denominator 0. In the second expression above, the denominator has no variables. Since $5 \neq 0$, the rational expression is always defined.

EXAMPLE 1: Finding Excluded Values

Determine the excluded value(s) for each expression.

a. $\frac{x-6}{x-9}$ **b.** $\frac{2x^2+3x+6}{x^2-8x+7}$ **c.** $\frac{x+7}{x^2+3x-9}$

Answers

a. $\frac{x-6}{x-9}$

1. The denominator cannot equal 0.

$x - 9 \neq 0$

$x \neq 9$

2. Solve.

$\frac{x-6}{x-9}$ is undefined when $x = 9$.

b. $\dfrac{2x^2 + 3x + 6}{x^2 - 8x + 7}$

1. The denominator cannot equal 0.

$x^2 - 8x + 7 \neq 0$

$(x - 7)(x - 1) \neq 0$

2. Solve by factoring.

$x - 7 \neq 0$ or $x - 1 \neq 0$

$x \neq 7, 1$

$\dfrac{2x^2 + 3x + 6}{x^2 - 8x + 7}$ is undefined when $x = 7, 1$.

---

c. $\dfrac{x + 7}{x^2 + 3x - 9}$

1. The denominator cannot equal 0.

$x^2 + 3x - 9 \neq 0$

$x \neq \dfrac{-3 \pm \sqrt{3^2 - 4(1)(-9)}}{2(1)} = \dfrac{-3 \pm \sqrt{45}}{2} = \dfrac{-3 \pm 3\sqrt{5}}{2}$

2. Since the trinomial cannot be factored, use the quadratic formula to solve.

$\dfrac{x + 7}{x^2 + 3x - 9}$ is undefined when $x = \dfrac{-3 \pm 3\sqrt{5}}{2} \approx 1.85, -4.85$.

SKILL CHECK **EXERCISES 3, 5** ☑

The rational expression $\dfrac{3x^2 + 4x - 5}{x^2 + 1}$ has no excluded values because there are no real-number solutions to $x^2 + 1 = 0$.

Rational expressions are found in almost every area of science. Medical doctors use rational expressions to model the persistence of prescribed drugs in the bloodstream. Mathematics gives us a glimpse of the incredible complexity and design found in our own bodies and points us to a Creator. Simplifying rational expressions makes real-world problems easier to solve. A rational expression is in simplest form when the numerator and denominator have no common factors other than 1.

EXAMPLE 2: Simplifying Rational Expressions

Simplify each expression. State the excluded values.

a. $\dfrac{180m^3n}{27mn^2}$ **b.** $\dfrac{3x + 6}{3}$ **c.** $\dfrac{x^2 + 4x + 4}{x^2 - 4}$

Answers

a. $\dfrac{180m^3n}{27mn^2} = \dfrac{\cancel{9} \cdot 20\cancel{m}m^2\cancel{n}}{\cancel{9} \cdot 3\cancel{m}\cancel{n}n} = \dfrac{20m^2}{3n}; m, n \neq 0$

Factor and cancel common factors.

---

b. $\dfrac{3x + 6}{3} = \dfrac{3(x + 2)}{3}$

1. Factor the numerator.

$= \dfrac{\cancel{3}(x + 2)}{\cancel{3}} = x + 2$

2. Cancel the common factor, 3.

There are no excluded values.

---

c. $\dfrac{x^2 + 4x + 4}{x^2 - 4} = \dfrac{(x + 2)(x + 2)}{(x - 2)(x + 2)}$

1. Factor each polynomial completely.

$= \dfrac{(x + 2)\cancel{(x + 2)}}{(x - 2)\cancel{(x + 2)}} = \dfrac{x + 2}{x - 2}; x \neq \pm 2$

2. Cancel the common factor, $x + 2$.

SKILL CHECK **EXERCISES 11, 17** ☑

AFTERSCHOOLHELP

Simplifying Rational Expressions

Notice in Examples 2a and 2c the excluded values in the original expression are different from the excluded values for the simplified expression ($n \neq 0$ for 2a and $x \neq 2$ for 2c). For this reason, excluded values for the original expression are often listed with the simplified rational expression.

EXAMPLE 3: Simplifying Rational Expressions

Simplify $\frac{4x^2 + 4x - 48}{3 - x}$, leaving the answer in factored form. State any excluded values.

Answer

$x \neq 3$

1. State any excluded values for the original expression.

$$\frac{4(x^2 + x - 12)}{3 - x} = \frac{4(x - 3)(x + 4)}{3 - x}$$

2. Factor the polynomial completely.

$$= \frac{4\cancel{(x - 3)}(x + 4)}{-1\cancel{(x - 3)}}$$

3. Recognize that $x - 3$ and $3 - x$ are opposites and factor -1 out of one of those factors so they can be canceled.

$$= -4(x + 4);\ x \neq 3$$

4. Simplify the result and state the excluded value.

SKILL CHECK **EXERCISE 21**

EXAMPLE 4: Applying Rational Expressions

Write a rational expression representing the portion of the square that is shaded. Then simplify and state the result as a percent (to the nearest tenth).

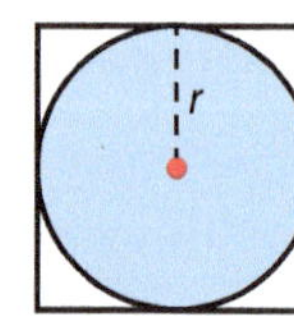

Answer

$$A_{\text{circle}} = \pi r^2$$
$$A_{\text{square}} = s^2 = (2r)^2 = 4r^2$$

1. Write expressions for the area of the circle and the square.

$$\frac{A_{\text{circle}}}{A_{\text{square}}} = \frac{\pi r^2}{4r^2}$$

2. Write a rational expression representing the percentage of area shaded.

$$= \frac{\pi \cancel{r^2}}{4\cancel{r^2}} = \frac{\pi}{4}$$

3. Cancel common factors.

$$\approx 0.785 = 78.5\%$$

4. Divide and convert to a percent.

SKILL CHECK **EXERCISE 29**

VOCABULARY

- excluded value
- rational expression

A. EXERCISES

State any excluded values for each rational expression.

1. $\frac{3x + 4}{x}$

2. $\frac{x^2 - 9}{x + 4}$

3. $\frac{a + 7}{a + 2}$

4. $\frac{b + 8}{b - 5}$

5. $\frac{x^2 + 3x - 2}{x^2 + 5x - 24}$

6. $\frac{y - 7}{y^2 + 7y - 18}$

7. $\frac{c + 4}{c^2 - 10c + 24}$

8. $\frac{z - 5}{z^3 - 3z^2 - 4z}$

Simplify, leaving each answer in factored form.

9. $\frac{3x^2}{15x}$

10. $\frac{3(a + 7)}{3}$

11. $\frac{24x^2y^3}{8xy}$

12. $\frac{2(y - 5)}{y - 5}$

13. $\frac{6x - 6}{6}$

14. $\frac{x^2 + x}{x}$

15. $\frac{4a^2 + 2a}{2a}$

16. $\frac{3c}{3c - 6c^2}$

B. EXERCISES

Simplify, leaving each answer in factored form. State any excluded values for *x*.

17. $\dfrac{x^2 - 4x - 45}{x^2 - 10x + 9}$

18. $\dfrac{3x^2 - 20x + 12}{3x^2 + x - 2}$

19. $\dfrac{x^3 - 3x^2 - 10x}{x^3 + 2x^2}$

20. $\dfrac{x^2 - y^2}{x + y}$

21. $\dfrac{y^2 - x^2}{x^2 - y^2}$

22. $\dfrac{3x^2 + 3xy}{x^2 - y^2}$

23. $\dfrac{x^2w^2 - 15xw^2 + 56w^2}{42x^2 - 6x^3}$

24. $\dfrac{121 - 9x^2}{6x^2 - 22x}$

25. $\dfrac{3x^5y - 3x^4y - 60x^3y}{9x^4 + 36x^3}$

26. $\dfrac{x^7y^2 - x^6y^2 - 6x^5y^2}{9x^3 - x^5}$

27. Suppose that the amount of a certain drug in the bloodstream (in milligrams) after *t* hours is given by the rational expression $\dfrac{5t^3 - 5t}{t^4 - 1}$. Simplify the expression and state the milligrams remaining after 12 hr.

28. **Essential Question:** Give one example of how a rational expression may be used in medicine.

MP7 Write a rational expression representing the portion of each figure that is shaded. State the result as a simplified expression and as a percent (to the nearest tenth).

29.

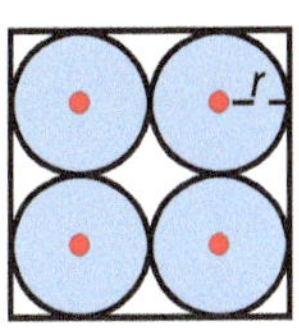

30.

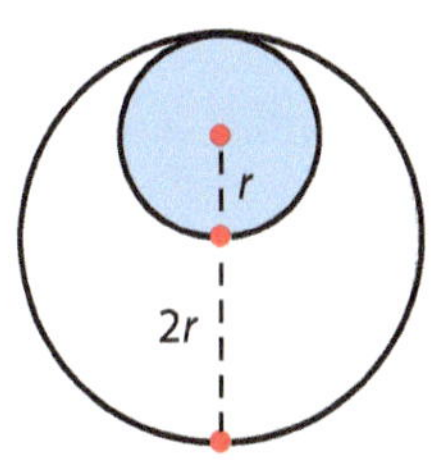

31.

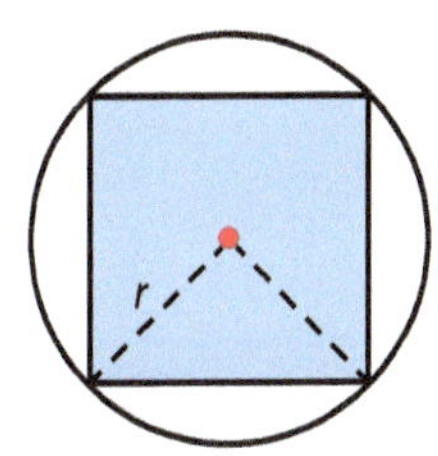

MP4 The ideal mechanical advantage (IMA) of a lever is given by the formula IMA = $\frac{l_e}{l_r}$, where l_e is the distance from the fulcrum to the effort and l_r is the distance from the fulcrum to the resistance.

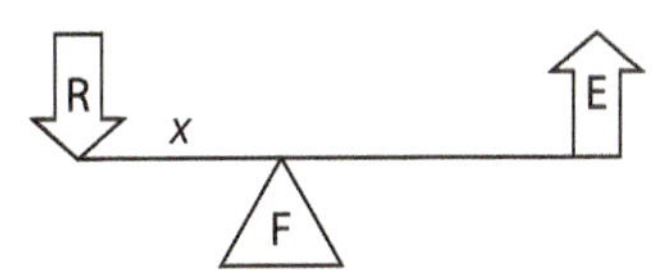

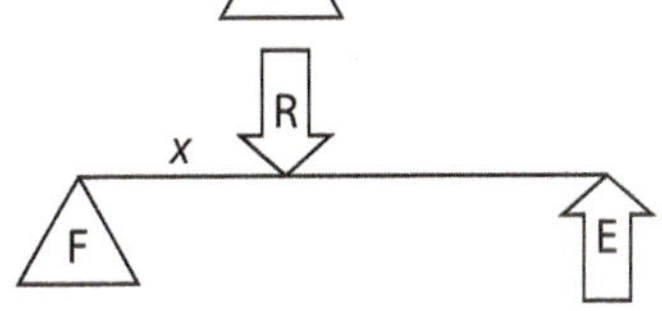

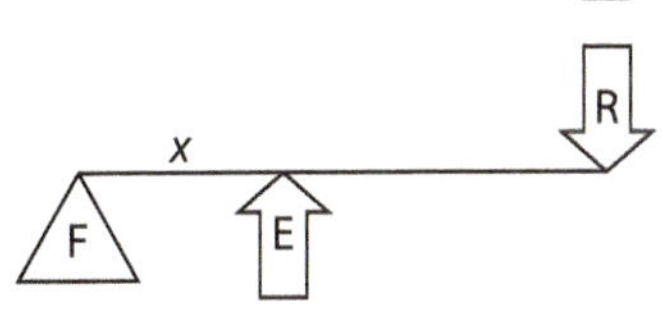

32. Write a rational expression representing the IMA of a 5 ft first-class lever if the fulcrum is placed *x* ft from the resistance.

33. Write a rational expression representing the IMA of a 5 ft second-class lever if the resistance is placed *x* ft from the fulcrum.

34. Write a rational expression representing the IMA of a 5 ft third-class lever if the effort is applied *x* ft from the fulcrum.

C. EXERCISES

Simplify, leaving each answer in factored form. State any excluded values.

35. $\dfrac{48x^2 + 34x - 5}{1 - 8x}$

36. $\dfrac{5xz + 5z}{10x^2 - 10}$

37. $\dfrac{x^4 - 5x^2 - 36}{x^3 - 3x^2 + 4x - 12}$

38. $\dfrac{x^5 + x^4 - 4x^3 - 4x^2}{x^5 + 4x^4 + 4x^3}$

39. **MP3** Explain why the expression $\dfrac{9x^2 - 16}{3x^2 + 4x}$ is not identical to the simplified expression $\dfrac{3x - 4}{x}$.

40. **MP7** Write a rational expression representing the portion of the figure that is shaded. State the result as a simplified expression and as a percent (to the nearest tenth).

41. **MP2** Write a rational expression representing the greatest portion of a cube that can be filled by a sphere. State the result as a simplified expression and as a percent (to the nearest tenth).

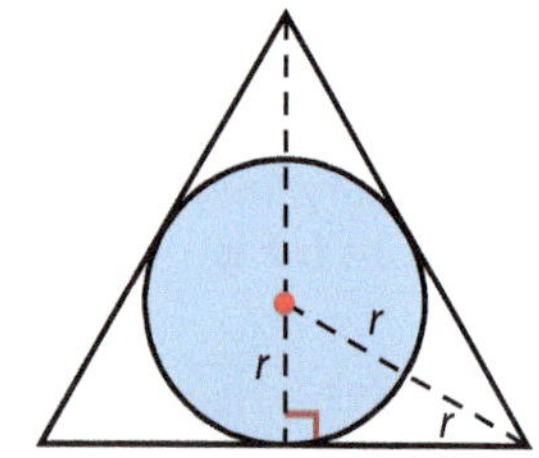

CUMULATIVE REVIEW

Find each product. [8.3]

42. $(2x - 5)(x + 4)$

43. $(x - 8)(3x - 5)$

Simplify each expression. [10.3]

44. $\frac{4}{\sqrt[3]{2}}$

45. $\frac{3x^2y\sqrt{20}}{9xy\sqrt{6}}$

Write each quadratic function in vertex form and state the vertex. [11.8]

46. $y = x^2 - 2x + 4$

47. $y = x^2 + 4x + 4$

48. $y = x^2 + 6x + 3$

49. $y = x^2 - 8x + 31$

50. Factor $4x^3 - 8x^2 + 9x - 18$ completely. **[9.5]**

A $(2x + 3)^2(x - 2)$

B $(4x^2 + 9)(x - 2)$

C $(2x - 3)^2(x + 2)$

D $(2x - 3)(2x + 3)(x + 2)$

E none of these

51. Which function has all real numbers greater than or equal to -3 as its domain? **[10.8]**

A $f(x) = \sqrt{x - 1} - 3$

B $g(x) = \sqrt{x - 3} - 2$

C $h(x) = \sqrt{x + 3} + 10$

D $j(x) = \sqrt{x + 5} + 3$

E none of these

MIND OVER MATH

One benefit of studying mathematics is the development of correct reasoning skills. Can you detect the faulty reasoning in the following proofs?

Proof: 1 = 2

$a = b$	**1.** Let $a = b$.
$ab = b^2$	**2.** Multiply both sides by b.
$ab - a^2 = b^2 - a^2$	**3.** Subtract a^2 from both sides.
$a(b - a) = (b + a)(b - a)$	**4.** Factor both sides.
$a = b + a$	**5.** Divide both sides by $(b - a)$.
$a = a + a$	**6.** Since $a = b$, substitute a for b.
$a = 2a$	**7.** Simplify the right side by adding like terms.
$1 = 2$	**8.** Divide both sides by a.

Proof: 4 = 8

$16 - 48 = 64 - 96$	**1.** Both sides of the equation equal -32.
$16 - 48 + 36 = 64 - 96 + 36$	**2.** Add 36 to both sides.
$4^2 - 2(4)(6) + 6^2 = 8^2 - 2(8)(6) + 6^2$	**3.** Factor the perfect square trinomials.
$(4 - 6)^2 = (8 - 6)^2$	
$4 - 6 = 8 - 6$	**4.** Take the square root of both sides.
$4 = 8$	**5.** Add 6 to both sides.

12.2 MULTIPLYING & DIVIDING RATIONAL EXPRESSIONS

What must we assume about any denominator when multiplying or dividing rational expressions?

After completing this section, I will be able to

- multiply rational expressions.
- divide rational expressions.

The same method used to multiply rational numbers is also used to multiply rational expressions.

A player's batting average is .314, which is calculated using the rational expression $A = \frac{H}{B}$, where H is the number of hits and B is the number of at bats.

EXAMPLE 1: Multiplying Rational Expressions

Simplify $\frac{3a^2c^2}{2b} \cdot \frac{10a}{9bc}$.

Answer

$\frac{\cancel{3}a^2\cancel{c}c}{\cancel{2}b} \cdot \frac{\overset{5}{\cancel{10}}a}{\underset{3}{\cancel{9}}b\cancel{c}}$ 1. Cancel common factors.

$= \frac{5a^3c}{3b^2}$ 2. Multiply.

SKILL CHECK EXERCISE 3

In Example 1 we assumed that $b \neq 0$ and $c \neq 0$ since the original expression would be undefined if $b = 0$ or $c = 0$. For the rest of this chapter, you may assume that no denominator of a rational expression has a value of 0.

MULTIPLYING RATIONAL EXPRESSIONS

1. Factor all algebraic quantities in the numerators and the denominators.
2. Cancel common factors appearing in both a numerator and a denominator.
3. Express the result as the product of the numerators divided by the product of the denominators.

EXAMPLE 2: Multiplying Rational Expressions

Simplify, leaving the answer in factored form.

a. $\dfrac{x^2+2xy+y^2}{2x+2y}\cdot\dfrac{18}{x^2-y^2}$ **b.** $\dfrac{x^3-x^2-72x}{3x^2}\cdot\dfrac{12x}{x^2+x-56}$

Answers

a. $\dfrac{(x+y)(x+y)}{2(x+y)}\cdot\dfrac{18}{(x+y)(x-y)}$

1. Factor the numerators and the denominators.

$=\dfrac{\cancel{(x+y)}\cancel{(x+y)}}{\cancel{2}\cancel{(x+y)}}\cdot\dfrac{\overset{9}{\cancel{18}}}{\cancel{(x+y)}(x-y)}$

2. Simplify by canceling common factors.

$=\dfrac{9}{x-y}$

3. The result is the product of the numerators divided by the product of the denominators.

b. $\dfrac{x(x-9)(x+8)}{3x^2}\cdot\dfrac{12x}{(x+8)(x-7)}$

1. Factor the numerators and the denominators.

$=\dfrac{\cancel{x}(x-9)\cancel{(x+8)}}{\cancel{3}\cancel{x}\cancel{x}}\cdot\dfrac{\overset{4}{\cancel{12}}\cancel{x}}{\cancel{(x+8)}(x-7)}$

2. Simplify by canceling common factors.

$=\dfrac{4(x-9)}{x-7}$

3. The result is the product of the numerators divided by the product of the denominators.

SKILL CHECK **EXERCISES 17, 19**

AfterSchoolHelp

Multiplying & Dividing Rational Expressions

To divide by a fraction, multiply by the reciprocal of the divisor.

$$\frac{1}{2}\div\frac{7}{4}=\frac{1}{\cancel{2}}\cdot\frac{\overset{2}{\cancel{4}}}{7}=\frac{2}{7}$$

This same procedure is used for dividing rational expressions.

EXAMPLE 3: Dividing Rational Expressions

Divide $\dfrac{x^2}{3y^2}$ by $\dfrac{x}{9y}$.

Answer

$\dfrac{x^2}{3y^2}\div\dfrac{x}{9y}=\dfrac{x^2}{3y^2}\cdot\dfrac{9y}{x}$

Multiply by the reciprocal of the divisor.

$=\dfrac{\cancel{x}x}{\cancel{3}\cancel{y}y}\cdot\dfrac{\overset{3}{\cancel{9}}\cancel{y}}{\cancel{x}}=\dfrac{3x}{y}$

SKILL CHECK **EXERCISE 13**

DIVIDING RATIONAL EXPRESSIONS

Multiply by the reciprocal of the divisor.

EXAMPLE 4: Dividing Rational Expressions

Simplify $\dfrac{x^2+3x+2}{x^2+x-2}\div\dfrac{x^2+2x+1}{x^2-1}$.

Answer

$\dfrac{x^2+3x+2}{x^2+x-2}\cdot\dfrac{x^2-1}{x^2+2x+1}$

1. Multiply by the reciprocal of the divisor.

$=\dfrac{(x+2)(x+1)}{(x+2)(x-1)}\cdot\dfrac{(x-1)(x+1)}{(x+1)(x+1)}$

2. Factor the numerators and the denominators.

$=\dfrac{\cancel{(x+2)}\cancel{(x+1)}}{\cancel{(x+2)}\cancel{(x-1)}}\cdot\dfrac{\cancel{(x-1)}\cancel{(x+1)}}{\cancel{(x+1)}\cancel{(x+1)}}=1$

3. Simplify by canceling common factors. Remember that $\frac{\cancel{a}}{\cancel{a}}=1$.

SKILL CHECK **EXERCISES 15, 29**

A. EXERCISES

Find each product.

1. $\frac{2}{5} \cdot \frac{19}{28}$

2. $\frac{16}{9} \cdot \frac{27}{8}$

3. $\frac{3a^2}{17b} \cdot \frac{34b^2}{6ab}$

4. $\frac{5x^2y}{18xy} \cdot \frac{3y^3}{9y}$

5. $\frac{x+2}{5} \cdot \frac{25}{x+2}$

6. $\frac{(x-3)(x+4)}{x-1} \cdot \frac{x-1}{(x-4)(x-3)}$

State the reciprocal of each expression.

7. $-\frac{2}{3}$

8. $\frac{32a^2b}{xy^3}$

9. $\frac{a^2-b^2}{a^2b}$

10. $x^2 - 2x - 3$

Find each quotient.

11. $\frac{4}{5} \div \frac{8}{35}$

12. $\frac{14}{9} \div \frac{28}{27}$

13. $\frac{8x^4y^2}{x^3y} \div \frac{24x}{7y}$

14. $\frac{a^2b^3}{x^5y^2} \div \frac{a^4b}{xy^3}$

15. $\frac{(x+y)^2}{x-y} \div \frac{(x+y)(x-y)}{(x-y)^2}$

16. $\frac{4(x+1)}{5(x-5)} \div \frac{3(x+1)}{x-5}$

B. EXERCISES

Simplify each expression, leaving each answer in factored form.

17. $\frac{x^2-5x-14}{8} \cdot \frac{4}{x-7}$

18. $\frac{x^2+3x-18}{x^2-x} \cdot \frac{x^3y}{xy-3y}$

19. $\frac{x^2+10x+16}{x-4} \cdot \frac{x^2-x-12}{x+2}$

20. $\frac{x^2-x-20}{x^2-6x+5} \cdot \frac{x+3}{x^2+7x+12}$

21. $\frac{2x^2-xy-y^2}{x^2-y^2} \cdot \frac{x^2+2xy+y^2}{3x^2-xy-4y^2}$

22. $\frac{x-2}{x^3-5x^2+6x} \cdot \frac{xy^2}{xy+4y}$

23. $\frac{x^2-y^2}{(x-y)^2} \div \frac{x^2-2xy+y^2}{x-y}$

24. $\frac{x^2+2x}{3x-9} \div \frac{x^3-9x}{x-3}$

25. $(x+1) \div \frac{x^2+2x+1}{x}$

26. $\frac{x^2-11x+18}{x-3} \div \frac{x^2+6x+8}{x+2}$

27. $\frac{x^2-xy-2y^2}{x^2-y^2} \div \frac{x^2-3xy+2y^2}{3x^2+4xy+y^2}$

28. $\frac{x^2+3x-54}{x^2+5x+6} \div \frac{x^2-12x+36}{x^2-3x-18}$

29. $\frac{a^2+ab-2b^2}{a^2-b^2} \cdot \frac{a^2-2ab-3b^2}{a^2-3ab}$

30. $\frac{3x^2-11xy-4y^2}{x^3y-4x^2y^2} \cdot \frac{x^2y-3xy^2}{3x+y}$

31. $\frac{a^2+7ab+10b^2}{a^2+ab-2b^2} \cdot \frac{3a^2-7ab+4b^2}{3a^2+11ab-20b^2}$

32. $\frac{3m^2-2mn-n^2}{m^3n^2} \div (m^2-2mn+n^2)$

33. $\frac{x^2-7x-18}{x+4} \cdot \frac{x^2-5x+6}{x-9} \cdot \frac{x+4}{x-3}$

34. $\frac{1}{2x^2+xy-y^2} \div \frac{x^2+2xy+y^2}{2x^2-3xy+y^2}$

35. $\frac{12x^2-29xy+14y^2}{x^2-xy-12y^2} \div \frac{3x^2+xy-2y^2}{x^2-2xy-8y^2}$

36. **MP3** When simplifying the expression $\frac{x^2+2xy-3y^2}{x^2-2xy-3y^2} \div \frac{x^2-y^2}{x^2-9y^2}$, Becky begins by factoring and rewriting the expression as $\frac{(x+3y)(x-y)}{(x-3y)(x+y)} \cdot \frac{(x-y)(x+y)}{(x-3y)(x+3y)}$. Then she cancels common factors to determine an answer: $\frac{\cancel{(x+3y)}(x-y)}{(x-3y)\cancel{(x+y)}} \cdot \frac{(x-y)\cancel{(x+y)}}{(x-3y)\cancel{(x+3y)}} = \frac{(x-y)^2}{(x-3y)^2}$. Describe and correct her error.

37. **Essential Question:** What must we assume when multiplying or dividing rational expressions?

C. EXERCISES

Simplify, leaving each answer in factored form.

38. $\frac{x^4 + x^3 - 6x^2}{xy^3 - 2y^3} \cdot \frac{x^2 + 2x - 15}{x^2 + 8x + 15} \div \frac{x^2}{y^3}$

39. $\frac{4ab^4}{a^2 + 8ab + 15b^2} \cdot \frac{a^2 + 3ab - 10b^2}{a + 7b} \div \frac{a^2b^3}{a^2 + 2ab - 3b^2}$

40. **MP1** $\frac{x^2 - 3x - 18}{x^2 - 2x - 48} \div \frac{x^2 - x - 12}{x^3 - 12x^2 + 36x} \cdot \frac{3x^2 - 6x - 144}{x^2 - 4x}$

41. $\frac{(3x + 1)(x - 1) + (5x - 3)(x + 2) - (19x - 12)}{x^2 - 1} \div \frac{4x^2 + 7x - 15}{x^2 + 4x + 3}$

42. **MP7** $\frac{(x - 1)^2 + (x - 1)(x + 3)}{(x - 1)^2 + (2x - 2)(x + 2)} \div \frac{(x - 2)^2 + (3x - 6)(x + 2)}{(x - 2)^2 + (x - 2)(x + 4)}$

CUMULATIVE REVIEW

Given the 2 points (6, −3) and (−2, −4), find the following. [10.5]

43. the midpoint of the line segment with the 2 points as endpoints

44. the distance between the points

Find the missing endpoint of a segment with the given endpoint and midpoint. [10.5]

45. $P_1\,(-10, 3)$ and $M\,(-2, -2)$

46. $P_1\,(5\sqrt{3}, 9)$ and $M\,(2\sqrt{3}, 5)$

Simplify each expression. [10.6]

47. $\frac{\sqrt{6}}{\sqrt{3} - \sqrt{2}}$

48. $(\sqrt{26} - 4\sqrt{2})(2\sqrt{13} + 5)$

Solve an equation modeling each problem. [10.7, 11.6]

49. Ten less than the square root of 1 more than twice a number is 1. Find the number.

50. A right triangle must have leg lengths that are 4 ft and 8 ft shorter than the hypotenuse, respectively. Find the measurements for the sides of the triangle.

51. Which equation has real solutions? **[11.2]**
   - A $6(x - 11)^2 = -24$
   - B $(x + 8)^2 + 6 = 1$
   - C $(x - 10)^2 - 10 = 0$
   - D $x^2 + 25 = 0$
   - E none of these

52. Which rational expression can be simplified further by canceling common factors? **[12.1]**
   - A $\frac{x^2 - 16}{(x + 4)^2}$
   - B $\frac{20 - 15x}{18 - 24x}$
   - C $\frac{45mpx}{28kyz}$
   - D $\frac{x^2 - 36}{x^2 - 9}$
   - E none of these

12.3 ADDING & SUBTRACTING EXPRESSIONS WITH COMMON DENOMINATORS

How are the rules for adding or subtracting rational expressions similar to adding or subtracting fractions?

After completing this section, I will be able to

- add rational expressions with common denominators.
- subtract rational expressions with common denominators.

Operations with rational expressions follow the same rules as operations with rational numbers. Expressions with a common denominator are added by writing the sum of the numerators over the common denominator.

The rational expression $\frac{1}{t}+\frac{2}{t}$ can be used to represent the rate at which Cynthia and her dad can complete their yardwork together.

EXAMPLE 1: Adding Rational Expressions

Simplify each expression.

a. $\frac{1}{t}+\frac{2}{t}$ **b.** $\frac{2a}{15}+\frac{8a}{15}$

Answers

a. $\frac{1}{t}+\frac{2}{t}=\frac{3}{t}$

b. $\frac{2a}{15}+\frac{8a}{15}=\frac{10a}{15}$

1. Write the sum of the numerators over the common denominator.

$=\frac{\cancel{10}^{2}a}{\cancel{15}_{3}}=\frac{2a}{3}$

2. When possible, simplify by canceling common factors.

SKILL CHECK **EXERCISE 1**

EXAMPLE 2: Adding Rational Expressions

Simplify each expression.

a. $\frac{2x+9}{x-4}+\frac{3x+6}{x-4}$ **b.** $\frac{3x^2+4x-8}{x+2}+\frac{x^2-3x-6}{x+2}$

Answers

a. $\frac{2x+9+3x+6}{x-4}=\frac{5x+15}{x-4}$

1. Add the numerators by combining like terms.

$=\frac{5(x+3)}{x-4}$

2. Factoring reveals that the expression cannot be simplified.

b. $\frac{3x^2+4x-8+x^2-3x-6}{x+2}=\frac{4x^2+x-14}{x+2}$

1. Add the numerators by combining like terms.

$=\frac{(4x-7)\cancel{(x+2)}}{\cancel{x+2}}=4x-7$

2. Factor and simplify by canceling common factors.

SKILL CHECK **EXERCISES 9, 27**

To subtract rational expressions, follow the procedure for subtracting rational numbers. When the expressions have a common denominator, subtract the numerators and write the result over the common denominator.

EXAMPLE 3: Subtracting Rational Expressions

Simplify each expression.

a. $\frac{2+x}{8x}-\frac{x-8}{8x}$ **b.** $\frac{3x^2+3x-20}{x^2+9x+20}-\frac{x^2+x+4}{x^2+9x+20}$

Answers

a. $\frac{(2+x)-(x-8)}{8x}=\frac{2+x-x+8}{8x}=\frac{10}{8x}$

1. Subtract the numerators.

$=\frac{\overset{5}{\cancel{10}}}{\underset{4}{\cancel{8}}x}=\frac{5}{4x}$

2. Simplify by canceling common factors.

b. $\frac{(3x^2+3x-20)-(x^2+x+4)}{x^2+9x+20}$

1. Subtract the numerators.

$=\frac{3x^2+3x-20-x^2-x-4}{x^2+9x+20}=\frac{2x^2+2x-24}{x^2+9x+20}$

$=\frac{2(x+4)(x-3)}{(x+5)(x+4)}$

2. Factor the numerator and the denominator.

$=\frac{2\cancel{(x+4)}(x-3)}{(x+5)\cancel{(x+4)}}=\frac{2(x-3)}{x+5}$

3. Simplify by canceling common factors.

SKILL CHECK **EXERCISES 5, 29**

AFTERSCHOOLHELP

Adding & Subtracting Expressions with Common Denominators

ADDING & SUBTRACTING RATIONAL EXPRESSIONS WITH COMMON DENOMINATORS

1. Add or subtract the numerators.
2. Simplify the resulting rational expression.

A rational number can be thought of as having 3 signs: 1 in the numerator, 1 in the denominator, and 1 preceding the fraction. The rational number $-\frac{4}{5}$ can be expressed as any 1 of these 3 equivalent expressions:

$$-\frac{4}{5}=\frac{-4}{5}=\frac{4}{-5}.$$

It is easier to simplify a problem like $\frac{3}{5}+\frac{2}{-5}$ if you use another form of the rational number.

This concept is especially helpful for simplifying rational expressions with denominators that have opposite binomial factors, such as $x-4$ and $4-x$. Factoring -1 from one of the binomials and using an equivalent rational expression allows you to combine expressions with opposite binomials in their denominators.

EXAMPLE 4: Subtracting Rational Expressions

Simplify $\frac{4x}{x-7}-\frac{3x}{7-x}$.

Answer

$\frac{4x}{x-7}-\frac{3x}{-1(x-7)}$

1. Recognize that the denominators are opposite binomials and factor -1 out of one of the denominators.

$=\frac{4x}{x-7}-\left(-\frac{3x}{x-7}\right)=\frac{4x}{x-7}+\frac{3x}{x-7}$

2. Use an equivalent expression to rewrite the subtraction as the addition of the opposite.

$=\frac{7x}{x-7}$

3. The result cannot be simplified.

SKILL CHECK **EXERCISES 15, 33**

A. EXERCISES

Perform the indicated operations and simplify.

1. $\frac{8}{x} + \frac{6}{x}$
2. $\frac{5}{x+2} + \frac{7}{x+2}$
3. $\frac{4x}{25} - \frac{9x}{25}$
4. $\frac{3x}{y^2} - \frac{5x}{y^2}$
5. $\frac{7}{2c} - \frac{9}{2c}$
6. $\frac{3x}{8} + \frac{5x}{8}$
7. $\frac{2x}{x-7} - \frac{14}{x-7}$
8. $\frac{8}{3x+2} + \frac{12x}{3x+2}$
9. $\frac{5x}{x^2+y} + \frac{4x}{x^2+y}$
10. $\frac{x-5}{x+2} - \frac{x+7}{x+2}$
11. $\frac{x+4}{7} - \frac{x-3}{7}$
12. $\frac{4a}{a-b} - \frac{4b}{a-b}$
13. $\frac{9x}{yz} + \frac{3y}{yz}$
14. $\frac{2x}{x+y} - \frac{2y}{x+y}$

B. EXERCISES

List all correct answers.

15. Which of the following are equivalent expressions to $\frac{3a}{a-b}$?

 A $\frac{3a}{b-a}$

 B $-\frac{3a}{b-a}$

 C $-\frac{-3a}{b-a}$

 D $\frac{-3a}{b-a}$

16. Which of the following are equivalent expressions to $\frac{y}{-x-y}$?

 A $\frac{-y}{x-y}$

 B $-\frac{y}{x+y}$

 C $\frac{-y}{y+x}$

 D $\frac{y}{y+x}$

Simplify the following expressions with opposite denominators.

17. $\frac{3}{m-4} + \frac{6}{4-m}$
18. $\frac{4}{x-2} - \frac{10}{2-x}$
19. $\frac{x}{x-y} + \frac{y}{y-x}$
20. $\frac{a}{b-a} - \frac{b}{a-b}$

Perform the indicated operations and simplify.

21. $\frac{6}{a+b} - \frac{3a}{a+b}$
22. $\frac{3a}{a^2-b^2} - \frac{3b}{a^2-b^2}$
23. $\frac{5}{a^2+3a-2} - \frac{7}{a^2+3a-2}$
24. $\frac{x+2}{x^2-3x-9} - \frac{x-4}{x^2-3x-9}$
25. $\frac{x+2}{x^2-4x-7} + \frac{x-9}{x^2-4x-7}$
26. $\frac{a-2}{a^2-3a+9} + \frac{a+4}{a^2-3a+9}$
27. $\frac{2m^2+4mn+n^2}{m^2+mn-2n^2} + \frac{m^2+3mn+n^2}{m^2+mn-2n^2}$
28. $\frac{5x^2+3x-9}{x^2+7x+10} + \frac{2x^2+8x+3}{x^2+7x+10}$
29. $\frac{3x^2+4x-10}{x^2+5x-36} - \frac{2x^2+x+18}{x^2+5x-36}$
30. $\frac{a^3}{a-1} - \frac{a}{a-1}$
31. $\frac{2x^2-6x+4}{x^2-8x+15} + \frac{x^2+2x-19}{x^2-8x+15}$
32. $\frac{x^2-8x+13}{x^2-4x-12} - \frac{-x^2-4x-1}{x^2-4x-12}$
33. $\frac{a}{a^2-b^2} - \frac{b}{b^2-a^2}$
34. $\frac{y}{x^2-y^2} + \frac{x}{y^2-x^2}$

35. **MP3** Marvin wrote the following solution: $\frac{3x}{x-5} - \frac{15}{5-x} = \frac{3x-15}{x-5} = \frac{3\cancel{(x-5)}}{\cancel{(x-5)}} = 3$. Describe and correct his error.

36. **Essential Question:** How are the rules for adding or subtracting rational expressions similar to adding or subtracting fractions?

C. EXERCISES

Given $f(x) = \frac{x^2-4}{x}$ and $g(x) = \frac{x+2}{x}$, find the simplified expression for each of the following.

37. $f(x) + g(x)$

38. $f(x) - g(x)$

39. $f(x) \cdot g(x)$

40. $f(x) \div g(x)$

Perform the indicated operations and simplify.

41. $\frac{-3x-3}{2x^2-5x-12} - \frac{5x^2-2x+1}{2x^2-5x-12} - \frac{x^2+6x-7}{2x^2-5x-12}$

42. **MP1** $\left(\frac{4x+5}{x-2} - \frac{x-4}{2-x}\right) \div \frac{x+3}{x-2}$

CUMULATIVE REVIEW

Simplify each expression. [8.2, 10.6, 12.1]

43. $(8x^2 + 3x - 5) + (2x^2 + 1)$

44. $\frac{4\sqrt{7}}{5-\sqrt{2}}$

45. $\frac{9 \pm 3\sqrt{7}}{6}$

46. $\frac{3x^2-15x}{x-5}$

Find each sum or difference. [1.1]

47. $\frac{4}{15} + \frac{3}{5}$

48. $\frac{13}{9} - \frac{8}{18}$

49. $\frac{3}{4} + \frac{4}{5}$

50. $\frac{3}{8} - \frac{7}{12}$

51. Choose the expression for b in $\frac{a}{b}$ that will result in the excluded values of $x = 3, -4$. **[12.1]**

A $x^2 + x + 12$

B $x^2 + x - 12$

C $x^2 - x + 12$

D $x^2 - x - 12$

E none of these

52. Simplify $\sqrt{x^3y} \cdot \sqrt{x^2y^3} \cdot \sqrt{xy^2}$. **[10.2]**

A x^2y^2

B x^3y^2

C x^2y^3

D x^3y^3

E none of these

12.4 ADDING & SUBTRACTING EXPRESSIONS WITH UNLIKE DENOMINATORS

How do I add or subtract rational expressions with unlike denominators?

After completing this section, I will be able to

- find the least common multiple of 2 algebraic expressions.
- add rational expressions with unlike denominators.
- subtract rational expressions with unlike denominators.

To add or subtract rational expressions with unlike denominators, you must first find a common denominator. The least common denominator (LCD) is the least common multiple (LCM) of the denominators. To find the LCM of 2 expressions, factor each expression completely and then find the product of the highest power of each factor.

The total time for a roundtrip canoe trip is the sum of 2 rational expressions: $t = \frac{d_{upstream}}{r_{upstream}} + \frac{d_{downstream}}{r_{downstream}}$.

LCM of Whole Numbers	
12	15
$2^2 \cdot 3$	$3 \cdot 5$
LCM = $2^2 \cdot 3 \cdot 5 = 60$	

LCM of Algebraic Expressions	
$9x^3y$	$3y^2z$
$3^2 \cdot x^3 \cdot y$	$3 \cdot y^2 \cdot z$
LCM = $3^2 \cdot x^3 \cdot y^2 \cdot z = 9x^3y^2z$	

EXAMPLE 1: Finding the LCM

Find the LCM of $x^4 - 3x^3 - 10x^2$ and $3x^3 - 30x^2 + 75x$.

Answer

$x^4 - 3x^3 - 10x^2$	$3x^3 - 30x^2 + 75x$	**1.** Factor each polynomial.
$= x^2(x^2 - 3x - 10)$	$= 3x(x^2 - 10x + 25)$	
$= x^2(x - 5)(x + 2)$	$= 3x(x - 5)^2$	
LCM = $3x^2(x - 5)^2(x + 2)$		**2.** The LCM is the product of the highest power of each factor.

SKILL CHECK EXERCISES 1, 5

EXAMPLE 2: Adding Rational Expressions

Simplify each expression.

a. $\frac{3}{2a}+\frac{5}{8}$ **b.** $\frac{2x-5}{3x}+\frac{x-1}{2x}+\frac{5x+6}{12x^2}$

Answers

a. $2a=2\cdot a; 8=2^3$
$\text{LCD}=2^3\cdot a=8a$

1. Find the LCD by determining the LCM of $2a$ and 8.

$\frac{4\cdot 3}{4\cdot 2a}+\frac{5a}{8a}$

2. Multiply the numerator and the denominator of each fraction by the factor that makes each denominator equal to the LCD.

$=\frac{12}{8a}+\frac{5a}{8a}=\frac{5a+12}{8a}$

3. Write the sum of the numerators over the LCD.

b. $3x=3\cdot x; 2x=2\cdot x; 12x^2=2^2\cdot 3\cdot x^2$
$\text{LCD}=2^2\cdot 3\cdot x^2=12x^2$

1. Find the LCD.

$\frac{4x(2x-5)}{4x(3x)}+\frac{6x(x-1)}{6x(2x)}+\frac{5x+6}{12x^2}$
$=\frac{8x^2-20x}{12x^2}+\frac{6x^2-6x}{12x^2}+\frac{5x+6}{12x^2}$

2. Multiply the numerator and the denominator of each fraction by the factor that makes each denominator equal to the LCD.

$=\frac{8x^2-20x+6x^2-6x+5x+6}{12x^2}$

3. Add the numerators by combining like terms.

$=\frac{14x^2-21x+6}{12x^2}$

4. The numerator cannot be factored, so the expression cannot be simplified.

SKILL CHECK **EXERCISES 7, 15**

EXAMPLE 3: Adding Rational Expressions

The expression $\frac{6}{r-1}$ represents the time the King family spent canoeing upstream, and $\frac{6}{r+1}$ represents the time they spent canoeing downstream. Find the simplified expression representing the time for the entire canoe trip.

Answer

$\frac{6}{r-1}+\frac{6}{r+1}$

1. The time for the entire trip will be the sum of the time spent canoeing upstream and downstream.

$r-1$ and $r+1$ have no common factors.
$\text{LCD}=(r-1)(r+1)$

2. Find the LCD.

$\frac{6(r+1)}{(r-1)(r+1)}+\frac{6(r-1)}{(r+1)(r-1)}$
$=\frac{6r+6}{(r-1)(r+1)}+\frac{6r-6}{(r-1)(r+1)}$

3. Multiply the numerator and the denominator of each fraction by the factor that makes each denominator equal to the LCD.

$=\frac{6r+6+6r-6}{(r-1)(r+1)}=\frac{12r}{(r-1)(r+1)}$

4. Write the sum of the numerators over the LCD.

The expression $\frac{12r}{(r-1)(r+1)}$ represents the total time the King family spent on their canoe trip.

SKILL CHECK **EXERCISE 13**

EXAMPLE 4: Subtracting Rational Expressions

Simplify each expression.

a. $\dfrac{x-3}{x+1} - \dfrac{x-4}{x+2}$ **b.** $\dfrac{b}{a^2+ab} - \dfrac{a}{ab+b^2}$

Answers

a. $x+1$ and $x+2$ have no common factors. $\text{LCD} = (x+1)(x+2)$

1. Find the LCD.

$$\frac{(x-3)(x+2)}{(x+1)(x+2)} - \frac{(x-4)(x+1)}{(x+2)(x+1)}$$
$$= \frac{x^2-x-6}{(x+1)(x+2)} - \frac{x^2-3x-4}{(x+1)(x+2)}$$

2. Multiply the numerator and the denominator of each fraction by the factor that makes each denominator equal to the LCD.

$$= \frac{x^2-x-6-x^2+3x+4}{(x+1)(x+2)}$$
$$= \frac{2x-2}{(x+1)(x+2)}$$

3. Subtract.

$$= \frac{2(x-1)}{(x+1)(x+2)}$$

4. Factoring the numerator shows that the expression cannot be simplified.

b. $\dfrac{b}{a(a+b)} - \dfrac{a}{b(a+b)}$

1. Factor the denominators to find the LCD, $ab(a+b)$.

$$= \frac{bb}{a(a+b)b} - \frac{aa}{b(a+b)a}$$
$$= \frac{b^2}{ab(a+b)} - \frac{a^2}{ab(a+b)}$$

2. Multiply the numerator and the denominator of each fraction by the factor that makes each denominator equal to the LCD.

$$= \frac{b^2-a^2}{ab(a+b)}$$

3. Subtract.

$$= \frac{(b-a)\cancel{(b+a)}}{ab\cancel{(a+b)}} = \frac{b-a}{ab}$$

4. Factor and simplify by canceling the common factors $a+b$ and $b+a$.

SKILL CHECK **EXERCISE 17**

ADDING & SUBTRACTING RATIONAL EXPRESSIONS WITH UNLIKE DENOMINATORS

1. Find the LCD of the rational expressions.
2. Make equivalent rational expressions that have the LCD.
3. Combine the numerators and place the result over the common denominator.
4. Simplify the resulting rational expression.

EXAMPLE 5: Rewriting Sums & Differences

AfterSchoolHelp

Adding & Subtracting Expressions with Unlike Denominators

Simplify $\frac{x+1}{2x-4}+\frac{x+1}{2x-x^2}$.

Answer

$$\frac{x+1}{2(x-2)}+\frac{x+1}{x(2-x)}$$

1. Factor the denominators to find the LCD.

$$=\frac{x+1}{2(x-2)}+\frac{x+1}{-x(x-2)}$$
$$=\frac{x+1}{2(x-2)}-\frac{x+1}{x(x-2)}$$

2. Since $x-2$ and $2-x$ are opposites, factor -1 out of the factor $2-x$ and restate as a subtraction problem.

$$=\frac{(x+1)x}{2(x-2)x}-\frac{2(x+1)}{2x(x-2)}$$
$$=\frac{x^2+x}{2x(x-2)}-\frac{2x+2}{2x(x-2)}$$

3. Multiply the numerator and the denominator of each fraction by the factor that makes each denominator equal to the LCD, $2x(x-2)$.

$$=\frac{x^2-x-2}{2x(x-2)}$$

4. Subtract.

$$=\frac{\cancel{(x-2)}(x+1)}{2x\cancel{(x-2)}}=\frac{x+1}{2x}$$

5. Factor and simplify by canceling common factors.

SKILL CHECK EXERCISE 21

A. EXERCISES

Find the LCM of each pair of expressions. Leave polynomials in factored form.

1. $6x$ and $8x^3$

2. $4ab^2$ and $12bc$

3. $2r+12$ and $3r+18$

4. $3x-1$ and $x-3$

5. a^2-b^2 and $a^2-2ab+b^2$

6. t^2-3t+2 and t^2-t-2

Perform the indicated operations and simplify.

7. $\frac{3x}{4}+\frac{9}{16}$

8. $\frac{x}{9}-\frac{4x}{7}$

9. $\frac{x+y}{5}-\frac{x}{15}$

10. $\frac{y-3}{18}+\frac{y+7}{4}$

11. $\frac{a^2+3a}{5}+\frac{a^2-9}{2}$

12. $\frac{4}{3x}-\frac{7}{21x^3}$

13. $\frac{3}{c}+\frac{4}{c+1}$

14. $\frac{9}{3m^3}+\frac{14}{8m^2}$

15. $\frac{5}{2x^3y^2}+\frac{10}{xy}$

16. $\frac{x^2-4}{x^2}+\frac{x-y}{x}$

17. $\frac{2b}{b+5}-\frac{3}{b-4}$

18. $\frac{g}{(g-2)^2}-\frac{3}{g-2}$

B. EXERCISES

Perform the indicated operations and simplify.

19. $\frac{2y}{(y-7)(y+4)}-\frac{9}{(y-7)^2}$

20. $\frac{7}{(x+2)(x-5)}+\frac{9}{(x+2)(x+3)}$

21. $\frac{x-3}{(x+2)(4-x)}-\frac{x+9}{(x+2)(x-4)}$

22. $\frac{2b-10}{(b-5)^2}+\frac{3b-5}{5-b}$

23. $\frac{-d}{2d+8}+\frac{2d}{3d-12}$

24. $\frac{1}{2x+8}-\frac{x+3}{3x+12}$

25. $\frac{1}{3x-9}+\frac{1}{4x+16}$

26. $\frac{5a}{a-1}-\frac{6a^2}{a^2-1}$

27. $\frac{3}{x^2-y^2}+\frac{5}{x+y}$

28. $\frac{10}{x^2-6x-27}+\frac{8}{x^2-3x-18}$

29. $\frac{3y}{y^2+7y+10}-\frac{2y^2}{y^2+10y+25}$

30. $\frac{m}{m^2-n^2}-\frac{n}{n^2-m^2}$

31. $\frac{x-3}{x+5}+\frac{2x-7}{x+3}+\frac{26}{x^2+8x+15}$

32. $\frac{x}{x+3}+x-\frac{x}{x+5}$

33. **Essential Question:** How do I add or subtract rational expressions with unlike denominators?

C. EXERCISES

Perform the indicated operations and simplify.

34. $\frac{a+3}{a+5}-\frac{a+2}{a-5}+\frac{a}{a^2-25}$

35. $\frac{x+3}{x^2+6x-16}+\frac{x-2}{x^2-6x+8}$

36. $\frac{3-x}{x^2-9}+\frac{x+2}{x^2+5x+6}-\frac{x+2}{4-x^2}$

37. $\frac{2}{x+1}+\frac{1}{x-1}+\frac{2}{1-x^2}$

38. **MP1** $\frac{7}{x-3}-\frac{5}{x+3}-\frac{4}{2x-1}$

39. **MP3** Describe the error in the work below and then complete the problem correctly.
$\frac{3x+5}{(x-2)(x-5)}-\frac{4x+1}{(2-x)(5-x)}=\frac{3x+5}{(x-2)(x-5)}+\frac{4x+1}{(x-2)(x-5)}$

40. **MP2** Simplify $\frac{x-1}{(x-3)(x-4)}-\frac{x-9}{(x-3)(4-x)}-\frac{x-3}{(3-x)(4-x)}+\frac{2-x}{(3-x)(x-4)}$.

CUMULATIVE REVIEW

Simplify each expression. [7.1, 10.1, 10.3, 12.2]

41. $(5x^{-3}y^2)^2(4x^2y^3)^{-1}$

42. $(-2)^{-3}(2a^{-2}b^2)^{-1}$

43. $\sqrt{50}$

44. $\sqrt{\frac{2}{5}}$

45. $\frac{x^2-1}{12}\cdot\frac{4}{x-1}$

46. $\frac{4a^2-4b^2}{a-b}\cdot\frac{1}{4}$

47. Angle 1 of a triangle is 10° more than angle 2. Angle 3 is 6° less than twice angle 2. Find the measure of each angle. **[2.2]**

48. A ladder must reach 20 ft up a telephone pole when it is positioned 6 ft from the base of the pole. How long does the ladder need to be? **[10.5]**

49. Which of the following expressions is not equivalent to $\frac{15}{8}\div\frac{9}{16}$? **[1.2]**

A $3\frac{2}{3}$

B $\frac{\frac{15}{8}}{\frac{9}{16}}$

C $\frac{15\cdot16}{8\cdot9}$

D $1\frac{7}{9}\cdot1\frac{7}{8}$

E none of these

50. Which quadratic function's vertex is also its only zero? **[11.9]**

A $f(x)=x^2-9$

B $g(x)=x^2+12x+36$

C $h(x)=x^2+1$

D $k(x)=(x-5)^2-1$

E none of these

MATH IN HISTORY
JOHN VON NEUMANN (1903–57)

John von Neumann, a Hungarian American mathematician, is well known in a variety of mathematical fields, especially computer science, game theory, and quantum mechanics. His discoveries helped bring modern society into both the atomic age and the computer age. Von Neumann was a child prodigy with a photographic memory. By age 10 he had completed college calculus. He earned a PhD in math at age 23 from the University of Budapest and had written ten major professional papers by age 25.

After teaching at two universities in Germany, von Neumann came to the United States as a visiting professor at Princeton. In 1933 he and Albert Einstein became two of the original faculty members of the Institute for Advanced Study at Princeton.

In 1937 von Neumann became an American citizen. During World War II he worked with the US War Department, studying submarine warfare, economic intelligence, and ordnance problems. His most famous work originated in secret meetings at Los Alamos, New Mexico, where scientists and mathematicians gathered to develop the atomic bomb. By creating a means of detonation by an inward blast, von Neumann expedited the development of the atomic bomb by a year.

12.5 MIXED & COMPLEX EXPRESSIONS

What accounts for the ability of math to model the physical universe?

After completing this section, I will be able to

- simplify mixed rational expressions.
- simplify complex rational expressions.
- apply what I am learning through math to praise the Creator.

The modeling ability of math has been described as mysterious, incredible, and even unreasonably effective. But for a believer, math is simply an attempt to discover and describe the design found in a created universe. We expect the mathematical structure found in human reasoning to mirror God's universe since both are reflections of their Creator. Even math should cause us to praise our God.

The formula $\phi = \frac{e}{\frac{v(v-1)}{2}}$ measures the connectivity relating the number of observed links (e) between nodes (v) to the number of possible links in a network.

The sum of a polynomial and a rational expression is a mixed rational expression. By adding the terms, you can write the sum as a single rational expression.

EXAMPLE 1: Simplifying a Mixed Expression

Simplify $4y + \frac{3}{y-2}$.

Answer

$$\frac{4y}{1} \cdot \frac{y-2}{y-2} + \frac{3}{y-2}$$

1. Multiply the numerator and the denominator of the first term by $y - 2$.

$$= \frac{4y^2 - 8y}{y-2} + \frac{3}{y-2}$$

2. Distribute.

$$= \frac{4y^2 - 8y + 3}{y-2}$$

3. Add the numerators.

$$= \frac{(2y-1)(2y-3)}{y-2}$$

4. Factoring reveals that no further simplifying can take place.

SKILL CHECK **EXERCISES 1, 5**

The division $\frac{1}{2} \div \frac{3}{4}$ can also be written as a fraction, $\frac{\frac{1}{2}}{\frac{3}{4}}$. A fraction containing a fraction in the numerator or denominator is called a complex fraction. A *complex rational expression* is a complex fraction that contains variables.

DEFINITION

A **complex rational expression** is an algebraic expression of the form $\frac{n}{d}$ in which the numerator, n, the denominator, d, or both are rational expressions.

Expressions containing fractions in the numerator or the denominator should always be simplified. Remembering that fractions are another way of writing a division problem can be helpful in this simplification.

EXAMPLE 2: Simplifying Complex Expressions

Simplify each expression.

a. $\dfrac{e}{\dfrac{v(v-1)}{2}}$ **b.** $\dfrac{\dfrac{x+2}{x^2}}{\dfrac{x+2}{x}}$

Answers

a. $\dfrac{e}{\dfrac{v(v-1)}{2}} = e \div \dfrac{v(v-1)}{2}$

1. Think of the complex fraction as a division problem.

$= e \cdot \dfrac{2}{v(v-1)}$

2. Multiply by the reciprocal of the denominator.

$= \dfrac{2e}{v(v-1)}$

3. Simplify.

b. $\dfrac{\dfrac{x+2}{x^2}}{\dfrac{x+2}{x}} = \dfrac{x+2}{x^2} \cdot \dfrac{x}{x+2}$

1. Express the division indicated by the complex fraction as multiplication of the numerator by the reciprocal of the denominator.

$= \dfrac{\cancel{x+2}}{xx} \cdot \dfrac{x}{\cancel{x+2}} = \dfrac{1}{x}$

2. Simplify.

SKILL CHECK **EXERCISES 11, 13**

Complex fractions can also be simplified by multiplying the numerator and the denominator by the LCD of all the fractions in both the numerator and the denominator. Each of the following examples shows 2 different methods of simplifying a complex expression. Choose whichever method seems easier to you.

SIMPLIFYING COMPLEX RATIONAL EXPRESSIONS	
Method 1	Multiply the numerator by the reciprocal of the denominator.
Method 2	Multiply the numerator and the denominator by their LCD.

EXAMPLE 3: Simplifying a Complex Expression

Simplify $\dfrac{\dfrac{x^2-9}{4x}}{\dfrac{x+3}{8x^2}}$.

Answer

Method 1

Multiply the numerator by the reciprocal of the denominator.

$\dfrac{x^2-9}{4x} \cdot \dfrac{8x^2}{x+3} = \dfrac{(x-3)\cancel{(x+3)}}{\cancel{4x}} \cdot \dfrac{\cancel{4x} \cdot 2x}{\cancel{x+3}}$

$= 2x(x-3) = 2x^2 - 6x$

Method 2

Multiply the numerator and the denominator by their LCD, $8x^2$.

$\dfrac{\dfrac{x^2-9}{4x} \cdot 8x^2}{\dfrac{x+3}{8x^2} \cdot 8x^2} = \dfrac{(x^2-9)2x}{x+3} = \dfrac{(x-3)\cancel{(x+3)}2x}{\cancel{x+3}}$

$= 2x(x-3) = 2x^2 - 6x$

SKILL CHECK **EXERCISE 29**

AFTERSCHOOLHELP

Mixed & Complex Expressions

The second method may be easier when the numerator or the denominator contains more than 1 term.

EXAMPLE 4: Simplifying a Complex Expression

Simplify $\dfrac{1+\frac{a^2}{b^2}}{1-\frac{a}{b}}$, leaving the answer in factored form.

Answer

Method 1

$$\frac{\frac{b^2}{b^2}+\frac{a^2}{b^2}}{\frac{b}{b}-\frac{a}{b}}=\frac{\frac{b^2+a^2}{b^2}}{\frac{b-a}{b}}$$

1. Since the fraction bar also acts as a grouping symbol, first combine the terms in the numerator and the denominator.

$$=\frac{b^2+a^2}{b\not{b}}\cdot\frac{\not{b}}{b-a}$$
$$=\frac{b^2+a^2}{b(b-a)}$$

2. Multiply by the reciprocal of the denominator.

Method 2

$$\frac{\left(1+\frac{a^2}{b^2}\right)b^2}{\left(1-\frac{a}{b}\right)b^2}$$

1. Multiply the numerator and the denominator by their LCD, b^2.

$$=\frac{b^2+\frac{a^2}{\not{b^2}}\cdot\not{b^2}}{b^2-\frac{a}{\not{b}}\cdot\not{b}b}$$
$$=\frac{b^2+a^2}{b^2-ab}$$

2. Distribute and simplify.

$$=\frac{b^2+a^2}{b(b-a)}$$

3. Factoring reveals that the expression cannot be simplified.

SKILL CHECK EXERCISE 25

VOCABULARY

- complex rational expression

A. EXERCISES

Simplify each mixed rational expression.

1. $2+\frac{1}{x}$
2. $y+\frac{1}{2}$
3. $2y+\frac{1}{y+1}$
4. $3x-\frac{1}{x-3}$
5. $2a-\frac{a}{b}$
6. $3b+\frac{a^2}{b-a}$

Simplify each complex rational expression.

7. $\dfrac{\frac{1}{4}}{\frac{5}{8}}$
8. $\dfrac{8\frac{4}{5}}{3\frac{3}{7}}$
9. $\dfrac{\frac{a}{b}}{\frac{c}{d}}$
10. $\dfrac{\frac{x}{y}}{z}$
11. $\dfrac{x}{\frac{y}{z}}$
12. $\dfrac{\frac{1}{r}}{\frac{1}{s}}$
13. $\dfrac{\frac{a^2}{d}}{\frac{ad}{c^3}}$
14. $\dfrac{x^2y^3}{\frac{x}{y}}$
15. $\dfrac{\frac{y}{c}}{c+y}$
16. $\dfrac{\frac{5}{x+y}}{x-y}$

B. EXERCISES

Simplify each mixed rational expression.

17. $x+2+\frac{1}{x}$
18. $y-1+\frac{1}{y-2}$
19. $2y+3-\frac{1}{y+1}$
20. $x+3-\frac{8x}{x-3}$
21. Given the general expression for a mixed number, $A\frac{b}{c}$, meaning $A+\frac{b}{c}$, write a general expression for the equivalent improper fraction.

22. Given the general expression for a mixed expression, $p(x) + \frac{n(x)}{d(x)}$, where $p(x)$, $n(x)$, and $d(x)$ are polynomials, write a general expression for the equivalent rational expression.

Simplify each complex rational expression, leaving the answer in factored form.

23. $\dfrac{x}{1 + \frac{1}{x}}$

24. $\dfrac{\frac{1}{m} + m}{\frac{1}{m} - m}$

25. $\dfrac{\frac{x}{y} + 1}{\frac{x}{y} - \frac{y}{x}}$

26. $\dfrac{\frac{1}{a+b}}{\frac{a+b}{a^2 - b^2}}$

27. $\dfrac{\frac{x+y}{x-y}}{x^2 - y^2}$

28. $\dfrac{x^2 + 2xy + y^2}{\frac{x+y}{x-y}}$

29. $\dfrac{\frac{3a}{2a^2 + 4a}}{\frac{6a}{a^2 + 2a}}$

30. $\dfrac{\frac{m+n}{m-n}}{\frac{1}{m^2 - n^2}}$

31. $\dfrac{\frac{4x^3 - 36x}{2x^3 - 2x^2 - 12x}}{x^2 + 5x + 6}$

32. $\dfrac{4x^3 - 36x}{\frac{2x^3 - 2x^2 - 12x}{x^2 + 5x + 6}}$

33. $\dfrac{\frac{3 - x}{x^2 + 7x + 10}}{\frac{x^2 - 9}{x^2 - 4x - 12}}$

34. **Essential Question:** Why does mathematics model our universe so well?

KEYWORD SEARCH

Doppler effect

MP4 The pitch of a car's horn as it approaches a listener is higher than when the car is stationary. As the car moves away from the listener, the pitch is lower. This Doppler effect can be modeled by the following equation for the perceived frequency: $h = \left(\frac{1}{1 - \frac{v}{s}}\right)f$, where v is the speed at which the sound's source is approaching, s is the speed of sound, and f is the frequency in hertz (cycles per second) of the sound emitted by the source.

A sonic boom is an extreme example of the Doppler effect.

35. Write the expression $\left(\frac{1}{1 - \frac{v}{s}}\right)f$ as a simplified rational expression.

36. Determine the frequency of the sound heard by a listener if a car's horn emits a frequency of 400 Hz and the car is approaching at 85 km/hr. Assume the speed of sound to be 1245 km/hr.

37. Determine the frequency of the sound heard by the listener as the car moves away at the same speed. (*Hint*: Use a negative value for speed.)

Graph theory is the study of relationships between a set of nodes (vertices) and how they are linked (connected). Applications of graph theory include transportation studies, computer networking, mapping DNA strands, and much more. In graph theory, a cycle is a series of nodes (v) connected by links (e) in which the first and last node in the path is the same. The formula $\alpha = \dfrac{e - v}{\frac{v(v-1)}{2} - (v - 1)}$ measures the number of cycles in a graph compared to the maximum number of possible cycles.

Alpha (α) values fall between 0 and 1, where values closer to 1 could reveal a high level of redundancy in a company's network.

38. Write the expression $\dfrac{e - v}{\frac{v(v-1)}{2} - (v - 1)}$ as a simplified rational expression.

39. Evaluate your simplified expression for α, where $e = 50$ and $v = 20$.

C. EXERCISES

Simplify.

40. $\dfrac{\frac{1}{2x}}{\frac{2x}{3}} \div \dfrac{\frac{3x}{4}}{\frac{4}{3} - x}$

41. MP1 $\dfrac{\frac{x}{y} - \frac{x-y}{x+y}}{\frac{y}{x} + \frac{x-y}{x+y}}$

42. MP4 Mr. Sung commutes *d* miles to and from work each day. His average speed for his drive to work early in the morning is 60 mi/hr, but during his drive home in rush-hour traffic his average speed is only 40 mi/hr. Using the fact that the average speed is the total distance divided by the total time, demonstrate that the average speed for his entire commute is 48 mi/hr, not 50 mi/hr.

43. MP8 Write a general expression for the average speed of a roundtrip in which the average speed for the drive to a destination is r_1 and the average speed for the return drive is r_2.

CUMULATIVE REVIEW

Simplify each expression and leave your answer in factored form. [12.1–12.4]

44. $\dfrac{(x-6)^2}{x(x-6)}$

45. $\dfrac{(a-11b)(a+7b)}{121b^2 - a^2}$

46. $\dfrac{(x+9)(x-4)}{(x+6)(x-4)} \cdot \dfrac{5x^2 + 30x}{x+9}$

47. $\dfrac{3x+1}{x^3 - 25x} \div \dfrac{2x(3x+1)}{x^2(5-x)}$

48. $-\dfrac{3a^2}{a^2 + 2ab - 3b^2} + \dfrac{4a^2 + 3ab}{a^2 + 2ab - 3b^2}$

49. $\dfrac{5a^2 + 2ab}{25a^2 - b^2} - \dfrac{b^2 - 2ab}{25a^2 - b^2}$

50. $\dfrac{x+1}{3x-12} - \dfrac{x-2}{4x - x^2}$

51. $\dfrac{6}{(3x-2)(x+4)} + \dfrac{3}{3x^2 - 11x + 6}$

52. Which expression represents the shaded portion of the figure? **[12.1]**

A $\frac{8}{\pi r}$

B $\frac{2}{\pi}$

C $\frac{\pi}{4}$

D $\frac{\pi}{8}$

E none of these

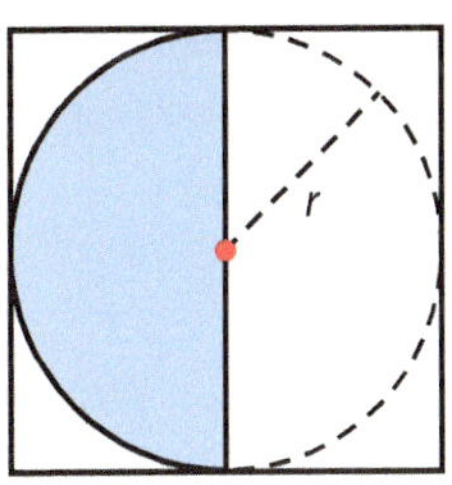

53. Which sum or difference is correctly paired with the LCD of its terms? **[12.4]**

A $\frac{3}{x-6} - \frac{x+1}{12-2x}$; LCD: $2(x-6)$

B $\frac{x+6}{x^2-1} + \frac{2x}{(x+1)(x+2)}$; LCD: $(x^2-1)(x+1)(x+2)$

C $\frac{y}{6x} + \frac{3x}{4y}$; LCD: $18xy$

D $\frac{5}{x^2-4} - \frac{x-8}{x^2+4x}$; LCD: $x(x^2-4)$

E none of these

12.6 SOLVING RATIONAL EQUATIONS

How can I solve an equation with a variable in the denominator?

After completing this section, I will be able to

- solve rational equations by using cross products.
- solve rational equations by clearing fractions.

DEFINITION

A **rational equation** is an equation containing rational expressions.

Rational equations can be solved by applying the Property of Cross Products if each side of the equation is a single rational expression.

The total resistance of a circuit with parallel resistors is a complex rational expression.

EXAMPLE 1: Using the Property of Cross Products

Solve $\frac{5}{x+1} = \frac{-2}{x-6}$.

Answer

$5(x-6) = -2(x+1)$ — 1. Apply the Property of Cross Products.

$5x - 30 = -2x - 2$ — 2. Solve.

$7x = 28$

$x = 4$

$\frac{5}{4+1} = \frac{-2}{4-6}$ — 3. Check your solution in the original equation.

SKILL CHECK **EXERCISE 11**

EXAMPLE 2: Using the Property of Cross Products

What number must be added to both the numerator and the denominator of $\frac{3}{5}$ to produce a fraction equivalent to $\frac{6}{7}$?

Answer

$\frac{3+n}{5+n} = \frac{6}{7}$ — 1. Write the appropriate equation, using n to represent the number added to the numerator and the denominator.

$7(3+n) = 6(5+n)$ — 2. Apply the Property of Cross Products.

$21 + 7n = 30 + 6n$ — 3. Solve.

$n = 9$

$\frac{3+9}{5+9} = \frac{12}{14} = \frac{6}{7}$ — 4. Check your solution in the original equation.

SKILL CHECK **EXERCISE 27**

When the rational equation is not a proportion, the fractions can be cleared by multiplying both sides of the equation by the LCD.

EXAMPLE 3: Solving by Clearing Fractions

Solve each equation.

a. $\frac{1}{R} = \frac{1}{4} + \frac{1}{8} + \frac{1}{8}$ **b.** $\frac{1}{2x} + \frac{5}{12} = -\frac{1}{3x}$

Answers

a. $$8R\left(\frac{1}{R}\right)=8R\left(\frac{1}{4}+\frac{1}{8}+\frac{1}{8}\right)$$
$$8=2R+R+R$$

1. Multiply both sides by the LCD, $8R$, to clear the fractions.

$$8=4R$$
$$R=2$$

2. Solve.

$$\frac{1}{2}=\frac{1}{4}\left(\frac{2}{2}\right)+\frac{1}{8}+\frac{1}{8}$$
$$\frac{1}{2}=\frac{2+1+1}{8}$$

3. Check your solution in the original equation.

b. $$12x\left(\frac{1}{2x}+\frac{5}{12}\right)=12x\left(-\frac{1}{3x}\right)$$
$$\overset{6}{\cancel{12x}}\left(\frac{1}{\cancel{2x}}\right)+\cancel{12}x\left(\frac{5}{\cancel{12}}\right)=\overset{4}{\cancel{12x}}\left(-\frac{1}{\cancel{3x}}\right)$$
$$6+5x=-4$$

1. Multiply both sides by the LCD, $12x$, to clear the fractions.

$$5x=-10$$
$$x=-2$$

2. Solve.

$$\frac{1}{2(-2)}+\frac{5}{12}=-\frac{1}{3(-2)}$$
$$-\frac{1}{4}+\frac{5}{12}=\frac{1}{6}$$
$$-\frac{3}{12}+\frac{5}{12}=\frac{2}{12}$$

3. Check your solution in the original equation.

SKILL CHECK **EXERCISE 15**

When you multiply by a variable expression to clear the denominators, the new equation may introduce solutions for which the original equation is undefined. You must check that your solution is not an excluded value for any of the expressions in the original rational equation. Any solution that does not check is extraneous.

AFTERSCHOOLHELP

Solving Rational Equations

EXAMPLE 4: Solving by Clearing Fractions

Solve $\frac{x-12}{x-10}=\frac{2}{x}-\frac{20}{x^2-10x}$.

Answer

$$\frac{x-12}{x-10}=\frac{2}{x}-\frac{20}{x(x-10)}$$

1. Factor the denominators to identify the LCD as $x(x-10)$.

$$x(x-10)\left(\frac{x-12}{x-10}\right)=x(x-10)\left(\frac{2}{x}-\frac{20}{x(x-10)}\right)$$
$$x\cancel{(x-10)}\left(\frac{x-12}{\cancel{x-10}}\right)=\cancel{x}(x-10)\left(\frac{2}{\cancel{x}}\right)-\cancel{x(x-10)}\left(\frac{20}{\cancel{x(x-10)}}\right)$$
$$x(x-12)=2(x-10)-20$$

2. Multiply both sides by the LCD to clear the fractions.

$$x^2-12x=2x-40$$
$$x^2-14x+40=0$$
$$(x-10)(x-4)=0$$
$$x-10=0 \quad \text{or} \quad x-4=0$$
$$x=10 \qquad\qquad x=4$$

3. Solve.

Since x cannot equal 0 or 10 in the original equation, $x=10$ is an extraneous solution.
Therefore, $x=4$ is the only solution to the equation.

4. Check that your solutions are not excluded values in the original equation.

SKILL CHECK **EXERCISE 21**

VOCABULARY

- rational equation

A. EXERCISES

Solve each equation.

1. $\frac{m}{3} = 9$

2. $\frac{x+1}{3} = 4$

3. $\frac{2}{x} = 8$

4. $\frac{8}{y} = 16$

5. $\frac{8}{x+2} = 3$

6. $\frac{10}{a-5} = 7$

7. $\frac{x+3}{x-2} = 1$

8. $\frac{x-5}{x-4} = 2$

9. $\frac{a-9}{15} = \frac{a-6}{5}$

10. $\frac{8}{2y} = \frac{4}{y-2}$

11. $\frac{1}{m-9} = \frac{2}{m+4}$

12. $\frac{2}{a-7} = \frac{9}{a+3}$

B. EXERCISES

Solve each equation.

13. $\frac{a+4}{11} - \frac{a+2}{3} = 7$

14. $\frac{4}{3x} + 7 = \frac{1}{6x}$

15. $\frac{1}{6} = \frac{1}{R} + \frac{1}{15}$

16. $\frac{x+5}{2x} - \frac{8x+3}{9x} = \frac{5}{18}$

17. $\frac{5}{x+9} - \frac{7}{x-8} = 0$

18. $\frac{a-3}{4} - \frac{5}{a+2} = 1$

19. $\frac{4}{a+3} + \frac{2}{a} = \frac{6}{a^2+3a}$

20. $\frac{1}{c+1} - \frac{2}{c^2-1} = \frac{1}{1-c}$

21. $\frac{x+3}{x-4} = \frac{4}{x} + \frac{28}{x^2-4x}$

22. $x - \frac{1}{1-x} = \frac{1}{x-1}$

23. $\frac{m+4}{m-9} = \frac{m-3}{m-6}$

24. $\frac{x+12}{x-2} - \frac{4}{x} = \frac{-4}{x^2-2x}$

25. The width of a rectangular picture is one-half its length. If its area is 648 in.^{2}, what are the dimensions of the picture?

26. Jack and Erin spent $\frac{1}{4}$ of their money on rides at the fair. They paid \$20 for food and transportation and returned with $\frac{4}{7}$ of their money. How much money did they take to the fair?

27. What number must be added to both the numerator and the denominator of $\frac{5}{11}$ to produce a fraction equivalent to $\frac{3}{4}$?

28. What number must be added to the numerator and subtracted from the denominator of $\frac{1}{2}$ to produce a fraction equivalent to $-\frac{8}{7}$?

29. One number is 6 less than another number. The quotient of the larger divided by the smaller is $\frac{5}{2}$. What are the 2 numbers?

30. Divide 84 into 2 parts that are in the ratio of 3 : 4.

31. **MP2** Petra has made 35 of 50 free-throw attempts in practice.

a. What is the least number of free throws she must make in a row so that her percentage of free throws made is at least 80%?

b. What is the greatest number of free throws she can miss in a row before her percentage of free throws made falls below 60%?

32. **Essential Question:** Explain how to solve a rational equation.

C. EXERCISES

Solve each equation.

33. $\frac{x^2}{9} = \frac{x}{3} + 4$

34. $\frac{x^2}{4} - \frac{3x}{8} = 2$

35. $\frac{4}{x^2 - 8x + 15} = \frac{1}{x^2 - 3x - 10}$

36. $\frac{1}{x^2 - 2x - 8} - \frac{4}{x^2 - x - 12} = \frac{2}{x^2 - 2x - 8}$

37. **MP4** In a circuit connecting 2 resistors in parallel, the total resistance R, in ohms, is $\frac{1}{R} = \frac{1}{R_1} + \frac{1}{R_2}$, where R_1 and R_2 represent the individual resistance of the 2 resistors. If $R = 6$ ohms and the individual resistance of R_1 is 16 ohms greater than R_2, find R_2. Assume positive resistance.

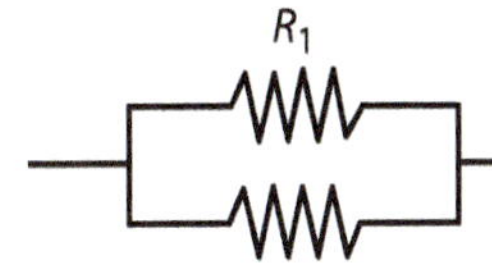

38. **MP8** Solve $\frac{1}{x^2 - a^2} - \frac{1}{x^2 - ax - bx + ab} = \frac{1}{(x - a)(x - b)}$ for x.

CUMULATIVE REVIEW

Solve each literal equation for the indicated variable. [2.3]

39. $f = \frac{1}{T}$ for T

40. $\frac{6yp - t}{q} = n$ for y

Solve each system. [6.4]

41. $4x + 3y = 9$
$3x - 7y = 16$

42. $y = \frac{3}{4}x - 3$
$3x - 4y = 12$

43. For which value(s) is $\frac{x}{3x + 5}$ undefined? **[12.1]**

Simplify each expression. [12.5]

44. $\frac{a}{\frac{a - b}{a + b}}$

45. $\frac{\frac{p - r}{p}}{p - \frac{r^2}{p}}$

46. $5c + \frac{b^2}{c - b}$

47. Which expression is equivalent to $\frac{2x - 4}{2 - x} - \frac{7x - 14}{x - 2}$? **[12.3]**

A $\frac{-5x + 10}{x - 2}$

B $\frac{9x + 10}{2 - x}$

C -9

D -5

E A and D

48. What is the LCM of $8xyz^2$, $2x^2y$, and $6x^3y^2z^5$? **[12.4]**

A $2x$

B $2xyz^2$

C $24x^3y^2z^5$

D $288x^6y^4z^7$

E none of these

TECHNOLOGY CORNER (TI-84 PLUS FAMILY)

Solving Rational Equations

Rational equations can be solved graphically using the TI-84 Plus. To begin solving the equation $\frac{5}{x+1} = \frac{-2}{x-6}$, press [y=] and enter the left side, $\frac{5}{x+1}$, as Y_1 and the right side, $\frac{-2}{x-6}$, as Y_2. Press [zoom] [6] to view the graphs in a standard 10 × 10 window.

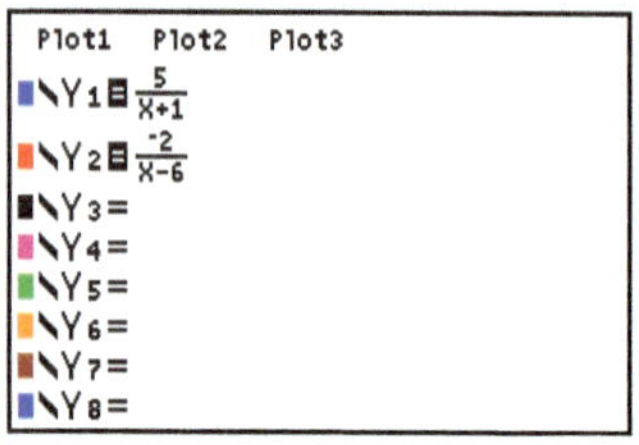

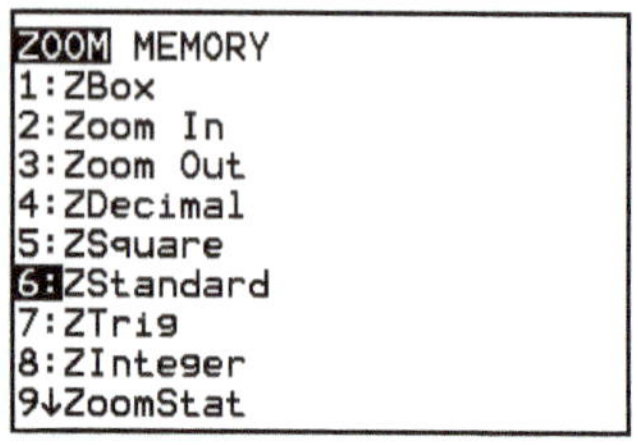

Select [calc] ([2nd] [trace]) and then choose option 5: intersect. Press [enter] three times, once to indicate the first equation's curve, once to indicate the second equation's curve, and once to accept the default value for the requested guess. The x-value of the intersection of the 2 graphs represents the solution to the equation.

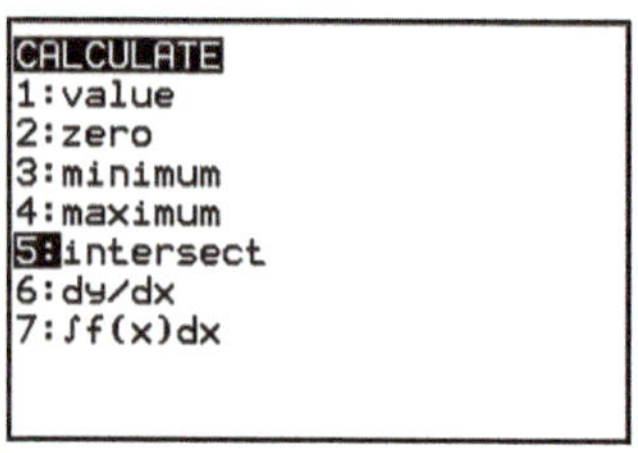

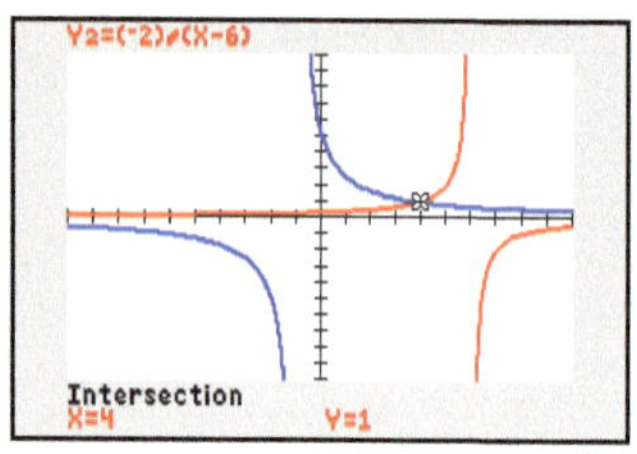

Now enter $\frac{1}{2x} + \frac{5}{12}$ as Y_1 and $-\frac{1}{3x}$ as Y_2. Select [calc] ([2nd] [trace]) and then choose option 5: intersect. Press the right arrow key [▶] until the cursor appears on the right side of the y-axis.

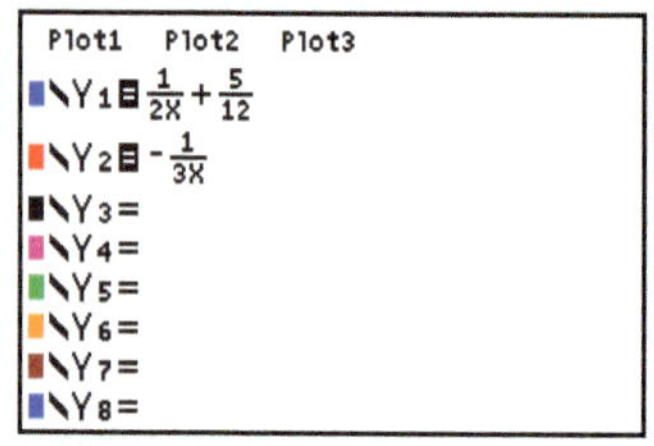

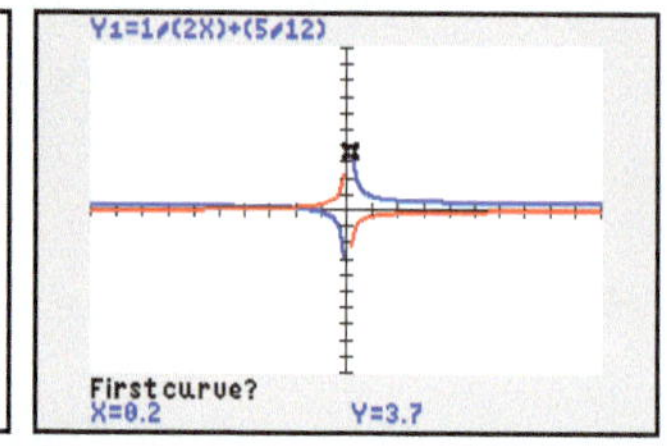

Pressing [enter] three times reveals an error message which indicates the right portions of the 2 graphs do not intersect. Press [enter] to clear the screen, then select [calc] ([2nd] [trace]) and choose option 5: intersect. Press the left arrow key [◀] until the cursor appears on the left side of the y-axis.

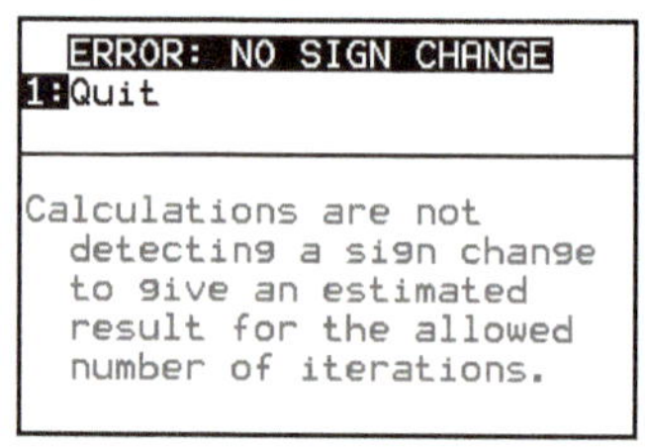

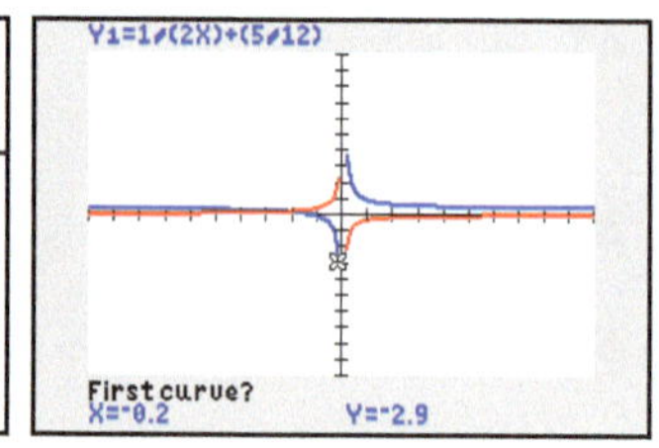

Then press [enter] three times to reveal the intersection of the 2 graphs. The x-value at the bottom of the screen will be the solution to the equation.

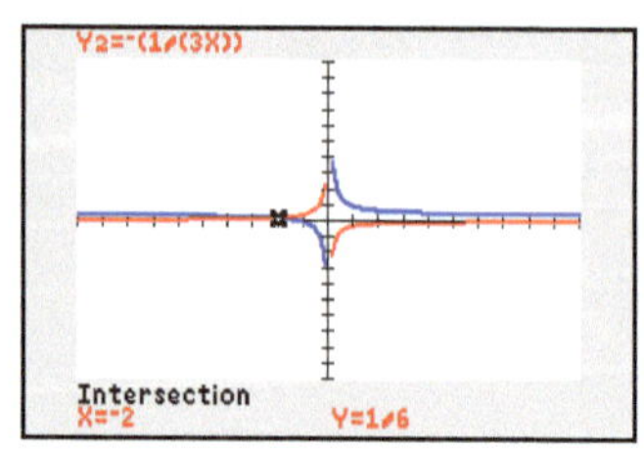

12.7 APPLYING RATIONAL EQUATIONS

How can rational equations be used to allocate resources in business?

After completing this section, I will be able to

- solve real-world rate problems by writing and solving rational equations.

The same table format used in Section 2.7 to organize information for distance problems can be applied to the problems in this section as well.

After filling in 2 of the columns with variable expressions and information from the problem, use $r \cdot t = d$ to derive the expressions in the third column. Then write an equation relating the quantities in the table and solve the problem.

EXAMPLE 1: Solving for Distance

Troy and Philip decided to spend the day canoeing on a river with a 3 mi/hr current. They began paddling up the river at 9:00 AM. After stopping on an island for 1 hr to eat lunch, they paddled back down the river and arrived back at their starting point at 3:00 PM. If they typically paddle at 5 mi/hr in still water, how far up the river did they travel? When did they stop for lunch?

Answer

Let d = the distance traveled up and down the river.

1. **Read** the problem carefully and assign a variable to represent the main unknown.

	r	t	d
Upstream	$5-3=2$	$\frac{d}{2}$	d
Downstream	$5+3=8$	$\frac{d}{8}$	d

2. **Plan** your solution by making a table.
   a. Fill in the distance and rate columns, using the information in the problem and the variable expression.
   b. Determine expressions for each time by solving $r \cdot t = d$ for $t = \frac{d}{r}$.

3. **Solve** an equation relating the times of the trips.

$$\frac{d}{2} + \frac{d}{8} = 5$$

   a. The total time spent traveling up and down the river was 5 hr.

$$8\left(\frac{d}{2}\right) + 8\left(\frac{d}{8}\right) = 8(5)$$
$$4d + d = 40$$
$$5d = 40$$
$$d = 8 \text{ mi up the river}$$

   b. Multiply each term by the LCD, 8, to clear the fractions.

$\frac{d}{2} = \frac{8}{2} = 4$ hr up the river, so they stopped for lunch at 1:00 PM.

   c. Use the expression in the table to determine the time spent traveling up the river and the time at which they stopped for lunch.

	r	t	d
Upstream	2	$\frac{8}{2}=4$	8
Downstream	8	$\frac{8}{8}=1$	8

4. **Check.** Determining the time spent traveling down the river to be 1 hr and adding the hour they spent eating lunch puts their return time at 3:00 PM.

SKILL CHECK **EXERCISES 1, 15**

These problems differ from the motion problems discussed in Section 2.7 only in that they often contain rational expressions.

EXAMPLE 2: Solving for Rate

When Lisa flies home for Christmas, her flight time is $12\frac{1}{2}$ hr shorter than the time it would take her to drive the 780 mi. If the plane's average speed is six times the average speed of her car, find both speeds.

Answer

Let r = the speed of the car and $6r$ = the speed of the plane.

1. **Read** the problem carefully and assign variable expressions to represent the main unknowns.

	r	t	d
By Car	r	$\frac{780}{r}$	780
By Plane	$6r$	$\frac{780}{6r}$	780

2. **Plan** your solution by making a table.
   a. Fill in the distance and rate columns, using the information in the problem and the variable expressions.
   b. Determine expressions for each time by solving $r \cdot t = d$ for $t = \frac{d}{r}$.

$$\frac{780}{r} = \frac{780}{6r} + \frac{25}{2}$$

$$6r\left(\frac{780}{r}\right) = 6r\left(\frac{780}{6r}\right) + 6r\left(\frac{25}{2}\right)$$

$$4680 = 780 + 75r$$

$$3900 = 75r$$

$$r = 52 \text{ mi/hr}$$

3. **Solve** an equation relating the times of the trips.
   a. The time by car is $12\frac{1}{2}$ hr more than the time by plane.
   b. Multiply each term by the LCD, $6r$, to clear the fractions.

$$6r = 6(52) = 312 \text{ mi/hr}$$

   c. Use the expression in the table to find the speed of the plane.

	r	t	d
By Car	52	$\frac{780}{52} = 15$	780
By Plane	312	$\frac{780}{312} = \frac{5}{2}$	780

4. **Check.** Filling in the values in your table shows that the 15 hr it takes to drive is $12\frac{1}{2}$ hr longer than the $2\frac{1}{2}$ hr it takes to fly.

SKILL CHECK **EXERCISES 3, 17**

Work is a God-given responsibility (Gen. 2:15; Exod. 20:9; 1 Thess. 4:11). Paul and his associates worked "night and day" so they would not be a burden to others. Paul reprimanded lazy people: "If any would not work, neither should he eat" (2 Thess. 3:8–10). All Christians have a responsibility to work hard because they serve not just their employer, but the Lord Jesus Himself (Rom. 12:11; Eph. 6:1–8; Col. 3:20–24).

A manager's ability to accurately predict the amount of time needed for his associates to complete a task allows him to maximize their efficiency without burdening them with unrealistic expectations. The work problems in this section can provide you with insights to everyday jobs.

Information from a work-rate problem can be organized in a modified table similar to the ones in Examples 1–2. Equations based on the information in the table can be set up and solved using the same techniques discussed in those examples.

EXAMPLE 3: Solving a Work-Rate Problem

The editorial manager knows that Emily can proofread a chapter in 60 min. Danny can proofread a similar chapter in 75 min. How long will it take them to proofread other similar chapters if they work together?

Answer

Let m = the number of minutes it should take for them to do the job together.

1. **Read** the problem carefully and assign a variable to represent the main unknown.

	Time (min)	Rate of Work
Emily	60	$\frac{1 \text{ chapter}}{60 \text{ min}}$
Danny	75	$\frac{1 \text{ chapter}}{75 \text{ min}}$
Together	m	$\frac{1 \text{ chapter}}{m \text{ min}}$

2. **Plan** by making a table listing each rate of work.

3. **Solve** an equation relating the work rates.

   a. The rate of working together is the sum of the individual rates.

$$\frac{1}{60} + \frac{1}{75} = \frac{1}{m}$$

   b. Multiply each term by the LCD, $300m$, to clear the fractions.

$$300m\left(\frac{1}{60}\right) + 300m\left(\frac{1}{75}\right) = 300m\left(\frac{1}{m}\right)$$
$$5m + 4m = 300$$
$$9m = 300$$
$$m = \frac{100}{3} = 33\tfrac{1}{3} \text{ min}$$

4. **Check** your answer in the original equation.

$$\frac{1}{60} + \frac{1}{75} = \frac{1}{\frac{100}{3}}$$
$$\overset{5}{\cancel{300}}\left(\frac{1}{\cancel{60}}\right) + \overset{4}{\cancel{300}}\left(\frac{1}{\cancel{75}}\right) = \overset{3}{\cancel{300}}\left(\frac{3}{\cancel{100}}\right)$$
$$5 + 4 = 9$$

SKILL CHECK **EXERCISE 13**

You should always check that your solutions are reasonable. A time of $33\frac{1}{3}$ min to complete the task together is between the 30 min and $37\frac{1}{2}$ min it would take Emily and Danny to do half the job individually. Notice also that the answer is not the average of these times.

EXAMPLE 4: Solving a Work-Rate Problem

AFTERSCHOOLHELP

Applying Rational Equations

Mr. Mathews owns a landscaping company and schedules the jobs for his company. He can complete the weekly maintenance for one of his corporate clients in 10 hr by himself or in 4 hr with his assistant, Ned. How long should he schedule for Ned to complete the job by himself?

Answer

1. **Read** the problem carefully and assign a variable to represent the main unknown.

Let h = the number of hours it should take for Ned to complete the job.

2. **Plan** by making a table listing each rate of work.

	Time (hr)	Rate of Work
Mr. Mathews	10	$\frac{1\text{ job}}{10\text{ hr}}$
Ned	h	$\frac{1\text{ job}}{h\text{ hr}}$
Together	4	$\frac{1\text{ job}}{4\text{ hr}}$

3. **Solve** an equation relating the work rates.

   a. The rate of working together is the sum of the individual rates.

$$\frac{1}{10}+\frac{1}{h}=\frac{1}{4}$$

   b. Multiply each term by the LCD, $20h$, to clear the fractions.

$$20h\left(\frac{1}{10}\right)+20h\left(\frac{1}{h}\right)=20h\left(\frac{1}{4}\right)$$
$$2h+20=5h$$
$$20=3h$$
$$h=\frac{20}{3}=6\frac{2}{3}\text{ hr}$$

4. **Check** your answer in the original equation.

$$\frac{1}{10}+\frac{1}{\frac{20}{3}}=\frac{1}{4}$$
$$\overset{2}{\cancel{20}}\left(\frac{1}{\cancel{10}}\right)+\cancel{20}\left(\frac{3}{\cancel{20}}\right)=\overset{5}{\cancel{20}}\left(\frac{1}{\cancel{4}}\right)$$
$$2+3=5$$

SKILL CHECK **EXERCISE 21** ☑

A. EXERCISES

Use the formula $r \cdot t = d$ to fill in the missing cells in each table.

1.

	r	t	d
Charissa	16		d
Sophia	20		d

2.

	r	t	d
Before Lunch	45		$x + 25$
After Lunch	55		x

3.

	r	t	d
Dion	x		2
Mitch	$x - 2$		2

4.

	r	t	d
Against Wind	$x - 5$		30
With Wind	$x + 5$		30

Make a table for each problem. Do not solve.

5. Mrs. Beck drove the 200 mi to her aunt's house at a speed 3 mi/hr faster than the speed at which she returned.

6. Bryce biked against a 5 mi/hr wind for 50 mi and then returned with the wind.

7. Pete takes 3 hr to mow the school's ball fields, and Adam takes 6 hr. How long will it take them working together?

8. Sheila and Keisha can each wash their parents' cars in 1 hr. How long will it take them working together?

Make a table and solve. Round to the nearest tenth if necessary.

9. At 7:00 AM Aiden started walking to school at a rate of 2.5 mi/hr. At 7:45 his brother William started riding a bicycle to school. If William rode at a rate of 10 mi/hr, how far did Aiden walk before William caught up with him?

10. The first bus for the senior trip left the school $\frac{1}{2}$ hr earlier than a second bus and traveled at an average speed of 45 mi/hr. When it finally departed, the second bus averaged 50 mi/hr. How long did it take for the second bus to catch up with the first bus?

11. Mr. O'Brian can plant his acreage in 74 hr. His neighbor can plant the same amount of acreage with his larger equipment in 50 hr. How long will it take to plant Mr. O'Brian's field if they work together?

12. Darla figured that making puppets for vacation Bible school would take her 4 hr. If she worked with her sister (who would take 6 hr to do the job alone), she could finish faster. If the girls work together, how long will it take them to finish the puppets?

13. The junior class is getting ready for the spring play. Landon is in charge of painting the sets, which will take 52 hr if he paints them by himself. Landon knows that Ethan is a good artist and could paint the sets in 36 hr. If he decides to ask Ethan to help him, how long will it take the two to paint the sets?

14. It takes 9 hr for one water pipe to fill a swimming pool. If another pipe is added and both pipes work together, the pool fills in only 3 hr. How long would it take the second pipe to fill the pool if it were the only one used?

B. EXERCISES

Solve. Round to the nearest tenth if necessary.

15. Two flights left Karr Airport at the same time. One plane flew 250 mi/hr. The other plane flew 375 mi/hr. The faster plane went 750 mi farther than the slower plane and traveled 1 hr longer. How far did each plane travel?

16. Mr. Truman drove to his cottage in the mountains at a rate of 45 mi/hr. On the way back he drove at a rate of 50 mi/hr but took a route 15 mi shorter than the route by which he traveled to the cottage. How many total miles did he travel if it took him $\frac{1}{2}$ hr longer to go than to return?

17. Gabriel walks $\frac{1}{10}$ as fast as a train travels. It takes 1.5 hr longer for him to walk 12 mi than for the train to travel 60 mi. Find the speed of each.

18. An airplane flew 700 mi with the wind in the same amount of time that it flew 625 mi against the wind. The average speed of the plane with no wind is 265 mi/hr. What was the wind speed?

19. Andrew and Emmy canoed a total of 14 mi upstream and back. The roundtrip took 8 hr. If they row at a rate of 4 mi/hr in still water, how fast was the current?

20. Hudson can load a truck in 2 hr. With Kevin's help, he can finish loading a truck in 45 min. How long does it take if Kevin is working alone?

21. A certain computer can grade 2000 objective-answer exams in 10 sec. A newer computer can grade the 2000 exams in less time. If the 2 computers together can grade 2000 exams in 3 sec, how long does it take the newer computer alone to do the grading?

22. If Reid, who takes twice as long as Silas to remove the tassels from an acre of corn, works with Silas, they can do an acre in 50 min. How long does it take each one to do the job alone?

23. Mr. Endicott, Mr. Bookout, and Mr. Piatt are going to paint a house together. Mr. Endicott can paint one side of the house in 4 hr. To paint an equal area, Mr. Bookout takes 3 hr and Mr. Piatt takes 2 hr. If the men work together, how long will it take them to paint one side of the house?

24. The Schultzes biked 15 mi from Brattleboro to Keene with a tailwind of 2 mi/hr. If the return trip against the wind took 2 hr longer, how fast would the family bike on a calm day?

25. A bicycle club took a weekend trip to a national forest and rode 60 mi at a certain average rate. On their return trip over the same route, they traveled 4 mi/hr faster and took 4 hr less. What were their average rates going and returning?

26. MP2 Logan and Conrad walked 30 min to the bicycle repair shop. They rode home in 45 min, taking a route that was 6 mi longer than their walk. If they rode 7 mi/hr faster than they walked, how fast did they walk and ride?

27. Hayley takes 3 hr longer than Lily to clean the house. If the girls work together, they can get the work done in 4 hr. How long does it take each girl working alone to clean the house?

28. Mr. Jansen can shingle a roof in 10 hr, and Mr. Doyle can do it in 8 hr. Mr. Davies has never done the job alone, but last time he helped with a similar job it took the 3 men $3\frac{1}{3}$ hr working together. How long would it take Mr. Davies alone?

29. How long will it take the Swansons to weed their vegetable garden if all 4 family members work together? They estimate that working alone would take Mr. Swanson 4 hr, Mrs. Swanson 4 hr, Brenda 6 hr, and Glen 7 hr.

30. At 1:00 PM Mrs. Fanelli tells Joey and Marcy that they can play miniature golf if they get the chores done (properly!). Joey usually takes 5 hr to do the chores, and Marcy does them in 4 hr. If they work together, how long will it take them to do the work? If the course closes at 5:00 PM and it takes 14 min to drive there, how long will they have to play?

31. **Essential Question:** How can rational equations help business managers optimize workflow?

C. EXERCISES

Solve. Round to the nearest tenth if necessary.

32. **MP1** A touring bus traveling to and from a historical site covered 200 mi in $4\frac{1}{2}$ hr. If the rate returning was 10 mi/hr faster than the rate going, find the rate each way. Also find the amount of time it took to reach the site.

33. **MP2** A machine at Eddie Sons light bulb factory produces metal bases for light bulbs. The company has a rush order that needs to be filled in 24 hr. After 7 hr the machine breaks down, and another machine is used to complete the job in 5 hr. It would have taken 15 hr to do the job using the first machine alone. How long would it have taken to do the job using only the second machine?

34. Penny takes 2 hr longer than Maggie to make a dress. Together they can finish a dress in 4 hr. How long does each girl take working alone?

35. **MP4** Mr. Desiatnikov drives a semitrailer. On one trip to a city 275 mi from his home, he traveled 30 min longer going than he did returning along the same route because of construction work. How long did it take him to make the trip each way if his rate going was 5 mi/hr slower than his rate returning? How fast did he travel each way?

CUMULATIVE REVIEW

Graph each quadratic function. [11.7–11.9]

36. $y = x^2 + 3$

37. $y = 2(x - 2)^2$

38. $y = x^2 - 4x + 7$

State the coordinates of the vertex of each parabola. [7.3]

39. $f(x) = (x - 12)^2 + 6$

40. $f(x) = -3(x + 8)^2 - 10$

41. Write a function rule that translates the function $y = x^2$ two units left and one unit down. **[7.3]**

42. Divide $x^2 + x + 4$ by $x - 1$. **[8.6]**

43. How many milliliters of a 28% potassium iodide solution and how many milliliters of a 60% potassium iodide solution should be combined by a chemist who needs 800 mL of a 40% solution? **[2.8]**

44. Which steps could be taken to solve $2x^2 - 7x = 3$ by completing the square? **[11.4]**

- **A** factor 2 from the left side; add $\left(\frac{3}{2}\right)^2$ to both sides
- **B** factor 2 from the left side; add $\left(-\frac{7}{2}\right)^2$ to both sides
- **C** divide both sides by 2; add $\left(\frac{3}{4}\right)^2$ to both sides
- **D** divide both sides by 2; add $\left(\frac{7}{4}\right)^2$ to both sides
- **E** none of these

45. What are the solutions to $x^2 - 8x = 7$? **[11.5]**

- **A** $-4 \pm \sqrt{46}$
- **B** $4 \pm \sqrt{23}$
- **C** $-7, -1$
- **D** $1, 7$
- **E** none of these

12.8 GRAPHING RATIONAL FUNCTIONS

How can I graph a rational function?

After completing this section, I will be able to

- define a rational function.
- graph simple rational functions.
- describe translations of the rational function $f(x) = \frac{a}{x}$.

Recall that a rational number is a number that can be written as a ratio of integers. Similarly, a *rational function* is a function that can be expressed as a ratio of polynomials.

Ohm's Law states that the relationship between the current (I), the voltage (V), and the resistance (R) of a circuit can be modeled by the function $I = \frac{V}{R}$.

DEFINITION

A **rational function** is a function expressed as a ratio of polynomials whose denominator is not 0.

You have already seen how the inverse variation $xy = 2$ can be written as $y = \frac{2}{x}$ or $f(x) = \frac{2}{x}$. This is an example of one of the simplest forms of a rational function, $f(x) = \frac{a}{x}$.

EXAMPLE 1: Graphing a Function of the Form $f(x) = \frac{a}{x}$

Graph $f(x) = \frac{2}{x}$.

Answer

$x \neq 0$

1. Determine the value for which the function is undefined.

2. Make a table of ordered pairs.

x	$f(x) = \frac{2}{x}$
−4	$-\frac{1}{2}$
−2	−1
−1	−2
$-\frac{1}{2}$	−4
0	undefined
$\frac{1}{2}$	4
1	2
2	1
4	$\frac{1}{2}$

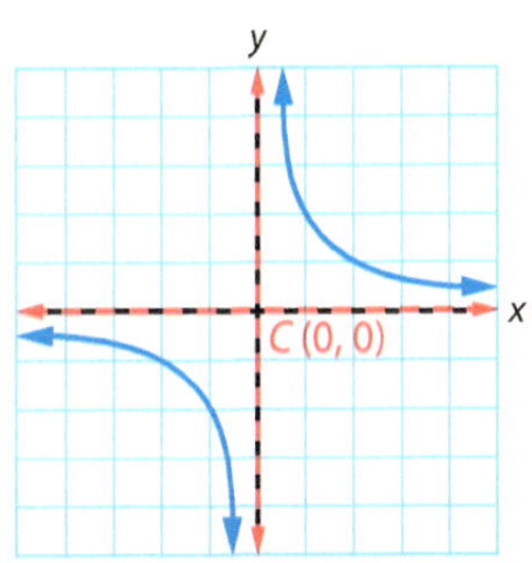

3. Graph the function.

SKILL CHECK **EXERCISE 24**

The graph in Example 1 is called a hyperbola. The ends of each branch of the hyperbola in Example 1 approach but never reach the x- and y-axes. Therefore, the x-axis is the horizontal asymptote, and the y-axis is the vertical asymptote in this graph. These asymptotes are indicated by dashed lines on the graph. The center of the hyperbola is at the intersection of its asymptotes, (0, 0).

Now consider a rational function of the form $f(x) = \frac{a}{x} + k$.

EXAMPLE 2: Graphing a Function of the Form $f(x) = \frac{a}{x} + k$

Graph $f(x) = \frac{2}{x} + 3$.

AFTERSCHOOLHELP

Graphing Rational Functions

Answer

Since the function is undefined when $x = 0$, $x = 0$ is the equation of the vertical asymptote.

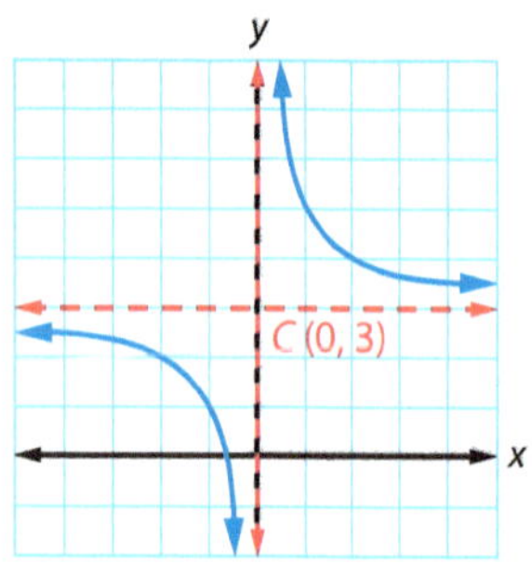

1. Determine the value for which the function is undefined.
2. Make a table of ordered pairs.

x	$\frac{2}{x} + 3$	$f(x)$
-4	$-\frac{1}{2} + 3$	$2\frac{1}{2}$
-2	$-1 + 3$	2
-1	$-2 + 3$	1
$-\frac{1}{2}$	$-4 + 3$	-1
0	undefined	
$\frac{1}{2}$	$4 + 3$	7
1	$2 + 3$	5
2	$1 + 3$	4
4	$\frac{1}{2} + 3$	$3\frac{1}{2}$

3. Graph the function.

SKILL CHECK **EXERCISE 25**

Observe that the graph of $f(x) = \frac{2}{x} + 3$ is the result of translating the graph of $f(x) = \frac{2}{x}$ three units up. The graph's center and horizontal asymptote have been moved up to (0, 3) and $y = 3$, while its vertical asymptote is still the y-axis.

In general, the graph of $f(x) = \frac{a}{x} + k$ is the result of translating the graph of $f(x) = \frac{a}{x}$, its parent function, k units up if k is positive and k units down if k is negative.

EXAMPLE 3: Graphing a Function of the Form $f(x) = -\frac{a}{x}$

Graph $g(x) = -\frac{6}{x}$.

Answer

Since the function is undefined when $x = 0$, $x = 0$ is the equation of the vertical asymptote.

1. Determine the value for which the function is undefined.

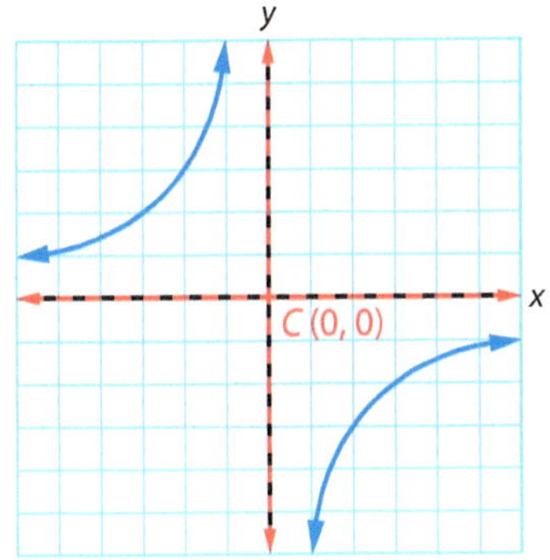

2. Make a table of ordered pairs.

x	$g(x) = \frac{-6}{x}$
−6	1
−4	$\frac{3}{2}$
−2	3
−1	6
0	undefined
1	−6
2	−3
4	$-\frac{3}{2}$
6	−1

3. Graph the function.

SKILL CHECK **EXERCISE 27**

In Example 3, when x is a very large positive value, $g(x)$ is negative and very close to 0. Likewise, for a very large negative value of x, $g(x)$ is positive and very close to 0. This indicates that the line $y = 0$ (the x-axis) is the horizontal asymptote. Since the vertical asymptote is $x = 0$, the graph's center is at (0, 0). Notice that the negative value of a causes the graph to be in the upper left and lower right quadrants formed by the asymptotes.

EXAMPLE 4: Graphing a Function of the Form $f(x) = -\frac{a}{x - h}$

Graph $h(x) = \frac{-6}{x - 2}$ and compare its graph to the graph of $g(x) = \frac{-6}{x}$.

Answer

Since the function is undefined when $x = 2$, $x = 2$ is the equation of the vertical asymptote.

1. Determine the value for which the function is undefined.
2. Draw the vertical asymptote.
3. Make a table of ordered pairs.

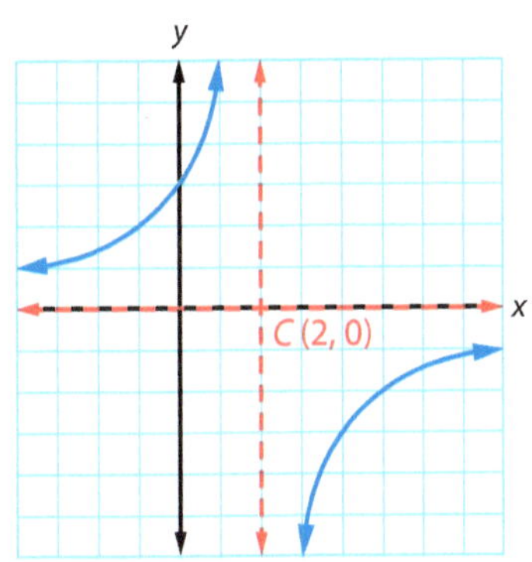

x	$\frac{-6}{x-2}$	$h(x)$
−4	$\frac{-6}{-4-2}$	1
−2	$\frac{-6}{-2-2}$	$\frac{3}{2}$
0	$\frac{-6}{0-2}$	3
1	$\frac{-6}{1-2}$	6
2	undefined	
3	$\frac{-6}{3-2}$	−6
4	$\frac{-6}{4-2}$	−3
6	$\frac{-6}{6-2}$	$-\frac{3}{2}$
8	$\frac{-6}{8-2}$	−1

4. Graph the function.

The graph of $h(x)$ is the result of translating the graph of $g(x)$ 2 units right.

5. Compare the graphs in Examples 3–4.

SKILL CHECK **EXERCISES 19, 26**

The graph of a function of the form $y = \frac{a}{x - h} + k$ is a translation of the base function $y = \frac{a}{x}$. It is a hyperbola with its center at (h, k), a vertical asymptote of $x = h$, and a horizontal asymptote of $y = k$. You will need to choose several points on either side of the hyperbola's center to determine its exact shape.

EXAMPLE 5: Graphing a Function of the Form $f(x) = \frac{a}{x - h} + k$

Given $f(x) = \frac{2}{x + 3} - 1$, identify the center of the hyperbola and both asymptotes. Then graph the function.

Answer

$f(x) = \frac{2}{x - (-3)} + (-1)$

$(h, k) = (-3, -1)$

vertical asymptote: $x = -3$

horizontal asymptote: $y = -1$

1. The graph of a function of the form $f(x) = \frac{a}{x - h} + k$ has its center at (h, k), a vertical asymptote of $x = h$, and a horizontal asymptote of $y = k$.
2. Draw the asymptotes and label the center.
3. Make a table of several convenient points on either side of the hyperbola's center to establish the general shape of the graph.

x	$\frac{2}{x+3} - 1$	$f(x)$
−7	$\frac{2}{-7+3} - 1$	$-\frac{3}{2}$
−5	$\frac{2}{-5+3} - 1$	−2
−4	$\frac{2}{-4+3} - 1$	−3
−3	undefined	
−2	$\frac{2}{-2+3} - 1$	1
−1	$\frac{2}{-1+3} - 1$	0
1	$\frac{2}{1+3} - 1$	$-\frac{1}{2}$

4. Graph the function.

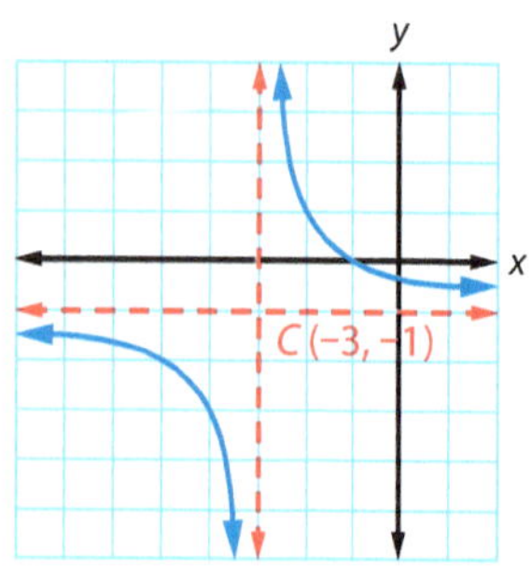

SKILL CHECK **EXERCISE 29**

The graph of $f(x) = \frac{2}{x + 3} - 1$ is the result of translating the graph of $f(x) = \frac{2}{x}$ three units left and one unit down. Notice the similarity to the translations of the absolute value, power, radical, and quadratic functions studied earlier.

VOCABULARY
- rational function

The table summarizes the types of functions studied in this course and their key characteristics.

FUNCTION TYPE	GENERAL FORM	VARIABLE	EXAMPLE	CHARACTERISTICS
Absolute Value (Section 4.7)	$f(x) = \lvert x - h \rvert + k$	within absolute value symbols	$f(x) = \lvert x - 3 \rvert + 4$	vertex: (h, k)
Linear (Section 5.3)	$f(x) = mx + b$	degree of 1	$f(x) = \frac{1}{2}x + 3$	y-intercept: $(0, b)$ slope: m
Power (Section 7.3)	$f(x) = ax^n$	the base of a power	$f(x) = 3(x - 2)^3 - 5$	a changes the graph's "steepness"
Exponential (Section 7.4)	$f(x) = ab^x$	the exponent of a power	$f(x) = 3 \cdot 2^x$	growth when $b > 1$ decay when $0 < b < 1$
Radical (Section 10.8)	$f(x) = \sqrt{x - h} + k$	in the radicand	$f(x) = \sqrt{x - 2} + 3$	"begins" at (h, k)
Quadratic (Section 11.8)	$f(x) = a(x - h)^2 + k$ $f(x) = ax^2 + bx + c$	degree of 2	$f(x) = (x - 5)^2 + 2$	vertex: (h, k)
Rational (Section 12.8)	$f(x) = \frac{a}{x - h} + k$	in the denominator	$f(x) = \frac{2}{x - 1} + 5$	center: (h, k) vertical asymptote: $x = h$ horizontal asymptote: $y = k$

A. EXERCISES

Classify each function.

1. $f(x) = 2x + 17$
2. $f(x) = (x + 12)^2 - 6$
3. $f(x) = |x + 7| - 21$
4. $f(x) = \sqrt{x + 7}$
5. $f(x) = 5^x$
6. $f(x) = \frac{1}{x}$
7. $f(x) = 5x - 1$
8. $f(x) = x^2 + 3x - 9$
9. $f(x) = x + |-5|$

A absolute value
B linear
C exponential
D radical
E quadratic
F rational

State the center and asymptotes for each function's graph.

10. $f(x) = \frac{2}{x} - 9$

11. $f(x) = \frac{3}{x}$

12. $f(x) = \frac{3}{x} + 7$

13. $f(x) = \frac{3}{x - 4}$

14. $f(x) = \frac{4}{x - 5} - 1$

15. $f(x) = \frac{2}{2x - 1} + 1$

16.

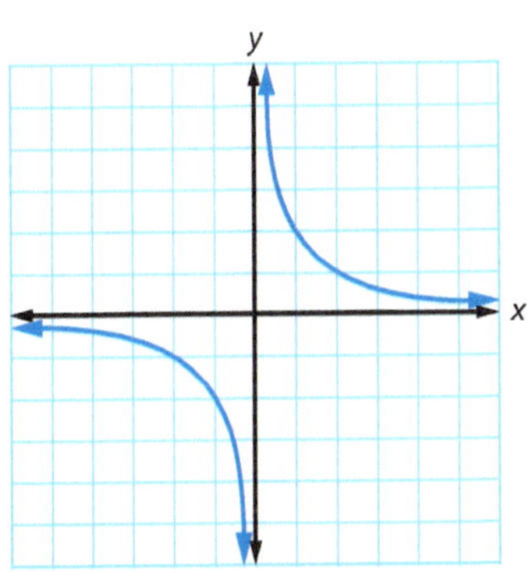

17.

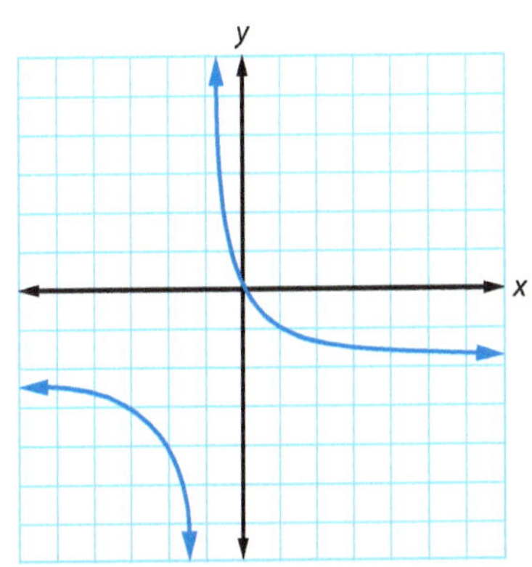

B. EXERCISES

Describe the translation(s) of $f(x) = \frac{1}{x}$ that produce the graph of each function.

18. $f(x) = \frac{1}{x - 10}$

19. $f(x) = \frac{1}{x + 14}$

20. $f(x) = \frac{1}{x} + 7$

21. $f(x) = \frac{1}{x} - 3$

22. $f(x) = \frac{1}{x + 4} - 9$

23. $f(x) = \frac{1}{x - 4} + 2$

Graph each rational function.

24. $f(x) = \frac{3}{x}$

25. $f(x) = \frac{3}{x} + 3$

26. $f(x) = \frac{-2}{x + 2}$

27. $f(x) = -\frac{1}{x} + 1$

28. $f(x) = \frac{3}{x + 2} + 2$

29. $f(x) = \frac{-1}{x - 4} - 2$

30. Why is the vertical asymptote not part of the solution of a rational function?

31. What does the horizontal asymptote of a rational function $f(x) = \frac{a}{x - h} + k$ represent?

32. **Essential Question:** How do you graph a rational function of the form $f(x) = \frac{a}{x - h} + k$?

Ohm's law states that current is directly proportional to the voltage and inversely proportional to the resistance in the circuit. This relationship can be modeled using the formula $I = \frac{V}{R}$, where I is the current in amps, V is the voltage in volts, and R is the resistance in ohms. If the voltage of a circuit is 120 V, the current is a rational function of the resistance in the circuit, $I = \frac{120}{R}$. Electrical engineers increase or decrease resistance in a circuit to maintain desirable current ranges.

33. **MP4** Graph the portion of the rational function $I = \frac{120}{R}$ that gives reasonable values for I and R.

Use the graph from exercise 33 to answer the following questions.

34. What happens in the circuit as the resistance increases?

35. What happens in the circuit as the resistance decreases?

36. What type of variation is illustrated by the graph?

C. EXERCISES

MP7 Use long division to express each rational function in $f(x) = \frac{a}{x-h} + k$ form; then graph the function.

37. $f(x) = \frac{2x-1}{x-1}$

38. $f(x) = \frac{3x+2}{x+2}$

39. **MP8** Consider a rational function of the form $f(x) = \frac{ax+b}{x+c}$, where a, b, and c are real numbers.

- **a.** Use long division to express this function in $f(x) = \frac{a}{x-h} + k$ form.
- **b.** Use the function from part a to verify your results from exercises 37–38.

40. **MP3** Answer the questions below to determine whether there is some value of x at which the graph of $y = \frac{a}{x-h} + k$ intersects its horizontal asymptote.

- **a.** Write the equation of the horizontal asymptote for this graph.
- **b.** Set the function and asymptote values equal and solve for the value of x at which the graph intersects the horizontal asymptote.
- **c.** Why are points on the horizontal asymptote not solutions to the rational function $y = \frac{a}{x-h} + k$?

41. Graph $f(x) = \frac{1}{x^2}$. Describe the characteristics of the graph compared to the graph of $f(x) = \frac{1}{x}$.

CUMULATIVE REVIEW

Solve each compound inequality. [3.3–3.4]

42. $7 + 5x \leq 9$ or $5x + 3 > 7$

43. $2x - 3 > 5$ and $5x - 1 < 9$

Given the 2 points (1, 3) and (−2, 5), find the following. [5.2, 10.5]

44. the slope of the line passing through the points

45. the distance between the points

Factor each expression. [9.5]

46. $7x^4 + 11x^2 + 4$

47. $3x^4 - 31x^2 + 36$

48. The square of a number reduced by twice the number is 15. Find the number. **[11.6]**

49. Solve $\frac{5x-9}{x-2} - 1 = \frac{13}{3} + \frac{1}{x+2}$. **[12.6]**

50. Which polynomial has $x + 3$ as a factor? **[9.4]**

- **A** $-x^2 + 3x$
- **B** $x^2 - 6$
- **C** $x^2 - 6x + 9$
- **D** $x^2 + 8x - 33$
- **E** none of these

51. For which value of m could the expression $\sqrt{m} + 2\sqrt{3}$ be written as a single term when simplified? **[10.4]**

- **A** 9
- **B** 24
- **C** 45
- **D** 48
- **E** none of these

APPLICATION PROBLEMS
ARITHMETIC & HARMONIC MEANS

Among the discoveries of Pythagoras and his students are the 3 classical Pythagorean means: the arithmetic, the harmonic, and the geometric. The arithmetic mean is the most common. Given n numbers, you can find this mean by dividing their sum by n. Less familiar are the harmonic and geometric means.

The harmonic mean of n numbers is found by dividing the number of values by the sum of their reciprocals. It can be used to find the average speed directly from several differing speeds when the distances traveled at each speed are the same. The geometric mean of n numbers is found by taking the nth root of their product. The geometric mean has applications in right triangles and is also used to find average growth rates. This section will focus on the difference between arithmetic and harmonic means.

Pythagoras is credited for relating arithmetic and harmonic means to the intervals that create harmonies that are pleasing to the ear. Tradition suggests that upon hearing the hammers of a blacksmith, Pythagoras became curious about their harmonious tones. The story suggests that the hammers had relative weights of 6, 8, 9, and 12. Inspired by the intervals he heard, Pythagoras applied the arithmetic and harmonic means to a string on a musical instrument by subdividing it into the same proportions as the hammers' weights.

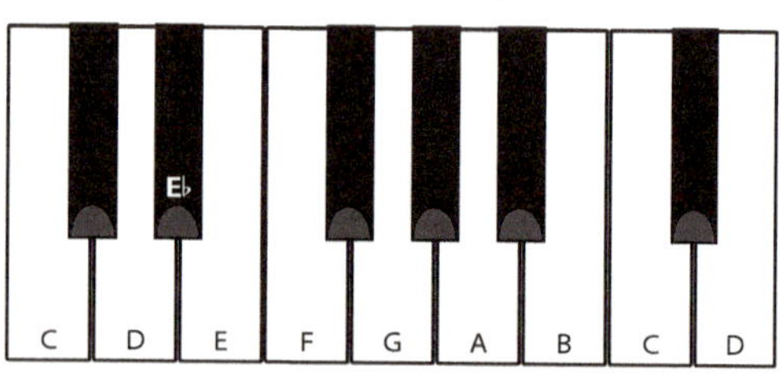

A musical interval is the distance in pitch between 2 notes. A scale in the key of C major is played on the white keys of a piano in succession from left to right. An interval of a 2nd can be heard by playing the first and second notes in the scale (C and D), an interval of a 3rd by playing the first and third notes (C and E), and so on up to an octave (equivalent to an interval of an 8th), which can be heard by playing 2 consecutive Cs.

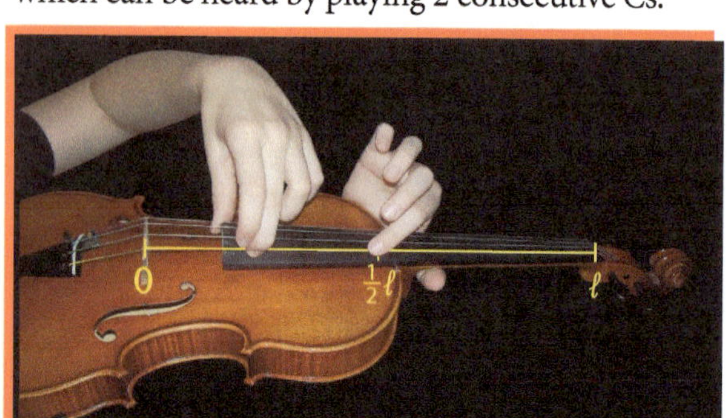

Pythagoras found that when a string was pressed at its midpoint, the plucked string produced a tone 1 octave higher than the tone produced by plucking the string without pressing it down. The arithmetic mean, A, of 0 and l (the original string length) predicts the length of string needed to produce a tone 1 octave higher.

$$A = \frac{a+b}{2} = \frac{0+l}{2} = \frac{l}{2} \text{ or } \frac{1}{2}l$$

1. Where should a string be pressed (in terms of l) to produce a tone that is 1 octave higher than the tone produced when the string is pressed at $\frac{1}{2}l$?
2. Describe how to produce a tone 3 octaves higher than the tone produced by the entire length of the string.

The string lengths that produce the intervals of a 4th and a 5th can be determined by using the arithmetic and harmonic means of $\frac{1}{2}l$ and l. The harmonic mean of 2 numbers, a and b, can be found by using the formula $H = \frac{2ab}{a+b}$.

3. Find the arithmetic mean of the string lengths for 2 notes 1 octave apart, $\frac{1}{2}l$ and l.
4. Find the harmonic mean of the string lengths for 2 notes 1 octave apart, $\frac{1}{2}l$ and l.
5. Copy the number line below, which models a string of length l.

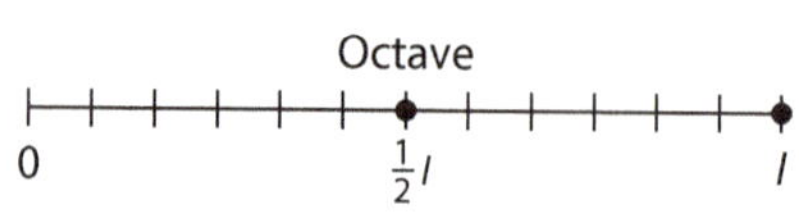

   a. On the number line, graph points representing the arithmetic and harmonic means found in exercises 3–4.
   b. Using the fact that an interval of a 5th is 1 step higher than an interval of a 4th, label the 2 points on your number line to show which length produces which interval.
6. Compare the hammer weights of 6, 8, and 9 to the heaviest weight of 12 by writing and simplifying 3 ratios. Label each ratio with the interval Pythagoras heard from each pair of hammers.

Mathematicians have continued to study musical intervals and have discovered mathematical relationships between several major and minor intervals. Just as the intervals of a 4th and a 5th are the arithmetic and harmonic means of 2 notes 1 octave apart, so the intervals of a major 3rd and a minor 3rd are the

arithmetic and harmonic means of 2 notes a 5th apart. As mentioned earlier, the interval of a major 3rd can be heard by playing the first and third notes of the C major scale (C and E). In a minor 3rd the second note is a half-step lower, so an E-flat would follow C instead of an E.

7. Find the arithmetic and harmonic means of the string lengths for 2 notes a 5th apart, $\frac{2}{3}l$ and l.

8. On the number line you created in exercise 5, graph and label additional points representing the intervals of a major 3rd and a minor 3rd.

9. Which mean (arithmetic or harmonic) produces a major 3rd? Which produces a minor 3rd?

It can be shown that $a : A = H : b$, where A is the arithmetic mean and H is the harmonic mean of 2 numbers a and b.

10. The intervals of a 4th and a 5th can be produced using $\frac{3}{4}l$ and $\frac{2}{3}l$, the arithmetic and harmonic means, respectively, of $\frac{1}{2}l$ and l. Verify the proportion $a : A = H : b$ for these values.

11. Using the formulas $A = \frac{a + b}{2}$ and $H = \frac{2ab}{a + b}$, verify the proportion $a : A = H : b$.

The average of several speeds can be found by using the arithmetic mean if the time traveled at each rate is the same. If the distance traveled at each rate is the same, then the harmonic mean of the individual rates provides the average rate of speed.

A missionary travels 50 mi to and from a village each week. He drives his vehicle part of the way, but he must hike the rest of the way into the mountains. It takes $2\frac{1}{2}$ hr to get to the village, but he returns home in just $1\frac{2}{3}$ hr.

12. To find the missing rates, fill in the table below using the formula $rt = d$.

	r (mi/hr)	*t* (hr)	*d* (mi)
Going to the Village		$\frac{5}{2}$	
Returning Home		$\frac{5}{3}$	

13. Find the arithmetic mean of the missionary's rates going to the village and returning.

14. Find the average rate using $rt = d$ and the total distance and time traveled.

15. The answers to exercises 13–14 are not the same. Use $H = \frac{2ab}{a + b}$ to find the harmonic mean of his going and returning rates. Verify that this answer gives the average rate for the entire trip.

To consider the reasonableness of building an airstrip for the delivery of supplies at a missionary compound, the mission board studies how long it takes the missionaries to travel to and from the nearest city. The harmonic mean of n numbers is found by dividing n by the sum of the numbers' reciprocals. Expressed as a formula, $H = \frac{n}{\frac{1}{x_1} + \frac{1}{x_2} + \dots + \frac{1}{x_n}}$, where each variable x_i is one of the numbers.

16. Show that $\frac{2}{\frac{1}{a} + \frac{1}{b}}$ is equivalent to $\frac{2ab}{a + b}$.

a. Starting with the first expression, find the LCD of the fractions in the denominator.

b. Multiply the numerator and the denominator of the complex fraction by the LCD and simplify.

17. Use $rt = d$ to find the average rate for each trip.

	r (mi/hr)	*t* (hr)	*d* (mi)
Trip 1		5	120
Return 1		$6\frac{2}{3}$	120
Trip 2		$7\frac{1}{2}$	120
Return 2		6	120

18. Find the harmonic mean of the 4 rates in the table above.

a. Substitute the individual rates into the formula for the harmonic mean.

b. Find the LCD of the fractions in the denominator.

c. Multiply the numerator and the denominator of the complex fraction by the LCD and simplify. State the rate to the nearest tenth.

12 CHAPTER REVIEW

Key Concepts

Simplifying Rational Expressions [12.1]

- A rational expression is a ratio of polynomials whose denominator is not 0.
- Rational expressions can be simplified by canceling factors that the numerator and denominator have in common.
- Excluded values are numbers that make the rational expression undefined.
- Identify excluded values by determining which x-values will make the denominator equal 0.

Multiplying & Dividing Rational Expressions [12.2]

- To multiply rational expressions:
  - factor all algebraic quantities in the numerators and the denominators,
  - cancel common factors appearing in both a numerator and a denominator, and
  - express the result as the product of the numerators divided by the product of the denominators.
- To divide rational expressions, multiply by the reciprocal of the divisor.

Adding & Subtracting Rational Expressions [12.3–12.4]

- To add or subtract rational expressions:
  - find the LCD if the denominators are different,
  - make equivalent rational expressions that have the LCD,
  - combine the numerators and place the result over the common denominator, and
  - simplify the resulting rational expression.

Mixed & Complex Expressions [12.5]

- A mixed rational expression is the sum of a polynomial and a rational expression.
- To simplify a mixed rational expression:
  - rename each part of the expression with a common denominator,
  - combine the expressions into a single expression, and
  - simplify the result.
- A complex rational expression is an algebraic expression of the form $\frac{n}{d}$ in which the numerator, n, the denominator, d, or both are rational expressions.
- To simplify a complex rational expression:
  - multiply the numerator by the reciprocal of the denominator, or
  - multiply the numerator and the denominator by their LCD.

Solving Rational Equations [12.6–12.7]

- Rational equations can be used to solve rate problems related to motion and work.
- To solve rational equations:
  - cross multiply if the equation is a proportion, or
  - multiply each term in the equation by the LCD to clear the fractions if the equation is not a proportion.

Graphing Rational Functions [12.8]

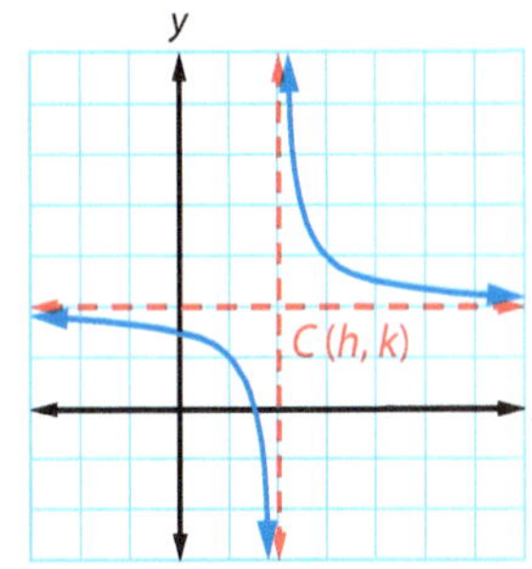

- Rational functions written in the form $f(x) = \frac{a}{x-h} + k$ reveal key characteristics.
  - center: (h, k)
  - vertical asymptote: $x = h$
  - horizontal asymptote: $y = k$
- Once the asymptotes have been established, the graph of a rational function can be easily determined by plotting a few points on either side of the vertical asymptote and connecting each set of points with smooth curves.

Biblical Worldview Shaping

Design through Algebra

How should we respond to the design we see in nature?

Mathematics helps us see and quantify design in the world. The more we recognize this design, the more we praise God for His wisdom and creativity.

EXERCISES

State any excluded values for each rational expression. [12.1]

1. $\frac{5}{x-3}$

2. $\frac{x-8}{x^2-4x-12}$

3. $\frac{x-6}{x^3-x}$

Simplify each expression. [12.1]

4. $\frac{12x+3}{8x+2}$

5. $\frac{3x+4}{6x^2-7x-20}$

6. $\frac{14-2x}{2x^2-22x+56}$

Find the LCM of each pair of polynomials. [12.4]

7. $8x^4$ and $28x^3y$

8. $4x-20$ and x^2-25

Simplify each expression, leaving each answer in factored form. [12.2–12.5]

9. $\frac{a}{\frac{b}{c}}$

10. $\frac{4x^3}{(x-5)^2} \cdot \frac{x-5}{2x}$

11. $\frac{2x+6}{3x+5} + \frac{x+4}{3x+5}$

12. $2c + \frac{d}{c}$

13. $\frac{56x^2-18}{7x-5} - \frac{7x^2+7}{7x-5}$

14. $\frac{(9x+2)(x-3)}{(2x-1)(x+4)} \div \frac{(9x+2)(x-6)}{(x-6)(2x-1)}$

15. $\frac{x^2+13x+22}{3} \cdot \frac{9}{x+2}$

16. $x + y - \frac{1}{x-y}$

17. $\frac{x}{x^2 - y^2} + \frac{y}{x^2 - y^2}$

18. $\frac{5x + 2}{x + 2} - \frac{x - 1}{x + 3}$

19. $\frac{\frac{x}{y} - 1}{\frac{y}{x} - 1}$

20. $\frac{5x^3}{3x^2 + x - 10} \div \frac{15x}{6x^2 + 13x + 2}$

21. $\frac{3}{1 - x} + \frac{7}{x - 1}$

22. $\frac{\frac{2x}{y} + 1}{\frac{2x + y}{y^3}}$

23. $\frac{xy}{5xy^3 - 11x^2} - \frac{y^3}{5y^5 - 11xy^2}$

24. $\frac{x^2 - 5x - 14}{2x + 3} \cdot \frac{2x^2 + x - 3}{x^2 - 7x}$

25. $\frac{4x^2 + 6x}{6x^2 - x - 15} - \frac{6x + 9}{15 + x - 6x^2}$

26. $\frac{\frac{x^2 - 4}{x}}{x + 2}$

27. $\frac{x + 2y}{27y} + \frac{x}{3}$

28. $\frac{x^3 + 4x}{3x^2 + 23x + 14} \div \frac{x^4 - 16}{x^2 + 5x - 14}$

29. $\frac{2}{xy} + \frac{3}{x^2 - 3xy} + \frac{1}{3y^2 - xy}$

30. $\frac{x + y}{x^2 + 2xy + y^2} + \frac{3}{x - y} - \frac{1}{x + y}$

Solve each equation. [12.6]

31. $\frac{x^2}{4} = \frac{13x}{8} + 3$

32. $\frac{6}{x - 3} = 3$

33. $\frac{x + 8}{2x - 6} = \frac{17}{23}$

34. $\frac{12x - 3}{x^2 + 4x - 5} - \frac{5}{x + 5} = \frac{6}{x - 1}$

35. $\frac{3}{8x} + \frac{1}{4} = \frac{5}{6x}$

36. $\frac{-5x + 8}{x^2 - 9} + \frac{5}{x + 3} = \frac{7x}{2x - 6}$

37. $\frac{2x - 2}{x^2 - 1} = \frac{x - 5}{x + 1}$

38. $\frac{6}{x^2 - x - 2} = \frac{-1}{x^2 + x - 6} - \frac{4}{x^2 + 4x + 3}$

Complete each table. Then write an equation modeling the data. Do not solve. [12.7]

39. trip going was 3 hr longer than trip returning

	r	*t*	*d*
Going	13		*d*
Returning	23		*d*

40. total time = 7 hr

	r	*t*	*d*
Day 1	$x + 3$		40
Day 2	x		50

41.

	Time (hr)	Rate of Work
First	5	
Second	2	
Together	x	

42.

	Time (hr)	Rate of Work
Seth	x	
Molly	$2x$	
Dominic	$x + 5$	
Together	8	

Solve. [12.7]

43. Dana has decorated the church for vacation Bible school in 5 hr, and Violet has done it in 3 hr. If they work together, how long will it take them?

44. Joan can wallpaper a living room in 9 hr. Kelly is inexperienced at wallpapering, but she wants to learn. If they work together on wallpapering the living room, it takes 6 hr. How long would it take Kelly to wallpaper the living room by herself?

45. Micah bikes 40 mi in the same time that Alexis drives 100 mi. If Alexis travels 12 mi/hr more than twice Micah's rate, how fast does each travel?

46. Gavin left early to drive to his parents' house and was able to drive 300 mi before lunch. Due to increased traffic after lunch, his average speed for the remaining 105 mi was 15 mi/hr less than his average speed before lunch. If he drove for a total of 9 hr, determine his average speed before lunch and after lunch. What was his average speed for the trip?

47. Write a rational expression representing the portion of the regular hexagon that is shaded. State the result as a simplified expression and as a percent (to the nearest tenth). **[12.1]**

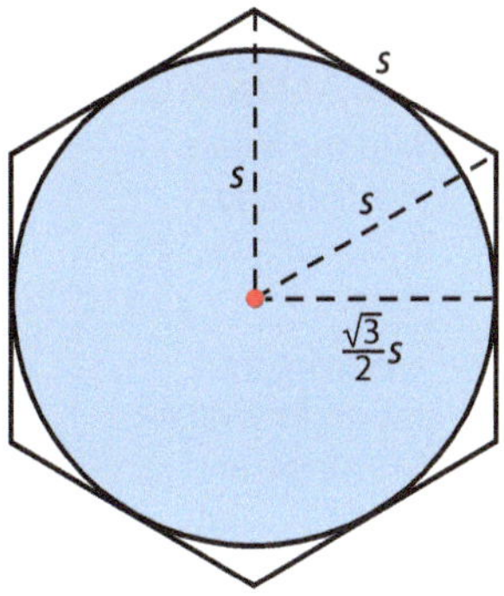

48. What number must be added to both the numerator and the denominator of $\frac{5}{8}$ to produce a fraction equivalent to $\frac{12}{13}$? **[12.6]**

State the center and asymptotes for each function's graph. [12.8]

49. $f(x) = \frac{12}{x+3}$

50. $f(x) = \frac{2}{x-3} + 4$

Graph each rational function. Be sure to include the asymptotes. [12.8]

51. $f(x) = -\frac{2}{x} + 1$

52. $f(x) = \frac{6}{x+2} - 3$

53. **BWS Essential Question:** What is the most appropriate response to the design we see and model through mathematics?

A praise for man's intellectual ability

B praise for the gift of math

C praise for the people who pioneered mathematical methods

D praise to God as Creator and Designer

A.1 STATISTICAL MEASURES

How can I describe typical values and the amount of variation found in a set of data?

After completing this section, I will be able to

- find the mean, median, and mode of a set of data.
- calculate the range, interquartile range, and mean absolute deviation of a set of data.

What is the typical number of roses for each yard in my neighborhood?

What is the average price for a pair of new running shoes? What is a typical low temperature in New York City for New Year's Eve? What number of patients suffering side effects from a medication would be unusual? Questions like these can be answered using **statistics**, the branch of mathematics that deals with collecting, organizing, analyzing, and reporting numerical information.

The collected information from recorded observations that can be measured or counted is called **data**. It is helpful to arrange the data in either ascending or descending order. There is a tendency for the numbers in the middle of an ordered data set to appear more frequently, especially when the data describes a naturally occurring phenomenon such as people's height, the weight of cows in a herd, or scores on a standardized test.

There are several ways to find a single number that represents a typical value for an entire set of data. The most common of these measures of central tendency are the *mean*, the *median*, and the *mode*.

DEFINITIONS

The **mean** is the average of the data, which is the sum of the values divided by the number of values. The mean is denoted $\bar{x}$ (pronounced "*x*-bar").

The **median** is the middle number in a set of ordered data or the average of the 2 middle numbers if there is an even number of values.

The **mode** is the value or values that occur most frequently in a set of data.

If no number occurs more than once, there is no mode. Data sets in which 2 values occur most frequently are called bimodal.

EXAMPLE 1: Finding Measures of Central Tendency

Find the mean, median, and mode from the set of data describing the number of correct answers on a 60-question test.

{58, 38, 60, 44, 45, 42, 49, 50, 41, 47}

Answer

38, 41, 42, 44, 45, 47, 49, 50, 58, 60

1. Arrange the data in ascending order.

The sum of the values is 474.

mean: $\frac{474}{10} = 47.4$

2. Divide the sum of the values by the number of values.

median: $\frac{45 + 47}{2} = 46$

3. Since there are 10 values, the median is the average of the fifth and sixth values.
38, 41, 42, 44, 45, 47, 49, 50, 58, 60

mode: none

4. All data values are different.

SKILL CHECK **EXERCISES 1–3**

EXAMPLE 2: Finding Measures of Central Tendency

Find the mean, median, and mode of the following low temperatures.

day	Mon	Tue	Wed	Thu	Fri	Sat	Sun
temperature (°F)	25°	21°	20°	30°	25°	22°	19°

Answer

30, 25, 25, 22, 21, 20, 19

1. Arrange the data in descending order.

mean: $\frac{30 + 25 + 25 + 22 + 21 + 20 + 19}{7}$

$= \frac{162}{7} \approx 23.1°F$

2. Divide the sum of the values by the number of values and round to the nearest tenth.

median: 22°F

3. Since there are 7 values, the median is the fourth value.
30, 25, 25, 22, 21, 20, 19

mode: 25°F

4. The value that occurs most often (twice) is 25°F.

SKILL CHECK **EXERCISES 4–6**

DETERMINING RANGE

Example 1:
60 − 38 = 22

Example 2:
30 − 19 = 11

Measures of variation describe how far the data spreads out in a data set. The *range*, the *interquartile range*, and the *mean absolute deviation* are commonly used measures of variation.

Since the **range** is the difference between the largest and smallest values in a set of data, 1 or more extreme values will cause a dramatic change in its value. To minimize the effect of extreme values, *quartiles* are used to separate a set of ordered data into 4 parts.

DEFINITIONS

Quartiles are values that divide the data into 4 equally sized groups.

The **lower quartile** (Q_1) is the median of the lower half of the data.

The **middle quartile** (Q_2) is the median of all the data.

The **upper quartile** (Q_3) is the median of the upper half of the data.

The **interquartile range** (IQR) is the difference of Q_3 and Q_1.

When the number of values is odd, the median is not included when finding Q_1 or Q_3.

EXAMPLE 3: Calculating the Interquartile Range

Calculate the interquartile range of the number of pages read by 11 students.

{12, 17, 22, 24, 25, 28, 29, 31, 32, 33, 35}

Answer

12, 17, 22, 24, 25, 28, 29, 31, 32, 33, 35

Q_2: 28	Q_2 is the median of the data set, the sixth value.
Q_1: 22	Q_1 is the median of the values below Q_2.
Q_3: 32	Q_3 is the median of the values above Q_2.
IQR: $32 - 22 = 10$	Find the interquartile range by subtracting Q_1 from Q_3.

SKILL CHECK **EXERCISE 14**

When the number of values is even, the smaller of the middle 2 values is included when finding Q_1, and the larger of the middle 2 values is included when finding Q_3.

EXAMPLE 4: Calculating the Interquartile Range

Calculate the interquartile range of the number of pages read by 12 students.

{16, 29, 29, 30, 35, 39, 40, 44, 46, 50, 51, 70}.

Answer

16, 29, 29, 30, 35, 39, 40, 44, 46, 50, 51, 70

Q_2: $\frac{39 + 40}{2} = 39.5$	Q_2 is the median of the data set, the average of the sixth and seventh values.
Q_1: $\frac{29 + 30}{2} = 29.5$	Q_1 is the median of the values below Q_2.
Q_3: $\frac{46 + 50}{2} = 48$	Q_3 is the median of the values above Q_2.
IQR: $48 - 29.5 = 18.5$	Find the interquartile range by subtracting Q_1 from Q_3.

SKILL CHECK **EXERCISE 19**

While all the data is included within the range, 50% of the data falls within the IQR.

DEFINITION

The **mean absolute deviation** (MAD) is the average of the distances of the values to the mean.

Use the following steps to calculate a data set's MAD.

1. Find the mean of the data.
2. Find the distance between each value and the mean.
3. Find the average of the distances.

EXAMPLE 5: Calculating the Mean Absolute Deviation

Find the MAD of the number of roses in a neighborhood's flower beds.

{12, 14, 22, 23, 26, 27, 30, 31, 32, 33}

Answer

mean: $\frac{12 + 14 + 22 + 23 + 26 + 27 + 30 + 31 + 32 + 33}{10}$

$= \frac{250}{10} = 25$

1. Find the mean of the data.

$|12 - 25| = 13$ $|27 - 25| = 2$

$|14 - 25| = 11$ $|30 - 25| = 5$

$|22 - 25| = 3$ $|31 - 25| = 6$

$|23 - 25| = 2$ $|32 - 25| = 7$

$|26 - 25| = 1$ $|33 - 25| = 8$

2. Find the distance between each value and the mean.

MAD: $\frac{13 + 11 + 3 + 2 + 1 + 2 + 5 + 6 + 7 + 8}{10} = \frac{58}{10} = 5.8$

3. Find the average of the distances.

SKILL CHECK **EXERCISES 20–21**

VOCABULARY

- data
- interquartile range
- lower quartile
- mean
- mean absolute deviation
- measures of variation
- median
- middle quartile
- mode
- quartile
- range
- statistics
- upper quartile

Calculating the mean and MAD in Example 5 allows us to say that a typical flower bed would have somewhere between 25 – 5.8 = 19.2 and 25 + 5.8 = 30.8 roses.

Finding and comparing the MADs of 2 similar data sets can illustrate how the MAD describes the variation within the data.

$A = \{3, 5, 7, 7, 8, 9, 10\}$

mean: $\frac{3 + 5 + 7 + 7 + 8 + 9 + 10}{7} = \frac{49}{7} = 7$

$|3 - 7| = 4$ $|8 - 7| = 1$
$|5 - 7| = 2$ $|9 - 7| = 2$
$|7 - 7| = 0$ $|10 - 7| = 3$
$|7 - 7| = 0$

MAD: $\frac{4 + 2 + 0 + 0 + 1 + 2 + 3}{7} = \frac{12}{7} \approx 1.7$

$B = \{1, 1, 3, 9, 10, 12, 13\}$

mean: $\frac{1 + 1 + 3 + 9 + 10 + 12 + 13}{7} = \frac{49}{7} = 7$

$|1 - 7| = 6$ $|10 - 7| = 3$
$|1 - 7| = 6$ $|12 - 7| = 5$
$|3 - 7| = 4$ $|13 - 7| = 6$
$|9 - 7| = 2$

MAD: $\frac{6 + 6 + 4 + 2 + 3 + 5 + 6}{7} = \frac{32}{7} \approx 4.6$

The MADs of the data sets indicate that the variability of set B is much higher than the variability of set A even though both sets have the same mean.

EXERCISES

Find each measure of central tendency for the given set of data. Round to the nearest tenth if necessary.

{7, 9, 7, 4, 7, 8, 10, 5, 5}

1. mean **2.** median **3.** mode

{45, 90, 86, 78, 91, 88, 79}

4. mean **5.** median **6.** mode

{25, 40, 19, 39, 18, 26}

7. mean **8.** median **9.** mode

Find each value for the given set of data.

{14, 5, 12, 0, 7, 16, 8, 25, 7}

10. range

11. Q_2

12. Q_1

13. Q_3

14. IQR

{514, 405, 312, 607, 715, 516, 803, 472, 425, 623}

15. range

16. Q_2

17. Q_1

18. Q_3

19. IQR

Calculate the mean and MAD of each set of data. Round to the nearest tenth if necessary.

hot dogs eaten: {75, 74, 72, 71, 70, 69, 68, 68, 62, 62, 59, 54}

20. mean

21. MAD

golf carts sold: {38, 87, 97, 112, 159, 202, 220, 257, 318, 410}

22. mean

23. MAD

websites built: {46, 52, 32, 33, 47, 56, 12, 42}

24. mean

25. MAD

books read: {8, 17, 34, 21, 12, 12, 15, 27, 11, 9, 10, 16}

26. mean

27. MAD

Leanna received the following test grades: 85, 90, 90, 95. Describe how each statistic changes if 3 bonus points are added to each test score.

28. mean

29. median

30. IQR

31. MAD

Last week Keith received homework grades of 80, 90, 90, and 100 and a 0 for a missed assignment. Describe how each statistic changes after the 0 is replaced with a 50 when he turns in the assignment late.

32. mean

33. median

34. IQR

35. MAD

36. **Essential Question:** How can I describe typical values and the amount of variation found in a set of data?

A.2 STANDARD DEVIATION

What does the standard deviation measure?

After completing this section, I will be able to

- define standard deviation.
- calculate the standard deviation of a data set.

The measures of variation discussed in Section A.1 are familiar ways to measure the spread of values in a data set. Another common measure of spread or deviation is called the *standard deviation*. A higher standard deviation indicates a greater spread in the data.

DEFINITION

The **standard deviation**, symbolized by σ (pronounced "sigma"), is the square root of the average of the squares of deviation.

FINDING THE STANDARD DEVIATION

1. Calculate the mean, $\overline{x}$.
2. Find the deviation of each data value, $x - \overline{x}$.
3. Find the square of the deviation for each data value, $(x - \overline{x})^2$.
4. Find the mean of the squared differences. This is called the **variance** (σ^2).
5. The square root of the variance is the standard deviation (σ).

EXAMPLE 1: Finding the Standard Deviation

Find the standard deviation of the data set.

{5, 8, 8, 9, 9, 10, 11, 12}

Answer

$$\bar{x} = \frac{5 + 8 + 8 + 9 + 9 + 10 + 11 + 12}{8} = \frac{72}{8} = 9$$

1. Calculate the mean, $\bar{x}$.
2. Find the deviation of each data value.
3. Find the square of the deviation for each data value.

x	$\bar{x}$	$(x - \bar{x})$	$(x - \bar{x})^2$
5	9	−4	16
8	9	−1	1
8	9	−1	1
9	9	0	0
9	9	0	0
10	9	1	1
11	9	2	4
12	9	3	9

$$\sigma^2 = \frac{16 + 1 + 1 + 0 + 0 + 1 + 4 + 9}{8} = \frac{32}{8} = 4$$

4. Find the mean of the squared differences (the variance).

$$\sigma = \sqrt{4} = 2$$

5. Take the square root of the variance.

The standard deviation is 2.

SKILL CHECK **EXERCISE 13** ☑

EXAMPLE 2: Finding the Standard Deviation

Find the standard deviation of the number of calories in 1 scoop (72 g) of ice cream.

Flavors (72 g)	Calories
chocolate chip	155
chocolate	156
coffee	170
rainbow	185
french vanilla	145
mint chocolate chip	172
rocky road	185

Answer

$$\bar{x} = \frac{155 + 156 + 170 + 185 + 145 + 172 + 185}{7} = \frac{1168}{7} \approx 166.86$$

1. Calculate the mean, $\bar{x}$.
2. Find the deviation of each data value.
3. Find the square of the deviation for each data value.

x	$\bar{x}$	$(x - \bar{x})$	$(x - \bar{x})^2$
155	166.86	−11.86	140.66
156	166.86	−10.86	117.94
170	166.86	3.14	9.86
185	166.86	18.14	329.06
145	166.86	−21.86	477.86
172	166.86	5.14	26.42
185	166.86	18.14	329.06

$$\sigma^2 = \frac{140.66 + 117.94 + 9.86 + 329.06 + 477.86 + 26.42 + 329.06}{7}$$

$$= \frac{1430.86}{7} \approx 204.41$$

$$\sigma = \sqrt{204.41} \approx 14.3$$

The standard deviation is about 14.3.

4. Find the mean of the squared differences (the variance).

5. Take the square root of the variance.

SKILL CHECK **EXERCISE 15**

The standard deviation is an important tool statisticians use to compare multiple data sets. The standard deviation may be interpreted differently depending on what the data set represents.

set *A*	set *B*
mean: 70.42	mean: 70.67
$\sigma \approx 4.65$	$\sigma \approx 1.84$

If the statistics represented by sets *A* and *B* are based on recent tests that 2 different classes took, then you might conclude that the class representing set *B* did better than set *A*. With a slightly higher mean and a relatively small standard deviation, this would indicate set *B* scores were more consistent than those for set *A*. However, you might interpret the statistics differently if the data represents the mean and the standard deviation of heights for 2 basketball teams. Since the 2 teams have roughly the same mean, but set *A* has a significantly higher standard deviation, you can conclude that the team representing set *A* has taller players, which may give them the advantage on the court.

VOCABULARY

- standard deviation
- variance

EXERCISES

1. The ______ (variance / standard deviation) is the mean of the squares of deviation.
2. The ______ (variance / standard deviation) is the square root of the mean of the squares of deviation.

Use the given set of data for exercises 3–7. Round to the nearest hundredth if necessary.

{7, 9, 7, 4, 7, 8, 10, 5, 5}

3. Calculate the mean.
4. Find the deviation for each data value.
5. Find the square for the deviation for each data value.
6. Calculate the variance.
7. Calculate the standard deviation.

Use the given data set and mean to calculate the standard deviation. Round to the nearest hundredth.

8. $\{45, 90, 86, 78, 91, 88, 79\}; \overline{x} = 79.57$
9. $\{25, 40, 19, 39, 18, 26\}; \overline{x} = 27.83$
10. $\{14, 5, 12, 0, 7, 16, 8, 25, 7\}; \overline{x} = 10.44$
11. $\{514, 405, 312, 607, 715, 516, 803, 472, 425, 623\}; \overline{x} = 539.2$

Use the given data set to find each statistic. Round to the nearest hundredth.

set *A* {28, 32, 36, 31, 40, 35, 32}

12. mean

13. standard deviation

set *B* {15, 22, 16, 24, 8, 17, 12}

14. mean

15. standard deviation

set *C* {1222, 1197, 1200, 921, 785, 704, 646}

16. mean

17. standard deviation

set *D* {266, 255, 264, 191, 185, 180, 172}

18. mean

19. standard deviation

20. Suppose the previous data sets *A* and *B* indicate the number of homes built by a contractor during 2 different 7 yr periods. Use the statistics found in exercises 12–15 to compare home building for the time periods.

21. Suppose the previous data sets *C* and *D* indicate the number of teenage passenger fatalities with a teen driver (set *C*) and a driver aged 30–59 (set *D*). Use the statistics found in exercises 16–19 to compare the drivers.

22. Essential Question: What does the standard deviation measure?

A.3 DATA DISPLAYS

How can I summarize and illustrate larger sets of data?

After completing this section, I will be able to

- represent sets of data with a dot plot and a box-and-whisker plot.
- analyze the information illustrated by a graph.
- summarize a large set of data by using a frequency distribution table or an interval frequency table.
- construct a histogram representing the data in an interval frequency table.

The **five-number summary** of a data set contains the lowest value, Q_1, Q_2, Q_3, and the highest value. A **box-and-whisker plot** uses these values to illustrate the median, interquartile range (IQR), and range.

After locating the values in the five-number summary on a number line, boxes are drawn from Q_1 to Q_2 and from Q_2 to Q_3 to represent the values within the IQR. Line segments (the whiskers) are drawn from the smallest value to Q_1 and from Q_3 to the largest value.

EXAMPLE 1: Constructing a Box-and-Whisker Plot

Construct a box-and-whisker plot to illustrate the five-number summary of the data: {27, 34, 29, 18, 28, 13, 26, 37, 33, 22}. Then calculate the range and IQR.

Answer

13, 18, 22, 26, 27, 28, 29, 33, 34, 37

1. Arrange the data in ascending order.

five-number summary:
13, 22, 27.5, 33, 37

2. Q_2 is the average of the fifth and sixth values. Q_1 is the third value, and Q_3 is the eighth value.

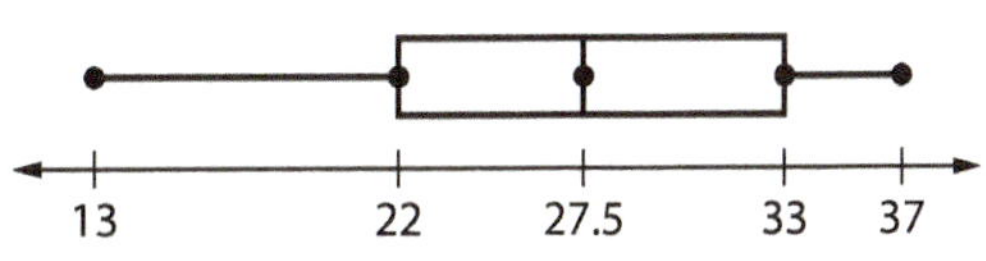

3. Construct a box-and-whisker plot.

range: 37 − 13 = 24
IQR: 33 − 22 = 11

4. Calculate the range and the IQR.

SKILL CHECK EXERCISE 5

EXAMPLE 2: Analyzing a Box-and-Whisker Plot

Use the box-and-whisker plot to determine the median and IQR for the times (in seconds) in a 400 m race.

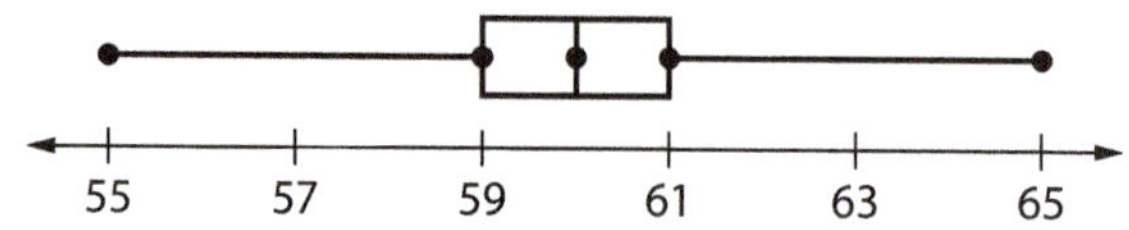

Answer

median: 60 sec

1. The median is the middle line in the large box.

IQR: $61 - 59 = 2$ sec

2. The IQR is the difference between the 2 ends of the large box.

SKILL CHECK **EXERCISES 7, 9**

A **dot plot** is often used to visualize the distribution of data when values occur repeatedly. After drawing a number line that includes each value found in the data, place a dot above the corresponding number each time the value appears in the data set.

Ages of Members in the Youth Group				
13	17	15	16	12
16	13	19	17	17
14	16	15	16	12
14	18	15	16	16

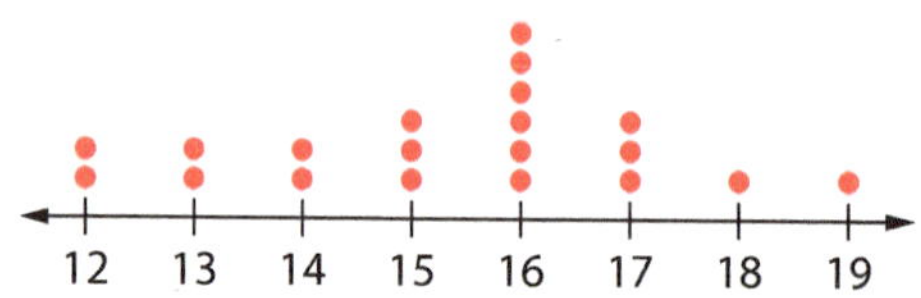

For a value with a decimal, such as 16.6, a dot would be placed above the 16.

EXAMPLE 3: Analyzing a Dot Plot

Use the dot plot shown above to determine the median and IQR for the data set.

Answer

median: $\frac{16 + 16}{2} = 16$

1. The median is the average of the 10th and 11th ages.

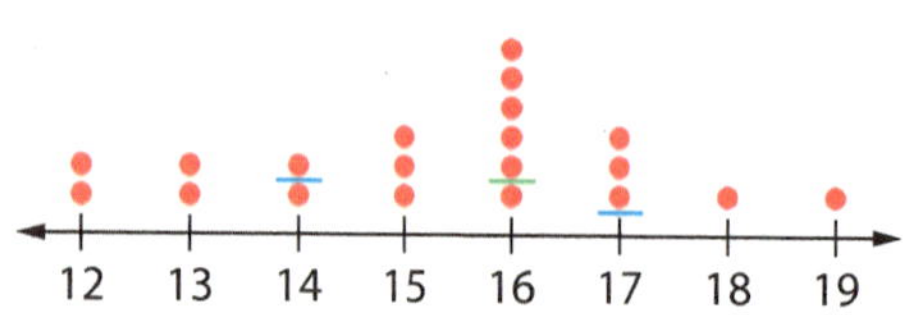

IQR $= 16.5 - 14 = 2.5$

2. The IQR is the difference between Q_1 and Q_3.

SKILL CHECK **EXERCISE 1**

A **frequency distribution table** is often used to organize larger data sets that contain repeated values. The frequency table below summarizes the results of an algebra test. The data column records each score, *d*, as an ordered list and is followed by a column for tallies to count the number of times each score appears. The number of tallies, *f*, is recorded in the frequency column. The product column contains the result of multiplying each score by its frequency, *df*. Totals of the frequency and product columns are used to find the mean of the data.

Frequency Distribution Table			
data (*d*)	**tally**	**frequency (*f*)**	**product (*df*)**
68	|	1	68
72	||	2	144
76	|||	3	228
80	|||	3	240
84	||	2	168
86	|	1	86
88	𝍸	5	440
92	||	2	184
96	|||	3	288
100	||	2	200
total		24	2046

Test Scores			
92	88	96	100
84	80	68	76
100	80	72	92
76	88	96	72
88	88	86	76
88	96	80	84

EXAMPLE 4: Finding Statistics with a Frequency Distribution Table

Use the frequency distribution table above to find the mean, median, mode, and range for the algebra test scores.

Answer

mean: $\frac{2046}{24} \approx 85.3$

1. Calculate the mean by dividing the sum of all the data (the total of the product column) by the number of scores (the total of the frequency column).

twelfth score: 86

thirteenth score: 88

median: $\frac{86 + 88}{2} = 87$

2. Count the tallies or add the frequencies to identify the twelfth and thirteenth scores. Then find their average.

mode: 88

3. Identify the number with the greatest frequency.

range: $100 - 68 = 32$

4. Calculate the range by subtracting the lowest data value from the highest.

SKILL CHECK **EXERCISE 21**

It is often helpful to group values from larger data sets in an **interval frequency table**. The table can then be used to produce reliable estimates of statistics that describe these large sets of data.

A **histogram** is used to illustrate the information found in an interval frequency table. The histogram's vertical bars represent the frequency of data in each interval. The bars are drawn without gaps to indicate that all numbers in the data set are accounted for by the intervals.

The range of the data in an interval frequency table can be estimated by subtracting the smallest value of the lowest interval from the largest value of the highest interval. An estimate for the range of ages in the illustrated data is 89 − 10 = 79.

To estimate the mean of data in intervals, find the product of each interval's midpoint and the frequency of values in that interval. The midpoint of each interval is the average of its beginning and ending values. Dividing the sum of the products by the sum of the frequencies produces an estimate of the mean.

Attendees	
age (interval)	**frequency (*f*)**
10–19	14
20–29	10
30–39	17
40–49	8
50–59	15
60–69	12
70–79	5
80–89	2
total	83

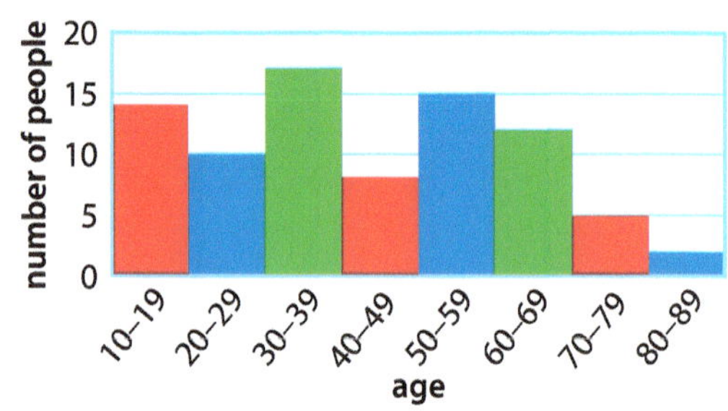

EXAMPLE 5: Estimating the Mean with an Interval Frequency Table

Complete midpoint and product columns for the previous interval frequency table. Then estimate the mean of the ages.

Answer

Attendees			
age (interval)	**midpoint (*m*)**	**frequency (*f*)**	**product (*mf*)**
10–19	14.5	14	203
20–29	24.5	10	245
30–39	34.5	17	586.5
40–49	44.5	8	356
50–59	54.5	15	817.5
60–69	64.5	12	774
70–79	74.5	5	372.5
80–89	84.5	2	169
total		83	3523.5

mean: $\frac{\text{sum of the products}}{\text{sum of the frequencies}} = \frac{3523.5}{83} \approx 42.5 \text{ yr}$

1. Find the midpoint of each age interval.
   $m_1 = \frac{10 + 19}{2} = \frac{29}{2} = 14.5$
   Repeat this process for the remaining intervals.
2. Find the product of each interval's midpoint and frequency.
   $m_1 f_1 = (14.5)(14) = 203$
   Repeat this process for the remaining intervals.
3. Find the sums of the frequency and the product columns.
4. Use these sums to calculate an estimate of the mean.

SKILL CHECK **EXERCISE 27**

VOCABULARY

- box-and-whisker plot
- dot plot
- five-number summary
- frequency distribution table
- histogram
- interval frequency table
- median interval
- modal interval

The table can also be used to identify the **modal interval,** the interval with the most data values.

In this case, the modal interval is 30–39 with a frequency of 17.

The **median interval** is the interval that contains the median. If n is the number of data values, the expression $\frac{n+1}{2}$ describes the position of the median. Since the table from Example 5 describes 83 values, the median is in position $\frac{n+1}{2} = \frac{83+1}{2} = \frac{84}{2} = 42$, the forty-second value. Since the first 3 intervals contain $14 + 10 + 17 = 41$ values, the forty-second data value must be in the next interval. Therefore, the median interval of the data is 40–49. If there were 84 values, then $\frac{n+1}{2} = \frac{84+1}{2} = \frac{85}{2} = 42.5$, which indicates that the median would be the average of the forty-second and the forty-third values.

EXERCISES

Determine the median and IQR from each dot plot.

1.

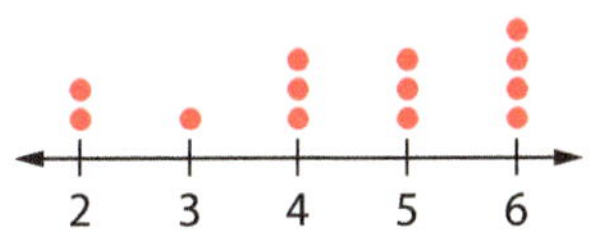

2.

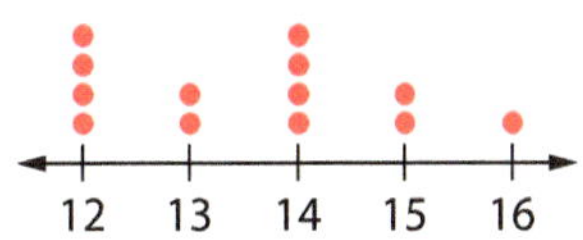

Make a dot plot to illustrate each set of data and give the mode and range for each.

3. {5.7, 5.1, 4.9, 6.3, 5.6, 6.4, 6.3, 6.3, 7.4, 8.1, 7.6, 4.7}

4. {2.3, 1.9, 2.1, 3.6, 2.4, 1.8, 3.2, 2.6, 4.0, 0.7, 2.9, 3.7, 2.8, 3.6}

Use the five-number summary 60, 71.5, 93.5, 100, 105 to complete exercises 5–7.

5. Create a box-and-whisker plot. **6.** Find the median. **7.** Determine the IQR.

Use the given box-and-whisker plot to complete exercises 8–10.

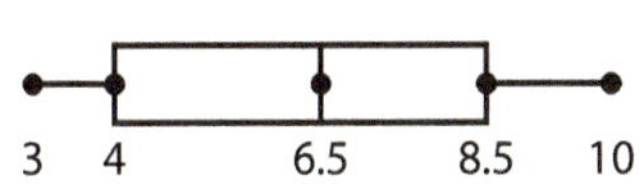

8. Write the five-number summary.

9. What is the median? **10.** Determine the range.

Use the data set {8, 10, 2, 10, 5, 9, 10, 4, 9} to complete exercises 11–13.

11. Find the five-number summary. **12.** Create a box-and-whisker plot. **13.** Determine the IQR.

Use the data set {33, 98, 84, 40, 86, 79, 80} to complete exercises 14–16.

14. Find the five-number summary. **15.** Create a box-and-whisker plot. **16.** Determine the IQR.

Make a frequency distribution table for each set of data.

17. {10, 10, 11, 12, 12, 12, 14, 15, 15, 15, 15, 17, 18, 18, 19, 20, 20, 20}

18. {3, 4, 5, 5, 6, 6, 7, 7, 7, 7, 8, 8, 9, 9, 9, 9, 10, 10, 10, 10}

19. {85, 80, 75, 100, 90, 75, 90, 100, 100, 95, 60, 75, 80, 85, 85, 80, 80}

20. {200, 250, 200, 225, 275, 275, 200, 200, 225, 225, 250, 250, 200, 200, 275}

Complete each frequency distribution table and use it to find the mean, median, mode, and range of the data. Round to the nearest tenth if necessary.

21.

<table>
<tr><th>Data (d)</th><th>Tally</th><th>Frequency (f)</th><th>Product (df)</th></tr>
<tr><td>100</td><td>|</td><td>1</td><td>100</td></tr>
<tr><td>96</td><td>||</td><td>2</td><td>192</td></tr>
<tr><td>92</td><td>||</td><td>2</td><td>184</td></tr>
<tr><td>88</td><td>|||</td><td>3</td><td>264</td></tr>
<tr><td>84</td><td>||||</td><td>4</td><td>336</td></tr>
<tr><td>80</td><td>|||</td><td>3</td><td>240</td></tr>
<tr><td>76</td><td>𝍸</td><td>5</td><td>380</td></tr>
<tr><td>72</td><td>|</td><td>1</td><td>72</td></tr>
<tr><td>total</td><td></td><td></td><td></td></tr>
</table>

22.

<table>
<tr><th>Data (d)</th><th>Tally</th><th>Frequency (f)</th><th>Product (df)</th></tr>
<tr><td>70</td><td>||||</td><td>4</td><td>280</td></tr>
<tr><td>68</td><td>𝍸</td><td>5</td><td>340</td></tr>
<tr><td>66</td><td>|||</td><td>3</td><td>198</td></tr>
<tr><td>64</td><td>||</td><td>2</td><td>128</td></tr>
<tr><td>62</td><td>||</td><td>2</td><td>124</td></tr>
<tr><td>60</td><td>|||</td><td>3</td><td>180</td></tr>
<tr><td>58</td><td>|</td><td>1</td><td>58</td></tr>
<tr><td>total</td><td></td><td></td><td></td></tr>
</table>

23.

<table>
<tr><th>Data (d)</th><th>Tally</th><th>Frequency (f)</th><th>Product (df)</th></tr>
<tr><td>40</td><td>|</td><td>1</td><td></td></tr>
<tr><td>50</td><td>||</td><td>2</td><td></td></tr>
<tr><td>60</td><td>|||</td><td>3</td><td></td></tr>
<tr><td>70</td><td>||||</td><td>4</td><td></td></tr>
<tr><td>80</td><td>|||</td><td>3</td><td></td></tr>
<tr><td>90</td><td>||||</td><td>4</td><td></td></tr>
<tr><td>100</td><td>||</td><td>2</td><td></td></tr>
<tr><td>total</td><td></td><td></td><td></td></tr>
</table>

24.

<table>
<tr><th>Data (d)</th><th>Tally</th><th>Frequency (f)</th><th>Product (df)</th></tr>
<tr><td>31</td><td>|</td><td>1</td><td></td></tr>
<tr><td>32</td><td>||</td><td>2</td><td></td></tr>
<tr><td>33</td><td>|</td><td>1</td><td></td></tr>
<tr><td>34</td><td>|||</td><td>3</td><td></td></tr>
<tr><td>35</td><td>||</td><td>2</td><td></td></tr>
<tr><td>36</td><td>|||</td><td>3</td><td></td></tr>
<tr><td>37</td><td>𝍸</td><td>5</td><td></td></tr>
<tr><td>38</td><td>||</td><td>2</td><td></td></tr>
<tr><td>39</td><td>|||</td><td>3</td><td></td></tr>
<tr><td>40</td><td>||</td><td>2</td><td></td></tr>
<tr><td>total</td><td></td><td></td><td></td></tr>
</table>

25.

<table>
<tr><th>Data (d)</th><th>Tally</th><th>Frequency (f)</th><th>Product (df)</th></tr>
<tr><td>210</td><td>||</td><td></td><td></td></tr>
<tr><td>200</td><td>||</td><td></td><td></td></tr>
<tr><td>195</td><td>|||</td><td></td><td></td></tr>
<tr><td>185</td><td>|||</td><td></td><td></td></tr>
<tr><td>170</td><td>||</td><td></td><td></td></tr>
<tr><td>160</td><td>𝍸</td><td></td><td></td></tr>
<tr><td>total</td><td></td><td></td><td></td></tr>
</table>

26.

<table>
<tr><th>Data (d)</th><th>Tally</th><th>Frequency (f)</th><th>Product (df)</th></tr>
<tr><td>21</td><td>|</td><td></td><td></td></tr>
<tr><td>18</td><td>|||</td><td></td><td></td></tr>
<tr><td>16</td><td>|</td><td></td><td></td></tr>
<tr><td>15</td><td>|</td><td></td><td></td></tr>
<tr><td>12</td><td>||</td><td></td><td></td></tr>
<tr><td>9</td><td>||||</td><td></td><td></td></tr>
<tr><td>8</td><td>||</td><td></td><td></td></tr>
<tr><td>total</td><td></td><td></td><td></td></tr>
</table>

For each set of data, make an interval frequency table with intervals of 0–9, 10–19, 20–29, etc. Then use the table to estimate the mean to the nearest tenth.

27. days of exercise in May:
{0, 5, 5, 5, 6, 6, 7, 11, 15, 18, 18, 18, 18, 19, 19, 20, 26, 28, 29}

28. apps on a smartphone:
{41, 45, 49, 51, 60, 64, 64, 67, 72, 79, 79, 83, 85, 89, 90, 95}

29. assembly steps completed on a CT scanner:
{12, 15, 30, 48, 49, 41, 17, 25, 26, 45, 48, 32, 36, 24}

For each set of data, make an interval frequency table with intervals of 0–19, 20–39, etc. Then use the table to estimate the mean to the nearest tenth.

30. push-ups: {20, 29, 51, 39, 43, 64, 23, 34, 28, 72, 42, 44, 18}

31. sit-ups: {62, 48, 63, 74, 28, 31, 27, 17, 19, 12, 10, 28, 37}

32. Make an interval frequency table for the set of data, {180, 201, 348, 302, 195, 190, 175, 160, 310, 379, 420, 490}, with intervals of 100–199, 200–299, etc. Use the table to estimate the mean to the nearest tenth.

Use the interval frequency table to estimate the range and mean. Then identify the median interval and the modal interval.

33.

Interval	Frequency
10–14	5
15–19	12
20–24	19
25–29	8
30–34	4
total	48

34.

Interval	Frequency
10–14	2
15–19	12
20–24	21
25–29	30
30–34	4
35–39	2
total	71

35. **Essential Question:** How can I summarize and illustrate larger sets of data?

A.4 RESIDUALS

How are residuals useful?

After completing this section, I will be able to

- define a residual.
- determine how well a trend line fits data by plotting residuals.
- define causation.
- identify correlation and causation in a given situation.

Recall that a trend line can be drawn on a scatter plot to model the data and assist in making predictions. This trend line is drawn so that it is as close as possible to all the points, ignoring any outliers. *Residuals* can be used to determine how well the trend line fits the data.

DEFINITION

A **residual** is the vertical difference between the trend line and an actual data point.

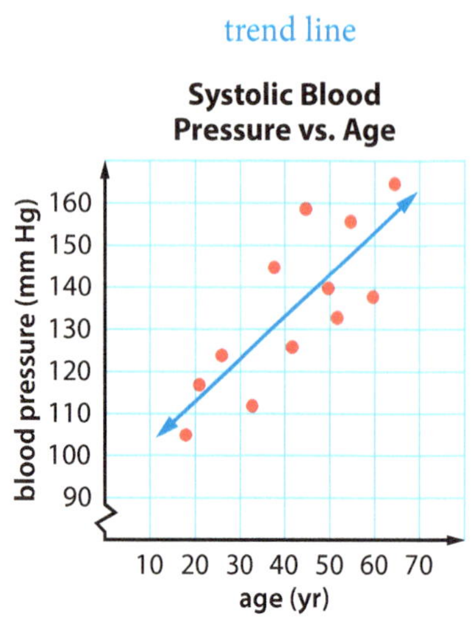

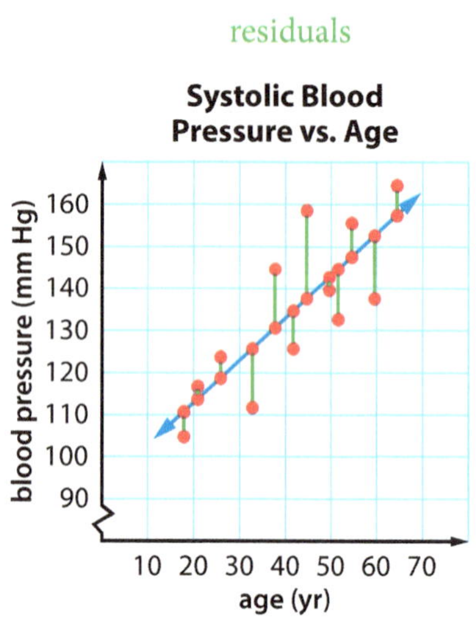

A residual can be positive, negative, or 0 depending on the location of the data point relative to the trend line. Graphing a scatter plot of the residuals, called a residual plot, can help determine how well the trend line fits the data. If the trend line is a good fit and the residual values are plotted, the values will be fairly small and somewhat evenly dispersed about the x-axis. If the line is not a good fit, the residual points will tend to be more spread out, or a pattern may emerge that may indicate a nonlinear correlation.

EXAMPLE 1: Graphing Residuals

Graph a scatter plot of residuals based on the table of values and the trend line $y = 2x - 5$.

x	1	2	3	4	5	6	7	8
y	−2	0	−1	4	4	7	8	12

Answer

x	y	2x − 5	Residual
1	−2	−3	−2 − (−3) = 1
2	0	−1	0 − (−1) = 1
3	−1	1	−1 − 1 = −2
4	4	3	4 − 3 = 1
5	4	5	4 − 5 = −1
6	7	7	7 − 7 = 0
7	8	9	8 − 9 = −1
8	12	11	12 − 11 = 1

1. Calculate the residuals.

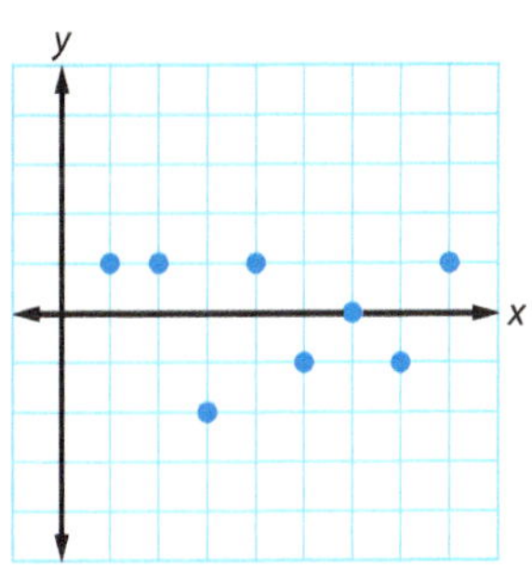

2. Plot the points (x, residual) to generate the scatter plot.

SKILL CHECK **EXERCISE 11**

Since the graph of residuals is close to the x-axis and well balanced between positive and negative values, the equation $y = 2x - 5$ is a good trend line for the table of values.

In Chapter 5 you learned to find statistical measures such as the line of best fit and the correlation coefficient by entering a table of values in the stat mode of your graphing calculator. This feature can also be used to help you graph residuals.

EXAMPLE 2: Graphing Residuals

The age (in years) and systolic blood pressure (in mm Hg) of 12 participants in a study are recorded in the table and displayed in the graph preceding Example 1. Use a graphing utility to construct a residual plot to determine whether the equation $y = x + 93$ is a good fit for the data.

age (yr)	18	21	26	33	38	42	45	50	52	55	60	65
blood pressure (mm Hg)	105	117	124	112	145	126	159	140	133	156	138	165

Answer

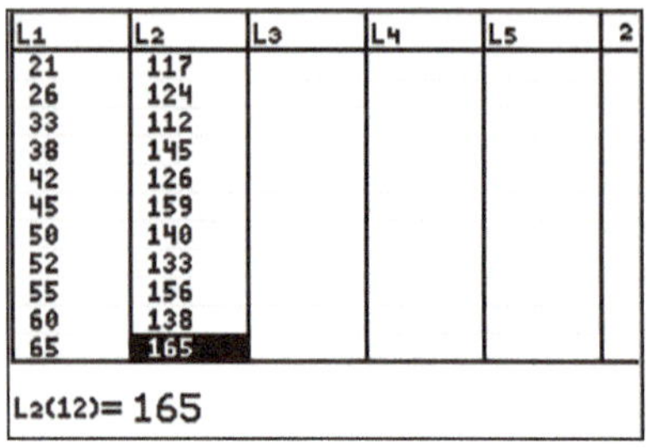

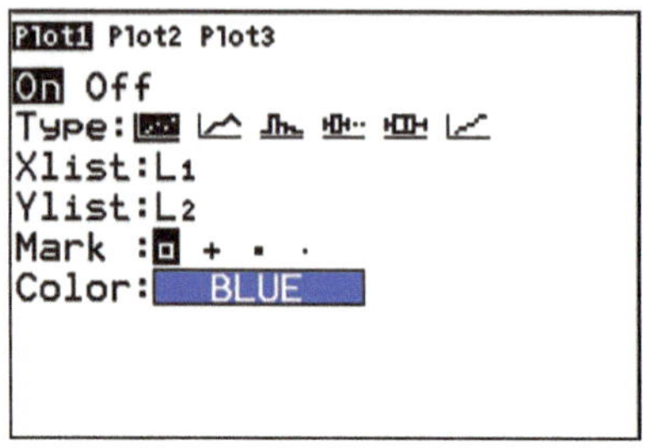

1. Enter the table of values in L_1 and L_2 in your graphing calculator. Then use your calculator to determine the line of best fit for L_1 and L_2. (If necessary, see the Chapter 5 Technology Corner for step-by-step instructions.)
2. Turn Plot1 on by pressing [2nd] [y=] and then press [enter] twice.

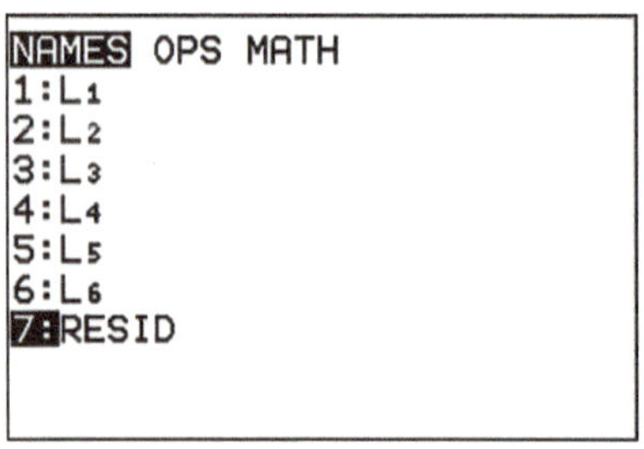

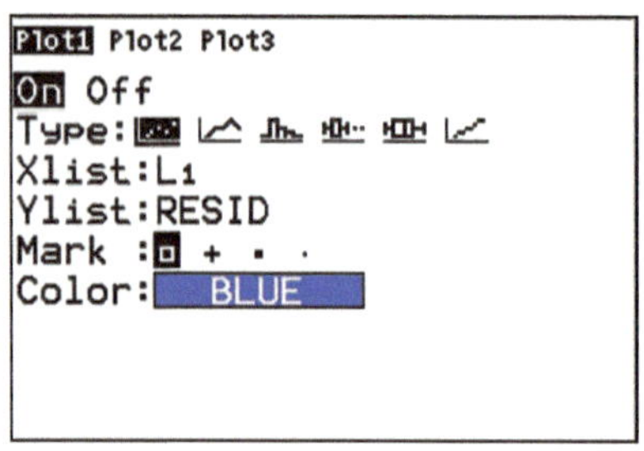

3. To plot residuals rather than the values from L_2, navigate down to Ylist and press [2nd] [stat] then select RESID.

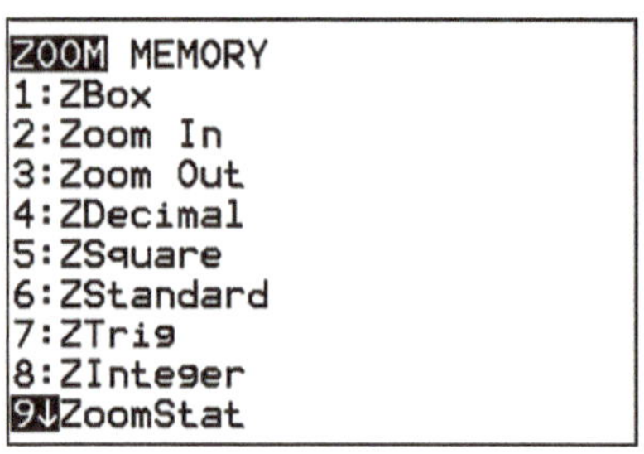

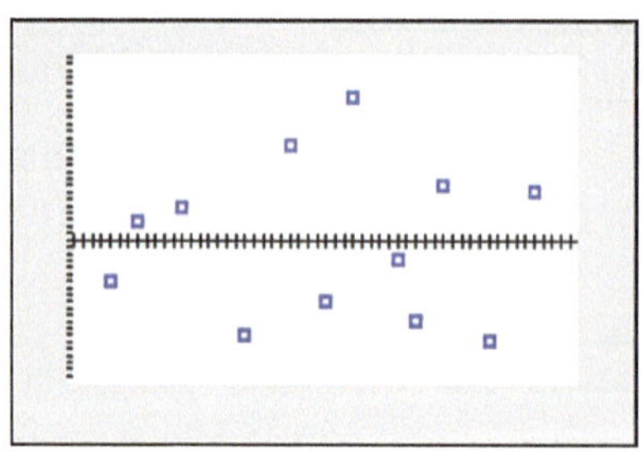

4. Press [zoom] [9] to select ZoomStat and the residual plot will automatically appear. (Note that you will receive an error message if you have not already used your calculator to determine the line of best fit.)

SKILL CHECK **EXERCISE 15**

Notice the residuals are not clustered as closely to the horizontal axis as in Example 1. However, the correlation coefficient (≈0.795) would indicate a fairly strong correlation between the data values given the line of best fit $y \approx 0.9942x + 93.1614$. Thus, the equation $y = x + 93$ would be a good fit for the data.

Statisticians are not only looking for correlations between data sets, but they are also interested in determining causation.

DEFINITION

Causation means that the change in one variable directly affects the change in another variable.

Causation implies a strong correlation between 2 variables, but a strong correlation does not imply causation. There may be a strong correlation between a person's age and high blood pressure, but a person's age does not cause high blood pressure. Identifying correlation and causation in situations is a valuable tool to applying statistics to our world.

EXAMPLE 3: Identifying Correlation & Causation

Determine whether there is correlation in each situation. If so, determine whether there is causation. Explain your reasoning.

a. The more a person exercises, the more calories he or she will burn.

b. A person with big feet will also wear large shirts.

Answers

a. There is a strong correlation between the amount of time spent exercising and the number of calories that will be burned. More time spent exercising will cause a person to burn more calories.

b. There is a strong correlation between people with big feet and people who wear large shirts. However, one does not necessarily cause the other to be true, so there is no causation.

SKILL CHECK **EXERCISE 21**

VOCABULARY

- causation
- residual

EXERCISES

Classify as true or false.

1. A strong correlation of events implies causation between them.
2. A residual plot with points clustered near the horizontal axis and on both sides of the axis indicates a good trend line.

Explain what the residual plot indicates about the trend line or data set.

3.

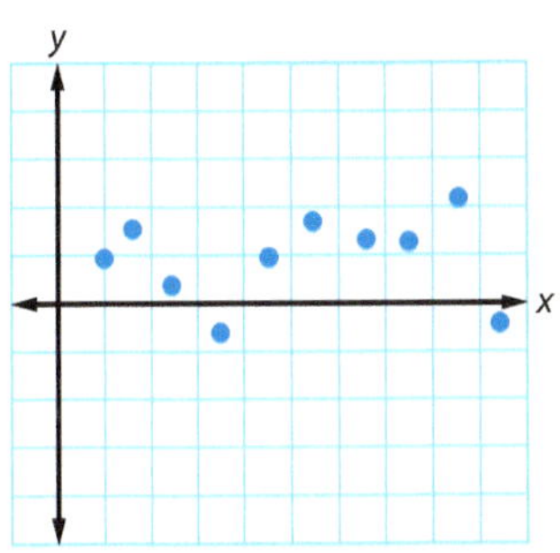

4.

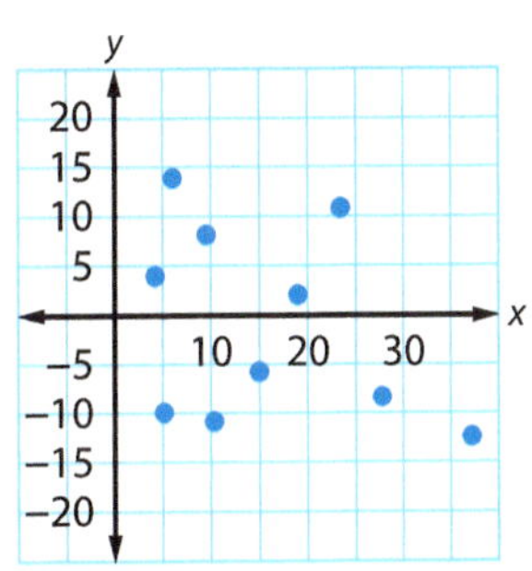

5.

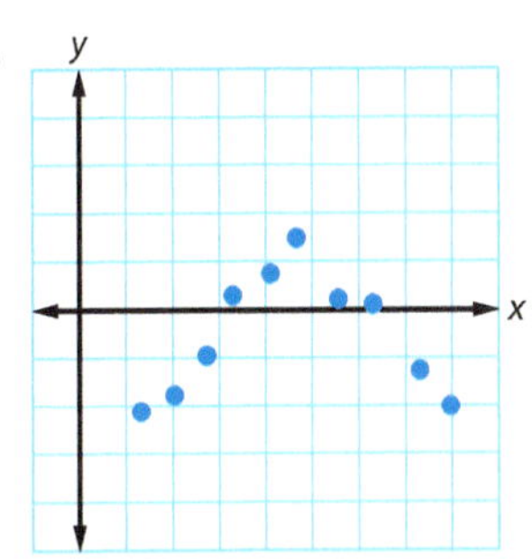

Complete the table to find the residuals algebraically.

6.

x	y	$f(x) = 8x - 3$	Residual
1	−2		
2	11		
3	24		
4	27		
5	32		

7.

x	y	$f(x) = -x + 4$	Residual
1	3.5		
3	0.9		
4	−1		
6	−1.7		
7	−3		

8.

x	y	$f(x) = 15 - \frac{1}{2}x$	Residual
4	10		
8	14		
12	10		
16	4		
20	8		

9.

x	y	$f(x) = 1.6x + 13$	Residual
−10	−1		
−5	12		
5	15		
10	31		
15	42		

MP3 Use residuals to determine whether the trend line equation is a good model for the table of values.

10. exercise 6: $y = 8x - 3$

11. exercise 7: $y = -x + 4$

12. exercise 8: $y = 15 - \frac{1}{2}x$

13. exercise 9: $y = 1.6x + 13$

MP5 Use technology to answer each question based on the table of values.

A ministry tracked how many of their podcasts were downloaded each month.

month	1	2	3	4	5	6	7	8	9	10	11	12
podcast downloads (thousands)	6.2	7.8	7	8.5	9.3	10.2	11.1	10.8	12.2	13.5	13.8	14.7

14. Find the line of best fit. State your answer to the nearest ten thousandth.

15. Graph the residuals.

16. Is the line of best fit a good model for the table of values? Explain.

A food scientist monitored how much yeast was present in a culture over time, measured in colony forming units (CFU) per milliliter of serum in which the yeast was kept.

hour	1	2	3	4	5	6	7	8	9	10	11	12
yeast CFU/mL (thousands)	50	62	85	117	160	220	375	550	600	620	630	633

17. Find the line of best fit. State your answer to the nearest ten thousandth.

18. Graph the residuals.

19. Is the line of best fit a good model for the table of values? Explain.

Identify the cause and the correlation(s) in each situation.

20. Jaimie has to replace her running shoes every 6 months because she jogs to stay in good physical condition.

    a. replace running shoes every 6 months

    b. jogs

    c. good physical condition

21. On a sunny day in June, Jeff ate freeze pops at the amusement park and got a sunburn.

    a. a sunny day

    b. ate freeze pops

    c. sunburn

Determine whether each statement reflects causation or correlation. Explain your reasoning.

22. People who smoke a lot are more likely to get lung cancer.

23. A windmill turns on a windy day.

24. Shark attacks increase when suntan lotion sales rise.

25. Describe a situation with a strong correlation that does not have a causal relationship.

26. **Essential Question:** How are residuals useful?

A.5 TWO-WAY FREQUENCY TABLES

How is a two-way frequency table useful?

After completing this section, I will be able to

- make and interpret a two-way frequency table.
- describe the association between the categories in a two-way frequency table.

Scatter plots are used to illustrate quantitative (numerical) bivariate data and describe the association of the 2 variables. Trend lines are then used to make predictions based on that data. **Categorical variables**, such as color or hobby, are used to describe qualitative data that is not measured numerically, such as blue, red, and green or quilting and woodworking.

DEFINITION

A **two-way frequency table** (or contingency table) displays the frequencies of 2 categorical variables from the same source.

A school is studying participation of its sixth-, seventh-, and eighth-grade students in the school's band. The possible values for the first category, grade level, are listed in columns, and the possibilities for the second category, participating or not, are listed in rows.

		Grade Level			
		6th	**7th**	**8th**	**total**
Band	**yes**	10	12	11	33
	no	13	16	14	43
	total	23	28	25	76

joint frequencies

marginal frequencies

grand total

The number representing each combination of categorical variables is a **joint frequency**. The sum of each row or column of joint frequencies is a **marginal frequency**. The sum of the marginal frequencies in the last row or the last column is the **grand total**. The grand total is also the sum of the joint frequencies.

EXAMPLE 1: Making & Interpreting a Two-Way Frequency Table

Students were asked whether they studied at least 5 min or less than 5 min for a quiz. Make a two-way frequency table to compare the time studied and quiz results. Then use the two-way frequency table to answer each question.

a. How many students studied at least 5 min but did not pass?

b. What percentage of those who studied for less than 5 min passed the quiz?

study time (≥5 or <5)	≥5	≥5	<5	≥5	<5	≥5	<5	≥5	<5	≥5	<5	≥5
result (pass or fail)	P	F	P	P	P	P	F	P	P	P	P	P

Answers

		Results		
		pass	fail	total
Study Time	<5	4	1	5
	≥5	6	1	7
	total	10	2	12

1. Count and record the number for each joint frequency.
2. Find the marginal frequencies by adding each row and column of joint frequencies.
3. Find and check the grand total. $10 + 2 = 12$ and $5 + 7 = 12$

a. There is just 1 student who studied for at least 5 min and did not pass.

b. $\frac{4}{5} = 0.8$; 80% of the students who studied less than 5 min passed the quiz.

SKILL CHECK **EXERCISE 5**

EXAMPLE 2: Completing & Interpreting a Two-Way Frequency Table

A survey of the favorite rides of 40 teens attending an amusement park indicated that 14 of the teens prefer rollercoasters. Nine boys and 5 of the 18 girls prefer water rides. Five boys prefer neither a rollercoaster nor a water ride. Complete a two-way frequency table describing the preferred type of rides for each teen and use the table to answer each question.

a. What percentage of the boys prefer rollercoasters?

b. What percentage of those who prefer rollercoasters are boys?

Answers

Ride Preferences	RC	WR	N	total
boys	8	9	5	22
girls	6	5	7	18
total	14	14	12	40

1. Fill in the known grand total, the joint frequencies for those preferring water rides or neither, and the marginal frequencies for teens preferring rollercoasters and total girls.
2. Find the unknown marginal frequencies.
   $40 - 18 = 22$; 22 boys attended.
   $9 + 5 = 14$; 14 teens prefer water rides.
   $40 - 14 - 14 = 12$; 12 teens prefer neither.
3. Find the unknown joint frequencies.
   $12 - 5 = 7$; 7 girls prefer neither.
   $22 - 5 - 9 = 8$; 8 boys prefer rollercoasters.
   $14 - 8 = 6$; 6 girls prefer rollercoasters.

a. $\frac{8}{22} \approx 0.364$; 36.4% of the boys prefer rollercoasters.

b. $\frac{8}{14} \approx 0.571$; 57.1% of those who prefer rollercoasters are boys.

SKILL CHECK **EXERCISES 8–9**

TIP

Rounding the relative frequencies in the table may cause a sum to appear to be inaccurate.

A **relative frequency** expresses the ratio of a joint or marginal frequency to a total, often as a percent. The association between the categorical variables can be described by examining the relative frequencies within each row or each column.

		Grade Level			
		6th	7th	8th	total
Band	yes	10	12	11	33
	no	13	16	14	43
	total	23	28	25	76

To make a table comparing relative frequencies for students in the band and those not in the band, divide each joint frequency by its row's total.

Grade Level

Band		6th	7th	8th	total
	yes	$\frac{10}{33} \approx 30\%$	$\frac{12}{33} \approx 36\%$	$\frac{11}{33} \approx 33\%$	$\frac{33}{33} = 100\%$
	no	$\frac{13}{43} \approx 30\%$	$\frac{16}{43} \approx 37\%$	$\frac{14}{43} \approx 33\%$	$\frac{43}{43} = 100\%$

Since there is practically no difference in the rows of relative frequencies, we can conclude there is no association between grade level and band participation. Greater differences in the relative frequencies indicate a greater association between the categorical variables.

EXAMPLE 3: Using Relative Frequencies to Determine Association

Make a table comparing the relative frequencies for each ride preference. Then use the relative frequencies to describe the association between gender and the preferred rides.

Ride Preferences

	RC	WR	N	total
boys	8	9	5	22
girls	6	5	7	18
total	14	14	12	40

Answer

1. Divide each frequency by its column's total.

Ride Preferences

	RC	WR	N
boys	$\frac{8}{14} \approx 57\%$	$\frac{9}{14} \approx 64\%$	$\frac{5}{12} \approx 42\%$
girls	$\frac{6}{14} \approx 43\%$	$\frac{5}{14} \approx 36\%$	$\frac{7}{12} \approx 58\%$
total	$\frac{14}{14} = 100\%$	$\frac{14}{14} = 100\%$	$\frac{12}{12} = 100\%$

2. The significant differences in the columns of relative frequencies indicate there is an association between a teen's gender and the preferred type of ride. It is likely that a surveyed teen preferring water rides is a boy, and it is somewhat likely that a surveyed teen preferring rollercoasters is a boy. It is somewhat likely that a surveyed teen preferring neither rollercoasters nor water rides is a girl.

SKILL CHECK **EXERCISE 21**

VOCABULARY

- categorical variable
- grand total
- joint frequency
- marginal frequency
- relative frequency
- two-way frequency table

EXERCISES

1. A ______ would be used to list frequencies of the soft drink preferences of different classifications of college students.
2. Each sum of a row or column of joint frequencies is a ______.
3. The sum of the row or the column of marginal frequencies is the ______.
4. Each ratio of a joint or marginal frequency to a total is a ______.

A survey of 20 middle-school students asked whether they preferred unit tests or daily quizzes and whether they preferred classroom or virtual instruction.

test or quiz	T	T	Q	Q	T	T	T	Q	Q	T
classroom or virtual	C	V	V	V	C	V	C	V	V	V

test or quiz	Q	T	T	T	Q	T	T	T	Q	Q
classroom or virtual	V	C	C	C	C	V	C	V	V	C

5. Complete the two-way frequency table to summarize the data.

		Assessment Preference		
		unit test	daily quizzes	total
Location	classroom			
	virtual			
	total			

6. How many students prefer daily quizzes and virtual instruction?
7. What percent of students prefer unit tests?

A survey of 14 employees were asked whether they prefer five 8 hr days or four 10 hr days of work per week and whether they preferred two 15 min breaks or three 10 min breaks each workday.

breaks (min)	15	15	10	15	10	10	15	15	10	15	10	15	15	15
workdays per week	4	4	4	4	5	4	5	5	4	5	4	4	5	5

8. Complete the two-way frequency table to summarize the data.

		Length of Breaks		
		10	15	total
Workdays	4			
	5			
	total			

9. What was the least popular combination?
10. What was the preferred number of working days?
11. The employer currently has a policy of a 5-day workweek and three 10 min breaks. What changes might the employer consider because of the data?

Use the two-way frequency table to find each relative frequency (to the nearest percent).

Favorite Sport to Watch on TV

	football	basketball	baseball	total
male	57	47	30	134
female	48	42	26	116
total	105	89	56	250

12. What percent of those surveyed prefer football?
13. What percent of surveyed females prefer baseball?
14. What percent of those who prefer baseball are female?
15. What percent of those surveyed prefer basketball?
16. What percent of surveyed females prefer football or basketball?
17. What percent of surveyed males do not prefer football?

During a tryout of 40 soccer players, 9 of the 25 girls play defense, and 7 boys play offense.

18. Construct a two-way frequency table summarizing the data.

19. How many boys play defense?

20. What percent of the players are boys?

21. Make a table listing relative frequencies for girls and boys.

22. What position is a girl more likely to play?

23. Make a table listing relative frequencies for each position.

24. Is a defender more likely to be a girl or a boy?

When testing a new treatment, researchers found that 65 of 89 patients who received the treatment improved and that 72 of 100 who did not receive the treatment did not improve.

25. Construct a two-way table summarizing the data.

26. Make a table comparing relative frequencies of receiving and not receiving the treatment. Describe the association between the treatment and improvement.

27. Make a table comparing relative frequencies of those who improved and those who did not. Describe the association between the treatment and improvement.

Jackson surveyed 75 students in his school about their participation in after-school activities and whether they worked a part-time job. Partial results are in the table.

After-School Activities

Part-Time Job	Bible club	sports	both	neither	total
yes		12		9	51
no	5		10		
total	15	18			75

28. Complete the two-way table.

29. Make a table comparing the relative frequencies of those with and without a part-time job. Describe any association between students having a part-time job and their participation in after-school activities.

30. Make a table comparing the relative frequencies of participation in after-school activities. Describe any association between students having a part-time job and their participation in after-school activities.

31. **Essential Question:** How is a two-way frequency table useful?

SELECTED ANSWERS

CHAPTER 1: EXPRESSIONS

Section 1.1

1. absolute value

3. adding **5.** 19 **7.** -21

9. 4 **11.** -1 **13.** -6

15. 15 **17.** -17 **19.** 6

21. -6 **23.** -624.17 **25.** 621.17

27. -6.646 **29.** -1 **31.** $-\frac{82}{21}$

33. $\frac{7}{9}$ **35.** $-\frac{18}{5}$ **37.** $-7\frac{1}{7}$

39. $17 - 5$; 12 knots

41. $14{,}494 - (-282)$; 14,776 ft

43. $20 + 3\frac{3}{4} + 30\frac{1}{2} + 7\frac{1}{6} + 20\frac{5}{6} + 7\frac{1}{6} + 25\frac{1}{4} - 5\frac{1}{2} - 7\frac{7}{12} - 8\frac{2}{3}$; $92\frac{11}{12}$ ft

45. Associative Property

49. -10.522 **51.** 28 **53a.** yes

Section 1.2

1. negative **3.** 4; Associative Property

5. -6 **7.** -48 **9.** $\frac{9}{10}$

11. 9 **13.** -90 **15.** $\frac{4}{9}$

17a. Commutative Property

17b. Multiplicative Inverse Property

17c. Multiplicative Identity Property

19. $\frac{1}{68}$ **21.** 32.838 **23.** $\frac{3}{8}$

25. $\frac{2}{3}$ **27.** $-\frac{12}{5}$ **29.** $\frac{5}{6}$

31. $-\frac{21}{4}$ **33.** $4\frac{4}{5}$ pt

35. Multiplicative Identity Property

37. $\frac{455}{4}$

39. No; $6 \div 3 = 2$, but $3 \div 6 = \frac{1}{2}$.

43. Commutative Property **45.** -15.03

47. 40 **49.** $\frac{3}{4}$ **51.** C

Section 1.3

1. $5^0 = 1$ by the definition of zero power.

3. A **5.** 64 **7.** -125

9. $-\frac{1}{27}$ **11.** 64 **13.** 2

15. $-\frac{1}{4}$ **17.** $(-12)^5$ **19.** 8^4

21. 1 **23.** 10^{50} **25.** 10^9

27. 4^7 **29.** $\frac{49}{100}$ **31.** -5

33. $\frac{16}{11}$ **35a.** Power Property

35b. Product Property **35c.** Quotient Property

37. $\frac{2^6}{5}$ **39.** $\frac{1}{2^{12}3^4}$

41. $y^5 \cdot y^0 = y^{5+0} = y^5$
According to the Multiplicative Identity Property, $y^5 \cdot 1 = y^5$.
The product is the same; thus, $y^0 = 1$.

43. $\frac{4}{5}$ **45.** 521 **47.** -27

49. -59 **51.** 1640 **53.** D

Section 1.4

1. divide 18 by 3, multiply by 9

3. evaluate the square root, divide by 5

5. 11 **7.** -11 **9.** 49

11. -144 **13.** 29 **15.** -3

17. $11^2 + (-5) + 4$; 120 **19.** $12 \cdot \frac{1}{6} \cdot 4 \cdot \frac{1}{8}$; 1

21. We can evaluate from left to right since multiplication is the only operation being performed.

23. -1 **25.** 0 **27.** -6

29. 8 **31.** $\frac{3}{2}$ **33.** -14

35. $5(6 + 7)$; 65 **37.** $20^2 - 13^2$; 231

39. $5 \cdot 6 + 20 \div 5$; 34 **41.** $\frac{1}{2}$

43. 162 **45.** 28 **47.** $\frac{5}{36}$

49. 8^{10} **51.** 28^8 **53.** A

Section 1.5

1. x; 3 and -6 **3.** r; π **5.** $\{-4, 6, 10, 20\}$

7. -2 **9.** $-\frac{14}{5}$ **11.** $8n$

13. $n + 24$ **15.** $\frac{8}{49}$ **17.** c^5

19. $\frac{7}{144}$ **21.** $h^3 + h^2$ **23.** $\frac{p+q+r}{3}$

25. $126 - n(n+1)$ **27.** $3(90 - a)$ **29.** $(180 - a) - a$

31. $25x + 90y$ **33.** \$37.92 **35.** \$10,395

37. 89.12 in.^{2} **39.** $-\frac{1219}{84}$ **41.** \$22.50

43. 150.72 ft^2

45. Associative Property of Multiplication

47. Commutative Property of Addition

49. $\frac{1}{3^{21}}$ **51.** 41 **53.** C

Section 1.6

1. 3, 4, –5; 4*y*, –5*y* **3.** 1, –2, –9; *xy*, –2*xy*, –9*xy*
5. $4x + 8y$ **7.** $5x^2 + 5xy$ **9.** $10a$
11. $16xy$ **13.** $3y^2 - 8y$ **15.** $6a + 9$
17. $x - 8y$ **19.** $2x$ **21.** $8x - 5y$
23. $5a - 4b$ **25.** $-5.94m + 1.6n$
27. $-\frac{6}{5}a^2 + \frac{12}{5}a - 3z + \frac{32}{5}$ **29.** $x^2 + 2xy + y^2$
31. $-4r^2 + 6r$ **33.** $-66x + 72$ **37.** $a^2 - 6$
39. $P = 2(l + w)$ **41.** $C = \frac{5}{9}F - \frac{160}{9}$ **43.** $t - 5$
45. $x^2 - 5x + 4$; –2
47. Multiplicative Identity Property
49. Distributive Property **51.** A

Application Problems

1. \$0.01/day; \$0.43/month; \$5.26/yr
3. \$0.10/day; \$3.02/month; \$36.79/yr
5. \$42.05/yr; \$84,100 **7a.** \$4.20/yr; \$26.28/yr
7b. ≈0.09 yr or 1.1 months
9. $\text{ROI} = \frac{\text{savings}}{\text{initial investment}} \times 100\%$
11. 16.4% **13a.** \$1700 **13b.** 3.4 yr

Chapter Review

1. –2 **3.** –1 **5.** $\frac{1}{6}$
7. $-\frac{11}{30}$ **9.** 0 **11.** $-\frac{2}{15}$
13. 14 **15.** $1\frac{13}{14}$ tn **17.** 8^{15}
19. $\frac{1}{3^{24}}$ **21.** $\frac{9}{16}$ **23.** $-\frac{5}{4}$
25. –8 **27.** $\frac{91}{120}$ **29.** –29
31. 15(–3) – 12; –57
33. 64 should be positive in step 2.
35. 1 **37.** $\sqrt{n} - 1$
39. $(180 - a) - (90 - a)$ **41.** 144 ft
43. $-3a + 6b + 15c - \frac{7}{2}d$ **45.** $2x$
47. $-3d + 10f$ **49.** $-8a - 3b$

CHAPTER 2: SOLVING EQUATIONS

Section 2.1

1. subtract 7; Subtraction Property of Equality
3. divide by 4; Division Property of Equality
5. $x = 9$ **7.** $y = -10$ **9.** $x = 41$
11. $c = -8$ **13.** $y = 30$
15. $n - 72 = 12$; $n = 84$ **17.** $6a = 84$; $a = 14$ yr old
19. $s + 15 = 21$; $s = 6$ in. **21.** $a = 0.49$
23. $y = 30$ **25.** $x = \frac{118}{35}$ **27.** $w = 30$
29. $y = -4.81$ **31.** $d = -\frac{77}{30}$ **33.** $4n = 20$; $n = 5$
35. $47 + 3n = 68$; $n = 7$ **37.** $6r = 37.80$; $r = \$6.30$
39. $310 + 5f = 395$; 17 fives **41.** $532 = 88t - 128$; 7.5 sec
43. 24 linear feet **45.** 216 ft by 108 ft by 266 ft
47. –144 **49.** $-\frac{17}{8}$ **51.** $6xy - 12y^2$
53. $6x^2 + 4x^2y$ **55.** C

Section 2.2

1. $a = 1$ **3.** $w = 9$ **5.** $n = -7$
7. $x = 2$ **9.** $x = 14.25$ **11.** $n = \frac{4}{3}$
13. $y = 30$ **15.** $b = -4$ **17.** $\frac{9}{11}$
19. $\frac{1354}{555}$ **21.** $k = 5$ **23.** $\mathbb{R}$; identity
25. no solution **27.** $t = -6$ **29.** $m = 7$
31. $b = \frac{16}{17}$
33. $n + (n + 1) + (n + 2) = 84$; 27, 28, and 29
35. $2n - 3 = 3n + 31$; –34
37. $4n = 3(n + 2) + 4$; 10 and 12
39. $2(3w) + 2w = 3w + 85$; 51 units by 17 units
41. r = ribbon length; $r + 2 = 2r - 6$; $r = 8$; desired length: 10 in.
43. $x - (85 - x) = 29$; 28 and 57
45. $-\frac{80}{9} = -8\frac{8}{9}$ **47.** 135 **49.** $\frac{9}{2} = 4.5$
51. \$1.20 **53.** C

Section 2.3

1. $x = \frac{a}{b}$ **3.** $x = r^2$ **5.** $b = y - mx$
7. $t = \frac{I}{Pr}$ **9.** $t = \frac{d}{r}$ **11.** $s = \frac{p}{3}$
13. $P = \frac{14r}{3x}$ **15.** $x = \frac{y - b}{m}$
17. $x_1 = \frac{mx_2 + y_1 - y_2}{m}$ **19.** $h = \frac{S - 2\pi r^2}{2\pi r}$
21. $x = -\frac{b}{2}$ **23.** $n = \frac{bx}{x - b}$
25. $v = \frac{d_f - d_i}{t}$; $v = 88$ ft/sec **27.** $h = \frac{S - cr}{c}$; $h = 9$ in.
29. A, B **31.** $p = \frac{mr + q}{2s - r}$
33. $a = \frac{c}{c - 2d + x}$ **35.** $m = \frac{E}{gh + \frac{1}{2}v^2}$; $m = 5$ kg
37. $x^4y^2 - x^2y^3$ **39.** $I = R - E$
41. $t = -2$ **43.** $x = 0$ **45.** E

Section 2.4

1. 4 to 5 **3.** 1 to 3
5. 17 points/game **7.** \$9.25/hr

9. $\frac{RS}{ZQ} = \frac{SB}{QM} = \frac{BR}{MZ}$

11. $\frac{P_{DEF}}{21} = \frac{15}{6}$; $P_{DEF} = 52.5$ units

13. 132 km
15. 53,600 g
17. \$18.25/yr
19. $80.\overline{6}$ ft/sec
21. $x = 3$
23. $n = 9$
25. $r = \frac{2}{3}$
27. \$190
29. 195 students
31. $DE = 3$ units; $BC = 3.75$ units
33. 6 : 5
35. 4 : 25
37. 64 ft^2
39. \$10
41. 4.9 units
43. $-\frac{9}{4}$
45. $x = -1.5$
47. $q = -6$
49. $h = \frac{2w + 440}{11}$
51. D

Section 2.5

1. 0.75; $\frac{3}{4}$
3. 25%; $\frac{1}{4}$
5. 0.6; 60%
7. 126
9. 82.76
11. 33,075
13. 25%
15. 650%
17. 1274
19. 40%; 60%
21. \$1500
25. 12.2%; Athletes
27. 22.1% increase; 83
29. subtract 39; 31.2% decrease
31. add 145.2; 419.2
33. 0.24; 0.54
35. \$36,000
37. \$3.50/lb; \$3.33/lb; \$3.42/lb
39. 125%
41. 20%
43. $x = \frac{2b + 3c}{ac}$
45. $x = \frac{21}{4} = 5.25$
47. $\frac{4}{6} = \frac{5}{x}$; $x = 7.5$ in.
49. No; the sides are not proportional since $\frac{4}{3} \neq \frac{6}{5}$.
51. B

Section 2.6

1. \$15; \$90
3. \$120; \$138
5. \$23.10; \$53.90
7. 6%; \$1800
9. \$4.00
11. \$13.00
13. \$132,050
15. \$300
17. \$5000
19. \$67.50; \$33.75
21. \$7.25; 64%
23. \$29.90; 90%
25. \$10.07; \$66.00
27. \$851
29. 16%
31. \$2500 at 5% and \$7500 at 6%
33. \$1800 for 9 months and \$2500 for 18 months
35. \$3500
37. 7 yr, 6 months
39. new car: \$20,000; used car: \$8500
41. 28
43. $\frac{19}{24}$
45. $-a^2b + 2ab^2$
47. 125 in.
49. B

Section 2.7

1. 54
3. $\frac{d}{19}$
5.

	r	t	d
Salvador	$r + 2$	30	$30(r + 2)$
Shayne	r	30	$30r$

7. $50t + 72t = 210$
9. $72t = 50t + 110$
11. $\frac{9}{2}r = 3r + 315$
13. $12(t - 1) = 9t$
15. $12(t - 1) = 9t + 3$
17. $\frac{3}{4}$ hr
19. 1 hr
21. 6 hr
23. up: 10 ft/sec; down: 80 ft/sec
25. 144 mi
27. 1 hr
29. Jay: ≈7.3 yd/sec; Zeke: 7 yd/sec in about 31.4 sec
31. 8.5 mi
33. $w = 4$
35. $y = 9$
37. $t = \frac{v_f - v_i}{a}$; $t = 15$ sec
39. \$25,500
41. E

Section 2.8

1. 7.2 oz
3. 80 L
5. $0.5x$
11. C
13. 50 gal; 4.92 gal; 9.84%
15. 7 dimes and 11 quarters
17. 2.4 lb of Pistachio Pleasure and 9.6 lb of Pistachio Surprise
19. 2.5 L
21. 7 dimes and 27 nickels
23. 3 tens, 4 fives, and 8 ones
25. 18 lb of hamburger and 12 lb of sausage
27. 5 gal
29. 66 dimes, 13 nickels, and 52 pennies
31. 11 quarters, 16 dimes, and 7 nickels
35. 2.4 oz
37. 20 pennies, 24 nickels, 15 dimes, and 10 quarters
39. 24 in.
41. 8 mi/hr
43. 12 games
45. The school exceeded its goal by 5 students.
47. A

Application Problems

3a. 346 gal
3b. 250 gal
5a. $\frac{3}{130}$
5b. $\frac{1}{60}$
5c. $\frac{1}{99}$
11. Answers are given in terms of g, the current price per gallon.
11a. $104.17g$
11b. $28.85g$
11c. $20.83g$
11d. $12.63g$
13. net cost: \$5000
payback period: $\approx \frac{5000}{20.83g}$
ROI: 12.5%
15. net cost: \$13,500
payback period: $\approx \frac{13{,}500}{20.83g}$
ROI: 4.6%
17. net cost: \$13,000
payback period: $\approx \frac{13{,}000}{104.17g}$
ROI: 24.0%

Chapter Review

1a. yes
1b. yes
3. $y = 8.32$
5. $a = 24$
7. $b = \frac{13}{8}$
9. $c = -20$

11. $x = \frac{6}{5}$

13. $25m = 5m + 300$; 15 lawns

15. $x = \frac{y-b}{m}$

17. $c = \frac{b}{a - bd}$

19. $v_2 = a(t_2 - t_1) + v_1$; $v_2 = 49$ m/sec

21. $27.\overline{7}$ m/sec

23. $x = 5$

25a. sample answer: $\frac{x}{17} = \frac{20}{25}$; $x = 13.6$ cm

25b. 56 cm; 134.4 cm^2

27. 150

29. 4% increase; 936

31. 120 people; 280 people

33. \$9240; \$9702

35. \$158.97; 77.9%

37. \$5000 at 4% and \$10,000 at 6%

39. 5760 yd ≈ 3.3 mi

41. 715 mi; 7 PM

43. 10 nickels, 5 dimes, and 20 quarters

45. 315 children and 735 adults

47. 4 L of water

CHAPTER 3: SOLVING INEQUALITIES

Section 3.1

1. $a = b$; $a > b$; $a < b$

3a. yes **3b.** no **3c.** no

5a. yes **5b.** yes **5c.** no

7.

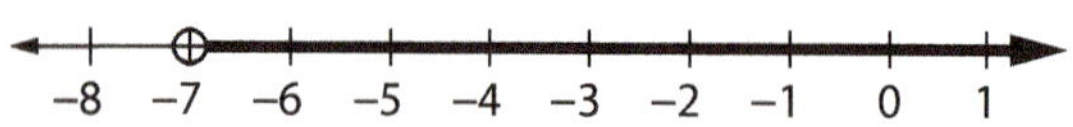

9.

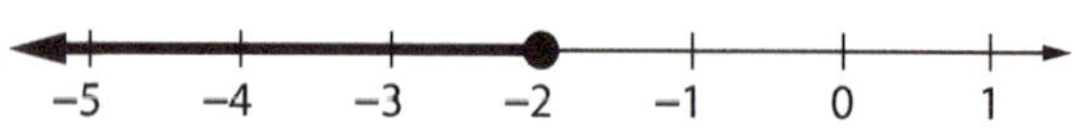

11.

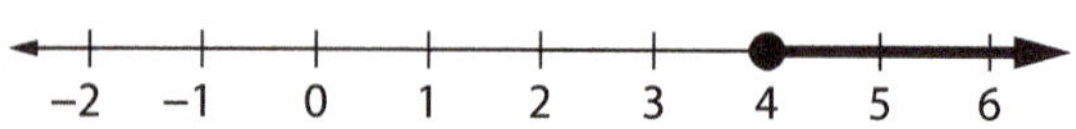

13. $y > 108$

15. $w \le -48$

17. $r < -6$

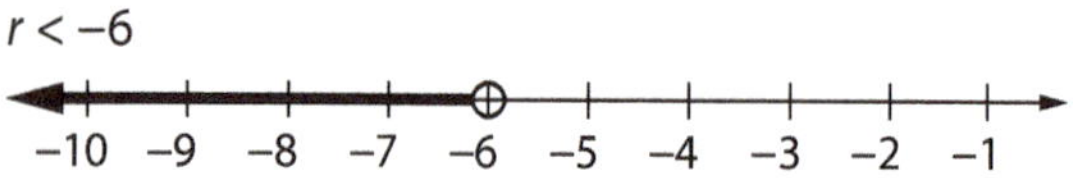

19. $x < -6$

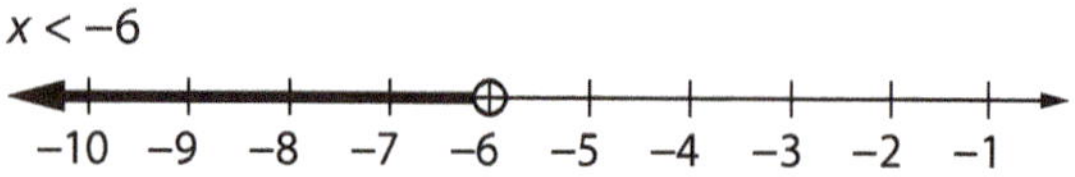

21. $x < 7$

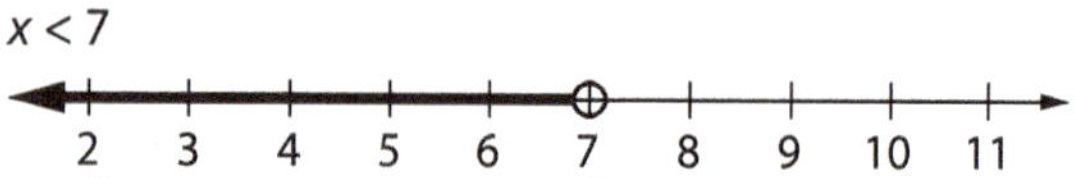

23. $n < \frac{3}{4}$

25. $w > -\frac{5}{3}$

27. any number less than 57

29. at most 5 teenagers

31. a maximum of 45.59 sec

33. above 15°C

35. Division Property of Inequality

37. The answer is $h \ge -\frac{5}{2}$.

39. $x < 0$

41. yes

43. $P = 2(l + w)$

45. Multiplication Property of Equality

47. Subtraction Property of Equality

49. 20 min

51. D

Section 3.2

1. $b < -130$ **3.** $c > 1$ **5.** $x \ge 8$

7. $m > -18$ **9.** $x \ne -55$; D **11.** $x \ge -2$; A

13. $x \ne 4$; E **15.** $z \le 30$ **17.** $s > 0$

19. $y \le 4$ **21.** $z \ge -3$ **23.** $x \ge 2$

25. $x > -12$ **27.** at least a 96

29. Jake: at most \$60; Naomi: at most \$90; Heston: at most \$180

31. whole milk: 20 calories/oz; skim milk: 10 calories/oz

33. at least \$2.5 million

35. $x < 2$

37. $q \ge 11$ **39.** \$833.34 **41.** $x = \frac{4}{3}$

43. $z = 5$ **45.** $x = c - a - b$ **47.** $a = 3$

49. E

Section 3.3

1. and

3.

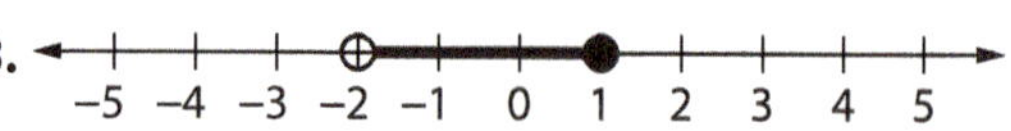

5.

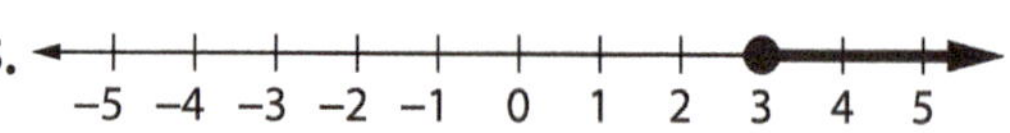

7.

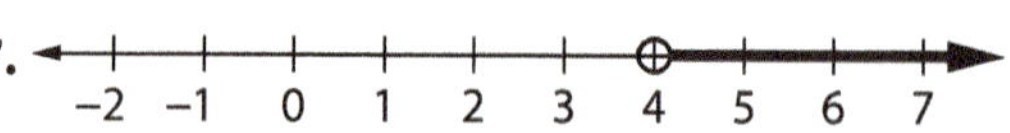

9.

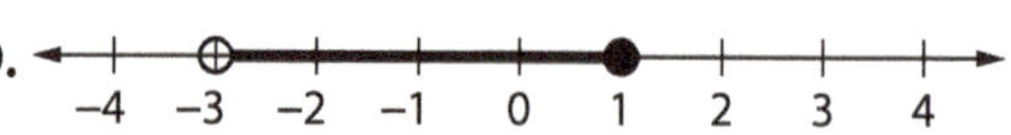

11. There are no points in the solution set.

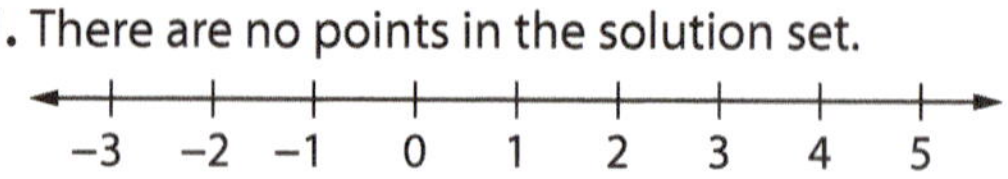

13. $-3 < x \le 1$ **15.** $-1 \le x < 6$ **17.** $-2 < x < 3$

19. $x < -12$ **21.** $\varnothing$ **23.** $-6 < x < 3$

25. $4 < x < 7$ **27.** $3 < x \le 5$

31. 200 m $< n <$ 1000 m

33. $18 \le n \le 28$

35. 136 mi/hr $\le w \le$ 155 mi/hr

37. $x < -4$

39. $\varnothing$

41. 20 ft $\le w \le$ 25 ft; 40 ft $\le l \le$ 50 ft

43. $\frac{65}{64}$ **45.** $x = 31.5$ **47.** $x = \frac{1}{12}$

49. $\frac{17}{6}$ **51.** B

Section 3.4

1. or

3.

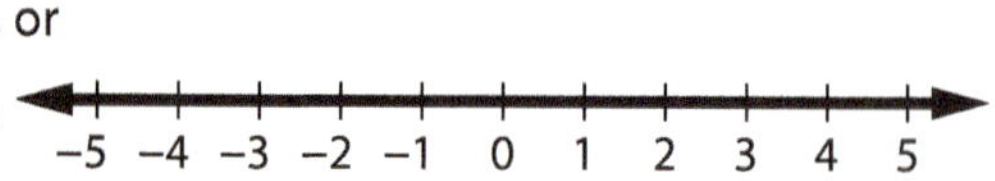

5.

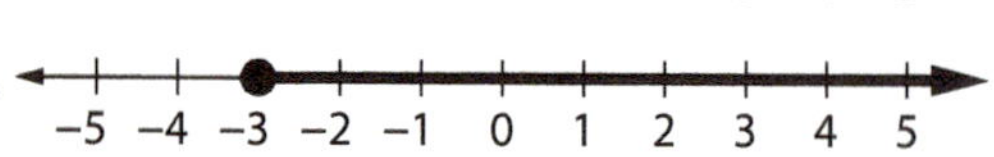

7. 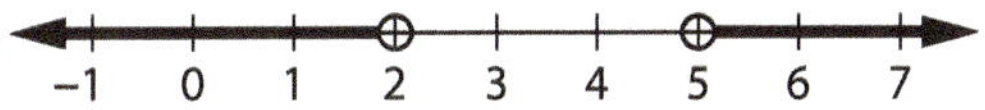

9. 

11. 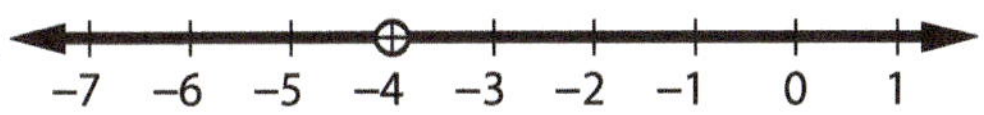

13. $\mathbb{R}$ **15.** $y \leq 0$ or $y > 5$ **17.** $b > 3$

19. $x < 6$ **21.** $x < \frac{5}{2}$ **23.** $\varnothing$

25. $\mathbb{R}$ **27.** $z \leq -\frac{5}{2}$ or $z > 3$

29. $t < 125°$ or $t > 160°$ **31.** $3 \cdot 10^{16}$ Hz $< f < 3 \cdot 10^{19}$ Hz

33. $w < 1 \cdot 10^{-8}$ m or $w > 3.80 \cdot 10^{-7}$ m

35. $x < -3$ or $x > 1$ **37.** $-5 < x \leq 0$ **41.** $x < \frac{21}{10}$

43. at least 111; at least 125 **45.** −16

47. −10 **49.** 81 **51.** 784

53. D

Section 3.5

1. $x = \pm 3$ **3.** $x = \pm 5$ **5.** $x = \pm\frac{8}{3}$

7. $\varnothing$ **9.** $y = \pm 43$ **11.** $\varnothing$

13. $y = -12.5, 4.5$ **15.** $x = -7, 5$ **17.** $y = -14, 19$

19. $y = -\frac{36}{5}, \frac{12}{5}$ **21.** $x = -\frac{7}{6}, 2$ **23.** $z = -\frac{8}{7}, 4$

25. $x = -\frac{41}{7}, 5$ **27.** $|x - 2| = 4$; −2 and 6

29. $|x + 4| = 8$; −12 and 4 **31.** $|x - 5| = 15$; −10 and 20

35. $|x + 7| = 3$ **37.** $|x - 6.5| = 13.5$

39. $\begin{cases} 2x - 3 \text{ if } x \geq \frac{3}{2} \\ -2x + 3 \text{ if } x < \frac{3}{2} \end{cases}$

41. \$258.75 **43.** \$60 **45.** \$320; \$4320

47. 33 lb of \$10 coffee and 17 lb of \$4 coffee

49. C

Section 3.6

1. conjunction; and

3. $x < -3$ or $x > 3$

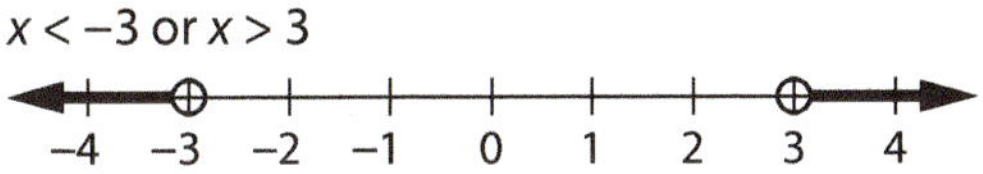

5. $-5 \leq x \leq 5$

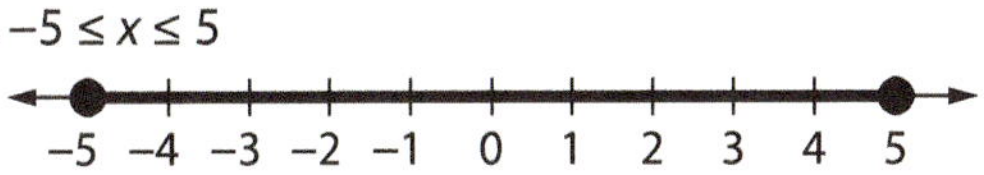

7. $x < -8$ or $x > 8$

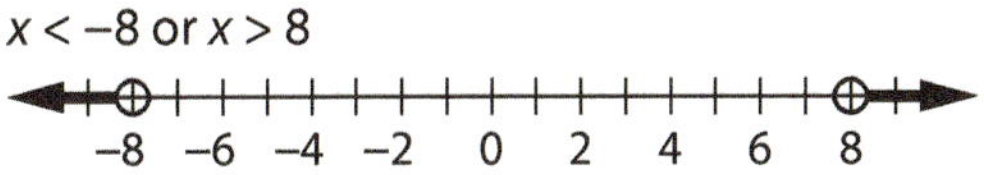

9. $-7 \leq x \leq 13$

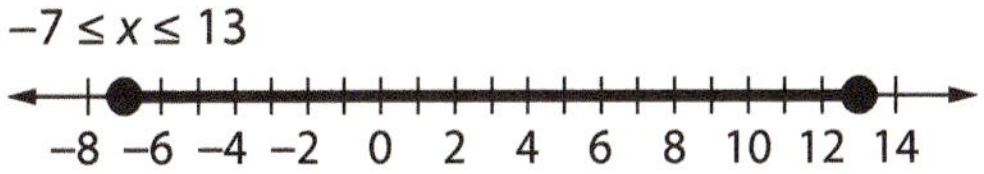

11. $|x| < 15$ **13.** $|y| \geq 19$ **17.** $\frac{1}{3} < x < \frac{19}{3}$

19. $\frac{9}{2} \leq x \leq \frac{33}{2}$ **21.** $-1 < x < 9$ **23.** $x < -3$ or $x > -1$

25. $-\frac{9}{2} < x < \frac{15}{2}$ **27.** $-\frac{4}{7} < x < \frac{16}{7}$ **29.** $|x - 5| \geq 2$

31. $|d - 5| \leq 0.05$; 4.95 mm $\leq d \leq$ 5.05 mm

33. $|t - 90| > 50$

35. $|v - 48| \leq 4$; 44% $\leq v \leq$ 52%

37. $\varnothing$ **39.** $z \neq \frac{1}{3}$ **41.** $-\frac{7}{2} \leq b \leq \frac{1}{2}$

43. $t > 101$ **45.** $r = \frac{2c - b}{4}$ **47.** 13.3%

49. \$7500 in the stock fund and \$2500 in bonds

51. A

Application Problems

1. interest: \$600; 1 payment of \$5600 at the end of 2 yr

3. amount loaned: \$5681.82; interest: \$681.82; 24 monthly payments of \$236.74

5. 19.6% **7.** 3.9% **9.** \$243.76

13. 4.47%; 4.28%; 4.4% quarterly

15. 3.82%; 3.84%; 3.8% semiannually

Chapter Review

1a. ≥ **1b.** ≤ **1c.** <

3a. 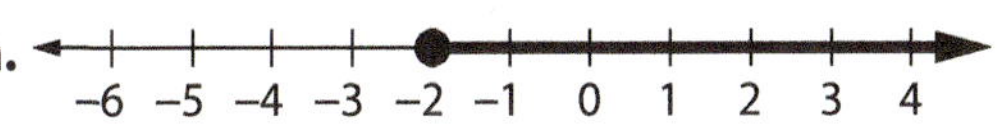

3b. 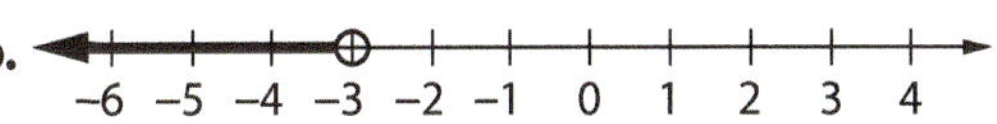

5. $y < -48$ **7.** $b \leq 27$ **9.** $x < \frac{17}{8}$

11. $x \geq 11$ **13.** $6x + 8 < -42$; −9

15. $36w + 12(2w) >$ \$480; Kimberly: more than \$8/hr; Samuel: more than \$16/hr

17. conjunction

19. $x > -1$

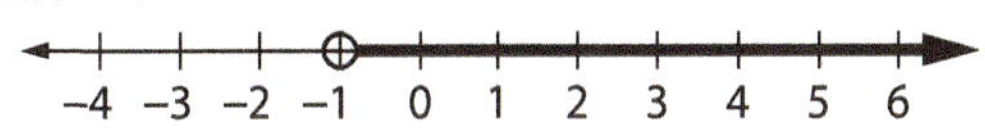

21. $3 < x < 7$

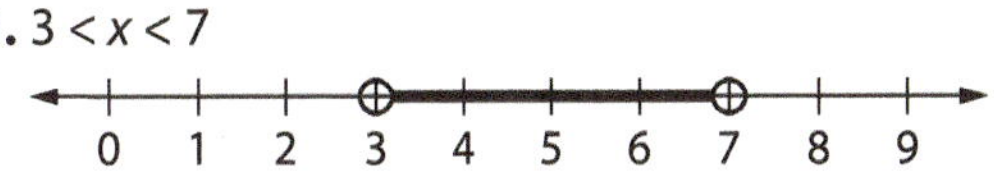

23. $x \neq 5$

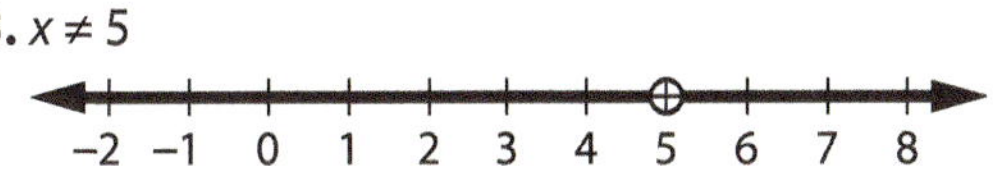

25. $x \leq -5$ or $x > 4$ **27.** $\varnothing$ **29.** $4 < x < 7$

31. $x \leq 0$ **33.** $x = \pm 25$ **35.** $x = -13, 7$

37a. $|x + 2| = 5$ **37b.** $|x + 3| = 4$ **37c.** $|x - 4| = 3$

39. $x < -10$ or $x > 4$

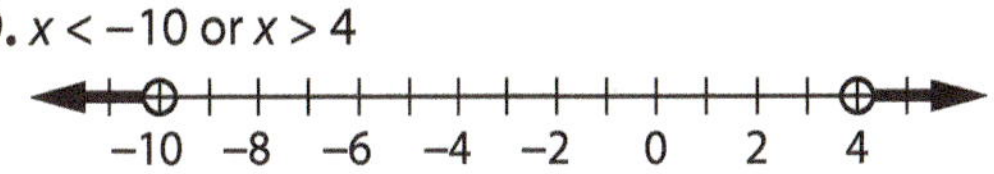

41. $-7 < x < 3$

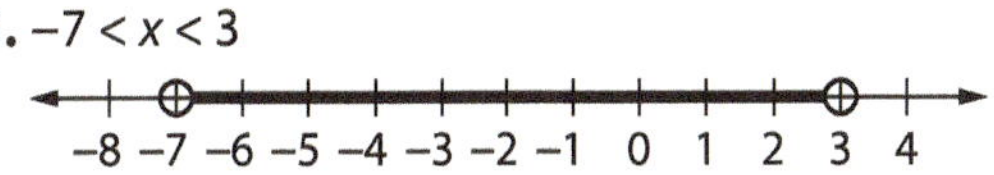

43. $|p - 73| > 14$; $p <$ \$59 or $p >$ \$87

CHAPTER 4: FUNCTIONS

Section 4.1

1. C **3.** no **5.** yes

7. $D = \{3\}$; $R = \{2, 3, 4, 5, 6\}$

9. $D = \left\{\frac{1}{2}, \frac{2}{3}, \frac{4}{5}, \frac{5}{6}\right\}$; $R = \left\{\frac{5}{9}, \frac{3}{5}, \frac{7}{8}\right\}$

11. yes

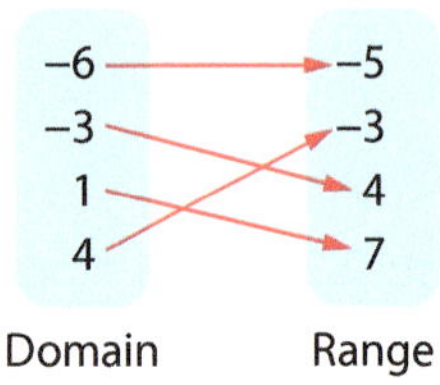

13. no

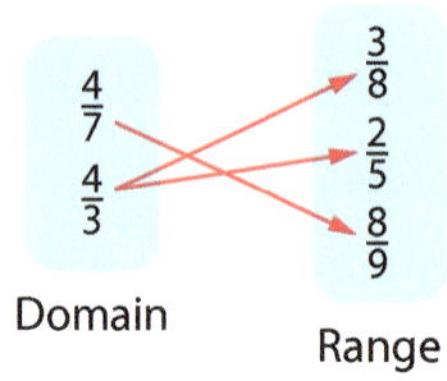

15. A **17.** B **19.** C

21. no

$\{(-2, 0), (-1, 4), (0, 3), (1, -1), (1, 4)\}$

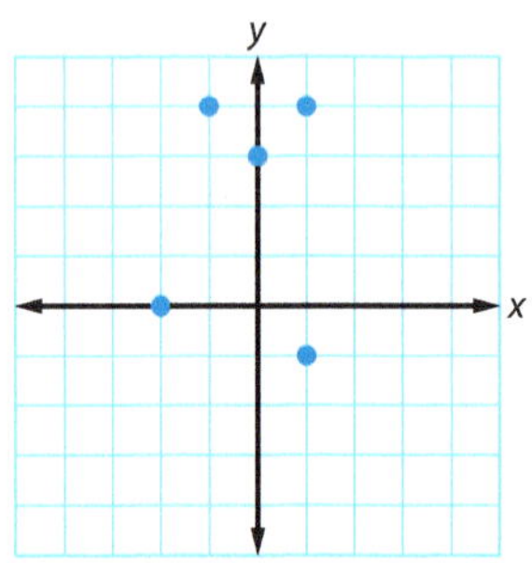

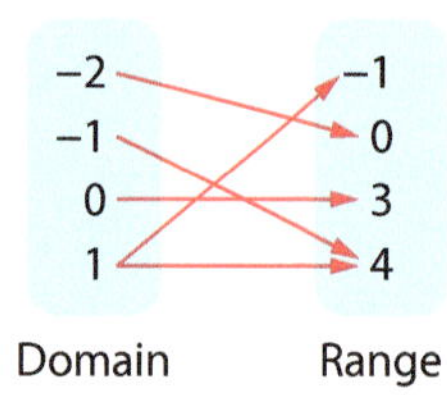

23. yes

$\{(-2, 1), (-1, 1), (1, 1)\}$

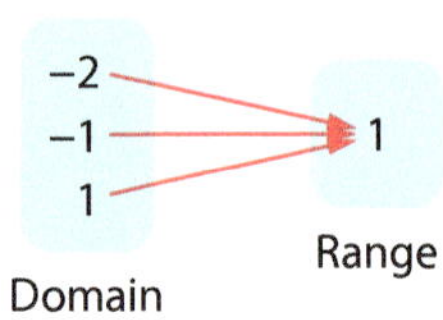

x	y
−2	1
−1	1
1	1

25. −4 **27.** −10 **33.** time; distance

35. age; height

37.

Resting Heart Rate vs. Daily Exercise

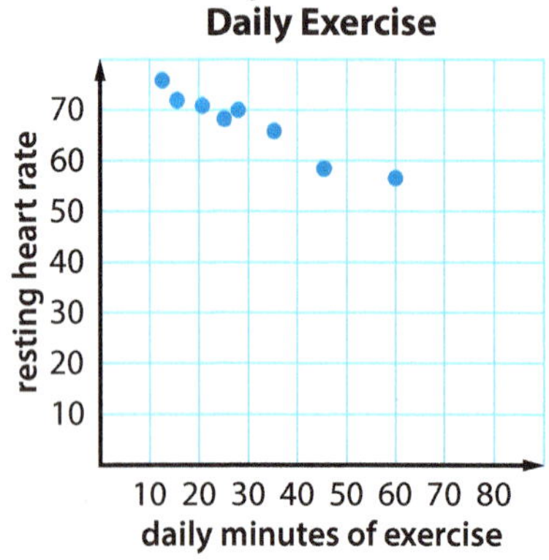

The resting heart rate decreases as exercise increases.

39. *o* **41.** *m*

45. Division Property of Equality

47. $x = 3$ **49.** $x = -5$ **51.** $2 \le x \le 8$

53. B

Section 4.2

1. yes **3.** no **5.** yes

7. $D = \{-4, -3, -1, 0\}$
$R = \{-3, -2, -1, 0\}$; no

9. $D = \{-4, -1, 0, 1, 2, 3, 4\}$
$R = \{1, 2\}$; yes

11. $D = \{x \mid -2 \le x \le 2\}$
$R = \{y \mid -4 \le y \le 4\}$; yes

13. $\{(-1, -3), (0, 0), (1, 3)\}$

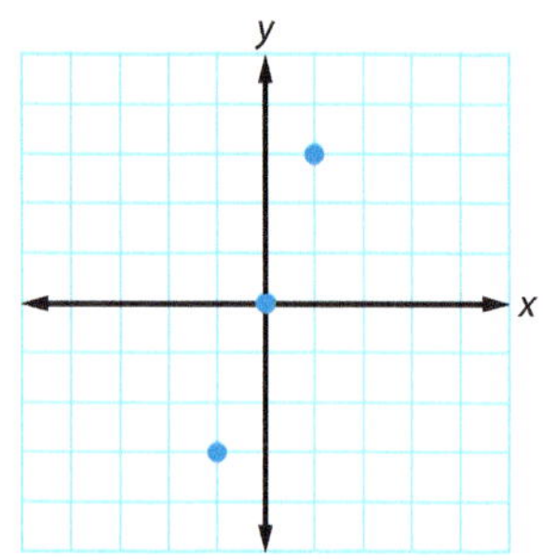

15. $\{(-2, -1), (0, 3), \left(\frac{1}{2}, 4\right), (1, 5)\}$

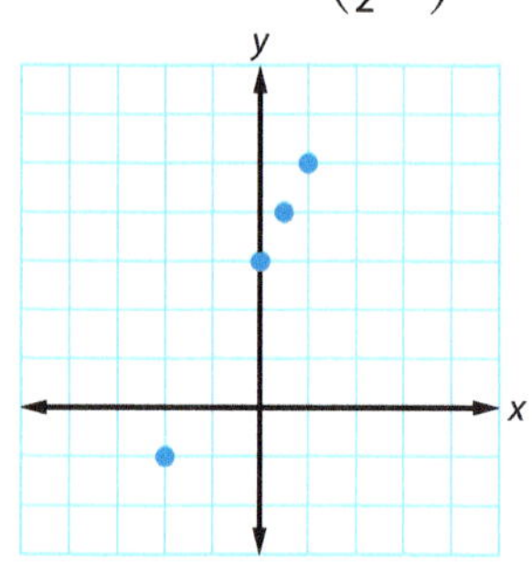

17.

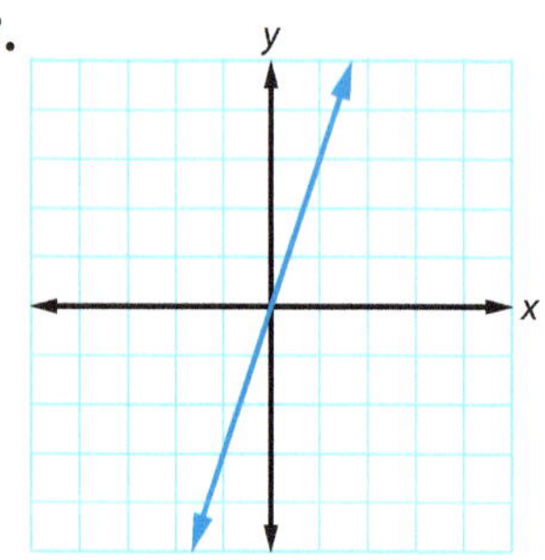

$D = \mathbb{R}$; $R = \mathbb{R}$

19.

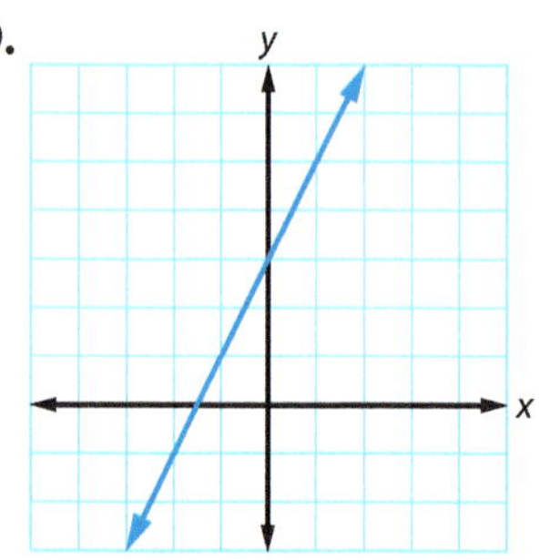

$D = \mathbb{R}$; $R = \mathbb{R}$

21. not a function

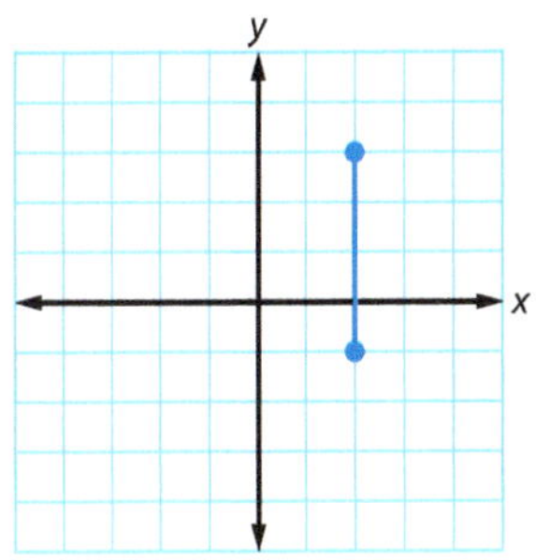

23. function

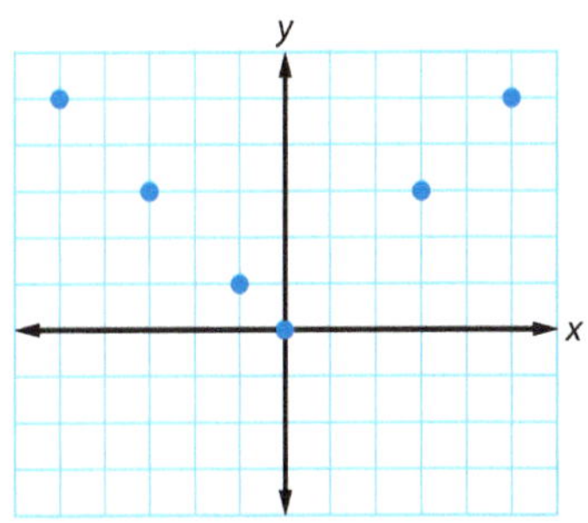

27.

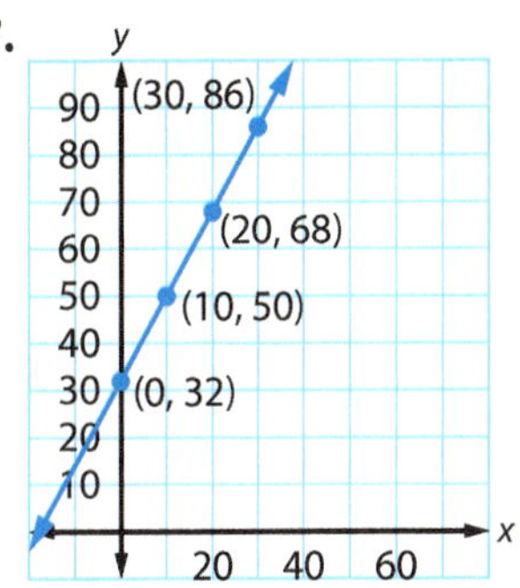

29. 260°C

31.

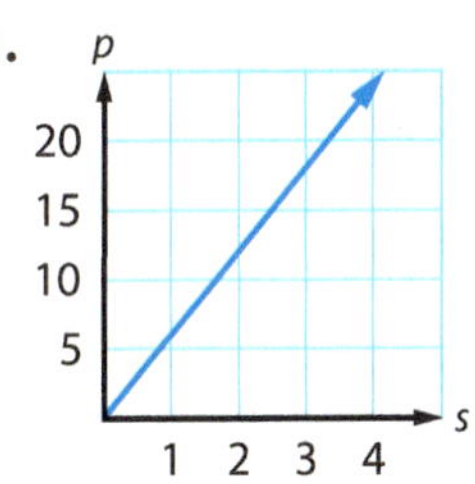

33. $D = \{x \mid x > 0\}$ **35.** \$3250 **37.** yes

39. no

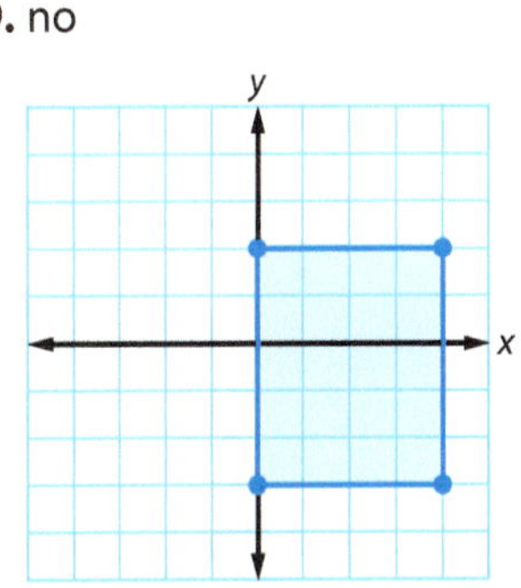

41.

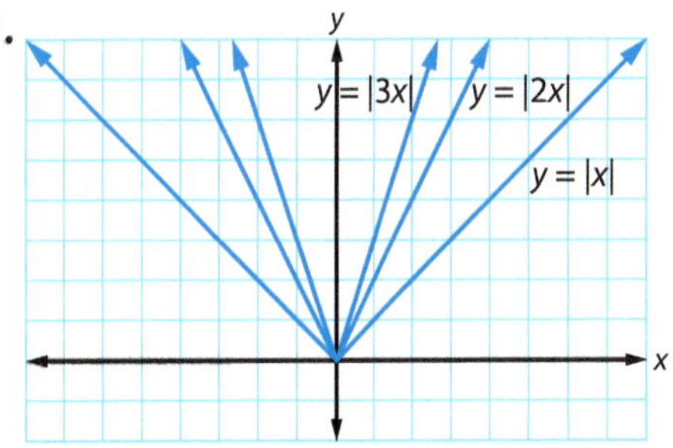

43. $x = \frac{114}{7}$ **45.** $x = 1.5$ **47.** $x = \frac{d - c}{3ab}$

49.

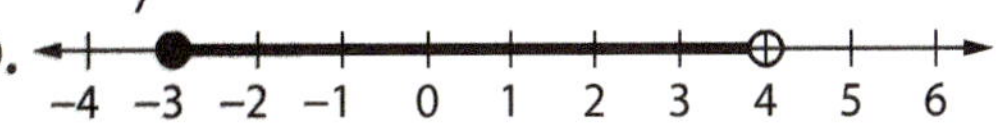

51. A

Section 4.3

1a. II **1b.** III **1c.** I

7.

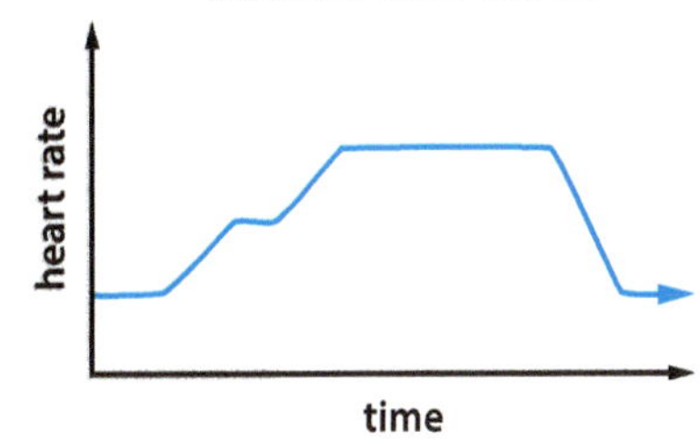

9.

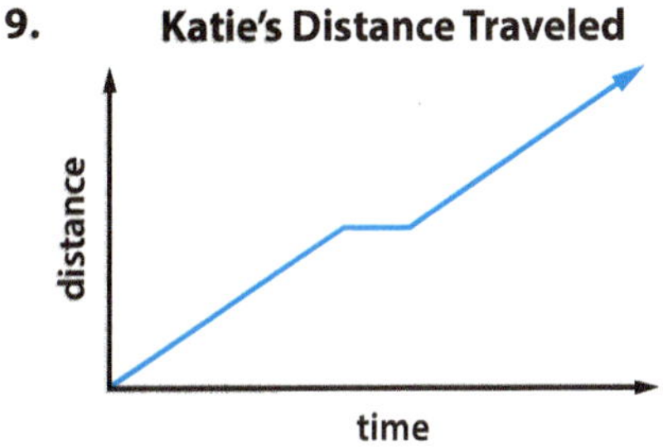

11.

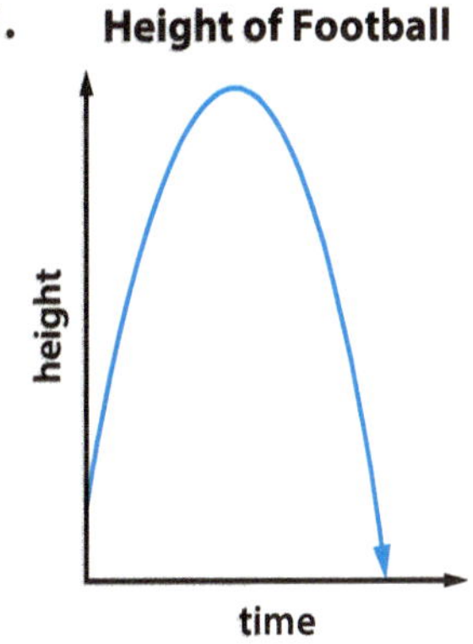

13.

15. \$30

17. **Calories Burned While Playing Basketball**

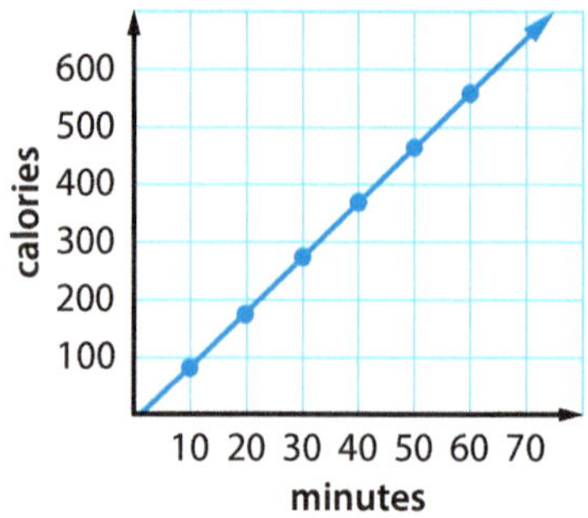

19. ≈16 min

21. Elevation; the boiling point is dependent on the elevation.

23. As the elevation increases, the boiling point of water decreases quickly at first and then not so quickly.

25. **Circle's Area vs. Diameter**

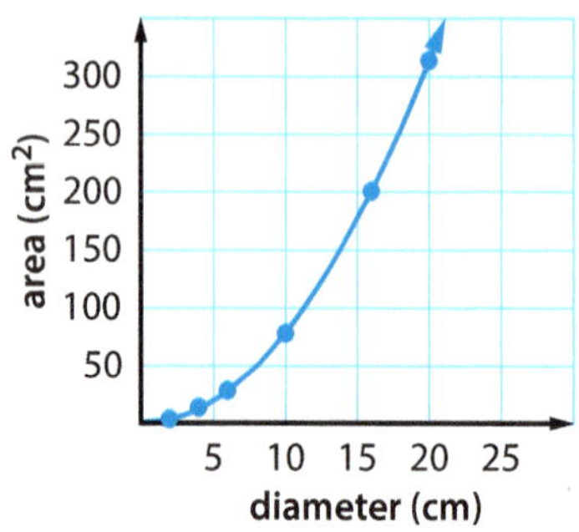

27. ≈20 cm^2

29. **Concentration of Medication over Time**

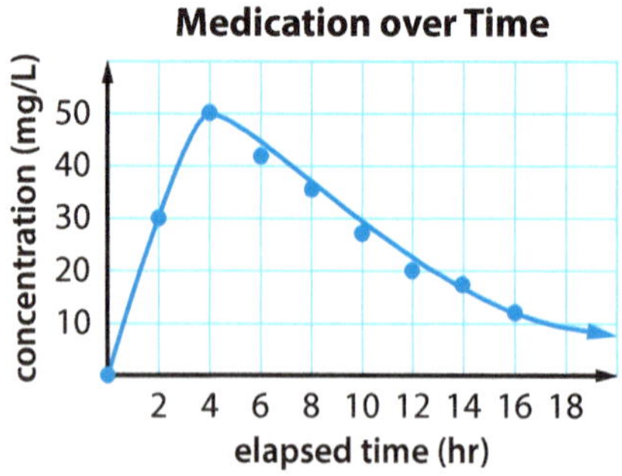

31. ≈31 mg/L

41. ≈23, 57, and 103 sec

43. $20 - 3x$

45. $x = -11, 6$

47. $x < -20$ or $x > 18.5$

49.

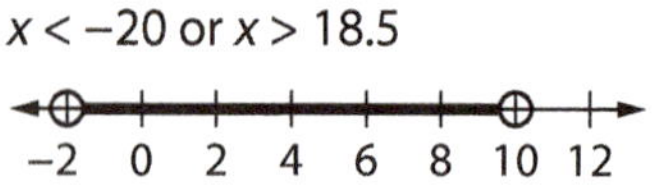

51. A

Section 4.4

1a. −8

1b. −2

1c. 7

3. $f = \{(-2, -6), (0, -4), (3, -1)\}$

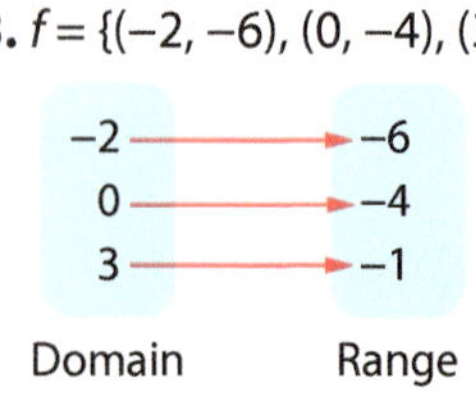

5.

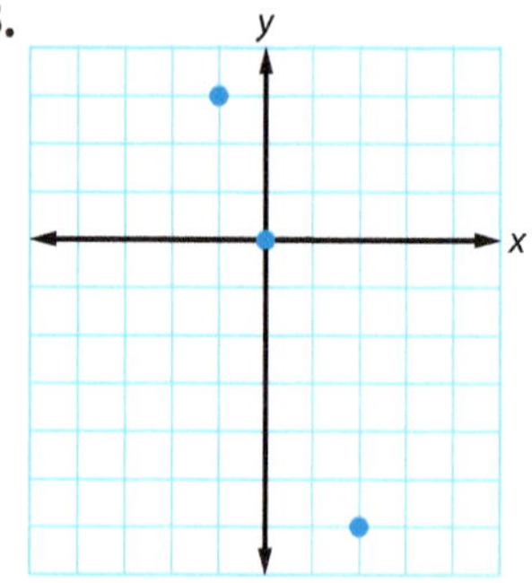

7.

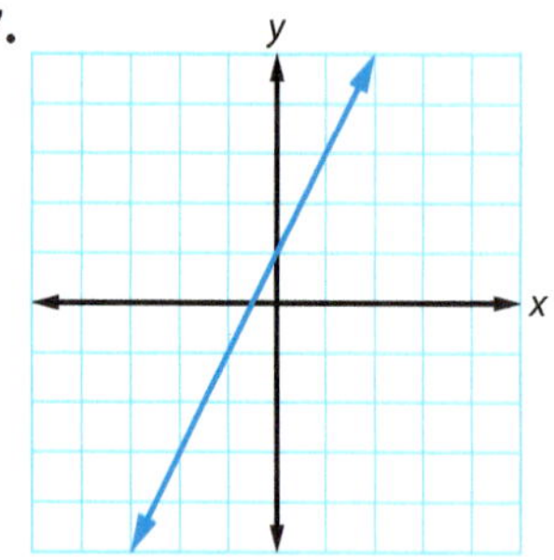

9. $f(x) = -4x$

11. $f(x) = x - 2$

13. $f(x) = 3x$

15. $f(x) = -x + 2$

17. $f(x) = x + 2$

19. $f(x) = \frac{1}{3}x$

21. $f(x) = 2x + 1$

23. $r(x) = 5x + 2$

25. $f(x) = 2x - 1$

27. $f(d) = 0.25d$

29. $N(a) = a + 2$

31. $P(c) = 0.25c$

33. $t(d) = \frac{d}{50}$

35. $A(r) = \frac{1}{2}\pi r^2$; ≈14.14 $in.^2$

37. $d(t) = 52t + 40$; 222 mi

41. $J(x) = 150 + 15x$; $C(x) = 200 + 12x$

Jamal's Earnings

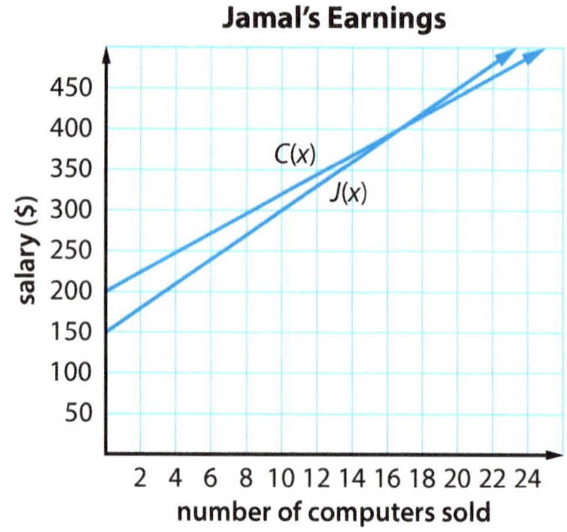

Java Jack's Electronics pays better if he sells at least 17 computers per week.

43. $\frac{9}{8}$ or $1\frac{1}{8}$

45. 42,000

47. $2x + 5$

49. $-\frac{16}{3}$

51. B

Section 4.5

1a. III

1b. I

1c. IV

1d. II

3. yes; $d = -0.5$; −1, −1.5, −2

5. no

7. yes; $d = 1$; 6, 7, 8

9. 7, 9, 11, 13

11. 27, 37, 47, 57

13.

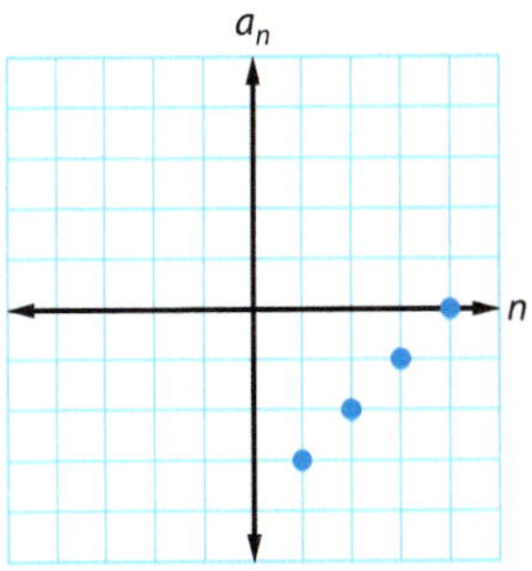

15.

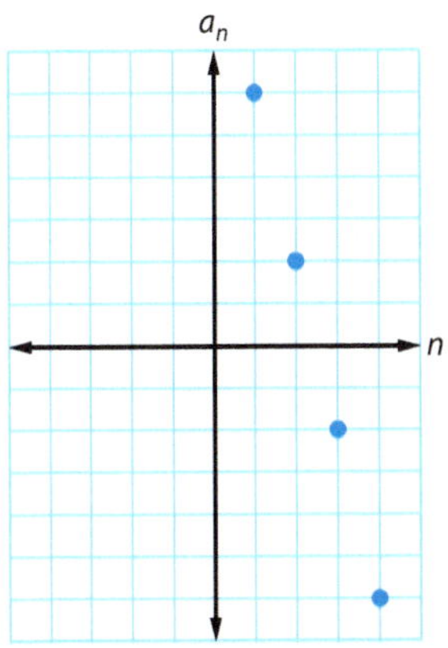

17. The proper order should be 25 – 39, 11 – 25, and –3 – 11, which gives a common difference of –14.

19.

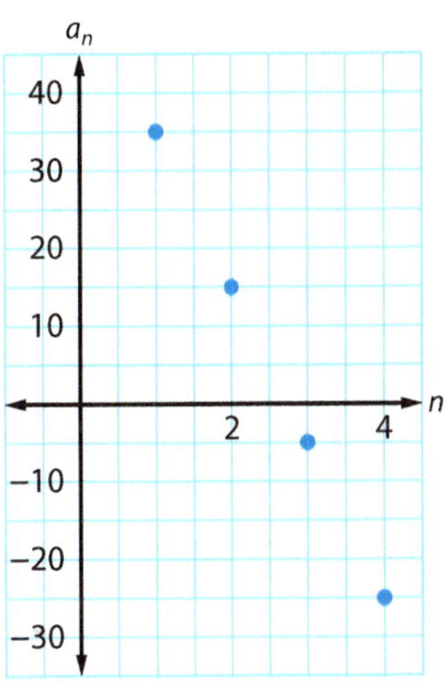

21.

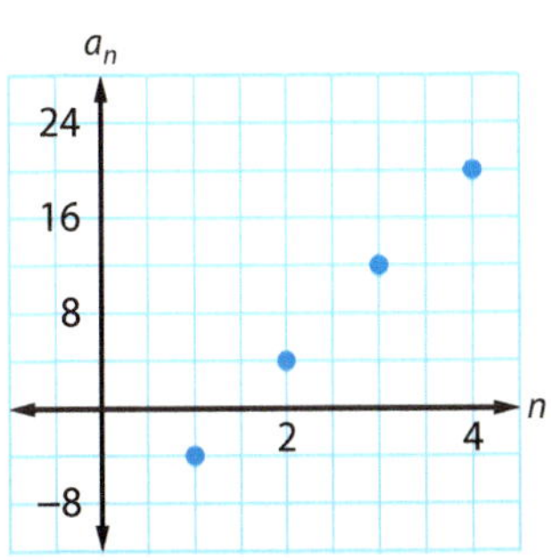

23. $a_n = a_{n-1} + 11; a_1 = -5$ **25.** $a_n = 2n + 1$

27. $a_n = 15n - 3; a_{52} = 777$

29. $a_n = -0.12n + 9.72; f(81) = 0$

31a. \$40 per guest; \$400 upfront

31b. $a_n = 40n + 400$ **31c.** \$4240 **35.** 23 houses

37. 1.75 **39.** $a_{23} = 106$ **41.** no

43. 7 **45.** 64 **47.** $x > 211$

49. $x \neq 13$ **51.** A

Section 4.6

1a. III **1b.** II **1c.** I

3. inverse **5.** direct **7.** neither

9. direct **11.** $k = 72; y = \frac{72}{x}$ **13.** $k = \frac{2}{3}; c = \frac{2}{3}v$

15. $k = 0.1; g = 0.1h$

17. direct; $y = 4x$

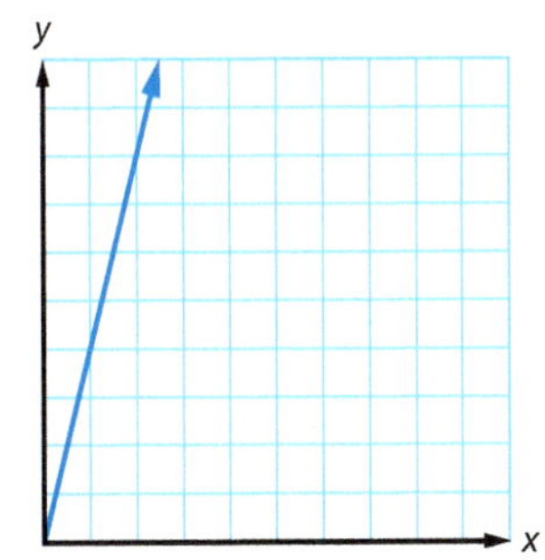

19. inverse; $y = \frac{10}{x}$

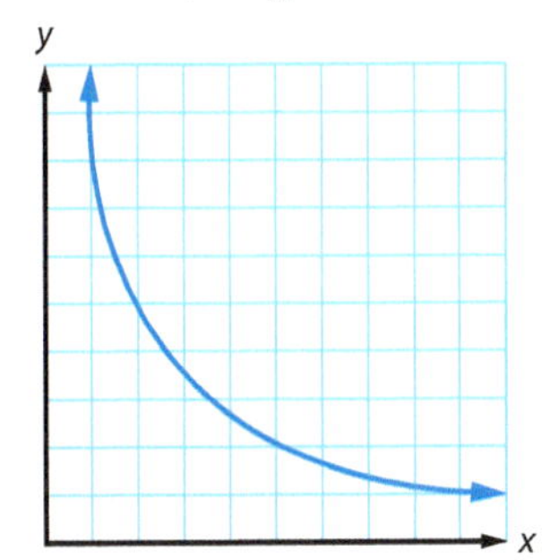

21. neither

23. inverse; $y = \frac{1.5}{x}$

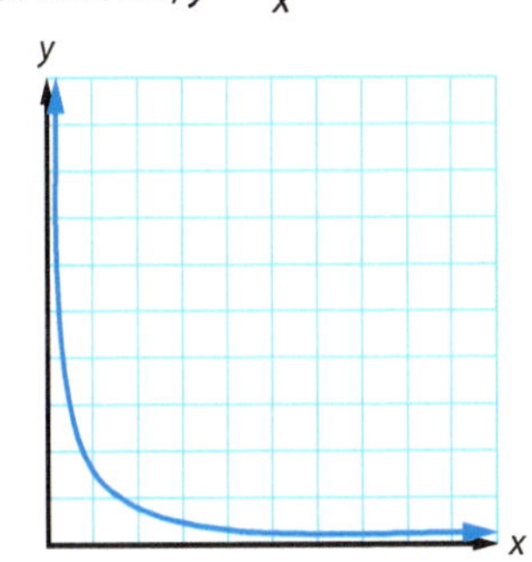

25. 1 **27.** 10.5

29. direct; $k = 0.433$ **31.** direct; $k = 0.012$

33. $d = 52t$; 130 mi **35.** $t = 0.07p; t = \$34.65$

37. $t = \frac{18}{p}$; 4.5 hr

39. neither

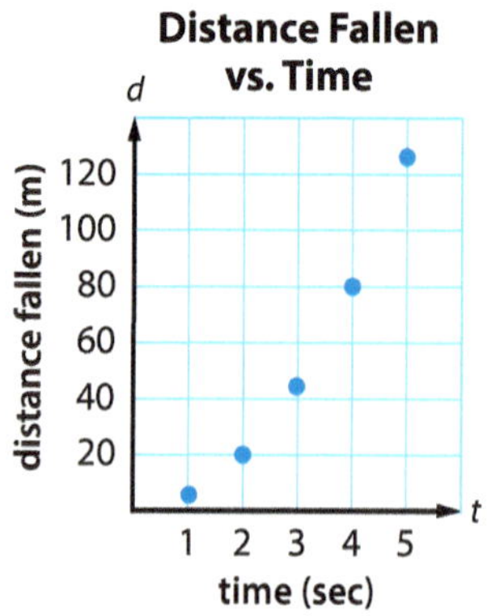

41. $d = 5t^2$ **41a.** 281.25 m **41b.** 10 sec

43. 113 in.^{3} **45.** $x = 82.69$ **47.** $a = -14$

49. 33%

51. just under 6.6 hr (≈6 hr 34 min)

53. B

Section 4.7

1. D **3.** A **5.** E

Table values will vary for exercises 7–13.

7.

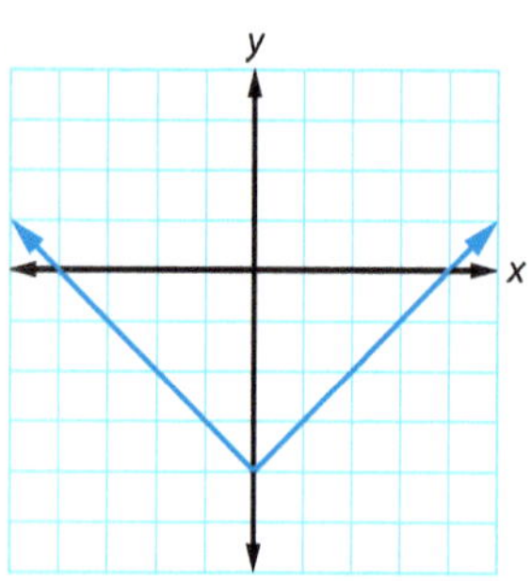

9.

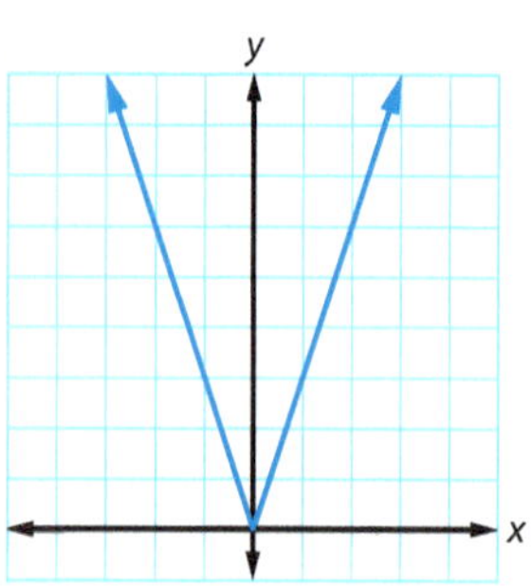

11.

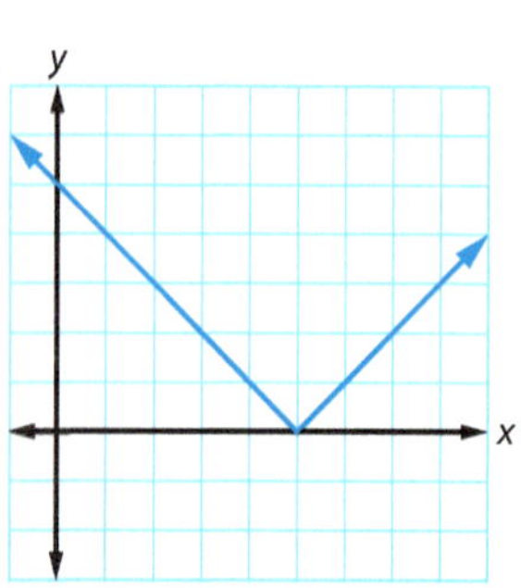

13.

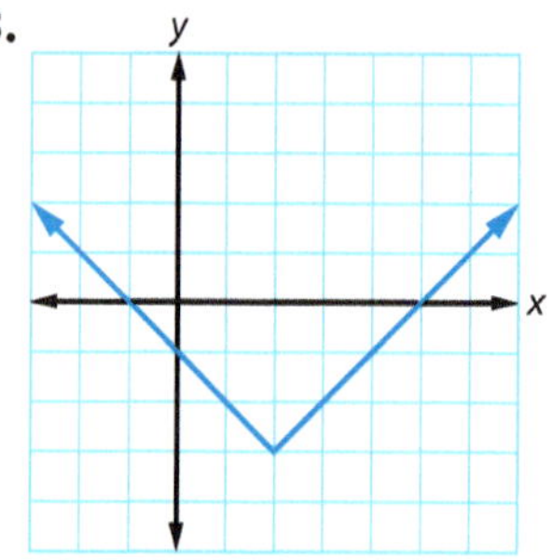

15. $f(x) = |x| - 2.5$ **17.** $f(x) = |x + 4|$

21. 2 units left, 6 units down

23. 1 unit right, 3 units up

25.

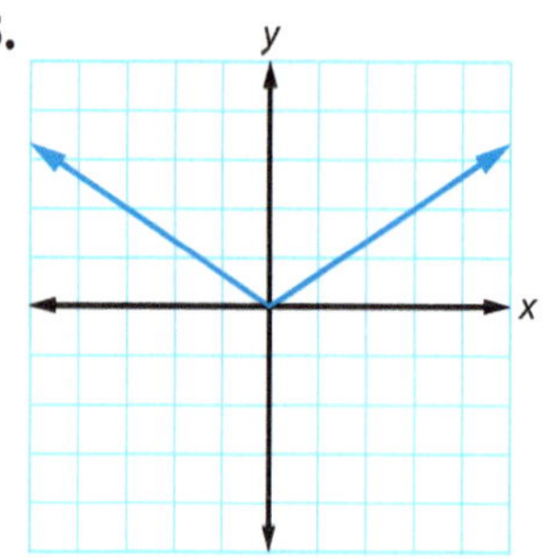

27.

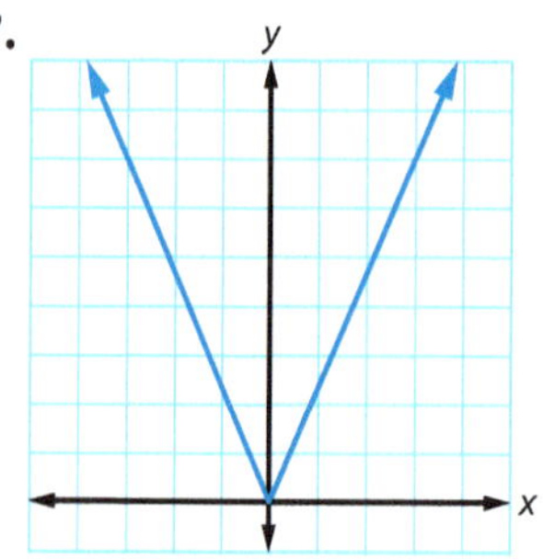

29. $f(x) = |x - 3| - 6$ **31.** $f(x) = \left|x + \frac{8}{9}\right| - \frac{2}{5}$

33.

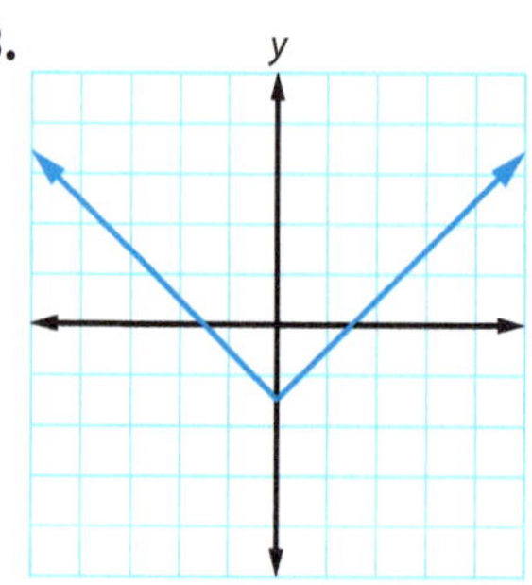

35.

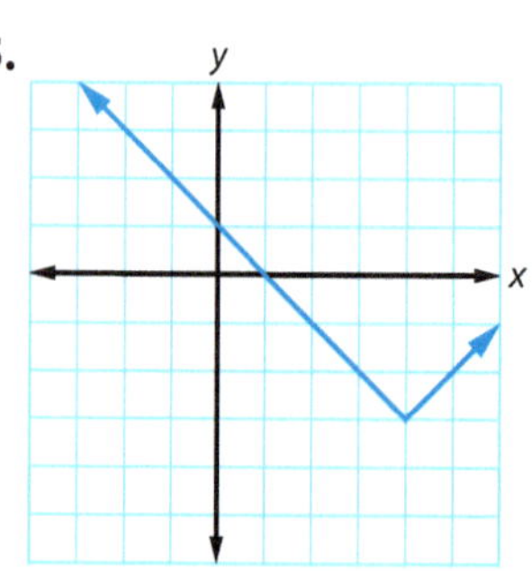

37.

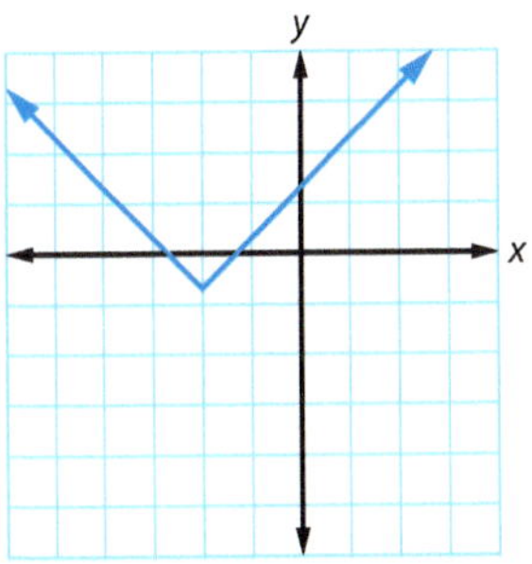

Table values will vary for exercises 39–43.

39.

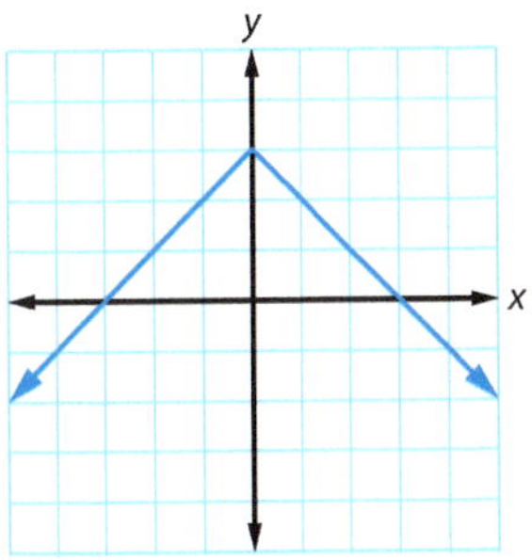

41.

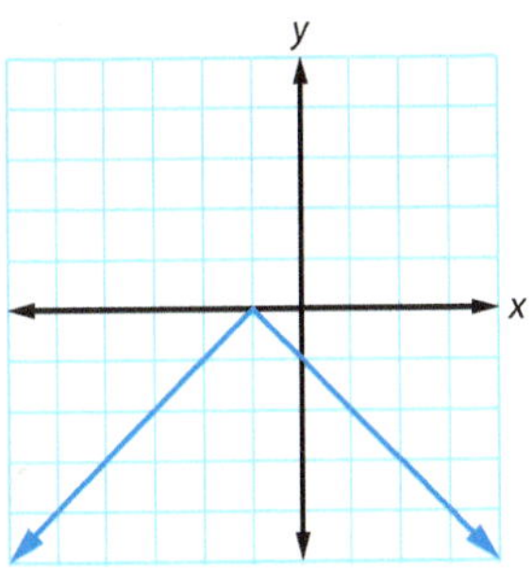

43.

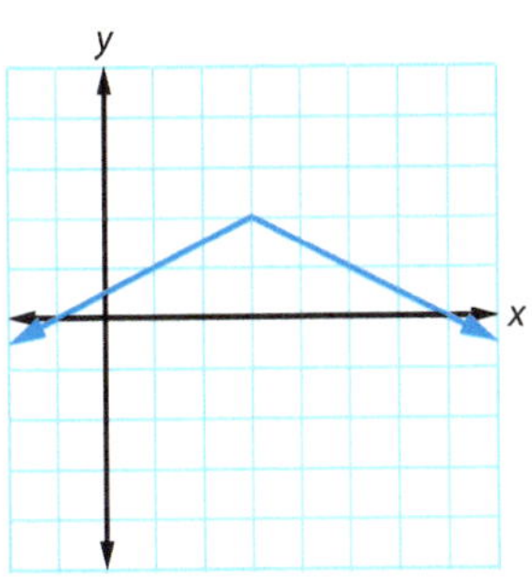

45. $x = y$, $x > y$, or $x < y$ **47.** $x > -17$

49. $x < -\frac{4}{7}$ **51.** $x < 10$ **53.** C

Application Problems

1. $\frac{4}{3}$ units **3a.** 64 **3b.** $\frac{1}{27}$ unit

3c. $\frac{64}{27}$ units

5a. There are four times as many segments.

5b. It is $\frac{1}{3}$ the length. **5c.** It is $\frac{4}{3}$ the length.

7a. segment length: $\left(\frac{1}{3}\right)^n$ or $\frac{1}{3^n}$

total length: $\left(\frac{4}{3}\right)^n$

7b. 10th iteration total length ≈ 17.76
100th iteration total length $\approx 3.12 \times 10^{12}$

9a. would continue to increase forever

9b. would get closer and closer to 0

9c. would continue to increase forever

Chapter Review

1. $D = \{-7, -3, -2, 1, 3, 5, 8\}$
$R = \{3, 4, 5, 6, 7, 8\}$

3. yes

5.

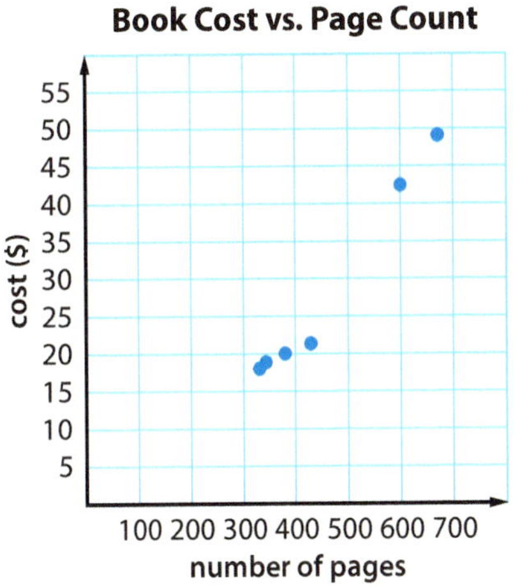

7. $D = \{-4, -1, 2, 5\}$
$R = \{-3, -1, 1, 3\}$; yes

9. $D = \{x \mid 0 \leq x \leq 4\}$
$R = \{y \mid -2 \leq y \leq 2\}$; no

11. yes

13. yes

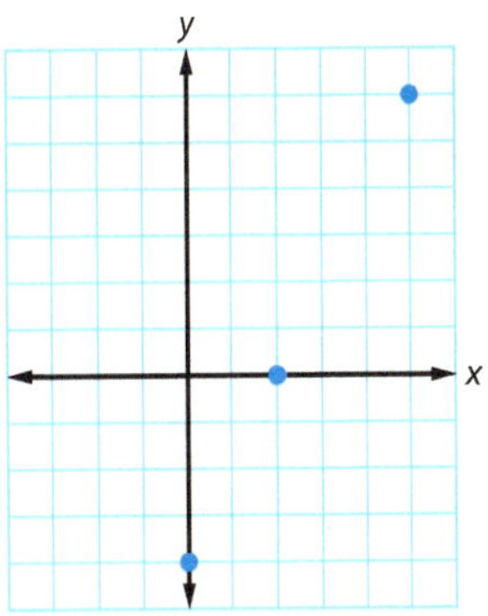

17.

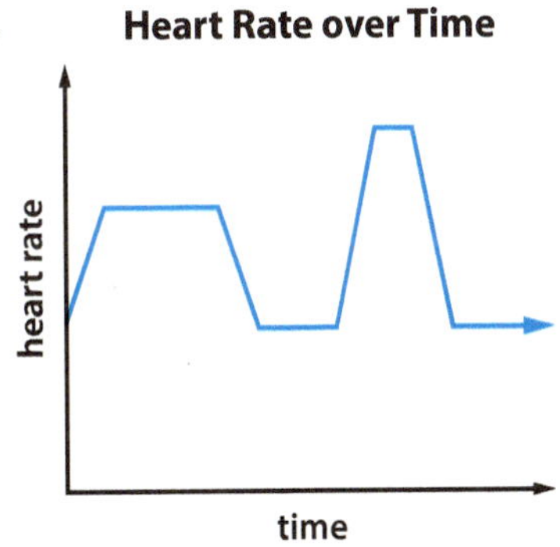

19. 9 gal **21a.** −3 **21b.** 5

21c. 123

21d.

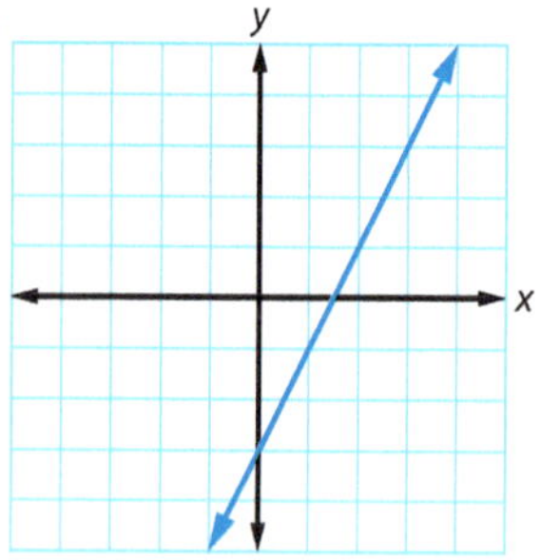

23. $g = \{(-2, 1), (-1, -2), (0, -3), (1, -2), (2, 1)\}$

25. $y = x + 3$

27. $h(x) = 3x - 2$

29.

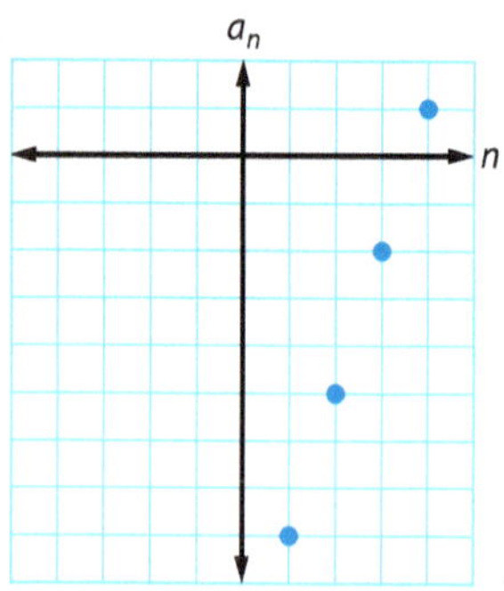

31. $a_1 = -6; a_n = a_{n-1} + 6$

33. $a_n = 6n - 14$

35. $a_{50} = 621; a_{89} = 1128$

37. 9 days

39. inverse; $f(x) = \frac{36}{x}$

41. 2.8 gal

43. $\{(-4, 4), (-2, 2), (0, 0), (2, 2), (4, 4)\}$

45a. $f(x) = |x| - 8$

45b. $f(x) = |x + 5.3|$

45c. $f(x) = |x - 4| + 11.5$

47.

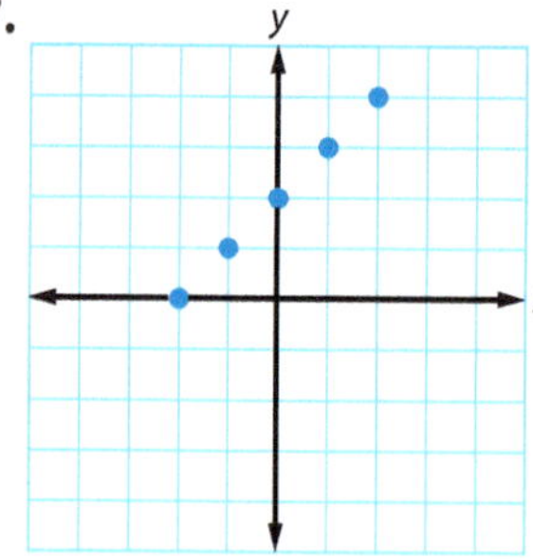

49.

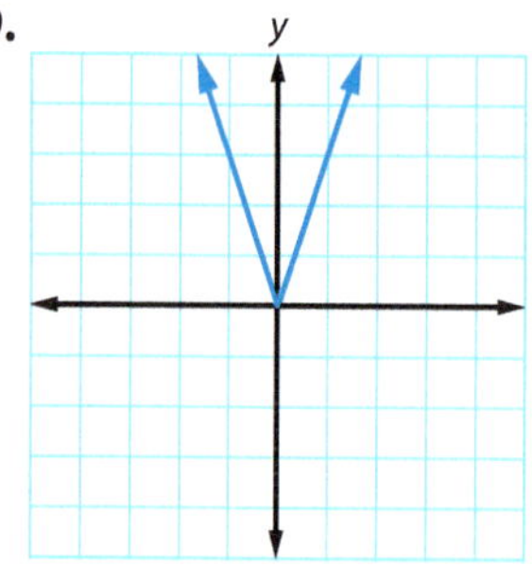

51. C

CHAPTER 5: LINEAR FUNCTIONS

Section 5.1

Table values will vary for exercises 1–7.

1.

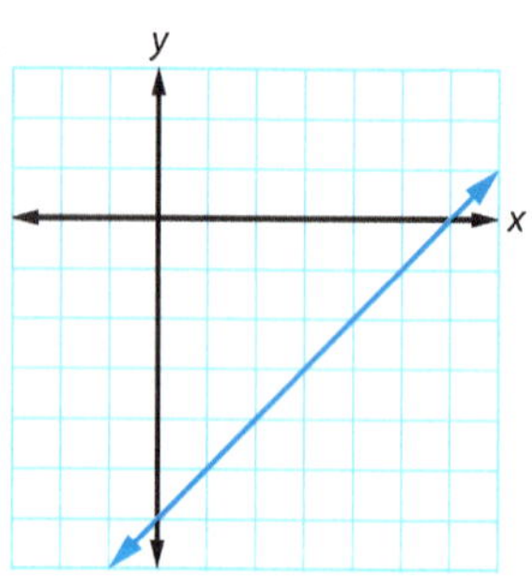

3.

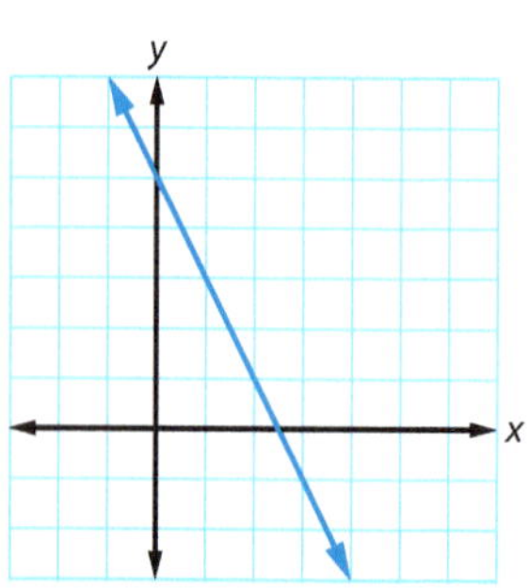

5.

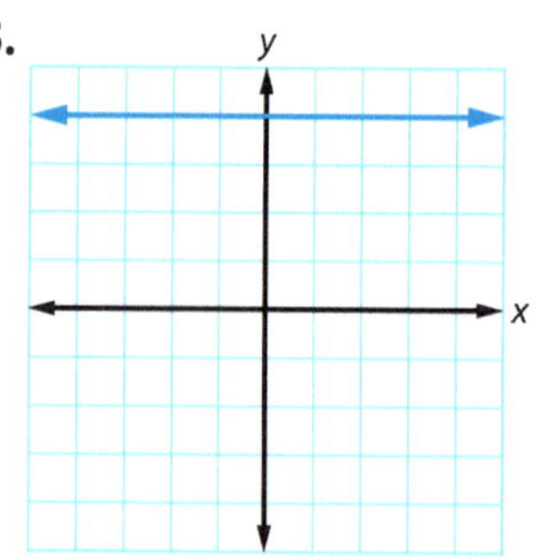

7.

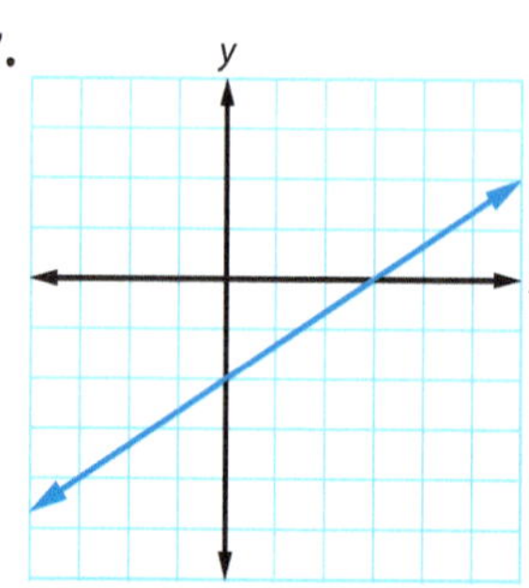

9. $x + y = 9$

11. $4x - 7y = 2$

13. $y = -5x + 20$

15. $y = 4x - 7$

17. (4, 0), (0, −2)

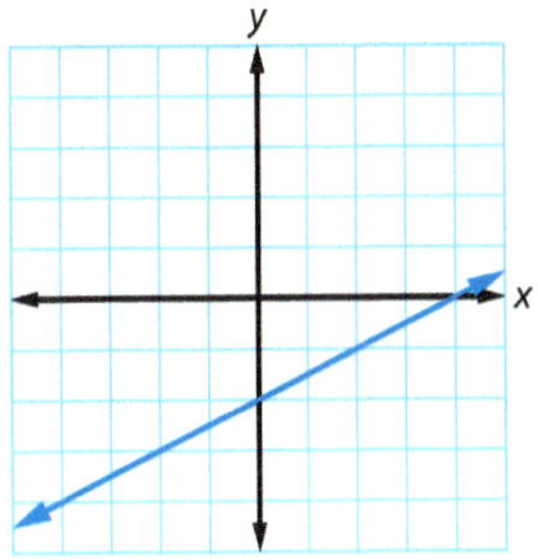

19. (4, 0), (0, 4)

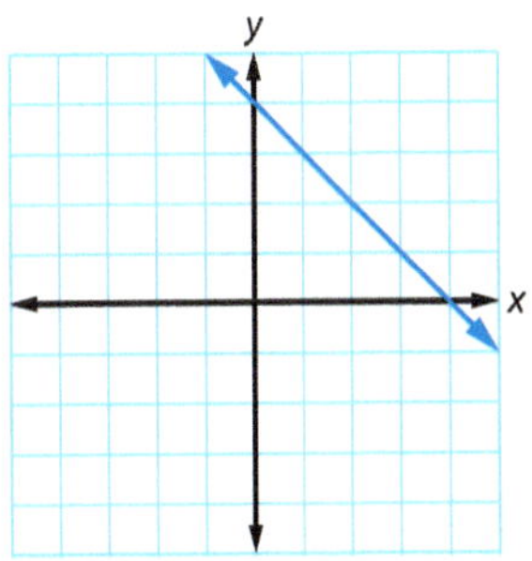

21. (−3, 0), $\left(0, \frac{3}{2}\right)$

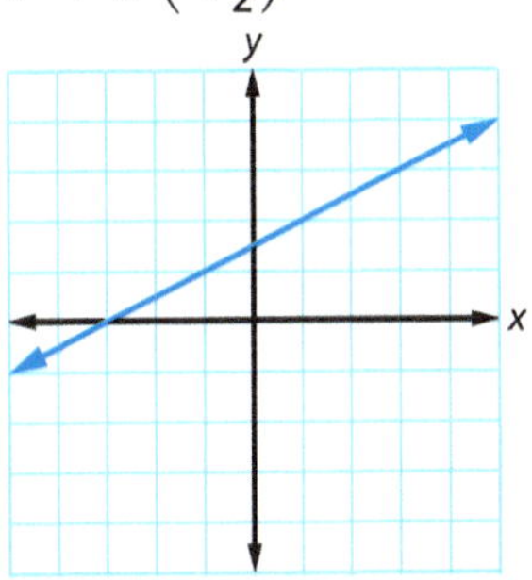

23. (2, 0), none

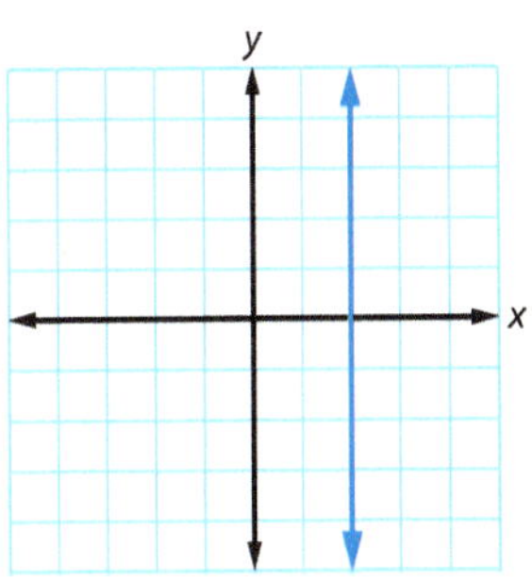

25.

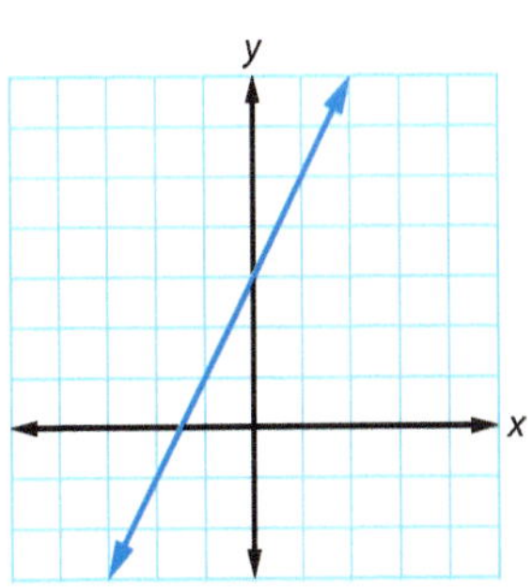

27.

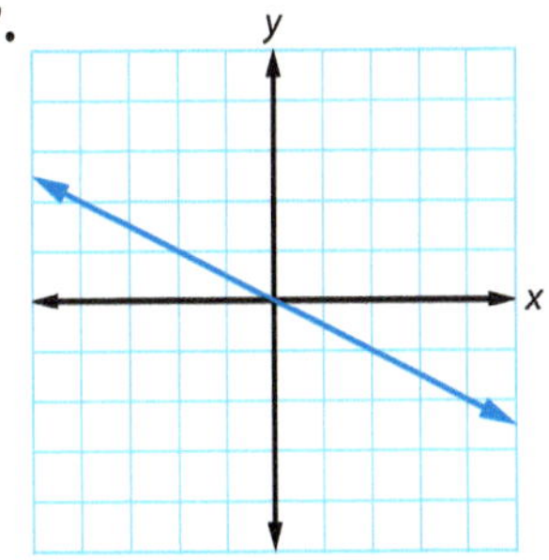

29.

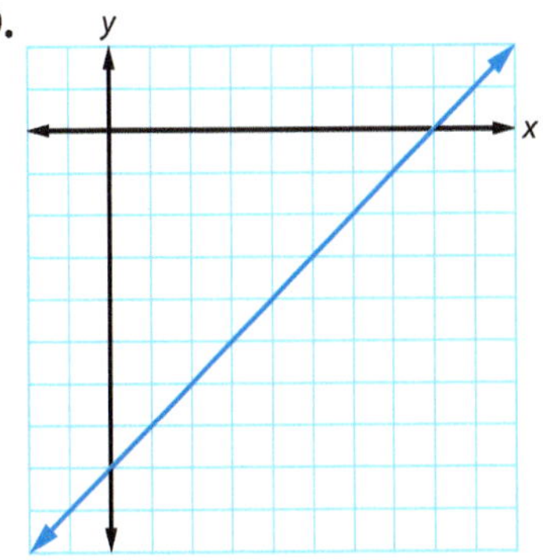

31.

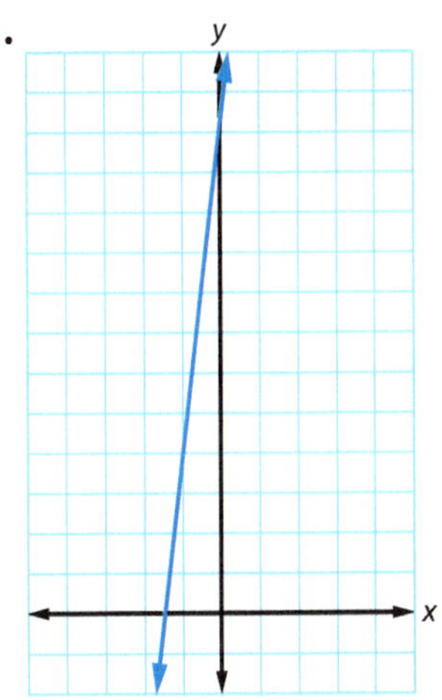

33. 0 **35.** horizontal **39.** (−4, 0), (0, −6)

41. In all the equations 1 is added to the *x*-term. The graphs all have a *y*-intercept of (0, 1).

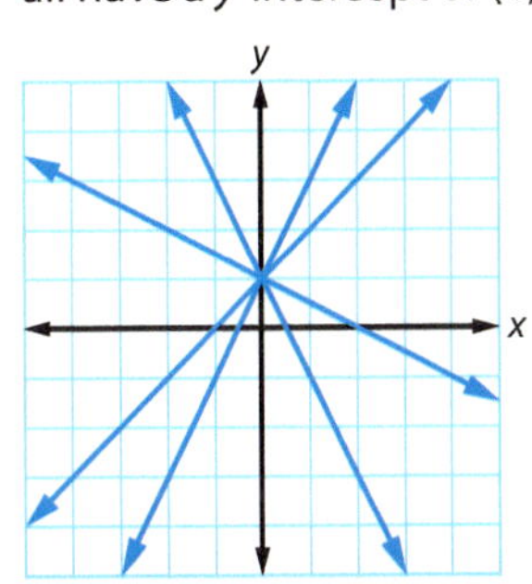

43. $l = \frac{S - b^2}{2b}$ **45.** 7.5 in. and 9 in. **47.** $-5 < a < -2$

49. $c \le \frac{2}{3}$ or $c \ge 4$ **51.** D

Section 5.2

1. *a*, *d* **3.** *e* **5.** *d*

7. $m = -\frac{2}{3}$ **9.** $m = \frac{1}{3}$ **11.** undefined

13. $m = 4$ **15.** $m = 0$ **17.** $m = 1$

19. $m = \frac{5}{2}$ **21.** undefined **23.** $m = 2.8$

25.

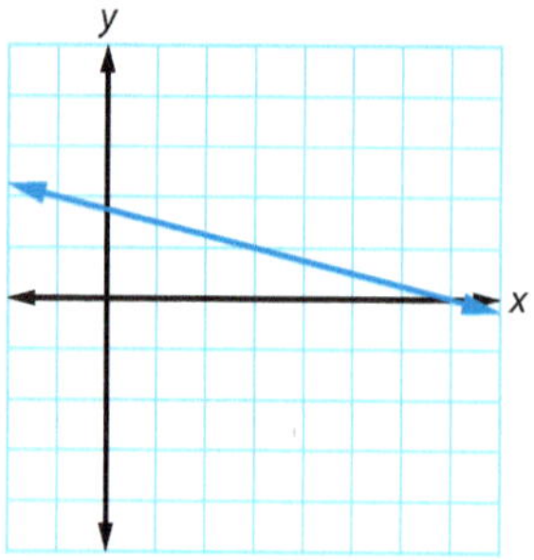

27.

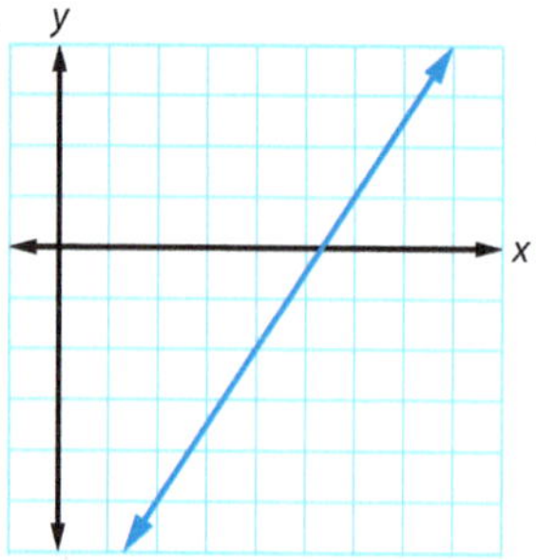

29a. 12 mi/hr **29b.** 20 mi/hr **29c.** 0 mi/hr

29d. stationary **31a.** 18 cm/yr **31b.** 8 cm/yr

31c. 5 cm/yr **31d.** 1 cm/yr **33.** $\frac{3}{25}; \frac{3}{20}$

35. $3\frac{3}{4}$ ft or 3.75 ft **37.** 27 in. **39.** 950.4 ft

41. 5.8 mi **43.** $70 **45.** $\frac{5}{6}$ qt = $0.8\overline{3}$ qt

47. 1 **49.** $x = -4, \frac{4}{5}$ **51.** D

Section 5.3

1. $m = 3$; (0, 4) **3.** $m = -1$; (0, 0) **5.** C

7. B

9.

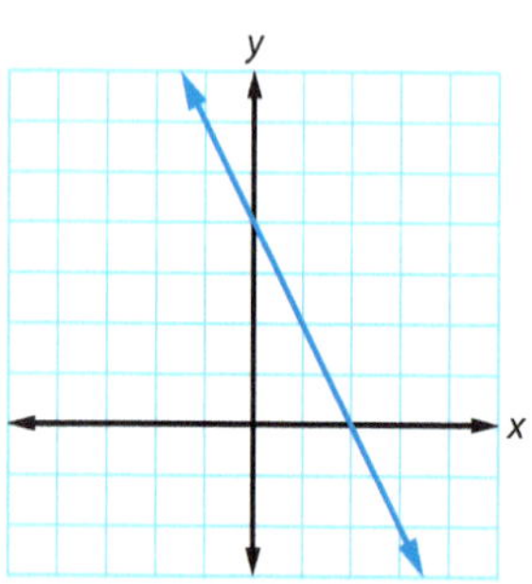

11.

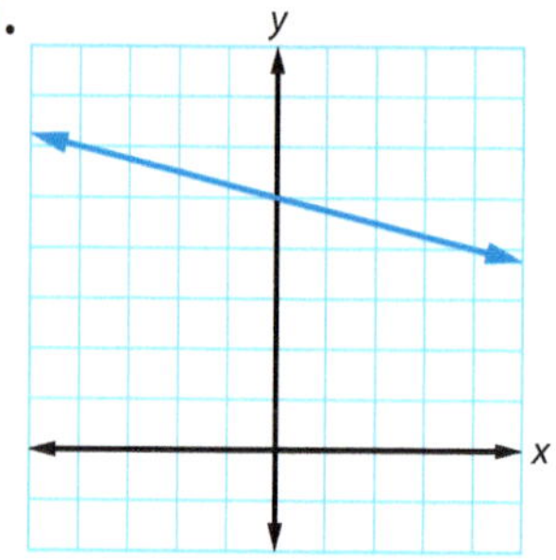

13.

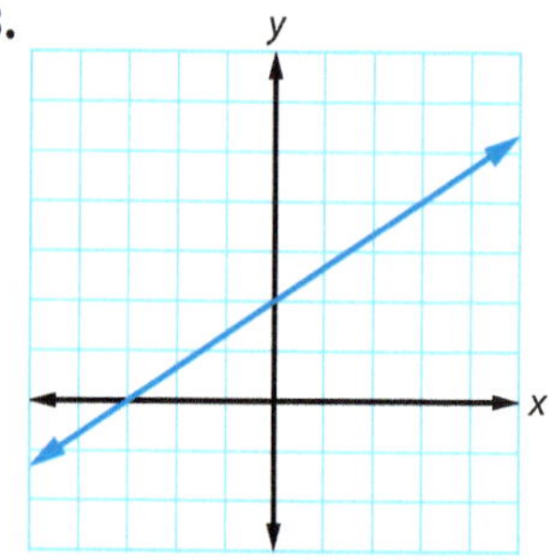

15. $y = -3x + 5$; $m = -3$; (0, 5) **17.** $y = -\frac{1}{7}x + 2$; $m = -\frac{1}{7}$; (0, 2)

19. $y = \frac{8}{3}x - 3$; $m = \frac{8}{3}$; (0, −3) **21.** $y = -dx + k$; $m = -d$; (0, *k*)

23. $y = -2x + 5$

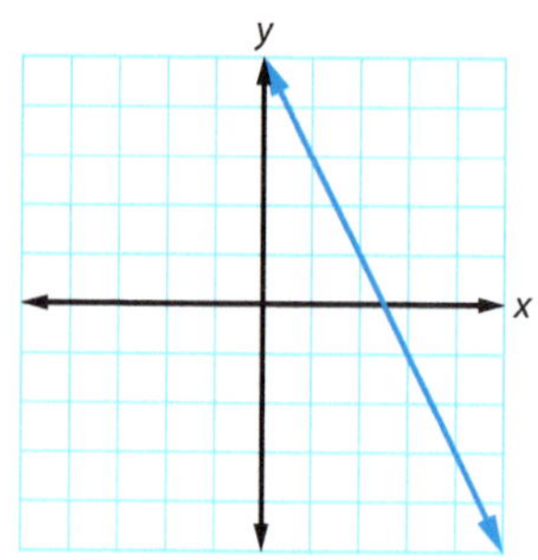

25. $y = -\frac{1}{2}x + 4$

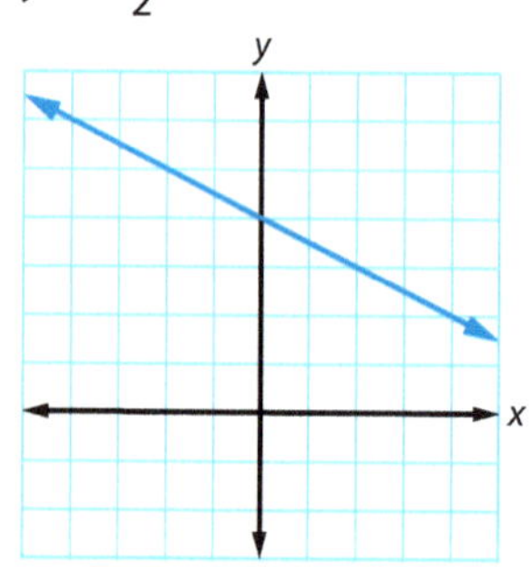

27. $y = -\frac{3}{4}x + 2$

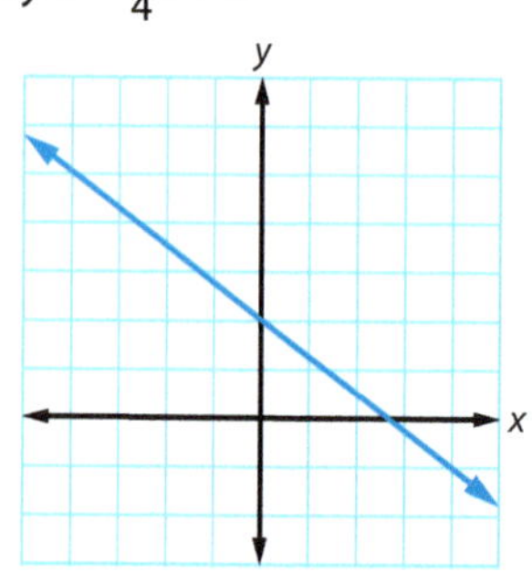

29. $y = \frac{1}{3}x + 3$

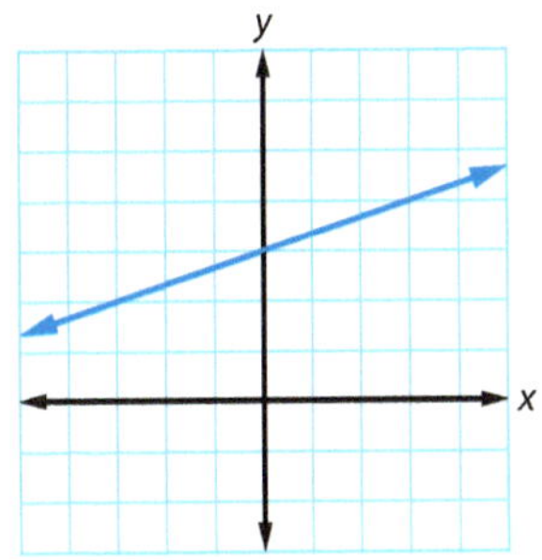

31. $y = 2x + 5$

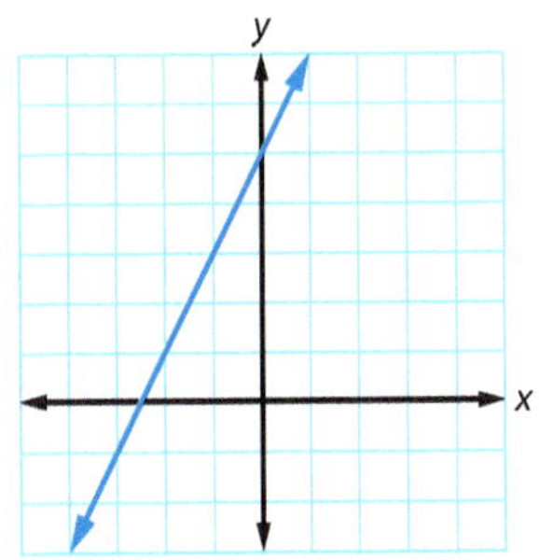

33. $f(x) = 18x + 226$; 586 students

35. $\frac{1}{55}$ gal/mi; $g(m) = -\frac{1}{55}m + 4.5$; 0.5 gal

37. $y = -\frac{A}{B}x + \frac{C}{B}$; $m = -\frac{A}{B}$; $\left(0, \frac{C}{B}\right)$

39. $y = -\frac{3}{2}x + \frac{13}{4}$

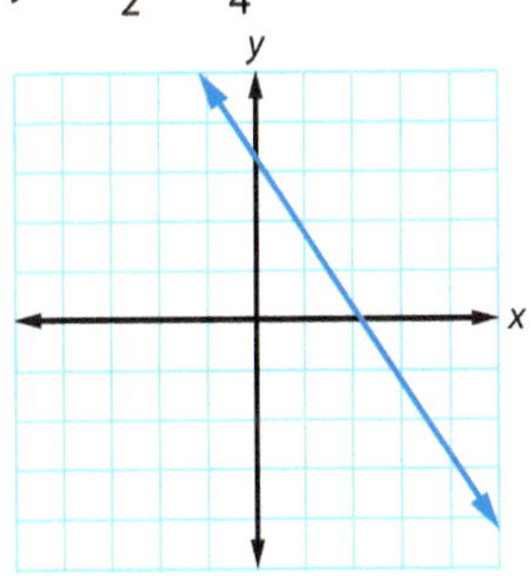

41. 1.34173 **43.** 1 **45.** $\frac{243}{1024}$

47. −4 **49.** B

Section 5.4

1. $y = -5x + 10$ **3.** $y = -\frac{1}{3}x + 3$ **5.** $y = 7$

7. $y = \frac{3}{2}x - 2$ **9.** $y = 3x - 1$ **11.** $m = -2$

15. $y = -3x + 1$ **17.** $y = -\frac{4}{5}x - 2$ **19.** $y = \frac{3}{2}x + 4$

21. $x = 2$ **23.** $y = -\frac{1}{4}x + \frac{25}{4}$ **25.** $y = \frac{1}{4}x + \frac{25}{4}$

27. $y = 2x + 1$ **29.** $y = -\frac{8}{13}x + \frac{2}{13}$

31. $y = \frac{12}{5}x + \frac{3}{5}$ or $y = 2.4x + 0.6$

33. $y = -\frac{1}{2}x + \frac{31}{2}$

35. $m = 10$ ft/sec; $f(x) = 10x + 300$

37. $f(x) = 1.25x - 231$; 499 hot lunches

41. $9x - 4y = 27$ **43.** $21x + 35y = 30$

45. (3, 0), (0, −4)

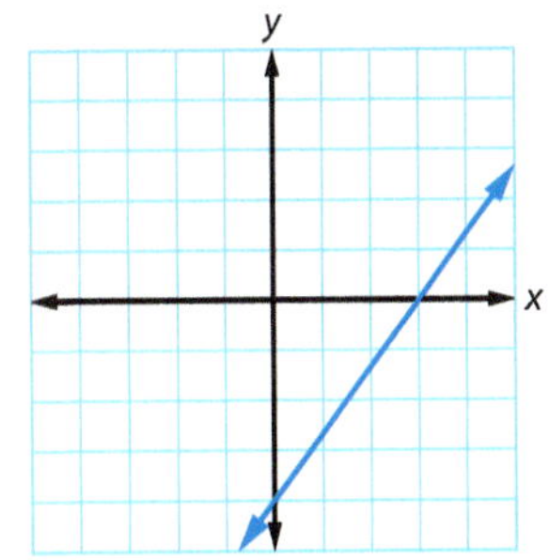

47. 0 **49.** $\frac{2}{3}$

51. $y = -\frac{2}{3}x + 5$; $m = -\frac{2}{3}$; (0, 5)

53. A

Section 5.5

1. $-\frac{1}{7}$ **3.** $\frac{5}{3}$ **5.** $-\frac{7}{4}$

7. $m = \frac{2}{3}$ **9.** $m = \frac{8}{5}$ **11.** $m = -5$

13. $m = -\frac{1}{4}$ **15.** $m = -\frac{9}{5}$ **17.** $y = -4x - 3$

19. $y = \frac{1}{9}x - \frac{71}{9}$ **21.** $y = x + 4$ **23.** $y = 5$

25. $y = \frac{7}{5}x - \frac{21}{5}$ **27.** neither **29.** perpendicular

31. $y = -x + 5$; $y = \frac{5}{3}x - 1$; neither

33. $y = 6x + 15$; $y = -\frac{1}{6}x + \frac{8}{3}$; perpendicular

37. $7x - y = 28$ **39.** −48

41. $m_{AC} = \frac{1}{7}$; $m_{BD} = -7$; perpendicular

43. 1 hr 55 min **45.** $-2 \le y \le 5$

47. $w \le -10$ or $w \ge -4$

49. $-36 < x < 24$

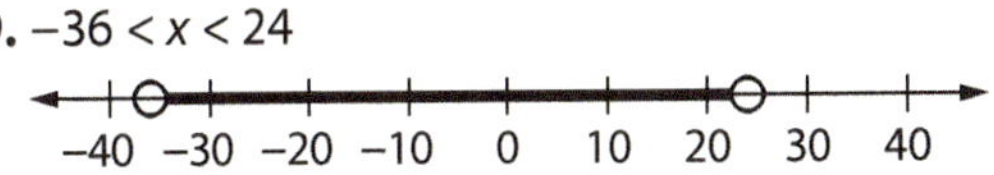

51. A

Section 5.6

1. E **3.** C **5.** A

7. $y = 2.5x + 25$ **9.** $y = 75$ **11.** strong negative

13. $y = 4.5$ **15.** yes

17. Answers will vary; $y = 0.8x + 18$.

19. 81.25 **21.** $y = 0.82x + 16.93$

23. strong positive **25.** 1,150,000

27. the population in 2020 **29.** $r = 0.99$; yes

33. 29.44 lb

35. Weak positive; predictions will most likely be inaccurate.

37. 29.48 lb; 23.20 in.

39. A curve; curves will vary.

41. $d = 0.0663v^2 + 16.01$

43. $x = 3.75$

45. $x = 16$

47. inverse

49. neither

51. A

Application Problems

1. 6 ft

3. 1 : 6

5. The bottom of the ladder could slide; the ladder could bend.

7. 28 in.

9. $h = 40$ ft; $w = 72$ ft 1 in.

11. 16

13. 9.375 in. or $9\frac{3}{8}$ in.; yes

15. 0.067; yes

Chapter Review

1. $y = 5x - 6$

3. Table values will vary.

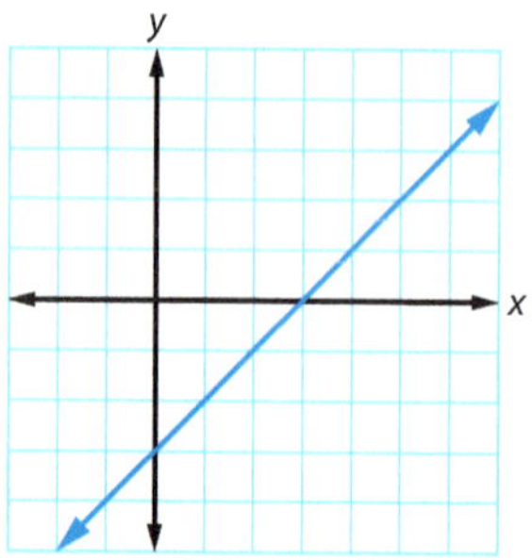

5. (4, 0), (0, −1)

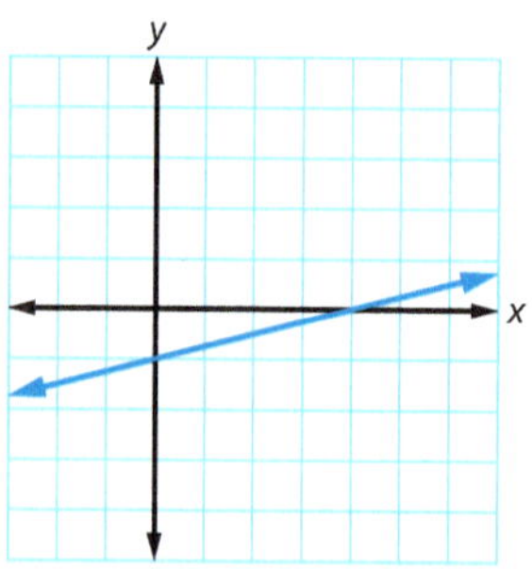

7. $m = \frac{2}{3}$

9. $m = \frac{7}{2}$

11. $m = 0$

13.

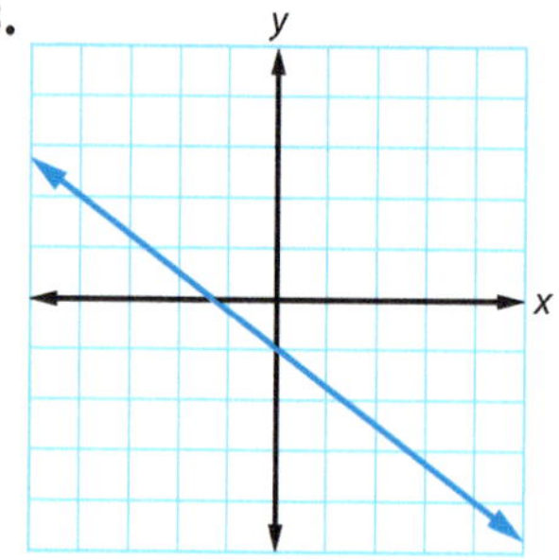

15. $m = \frac{5}{6}$; (0, −3)

17.

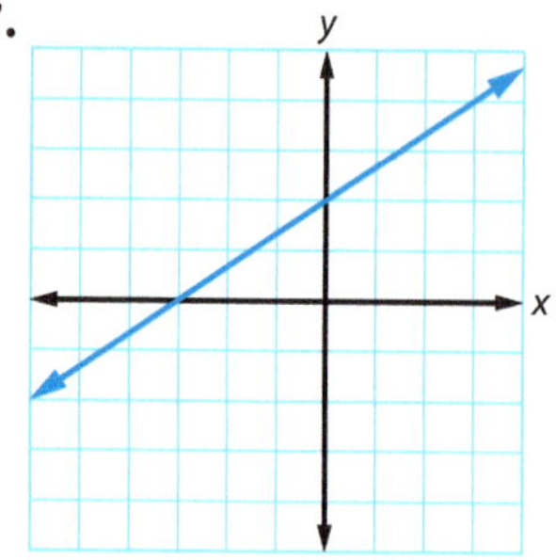

19. $f(x) = -\frac{5}{4}x + 2$

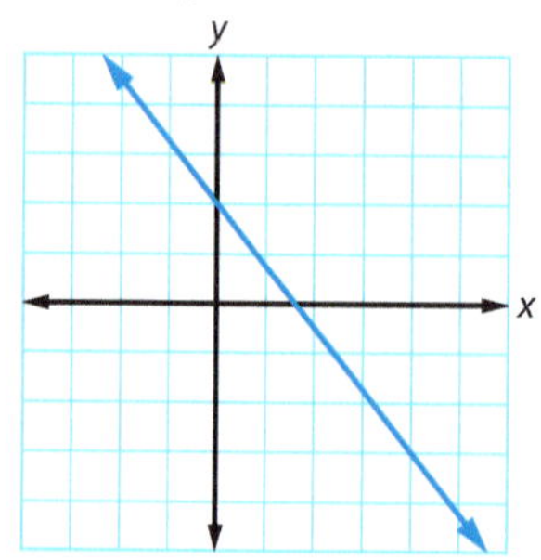

21. $f(x) = \frac{1}{4}x + 8$; 22 lb

23. $y = 2x + 3$

25. $y = \frac{3}{2}x + \frac{5}{2}$

27. $y = -\frac{1}{2}x + 4$

29. $y = -3$

31. $y = \frac{2}{7}x + \frac{22}{7}$

33. $3x - 2y = 6$

35. $\frac{1}{2}$ and $-\frac{8}{3}$

37. neither

39. $m_c = 2$, $m_d = -\frac{1}{2}$; perpendicular

41. strong positive

43. 60.4 kg

45. C

47. the census

CHAPTER 6: LINEAR SYSTEMS

Section 6.1

1. (1, 5)

3. ∅

5. yes

7. no

9. no

11. no

13. yes

15. (−6, −7)

17. (1, 1)

19. (3, −5)

21. (−2, 6)

23. (5, 7)

25. (2, 4)

27. (−3, −1)

29. all points on $y = -\frac{x}{2} + 4$

31. ∅

33a. $x + y = -7$ and $x - y = 3$

33b. (−2, −5)

35a. $r(x) = 30x$

35b. $e(x) = 20x + 120$

35c. 12 jackets

39. about 233 mi; Great Haul

41. $\frac{37}{15}$

43. $x = \frac{-4y + 5}{3}$

45. $x \geq -3$

47. $\mathbb{R}$

49. C, D

Section 6.2

1. A

3. C

5. C

7. A

9. (7, 21)

11. (16, 12)

13. (9, 2)

15. (9, −1)

17. (−2, 8)

19. (10, 5)

21. $\left(-\frac{9}{26}, -\frac{25}{26}\right)$

23. $\left(3, \frac{1}{5}\right)$

25. $\left(\frac{1}{6}, \frac{7}{10}\right)$

27. $\left(\frac{1}{3}, \frac{5}{6}\right)$

29. It eliminates one of the variables.

31. 19 ft by 26 ft

33. 5 apple and 7 pecan

35. down payment: \$800; bank loan: \$4000

37. $\left(\frac{1}{2}, \frac{3}{4}\right)$ or (0.5, 0.75)

39. (1, 4) and (−1, 4)

41. $\left(\frac{28}{5}, \frac{14}{5}\right)$

43. $x = 3, 5$

45. $x = -\frac{10}{3}, 6$

47. ∅

49. $f(x) = 2x + 1$

51. A

Section 6.3

1. **a)** 2nd equation by −7
   b) just subtract

3. **a)** 2nd equation by −2
   b) just subtract

5. (1, 4)

7. (−3, −1)

9. (−3, 4)

11. **a)** 1st equation by −2
    b) 1st equation by 5 and 2nd equation by 3

13. **a)** 2nd equation by −6
    b) 1st equation by 3 and 2nd equation by 11

15. $\left(4, \frac{5}{2}\right)$

17. (4, 7)

19. (−1, −3)

21. (6, −6)

23. $\left(\frac{5}{2}, \frac{9}{2}\right)$

25. $\left(\frac{19}{8}, \frac{1}{16}\right)$

27. $\left(\frac{31}{17}, \frac{4}{17}\right)$

29. (2, 3)

31. 365 and 256

33. 75 adult and 170 student tickets

35. 18 softballs and 25 baseballs

37. $\left(\frac{41}{18}, \frac{67}{30}\right)$

39. $\frac{24}{7}$ and $\frac{8}{7}$

41. $7y + 14$

43. $x^2 + 2x + 1$

45. $8x + 17$

47. 2

49. A

Section 6.4

1. B

3. C

5. A

7. A, D

9. C

11. A, B

13. (−3, −3)

15. (3, −5)

17. (2, 0)

19. ∅; inconsistent

21. (2, 4); consistent independent

23. $y = -\frac{4}{3}x + \frac{7}{3}$; consistent dependent

25. ∅; inconsistent

27. $\left(\frac{41}{7}, -\frac{3}{7}\right)$; consistent independent

29. ∅; inconsistent

31. $y = \frac{5}{6}x - \frac{7}{6}$; consistent dependent

35. The equations have the same slope and the same *y*-intercept.

37. The lines are straight and coplanar.

39. books: \$4 each; DVDs: \$12 each

41. Each equation has a fractional *y*-intercept.

43. 18 credit hours; if he takes more than 18 hours per semester

45. They both have a *y*-intercept of 0 or they both go through the origin.

47. $p(d) = 0.445d + 14.7$

49. 1086.0 atm

51. 73.7 atm

53. A, D

Section 6.5

1. $12 - w$

3. $4 + c$

5. $500 - w$

7a. b = the rate with no wind

7b. $1.125(b + 8) = 36$ or $2.25(b - 8) = 36$

7c. There is only 1 unknown.

9.

	r	*t*	*d*
Upstream	$b - 3$	6	$6(b - 3)$
Downstream	$b + 3$	4	$4(b + 3)$

11. x = the rate with no wind; w = the wind speed

13. $2(x + w) + 5(x - w) = 150$ or $7x - 3w = 150$

15. x = the rate with no wind; w = the wind speed

17. $2(x + w) = 3(x - w)$

19. 4 hr

21. plane: 275 knots; wind: 25 knots

23. southbound train: 65 mi/hr; northbound train: 55 mi/hr

25. 24 mi/hr

27. first car: 65 mi/hr; second car: 59 mi/hr

29. express train: 59.5 mi/hr; local train: 45.5 mi/hr

31. When both the normal rate and the rate of wind or current are unknown, you can create a system of equations to solve for both.

33. 2 hr

35. Feusses: 45 mi/hr; Sallés: 48 mi/hr

37. (5, 4)

39. ∅

41. (17, 19); consistent independent

43. $y = \frac{3}{7}x - \frac{5}{7}$; consistent dependent

45. A

Section 6.6

1. \$40

3. 82 bags

5. $30x$

7. 28% salt

9. $0.12x$ gal of acid

11. 0%

13. 18 oz of nuts and 30 oz of raisins

15. 1.5 lb of caramels and 3.5 lb of butterscotch candies

17. 8 roses and 6 carnations

19. 12 gal of the 30% salt solution and 28 gal of pure water

21. 8 g of cereal and 4 g of high-protein grain

23. solution I: 15%; solution II: 35%

25. \$4200 at 10% and \$5800 at 9.5%

27. 116.7 gal of type A and 83.3 gal of type B

29. 10.7 gal of cream and 49.3 gal of milk

31. 26.6 lb of A and 58.4 lb of B

35. 6426 in district 12 and 3827 in district 13

37. corn: \$3.30/bu; beans: \$9.60/bu

39. 3^7 **41.** $\frac{1}{2^4 \cdot 3^4 \cdot 5^7}$ **43a.** no

43b. no **43c.** yes **45.** $x \le 3$

47. E

Section 6.7

1. solid **3.** dashed **5.** *D*

7. *M* **9.** B **11.** A

13. B **15.** A

17.

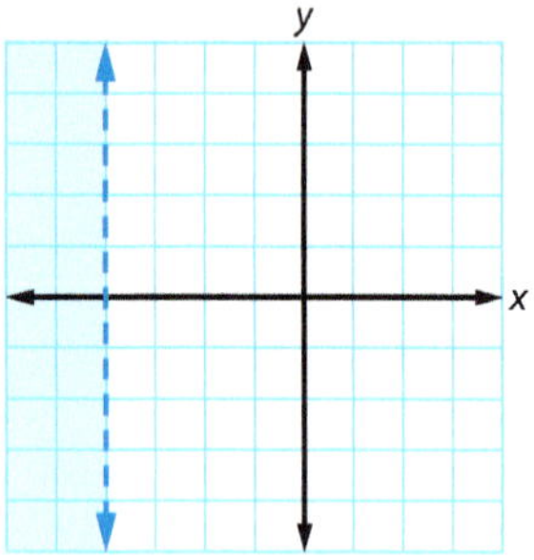

19.

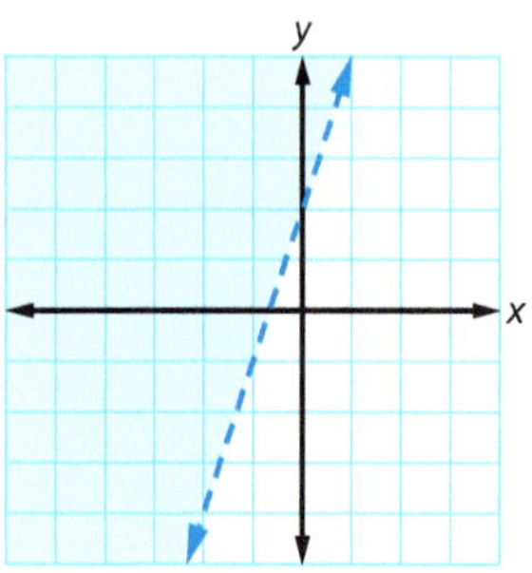

21.

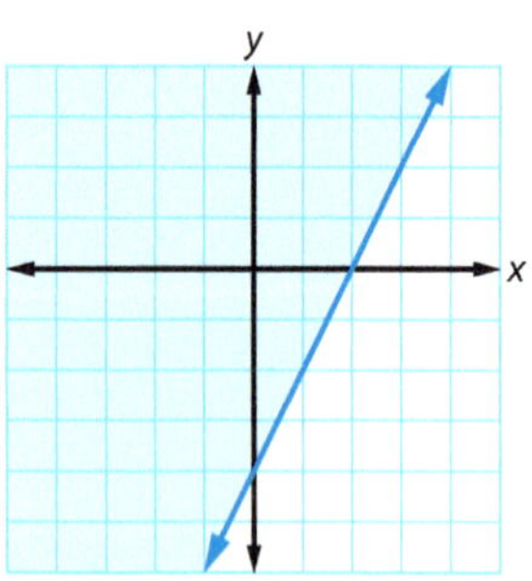

23.

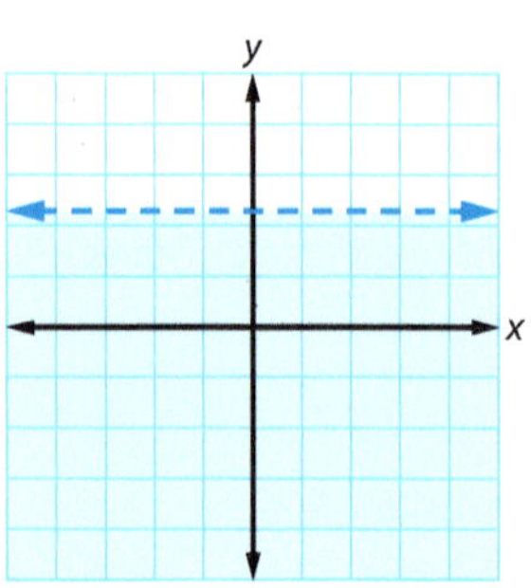

25. $y \ge 2x - 6$ **27.** $y > -2x + 1$ **31.** yes; no

33. $y \le 5x$

35. $x + 3y \le 12$; answers will vary.

37. $25x + 5y \ge 325$; 9 computers

39.

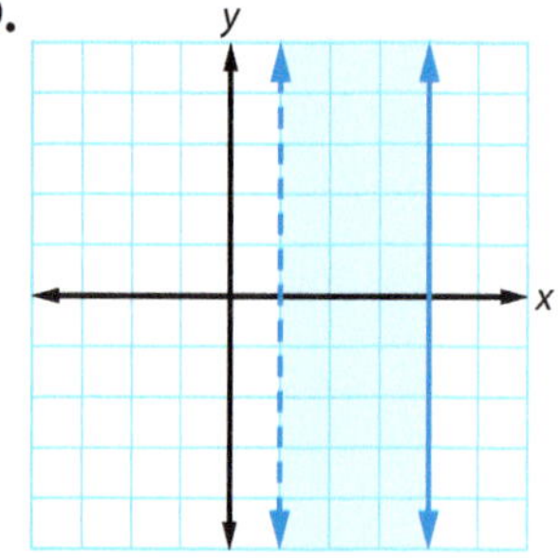

41. She could buy 12 lb of nuts and 4 lb of candy, 8 lb of each, or 9 lb of nuts and 6 lb of candy.

43. $x = \frac{136}{27} = 5.\overline{037}$ **45.** 14,720 bu

47. 222 ft **49.** $(-6, 6)$ **51.** B, C

Section 6.8

1. I **3.** II **5.** yes

7. no

Points will vary for exercises 9–13.

9.

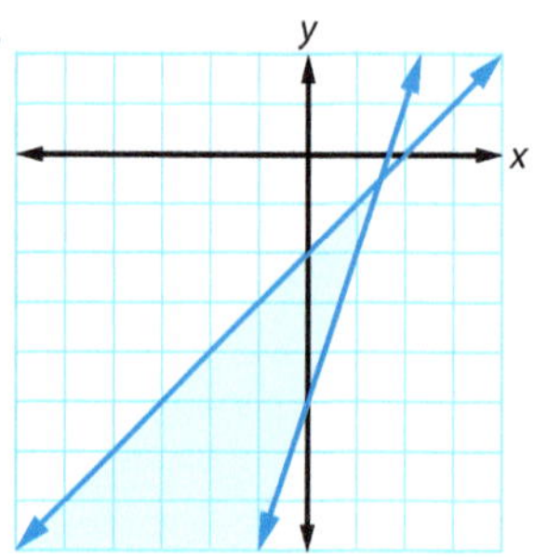

11.

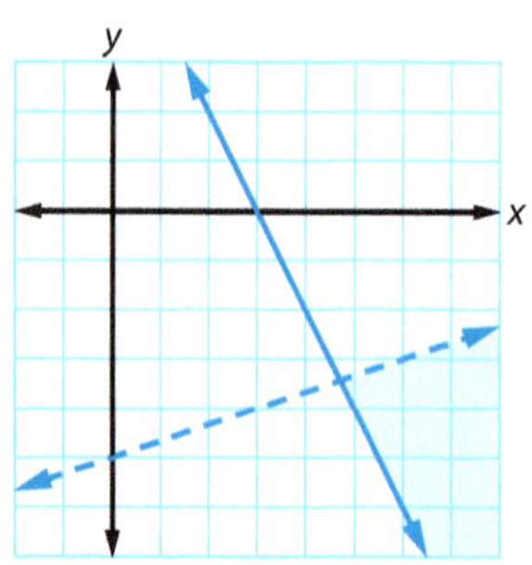

13.

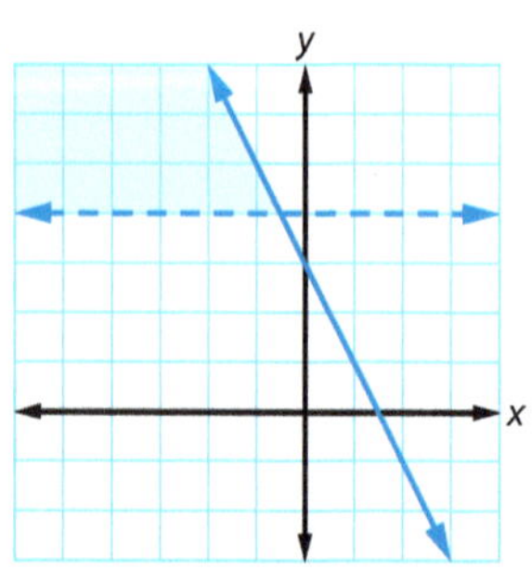

15. $y \le 2$ and $y \le -\frac{1}{3}x + 1$

17. $y > \frac{1}{3}x + 4$ and $y \le 2x + 2$

19.

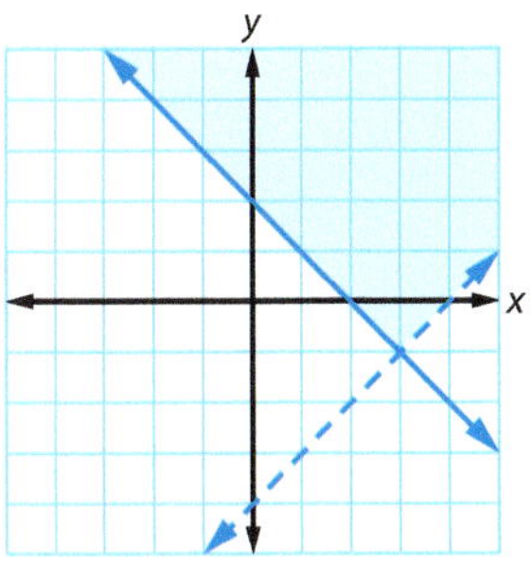

21.

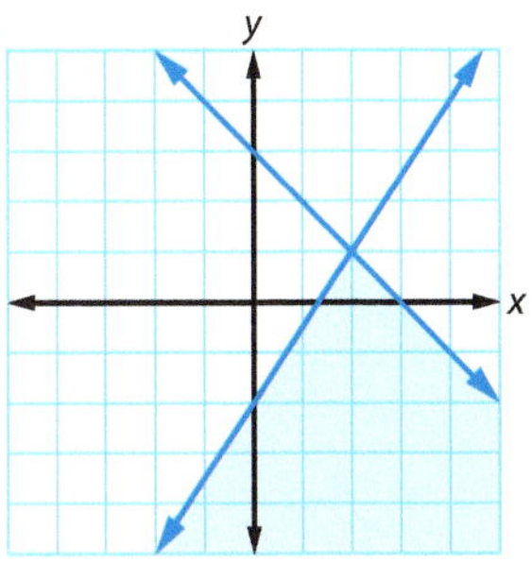

23.

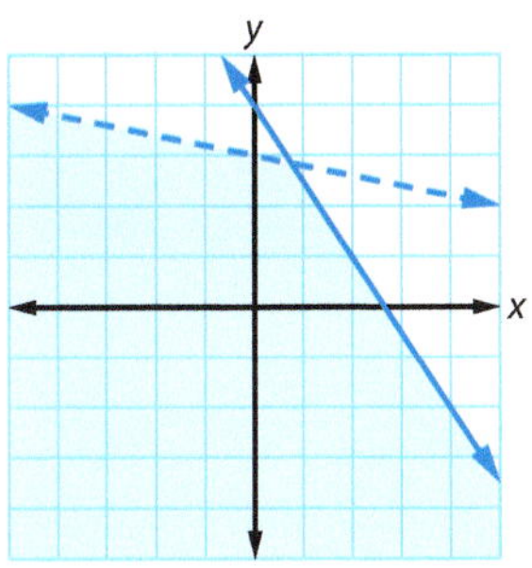

25.

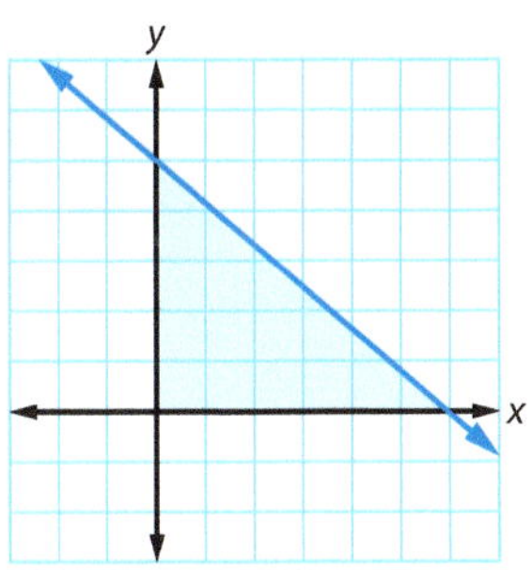

27.

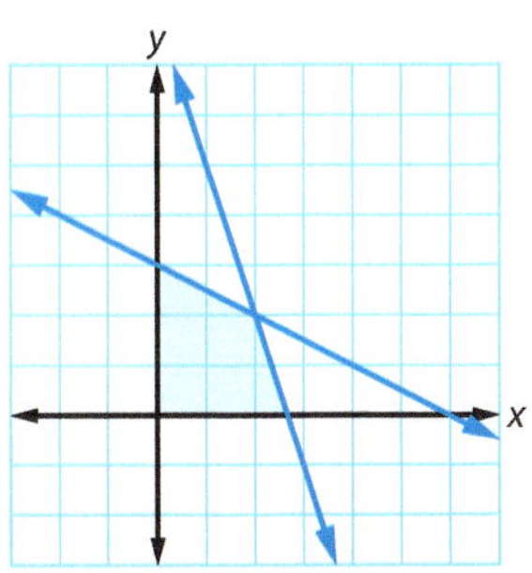

29. $y \geq -3x - 2$ and $y < -3x + 5$

31. $y \geq -3$, $x < 2$, and $y < 2x + 1$

33a. $12x + 7y \leq 70$, $8x + 12y \geq 72$, $y \geq 0$, and $x \geq 0$

33b.

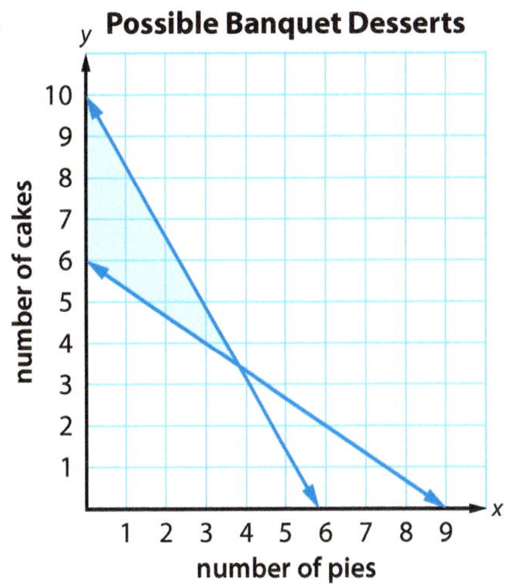

33c–d.

Chocolate Silk	Lemon Chiffon	Number of Servings	Cost
2	5	76	$59
2	6	88	$66
3	4	72	$64

37. The shared dashed line belongs to neither solution set and there is no intersection of the shaded regions.

39. $y < -\frac{3}{4}x + \frac{13}{4}$, $y > \frac{1}{4}x + \frac{9}{4}$, $x > -5$, and $x < -1$

41. (11, 19) **43.** (5, −8) **45.** (4, −5)

47. 30 km/hr **49.** B

Application Problems

1.

Year	Mexico (in millions)	Colombia (in millions)
1950	26	12
1960	35	16
1970	52	21
1980	69	27
1990	83	33
2000	100	40
2010	111	46
2020	126	51

5. Record Holders: $r(x) = 0.18x + 50$; Digital Dzign: $c(x) = 0.2x$

7. Record Holders; $1850

9. 200 CDs per person

11. $400{,}000x + 4300 = 70{,}000$; $x \approx \$0.164$ per CD

13. $16.5x \leq 2000$; $x \leq 121.\overline{21}$; 121 boxes

15. 1700 lb

17. $\frac{0.9}{16}x + 6 \le 25$ or $0.05625x + 6 \le 25$

19. $x \le 337.78$ and $x \le 816$; 337 CDs

Chapter Review

1. yes
3. yes
5. (3, −5)
7. (4, 0)
9. ∅
11. (−2, 5)
13. −9 and −14
15. 33 people
17. (1, 1)
19. (2, 4)
21. (−4, 19)
23. $\left(\frac{4}{3}, -\frac{1}{3}\right)$
25. ∅
27. They are parallel.
29. They intersect at their *y*-intercept.
31. no solutions; inconsistent
33. infinite number of solutions; consistent dependent
35. 1 solution; consistent independent
37. first car: about 2.04 hr or 2 hr 2 min; second car: 1.79 hr or 1 hr 47 min
39. plane speed: 260 mi/hr; wind: 60 mi/hr
41. 3.3 gal of the 35% salt solution and 1.7 gal of pure water
43. dashed; above
45.

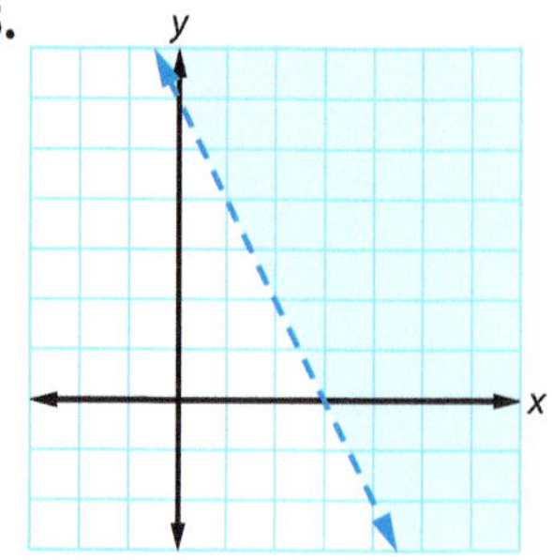

47.

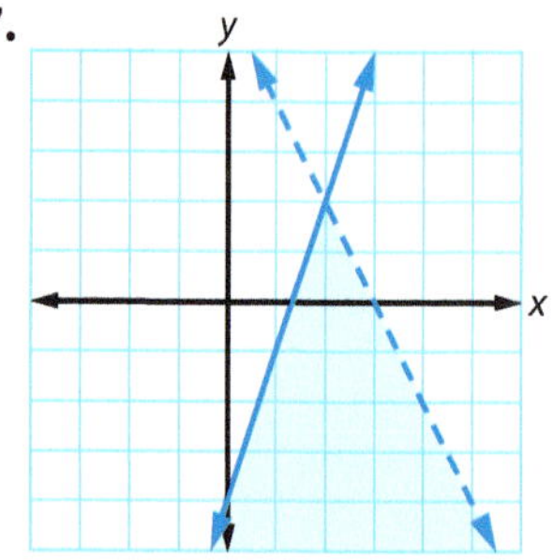

49. $y \le -\frac{5}{2}x + 4$ and $y > -x - 5$
51. $x \le 12$, $y \le 15$, $x + y \le 20$; 9 quail

CHAPTER 7: EXPONENTS

Section 7.1

1. product; powers **3.** x^8

5. $12w^7$ **7.** $-\frac{3ab^5}{4}$ **9.** a^{12}

11. $4b^4c^2d^{10}$ **13.** $\frac{y}{x}$ **15.** $-12a^4b^2c^2$

17. $8x^5y^{-9}$ **19.** $500x^{11}$ **21.** $\frac{19x^4y^6}{z}$

23. $-\frac{a^2bc^3}{3}$ **25.** $\frac{y^9}{9x^4}$ **29.** $\frac{1}{6}h^6$

31. $\frac{r}{2s^2}$ **33.** $27a^9b^3$ **35.** a^2

37. y^{5z+3} **39.** $\frac{3}{m^4n^2} + 4m^3$ **41.** $5w^3z^2 - w^2z^3$

45. $x^3 + 7x^2$

47. $D = \{-2, 4, 8, 9\}$; $R = \{4, 8, 9\}$; yes

49. rate; time **51.** $t(r) = \frac{300}{r}$ **53.** A

Section 7.2

1. 3.9×10^8 **3.** 5.2×10^{-5} **5.** 1.36×10^8

7. 2.356×10^{-2} **9.** 254,000,000 **11.** 0.000153

13. 6×10^7 **15.** 8.1×10^{21} **17.** 5×10^6

19. 9.5×10^3 **21.** 2.77×10^9 **23.** 3.01×10^3

25. 5.625×10^{-29} **27.** 2.73×10^7

29. capillary: 8×10^{-6} m; aorta: 2.5×10^{-2} m

31. 7.94×10^8 mi

33. 2.25×10^8 m; 2.25×10^5 km

37. 1.5×10^{11} **39.** 2×10^6 **41.** 10^9

43. 189×10^9; 189 billion **45.** 100×10^{-6}; 100 millionths

47. $f(x) = |x - 2| + 11$ **49.** $f(x) = \left|x + \frac{3}{2}\right| - \frac{5}{9}$

51. $4x - y = -10$ **53.** $(2, -2)$ **55.** E

Section 7.3

1. vertex **3.** 5 units up **5.** 7 units left

7. 5 units right, 6 units up **9.** h units right, k units up

11. $(3, 1)$ **13.** $(-8, 6)$ **15.** $(0, 19)$

17. $(7, 0)$ **19.** $(h, -k)$

Table values will vary for exercises 21–25.

21. $(0, 3)$

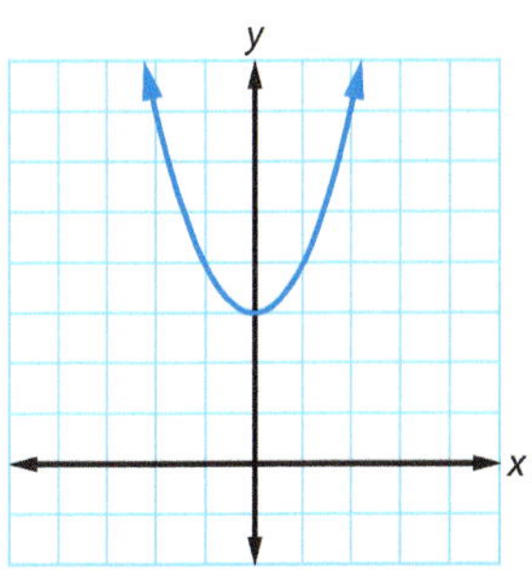

23. $(-2, 0)$

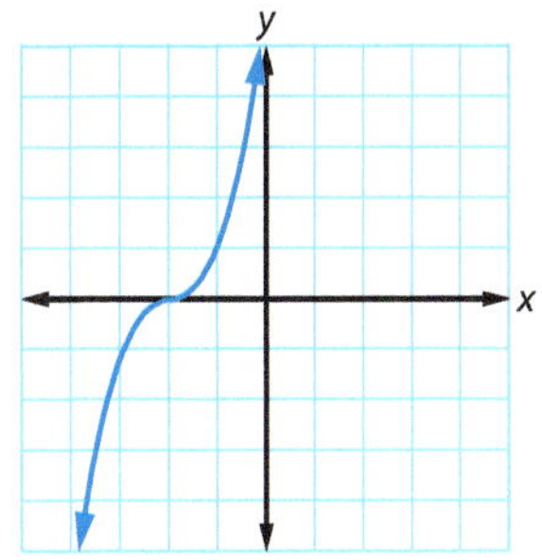

25. $(4, 1)$

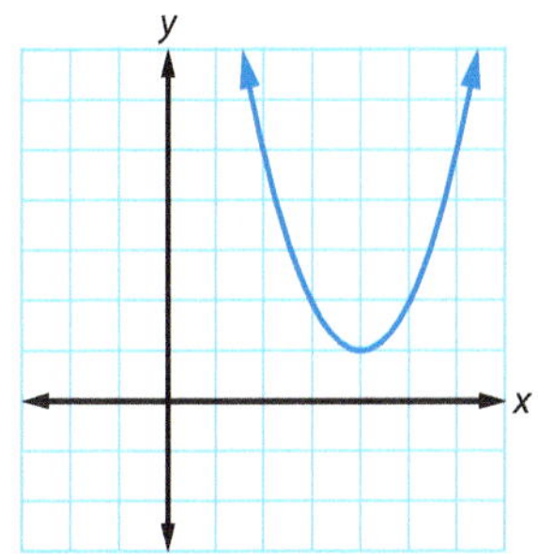

27. $f(x) = (x - 10)^2$ **29.** $f(x) = (x + 3)^2 - 12$

31. $f(x) = (x - 2)^2 + 1$ **33.** $y = x^2 + 2$

35. $y = (x - 2)^2 - 2$

39. Tables values will vary.

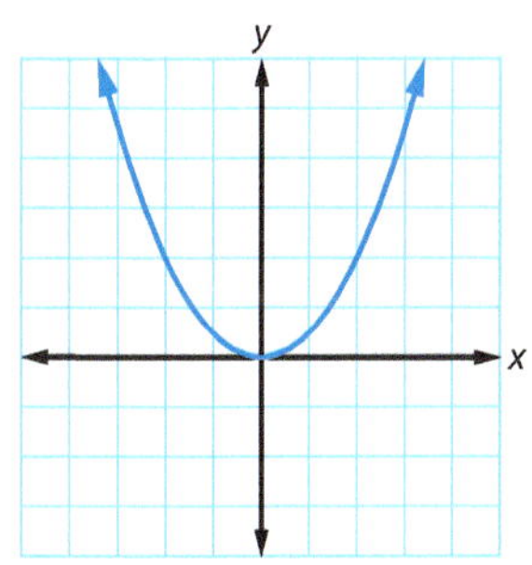

41. It gets steeper. **43.** It turns upside down.

45. $g(x) = 4x - 3$ **47.** e **49.** c

51.

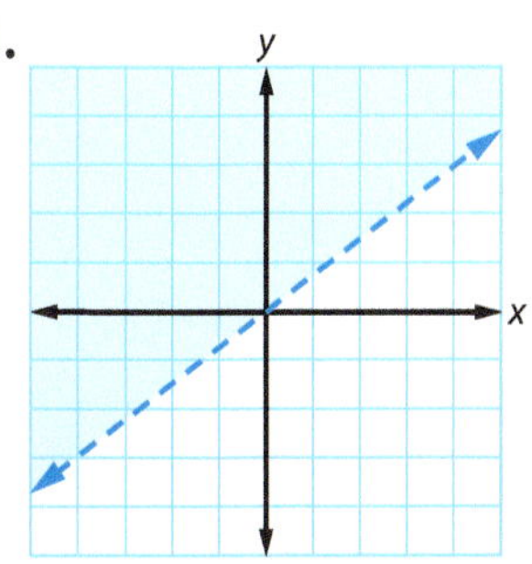

53. E

Section 7.4

1a. II **1b.** I **1c.** III

3. $\frac{1}{32}$ **5.** 192

7.

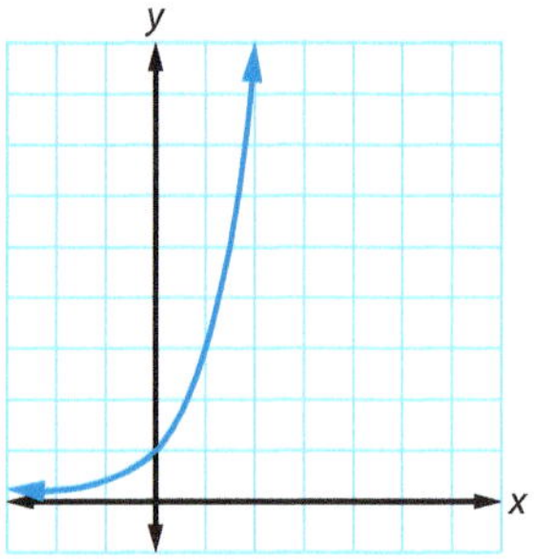

9.

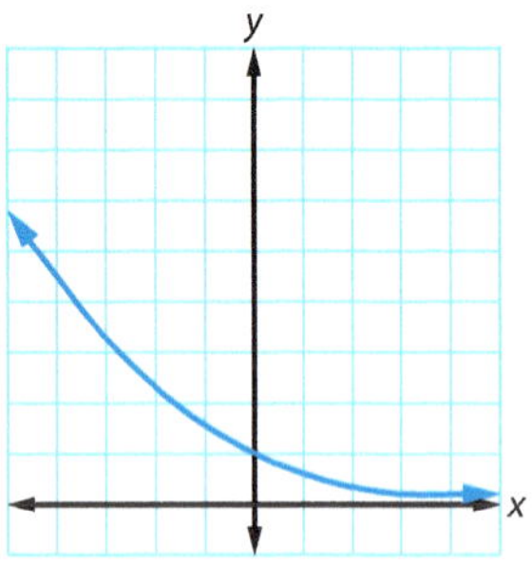

11.

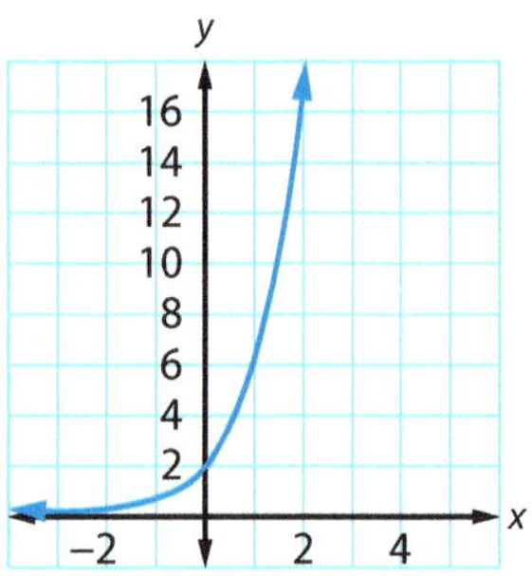

13.

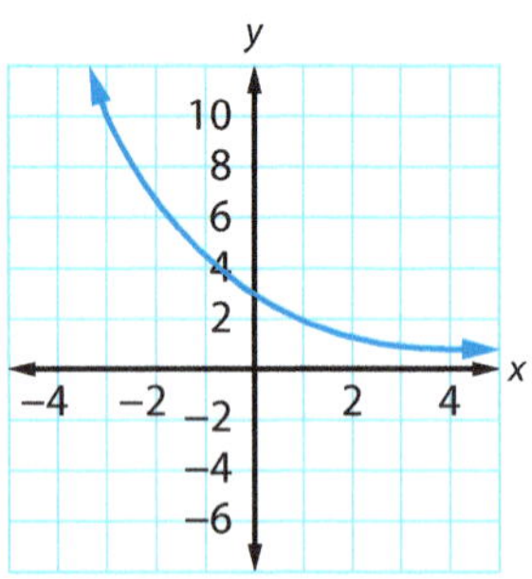

15. E

17. As x increases, y decreases.

19. $y = 1.25^x$; 156% **21.** $y = 0.75^x$; 32% **23a.** $1020.00

23b. $1040.40 **23c.** $1218.99 **23d.** $1640.61

23e. $2691.59

25.

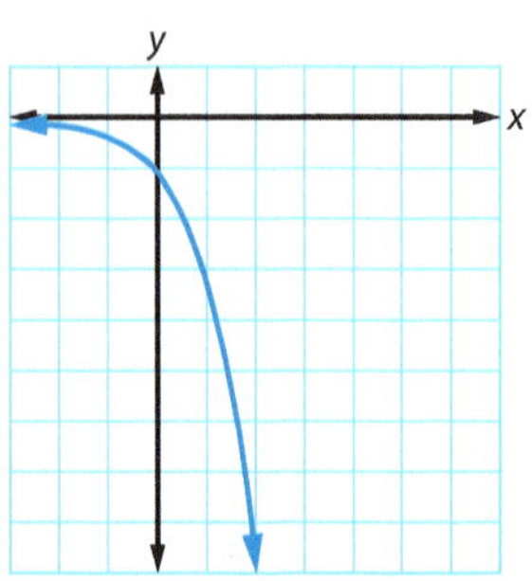

27.

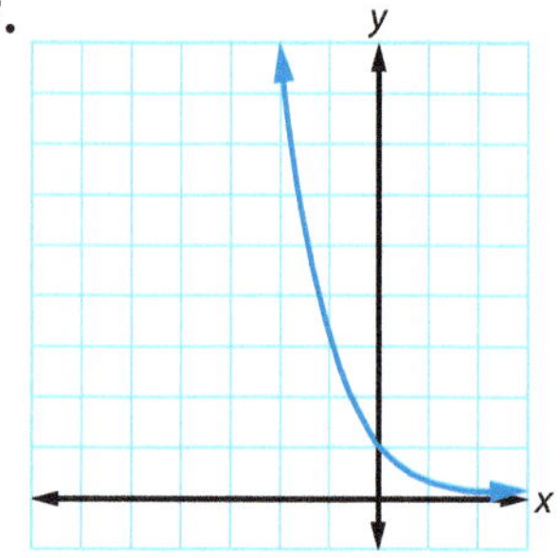

29. Answers will vary. Sample answer: $y = b^x$ and $y = -b^x$ are mirror images (reflections) of each other across the x-axis.

31.

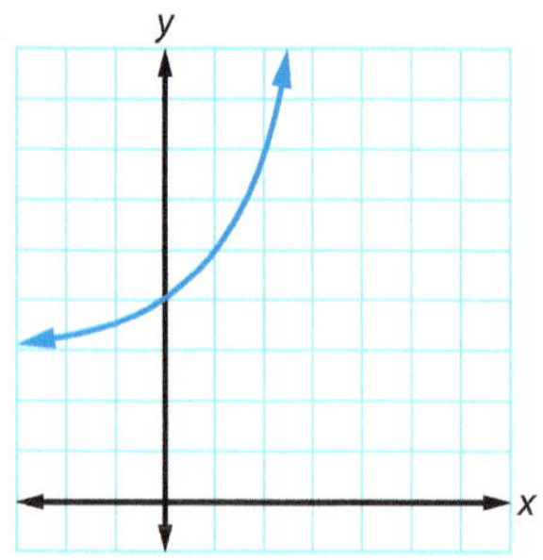

33. *a* **35.** *d* **39.** $10,737,418.24

41. No; if $0 < b < 1$, then the function decreases.

43. rs^3 **45.** $\frac{c}{a^3b^2}$ **47.** 7.5×10^{10}

49. 1.5×10^{-2} m **51.** A

Section 7.5

1. 35; 5%; 1.05; growth **3.** 1; −30%; $\frac{7}{10}$; decay

5. $f(t) = 10(2.5)^t$; growth

7. $f(t) = 81\left(\frac{3}{2}\right)^t$ or $f(t) = 81(1.5)^t$; growth

9. $f(t) = 75(0.5)^t$ or $f(t) = 75\left(\frac{1}{2}\right)^t$; decay

11. $y = 800{,}000(0.9)^x$ **13.** $y = 1500(2)^x$

15. 10 periods; 6% **17.** 40 periods; 1.5%

19. 3650 periods; ≈0.016% **21.** $5000(1.02)^{20}$

23. $37{,}855(1.00\overline{3})^{360}$

25. $f(q) = 10{,}000(1.015)^q$; $f(48) \approx \$20{,}434.78$

27. $f(h) = 2000(1.03)^h$; $f(40) \approx \$6524.08$

29. $f(n) = 5000(1.045)^n$; $f(6.5) \approx \$6656.19$

31a. $f(x) = 500(0.9)^x$ **31b.** 328 mg

31c.

Medication Levels

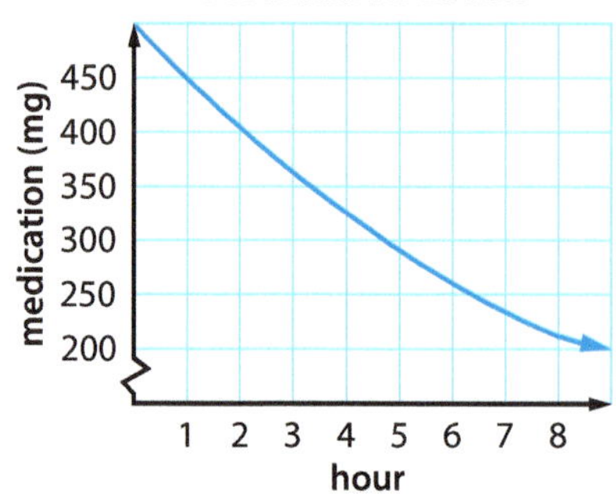

33a. $f(p) = 500(2)^p$ **33b.** 32,000 bacteria

33c. 2.4×10^{24} bacteria **35a.** $f(m) = 2(2)^{\frac{m}{3}}$

35b. $f(n) = 2(2)^{4n}$ or $f(n) = 2(16)^n$

35c. 8192 squirrels **37.** $b > 1$ **39.** $b \leq 0; b = 1$

41. no **43.** $y < -\frac{1}{3}x + 3$ and $y \leq x + 4$

45. $y = -2x + 3$; consistent dependent

47. ∅; inconsistent **49.** E

Section 7.6

1. false **3.** 4 **5.** $\frac{1}{5}$

7. geometric **9.** neither **11.** arithmetic

13. 50, 250, 1250, 6250 **15.** 8, 80, 800, 8000

17.

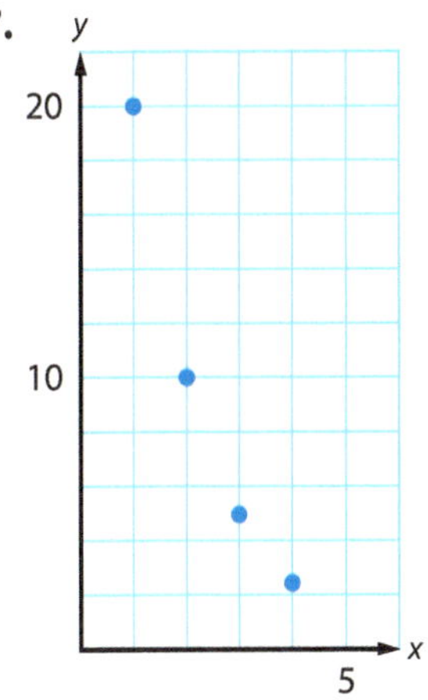

19.

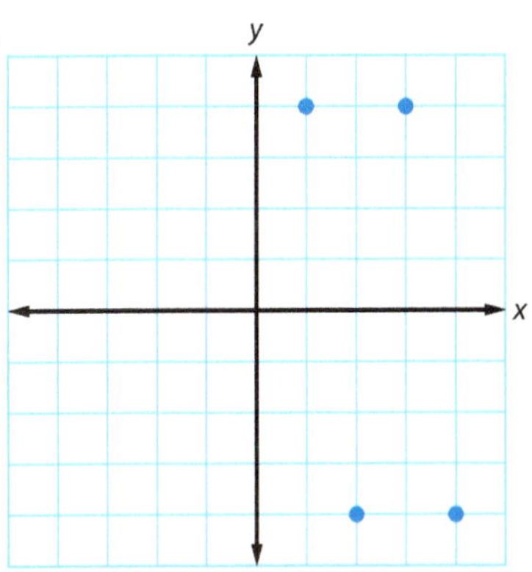

21. $a_n = -2a_{n-1}; a_1 = 3$ **23.** $a_n = 2\left(\frac{1}{2}\right)^{n-1}$

25. $a_n = -15(2)^{n-1}; a_{14} = -122{,}880$

27. $a_n = 500\left(-\frac{3}{5}\right)^{n-1}; a_7 = 23.328$

29. $a_n = 256\left(\frac{3}{2}\right)^{n-1}; a_9 = 6561$

31. The common ratio is 3. **33a.** 6 ft, 3.6 ft, 2.16 ft

33b. $a_n = 0.6a_{n-1}; a_1 = 6$ ft **33c.** 5th bounce

35a. 2.3×10^{10} represents the number of transistors in 2020. 1.41 represents the yearly growth factor.

35b. 3.59×10^{11} **43.** $y = 2x + 8$

45. boat: 25 mi/hr; current: 5 mi/hr

47. $7500 at 6% and $2500 at 2%

49. $\frac{9x^8}{y^{11}}$ **51.** C

Application Problems

3. 0.85 sec **5.** 0.02 sec **7.** 1×10^{21}

9a. $\approx 2.14 \times 10^{13}$ bytes **9b.** ≈335 phones

11. $4.602 \times 10^9 = 4{,}602{,}000{,}000$ users

13. $1.25 \times 10^3 = 1250$ messages per second

Chapter Review

1. $-18m^3n^{12}$ **3.** $4m^2np^4$ **5.** y^{2a}

7a. $-3m^{-4}r^5s^{-6}$ **7b.** $-\frac{3}{4}x^{-3}y^{-2}$ **9a.** 3.19×10^{-4}

9b. 3.2×10^4 **11.** 3×10^2 **13.** 8.3×10^3

15. 1×10^{-10} m **17.** 2 units down **19.** (1, 0)

21. (−3, 5) **23.** $f(x) = (x + 5)^2 + 3$

25. $y = (x + 1)^3 - 2$ **27a.** exponential **27b.** power

Table values will vary for exercises 29–31.

29.

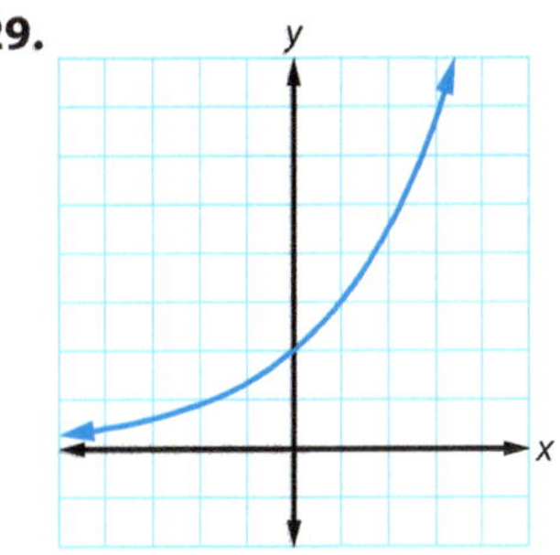

31.

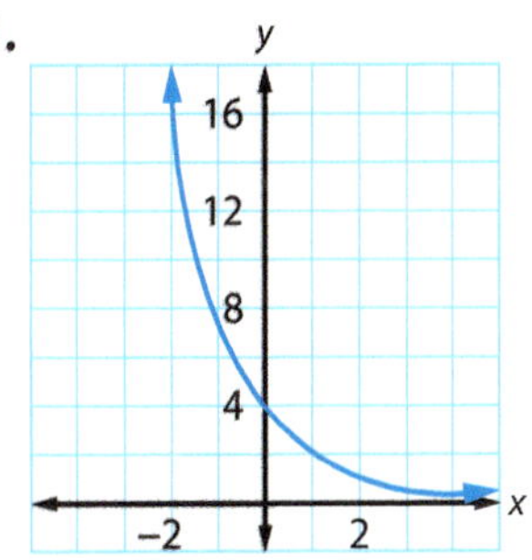

33. $0 < b < 1$ **35.** $y = 1.2^x$; 144%

37. 1; 50%; 1.5; growth

39. 40 periods; 0.75%; $6741.74

41a. 200%; 3 **41b.** $f(d) = 500(3)^d$

41c. 1,093,500 mosquitoes

43. An arithmetic sequence has a common difference between consecutive terms, but a geometric sequence has a common ratio between terms.

45. 5, −10, 20, −40, 80, −160

47. $a_n = 3a_{n-1}, a_1 = -2; a_n = -2(3)^{n-1}; -39{,}366$

49a. 36 ft, 21.6 ft, 12.96 ft **49b.** $a_n = 36(0.6)^{n-1}$

49c. ≈0.36 ft

CHAPTER 8: POLYNOMIAL OPERATIONS

Section 8.1

1. 7 **3.** 0 **5.** binomial; 8

7. polynomial; 3 **9.** not a polynomial

11. binomial; 3 **13.** trinomial; 5

15. $-2x^2 + 3x + 5$; 2; trinomial

17. $-6x^5 + \frac{1}{3}x^2 - 4x + 12$; 5; polynomial

19. $6x^4 - 7x^2y + 2xy$; 4; trinomial

21. 133 **23.** −3 **25.** −23

27. 9 **29.** −33 **31.** −3240

35. 1.15 **37.** $n + 1$ **39.** C

41. D **43.** B **45.** 3%

47. B

Section 8.2

1. $19m - 6$ **3.** $7m^2 + 10m - 5$ **5.** $3a + 8$

7. $-3x + 10y - 6$ **9.** $9x + 3y$

11. $5x^3 + 9x^2 + 2x - y + 15$ **13.** $x + 2y - 14$

15. $-a^3 + 4b^2 + 9c$ **17.** $-8y^2 + 7y + 11$ **19.** $-9x^2 + 10x + 3$

21. $\frac{7}{3}x - \frac{3}{5}y + \frac{8}{7}$ **23.** $a^3b - 11a^2b^3 - 4ab - 9$

25. $12a^2 - 26ab - 9b^2$ **27.** $-4x + 7y + 47$

29. $14x + 3$ **31.** $3x^2 + 8x + 13$ **33.** $2x + 1$

37. $\frac{1}{6}x^2 + \frac{1}{8}y^2$ **39.** $7n + 27$

41. $8 - 3 = 5$, but $3 - 8 = -5$. **43.** If $a = b$, then $a + c = b + c$.

45. $15x^5$ **47.** x^2y **49.** C

Section 8.3

1. $x(x + 3) = x^2 + 3x$

3. $(x + 4)(2x + 1) = 2x^2 + 9x + 4$

5. $-20ab - 40b^2$ **7.** $9x^2 - 3x - 30$

9. $12x^3 + 12x^2y - 15xy^2 - 9y^3$ **11.** $-56k^3 - 63k^2 - 42k$

13. $12a^6 + 15a^3 - 54a^2$ **15.** $y^2 - 7y + 12$

17. $c^2 + 5c - 24$ **19.** $-24x^5y - 56x^4y^2 - 80x^3y$

21. $4x^3 + 2x^2y + 8x$ **23.** $-8b^4 + 24b^3 + 16b^2 - 48b$

25. $-12x^3y - 9x^2y^2 + 16xy^2 + 12y^3$

27. $6x^3 - 3x^2 - 6x + 48$ **29.** $x^3 - 27$

31. $6x^4 - 29x^3 + 48x^2 - 29x - 28$

33. $x^3 - 15x^2 + 71x - 105$ **35.** $2x^3 + 13x^2 + 30x + 40$

37. $2b^2 + 15b + 7$ **39.** $-x^2 + 5x - 26$ **43.** $x^2 - y^2$

45. $x = \frac{2}{3}, 4$ **47.** $(4, -5)$ **49.** 1.53×10^6

51. 1.0×10^8 mi **53.** B

Section 8.4

1. $x^2 + 7x + 10$ **3.** $b^2 - 7b + 12$ **5.** $v^2 - 4v - 5$

7. $t^2 - 13t + 42$ **9.** $y^2 - 25$ **11.** $a^2 - 49$

13. $c^2 - 2c + 1$ **15.** $x^2 + 6x + 9$ **17.** $15z^2 - 8z - 12$

19. $a^2 + 12a + 36$ **21.** $49g^2 + 140g + 100$

23. $10z^2 - 23z - 42$ **25.** $9h^2 - 16$ **27.** $6v^2 - 24$

29. $3x^4 + 3x^2 - 6$ **31.** $6b^6 + 17b^4 + 12b^2$

33. $20x^6 - 24x^5 + 10x^3 - 12x^2$ **35.** $x^2 + 13x + 42$

37. $144x^2 + 82x + 8$ **39a.** $15x^2 + 96x + 144$

39b. $x = 4$; 32 ft^2 **39c.** 20 ft × 12 ft **41.** $x^2 + 46$

43. $5x^2 + 3x - 1$ **45.** binomial; 6 **47.** $17a - 17b + 1$

49. $25f^2g + 10fg^2 - 11fg + 7$ **51.** $16x^2 - 4x - 42$

53. D

Section 8.5

1. $2ab$ **3.** − **5.** $y^2 - 8y + 16$

7. $4a^2 + 12a + 9$ **9.** $x^2 - 18xy + 81y^2$

11. $y^2 - 64$ **13.** $25z^2 - 9$ **15.** $4d^2 - 121$

17. $(30 - 3)(30 + 3) = 891$ **19.** $(30 - 1)^2 = 841$

21. $(50 + 2)^2 = 2704$ **23.** $16x^2 - 9y^2$

25. $400x^2 - 81y^2$ **27.** $144a^2 + 8a + \frac{1}{9}$

29. $0.64x^2 - 17.6xy + 121y^2$ **31.** $\frac{100}{9}x^4 - 10x^2y + \frac{9}{4}y^2$

33. $-4b + 8$

35. The correct answer is $x^2 - 2x + 1$.

37. $y^3 + 2y^2 - 36y - 72$ **43.** 1

45. $x = \frac{25}{8}$ **47.** $y = \frac{101}{4}$ **49.** (0, 3)

51. B

Section 8.6

1. $x + 5$ **3.** $4x^2 - 1$ **5.** $x^2 + 5$

7. $a - 6b$ **9.** $8a^2 - 4a + 3b$

11. $-\frac{15}{2}x^2 + \frac{3}{2}x - \frac{27}{2}$ **13.** $3n^2 + 5n - 1$

15. $\frac{1}{3}x^3 + 3x^2 - \frac{5}{3}$ **17.** $x - 2$ **19.** $3x + 1$

21. $a + 10 + \frac{72}{a - 6}$ **23.** $24x - 11 + \frac{2}{x + 1}$

25. $3y - 5$

27. The quotient is $a^2 + a + 7 - \frac{7}{a - 5}$.

29. lower **31.** $x - 4 + \frac{14}{x + 3}$

33. $x^2 + 2x + 11 + \frac{44}{x - 4}$ **35.** $x^2 + 3x + 9$

37. $x^2 - 12 + \frac{64}{x^2 + 4}$ **39.** $2x + 2$

41. $x^3 - 4x - 2 + \frac{16x + 14}{x^2 + 4}$ **43.** $2x + 2$

45. $-\frac{5}{2}$ **47.** −4 **49.** $y = 2x$

51. A **53.** A **55.** B

Application Problems

1. 113,755 **5.** 19,142,046 **7.** 371,425,443

9. 2160 **11.** 1.3%; 1%

Chapter Review

1. binomial; 1 **3.** trinomial; 3 **5.** 36

7. 12 **9.** $9x^2 + 3x - 14$

11. $-a^2 + 6ab - 13b^2$ **13.** $13x^2 - x + y$

15. $3x^3 - 15x^2 + 12x$ **17.** $-6a^7b^2c^2 - 21a^5b^2c^7$

19. $x^3 + 8$ **21.** $x^2 - 2x - 24$ **23.** $x^2 - 20x + 99$

25. $6x^2 + x - 35$ **27.** $12x^2 + 32x + 20$ **29.** $y^2 - 36$

31. $x^2 - 4xy + 4y^2$ **33.** $16a^2 - 40ab + 25b^2$

35. $144c^{10} + 360c^5d^4 + 225d^8$ **37.** $5x - 7$

39. $4xyz + 6y^3$ **41.** $6x + 3$ **43.** $x + 2 + \frac{9}{x-2}$

45. $x - 1$ **47.** $x^3 - 2x^2 + 4x - 3$

49. $11x^2 + 12x + 3$ **51a.** $-2x + 11$ **51b.** $-2x + 8.5$

51c. $4x^2 - 39x + 93.5$ **51d.** $4x^3 - 39x^2 + 93.5x$

CHAPTER 9: FACTORING POLYNOMIALS

Section 9.1

1. 15 **3.** $8x$ **5.** x^2y^2

7. $3(x + y)$ **9.** $8(x - 4y)$ **11.** $a(a - 1)$

13. $7(x^2 - 7xy + 4y^2)$ **15.** prime

17. $3ab(2a + b)$ **19.** $13a^2b(7ab + 1)$

21. prime **23.** $14x(4x + 3y - 2)$

25. $7a(a^4 - 12a^2 + 3)$ **27.** $(x - 2)(x + 8)$

29. $(b + c)(5 + a)$ **31a.** the Distributive Property

33. 12 children; 9 rows **35.** $3x - 1$

37. $3x^2y^4z^3(2x^2y - 9xy^2z + 6z^2)$

39. $8z(x + 1)(3x + 7y)$

41. $V = \pi r_1^2H - \pi r_2^2H; \pi H(r_1^2 - r_2^2)$

43. $D = \{-3, -1, 2, 4\}$ **45.** yes

47. $-3, -1, 2, 4$ **49.** $3x^2 + 15x - 42$ **51.** A

Section 9.2

1. C **3.** A **5.** $(a + 3)(a + 5)$

7. $(a - 3)(a - 1)$ **9.** $(b + 6)(b - 5)$ **11.** $(a + 4)(a - 2)$

13. prime **15.** $(x - 6)(x - 5)$ **17.** $(y + 5)(y + 2)$

19. prime **21.** $(x + 4)(x + 14)$ **23.** $(x - 3)(x + 18)$

25. $3(x - 9)(x + 1)$ **27.** $3(x + 6)(x + 2)$ **29.** $x^2(x + 8)(x - 2)$

31. $(x + 3y)^2$ **33.** $(x - 3y)(x - 15y)$

35. prime

37. The correct answer is $(x - 12)(x + 2)$.

39. $(x - 4)(x - 14)$; 10 units **41.** $-ax(x + 3)(x - 48)$

43. $(a^3 - 9b^2)(a^3 + 13b^2)$ **45.** 8

47. 6 **49.** $\frac{2}{3}x^2 - 3xyz + \frac{8}{9}x$ **51.** 62 mi/hr

53. B

Section 9.3

1. $(2x + 3)(x + 4)$ **3.** $(x - 4)(3x - 2)$ **5.** $(3x - 7)(x + 4)$

7. $(3x + 4)(2x - 1)$ **9.** $(3c + 2)(5c + 3)$

11. $(2a + 1)(3a - 4)$ **13.** $(5x + 1)(x - 3)$

15. $(5x - 7y)(x + 3y)$ **17.** $3(2x + 5)(x - 2)$

19. $8(2a - b)(a + 3b)$ **21.** $(a - 4b)(a + 7b)$

23. $-(x - 14y)(x + 7y)$ **25.** $8(a + 1)(a - 14)$

27. $17xy(3x - 5)(x + 1)$ **29.** $-2(2x - 3)(7x + 4)$

31. $3(x - 11y)(x + 14y)$ **35.** $(3x + 4)(5x + 4)$

37. 5 : 3 **41.** $x = 9$ **43.** (6, 2)

45. 1.225×10^{-21} **47.** 20 hr **49.** D

Section 9.4

1. +, − or −, + **3.** 2, 2 **5.** $6a$, 7

7. Yes; $3(x)$ doubled $= 6x$.

9. No; x^2y is not a perfect square.

11. No; $2x(5)$ doubled $= 20x$. **13.** 8 is not a perfect square.

15. $(ax - y)(ax + y)$ **17.** $(x - 8)^2$ **19.** $(4a + x)^2$

21. $(a - 7)^2$ **23.** $3(x - 7)(x + 7)$ **25.** $(6x + 1)^2$

27. $a^2(a - b)(a + b)$ **29.** $(3x^4 + 5z)^2$

31. $x^5(7 - 3x)(7 + 3x)$ **33.** prime

35. $3(7a^5 - 13b^4)(7a^5 + 13b^4)$ **39.** $3xy(3x^2 - 4y)^2$

41a. $a^3 + b^3$ **43.** $-5x(8y^2 - y - 4)$

45. $(y - 5)(y - 6)$ **47.** $(2x + 3)(4x + 1)$

49. $(6r - p)(r + 5p)$ **51.** D

Section 9.5

1a. II **1b.** I **1c.** IV

1d. III **3.** $x(x - 1)^2$

5. $2a^6(2a - 1)(2a + 1)$ **7.** $a^3(3a^2 - 4)$

9. $(m^5 + 1)^2$ **11.** $(4 - m)(b + c)$ **13.** $(x + 2)(t - 3w)$

15. $(m + n)(2a + b)$ **17.** $3x(x - 4)(x - 7)$

19. $(ax - 3)(ax + 3)(a^2x^2 + 9)$ **21.** $5(3y + 4)(2y - 5)$

23. $2(5x + 2)(7x - 6)$ **25.** $(x + 3)(y + 5)$

27. $(x - 3y)(x - 2)$ **29.** $(x - y)(x + y)(2a + b)$

31. $(a - 3)(a + 3)(a + 2)$ **33.** $3(x - 4)(x + 4)(x - 2)$

35. $(x - 2)(x + 3)(x - 4)$ **37.** $(y + 1)(y - 3)(2y - 5)$

39. $(x + y - 3)(x + y + 3)$

41. plane: 183 mi/hr; wind: 8 mi/hr

43. 1.875 gal of the 40% saline solution and 1.125 gal of water

45. $f(x) = (x + 6)^3 - 5$ **47.** $f(x) = (x + 11)^2 + 8$

49. D

Application Problems

1. about 12 km; about 1 km

3.

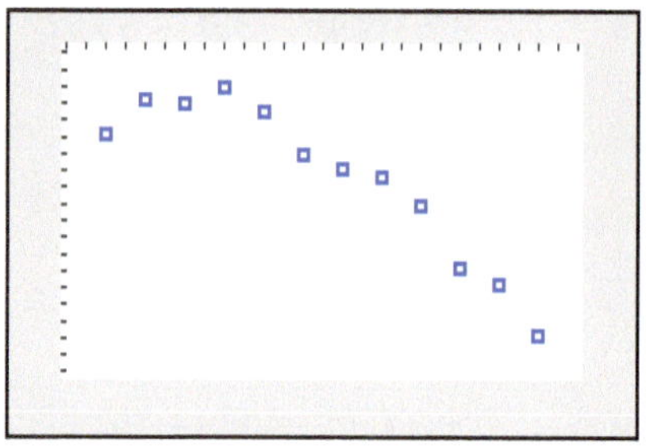

5.

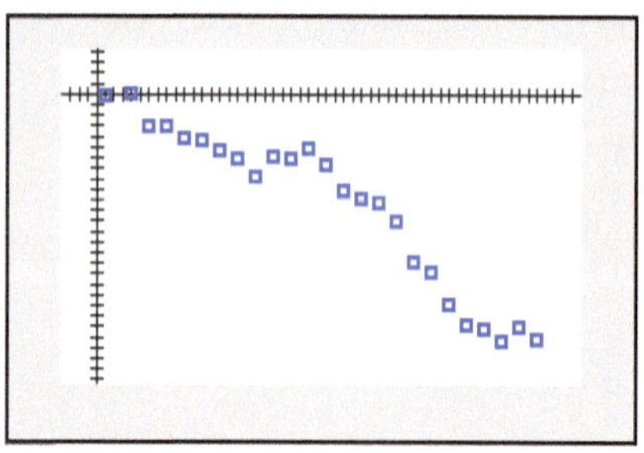

7. 17,952 days ≈ 49 yr

9a. $y \approx -3.661x^2 + 30.863x + 196.339$

9b.

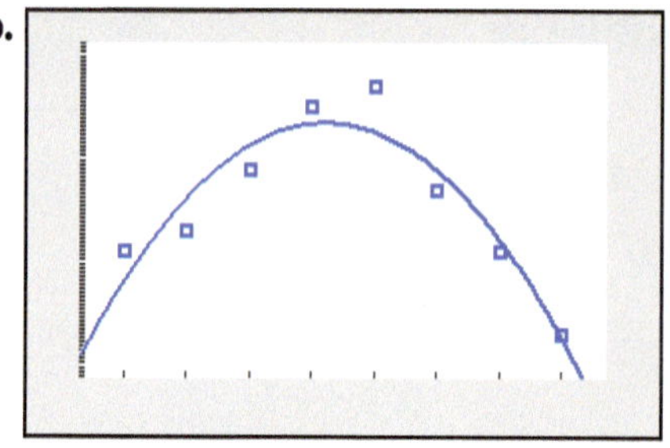

11a. ≈0.44 = 44% grade **11b.** ≈0.20 = 20% grade

Chapter Review

1. $r(3r + 5)$ **3.** $4x^2(x - 2)$ **5.** $(y - 7)(y - 8)$

7. $(2d + 7)(d + 1)$ **9.** $(x + 7)(x + 2)$

11. $-(5b + 3)(3b - 2)$ **13.** $y(x + 8)(x - 7)$

15. $2(9x - 5)(x - 5)$ **17.** $(a - 3)(a + 3)$

19. $(3x - 5)^2$ **21.** $2(x - 9)(x - 6)$

23. $(y + 5z)(y + 12z)$ **25.** $(2 - x)(9x - 25)$

27. prime **29.** $5(b - 5)^2$ **31.** $(2a - 3)(a + 6)$

33. prime **35.** $(3a - 7)(a - 1)$

37. $2(3y - 5)(2y + 7)$ **39.** $(2x^2 - 3)(5x + 7)$

41. $(w^3 + 3z^2)^2$

43. $(2a - 1)(2a + 1)(4a^2 + 1)(16a^4 + 1)$

45. $(x - 1)(xy + x + 1)$ **47.** $h = -(16t + 1)(t - 4)$

49. 35 in.^{2} **51.** B

CHAPTER 10: RADICALS

Section 10.1

1. $5^{\frac{1}{2}}$ **3.** $3^{\frac{1}{5}}x^{\frac{2}{5}}y^{\frac{3}{5}}$ **5.** $\sqrt{3y}$

7. $\sqrt[9]{16y^3z^6}$ **9.** $2\sqrt{3}$ **11.** $3\sqrt{11}$

13. $2\sqrt[3]{4}$ **15.** $10\sqrt{2}$ **17.** $4\sqrt[3]{5}$

19. cannot be simplified **21.** x^4

23. $|y|^5$ **25.** $3|x|\sqrt{2y}$ **27.** $3x^2|yz^3|$

29. $\frac{|x^3y^3|}{10}$ **31.** $x|y|z\sqrt[4]{10xy^2z^3}$

35. 240 mi/hr **37.** 60 mi/hr **39.** 3125

41. cannot be simplified **43.** 5400

45. $-10a^3b^4c^6$ **47.** $32x^8$

49. $(2x - 17)(2x + 17)$ **51.** $(m + 14)^2$

53. C

Section 10.2

1. $\sqrt{15}$ **3.** $2\sqrt{6}$ **5.** $2\sqrt{15}$

7. $-60\sqrt{2}$ **9.** $14\sqrt{5}$ **11.** $\sqrt{590}$

13. $6\sqrt[3]{10}$ **15.** $4\sqrt[4]{8}$ **17.** $xy^2\sqrt{6x}$

19. x^2y^2 **21.** $6x^4y^2\sqrt{35x}$ **23.** $5ac^3\sqrt[3]{2a^2b^2}$

25. $\sqrt[6]{243}$ **27.** $3\sqrt[6]{3}$ **29.** $70\sqrt{2}$ ft^2

31. $3x^5$ units^3 **33.** $8\sqrt{14}$ m^2 **35.** $57\sqrt[3]{3}$

37. $x\sqrt[20]{512x^7}$ **39a.** $\sqrt[mn]{a^{m+n}}$ **41.** $-\frac{r^2s}{7}$

43. $\frac{x^{10}}{4}$ **45.** $n^3 + 5n - 1$ **47.** $a^{\frac{7}{8}}b^{\frac{1}{2}}c^{\frac{1}{4}}$

49. C

Section 10.3

1. $\frac{2}{5}$ **3.** $\frac{\sqrt{5}}{2}$ **5.** $\frac{\sqrt{3}}{2}$

7. $\frac{\sqrt{3}}{15}$ **9.** $\frac{3\sqrt{13}}{13}$ **11.** $\frac{\sqrt{6}}{6}$

13. $\frac{\sqrt{6}}{3}$ **15.** $\sqrt{3}$ **17.** $\frac{\sqrt[3]{4}}{2}$

19. $\frac{\sqrt[3]{2}}{2}$ **21.** $\frac{3\sqrt[3]{25}}{4}$ **23.** $\frac{\sqrt{2x}}{2xy}$

25. $\frac{3z^2\sqrt{35txz}}{t}$ **27.** $\frac{y\sqrt[3]{3x^2}}{x}$

29. No; the correct answer is $\frac{\sqrt{6}}{2}$.

33. $3\sqrt{10}$ cm **35.** $\frac{\sqrt{14}}{2} \approx 1.9$ m^2 **37.** $\frac{2\sqrt{2b}}{a^2b^3}$

39. $\frac{\sqrt[4]{14x}}{3x}$

41.

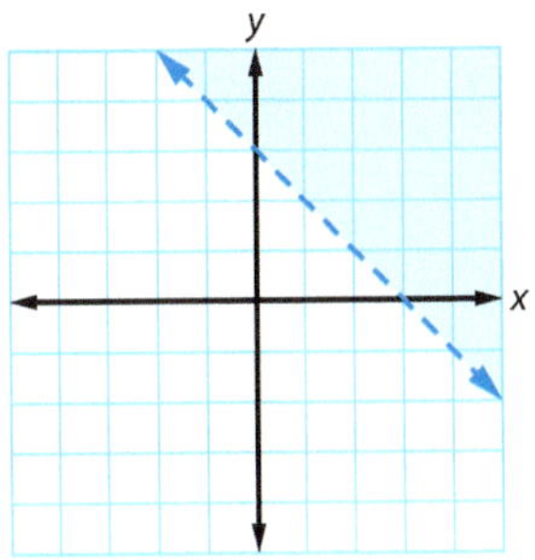

43. $4x^2 - 4xy + y$ **45.** $2x^3y - 11x^2y^3 - 4xy - 15$

47. $2(b - 7)^2$ **49.** E

Section 10.4

1. radicand; index **3.** $6\sqrt{3}$ **5.** $2\sqrt{3} + 4\sqrt{2}$

7. $5\sqrt{2}$ **9.** $4\sqrt{3}$ **11.** $3\sqrt{2}$

13. $-\sqrt[3]{4}$ **15.** 5; 12; 17; 13

17. $3\sqrt{5}$; $5\sqrt{3}$; $3\sqrt{5} + 5\sqrt{3}$; $2\sqrt{30}$

21. 0 **23.** $2\sqrt[3]{5} - 5\sqrt[3]{7}$ **25.** $5\sqrt[3]{4} + 16$

27. $3\sqrt{ab} + 4\sqrt{a}$ **29.** $5a\sqrt{2a} - 3\sqrt{6}$ **31.** $\frac{5x}{12}$

33. $-2a\sqrt[3]{3} + \sqrt{3}$ **35.** $16x\sqrt{5} + 48y\sqrt{3}$ units

37. $2\sqrt{3} + 9$ units^2 **39.** $-5xy\sqrt[3]{9x^2y^2}$ **41.** $2a\sqrt{ab}$

43. $x = 14$ **45.** $x - 7$ **47.** $(x - 16)^2$

49. $(x - 4)(x - 3)(x + 3)$ **51.** E

Section 10.5

1. 5 **3.** $\sqrt{41}$ **5.** $2\sqrt{15}$

7. $4\sqrt{2}$ **9.** yes **11.** no

13. (4, 5) **15.** 5 **17.** 13

19. $\sqrt{71}$ **21.** $c = 4$ **23.** $c = 2\sqrt{6}$

25. (−10, −14) **27.** $(\sqrt{2}, -7\sqrt{5})$

29. $\sqrt{10} + \sqrt{17} + 5$ **31.** $2\sqrt{5} + 2\sqrt{17} + \sqrt{10}$

33. $\sqrt{0.45} \approx 0.67$ mi **35.** $\sqrt{157} \approx 12.5$ nautical miles

37. 15 ft **39a.** $\sqrt{l^2 + w^2}$

39b. $\sqrt{l^2 + w^2 + h^2}$ **41.** (44, 240, 244)

43. $5|x|^3\sqrt{2}$ **45.** $6\sqrt{5}$ **47.** $\frac{5\sqrt{3xy}}{y}$

49. $4x\sqrt[3]{2x}$ **51.** A

Section 10.6

1. $\sqrt{6} - \sqrt{3}$ **3.** $7\sqrt{3} + \sqrt{14}$ **5.** 1

7. $17 - 2\sqrt{70}$ **9.** $71 + 16\sqrt{7}$

11. $12 + 6\sqrt{5} - 4\sqrt{10} - 10\sqrt{2}$ **13.** $-3 - \sqrt{5}$

15. $2 - \sqrt{3}$ **17.** $\sqrt{33} - \sqrt{22}$ **19.** $\sqrt{2} + 1$

21. $-4\sqrt{2} - 4$

23. $5\sqrt{105} + 10\sqrt{7} + 25\sqrt{3} + 10\sqrt{5}$

25. $x^2\sqrt{3} + 2\sqrt{3x} - 5x^2 - 10\sqrt{x}$

27. $5\sqrt{3} - 6\sqrt{15} - 5\sqrt{2} + 6\sqrt{10}$

29. $12\sqrt{5}$ **31.** $\sqrt{3} + 1$ **33.** $\frac{-5\sqrt{2} - 5\sqrt{6}}{4}$

35. $4\sqrt{7} - 4\sqrt{3}$ **37.** $42\sqrt{10} + 12\sqrt{30}$ units^2

39. $\frac{-\sqrt{6} - 2\sqrt{5} - \sqrt{15} - 5\sqrt{2}}{7}$

41. $x^2 - 22x\sqrt{x} + 177x - 616\sqrt{x} + 784$

43. $3\sqrt{7}$ **45.** $2\sqrt{x - 3}$ **47.** 1.5×10^2

49. $16x^2 - 56x + 49$ **51.** $2(a - 11b)(a + 2b)$

53. D

Section 10.7

1. $x = 16$ **3.** $x = 62$ **5.** $x = 36$

7. $x = -5$ **9.** $x = -1$ **11.** $x = 100$

13. no solution **15.** $y = 124$ **17.** $x = 19$

19. $x = \frac{73}{16}$ **21.** $\sqrt{2x} = 12$; 72 **23.** 180 ft

25. yes **27.** 24.8 cm **29.** $l = \frac{gT^2}{4\pi^2}$

31. 968 ft **33.** $d = \frac{484V^2}{225g}$ **35.** $x = \frac{1}{2}$

37. $x = 1$ **39.** $x = 51$

41. $f(x) = (x + 3)^2 + 2$ **43.** $x^2 - 5x + 2$

45. $(x + 3)(x - 3)$ **47.** $\frac{7\sqrt{3}}{10}$ **49.** B

Section 10.8

1. $x \geq 0$ **3.** $x \geq 0$ **5.** $x \geq 5$

7. $x \geq -2$

Table values will vary for exercises 9–15.

9.

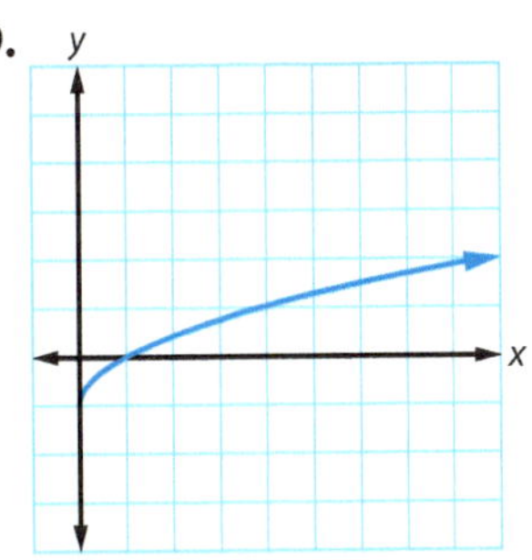

11.

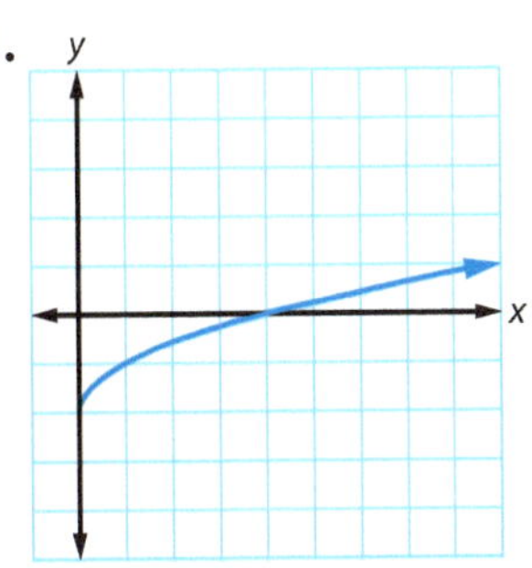

13.

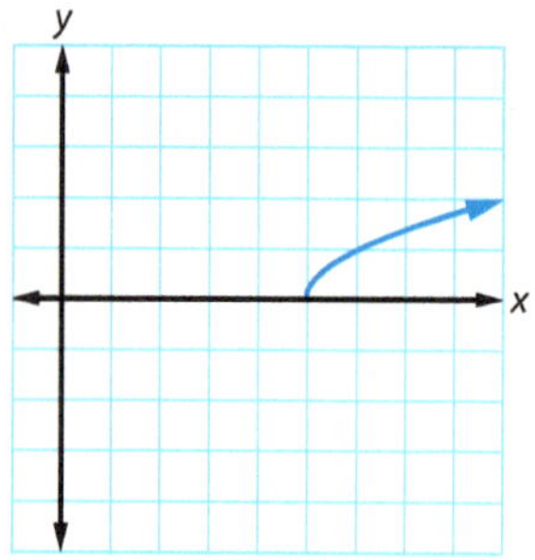

15.

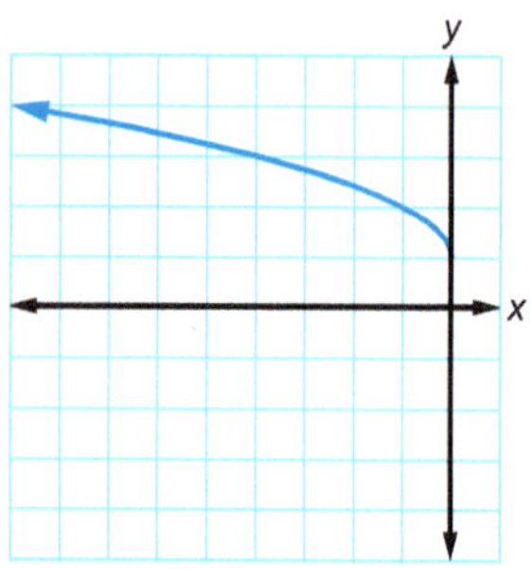

17. F **19.** A **21.** C

23. $f(x) = \sqrt{x} + 1$ **25.** $f(x) = \sqrt{x - 2} - 2$

27. $f(x) = \sqrt{-x} + 4$

29.

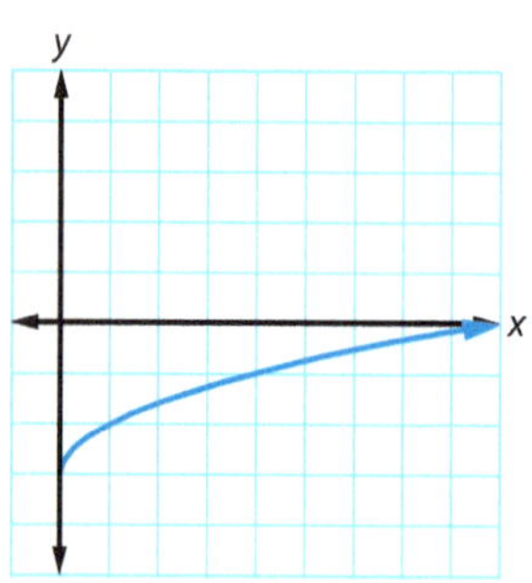

31.

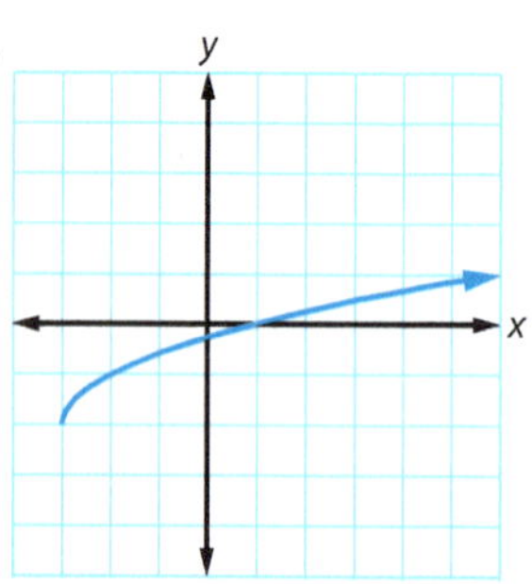

33.

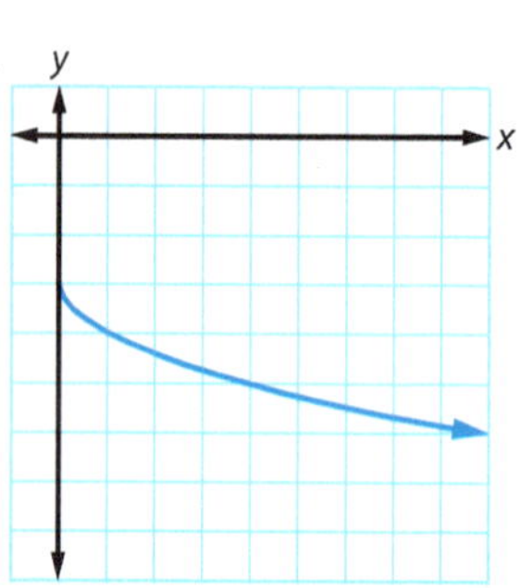

37. $\mathbb{R}$ **39.** $y = -\sqrt{x - 3} + 2$

41. $2\sqrt{13} \approx 7$ units **43.** $2\sqrt{5}$; (7, −2)

45. $x(x - 7)$ **47.** $(2v - 3w)(2v + 3w)$

49. C

Application Problems

1. 1.618034

3. a number that cannot be expressed as a ratio of integers

5a. 1.618; 1.618 **5b.** yes **7a.** $\frac{3}{2}$; 1.5

7b. $\frac{5}{3}$; 1.667 **7c.** $\frac{8}{5}$; 1.6 **7d.** $\frac{13}{8}$; 1.625

9. 0.618034 **11.** 1.616 **13.** 1

Chapter Review

1. $2^{\frac{1}{2}}x^{\frac{3}{2}}$ **3.** $a\sqrt[3]{9a^2}$ **5.** −2

7. $2\sqrt{21}$ **9.** $2|a|b^2\sqrt{3a}$ **11.** $3\sqrt[3]{7}$

13. $6ab^2\sqrt{35a}$ **15.** $\sqrt[6]{3125}$ **17.** $\frac{\sqrt[3]{3}}{3}$

19. $\frac{x^4}{2y^3}$ **21.** $5\sqrt{2} + 2\sqrt{3}$

23. $6a\sqrt{2a} - 3a^2\sqrt{2a}$ **25.** $2g\sqrt{5} - 15\sqrt{6g}$

27. $53 - 10\sqrt{6}$ **29.** $\frac{18 - 6\sqrt{2}}{7}$ **31.** $-\frac{3 + 3\sqrt{6}}{5}$

33. 40 cm^3 **35.** $2\sqrt{34} \approx 11.66$ units

37. $9\sqrt{2}$ **39.** $P_2\left(-\sqrt{5}, -6\sqrt{10}\right)$

41. $x = 620$ **43.** $x = -3$ **45.** 65.7 ft

Table values will vary for exercises 47–49.

47. $x \geq 0$

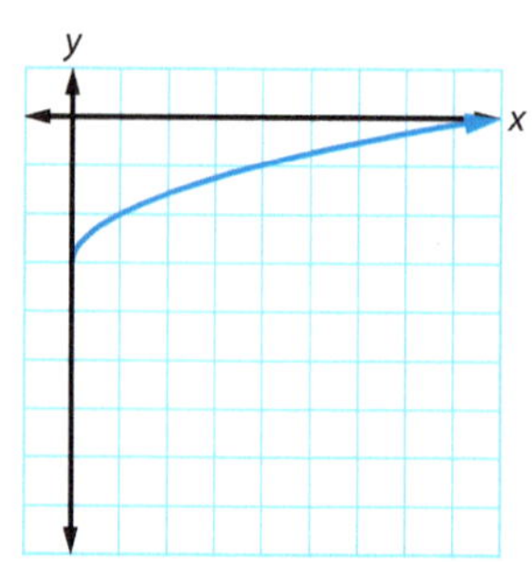

49. $x \leq 0$

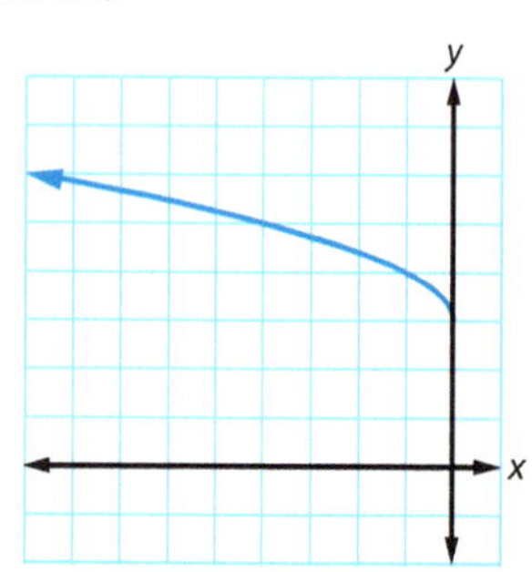

51. $f(x) = \sqrt{x + 5} - 3$

CHAPTER 11: QUADRATIC EQUATIONS & FUNCTIONS

Section 11.1

1. $x = 2, -4$ **3.** $z = 0, -7$ **5.** $x = 0, -1, 10$ **7.** $b = -\frac{1}{4}, -\frac{2}{3}$ **9.** $y = -3$ **11.** $x = \pm 3$ **13.** $a = -9, 4$ **15.** $x = 0, 1$ **17.** $c = 0, 7$ **19.** $x = 2, 8$ **21.** $p = -4, 6$ **23.** $y = \pm 13$ **25.** $x = \pm\frac{3}{2}$ **27.** $x = -3, 24$ **29.** $t = -\frac{3}{2}, \frac{4}{5}$ **31.** $r = -2, 0, 1$ **33.** The answer is $x = -1, 3$. **35.** 15 and 16 or −16 and −15 **37.** 8 ft; 20 ft × 28 ft **39.** 14 **41.** $x = -\frac{6}{5}, 0, \frac{7}{3}$ **43.** $k = 10, 22$ **45.** $k = 1, 4, 11$ **47.** $2\sqrt{33}$ **49.** $b = 20$ **51.** E

Section 11.2

1. $x = \pm 5$ **3.** $x = \pm\sqrt{6}$ **5.** $x = \pm 10$ **7.** $x = \pm 6$ **9.** no real solution **11.** $x = \pm\sqrt{17}$ **13.** $x = \pm 3\sqrt{3}$ **15.** $x = \pm\frac{5\sqrt{2}}{2}$ **17.** $x = 1, -9$ **19.** $x = -2 \pm 2\sqrt{3}$ **21.** $x = \frac{2}{3}, -\frac{2}{9}$ **23.** $x = \frac{10}{3}, \frac{2}{3}$ **25.** 2 real solutions; $x = 8, -14$ **27.** no real solution **29.** 1 real solution; $x = 0$ **31.** 12 ft **33a.** 0 **33b.** $t = \sqrt{\frac{h_i}{16}}$ or $t = \frac{\sqrt{h_i}}{4}$ **33c.** 47 sec **35.** 2.7 sec **37.** 372 ft **39.** $x = \frac{5 \pm 4\sqrt{2}}{2}$ **41a.** $d = |h_c - h_i|$ or $d = |h_i - h_c|$ **41b.** $h_c = -\frac{1}{2}gt^2 + h_i$ **41c.** $h_c = -\frac{1}{2}(32)t^2 + h_i = -16t^2 + h_i$ **41d.** 9.8 m/sec^2; $h_c = -4.9t^2 + h_i$ **43.** $\frac{5}{12}$ **45.** $\left(\frac{1}{2} \cdot \frac{3}{8}\right)^2; \frac{9}{256}$ **47.** $x = \pm\frac{8}{3}$ **49.** $x = -\frac{1}{2}, 0$ **51.** C

Section 11.3

1. 16 **3.** 64 **5.** 1 **7.** $\frac{9}{4}$ **9.** $\frac{121}{4}$ **11.** $\frac{9}{16}$ **13.** $\frac{25}{121}$ **15.** $a = 3, -5$ **17.** $x = 1, -5$ **19.** $x = 12, -2$ **21.** $x = 11, -9$ **23.** $m = \frac{4}{3}, 0$ **25.** $x = 1, -2$ **27.** $x = -10 \pm \sqrt{85} \approx -0.78, -19.22$ **29.** $x = 9 \pm 6\sqrt{2} \approx 17.49, 0.51$ **31.** $x = 4 \pm 2\sqrt{3} \approx 7.46, 0.54$ **33.** $x = 2 \pm \sqrt{7} \approx 4.65, -0.65$ **37.** 10 in. on each side **39.** $x = \frac{-3 \pm \sqrt{5}}{2}$ **41.** $t = \frac{-1 \pm \sqrt{17}}{2}$ **43.** $x^2 + 3x - 5$ **45.** $x^2 - 6x + 2$ **47.** $x = 81$ **49.** $x = \frac{25}{4}$ **51.** C

Section 11.4

1. $x = -3 \pm \sqrt{11}$ **3.** $a = 8, 0$ **5.** $y = -4 \pm \sqrt{19}$ **7.** no real solution **9.** $x = 4 \pm 2\sqrt{5}$ **11.** $y = 1, -\frac{3}{2}$ **13.** $c = \frac{5}{2}, -\frac{1}{2}$ **15.** $x = 1, -\frac{1}{2}$ **17.** $g = \frac{1 \pm \sqrt{17}}{4}$ **19.** no real solution **21.** $k = 2, -\frac{4}{3}$ **23.** $x = -\frac{1}{2}, -1$ **25.** $m = \frac{1 \pm 2\sqrt{3}}{2}$ **27.** $y = \frac{-12 \pm \sqrt{142}}{2}$ **29.** 11 and 21 **31.** 32 ft × 49 ft **33.** 1.5 sec **35.** 13 and 25 **37.** opponent's 13.5 yd line **39.** $x = \frac{-12 \pm \sqrt{139}}{5}$ **41.** $y \approx 0.56, -3.56$; $x \approx \pm 0.75$ **43.** $\frac{c^4}{3a^2}$ **45.** $4x^2 + 22x + 30$ **47.** $4\sqrt[4]{6}$ **49.** $\frac{5 + \sqrt{5}}{2}$ **51.** A

Section 11.5

1. $a = 1, b = 6, c = 5$ **3.** $a = 1, b = 5, c = 6$ **5.** $a = 1, b = 0, c = -6$ **7.** $x = -2, -4$ **9.** $x = -3, -5$ **11.** $a = \pm 3$ **13.** $c = \frac{3}{2}, -1$ **15.** $y = 3$ **17.** $x = 4, \frac{1}{2}$ **19.** $y = \frac{5 \pm \sqrt{17}}{4}$ **21.** $x = \pm\frac{\sqrt{21}}{3}$ **23.** $x = \frac{5 \pm \sqrt{105}}{4}$ **25.** $m = -\frac{2}{7}, -1$ **27.** $k = \frac{1 \pm \sqrt{31}}{3}$ **29.** $x = \frac{7}{2}, -\frac{1}{4}$ **31.** $x \approx 2.39, 0.28$ **33.** $2 \pm \sqrt{3}$ **35.** ≈4.6 in. **39.** $x \approx 1.65, -0.45$ **41.** 108.4 ft **43.** yes, at about 44 ft and 76 ft **45.** irrational **47.** rational **49.** irrational **51.** $x = 9$ **53.** E

Section 11.6

1. A **3.** B **5.** D **7.** 196; A **9.** 0; C **11.** −32; D **13.** $a = \frac{-3 \pm \sqrt{5}}{2}$; irrational **15.** $x = \frac{7 \pm \sqrt{33}}{4}$; irrational **17.** $x^2 + 4x - 32 = 0$ **19.** $8x^2 - 10x + 3 = 0$ **21.** $y = 2, \frac{1}{6}$ **23.** $a = -\frac{4}{3}$ **25.** $x = -4 \pm 3\sqrt{2}$ **27.** $a = -22 \pm \sqrt{482}$ **29.** $x = \frac{8}{3}, -\frac{3}{2}$ **31.** no real solution **35.** 17.9 in., 14.9 in., and 9.9 in. **37.** $y = -\frac{7}{6}$ **39.** $x \approx 7.110, -2.110$ **41.** $9x^2 + 18x + 4 = 0$ **43.** 9 units up

45. 6 units right **47.** $x = \frac{14}{5}$ **49.** ≈8.9 ft

51. A

Section 11.7

1. 1, 0, 4, 16 **3.** −3, −4, 0, 12 **5.** −2, 0, −8, −32

7. −4, −7, 5, 41

9. up; (0, 0); minimum; vertical stretch

11. down; (0, 4); maximum; same

13.

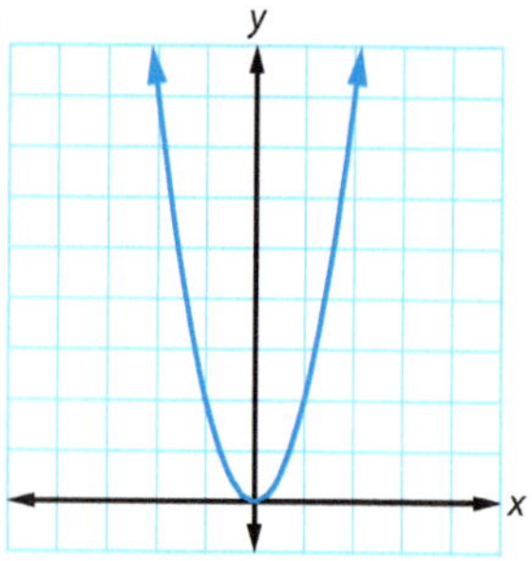

15.

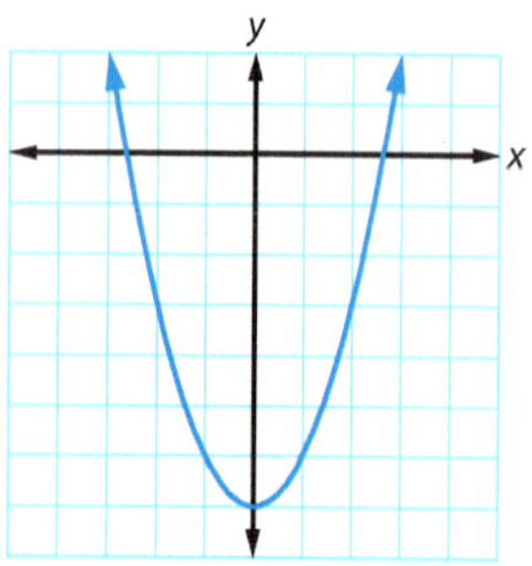

17.

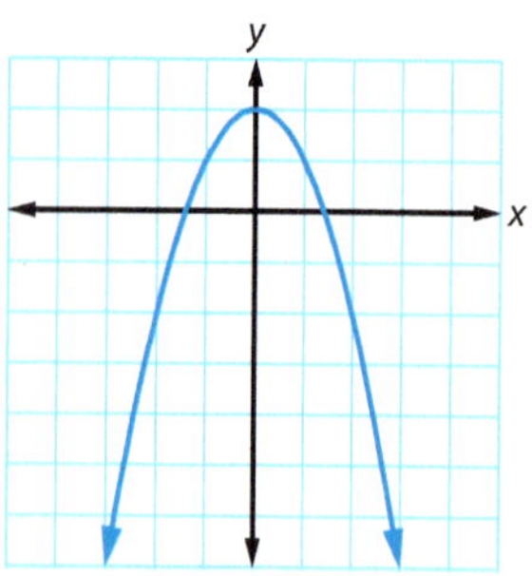

19.

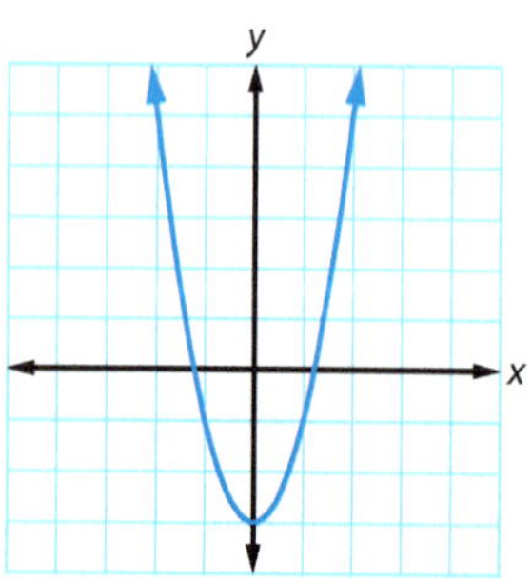

21.

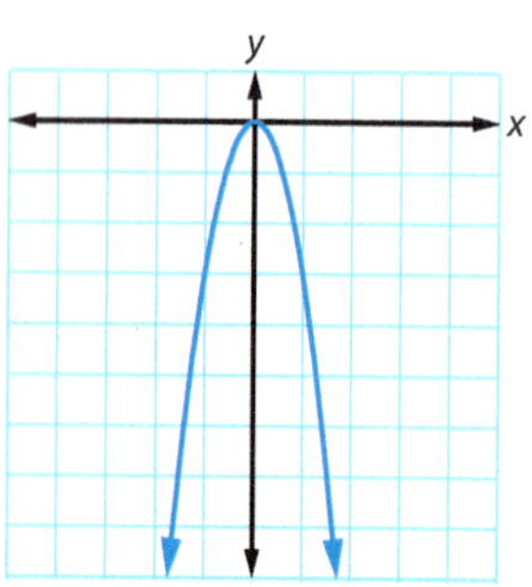

23.

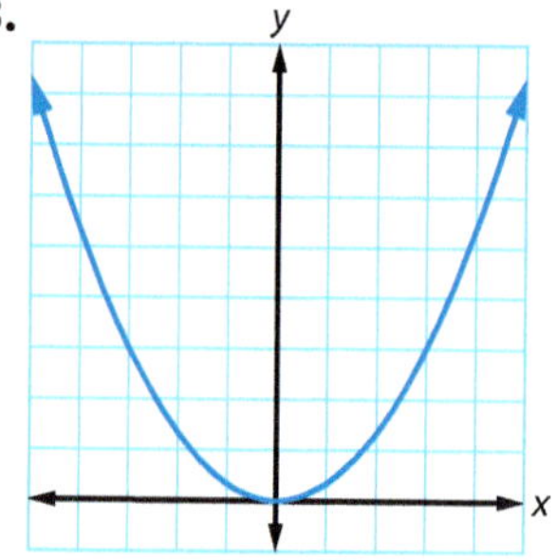

25.

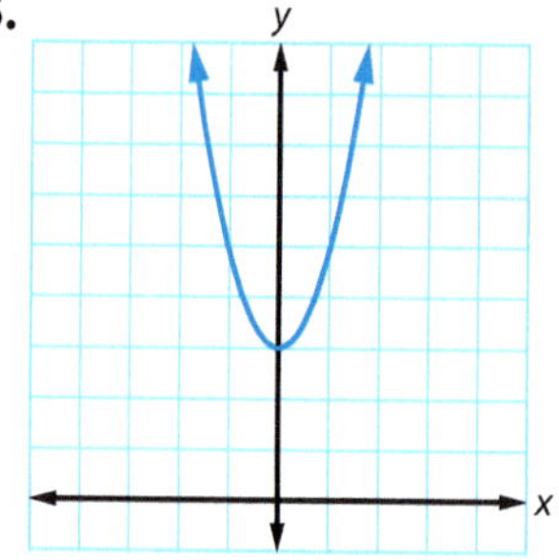

27.

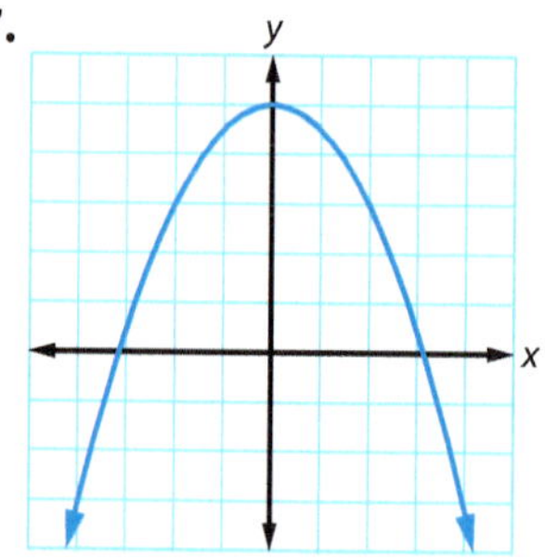

31. origin or (0, 0) **33.** along the *y*-axis; $x = 0$

35. I and II

37a. (0, 0.3), (±1, 0.5), (±2, 1.1), and (±3, 2.1)

37b.

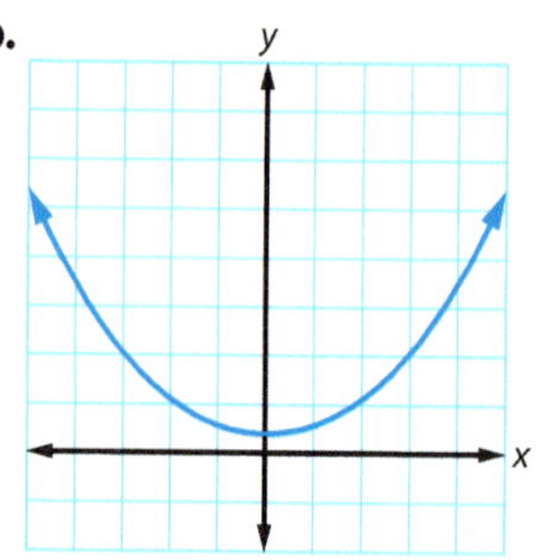

37c. (0, 0.3); minimum **37d.** $x = 0$

39. $f(x) = 3x^2 - 5$ **41.** $f(x) = -\frac{3}{2}x^2 + 15$

43. (5, −2) **45.** (−8, −15) **47.** $\frac{\sqrt{13}}{26}$

49. $\sqrt{3}$ **51.** A

Section 11.8

1. (2, 9); $x = 2$; 1; up; same

3. (2, −4); $x = 2$; 3; up; vertical stretch

5. (−1, 0); $x = -1$; −5; down; vertical stretch

7. (0, 7); $x = 0$; $-\frac{7}{4}$; down; vertical stretch

9. (4, −7); $x = 4$; 1; up; same

11.

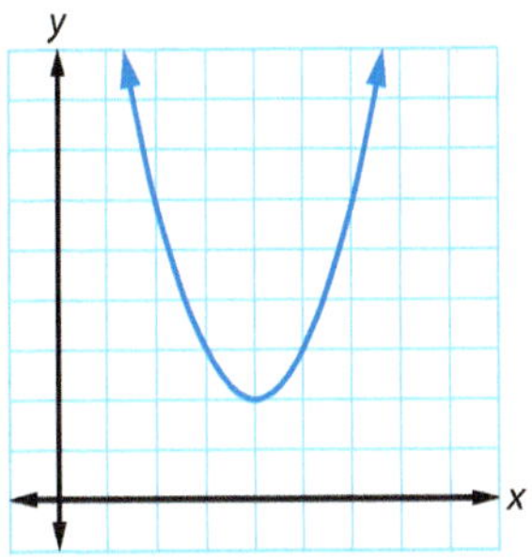

13.

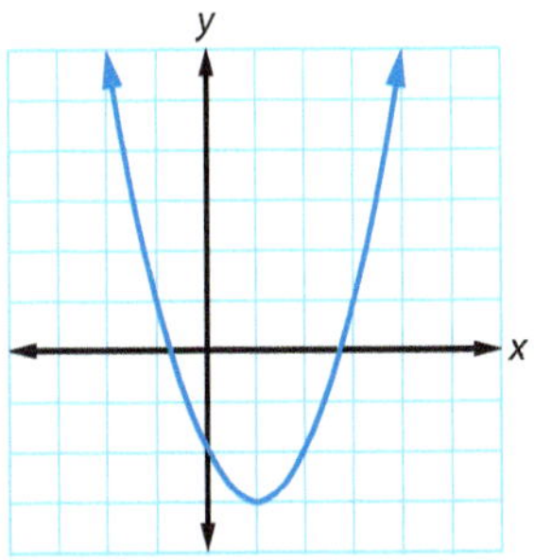

15.

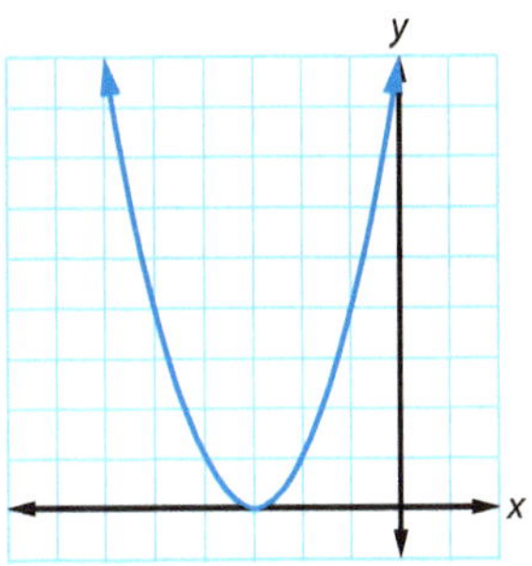

17.

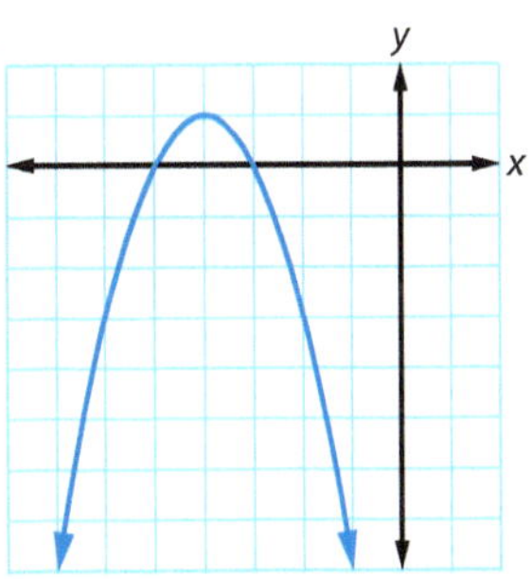

19.

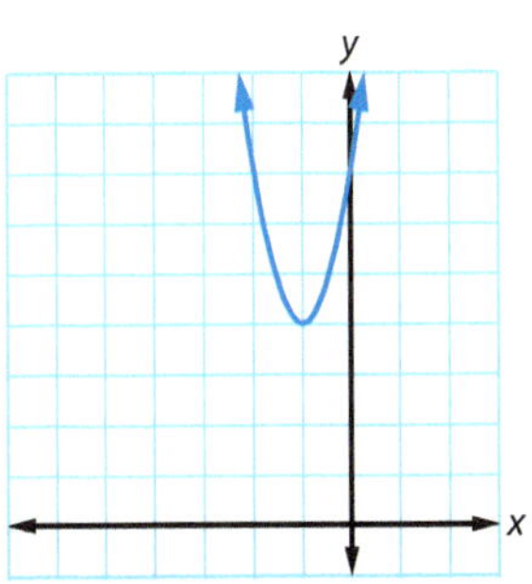

21. $f(x) = (x + 2)^2 - 16$; $(-2, -16)$

23. $f(x) = (x + 7)^2 - 9$; $(-7, -9)$

25. $(-5, -1)$

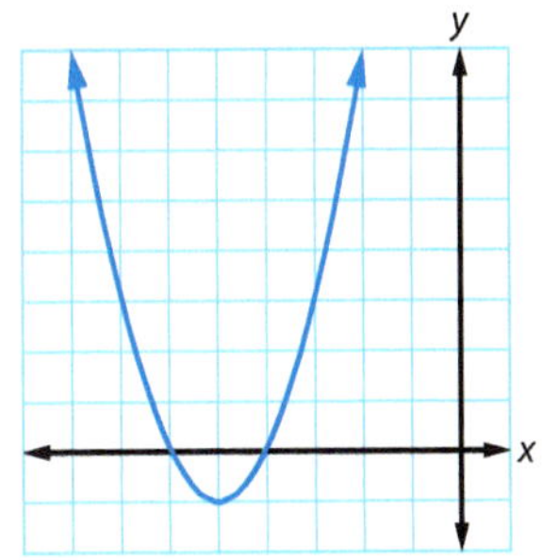

27. $(-4, 4)$

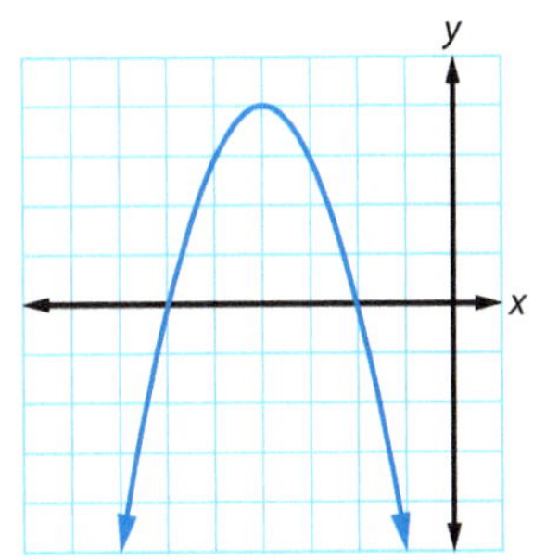

29. $(3, -7)$

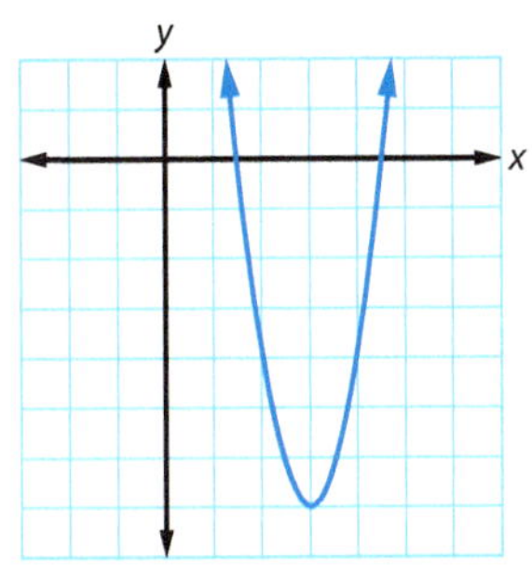

31a. \$1750 lost; \$125 gained **31b.** 200 items

31c. \$750 **33.** 74.0625 ft; 2.0625 sec

35a. $90 - 2x$

35b. $f(x) = x(90 - 2x)$ or $f(x) = -2x^2 + 90x$

35c. (22.5, 1012.5) **35d.** 22.5 ft × 45 ft **35e.** 1012.5 ft^2

37. $(2, -3)$

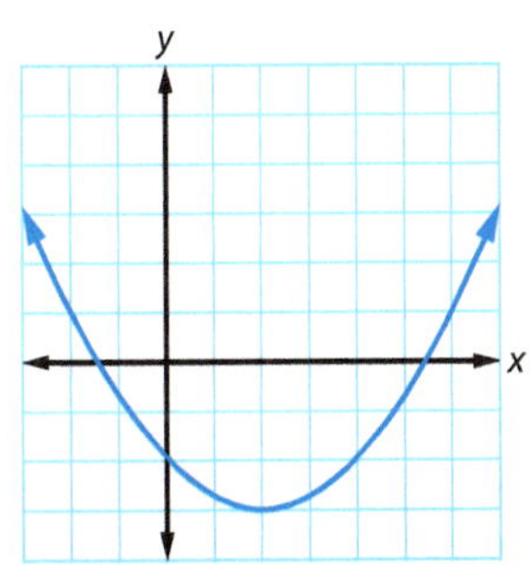

39. 1326.5 ft; 8 sec **41.** $\left(\frac{1}{2}, 2\right)$ **43.** $x^5 - 1024$

45. $x = 7, -15$ **47.** $x = 1, \frac{1}{2}$ **49.** A

Section 11.9

1. (0, 0) **3.** (0, −5) **5.** (0, 1)

7. (0, 8) **9.** (0, −5) **11.** −2, 4

13. 3 **15.** no real zero **17.** 1, −3

19. $-3 \pm \sqrt{17}$ **21.** (0, c)

23. the x-axis or $y = 0$

27. (0, −4); −2, 2; (0, −4)

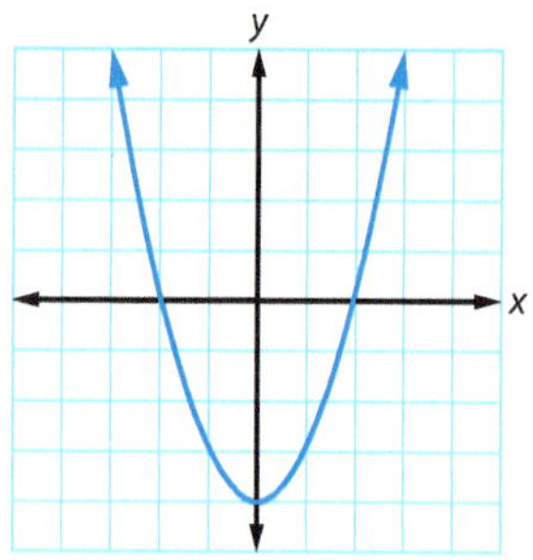

29. (0, 8); 2, 4; (3, −1)

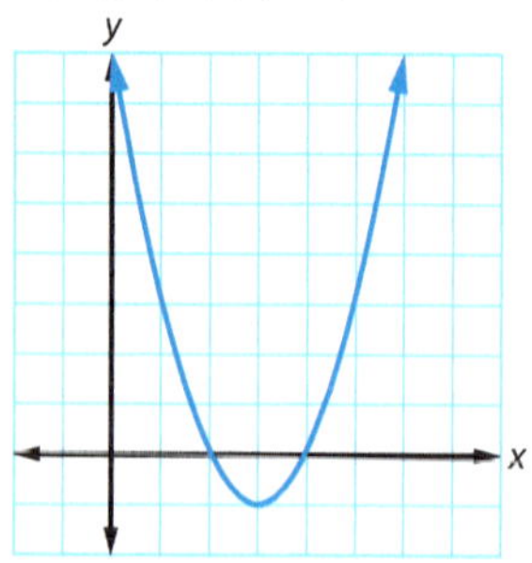

31. (0, 1); $-2 \pm \sqrt{3}$; (−2, −3)

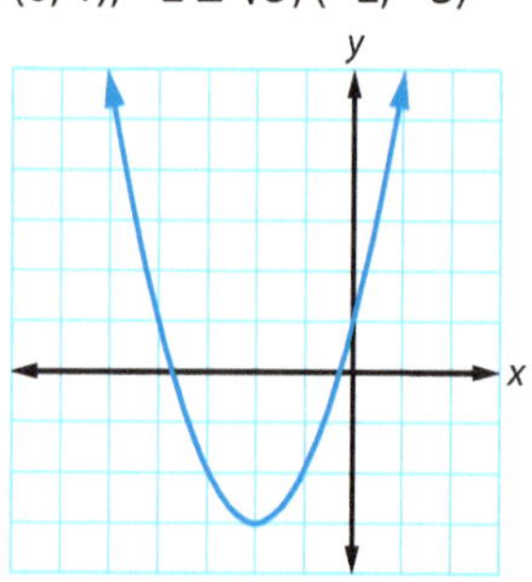

33. (0, −4); $2 \pm 2\sqrt{2}$; (2, −8)

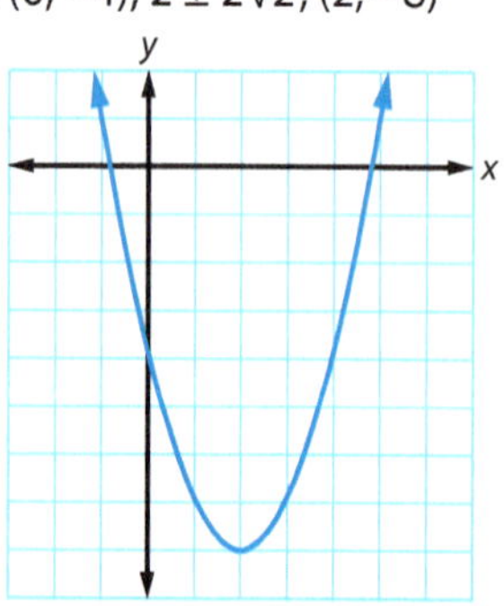

35. (0, 9.25); no real zero; (5, 3)

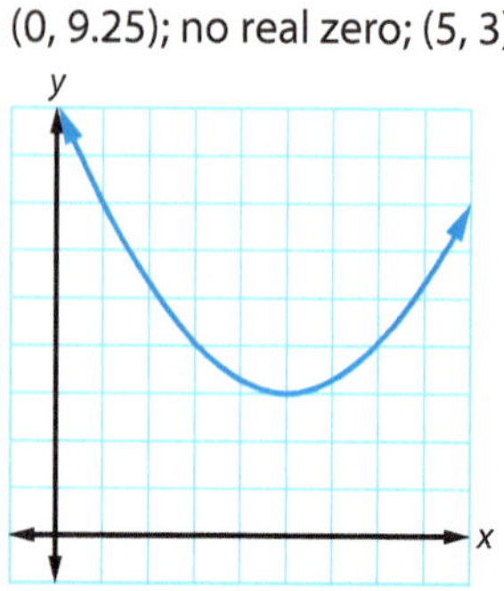

37a. 70 items **37b.** $800 **37c.** never

39a. $s = \frac{x}{4}$; $r = \frac{x}{2\pi}$ **39b.** $A_s = \frac{x^2}{16}$; $A_c = \frac{x^2}{4\pi}$

39c. the area of the circle **41.** $2(x + 7)$

43. $(x + 3)(x + 2)$ **45.** $x = -4, 0$ **47.** $x = 4 \pm 2\sqrt{3}$

49. C

Application Problems

1. ≈1.5 ft **3a.** 65 ft **3b.** 2.0 sec

5. 44 ft × 54 ft **7a.** 1521 ft² **7b.** 28%

9. $x = 10$; $y = 3$ **11.** $f(x) = -1.44x^2 + 7.2x$

Chapter Review

1. $x = 2, -3$ **3.** $a = 0, \frac{1}{4}, -3$ **5.** $x = -1, 7$

7. $x = \pm\sqrt{6}$ **9.** $x = 4 \pm \sqrt{11}$

11a. 2 real solutions **11b.** no real solutions

11c. 1 real solution **11d.** 2 real solutions

13. $x = \frac{-5 \pm \sqrt{37}}{2}$ **15.** $x = 5 \pm \frac{\sqrt{110}}{2}$ **17.** $x = 7, -3$

19. $x = \frac{1 \pm \sqrt{13}}{4}$ **21.** 36; 2 rational **23.** 57; 2 irrational

25. $a = \pm 4$ **27.** no real solution **29.** $x = 0, \frac{8}{5}$

31. $x = -3$ **33.** $10x^2 + x - 3 = 0$ **35.** 4.75 sec

37. 70 boxes; $960

39. (0, −6); $\pm\sqrt{6}$; $x = 0$

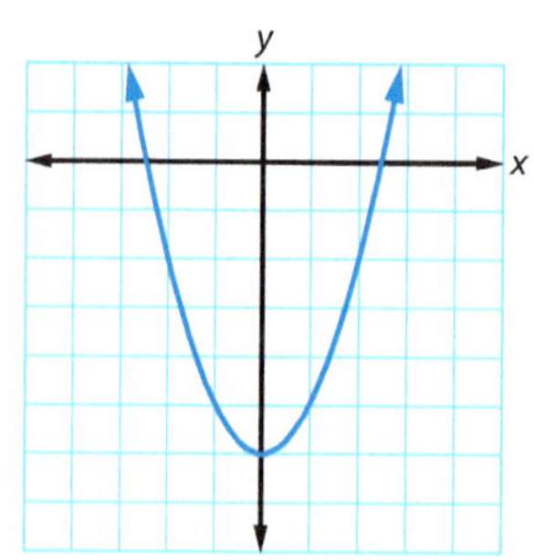

41. (1, 4); −1, 3; $x = 1$

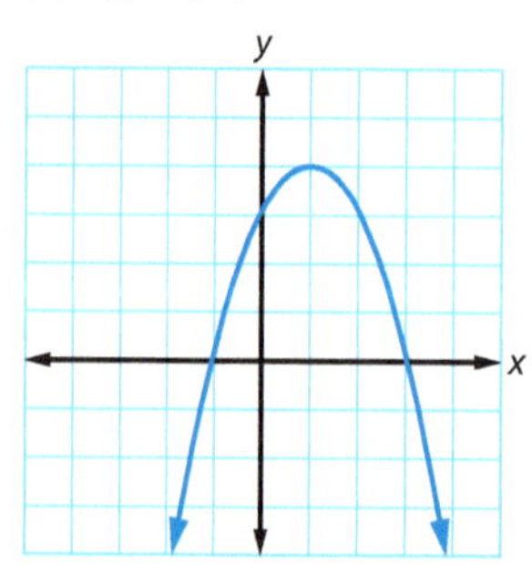

43. 2, 4 **45.** $x = 3$ **47.** 1, 3

49. Yes; the graph opens upward from the minimum of (2, −2), so it crosses every y-value greater than −2.

CHAPTER 12: RATIONAL EXPRESSIONS & EQUATIONS

Section 12.1

1. $x \neq 0$ **3.** $a \neq -2$ **5.** $x \neq -8, 3$

7. $c \neq 4, 6$ **9.** $\frac{x}{5}$ **11.** $3xy^2$

13. $x - 1$ **15.** $2a + 1$ **17.** $\frac{x+5}{x-1}; x \neq 1, 9$

19. $\frac{x-5}{x}; x \neq -2, 0$ **21.** $-1; x \neq \pm y$

23. $-\frac{w^2(x-8)}{6x^2}; x \neq 0, 7$ **25.** $\frac{y(x-5)}{3}; x \neq -4, 0$

27. $\frac{5t}{t^2+1}$; ≈0.41 mg **29.** $\frac{\pi}{4} \approx 78.5\%$

31. $\frac{2}{\pi} \approx 63.7\%$ **33.** $\frac{5}{x}$ **35.** $-6x - 5; x \neq \frac{1}{8}$

37. $x + 3; x \neq 3$

39. The simplified expression has only 1 excluded value ($x \neq 0$), but the original expression has 2 excluded values $\left(x \neq -\frac{4}{3}, 0\right)$.

41. $\frac{\pi}{6} \approx 52.4\%$ **43.** $3x^2 - 29x + 40$ **45.** $\frac{x\sqrt{30}}{9}$

47. $y = (x+2)^2; (-2, 0)$ **49.** $y = (x-4)^2 + 15; (4, 15)$

51. C

Section 12.2

1. $\frac{19}{70}$ **3.** a **5.** 5

7. $-\frac{3}{2}$ **9.** $\frac{a^2b}{a^2-b^2}$ **11.** $\frac{7}{2}$

13. $\frac{7y^2}{3}$ **15.** $x + y$ **17.** $\frac{x+2}{2}$

19. $(x+8)(x+3)$ **21.** $\frac{2x+y}{3x-4y}$ **23.** $\frac{x+y}{(x-y)^2}$

25. $\frac{x}{x+1}$ **27.** $\frac{(3x+y)(x+y)}{(x-y)^2}$ **29.** $\frac{a+2b}{a}$

31. 1 **33.** $(x+2)(x-2)$

35. $\frac{(4x-7y)(x+2y)}{(x+3y)(x+y)}$ **39.** $\frac{4b(a-b)(a-2b)}{a(a+7b)}$

41. $\frac{2x-1}{x-1}$ **43.** $(2, -3.5)$ **45.** $(6, -7)$

47. $3\sqrt{2} + 2\sqrt{3}$ **49.** 60 **51.** C

Section 12.3

1. $\frac{14}{x}$ **3.** $-\frac{x}{5}$ **5.** $-\frac{1}{c}$

7. 2 **9.** $\frac{9x}{x^2+y}$ **11.** 1

13. $\frac{3(3x+y)}{yz}$ **15.** B, D **17.** $-\frac{3}{m-4}$

19. 1 **21.** $\frac{3(2-a)}{a+b}$ **23.** $-\frac{2}{a^2+3a-2}$

25. $\frac{2x-7}{x^2-4x-7}$ **27.** $\frac{3m+n}{m-n}$ **29.** $\frac{x+7}{x+9}$

31. $\frac{3x+5}{x-5}$ **33.** $\frac{1}{a-b}$

35. The answer is $\frac{3(x+5)}{x-5}$. **37.** $\frac{x^2+x-2}{x}$

39. $\frac{x^3+2x^2-4x-8}{x^2}$ **41.** $\frac{-3x+1}{x-4}$ or $-\frac{3x-1}{x-4}$

43. $10x^2 + 3x - 4$ **45.** $\frac{3 \pm \sqrt{7}}{2}$ **47.** $\frac{13}{15}$

49. $1\frac{11}{20}$ **51.** B

Section 12.4

1. $24x^3$ **3.** $6(r+6)$ **5.** $(a+b)(a-b)^2$

7. $\frac{12x+9}{16}$ **9.** $\frac{2x+3y}{15}$ **11.** $\frac{7a^2+6a-45}{10}$

13. $\frac{7c+3}{c(c+1)}$ **15.** $\frac{20x^2y+5}{2x^3y^2}$ **17.** $\frac{2b^2-11b-15}{(b-4)(b+5)}$

19. $\frac{2y^2-23y-36}{(y-7)^2(y+4)}$ **21.** $\frac{-2(x+3)}{(x+2)(x-4)}$ **23.** $\frac{d(d+28)}{6(d+4)(d-4)}$

25. $\frac{7(x+1)}{12(x-3)(x+4)}$ **27.** $\frac{5x-5y+3}{(x+y)(x-y)}$

29. $-\frac{y(2y^2+y-15)}{(y+2)(y+5)^2}$ **31.** $\frac{3(x-2)}{x+5}$

35. $\frac{2x^2+5x-28}{(x+8)(x-4)(x-2)}$ **37.** $\frac{3}{x+1}$

39. $\frac{-x+4}{(x-2)(x-5)}$ **41.** $\frac{25y}{4x^8}$ **43.** $5\sqrt{2}$

45. $\frac{x+1}{3}$ **47.** 44°, 54°, and 82° **49.** A

Section 12.5

1. $\frac{2x+1}{x}$ **3.** $\frac{2y^2+2y+1}{y+1}$

5. $\frac{2ab-a}{b}$ or $\frac{a(2b-1)}{b}$ **7.** $\frac{2}{5}$

9. $\frac{ad}{bc}$ **11.** $\frac{xz}{y}$ **13.** $\frac{ac^3}{d^2}$

15. $\frac{y}{c^2+cy}$ **17.** $\frac{x^2+2x+1}{x}$ or $\frac{(x+1)^2}{x}$

19. $\frac{2y^2+5y+2}{y+1}$ or $\frac{(2y+1)(y+2)}{y+1}$

21. $\frac{Ac+b}{c}$ **23.** $\frac{x^2}{x+1}$ **25.** $\frac{x}{x-y}$

27. $\frac{1}{(x-y)^2}$ **29.** $\frac{1}{4}$ **31.** $\frac{2}{(x+2)^2}$

33. $\frac{-x+6}{(x+5)(x+3)}$ **35.** $\frac{sf}{s-v}$ **37.** ≈374 Hz

39. $\frac{10}{57} \approx 0.175$ **41.** $\frac{x}{y}$ **43.** $\frac{2r_1r_2}{r_1+r_2}$

45. $-\frac{a+7b}{a+11b}$ **47.** $-\frac{1}{2(x+5)}$ **49.** $\frac{a+b}{5a+b}$

51. $\frac{3}{(x+4)(x-3)}$ **53.** A

Section 12.6

1. $m = 27$ **3.** $x = \frac{1}{4}$ **5.** $x = \frac{2}{3}$

7. no solution **9.** $a = \frac{9}{2}$ **11.** $m = 22$

13. $a = -\frac{241}{8}$ **15.** $R = 10$ **17.** $x = -\frac{103}{2}$

19. no solution **21.** $x = -3$ **23.** $m = \frac{51}{10}$

25. 36 in. × 18 in. **27.** 13 **29.** 4 and 10

31a. 25 made free throws **31b.** 8 missed free throws

33. $x = \frac{3 \pm 3\sqrt{17}}{2}$ **35.** $x = -\frac{11}{3}$ **37.** $R_2 = 8$ ohms

39. $T = \frac{1}{f}$ **41.** (3, −1) **43.** $x = -\frac{5}{3}$

45. $\frac{1}{p + r}$ **47.** C

Section 12.7

1.

	r	t	d
Charissa	16	$\frac{d}{16}$	d
Sophia	20	$\frac{d}{20}$	d

3.

	r	t	d
Dion	x	$\frac{2}{x}$	2
Mitch	$x - 2$	$\frac{2}{x - 2}$	2

5.

	r	t	d
Going	$r + 3$	$\frac{200}{r + 3}$	200
Returning	r	$\frac{200}{r}$	200

7.

	Time (hr)	**Rate of Work**
Pete	3	$\frac{1}{3}$
Adam	6	$\frac{1}{6}$
Together	x	$\frac{1}{x}$

9. 2.5 mi **11.** 29.8 hr **13.** 21.3 hr

15. slower plane: 750 mi; faster plane: 1500 mi

17. Gabriel: 4 mi/hr; train: 40 mi/hr

19. 3 mi/hr **21.** 4.3 sec **23.** 0.9 hr

25. going: 6 mi/hr; returning: 10 mi/hr

27. Lily: 6.8 hr; Hayley: 9.8 hr

29. 1.2 hr **33.** 9.4 hr

35. going: 5.5 hr, 50 mi/hr; returning: 5 hr, 55 mi/hr

37.

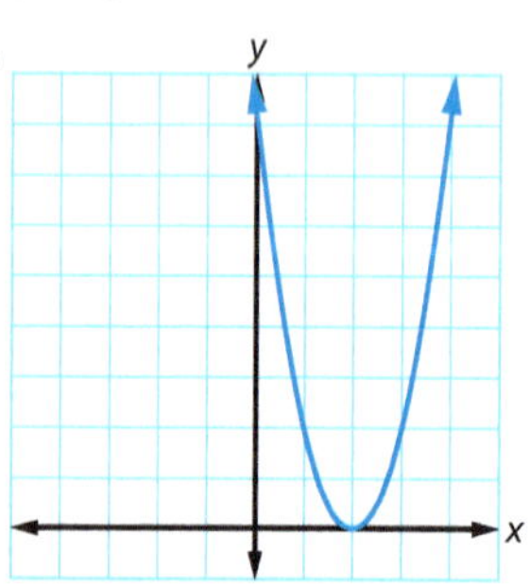

39. (12, 6) **41.** $f(x) = (x + 2)^2 - 1$

43. 500 mL of the 28% solution and 300 mL of the 60% solution

45. B

Section 12.8

1. B **3.** A **5.** C

7. B **9.** B

11. (0, 0); $x = 0$; $y = 0$ **13.** (4, 0); $x = 4$; $y = 0$

15. $\left(\frac{1}{2}, 1\right)$; $x = \frac{1}{2}$; $y = 1$ **17.** (−1, −2); $x = -1$; $y = -2$

19. 14 units left **21.** 3 units down

23. 4 units right and 2 units up

25.

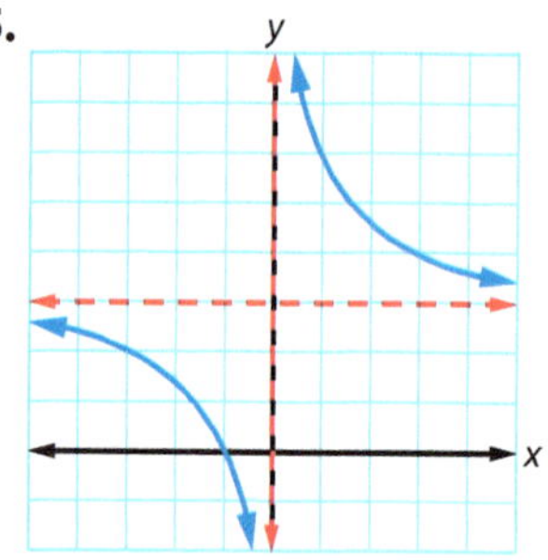

27.

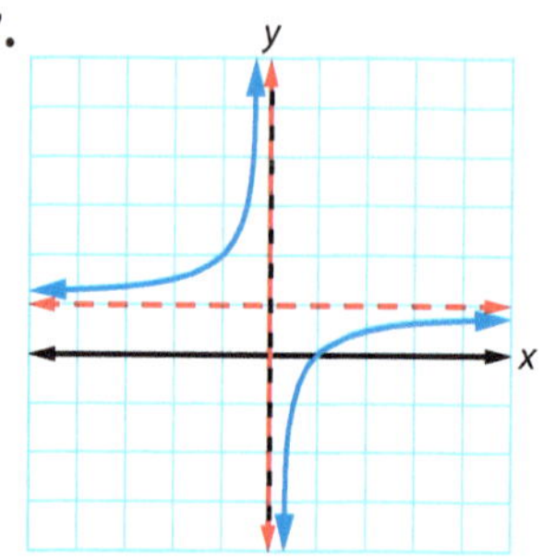

29.

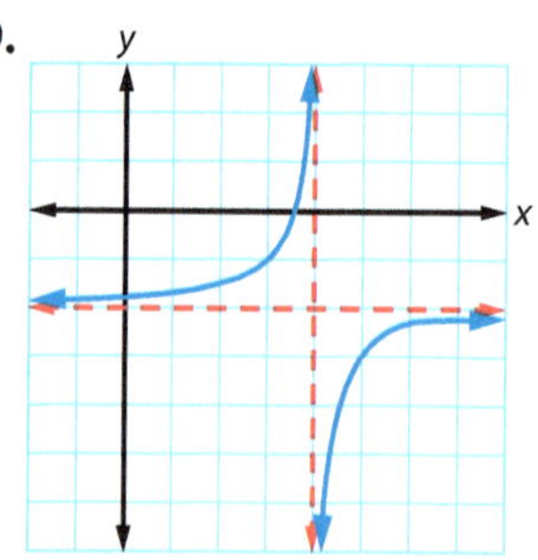

31. a line that the solutions approach but never reach

33.

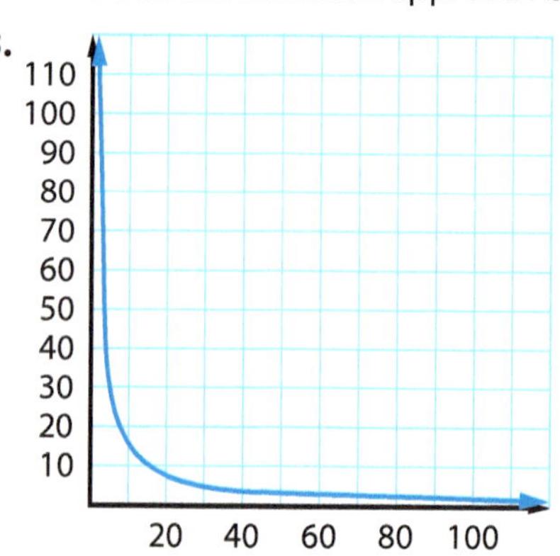

35. The current increases. **37.** $f(x) = \frac{1}{x - 1} + 2$

39a. $f(x) = \frac{b - ac}{x + c} + a$

41.

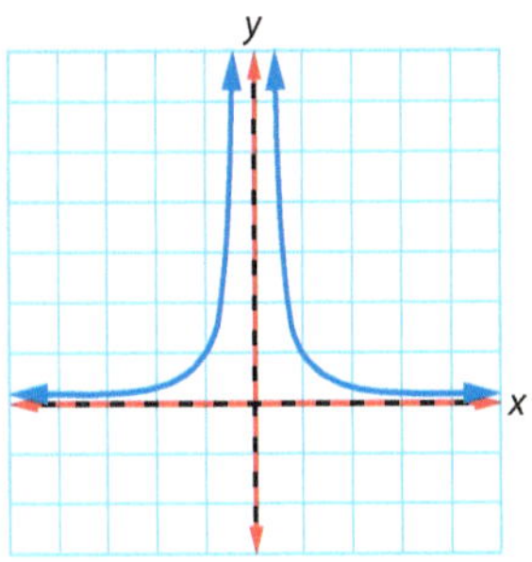

43. $\varnothing$ **45.** $\sqrt{13}$

47. $(x - 3)(x + 3)(3x^2 - 4)$

49. $x = \pm 4$ **51.** D

Application Problems

1. $\frac{1}{4}l$ **3.** $\frac{3}{4}l$

5.

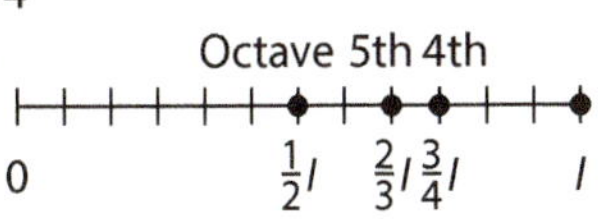

7. $\frac{5}{6}l; \frac{4}{5}l$ **9.** harmonic; arithmetic

13. 25 mi/hr **15.** 24 mi/hr

17.

	r (mi/hr)	*t* (hr)	*d* (mi)
Trip 1	24	5	120
Return 1	18	$6\frac{2}{3}$	120
Trip 2	16	$7\frac{1}{2}$	120
Return 2	20	6	120

Chapter Review

1. $x \neq 3$ **3.** $x \neq 0, \pm 1$ **5.** $\frac{1}{2x - 5}$

7. $56x^4y$ **9.** $\frac{ac}{b}$ **11.** $\frac{3x + 10}{3x + 5}$

13. $7x + 5$ **15.** $3(x + 11)$ **17.** $\frac{1}{x - y}$

19. $-\frac{x}{y}$ **21.** $\frac{4}{x - 1}$ **23.** 0

25. $\frac{2x + 3}{3x - 5}$ **27.** $\frac{x + 9xy + 2y}{27y}$ **29.** $\frac{1}{xy}$

31. $x = -\frac{3}{2}, 8$ **33.** $x = 26$ **35.** $x = \frac{11}{6}$

37. $x = 7$ **39.** $\frac{d}{13} = \frac{d}{23} + 3$ **41.** $\frac{1}{5} + \frac{1}{2} = \frac{1}{x}$

43. $1\frac{7}{8}$ hr

45. Micah: 24 mi/hr; Alexis: 60 mi/hr

47. $\frac{\sqrt{3}\pi}{6} \approx 90.6\%$ **49.** $(-3, 0); x = -3; y = 0$

51.

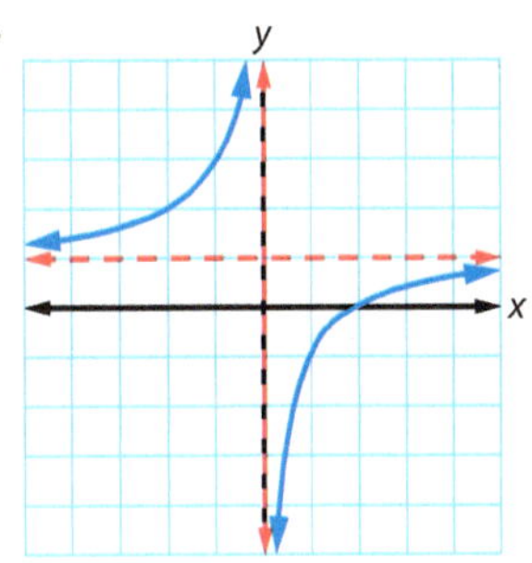

53. D

APPENDIX LESSONS

Section A.1

1. 6.9 **3.** 7 **5.** 86

7. 27.8 **9.** none **11.** 8

13. 15 **15.** 491 **17.** 425

19. 198 **21.** 5.2 **23.** 91.4

25. 10.75 **27.** 5.8 **29.** increases by 3

31. no change **33.** no change

35. decreases by 15.2

Section A.2

1. variance **3.** 6.89

5. 0.01, 4.45, 0.01, 8.35, 0.01, 1.23, 9.67, 3.57, 3.57

7. 1.85 **9.** 8.74 **11.** 141.59

13. 3.62 **15.** 5.09 **17.** 232.50

19. 39.90

21. On average, there were many more fatalities involving teen drivers, and the number of fatalities involving teen drivers also varied more.

Section A.3

1. 5; 2.5 **3.** mode: 6.3; range: 3.4

5.

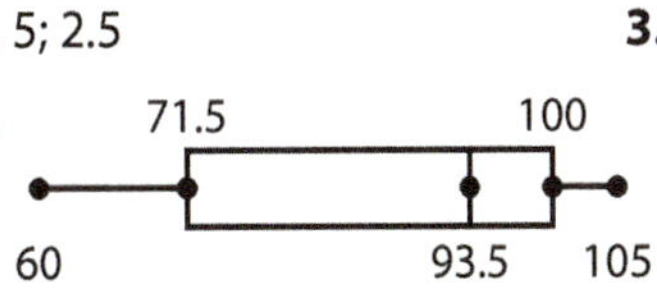

7. 28.5 **9.** 6.5 **11.** 2, 4.5, 9, 10, 10

13. 5.5

15.

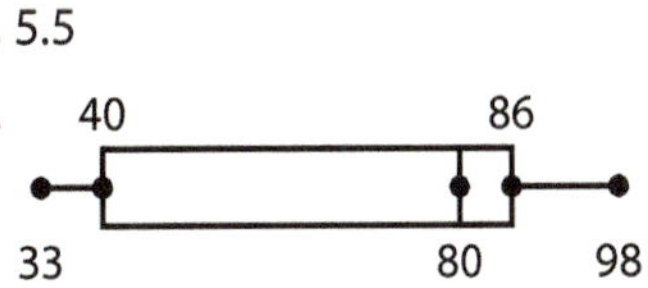

17.

Data	Tally	*f*	*df*
10	||	2	20
11	|	1	11
12	|||	3	36
14	|	1	14
15	||||	4	60
17	|	1	17
18	||	2	36
19	|	1	19
20	|||	3	60
total		18	273

19.

Data	Tally	*f*	*df*
60	|	1	60
75	|||	3	225
80	||||	4	320
85	|||	3	255
90	||	2	180
95	|	1	95
100	|||	3	300
total		17	1435

21. mean: 84.2; median: 84; mode: 76; range: 28

23. mean: 73.7; median: 70; mode: 70 and 90; range: 60

25. mean: 182.4; median: 185; mode: 160; range: 50

27. 12.9 days **29.** 31.6 steps **31.** 34.1 sit-ups

33. range: 24; mean: 21.4; median interval: 20–24; modal interval: 20–24

Section A.4

1. false

3. The trend line is not a good fit since the residuals are not dispersed evenly above and below the horizontal axis.

5. There is a discernable pattern which may indicate that a nonlinear correlation exists in the original data set.

11. The residuals are clustered close to the horizontal axis and evenly dispersed about it. It appears $y = -x + 4$ models the data well.

13. The residuals are not evenly dispersed about the horizontal axis. It appears $y = 1.6x + 13$ does not model the data well.

15.

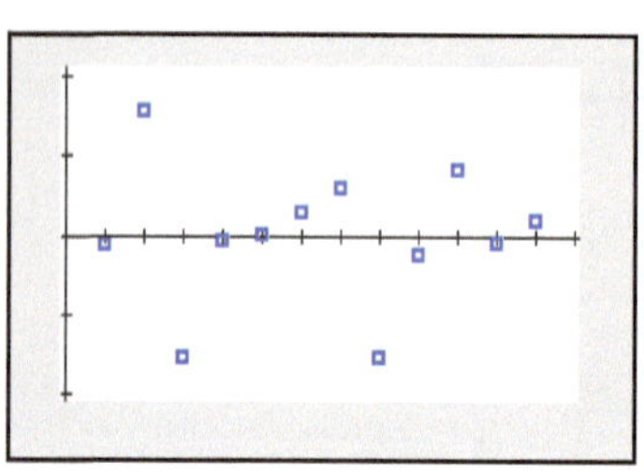

17. $y = 66.4685x - 90.2121$

19. No; the residuals appear to form a nonlinear pattern about the horizontal axis, suggesting that the data are not linear and that the line of best fit does not model the data well.

21a. cause **21b.** correlation **21c.** correlation

23. Causation; the wind is the cause of the windmill turning.

25. Answers will vary. Sample answer: Mark applied suntan lotion and is wearing sunglasses.

Section A.5

1. two-way frequency table

3. grand total

5.

		Assessment Preference		
		unit test	**daily quizzes**	**total**
Location	**classroom**	7	2	9
	virtual	5	6	11
	total	12	8	20

7. 60%

9. 5-day workweeks with three 10 min breaks

11. The data suggests that two 15 min breaks would be more desirable. While there is also support for a 5-day workweek, a significant number prefer a 4-day workweek.

13. 22% **15.** 36% **17.** 57%

19. 8

21.

	Position		
	offense	**defense**	**total**
girl	64%	36%	100%
boy	47%	53%	100%

23.

	Position	
	offense	**defense**
girl	70%	53%
boy	30%	47%
total	100%	100%

25.

		Treatment		
		yes	**no**	**total**
Improved	**yes**	65	28	93
	no	24	72	96
	total	89	100	189

27. There is an association between treatment and improvement. Those who improved were more likely to have been treated, and those who did not improve were less likely to have been treated.

29. There is very little association between having a part-time job and participation in after-school activities. Those with a part-time job were slightly more likely to participate in neither Bible club nor sports.

GLOSSARY

absolute value (p. 4) The number of units (distance) between 0 and the location of the number on a number line, denoted $|x|$. Formally, $|x| = x$ if $x \geq 0$ and $|x| = -x$ if $x < 0$.

absolute value equation (p. 135) An equation whose variable expression is contained within the absolute value symbols.

absolute value function (p. 200) A function containing an absolute value expression.

algebraic expression (p. 29) A mathematical expression containing numbers, variables, and operational signs.

arithmetic mean (p. 608) The sum of n numbers divided by n. Commonly called the *mean* or *average*.

arithmetic sequence (p. 185) A sequence in which the difference between any 2 consecutive terms is a constant.

asymptote (p. 352) A line that a graph approaches but never intersects.

base of an exponential expression (p. 18) A number, variable, or expression that is raised to a power indicated by an exponent (e.g., in the expression a^3, a is the base).

binomial (p. 380) A polynomial with exactly 2 terms.

boundary line (p. 307) A line separating portions of an inequality graph into half-planes on a coordinate plane.

box-and-whisker plot (p. 623) A diagram illustrating the median, quartiles, and extremes of a data set.

categorical variable (p. 636) A variable that describes qualitative data that is not measured numerically, such as blue, red, and green or male and female.

causation (p. 632) The change in one variable directly affects the change in another variable.

coefficient (p. 36) The numerical factor accompanying the variables of a term.

common difference (p. 185) The constant describing the difference between any 2 consecutive terms in an arithmetic sequence.

common ratio (p. 365) The constant describing the quotient between any 2 consecutive terms in a geometric sequence.

completing the square (p. 518) A method of solving quadratic equations by obtaining a perfect square trinomial.

complex fraction (p. 13) A fraction whose numerator and/or denominator contains a fraction.

complex rational expression (p. 583) An algebraic expression of the form $\frac{n}{d}$ in which the numerator, n, the denominator, d, or both are rational expressions.

compound sentence (p. 123) Two or more equations or inequalities connected by using the word *and* or the word *or*.

conjugates (p. 400) Two algebraic or numerical expressions in the form $a + b$ and $a - b$.

conjunction (p. 123) A compound sentence consisting of mathematical statements connected by the word *and*, referring to the intersection of the graphs.

consistent system (p. 287) A system of equations with 1 or more solutions.

constant (p. 29) A symbol that represents a fixed number.

constant of variation (proportionality) (p. 192) The constant relating 2 variables in a direct or inverse variation (e.g., in the expressions $y = kx$ and $yx = k$, k is the constant of variation).

cube root (p. 21) One of a number's 3 equal factors.

data (p. 614) Recorded observations that can be measured or counted.

decay factor (p. 358) The quantity $(1 - r)$ in an exponential decay function $f(t) = A(1 - r)^t$.

decay rate (p. 358) The value of r in the exponential decay function $f(t) = A(1 - r)^t$.

deductive reasoning (p. 140) Reasoning from a general premise to a specific conclusion.

degree of a polynomial (p. 381) The highest degree of any term in the polynomial.

degree of a term (p. 380) The sum of the exponents of the variables of a term.

dependent system (p. 287) A consistent system of equations with an infinite number of solutions. The graphs coincide because the equations are equivalent.

dependent variable (p. 157) A variable that represents the values of the range of a relation or function.

depreciation (p. 360) The decrease in value of an item by the same percent each year.

direct variation (proportion) (p. 192) A function in which the ratio of variables is a nonzero constant, k. That is, $\frac{y}{x} = k$, where $k \neq 0$.

discriminant (p. 533) The value of the radicand in the quadratic formula, $b^2 - 4ac$, that can be used to identify the number and type of solutions to the equation.

disjunction (p. 129) A compound sentence consisting of mathematical statements connected by the word *or*, referring to the union of the separate graphs.

division (p. 12) Multiplication by the reciprocal of the divisor.

domain of a relation (p. 154) The set of first elements (x-coordinates) of the ordered pairs.

domain of a variable (p. 29) The set of numbers the variable may represent.

dot plot (p. 624) A graphical representation of a data set where for each number in the set a dot is placed above the corresponding value on a number line.

double root (p. 550) A single root that results from 2 repeated factors in a quadratic function.

element (p. 18) A distinct object or member of a set.

elimination method (p. 279) Solving a system of equations by adding or subtracting multiples of the equations to eliminate one of the variables.

empty set (p. 124) A set containing no elements, denoted ∅.

equation (p. 46) A mathematical sentence stating that 2 expressions are equal.

excluded value (p. 564) A number that makes a rational expression undefined.

explicit formula (p. 187, 366) A formula used to find the *n*th term in a sequence.

exponent (p. 18) The number of times a base occurs as a factor, indicated by a superscript numeral to the right of the base (e.g., in the expression a^3, 3 is the exponent).

exponential decay (p. 358) A quantity exhibits exponential decay if it decreases by the same percent during each unit of time.

exponential form (p. 18) A simplified form of writing repeated multiplication (e.g., 5^3). Also called *exponential notation*.

exponential function (p. 351) A function of the form $f(x) = ab^x$, where $b > 0$ and $b \neq 1$.

exponential growth (p. 358) A quantity exhibits exponential growth if it increases by the same percent during each unit of time.

extraneous solution (p. 487) A derived solution that is invalid because it does not make the original equation true.

extrapolation (p. 249) The process of making predictions outside the range of existing data points.

factoring (p. 418) The process of determining the individual factors of a product.

five-number summary (p. 623) The lowest value, Q_1, Q_2, Q_3, and the highest value of a set of data.

FOIL method (p. 394) A method of multiplying 2 binomials by finding the products of the first, outer, inner, and last terms and then combining like terms.

formula (p. 31) An equation that expresses the relationship between related quantities.

fractals (p. 208) Rough or fragmented geometric shapes that are repeated at an increasingly smaller scale.

frequency distribution table (p. 625) A table used to organize a large set of data.

function (p. 155) A relation in which each element in the domain is paired with exactly 1 element in the range.

function form of a linear equation (p. 231) A linear equation in the form $f(x) = mx + b$, where m is the slope of the line and $(0, b)$ is the y-intercept.

function notation (p. 157) A means of expressing the dependence of one variable upon another (e.g., $y = x + 5$ is expressed as $f(x) = x + 5$ in function notation).

function rule (p. 177) An equation that pairs each member of the domain with an element of the range.

geometric sequence (p. 365) A sequence in which the quotient between any 2 consecutive terms is a constant.

golden ratio (p. 496) The irrational mathematical constant $\frac{1+\sqrt{5}}{2} \approx 1.6180339887$, represented by the lowercase Greek letter phi (ϕ).

grand total (p. 636) The sum of the joint frequencies in a two-way table.

greatest common factor (GCF) (p. 418) The largest positive integer that evenly divides 2 or more given integers. The GCF of a polynomial is the product of all factors shared by every term of the polynomial.

growth factor (p. 358) The quantity $(1 + r)$ in an exponential growth function $f(t) = A(1 + r)^t$.

growth rate (p. 358) The value of r in an exponential growth function $f(t) = A(1 + r)^t$.

half-plane (p. 307) The portion on either side of the boundary line of an inequality graphed in the coordinate plane.

harmonic mean (p. 608) The number of values divided by the sum of their reciprocals. The harmonic mean of n numbers can be found using the formula $H = \frac{n}{\frac{1}{x_1} + \frac{1}{x_2} + \ldots + \frac{1}{x_n}}$. When there are just 2 values, a and b, this simplifies to $H = \frac{2ab}{a+b}$.

histogram (p. 626) A bar graph representing equal intervals with no space between the bars.

hypotenuse (p. 474) The side of a right triangle opposite the right angle.

inconsistent system (p. 287) A system of equations with no solution.

independent system (p. 287) A consistent system of equations with only 1 solution.

independent variable (p. 157) A variable that represents the values of the domain.

index (p. 459) A superscript number within the radical sign that indicates the root (e.g., in $\sqrt[3]{x}$, the index is 3). If no index is given, the index is understood to be 2.

inequality (p. 112) A mathematical sentence stating that 2 quantities are not always equal.

intercept (p. 220) A point where a graph intersects an axis.

interpolation (p. 249) The process of making estimates within the range of existing data points.

interquartile range (IQR) (p. 615) The difference of Q_3 and Q_1 from a set of data.

interval frequency table (p. 626) A table used to organize large sets of data with many different values.

inverse operation (p. 46) A mathematical operation that reverses another operation, such as addition and subtraction or multiplication and division.

inverse variation (proportion) (p. 194) A relation in which the product of the 2 variables is a nonzero constant, k. That is, $xy = k$, where $k \neq 0$.

joint frequency (p. 636) The number representing each combination of categorical variables.

like radicals (p. 469) Radicals that have the same radicand and the same index.

like terms (p. 35) Terms that contain an identical variable or variables, all raised to the same respective powers.

linear equation (p. 218) An equation whose graph is a straight line. The standard form of a linear equation is $Ax + By = C$, where A, B, and C are real numbers.

linear function (p. 218) A function whose graph is a straight line and is described by the equation $f(x) = mx + b$, where m and b are real numbers.

line of best fit (p. 248) The trend line that best illustrates the relationship between data sets.

line of symmetry (p. 539) A line that divides a figure into 2 identical halves.

literal equation (p. 60) An equation containing more than one variable.

lower quartile (p. 615) The median of the lower half of a set of data.

marginal frequency (p. 636) The sum of each row or column of joint frequencies in a two-way table.

mathematical modeling (p. 169) Using mathematical language and representations such as functions and graphs to represent relationships between quantities related to the world around us.

mathematical property (p. 5) An equation or statement that is true for any value of the variable. Also called an *identity*.

maximum (p. 539) The highest point on a graph, having the greatest y-value of any point on the graph.

mean (p. 614) The sum of the values divided by the number of values in a data set.

mean absolute deviation (MAD) (p. 616) The average of the distances from each value in a data set to the mean.

median (p. 614) The middle number in a set of ordered data or the average of the 2 middle numbers if there is an even number of values.

median interval (p. 627) The interval of a frequency interval table that contains the median.

middle quartile (p. 615) The median of all the data in a set.

minimum (p. 539) The lowest point on a graph, having the smallest y-value of any point on the graph.

modal interval (p. 627) The interval of a frequency interval table that contains the most data.

mode (p. 614) The value or values that occur most frequently in a set of data.

monomial (p. 380) A polynomial with exactly 1 term.

multiplicative identity (p. 11) The number 1; the value that when multiplied by another number yields the same number.

multiplicative inverse (p. 12) The reciprocal of a, denoted $\frac{1}{a}$.

negative exponent (p. 20) For any nonzero number x and integer a, $x^{-a} = \frac{1}{x^a}$.

parabola (p. 344) The curved graph of a quadratic function.

parallel lines (p. 243) Lines in the same plane that have no points in common and have equal slopes.

parent function (p. 201) The simplest form of a function in a family of functions.

percent (p. 73) The ratio of a number to one hundred.

perfect square trinomial (p. 437) A trinomial that is the square of a binomial.

perpendicular lines (p. 244) Lines that meet at right angles (90°) and have slopes that are negative reciprocals of each other.

point of inflection (p. 347) The point at which the 2 branches of a cubic function meet.

point-slope form (p. 237) A linear equation in the form $y - y_1 = m(x - x_1)$, where m is the slope of the line and (x_1, y_1) is a given point.

polynomial (p. 380) An expression that can be written as a sum of terms consisting of constants, variables with nonnegative integral exponents, or products of a constant and 1 or more such variables.

power function (p. 344) A function of the form $y = ax^n$, where the base x is a variable and the exponent n and coefficient a are constants.

principal (p. 83) The amount of money initially invested or borrowed.

principal root (p. 21) The positive root of a number (e.g., $\sqrt{9} = 3$).

proportion (p. 66) A statement of equality between 2 ratios.

quadratic equation (p. 506) A second-degree equation that can be written in the standard form $ax^2 + bx + c = 0$, where a, b, and c are real numbers and $a \neq 0$.

quadratic function (p. 538) A function that can be written in the standard form $f(x) = ax^2 + bx + c$, where a, b, and c are real numbers and $a \neq 0$.

quartile (p. 615) Any of 3 values that divide a data set into 4 equally sized groups.

radical (p. 452) An expression in the form $\sqrt[n]{x}$.

radical equation (p. 486) An equation that contains a variable in a radicand.

radical expression (p. 482) An algebraic expression containing at least 1 radical.

radical function (p. 490) A function with the independent variable in the radicand.

radical sign (p. 21, 452) The symbol $\sqrt{\ }$ or $\sqrt[n]{\ }$, used to indicate a square root or nth root. When n is even, the principal (positive) root is indicated.

radicand (p. 452) The number under the radical sign (e.g., in $\sqrt[n]{3x}$, $3x$ is the radicand).

range of a data set (p. 615) The difference between the largest and smallest values in a set of data.

range of a relation (p. 154) The set of second elements (y-coordinates) of the ordered pairs.

rate (p. 65) A ratio comparing 2 quantities with different units.

rate of change (p. 227) The comparison of 2 changing quantities, such as time and distance or rise and run on a graph.

ratio (p. 65) A comparison of 2 numbers by division, often expressed as a fraction or in the form $x : y$.
rational equation (p. 588) An equation containing rational expressions.
rational exponent (p. 452) For any real number x and integers a and b, where $b > 1$, the exponential expression $x^{\frac{a}{b}}$ is equivalent to the radical $\sqrt[b]{x^a}$ or $(\sqrt[b]{x})^a$.
rational expression (p. 564) A ratio of polynomials whose denominator is not 0.
rational function (p. 601) A function expressed as a ratio of polynomials whose denominator is not 0.
rationalizing the denominator (p. 466) The process of eliminating any radicals from the denominator of a fraction.
rational number (p. 6) A number that can be expressed as a ratio of 2 integers $\frac{a}{b}$, where $b \neq 0$.
reciprocals (p. 12) Two numbers whose product is 1 (e.g., 3 and $\frac{1}{3}$ are reciprocals). Also called *multiplicative inverses.*
recursive formula (p. 186, 365) A formula that applies specific criteria to a given term in a sequence to generate consecutive terms.
relation (p. 154) Any set of ordered pairs.
relative frequency (p. 637) The ratio of a joint or marginal frequency to a total, often as a percent.
relatively prime numbers (p. 418) Two numbers or terms whose greatest common factor is 1.
residual (p. 630) The vertical difference between the trend line and an actual data point.
root (p. 549) A value of x at which the graph of the function crosses the x-axis. It is the x-coordinate of an x-intercept of the graph. Also called a *zero* of the function.
scientific notation (p. 337) A number expressed as a product in the form $a \times 10^n$, where $1 \leq a < 10$.
sequence (p. 185, 365) An ordered list of numbers.
similar figures (p. 68) Figures that have the same shape and proportional corresponding parts.
similar polygons (p. 68) Polygons in which corresponding angles are congruent and corresponding sides are proportional.
slope-intercept form (p. 231) A linear equation in the form $y = mx + b$, where m is the slope of the line and $(0, b)$ is the y-intercept.
slope of a line (p. 224) The ratio of the rise (vertical change) to the run (horizontal change). Also called the *rate of change.*
solution (p. 46) The value(s) that make a mathematical equation true.
solution to a system of equations (p. 264) Any ordered pair that makes all the equations in a system true.
square root (p. 21) One of a number's 2 equal factors.
standard deviation (p. 619) The square root of the average of the squares of deviation.
standard form of a polynomial (p. 381) A polynomial whose terms are arranged in descending order of the degrees of one of the variables.
statistics (p. 614) The branch of mathematics that deals with collecting, organizing, analyzing, and reporting numerical information.
substitution method (p. 272) Solving a system of equations by first solving an equation for one variable and then replacing that variable in the other equation, resulting in a single equation with 1 variable.
system of equations (p. 264) Two or more equations with 2 or more variables to be solved simultaneously.
term (p. 35) A constant, a variable, or the product or quotient of a constant and 1 or more variables. In algebraic expressions, terms are separated by + or −.
term in a sequence (p. 185, 365) Each value in an arithmetic or geometric sequence.
theorem (p. 474) A statement that has been proven to be true.
translation (p. 201) The movement of a parent function in the coordinate plane.
trend line (p. 248) A line that estimates the relationship between data sets.
trinomial (p. 380) A polynomial with exactly 3 terms.
two-way frequency table (p. 636) Displays the frequencies of 2 categorical variables from the same source.
unit multiplier (p. 66) A fraction equal to 1 with different units representing the same quantity in the numerator and the denominator.
unit rate (p. 65) A rate in which the denominator equals 1.
upper quartile (p. 615) The median of the upper half of the data.
variable (p. 29) A symbol used to represent an unknown quantity.
variance (p. 619) The mean of the squared differences in a data set.
vertex (p. 201, 344) The turning point of a graph at which it reaches its highest or lowest point.
vertex form (p. 347) The equation of a parabola of the form $y = (x - h)^2 + k$, where h and k are real numbers taken from the vertex (h, k).
vertical line test (p. 161) The method of moving a vertical line across a graph of a relation to determine whether the relation is a function.
***x*-intercept** (p. 220) The point at which the graph of an equation crosses, or intersects, the x-axis.
***y*-intercept** (p. 220) The point at which the graph of an equation crosses, or intersects, the y-axis.
zero of a function (p. 549) A value of x such that $f(x) = 0$. It is the x-coordinate of any point at which a graph of the function intersects the x-axis. Also called a *root* of the function.
zero power (p. 20) Any nonzero number to the zero power equals 1. That is, when $x \neq 0$, $x^0 = 1$.

INDEX

PHOTO CREDITS

Key: (t) top; (c) center; (b) bottom; (l) left; (r) right; (i) inset; (bg) background

Front Matter

viiit Apic/Hulton Archive via Getty Images; **viii**b Ground Picture/Shutterstock.com; **ix**t Pixels Hunter/Shutterstock.com; **ix**bl "Portrait of Sir Isaac Newton, 1689" by Godfrey Kneller/ Wikimedia Commons/Public Domain; **ix**br Andrey_Popov/Shutterstock.com; **x** Klaus Vedfelt/ DigitalVision via Getty Images; **xi** Heritage Images/Hulton Archive via Getty Images

Chapter 1

2–3 ©Dimitry Kalinovsky/Dreamstime.com; **4** xPACIFICA/Alamy Stock Photo; **7** dowell/ Moment via Getty Images; **10** antoniodiaz/ Shutterstock.com; **13** skynesher/E+ via Getty Images; **17**bg NIKCOA/Shutterstock.com; **17**i Pratchaya.Lee/Shutterstock.com; **18** chuyuss/ Shutterstock.com; **29** Stephen Kiers/Shutterstock .com; **32** cjp/E+ via Getty Images; **35** Nikada/E+ via Getty Images; **39** BM_27/Shutterstock.com; **40** Magsi/Shutterstock.com

Chapter 2

44–45 Avigator Fortuner/Shutterstock.com; **46** Joseph Sohm/Shutterstock.com; **49** Lynda McFaul/ Shutterstock.com; **54** Andrii_K/Shutterstock.com; **57** Dziurek/Shutterstock.com; **60** Lisa F. Young/ Shutterstock.com; **62** kali9/E+ via Getty Images; **65** bibiphoto/Shutterstock.com; **66** Hamara/ Shutterstock.com; **72** "Portrait of Sir Isaac Newton, 1689" by Godfrey Kneller/Wikimedia Commons/Public Domain; **73** katatonia82/ Shutterstock.com; **75**t J.Gatherum/Shutterstock .com; **75**b Nolte Lourens/Shutterstock.com; **76** Studio Romantic/Shutterstock.com; **78** BJU Photo Services/Bradley Allweil; **80** izusek/ iStock/Getty Images Plus via Getty Images; **81** BearFotos/Shutterstock.com; **82** RossHelen/ Shutterstock.com; **83** Sorbis/Shutterstock.com; **88** Gorodenkoff/Shutterstock.com; **90**t Tomas Kulaja/ Shutterstock.com; **90**b Blue Titan/Shutterstock .com; **92** Jacek Chabraszewski/Shutterstock.com; **94** NDAB Creativity/Shutterstock.com; **95** Anton Havelaar/Shutterstock.com; **96** A. Zhuravleva/ Shutterstock.com; **98** Tamakhin Mykhailo/ Shutterstock.com; **100** Krisana Antharith/ Shutterstock.com; **103** MAD_Production/ Shutterstock.com

Chapter 3

110–11 Pixels Hunter/Shutterstock.com; **112** Virrage Images/Shutterstock.com; **119** YinYang/ iStock/Getty Images Plus via Getty Images; **123** mapman/Shutterstock.com; **126** SDI Productions/ E+ via Getty Images; **129**, **131** Heritage Images/Hulton Archive via Getty Images; **134** Andrey_Popov/Shutterstock.com; **135** muratart/ Shutterstock.com; **140** Cat Box/Shutterstock.com; **142**t Megan Czarnocki/Shutterstock.com; **142**b NDAB Creativity/Shutterstock.com; **146** Klaus Vedfelt/DigitalVision via Getty Images

Chapter 4

152–53 asbestos_cemento/Shutterstock.com; **154** Adisa/Shutterstock.com; **157** dowell/Moment via Getty Images; **161** Michael Mitchell/Shutterstock .com; **166** Andrey_Popov/Shutterstock.com; **169** FedBul/Shutterstock.com; **171** 2happy/ Shutterstock.com; **172** Kilito Chan/Moment via Getty Images; **173** (all) BJU Photo Services; **180** Robin Craig/Shutterstock.com; **184** "Frans Hals - Portret van René Descartes" by Frans Hals/Wikimedia Commons/Public Domain; **185** mitifoto/Shutterstock.com; **188** muzsy/ Shutterstock.com; **192** Have a nice day Photo/ Shutterstock.com; **194**, **198** ttsz/iStock/Getty Images Plus via Getty Images; **195** Ground Picture/Shutterstock.com; **200** artpartner-images/The Image Bank via Getty Images; **208** Andrij Vatsyk/Shutterstock.com; **209**t yanikap/Shutterstock.com; **209**b Alexey Kljatov/ Shutterstock.com; **210** Steve Mann/Shutterstock .com

Chapter 5

216–17 © Pakhnyushchyy | Dreamstime.com; **218** serato/Shutterstock.com; **224** Mystic Stock Photography/Shutterstock.com; **230** NirutiStock/ iStock/Getty Images Plus via Getty Images; **231** Lukas Gojda/Shutterstock.com; **236** Paul Brady Photography/Shutterstock.com; **242** CrispyPork/iStock/Getty Images Plus via Getty Images; **243** AleksSafronov/Shutterstock.com; **248** industryviews/Shutterstock.com; **255** Matej Kastelic/500px Plus via Getty Images

Chapter 6

262–63 Peppy Graphics/Shutterstock.com; **264** Kzenon/Shutterstock.com; **272** Anjelika Gretskaiai/Moment via Getty Images; **277**

Billion Photos/Shutterstock.com; **279** Richie Chan/Shutterstock.com; **282** ChameleonsEye/ Shutterstock.com; **284** Pavel Ignatov/Shutterstock .com; **285** Cast of Thousands/Shutterstock.com; **292** Luca Oleastri/Alamy Stock Photo; **293** "Carl Friedrich Gauss 1840 by Jensen" by Christian Albrecht Jensen/Wikimedia Commons/Public Domain; **294** Jerry Monkman/Cavan via Getty Images; **295** pixdeluxe/E+ via Getty Images; **297** Denis Belitsky/Shutterstock.com; **298** MeskPhotography/Shutterstock.com; **301** Danil Bukharov/Shutterstock.com; **307** BJU Photo Services/Marissa Ellerbrock; **312** Dean Drobot/ Shutterstock.com; **314** Fotografia de eLuVe/ Moment via Getty Images; **316** Ovchinnikova Irina/ Shutterstock.com; **317** Melanie Cooper Studios/ Shutterstock.com; **321**t David Huamani Bedoya/ Shutterstock.com; **321**b Courtesy of Mark Ring/ Comforting Mercies Ministries; **322** Monticello/ iStock/Getty Images Plus via Getty Images

Chapter 7

328–29 NicoElNino/Shutterstock.com; **330** Sean Anthony Eddy/E+ via Getty Images; **337** Zakharchuk/Shutterstock.com; **340** DETLEV VAN RAVENSWAAY/Science Source; **341** Phonlamai Photo/Shutterstock.com; **344** Corel Corporation; **351** Dragon Images/Shutterstock.com; **352** A_ stockphoto/Shutterstock.com; **354** Raycat/E+ via Getty Images; **357** DC Studios/Shutterstock.com; **358** (all) John Lucas-South Carolina Department of Natural Resources; **359** Arpon Pongkasetkam/ Shutterstock.com; **360** djile/Shutterstock.com; **363** Callista Images/Image Source via Getty Images; **365** Vad-Len/Shutterstock.com; **368** ninjaMonkeyStudio/E+ via Getty Images; **371** Peter Photo123/Shutterstock.com

Chapter 8

378–79 Alexander Image/Shutterstock.com; **380** redbrickstock.com/Alamy Stock Photo; **384** Audio und werbung/Shutterstock.com; **389** fotoVoyager/ E+ via Getty Images; **394** OlafSpeier/iStock/Getty Images Plus via Getty Images; **396** BJU Photo Services/John Smoker; **398** "Commodore Grace M. Hopper, USN (covered)" by James S. Davis/ Wikimedia Commons/Public Domain; **399** R.M. Nunes/iStock/Getty Images Plus via Getty Images; **404** Allen Baxter/The Image Bank via Getty Images; **410** Dan J Rozak/iStock/Getty Images Plus via Getty Images

Chapter 9

416–17 Santiago Lecue/EyeEm via Getty Images; **418** Sussi HJ/Shutterstock.com; **423, 427** Andrew Mayovskyy/Shutterstock.com; **430** Vasyl Shulga/ Shutterstock.com; **433** MaksymA/Shutterstock .com; **436** SlobodanMiljevic/iStock/Getty Images Plus via Getty Images; **440** Studio Romantic/ Shutterstock.com; **441** grafvision/Shutterstock.com; **445**t brandstaetter images/Hulton Archive via Getty Images; **445**b Sean Gallup/Getty Images News via Getty Images; **446** (all) The Glacier Photograph Collection, National Snow and Ice Data Center/ World Data Center for Glaciology; **447** JEAN-PIERRE CLATOT/AFP via Getty Images

Chapter 10

450–51 Christian Offenberg/Alamy Stock Photo; **452** Marti Bug Catcher/Shutterstock.com; **456** (all) NASA; **459** melnikof/Shutterstock.com; **465** Lordn/ Shutterstock.com; **468** AldoMurillo/iStock/Getty Images Plus via Getty Images; **469** © Vyacheslav Bukhal | Dreamstime.com; **473**t "Johann Bernoulli2" by Johann Rudolf Huber/Wikimedia Commons/Public Domain; **473**c "Jakob Bernoulli" by Niklaus Bernoulli/Wikimedia Commons/Public Domain; **473**b "Porträt_des_Daniel_Bernoulli_-_ edit1" by Bammesk/Wikimedia Commons/CC-BY-SA-4.0; **474** phil_berry/Shutterstock.com; **482** bluedog studio/Shutterstock.com; **486** BanksPhoto/ iStock/Getty Images Plus via Getty Images; **490** isak55/Shutterstock.com; **496**t Min C. Chiu/ Shutterstock.com; **496**b DEA/ICAS94/DeAgostini Editorial via Getty Images; **497** BJU Photo Services/ Rita Mitchell; **503** Edward Haylan/Shutterstock .com

Chapter 11

504–5 Peter Ravallo/Alamy Stock Photo; **506** Chlorophylle Photography/Shutterstock .com; **511** joyfull/Shutterstock.com; **513** mooninblack/Shutterstock.com; **514** "Galileo-sustermans4" by Justus Sustermans/Wikimedia Commons/Public Domain; **517** wavebreakmedia/ Shutterstock.com; **520** Air Images/Shutterstock .com; **523, 524** Conor P. Fitzgerald/Shutterstock .com; **527** Poznyakov/Shutterstock.com; **528** Danil Evskyi/Shutterstock.com; **533** myphotobank .com.au/Shutterstock.com; **535** FOTOGRIN/ Shutterstock.com; **536** RAYphotographer/ Shutterstock.com; **538** Image Professionals GmbH/ Alamy Stock Photo; **543** Alexandra Robins/ Moment via Getty Images; **546** ronstik/iStock/ Getty Images Plus via Getty Images; **548** Wead/ Shutterstock.com; **549** Unwind/Shutterstock .com; **553** Bet_Noire/iStock/Getty Images Plus via Getty Images; **554**t VW Pics/Universal Images Group Editorial via Getty Images; **554**b Milkovasa/

Shutterstock.com; **555** ©Psalm113.Image from BigStockPhoto.com

Chapter 12

562–63 Michal Bednarek/Shutterstock.com; **564** Alexxxey/Shutterstock.com; **569** zsolt_uveges/ Shutterstock.com; **573** Air Images/Shutterstock .com; **577** © Motionsense | Dreamstime.com; **582** Photo Quest/Archive Photos via Getty Images; **583** metamorworks/Shutterstock.com; **586**t Katerina_S/ Shutterstock.com; **586**b PopTika/Shutterstock .com; **588** ACTS-DATA STOCK/Shutterstock.com; **590** Boris Riaposov/Shutterstock.com; **593** Virrage Images/Shutterstock.com; **594** gece33/iStock/ Getty Images Plus via Getty Images; **599** Parilov/ Shutterstock.com; **601** urbans/Shutterstock.com; **608** BJU Photo Services/Mark Wetzel; **609** Beyla Balla/Shutterstock.com

Appendix

614 Gary Matuschka/Shutterstock.com; **619** Eternity in an Instant/Stone via Getty Images; **623** NicoElNino/Alamy Stock Photo; **630** pixelheadphoto digitalskillet/Shutterstock.com; **636** SIHASAKPRACHUM/Shutterstock.com

QUICK REFERENCE

Properties of Addition & Multiplication		Properties of Exponents & Radicals		
Commutative		**Definitions**		
$a + b = b + a$	$ab = ba$	$x^0 = 1\ (x \neq 0)$	$x^{-n} = \frac{1}{x^n}$	$\sqrt{x} = x^{\frac{1}{2}}$
Associative		**Product of Powers**		
$(a + b) + c = a + (b + c)$	$(ab)c = a(bc)$	$x^a \cdot x^b = x^{a+b}$		
Identity		**Power of a Power**		
$a + 0 = 0 + a = a$	$a \cdot 1 = 1 \cdot a = a$	$(x^a)^b = x^{ab}$		
Inverse		**Quotient of Powers**		
$a + (-a) = -a + a = 0$	$a \cdot \frac{1}{a} = \frac{1}{a} \cdot a = 1$	$\frac{x^a}{x^b} = x^{a-b}$		
Distributive		**Powers of Products & Quotients**		
$a(b + c) = ab + ac$ $a(b - c) = ab - ac$		$(x \cdot y)^a = x^a \cdot y^a$ and $\left(\frac{x}{y}\right)^a = \frac{x^a}{y^a}\ (y \neq 0)$		
	Zero	**Product of Radicals**		
	$a \cdot 0 = 0$	For $x \geq 0$ and $y \geq 0$, $\sqrt[n]{xy} = \sqrt[n]{x} \cdot \sqrt[n]{y}$.		
	Zero Product	**Quotient of Radicals**		
	If $pq = 0$, then $p = 0$ or $q = 0$.	For $x \geq 0$ and $y > 0$, $\sqrt[n]{\frac{x}{y}} = \frac{\sqrt[n]{x}}{\sqrt[n]{y}}$.		

Properties of Equality			
Addition	If $a = b$, then $a + c = b + c$.	**Subtraction**	If $a = b$, then $a - c = b - c$.
Multiplication	If $a = b$, then $ac = bc$.	**Division**	If $a = b$ and $c \neq 0$, then $\frac{a}{c} = \frac{b}{c}$.
Symmetric	If $a = b$, then $b = a$.	**Substitution**	If $a = b$, replacing a with b in any expression produces an equivalent expression.
Properties of Inequality			
Addition	If $a < b$, then $a + c < b + c$.	**Subtraction**	If $a < b$, then $a - c < b - c$.
Multiplication	If $a < b$ and $c > 0$, then $ac < bc$, but if $c < 0$, then $ac > bc$.	**Division**	If $a < b$ and $c > 0$, then $\frac{a}{c} < \frac{b}{c}$, but if $c < 0$, then $\frac{a}{c} > \frac{b}{c}$.

Area, Surface Area, Perimeter & Volume Formulas		
Circle	$A = \pi r^2$	$C = 2\pi r$ or $C = \pi d$
Parallelogram	$A = bh$	$P = 2a + 2b$
Rectangle	$A = lw$	$P = 2l + 2w$
Square	$A = s^2$	$P = 4s$
Trapezoid	$A = \frac{1}{2}(b_1 + b_2)h$	
Triangle	$A = \frac{1}{2}bh$	$P = a + b + c$
Cube	$S = 6s^2$	$V = s^3$
Rectangular Prism	$S = PH + 2B$	$V = lwh$ or $V = BH$
Right Circular Cylinder	$S = 2\pi rH + 2\pi r^2$	$V = \pi r^2 H$
Sphere	$S = 4\pi r^2$	$V = \frac{4}{3}\pi r^3$

Triangles		
Sum of the Angles of a Triangle	$m\angle A + m\angle B + m\angle C = 180°$	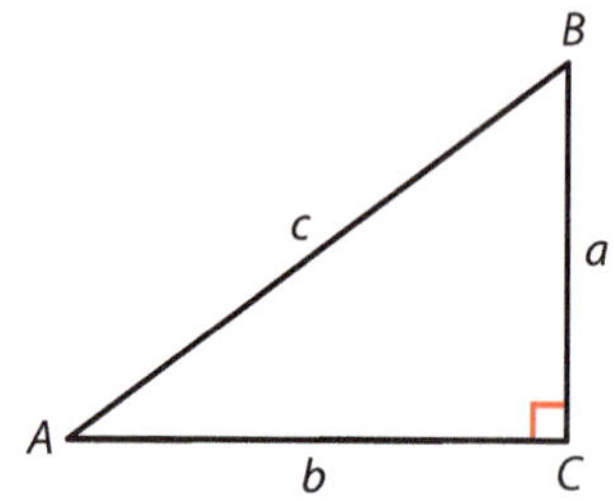
Heron's Formula	$A = \sqrt{s(s-a)(s-b)(s-c)}$, where $s = \frac{a+b+c}{2}$	
Pythagorean Theorem	In right triangle ABC, $a^2 + b^2 = c^2$.	

Formulas		
Distance & Midpoint	**On a Number Line**	**In the Coordinate Plane**
Distance between 2 Points	$d = \lvert b - a \rvert$	$d = \sqrt{(x_2 - x_1)^2 + (y_2 - y_1)^2}$
Midpoint of 2 Points	$m = \frac{a+b}{2}$	$M\left(\frac{x_1 + x_2}{2}, \frac{y_1 + y_2}{2}\right)$
Sequences	**Arithmetic**	**Geometric**
Recursive Formula	$a_n = a_{n-1} + d$	$a_n = ra_{n-1}$
Explicit Formula	$a_n = a_1 + (n-1)d$	$a_n = a_1 r^{n-1}$

Linear Equations	
Standard Form	$Ax + By = C$
Slope-Intercept Form	$y = mx + b$
Point-Slope Form	$y - y_1 = m(x - x_1)$

Other Formulas		
Slope	$m = \frac{\text{rise}}{\text{run}} = \frac{y_2 - y_1}{x_2 - x_1} = \frac{\Delta y}{\Delta x} = \frac{\text{change in } y}{\text{change in } x}$	
Distance	*distance* = *rate* × *time*	$d = rt$
Percent Change	$\frac{\text{amount of change}}{\text{original amount}} \times 100\%$	
Simple Interest	*interest* = *principal* × *rate* × *time*	$I = Prt$
Quadratic Formula	Given $ax^2 + bx + c = 0$, where $a \neq 0$, $x = \frac{-b \pm \sqrt{b^2 - 4ac}}{2a}$ and the discriminant $D = b^2 - 4ac$.	
Exponential Growth	$f(t) = A(1 + r)^t$	
Exponential Decay	$f(t) = A(1 - r)^t$	

Conversions		
1 inch = 2.54 centimeters	1 mile = 5280 feet	1 meter = 100 centimeters
1 meter ≈ 3.28 feet	1 nautical mile ≈ 1.15 statutory miles	1 kilogram ≈ 2.20 pounds
1 mile ≈ 1.61 kilometers	1 kilometer = 1000 meters	

SYMBOLS

$+$	addition, positive
$-$	subtraction, negative, opposite
$\times, \cdot$	multiplication
$\div$	division
$\frac{p}{q}$	division, fraction
$\pm$	plus or minus
$\lvert x \rvert$	absolute value of *x*
$\%$	percent
$0.\overline{3}$	repeating decimal
$f(x)$	function of *x*
$\ldots$	ellipses (and so on)
$\{\ \}$	set braces
$\{x \mid \ldots\}$	the set of *x* such that . . .
$\in$	element of
$\varnothing$	empty (null) set
$\mathbb{R}$	real numbers
$\mathbb{Q}'$	irrational numbers
$\mathbb{Q}$	rational numbers
$\mathbb{Z}$	integers
$\mathbb{W}$	whole numbers
$\mathbb{N}$	natural numbers
$=$	equal to
$\neq$	not equal to
$\approx$	approximately equal to
$\sim$	similar to
$>$	greater than
$\ngtr$	not greater than
$\geq$	greater than or equal to
$\ngeq$	not greater than or equal to
$<$	less than
$\nless$	not less than
$\leq$	less than or equal to
$\nleq$	not less than or equal to
$\sqrt{\ }$	square root
$\sqrt[3]{\ }$	cube root
$\sqrt[n]{\ }$	*n*th root
$^\circ$	degree(s)
$\triangle$	triangle
π	pi
Δ	delta
ϕ	phi
σ	sigma